COUPLED OCEAN - ATMOSPHERE MODELS

Elsevier Oceanography Series, 40

COUPLED OCEAN-ATMOSPHERE MODELS

Edited by

J.C.J. NIHOUL

University of Liège, B5 Sart Tilman, B-4000 Liège, Belgium

ELSEVIER

Amsterdam — Oxford — New York — Tokyo 1985

ELSEVIER SCIENCE PUBLISHERS B.V.
Molenwerf 1
P.O. Box 211, 1000 AE Amsterdam, The Netherlands

Distributors for the United States and Canada:

ELSEVIER SCIENCE PUBLISHING COMPANY INC.
52, Vanderbilt Avenue
New York, NY 10017, U.S.A.

Library of Congress Cataloging in Publication Data
Main entry under title:

Coupled ocean-atmosphere models.

   (Elsevier oceanography series ; 40)
   Papers presented at the 16th International Liège
Colloquium on Ocean Hydrodynamics, held in 1984.
   Includes bibliographies.
   1. Ocean-atmosphere interaction--Mathematical models
--Congresses. I. Nihoul, Jacques C. J. II. Inter-
national Liège Colloquium on Ocean Hydrodynamics (16th :
1984) III. Series.
GC190.5.C68 1985   551.5   85-10294
ISBN 0-444-42486-5 (U.S.)

ISBN 0-444-42486-5 (Vol. 40)
ISBN 0-444-41623-4 (Series)

Printed in The Netherlands

LIST OF CONTRIBUTORS

Anderson, D.L.T., Department of Atmospheric Physics, University of Oxford, Parks Road, Oxford OX1 3PU, Great Britain.

Arkin, P.A., Climate Analysis Center, NMC, NWS, NOAA, World Weather Building, Washington, DC 20233, U.S.A.

Blackmon, M.L., National Center for Atmospheric Research, P.O. Box 3000, Boulder, CO 80307, U.S.A.

Boer, G.J., Canadian Climate Centre (CCRN), 4905 Dufferin Street, Downsview, Ont. M3H 5T4, Canada.

Bottomley, Dynamical Climatology Branch, Meteorology Office, London Road, Bracknell, Berkshire RG12 2SZ, Great Britain.

Bryan, K., Geophysical Fluid Dynamics Laboratory, Princeton University, P.O. Box 307, Princeton, NJ 08542, U.S.A.

Busalacchi, A.J., NASA/Goddard Space Flight Center, Goddard Laboratory for Atmospheric Sciences, Greenbelt, MD 20771, U.S.A.

Bye, J.A.T., School of Earth Sciences, The Flinders University of South Australia, Bedford Park, S.A. 5042, Australia.

Chao Jih-Ping, Institute of Atmospheric Sciences, Academia Sinica, Beijing, People's Republic of China.

Chi-San Liou, Department of Meteorology, Naval Postgraduate School, Monterey, CA 93943, U.S.A.

Crépon, M.L., Muséum National d'Histoire Naturelle, Laboratoire d'Océanographie Physique, 43–45 Rue Cuvier, F-75231 Paris Cédex 05, France.

Cubasch, U., European Centre for Medium Range Weather Forecasts (ECMWF), Shinfield Park, Reading, Berkshire RG2 9AX, Great Britain.

Cushman-Roisin, B., Mesoscale Air–Sea Interaction Group, Meteorology Annex, Florida State University, Tallahassee, FL 32306, U.S.A.

Davey, M.K., Department of Applied Mathematics and Theoretical Physics, University of Cambridge, Silver Street, Cambridge CB3 9EW, Great Britain.

Deque, M. Centre National de Recherches Météorologique, Avenue Coriolis, 30957 Toulouse, France

Elsberry, R.L., Department of Meteorology, Naval Postgraduate School, Monterey, CA 93943, U.S.A.

Esbensen, S.K., Department of Atmospheric Sciences, Oregon State University, Corvallis, OR 97331, U.S.A.

Fennessy, M.J., Center for Ocean–Land–Atmosphere Interactions, Department of Meteorology, University of Maryland, College Park, MD 20742, U.S.A.

Frankignoul, C., Laboratoire de Physique et Chimie Marines, Université Pierre et Marie Curie, Tour 24–25, 4 Place Jussieu, 75230 Paris Cédex 05, France.

Gates, W.L., Department of Atmospheric Sciences, Oregon State University, Corvallis, OR 97331, U.S.A.

Gill, A.E., Hooke Institute, Clarendon Laboratory, Oxford OX1 3PU, Great Britain.

Gordon, C., Dynamical Climatology Branch, Meteorology Office, London Road, Bracknell, Berkshire RG12 2SZ, Great Britain.

Guo-Yu-Fu, Institute of Atmospheric Sciences, Academia Sinica, Beijing, People's Republic of China.

Han, Y.-J., Department of Atmospheric Sciences, Oregon State University, Corvallis, OR 97331, U.S.A.

Hirst, A.C., Department of Meteorology, University of Wisconsin, 1225 West Dayton Street, Madison, WI 53706, U.S.A.

Killworth, P.D., Department of Applied Mathematics and Theoretical Physics, University of Cambridge, Silver Street, Cambridge CB3 9EW, Great Britain.

Kraus, E.B., Cooperative Institute for Research in Environmental Sciences, University of Colorado, Boulder, CO 80309, U.S.A.

Krishnamurti, T.N., Department of Meteorology, Florida State University, Tallahassee, FL 32306, U.S.A.

Kruse, H., Max Planck Institut für Meteorologie, Bundesstrase 55, D-2000 Hamburg 13, F.R.G.

Latif, M., Max Planck Institut für Meteorologie, Bundesstrasse 55, D-2000 Hamburg 13, F.R.G.

Lau, K.-M. Goddard Laboratory for Atmospheric Sciences, NASA/Goddard Space Flight Center, Greenbelt, MD 20771, U.S.A.

Lau, N.-C., Geophysical Fluid Dynamics Laboratory/NOAA, Princeton University, P.O. Box 308, Princeton, NJ 08542, U.S.A.

Le Treut, H., Laboratoire de Météorologie Dynamique, 24 Rue Lhomond, 75231 Paris Cédex 05, France.

Luther, M.E., Mesoscale Air–Sea Interaction Group, Meteorology Annex, Florida State University, Tallahassee, FL 32306, U.S.A.

Maher, M.A.C., U.S. Department of Commerce/NOAA, Geophysical Fluid Dynamics Laboratory, Princeton University, P.O. Box 308, Princeton, NJ 08542, U.S.A.

Maier-Reimer, E., Max Planck Institut für Meteorologie, Bundesstrasse 55, D-2000 Hamburg 13, F.R.G.

Manabe, S., Geophysical Fluid Dynamics Laboratory, Princeton University, P.O. Box 307, Princeton, NJ 08542, U.S.A.

Marx, L., Center for Ocean–Land–Atmosphere Interactions, Department of Meteorology, University of Maryland, College Park, MD 20742, U.S.A.

McCreary Jr., J.P., Nova University Oceanographic Center, 8000 North Ocean Drive, Dania, FL 33004, U.S.A.

Meng, A.H., Mesoscale Air–Sea Interaction Group, Meteorology Annex, Florida State University, Tallahassee, FL 32306, U.S.A.

Michaud, R., Laboratoire de Météorologie Dynamique du CNRS, Ecole Normale Supérieure, 24 Rue Lhomond, F-75231 Paris Cédex 05, France.

Mitchell, J.F.B., Meteorological Office, London Road, Bracknell, Berkshire RG12 2SZ, Great Britain.

Muir, D., Department of Mathematics, Exeter University, Exeter EX4 4QE, Great Britain.

Murdoch, N., Department of Mathematics, Exeter University, Exeter EX4 4QE, Great Britain.

Newman, M.R., Meteorological Office, London Road, Bracknell, Berkshire RG12 2SZ, Great Britain.

Nihoul, J.C.J., University of Liège, B5 Sart Tilman, B-4000 Liège, Belgium.

O'Brien, J.J., Mesoscale Air–Sea Interaction Group, Meteorology Annex, Florida State University, Tallahassee, FL 32306, U.S.A.

Olbers, D.J., Max Planck Institut für Meteorologie, Bundesstrasse 55, D-2000 Hamburg 13, F.R.G.

Oort, A.H., U.S. Department of Commerce/NOAA, Geophysical Fluid Dynamics Laboratory, Princeton University, P.O. Box 308, Princeton, NJ 08542, U.S.A.

Palmer, T.N., Meteorological Office, London Road, Bracknell, Berkshire RG12 2SZ, Great Britain.

Philander, S.G.H., Geophysical Fluid Dynamics Laboratory, Princeton University, P.O. Box 308, Princeton, NJ 08542, U.S.A.

Picaut, J., Laboratoire d'Océanographie Physique, Université de Bretagne Occidentale, 29200 Brest, France.

Price, C., Meteorological Office, London Road, Bracknell, Berkshire RG12 2SZ, Great Britain.

Ranelli, P.H., Department of Meteorology, Naval Postgraduate School, Monterey, CA 93943, U.S.A.

Rasmusson, E.M., Climate Analysis Center, NMC, NWS, NOAA, World Weather Building, Washington, DC 20233, U.S.A.

Rowe, M.A., Department of Oceanography, University of Southampton, Southampton SO9 5NH, Great Britain.

Royer, J.F., Centre National de Recherches Météorologiques, Avenue Coriolis, 30957 Toulouse, France.

Sadourny, R., Laboratoire de Météorologie Dynamique de CNRS, Ecole Normale Supérieure, 24 Rue Lhomond, F-75231 Paris Cédex 05, France.

Sandgathe, S.A., Department of Meteorology, Naval Postgraduate School, Monterey, CA 93943, U.S.A.

Sarkisyan, A.S., Department of Numerical Mathematics, USSR Academy of Sciences, Gorkistreet 11, Moscow 103009, U.S.S.R.

Schlesinger, M.E., Department of Atmospheric Sciences, Oregon State University, Corvallis, OR 97331, U.S.A.

Schopf, P.S., Goddard Laboratory for Atmospheric Sciences, NASA/Goddard Space Flight Center, Greenbelt, MD 20771, U.S.A.

Seigel, A.D., Geophysical Fluid Dynamics Laboratory, Princeton University, P.O. Box 308, Princeton, NJ 08542, U.S.A.

Servain, J., Laboratoire d'Océanographie Physique, Université de Bretagne Occidentale, 29200 Brest, France.

Shukla, J., Center for Ocean–Land–Atmosphere Interactions, Department of Meteorology, University of Maryland, College Park, MD 20742, U.S.A.

Simonot, J.Y., Muséum National d'Histoire Naturelle, Laboratoire d'Océanographie Physique, 43–45 Rue Cuvier, F-75231 Paris Cédex 05, France.

Storey, A.M., Meteorological Office, London Road, Bracknell, Berkshire RG12 2SZ, Great Britain.

Suarez, M.J., Code 911, Global Modeling and Simulation Branch, NASA/Goddard Space Flight Center, Greenbelt, MD 20771, U.S.A.

Taylor, A.H., Natural Environment Research Council, Institute for Marine Environmental Research, Prospect Place, The Hoe, Plymouth PL1 2DH, Great Britain.

Tourre, Y., Laboratoire de Physique et Dynamique de l'Atmosphère/ORSTOM, Université Pierre et Marie Curie, Place Jussieu 4, 75230 Paris Cédex 05, France.

Von Storch, H., Meteorologisches Institut der Universität Hamburg, Bundesstrasse 55, D-2000 Hamburg 13, F.R.G.

Wang Xiao-Xi, Institute of Atmospheric Sciences, Academia Sinica, Beijing, People's Republic of China.

Wells, N.C., Department of Oceanography, University of Southampton, Southampton SO9 5NH, Great Britain.

Wilson, C.A., Meteorological Office, London Road, Bracknell, Berkshire RG12 2SZ, Great Britain.

Woods, J.D., Institut für Meereskunde, Universität Kiel, Düsternbrooker Weg 20, D-2300 Kiel 1, F.R.G.

Yamagata, T., Research Institute for Applied Mechanics, Kyushu University 87, Kasuga 816, Japan.

PREFACE

The International Liège Colloquia on Ocean Hydrodynamics are organized annually. Their topics differ from one year to another and try to address, as much as possible, recent problems and incentive new subjects in physical oceanography.

Assembling a group of active and eminent scientists from different countries and often different disciplines, they provide a forum for discussion and foster a mutually beneficial exchange of information opening on to a survey of major recent discoveries, essential mechanisms, impelling question-marks and valuable recommendations for future research.

The Scientific Organizing Committee and all the participants wish to express their gratitude to the Belgian Minister of Education, the National Science Foundation of Belgium, the University of Liège, the Intergovernmental Oceanographic Commission and the Division of Marine Sciences (UNESCO), the Committee on Climate Changes and the Ocean (SCOR), the WMO/ICSU Joint Scientific Committee for the World Climate Research Program, and the Office of Naval Research for their most valuable support.

The editor is deeply indebted to Dr. David Anderson who so efficiently directed the work of the Scientific Organizing Committee and whose unwearying dedication to the success of the Colloquium made the publication of this book possible. Dr. Jamart's help in editing these Proceedings is also acknowledged.

JACQUES C.J. NIHOUL (Editor)

CONTENTS

XIV

XX

CHAPTER 1

A COUPLED OCEAN–ATMOSPHERE AND THE RESPONSE TO INCREASING ATMOSPHERIC CO_2*

KIRK BRYAN and SYUKURO MANABE

PLAN OF THE NUMERICAL EXPERIMENTS

The exchanges of heat and carbon between the ocean and atmosphere are crucially important in determining the climatic response to increasing anthropogenic CO_2 in the atmosphere. Subject to an uncertainty caused by a possible growth or decay of the biospheric carbon reservoir, the oceans are estimated to absorb enough CO_2 to reduce the anthropogenic build-up in the atmosphere by as much as 50%. *Sensitivity* studies such as those carried out by Manabe and Wetherald (1975), Manabe and Stouffer (1980), and many others, estimate the equilibrium response to an increase in atmospheric CO_2. The actual timing of the outset of climate change, however, will be controlled by heat exchange with the ocean, which has a heat capacity orders of magnitude larger than the atmosphere. The present study is concerned with the physical processes involved in this transient response.

It is impossible to use historical data to estimate the response time of the coupled ocean–atmosphere system, since the rapid increase of greenhouse gases is an un-precedented event. Transient tracers produced by the bomb tests of the late fifties and early sixties provide one source of information on downward pathways from the ocean surface to deeper layers. The downward movement of these tracers offers an interesting analogue to the penetration of a heat anomaly originating at the surface, with one im-portant difference. The tracers are neutrally buoyant, while a heat anomaly will change the ocean circulation as it is carried downward in the main thermocline. The question then arises as to whether data from the transient tracers (Ostland et al., 1976) can be a quantitatively useful analogue. One of the motivations for this study is to investigate this important question.

At this point, it is crucially important to identify all the significant processes in the transient response of climate to increasing CO_2. For this reason, we have chosen to use models of the ocean and atmosphere that are quite complete physically, but other aspects such as the geometry of continents and oceans are deliberately kept as simple as possible. The actual configuration is shown in Fig. 1. Computations are carried out for a domain spanning $120°$ of longitude, and extending from equator to pole. Cyclic boundary conditions are assumed in the zonal direction, and mirror symmetry is assumed across the equator. The ocean is confined to a $60°$ longitude sector which extends from equator to pole. Our geometry is equivalent to both hemispheres having three identical continents

---

*Some of this material has been published elsewhere. The reader is referred to Bryan et al. (1982) and Spelman and Manabe (1984).

2

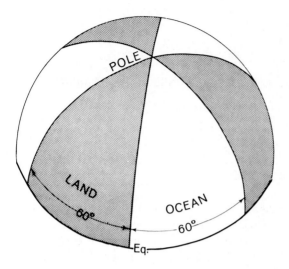

Fig. 1. The simplified geometry of the model. Cyclic symmetry is assumed over a 120° longitude sector. Mirror symmetry is assumed across the equator.

about the size of North America. The three continents are separated by three identical oceans with approximately the width of the North Atlantic.

The atmospheric component of the model is similar to that used in many previous CO_2/climate studies (e.g., Manabe and Stouffer, 1980). Essentially the model is an advanced numerical weather prediction model with the more detailed representation of radiation and boundary layer processes needed for climatic purposes. Clouds are not predicted by the model, but specified from data. This may have an important bearing on the sensitivity of the model (Hansen et al., 1984). We will return to this point in the final section. The atmospheric model is based on a spherical harmonic representation of variables with a resolution of approximately 400 × 400 km in the horizontal plane, and nine vertical levels. The ocean model has the equivalent horizontal resolution but twice the number of grid points in the zonal direction and twelve levels in the vertical. The horizontal resolution in the ocean model is minimal in its ability to resolve ocean currents and should certainly be improved in future studies. Snow over land surfaces, and an ice pack over the sea are included. The sea-ice parameterization was originally specified in Bryan (1969). Snow and ice permit an important albedo feedback which amplifies external influences.

The experimental procedure is illustrated in Fig. 2. Equilibrium climates are calculated for the coupled model corresponding to a normal concentration of atmospheric CO_2, and for the case of four times the normal concentration of atmospheric CO_2. The ordinate in the diagram is the globally averaged sea-surface temperature. The two climate equilibria are illustrated by two horizontal lines. The abscissa represents the time after "switch on"; that is, the time elapsed after the climate corresponding to a normal level of CO_2 is perturbed by impulsively quadrupling the CO_2 level in the model atmosphere. Our study is intended to study the processes that govern the speed and character of the response shown schematically by the dashed line in Fig. 2. Will the delay caused by the ocean's capacity be measured in years, decades, or centuries?

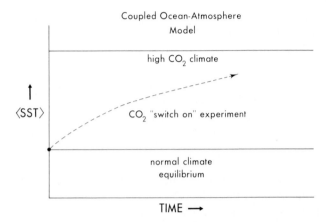

Fig. 2. A schematic diagram of the numerical experiments. The ordinate is the globally averaged sea-surface temperature. The horizontal lines represent climatic equilibria for normal and high atmospheric CO_2. The dashed line represents the "switch-on" experiment.

THE EQUILIBRIUM CALCULATIONS

For fixed lower boundary conditions the calculation of climate equilibrium for an atmospheric model without seasonal variations is relatively simple. Since the response of the atmosphere is only a few months, climatic equilibrium can be reached by simply extending a numerical integration with respect to time over a period of six months to a year. On the other hand, the ocean has a heat capacity orders of magnitude greater than the atmosphere and its thermal relaxation time is correspondingly greater. To integrate a coupled model to climatic equilibrium by straightforward time-stepping would require numerical integration over a period of at least one thousand years. This is clearly not a feasible procedure. To circumvent this difficulty, another method for reaching climatic equilibrium has been devised (Manabe and Bryan, 1969; Bryan et al., 1975; Bryan, 1984). Essentially the method takes into account the fact that different elements of the climate system have very different time scales. By decoupling "slow" and "fast" time-scale processes, each component can be adjusted according to its own natural time scale. Since systems with slow time scales can be integrated with much longer time steps than systems with fast time scales, decoupling allows a much more efficient approach to equilibrium. Once equilibrium is achieved, response experiments are carried out with a fully synchronous version of the coupled model to avoid any possible distortion.

To test the degree of equilibrium achieved, the coupled model was integrated over fifty years in a control run without any change in the atmospheric CO_2 concentration. The net exchange of heat between the ocean and atmosphere was carefully monitored. It was found that the average exchange of heat was less than $0.4 \, W \, m^{-2}$. At this rate of heating, 300 years would be required to change the upper kilometer of the water column by one degree. This is a very small drift compared to the signal imposed by our "switch-on" experiment.

4

RESULTS OF THE "SWITCH-ON" EXPERIMENT

The result of the "switch-on" experiment is shown in Fig. 3 for the sea surface temperature and the air temperature over both land and sea. The normalized response, R, is:

$$R = \frac{T - T_0}{T_\infty - T_0} \qquad (1)$$

where T is the temperature as a function of time after "switch on", T_0 is the initial temperature, and T_∞ is the final equilibrium temperature. The denominator in eqn. (1) is the sensitivity to a change in CO_2, and the numerator is the time-dependent response. The advantage of this normalization is that it allows a direct interpretation of the transient experiment in terms of previous sensitivity studies which evaluate equilibrium response. An important finding of the CO_2 sensitivity studies (Manabe and Wetherald, 1975; Manabe and Stouffer, 1980; Hansen et al., 1984) is a tendency for polar amplification in CO_2-induced warming. As we examine the zonally averaged response in Fig. 3b it can be seen that shortly after "switch on" R is greater at low latitudes. As time goes on, the latitudinal gradient of R gradually diminishes, almost disappearing at 25 years. This means that the transient response at 25 years has nearly the same latitudinal structure as the

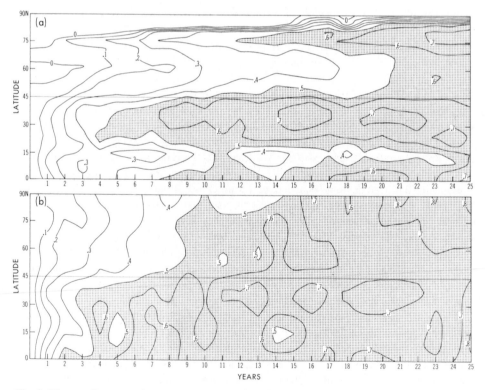

Fig. 3. The zonally averaged normalized response based on one year averages. The ordinate is latitude and the abscissa is time after "switch on". (a) The sea-surface temperature; (b) air temperature over both land and sea.

equilibrium response, including the polar amplification noted earlier. It also suggests that the pattern of transient response to a slow build-up of CO_2 will be similar to the patterns found in sensitivity studies, as long as the time scale of CO_2 increase is longer than 20–30 years.

As mentioned previously, transient tracers provide the only observed analogue for the downward penetration of a large surface heat anomaly caused by a build-up of atmospheric CO_2. Simple one-dimensional models (Oeschger et al., 1975) have been fitted to transient tracer data and used to estimate the delayed response of a coupled ocean–atmosphere model (Cess and Goldenberg, 1981). Our model allows us to actually compare the downward penetration of a tracer and the downward penetration of a heat anomaly. In the control run for normal atmospheric CO_2 a uniform tracer is "switched on" at the upper boundary. The scale depth, d, of penetration may be defined as follows:

$$d = \frac{1}{\mu_s} \int_{-H}^{0} \mu \, dz$$

where μ is the horizontally averaged value of the tracer as a function of depth, and μ_s is the corresponding surface value. H is the total depth of the ocean, independent of position.

In Fig. 4 the penetration depth of the tracer for the control run is shown as a function of time after the tracer is "switched on" at the surface. Tracer input is confined to the top level of the ocean model. Another curve shows the corresponding penetration of a heat anomaly when atmospheric CO_2 is "switched on". Intuition would suggest that the buoyancy associated with a heat anomaly would slow down penetration relative to a passive tracer. Such a feedback has been suggested by Harvey and Schneider (1985), who have incorporated it in a simple, one-dimensional ocean–atmosphere model. In our present calculations the feedback is in the reverse sense. Surface heating causes a greater penetration of a heat anomaly than would be expected for a passive tracer. Vertical transport takes place in three ways: advection, convection and diffusion. If the model

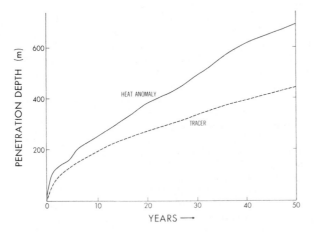

Fig. 4. The penetration depth in meters as a function of time after "switch on". The tracer experiment corresponds to normal atmospheric CO_2.

response was linear, a temperature anomaly would be carried downward by these three types of vertical transport in exactly the same way as a transient tracer introduced at the surface. The nonlinear effect is mainly due to the suppression of convection by the temperature anomaly. In the model, diffusion is independent of changes in the density structure and analysis shows that the vertical advection is also relatively independent of small changes in stratification. Since convection at high latitudes ordinarily transfers heat upward, the suppression of convection traps heat below the surface that has been transported to the subArctic gyre from lower latitudes. Thus, the heat anomaly is carried downward by the same linear processes which carry a tracer downward. Superimposed is another mechanism which involves the suppression of convection which in the undisturbed state cools subsurface waters in the subArctic gyre.

This finding provides an example of the unexpected results which can be obtained in the study of coupled models. Recently Hansen et al. (1984) have pointed out that cloud feedback may greatly increase climate sensitivity to increasing atmospheric CO_2. If this is true, the time scale of climatic response will also be strongly influenced, and the results given here which neglect cloud feedback may be underestimated. This is just one of the many interdisciplinary problems that are awaiting investigation by coupled models of the atmosphere and ocean.

REFERENCES

Bryan, K., 1969. Climate and the ocean circulation, III. The ocean model. Mon. Weather Rev., 97: 807–826.
Bryan, K., 1984. Accelerating the convergence to equilibrium of ocean-climate models. J. Phys. Oceanogr., 14: 666–673.
Bryan, K., Manabe, S. and Pacanowski, R.L., 1975. A global ocean–atmosphere model. Part II. The oceanic circulation. J. Phys. Oceanogr., 5: 30–46.
Bryan, K., Komro, F.G., Manabe, S. and Spelman, M.J., 1982. Transient climate response to increasing atmospheric carbon dioxide. Science, 215: 56–58.
Cess, R.D. and Goldenberg, S.D., 1981. The effect of ocean heat capacity upon global warming due to increasing atmospheric carbon dioxide. J. Geophys. Res., 86: 498–502.
Hansen, J., Lacis, A., Rind, D., Russel, G., Stone, P., Fung, I., Ruedy, R. and Lerner, J., 1984. Climate Sensitivity: Analysis of Feedback Mechanisms. Monogr. 29, Maurice Ewing Vol. 5, Am. Geophys. Union, Washington, D.C., 368 pp.
Harvey, L.B.D. and Schneider, S.H., 1985. Transient climate response to external forcing on 10^0–10^4 year time-scales. I. Experiments with globally averaged, coupled atmosphere and ocean energy balanced models. J. Geophys. Res., in press.
Manabe, S. and Bryan, K., 1969. Climate calculations with a combined ocean–atmosphere model. J. Atmos. Res., 26: 786–789.
Manabe, S. and Stouffer, R.J., 1980. Sensitivity of a global climate model to an increase of CO_2 concentrations in the atmosphere. J. Geophys. Res., 85: 5529–5554.
Manabe, S. and Wetherald, R.T., 1975. The effect of doubling the CO_2 concentration on the climate of a general circulation model. J. Atmos. Sci., 32: 3–15.
Oeschger, H., Siegerthaler, H., Schotterer, U. and Gugelmann, A., 1975. A box diffusion model to study the carbon dioxide exchange in nature. Tellus, 27: 168–192.
Ostland, H.G., Dorsey, H.G., Rooth, C.G.H. and Brescher, R., 1976. GEOSECS Atlantic radiocarbon and tritium results. Data Rep. No. 5, Rosenstiel School Mar. Atmos. Sci., Univ. of Miami, Miami, Fla.
Spelman, M.J. and Manabe, S., 1984. Influence of oceanic heat transport upon the sensitivity of a model climate. J. Geophys. Res., 89: 571–586.

CHAPTER 2

MODELLING THE ATMOSPHERIC RESPONSE TO THE 1982/83 EL NIÑO

G.J. BOER

ABSTRACT

The atmospheric response to the observed 1982/83 El Niño sea-surface temperature anomaly is simulated using the Canadian Climate Centre general circulation model. Five independent evolutions of the atmospheric response are obtained. The results for the December–February season are pooled together and the resulting anomalies from model climatology, which is based on a ten year simulation, are analysed and compared with observations.

The effects of the sea-surface temperature anomaly are clearly seen in the tropical region and the results are interpreted in terms of anomalies in transports of moisture and energy. The tropical response is apparently relatively straightforward and agrees with the qualitative ideas of expected behaviour.

At extratropical latitudes, the model response to this extreme El Niño differs from the typical or composite response as delineated by Pan and Oort for instance. It is argued that, despite the rather modest statistical significance of the result as measured by the univariate t-test, the model's extratropical response to this extreme El Niño resembles the observed anomaly of January 1982 while both differ from the composite or typical El Niño response.

INTRODUCTION

The 1982/83 Northern Hemisphere winter was unusual in many respects. This is usually ascribed to anomalously warm ocean-surface temperatures which occurred at the time in the equatorial eastern Pacific. Despite the general acceptance of the "El Niño" as the cause of the observed anomalies in the winter's climate, the justification for this conclusion, especially at extratropical latitudes, is not strong.

General circulation models provide one method of understanding the response of the (model) atmosphere to an imposed equatorial sea-surface temperature anomaly. A model simulation with and without an anomaly provides information on the behaviour of the model atmosphere which may be compared to that of the real atmosphere. Since the model may be integrated a number of times, several 82/83 style El Niño cases may be simulated and the features common to the simulations compared with what occurred.

What follows are preliminary results of an experiment aimed at simulating the atmospheric response to the 82/83 El Niño. Much more analysis is required to fully understand the model's response and how it relates to that of the atmosphere. The results are discussed in terms of the divergent and rotational components of the flow. The divergent aspect of the flow aids the understanding of the tropical response in terms of the moisture and energy balances while the rotational aspect illustrates important tropical and extratropical responses.

8

THE MODEL

The model used for these integrations is the Canadian Climate Centre general circulation model as described in Boer et al. (1984a, b). It is a spectral T20 model with ten levels in the vertical. It includes annual and diurnal cycles but clouds are specified as are ocean-surface temperatures.

The model has been integrated for ten annual cycles with climatologically specified sea-surface temperatures. The resulting simulated observations have been used to compile the climatology of the model to which the results of the "El Niño" simulation are compared.

THE EXPERIMENT

The observed monthly mean tropical sea-surface temperatures for the belt 60°N–40°S prepared by the NOAA Climate Analysis Centre for the 82/83 El Niño event are used in the simulation. Using initial conditions from 1 June for five of the individual years of the 10-year climatological run, the model is integrated from 1 June to 31 May with the 82/83 observed SST's.

The resulting five simulated cases are analysed for the December–February season in terms of differences or anomalies from the climatological state. The December–February sea-surface temperature anomaly obtained by subtracting the model climatological sea-surface temperatures from the observed 82/83 values is shown in Fig. 1. Note that the observed sea-surface temperatures are used in the simulation. Note also that the resulting SST's include anomalies in excess of 1°C in parts of the Atlantic and Indian Oceans.

Features of the simulation include the simulation of the seasonal cycle (i.e. not perpetual January), and the use of model generated rather than of observed initial conditions (the response of interest is not influenced by initial conditions which may be, in any case, inconsistent with model climatology).

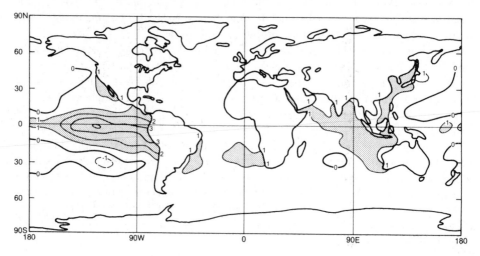

Fig. 1. Sea-surface temperature anomaly for December–February as used in the simulation. Units °C.

THE DIVERGENT COMPONENT

Model climatology

The vertically integrated moisture and energy budgets in the tropical region may be considered in terms of the divergent component of the flow in the following way. Consider the prototype equation:

$$\frac{\partial X}{\partial t} + \nabla \cdot XV + \frac{\partial}{\partial p} X\omega = \hat{S}$$

where X represents moisture, energy, or other atmospheric variable and $\hat{S}$ is the associated source/sink term. The vertically integrated and time-averaged form of this equation may be written generally as:

$$\nabla \cdot F = S$$

where:

$$F = \int_0^{\overline{p_s}} XV(\mathrm{d}p/g) \quad \text{and:}$$

$$S = \int_0^{\overline{p_s}} \hat{S}(\mathrm{d}p/g)$$

The flux vector may be decomposed into its rotational and divergent parts with associated stream function and velocity potential:

$$F = F_R + F_D = k \times \nabla\psi + \nabla\chi$$

The divergent component connects sources and sinks in the flow:

$$\nabla \cdot F_D = S$$

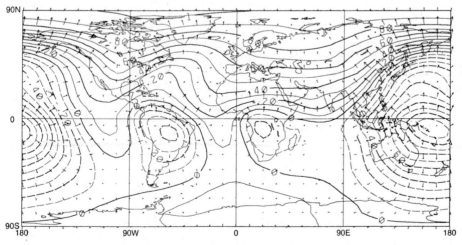

Fig. 2. Divergent part of the energy flux vector E_D and the associated potential function for December–February model climatology. Units $10^6 \ \mathrm{W \, m^{-1}}$ and $10^{15} \ \mathrm{W}$, respectively.

The flux of energy in the atmosphere (not including latent energy associated with water vapour) is:

$$E = \int_0^{\overline{p_s}} (C_p T + \phi) V (dp/g) \tag{1}$$

where the small contribution from kinetic energy is neglected. The divergent component of this vector and the associated potential function are shown in Fig. 2 for the model climatology. This diagram indicates that the flow of energy out of the source regions in the tropics is directed primarily poleward and toward the cold interiors of the continents in the Northern Hemisphere and toward regions of cold sea-surface temperatures such as occur in the eastern tropical Pacific in the absence of an El Niño. In the tropics at least, this net energy flux vector is closely associated with the divergent component of the flow in the upper troposphere as can be seen from Fig. 3. This is explained by the dominance of the potential energy in the integral 1 at these levels (Boer and Sargent, 1984).

The source of energy in the tropics comes largely from latent heat release as may be inferred from Fig. 4 in which the divergent component of the vertically integrated moisture flux vector:

$$Q = \int_0^{\overline{p_s}} q V (dp/g)$$

is shown. The divergence of the flux:

$$\nabla \cdot Q_D = E - P$$

is the evaporation minus the precipitation and sink regions of atmospheric moisture are closely related to source regions of atmospheric energy in the tropics.

The divergent part of the moisture flux vector is related to the divergent component of the flow at low levels in the tropical troposphere since moisture is concentrated at these levels.

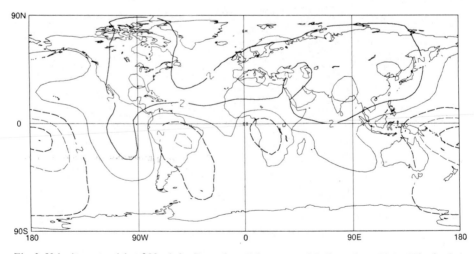

Fig. 3. Velocity potential at 200 mb for December–February model climatology. Units 10^6 m^2 s^{-1}.

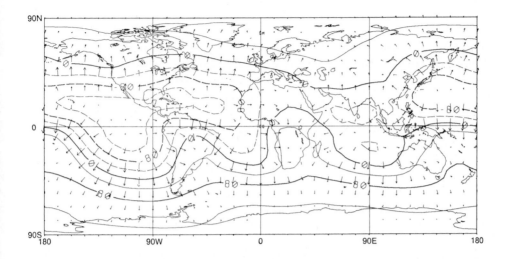

Fig. 4. Divergent part of the moisture flux vector Q_D and the associated potential function for December–February model climatology. Units kg m^{-1} s^{-1} and 10^6 kg s^{-1}.

The resulting picture in the tropics is of the convergence of moisture associated with the divergent component of the low-level flow into tropical regions of precipitation. This is accompanied also by a convergence of internal energy. The associated release of latent heat and rising motion convert this energy into potential energy so that there is a net energy divergence from these regions in conjunction with the upper tropospheric divergent flow. These divergent flow components comprise the Hadley and Walker circulations in the tropics. It is this balance which is disturbed by the El Niño.

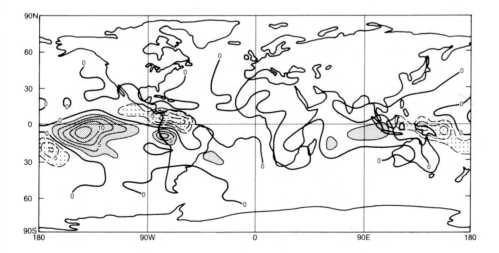

Fig. 5. Precipitation anomaly for December–February associated with the model response to the El Niño sea-surface temperature anomaly. Units 10^{-5} kg m^{-2} s^{-1} (10^{-5} kg m^{-2} s^{-1} = 0.864 mm per day).

12

The El Niño case

One of the most striking of the El Niño effects in the tropical region is the alteration of the precipitation pattern. The model precipitation anomaly is shown in Fig. 5 where an increase in the eastern and decrease in the western Pacific are seen. Note also the increase in the Indian Ocean region in response to the sea-surface temperature anomaly there. This precipitation anomaly agrees well with that implied by the outgoing long-wave radiation anomaly which is related to tropical precipitation (Arkin et al., 1983) and shown in Fig. 6.

The anomaly in precipitation develops in response to an anomaly in the flow and in the transport of moisture. The anomaly in the divergent moisture flux vector is shown in Fig. 7 and clearly indicates the change in the divergent flow.

As discussed above, the anomaly in the divergent moisture flux vector and hence in precipitation affects the distribution of sources and sinks of energy for the atmosphere

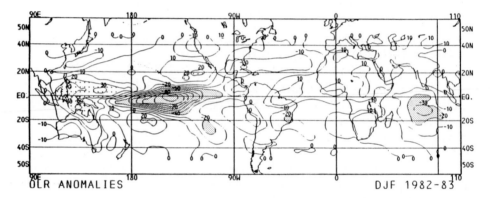

Fig. 6. OLR anomaly for December–February 82/83 from Arkin et al. (1983). Note that this chart is centered at 90°W rather than at 0.

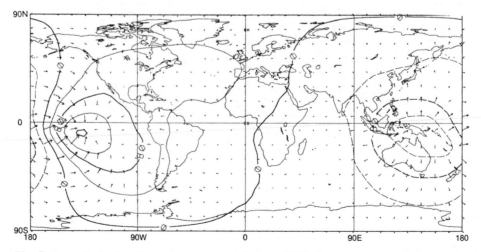

Fig. 7. Anomaly in the divergent component of the moisture flux vector Q_D and the associated potential function for December–February from the model. Units $kg\,m^{-1}\,s^{-1}$ and $10^6\,kg\,s^{-1}$.

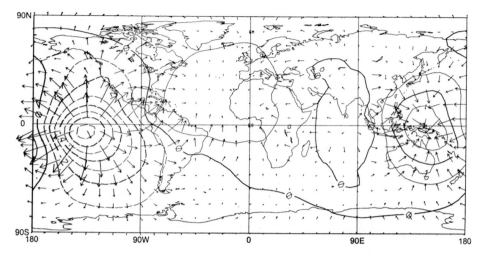

Fig. 8. Anomaly in the divergent component of the energy flux vector E_D and the associated potential function for December–February from the model. Units $10^6 \, \mathrm{W \, m^{-1}}$ and $10^{14} \, \mathrm{W}$.

and will be reflected in anomalous energy transport. This is shown in Fig. 8. Taken together these diagrams indicate the "anti-Walker" component of the flow induced by the sea-surface temperature anomaly as well as the modification to the Hadley circulation. These divergent components of the flow and of the transport vectors form a consistent picture which agrees essentially with the ideas of Bjerknes (1969). Some of these effects are discussed by the author in a previous data study (Newell et al., 1974, chapter 9).

THE ROTATIONAL COMPONENT

The dominant feature of the rotational component of the flow at tropical latitudes is an anticyclonic dipole at 200 mb associated with the region of divergent mean flow of

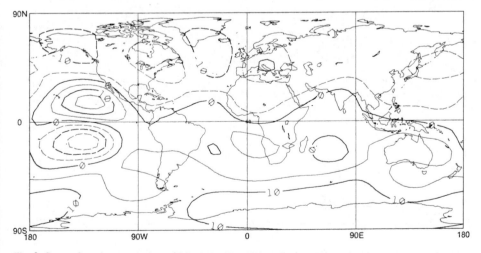

Fig. 9. Streamfunction anomaly at 200 mb for December–February from the model. Units $10^6 \, \mathrm{m^2 \, s^{-1}}$.

14

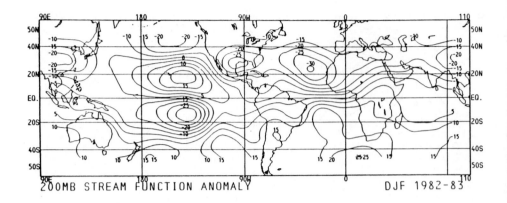

Fig. 10. Observed streamfunction anomaly at 200 mb for December–February 82/83 from Arkin et al. (1983). Units 10^6 m^2 s^{-1}. Note that this chart is centred at 90°W rather than at 0.

mass at that level and of vertically integrated energy. This feature is well reproduced in the simulation (Fig. 9) as compared to the observed values of Fig. 10. The associated equatorial region of strong flow near Africa is also present although weaker than observed, as is the cyclonic flow region in the western Pacific. It must be recalled that the mean of five simulations of the 82/83 El Niño is being compared to a single observed case. Presumably the atmosphere would react somewhat differently to the same El Niño if it were to occur again so that it is difficult to be specific as to just how similar the five experiment composites should be to the 82/83 case. Presumably the strong features directly associated with the region of anomalous sea-surface temperature would be a constant of similar El Niños.

The extratropical response is apparently equivalent barotropic in nature as the same patterns are seen at the surface and in the troposphere. For extratropical regions, the 500 mb height field is approximately a streamline field of the flow. The observed 82/83

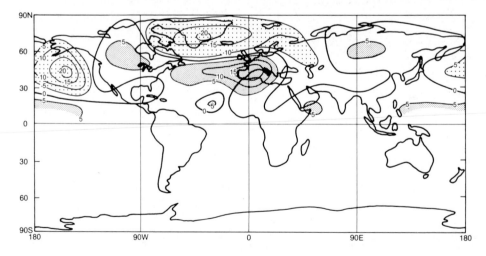

Fig. 11. Observed anomaly in the 500 mb height field for January 1983. Units dm.

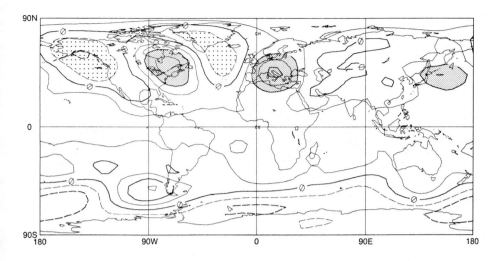

Fig. 12. Anomaly in the 500 mb geopotential field ϕ for December–February from the model. Units 10^2 m² s⁻². Note that the contour interval and shading differ from the previous diagram.

anomaly in the 500 mb height field is shown in Fig. 11 and that of the model in Fig. 12. For comparison the 1000 mb observed composite anomaly pattern from Pan and Oort (1983) is shown in Fig. 13. Since the flow is equivalent barotropic this pattern is also appropriate to 500 mb although with different amplitude.

It is notable that the 82/83 anomaly, both as observed and as modelled, is quite different from the "P & O" pattern of Fig. 13. In particular, the north–south pattern of high and low pressure in the eastern Atlantic is replaced by a strong anomaly pattern of opposite sign.

Apparently the strong El Niño of 82/83 is not typical of the composite anomaly of the

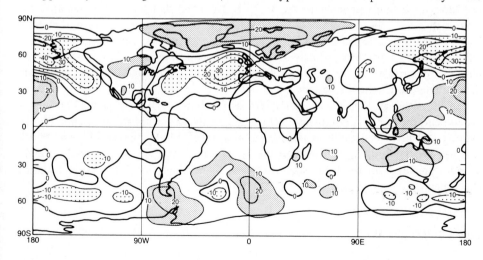

Fig. 13. Composite 1000 mb height anomaly pattern for December–February from Pan and Oort (1983). Units m.

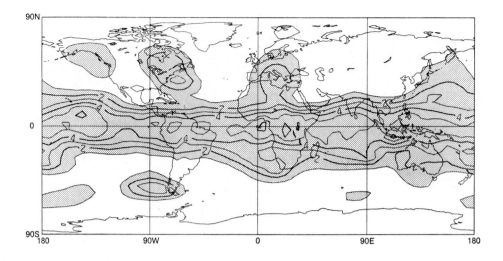

Fig. 14. The "reliability ratio" for the univariate t-test statistic given by $R = |t|/t_{0.5}$, that is by the ratio of the t statistic to the 5% significance value. Shaded areas indicate significant values for which $R > 1$.

Pan and Oort study either for the atmosphere or for the model. Other simulation studies performed with the model give a pattern consistent with the Pan and Oort pattern when the climatological composite sea-surface anomaly pattern of Rasmusson and Carpenter (1982) is used or even when twice this anomaly is used.

Thus not only is the observed flow for the 82/83 El Niño different from the "composite" case, the model results differ as well. The agreement of the simulated with the observed 82/83 case as shown in Figs. 11 and 12 is surprisingly good although the simulated amplitudes are low. The conclusion is that both the atmosphere and the model responded in a similar manner to the 82/83 El Niño and that this response is different from the weaker "climatological" El Niño case. Apparently the response is not, therefore, a simple one where the Pan and Oort pattern is seen with increased amplitude for increasing sea-surface temperature anomaly. Of course the sea-surface temperature anomaly does not directly measure the position and magnitude of the forcing and this must be kept in mind in interpreting the simulations.

Finally, although the number of degrees of freedom is relatively large (13) for this experiment, the results of a statistical test using the univariate t statistic as given in Fig. 14 indicate that model variability is relatively large compared to the imposed El Niño signal. It is somewhat surprising perhaps, that the "significant" areas are no larger and certainly do not compare with the large values obtained by Blackmon et al. (1983) in a related experiment.

CONCLUDING REMARKS

The 82/83 "El Niño" winter is simulated using the CCC GCM with observed SST's as lower boundary conditions over the tropical oceans. The divergent component of the flow in the tropics reacts in a direct way to the anomalous sea-surface temperatures resulting in an increase of precipitation in the eastern tropical Pacific and a compensating decrease in the western Pacific similar to that which was observed. This precipitation change is associated with enhanced convergence of moisture into the region of enhanced precipitation and divergence of moisture from the regions of decreased precipitation. As well, the region of enhanced precipitation is a region of anomalous export of atmospheric energy while the reverse is the case on the other side of the Pacific. These changes are associated in a direct way with changes in the Walker and Hadley circulations.

Some aspects of the rotational flow in the tropics agree with observations notably the anticylonic dipole at 200 mb over the region of enhanced precipitation. In the extra-tropical region there is some agreement between the 500 mb height anomalies which are modelled and observed. The notable result is that both differ in a major way from the Pan and Oort composite pattern indicating the different response in both model and atmosphere to the very strong El Niño forcing of 1982/83.

ACKNOWLEDGEMENTS

Mike Lazare set up the modified sea-surface temperatures and ran the simulations. The results could not have been obtained without his help. Lynda Smith produced the manuscript and Brian Taylor aided with the figures.

REFERENCES

Arkin, P.A., Kopman, J.D. and Reynolds, R.W., 1983. 1982–1983 El Niño/Southern Oscillation Event Quick Look Atlas. NOAA/National Weather Service, Washington, D.C.

Bjerknes, J., 1969. Atmospheric teleconnections from the equatorial Pacific. Mon. Weather Rev., 97: 163–172.

Blackmon, M.L., Geisler, J.E. and Pitcher, E.J., 1983. A general circulation model study of January climate anomaly patterns associated with interannual variation of Equatorial Pacific Sea surface temperatures. J. Atmos. Sci., 40(6): 1410–1425.

Boer, G.J. and Sargent, N.E., 1984. Vertically integrated budgets of mass and energy for the globe. Submitted for publication.

Boer, G.J., McFarlane, N.A., Laprise, R., Henderson, J.D. and Blanchet, J.-P., 1984a. The Canadian Climate Centre spectral atmospheric general circulation model. Atmos.-Ocean, 22(4): 397–429.

Boer, G.J., McFarlane, N.A. and Laprise, R., 1984b. The climatology of the Canadian Climate Centre GCM as obtained from a five year simulation. Atmos.-Ocean, 22(4): 430–473.

Newell, R.E., Kidson, J.W., Vincent, D.G. and Boer, G.J., 1974. The General Circulation of the Tropical Atmosphere and Interactions with Extratropical Latitudes, Vol. 2. MIT Press, Cambridge, Mass., 371 pp.

Pan, Y.-H. and Oort, A.H., 1983. Global climate variations connected with sea surface temperature anomalies in the Eastern Equatorial Pacific Ocean for the 1958–73 period. Mon. Weather Rev., 111: 1244–1258.

Rasmusson, E.M. and Carpenter, T.H., 1982. Variations in tropical sea surface temperature and surface wind fields associated with the Southern Oscillation/El Niño. Mon. Weather Rev., 110: 354–384.

CHAPTER 3

SENSITIVITY OF JANUARY CLIMATE RESPONSE TO THE POSITION OF PACIFIC SEA-SURFACE TEMPERATURE ANOMALIES

MAURICE L. BLACKMON

This paper summarizes the results of a set of equatorial Pacific sea-surface temperature (SST) anomaly experiments which were run using the Community Climate Model (CCM) developed at the National Center for Atmospheric Research. The CCM is a spectral general circulation model of moderately coarse resolution with nine vertical levels and rhomboidal truncation with fifteen waves horizontally. Boundary conditions are quite simple, and in particular, the sea-surface temperatures are held constant throughout each experiment. For details of the model formulation and documentation of the model performance see Pitcher et al. (1983).

In each of the experiments discussed below, the model has been run in a perpetual January mode for 1200 days with a different SST distribution. From each of these runs we skip 60 days, average data for 90 days, skip 60 days, average for 90 days, etc., and therefore have an eight-sample ensemble of 90-day averaged data. Each sample within each run is presumed to be independent of the others because of the gap of 60 days between any two samples. We calculate the difference of the ensemble means (720-day averages) between any of the sensitivity experiments and the model climatology. We can also calculate the standard deviation of the 90-day means and the Student t-statistic. For the number of samples in our calculation, values of t of 2.15 (2.98) are significant at the 5% (1%) level.

In Fig. 1a is shown the SST anomaly used for the "twice-as-warm" case. This anomaly is twice as large as that presented in Rasmusson and Carpenter (1982) as the composite of the mature phase of six El Niños. Blackmon et al. (1983) found that this anomaly produced realistic climate anomalies and offered a rationale for using the "twice-as-warm" forcing, rather than the SST anomaly of Rasmusson and Carpenter. The total SST distribution, climatology plus anomaly, is shown in Fig. 1b. Note the area of heavy precipitation near 165°W, shown in Fig. 1c, where the SST distribution is a maximum. Figure 1d shows the precipitation anomaly (experiment minus control) which has a maximum of 8.7 mm per day near 165°W. In Blackmon et al. (1983) a variety of fields were examined and both the tropical and mid-latitude responses to the anomalous forcing shown in Fig. 1 were discussed. Here we will concentrate on the 200-mb geopotential height field only, but mention other fields in passing.

Figure 2a shows the 200 mb height anomaly. Near 165°W we see a pair of anticyclones straddling the equator. The rest of the tropics also shows positive anomalies, although not so large as those near the SST anomaly. Figure 2b shows the t-statistic. In the tropics we see values of $t = 9$ or 10 near the SST anomaly and large areas of $t > 3$. All this is highly significant.

20

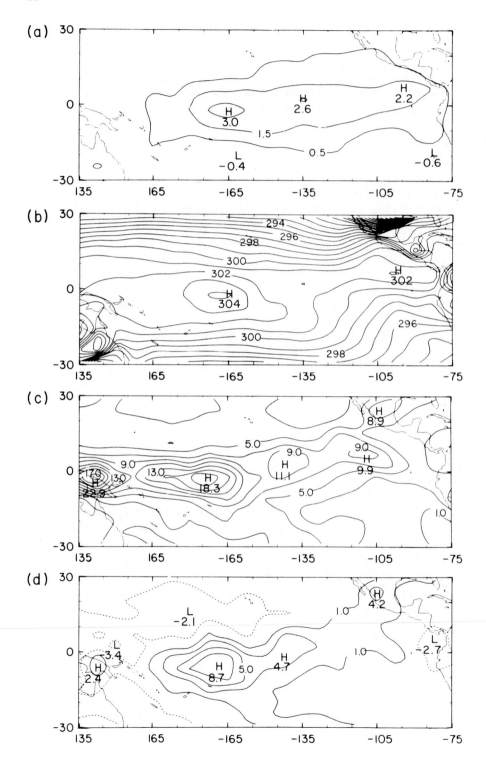

21

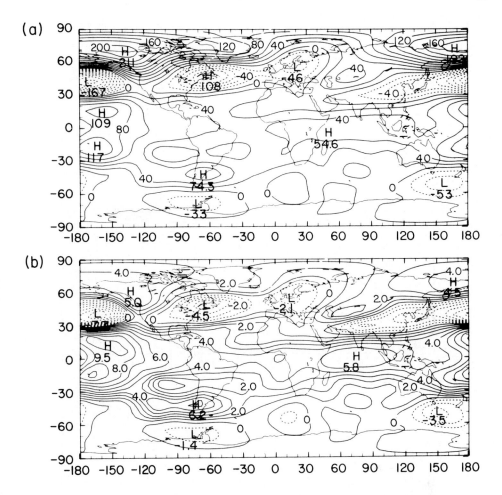

Fig. 2. (a) 200-mb height anomaly for the "twice-as-warm" case. Contour interval 20 m. (b) t-statistic for 200-mb height. Contour interval 1.

In the Northern Hemisphere mid-latitudes we see a wave pattern starting near 160°W and arcing across North America. This is the model version of the Pacific/North American pattern discussed by Wallace and Gutzler (1981) and Horel and Wallace (1981). The amplitude of the anomaly is substantial and the region where the t-statistic (Fig. 2b) is greater in absolute value than three is sizeable. The mid-latitude response is also highly significant, but only over the Pacific and North America. The response over Eurasia is less significant.

Fig. 1. (a) Sea-surface temperature anomaly for the "twice-as-warm" composite case of Blackmon et al. (1983). (b) Total sea-surface temperature distribution for the "twice-as-warm" case. (c) Total precipitation in the "twice-as-warm" case. Contour interval 2 mm per day. (d) Precipitation anomaly in the "twice-as-warm" case. Contour interval 2 mm per day.

22

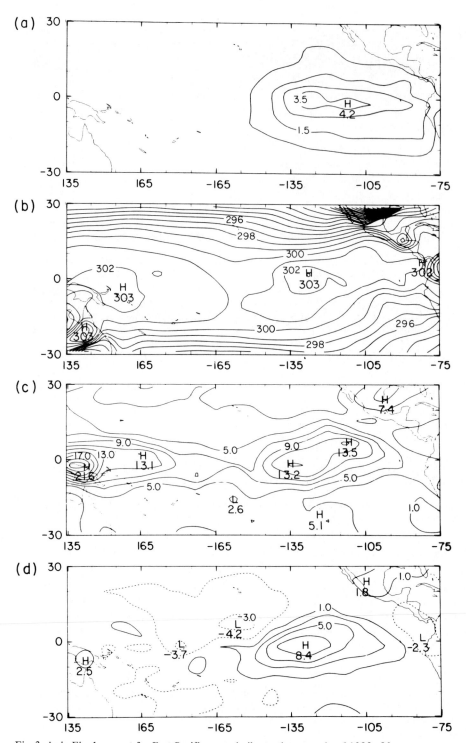

Fig. 3. As in Fig. 1, except for East Pacific case, similar to the anomaly of 1982–83.

In Blackmon et al. (1983), the climate anomalies for other fields are discussed as well. The sea-level pressure field shows a well-developed Southern Oscillation signature, with a low-pressure anomaly from the dateline to South America and a high-pressure anomaly from the dateline to the Indian Ocean. The 200-mb zonal wind in the tropics has an easterly anomaly to the west of, and westerly anomaly to the east of, the SST anomaly. The mid-latitude Pacific jet is shifted to the south by about $10°$ lat. The lower troposphere temperature fields show significantly colder than normal temperatures over the eastern United States and Canada. (The temperature anomaly fields have their maximum amplitude some $5°-10°$ lat. farther north than those in the observations.) The 10-mb height and wind fields also show significant anomalies, with higher than normal heights and weaker than normal zonal winds near the North Pole associated with warm SST anomalies. All of these features are in agreement with observations.

An important result in the "twice-as-warm" case is that each 90-day sample in the ensemble shows the same basic response. There is some variation in the strength of the wave pattern and some variation in the position of the nodes in the mid-latitude wave

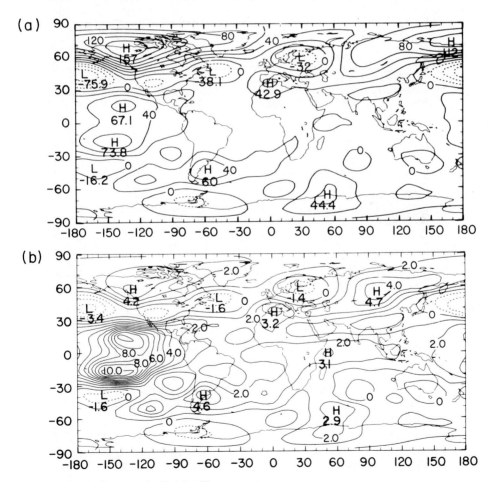

Fig. 4. As in Fig. 2, except for East Pacific case.

24

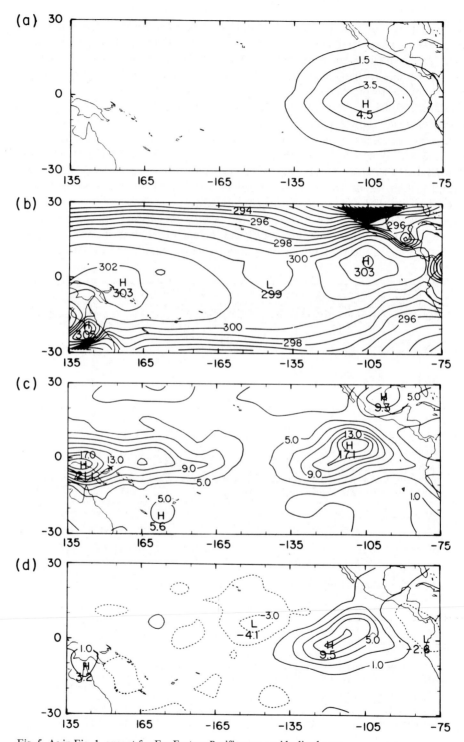

Fig. 5. As in Fig. 1, except for Far Eastern Pacific case, an idealized case.

pattern. However, there is always a positive anomaly over western North America and negative anomalies over the North Pacific and eastern North America. The consistent appearance of this Pacific/North American anomaly is not in agreement with observational work and is not consistent with the observed anomaly for the winter 1982–83.

One significant difference between the 1982–83 SST anomaly and the "twice-as-warm" anomaly shown in Fig. 1 is that the 1982–83 SST anomaly had a maximum near $110°–135°W$. The anomaly in Fig. 3a was taken from a preliminary analysis by the Climate Analysis Center for December 1982. The total SST distribution is shown in Fig. 3b. Because of the decreasing SST from west to east along the equator, the total SST distribution has a maximum a little to the west of the maximum anomaly. The rainfall anomaly (Fig. 3d) is maximum near the SST maximum.

In Fig. 4a is shown the 200-mb height anomaly for this case, called the East Pacific case. The tropical response has a maximum near $135°W$, on either side of the maximum precipitation anomaly. It is also a little smaller than that of the first case shown in Fig. 2.

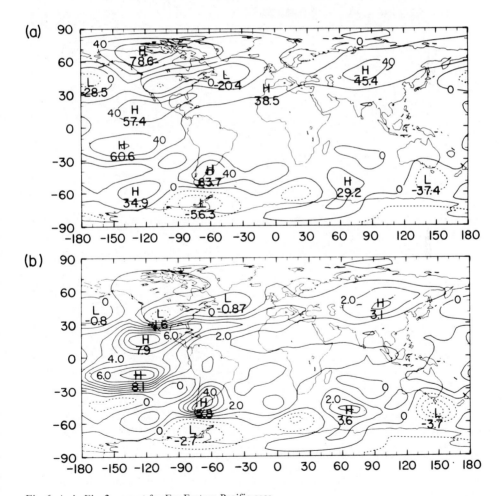

Fig. 6. As in Fig. 2, except for Far Eastern Pacific case.

The map of t-statistic (Fig. 4b) shows that the tropical response is still highly significant. The mid-latitude response is different in character from the tropical response in that it is still in the same geographical location as in the "twice-as-warm" case. It is weaker in amplitude and the map of t-statistic shows that the anomaly pattern is marginally significant.

Yet another experiment, called the Far East Pacific case, was run with the SST anomaly shown in Fig. 5a. The SST distribution, precipitation distribution and precipitation anomaly are shown in Figs. 5b, c and d, respectively. In Fig. 6a we see the resulting 200-mb height anomaly. Again the tropical response has moved to the east, compared to the two previous cases, but this time the maximum tropical anomaly is not exactly straddling the maximum precipitation anomaly. The mid-latitude response is again the usual Pacific/

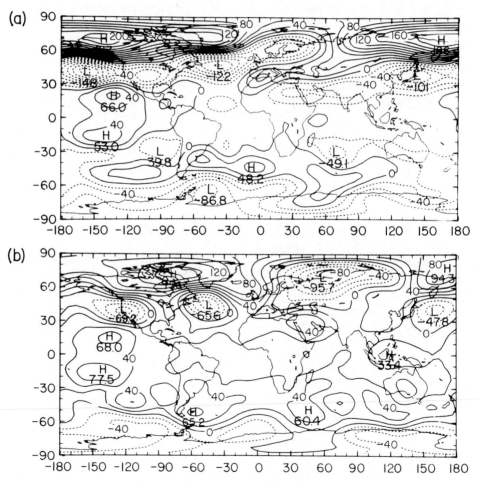

Fig. 7. Two examples of 90-day averaged 200-mb height anomalies from the East Pacific case. Panel (a) shows a response similar to the long-term average response while panel (b) shows a response almost in quadrature with the long-term average. The anomaly shown in panel (b) has some common features with the observed anomaly of 1982–83.

North American anomaly. It is still geographically fixed and the long-term average amplitude is reduced from the previous two cases. The map of *t*-statistic shows that the response is significant in the tropics, but not significant in mid-latitudes.

Examining each 90-day anomaly for the latter two cases, we find that some samples have strong Pacific/North American anomalies. Others have distorted anomalies or even anomalies with opposite signs to the long-term average. This is in contrast to the consistent response in the "twice-as-warm" case. Figure 7 shows two 90-day average anomalies from the East Pacific experiment. In the first example, Fig. 7a, the pattern resembles the long-term average anomaly shown in Fig. 4a. In Fig. 7b, on the other hand, we see a pattern with a negative anomaly centered on the west coast of North America and a positive anomaly over Hudson Bay. This second pattern has some features of the observed anomaly pattern for 1982–83. However, the most likely pattern to occur, as shown in Fig. 4a, is the Pacific/North American pattern.

These results seem most consistent with the idea that mid-latitude anomalies are a reflection of the consistent stimulation of a barotropically unstable mode. This mode converts energy from the mean Pacific jet into the anomalous eddies (see Simmons et al., 1983, for a complete discussion). The fact that the mid-latitude response is fixed geographically is consistent with the unstable mode concept and inconsistent with the idea of wave trains emanating directly from a tropical source, as proposed, for example, by Hoskins and Karoly (1981). Individual 90-day samples from the East Pacific and Far East Pacific experiments look as if they could be wave trains coming from the tropics. Figure 7b is one example. However, these results suggest that, over the long term, the stimulation of the unstable mode dominates. For a more detailed discussion see Geisler et al. (1985).

REFERENCES

Blackmon, M.L., Geisler, J.E. and Pitcher, E.J., 1983. A general circulation model study of January climate anomaly patterns associated with interannual variation of equatorial Pacific sea surface temperatures. J. Atmos. Sci., 40: 1410–1425.

Geisler, J.E., Blackmon, M.L., Bates, G.T. and Muñoz, S., 1985. Sensitivity of January climate response to the magnitude and position of a warm equatorial Pacific sea surface temperature anomaly. J. Atmos. Sci., 42, in press.

Horel, J.D. and Wallace, J.M., 1981. Planetary-scale atmospheric phenomena associated with the Southern Oscillation. Mon. Weather Rev., 109: 813–829.

Hoskins, B.J. and Karoly, D.J., 1981. The steady linear response of a spherical atmosphere to thermal and orographic forcing. J. Atmos. Sci., 38: 1179–1196.

Pitcher, E.J., Malone, R.C., Ramanathan, V., Blackmon, M.L., Puri, K. and Bourke, W., 1983. January and July simulations with a spectral general circulation model. J. Atmos. Sci., 40: 580–604.

Rasmusson, E.M. and Carpenter, T.H., 1982. Variations in tropical sea surface temperature and surface wind fields associated with the Southern Oscillation/El Niño. Mon. Weather Rev., 110: 354–384.

Simmons, A.J., Wallace, J.M. and Branstator, G., 1983. Barotropic wave propagation and instability, and atmospheric teleconnection patterns. J. Atmos. Sci., 40: 1363–1392.

Wallace, J.M. and Gutzler, D.S., 1981. Teleconnections in the geopotential height field during the Northern Hemisphere winter. Mon. Weather Rev., 109: 784–812.

CHAPTER 4

SUBSEASONAL SCALE OSCILLATION, BIMODAL CLIMATIC STATE AND THE EL NIÑO/SOUTHERN OSCILLATION

K.-M. LAU

ABSTRACT

In this paper, we present some observational evidence showing the presence of a low-frequency oscillation in tropical convection that may be related to the 40–50-day tropical wind oscillation revealed by earlier studies (e.g. Madden and Julian, 1972). We suggest that this oscillation is closely related to a bimodal climate state exhibited by the tropical ocean–atmosphere system, with the El Niño/Southern Oscillation (ENSO) occurring as an amplification of one of these climatic modes through large scale air–sea interaction. To illustrate the stochastic nature of this interaction, a simple model of a randomly forced nonlinear oscillator is constructed. We find that there is a remarkable resemblance between the long-term temporal variability of the model output and that of the Southern Oscillation. This resemblance suggests that the episodic occurrence of ENSO may be a manifestation of an instability in the tropical ocean–atmosphere system triggered by stochastic forcing and modulated by the deterministic component (i.e. the seasonal cycle) of the coupled system.

INTRODUCTION

Recent studies indicate the existence of a bimodal climatic state in the Pacific tropical ocean–atmospheric system in connection with the occurrence of large positive sea-surface temperature (SST) anomaly (El Niño) and large negative SST anomaly (anti-El Niño) conditions (Meyers, 1982; Wyrtki, 1982; Barnett, 1983; Lau and Chan, 1983a, b). While the duration (or residence time) of each of the climatic mode is quite uniform in the order of 18–24 months, its occurrence appears to be rather episodic with a temporal separation ranging from 2 to 7 years. An intriguing feature concerning the El Niño/ Southern Oscillation (ENSO) is that although the occurrence of an ENSO is quite random, the onset of each event appears to coincide with specific phases of the annual cycle (Wyrtki, 1975; Rasmusson and Carpenter, 1982). This suggests that the annual cycle plays an essential role in the timing of the onset of the ENSO. The abruptness of the onset, and the apparent strong tendency of the oceanic and atmospheric components in mutually adjusting and amplifying each other during the course of the ENSO, suggest the presence of an instability in the tropical ocean–atmosphere system. Such an instability has been explored in a number of recent theoretical and numerical modelling studies (Lau, 1981, 1982; McCreary, 1983; Philander et al., 1984; McCreary and Anderson, 1984, 1985; and others). While the details are still being investigated it is now generally agreed that such an instability exists and is responsible for the rapid onset of an ENSO. The above discussion suggests that the long-term variability of the ENSO possesses a strong stochastic component which is modulated by the seasonal cycle and possibly by inherent instabilities in the coupled system.

In this paper, we first discuss some observational evidence that further supports the notion of multiple climatic states (equilibria) in the tropical ocean–atmosphere system. Then we use a simple analog model, in the form of a nonlinear mechanical oscillator, to demonstrate a possible way in which the stochastic and deterministic components of the tropical air–sea system may interact to produce episodic coupled oscillations. Preliminary model results show a surprising resemblance to the characteristics of the Southern Oscillation. A full discussion of the model results, including the formulation of a stochastic-dynamical theory of ENSO, will be reported elsewhere (Lau, 1985).

OBSERVATIONS

Observational evidence suggesting possible relationships between subseasonal and ENSO-scale fluctuations are presented in this section. The primary data used are NOAA global outgoing longwave radiation (OLR) fluxes on a $2.5° \times 2.5°$ grid from 1974–83. For this study, we only focus on the tropics, where OLR is known to be a good indicator of convective cloudiness and precipitation (Lau and Chan, 1983a, b). Monthly and five-day mean data are used, and the climatological seasonal cycle is removed from both data sets.

Dipolar oscillation in tropical convection

Figure 1 shows the time–longitude section of monthly OLR averaged between 10°S and the equator from 60°E to 60°W for the entire data period. Here, large negative deviations of OLR are due to anomalously cold cloud top temperatures associated with deep convection and enhanced precipitation, and the reverse for positive deviations. In the following discussion, negative (positive) anomalies will be used to mean wet (dry) condition in a relative sense. There are two ENSO events during the data period, one in 1976–77 and the other in 1982–83. During these ENSO's, the most obvious feature is the large amplitude of a dipole-like anomaly corresponding to wet conditions over the equatorial central Pacific (170°E–140°W) and dry condition over the maritime continent of Indonesia and New Guinea (120°–150°E). Also apparent are the fluctuations in the form of a reversal of the polarity of this dipole anomaly on shorter (or subseasonal) time scales. Interspersed with these subseasonal fluctuations, the dipole anomaly during 1974–75 was completely reversed from those typically observed during ENSO, suggesting an "anti-ENSO" condition was prevailing prior to the 1976–77 event. During the 1976–77 winter, the subseasonal fluctuations near the dateline (represented by two reversals in the sign of the anomaly within a 4–5 month duration) was most apparent before the atmosphere settled down to the amplified dipole state of 1977–78. The period 1979–81 saw a succession of these small-scale reversals prior to the 1982–83 ENSO, which begins around June 1982 following a substantial anti-ENSO anomaly between September and December of 1981.

In order to focus on these subseasonal oscillations, we have further removed the inter-annual variability of the OLR by subtracting the individual seasonal mean trend from the daily data and used a 5-day running mean on the remaining data. The detailed analyses of these cloud fluctuations and their connection with the 40–50-day waves are reported

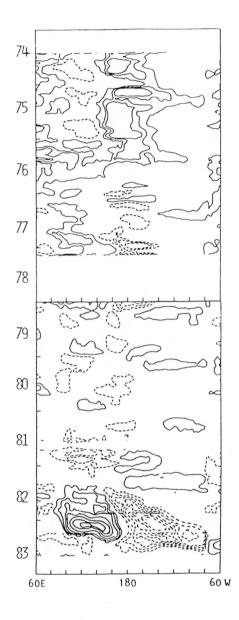

Fig. 1. Time–longitude section of monthly anomalous *OLR* averaged between 10°S and the equator. Contour interval 10 W m^{-2}. Negative contours indicating enhanced convection are dashed. The zero isopleths have been removed for clarity.

in Lau and Chan (1985). One of the most interesting results from the analyses can be seen in the Hövmöller-type diagram shown in Fig. 2. An oscillation exists with a "period" of the order ~30–60 days, and shows a preferred eastward migration along the equator from the Indian Ocean to the central Pacific. Although the "periodicity" of the oscillation varies substantially from year to year and appears to be quite ill-defined for some years,

32

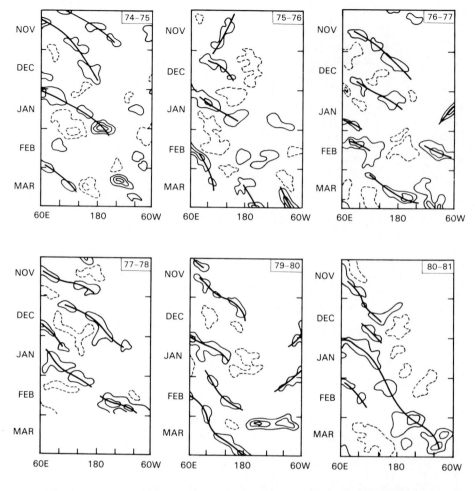

Fig. 2. Time–longitude sections of 5-day averaged *OLR* between 10°S and the equator showing pulse-like perturbations originating from the Indian Ocean. The speed of the eastward propagating signal (~3° longitude per day) is estimated from the slope of the heavy solid lines. Contour interval is $10\,W\,m^{-2}$.

we shall keep the "40–50-day" terminology for convenience. In view of the large variability in the observed "periods" and hence the difficulty in interpreting these fluctuations as regular wave phenomena, we offer in the following section an alternate interpretation of this phenomenon in terms of the notion of multiple equilibria in the tropical ocean–atmosphere system.

Fig. 3. Potential density functions (see text for explanation) of *OLR* for selected locations along the equator. Abscissa and ordinate are normalized values and its 5-day tendency, respectively. Contour unit is only relative with interval of five units.

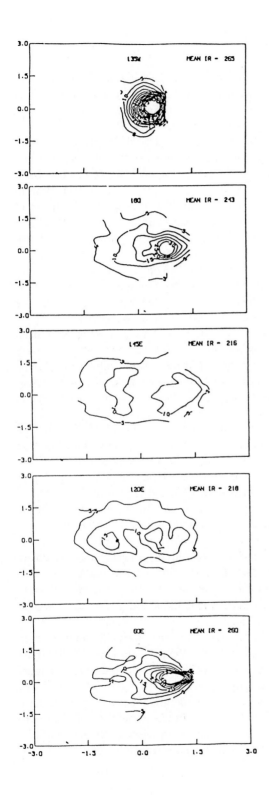

Multiple-equilibria: Bimodal climatic state

To examine the possible existence of a bimodal state in the subseasonal time scale, we have constructed the distribution of the 5-day mean *OLR* fluctuation as a function of its magnitude and tendency $(\partial\, OLR)/\partial t$ normalized by the corresponding standard deviation at different locations along the equator over the Pacific. Scatter diagrams are first constructed with magnitude on the X-axis and tendency on the Y-axis. Isopleths of constant densities are then obtained by counting the total number of data points that fall into intervals of 0.5 units in the X- and Y-direction. Figure 3 shows plots of these isopleths for the labeled longitudes. Hereafter, we shall refer to the quantity plotted as the potential density function (PDF). Each point on the PDF represents a climatic state whose stability is measured by the distance in the Y-direction from the X-axis. Stable climatic states or multiple equilibria can be identified by a clear separation of the relative maxima in the PDF close to the X-axis, where the tendency is small, i.e. stable states.

For the *OLR*-anomaly over the eastern Pacific (135°W), there is a single maximum concentrated near +0.5 on the X-axis. It follows that the atmospheric system there is dominated by a single equilibrium state. The mean value of $265\,\mathrm{W\,m^{-2}}$ suggests that the prevailing condition is either low-cloudiness or clear-sky (a value of less than $240\,\mathrm{W\,m^{-2}}$ in the tropics is generally associated with deep convection). Occasional outbreak of deep convection does occur over this region as indicated by the presence of points with negative values and large tendencies although their occurrence is too infrequent to show up in the PDF. At the dateline (180°), the dry equilibrium state is still quite well defined. Here, the tendency for frequent outbreak of deep convection is greater as can be seen in the broadening of the PDF towards negative values. This results in the dry state being found further away ($X = 1.0$) from the origin than that found at 135°W. The climato-logical value of $243\,\mathrm{W\,m^{-2}}$ implies a background only marginal in deep convection. Since the dateline lies near the boundary between the wet zone to the west and the dry zone to the east along the equator, the PDF here is likely to reflect an alternation of cloudy and clear conditions, associated with the transient fluctuation of the Pacific Walker circulation. The more diffuse contours due to deep convection (negative *OLR*) contrasting with the more localized contours due to the absence of convection (positive *OLR*) gives rise to the somewhat asymmetric distribution with respect to positive and negative anomalies.

The PDF's over the western Pacific (145°E) and Indonesia (120°E) are markedly different from the previous two. The fact that there are two relative maxima near the X-axis is indicative of the presence of a bimodal climatic state. The climatological mean *OLR* of $216-218\,\mathrm{W\,m^{-2}}$ suggests a cloudy background and accounts for the somewhat symmetric appearance of the PDF's in these locations.

The bimodal signal diminishes gradually further westward towards the Indian Ocean (60°E), where the PDF is structurally similar to that at the dateline. Thus the longitudes of 60° and 180°E appear to mark the east–west limit of the bimodal signal. Similar analyses applied to different latitudes indicate the signal is strongest within this east–west limit between 10°S and 5°N, suggesting that the structure of these climatic modes is geographically fixed. The reversal in the monthly anomalies in convection over Indonesia and the dateline respectively (Fig. 1) can now be viewed in terms of the recurrence of these intrinsic climatic modes which sometimes may intensify into an ENSO.

AIR–SEA INTERACTION

Although the origin of the 40–50 day oscillation is still not clear, it is interesting to note that evidences of oscillations of similar periods have been found in the surface-wind-driven current (McPhaden, 1982) but not in SST in the Indian Ocean. Atmospheric and oceanic observations of ENSO so far indicated that enhanced convection over the maritime continent is associated with a strong Walker circulation, a strong east–west SST gradient, and reduced convection over the equatorial central Pacific. On the other hand, enhanced convection over the equatorial central Pacific is associated with a weak Walker circulation and a reduced east–west SST gradient. The two atmospheric conditions mentioned above coincide well with the two opposite polarities of the dipolar convection anomaly in the 40–50-day oscillation. The main difference is that the ENSO is known to be a strongly coupled ocean–atmosphere phenomenon, whereas for the 40–50-day oscillation, the role of the ocean has yet to be identified.

The presence of coupled ocean–atmosphere modes, as suggested by many studies of ENSO, is supported by recent theoretical and modelling studies. Lau (1981) showed that large-scale coupled Kelvin wavelike oscillations in the ocean and atmosphere can become unstable under strong air–sea coupling conditions. Philander et al. (1984) suggested that the migration and amplification of warm SSTA and atmospheric convergence during ENSO is a result of unstable air–sea interaction. McCreary and Anderson (1984a, b) demonstrated that the interaction between the ocean and a two-state atmosphere can lead to successive generation of El Niño-like events in the ocean. These studies and the observational evidence discussed above have motivated the author into the following hypothesis regarding the 40–50-day oscillation and the ENSO.

A HYPOTHESIS

During the phase of the 40–50-day cycle when enhanced convection occurs over the equatorial central Pacific, a surface, westerly inflow will be driven to the west of the convective source (Gill, 1980; Lau and Lim, 1982) thus providing a condition favourable for the development of ENSO oceanic anomaly by remote forcing (cf. Wyrtki, 1975; Busalacchi et al., 1983). It is possible that an ENSO may be the result of an amplification of this phase of the 40–50-day wave. We hypothesize that such an amplification can be achieved through the triggering of an instability in the ocean–atmosphere system.

While the 40–50-day oscillations are present all year around (Madden and Julian, 1972), it is known that the large anomalies in SST during ENSO begin in December to February. To account for this phase-locking behavior, the seasonal cycle of the mean state of the ocean–atmosphere system (e.g., trade wind and SST) must be changing in such a way that instability is favourably triggered only during a particular season. Since the instability involves an interaction between the seasonal mean and the subseasonal-scale fluctuations, the process is basically nonlinear. Further, there is also the possibility that large fluctuations in the amplitude of the 40–50-day oscillation could trigger the instability during other times of the year. Thus, the nonlinear interaction between an instability mechanism and stochastic forcings in an ocean–atmosphere system may also provide a plausible explanation to the aperiodic occurrence of ENSO and the large variability between individual events.

We should stress that the stochastic forcing that we referred to above is not only restricted to the 40–50-day oscillation. The observation that the 40–50-day oscillation appears to be an ENSO in miniature only suggests that it is perhaps the "most" probable trigger for the above type of instability. Indeed we suggest that in the time scale of the ENSO, all subseasonal variability including the 40–50-day waves can be considered as white noise (cf. Hasselmann, 1976). These ideas are explored in the next section with a very simple model of a nonlinear oscillator.

A SIMPLE NONLINEAR OSCILLATOR RELEVANT TO ENSO

In this section we discuss a nonlinear oscillator model that provides a useful mechanical analog to the ENSO climate system. To assist in the comparison of model with observations, we put in brackets next to the component of the oscillator, the corresponding feature identifiable in the tropical ocean–atmosphere system described in the previous sections.

Consider the mechanical system depicted in Fig. 4. A spring having a mass M attached to its end is freely suspended from a rigid support which can take on any of the positions C (mean state), and two fixed positions K_1 and K_2 (multiple-equilibria). The system composed of the spring, the support and the mass (climate system) is allowed to oscillate with its natural frequency ω_0 (annual cycle) such that on its upward motion, when the mass M passes through the threshold position X_1, an electronic device will cause the support to move rapidly into position K_1 (instability triggering). However, this position is quasi-stable so that the support position, if undisturbed, will relax back to C in a time τ_1 (residence time). Similarly, on its downward motion, when the mass passes through a second threshold X_2, a trigger will be set off to send the support to the position K_2, which will again automatically relax back to the mean position in time τ_2. Thus the mass M will either oscillate with its natural frequency (when the amplitude is small such that the

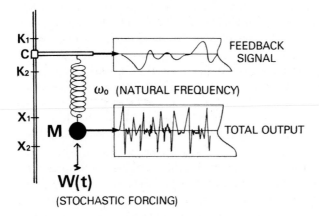

Fig. 4. A mechanical nonlinear oscillator as an analog to the ENSO climate system (see text for explanation).

oscillation takes place between the positions X_1 and X_2) or oscillate with its natural frequency but modulated by impulse-like transitions towards and away from the positions K_1 and K_2. Given the positions of X_1, X_2, K_1, and K_2 and the amplitude of the natural oscillation, the system is completely deterministic and exhibits limit-cycle oscillations. However, when M is also subjected to a random perturbation $W(t)$ of sufficiently large amplitude, the regular oscillations will be disrupted by the random forcing which causes the support position to flip between K_1 and K_2 "almost randomly" (stochastic transition). Figure 4 also shows the schematic drawing of the time traces of two pointers attached to the mass M and the support C when the modulated oscillation occurs. The former represents the total output of the system and the latter the feedback signal (climate anomaly) arising from the interaction of the natural oscillation, the stochastic forcing and the transition between equilibrium positions.

Mathematically the position of the mass M can be expressed as:

$$X(t) = A \cos \omega_0 t + \epsilon W(t) + \sum_{j=1}^{2} \sum_{i=1}^{I_j(t)} (K_j - C)\phi(t - t_{ij}) \tag{1}$$

where A and ϵ are respectively the amplitude of the natural oscillation, and the random process $W(t)$ represented by Gaussian white noise. K_1, K_2 and C represent the three possible equilibrium positions of the mass M and:

$$\phi = t e^{-t/\tau}[(\tfrac{1}{2})\Gamma(\tfrac{3}{2})\tau^3]^{-1/2} \tag{2}$$

ϕ is the impulse-like function with unit variance. In eqn. (2), Γ is the gamma function and the residence time τ is chosen to be the same for the two positions K_1 and K_2; t_{ij} is the time of occurrence of the ith impulse for the jth state determined in the manner described above; $I_j(t)$ is the total number of impulses for the jth state that have occurred up to time t and is a function of the positions $X(t)$ and the thresholds X_1 and X_2. Equation (1) is a nonlinear integral-type equation in $X(t)$ which can be solved very efficiently by numerical methods (Morse and Feshbach, 1953).

By dividing the natural period $(2\pi \omega_0^{-1})$ into twelve time intervals each equivalent to one month, we have integrated eqn. (1) for many thousands of years (12 units = 1 yr) in 50-year segments. The detailed analysis of the general system similar to eqn. (1) will be presented elsewhere. Here, we only provide one example of an output of this model to compare with the observed Southern Oscillation Index (SOI, Tahiti minus Darwin surface pressure).

Figure 5 shows five sample time plots of the model-generated anomalies (annual cycle removed) form randomly selected 50-year segments and the actual SOI from 1935–82. The SOI time series shown have been normalized with respect to the peak amplitude of the annual cycle. A model ENSO is defined to be a negative anomaly having a peak amplitude greater than unity with the negative anomaly lasting over at least one year. All the five sample-time series show 8–12 ENSO's occurring over the 50-year period, comparing well with the 11 ENSO's that actually occurred from 1935 to 83. If we define a "super-ENSO" as one that has either peak amplitude larger than 2.5 times that of the seasonal cycle and/or that lasts for over two years, we can see that two such super-ENSO's have occurred during the last 50-years (i.e., 1940–42 and 1982–83). In the sample series, one to two super-ENSO's are found to occur within each 50-year segment. Averaged over

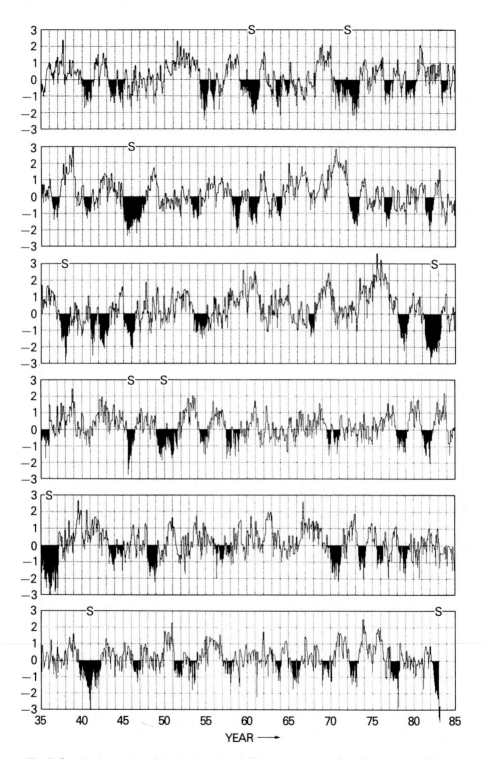

Fig. 5. Sample-time series of randomly selected 50-year segments of oscillator output. Bottom panel shows actual SOI variation from 1935–83. ENSO events are shaded and *S* denotes super-events (see text for definition).

the five 50-year periods, the model yields a probability of occurrence of super-ENSO's of about one in every 30–40 years.

While the oscillator output looks similar to observation, we should not, however, belabour with the specific justification as to what exact parameter(s) in the observation to identify with the variable $X(t)$. The choice of the SOI here is a pure matter of convenience. Instead, we should look upon eqn. (1) as representing some general characteristics of a stochastic-dynamical system. Since any system that has the basic ingredients contained in eqn. (1) will behave statistically the same, eqn. (1) provides little information on the exact details of the nature of the interaction. Hence, the interpretation of the above system in terms of *specific* interaction mechanism(s) should not be emphasized. For the ENSO system, it is natural to think in terms of the interaction between the ocean and the atmosphere. Nonetheless, we can relate specific mechanisms to the above mentioned only after they have been identified elsewhere by more detailed process studies (e.g., Lau, 1979, 1981; Philander et al., 1984; McCreary and Anderson, 1984).

CONCLUSION

We have presented some evidence of the presence of a bimodal climatic state in the tropical ocean–atmosphere system over the Pacific. Our results suggest that these modal states are realizable both in the subseasonal (in particular, the 40–50-day) time scale and the interannual time scales related to ENSO. We further postulated that the interaction between stochastic forcing represented by subseasonal transients (including but not exclusively the 40–50-day oscillation) and the deterministic forcing represented by the annual cycle could lead to episodic occurrence of large anomalies resembling ENSO events. By drawing an analogy between the behavior of a stochastically forced non-linear oscillator and the Southern Oscillation, we speculate that in the time scale of ENSO, the tropical ocean–atmosphere system may indeed be described by such a stochastic-dynamical climate system. Hopefully our result will stimulate new ideas and research in this area.

ACKNOWLEDGMENT

The author wishes to thank Dr. J. McCreary for his useful comments which helped to clarify and improve the manuscript. This research was supported in part by the NASA Global Scale Atmospheric Process Program.

REFERENCES

Barnett, T.P., 1983. Interaction of the monsoon and the Pacific trade wind system in interannual time scales. Part I: The equatorial zone. Mon. Weather Rev., 111: 756–773.
Busalacchi, A.J., Takenchi, K. and O'Brien, J.J., 1983. Interannual variability of the equatorial Pacific – revisited. J. Geophys. Res., 88: 7551–7562.

Gill, A.E., 1980. Some simple solutions to heat induced tropical circulation. Q.J. R. Meteorol. Soc., 106: 447–462.

Hasselmann, K., 1976. Stochastic climate model, Part I: Theory. Tellus, 28: 473–484.

Lau, K.-M., 1979. A numerical study of tropical large scale air–sea interaction. J. Atmos. Sci., 36: 1467–1489.

Lau, K.-M., 1981. Oscillations in a simple equatorial climate system. J. Atmos. Sci., 38: 248–261.

Lau, K.-M., 1982. A simple model of atmosphere–ocean interaction during ENSO. Trop. Ocean-Atmos. Newsl., 13: 1–2.

Lau, K.-M., 1985. Elements of a stochastic-dynamical theory of the temporal variability of ENSO. J. Atmos. Sci., in press.

Lau, K.-M. and Chan, P.H., 1983a. Short-term climate variability and atmospheric teleconnection from satellite-observed outgoing longwave radiation. I: Simultaneous relationships. J. Atmos. Sci., 40: 2735–2750.

Lau, K.-M. and Chan, P.H., 1983b. Short-term climate variability and atmospheric teleconnection from satellite observed outgoing longwave radiation. II: Lagged correlations. J. Atmos. Sci., 40: 2751–2767.

Lau, K.-M. and Chan, P.H., 1985. Aspects of the 40–50 day oscillation during northern winter as inferred from outgoing longwave radiation. Mon. Weather Rev. (submitted).

Lau, K.-M. and Lim, H., 1982. Thermally driven motions in an equatorial β-plane: Hadley and Walker circulations during the winter monsoon. Mon. Weather Rev., 110: 336–353.

Madden, R. and Julian, P.R., 1972. Description of global scale circulation cells in the tropics with a 40–50 day period. J. Atmos. Sci., 29: 1109–1123.

McCreary, J.P., 1983. A model of tropical ocean–atmosphere interaction. Mon. Weather Rev., 111: 370–389.

McCreary, J.P. and Anderson, D.L.T., 1984. A model of the El Niño and the Southern Oscillation. Mon. Weather Rev., 112: 934–946.

McCreary, J.P. and Anderson, D.L.T., 1985. Simple models of the El Niño and the Southern Oscillation. In: J.C.J. Nihoul (Editor), Coupled Ocean–Atmosphere Models. (Elsevier Oceanography Series, 40) Elsevier, Amsterdam, pp. 345–370 (this volume).

McPhaden, M., 1982. Variability in the central equatorial Indian Ocean. Part I: Ocean Dynamics. J. Mar. Res., 40: 157–176.

Meyers, G., 1982. Interannual variation in sea level near Truk island – a bimodal seasonal cycle. J. Phys. Oceanogr., 12: 1161–1168.

Morse, P.M. and Feshbach, H., 1953. Methods in Theoretical Physics, Part I. McGraw-Hill, New York, N.Y., 996 pp.

Philander, S.G.H., Yamagata, T. and Pacanowski, R.C., 1984. Unstable air–sea interaction in the tropics. J. Atmos. Sci., 41: 603–613.

Rasmusson, E.N. and Carpenter, T.H., 1982. Variations in tropical sea surface temperature and surface wind fields associated with the Southern Oscillation/El Niño. Mon. Weather Rev., 110: 354–384.

Wyrtki, K., 1975. El Niño – the dynamical response of the equatorial Pacific Ocean to atmospheric forcing. J. Phys. Oceanogr., 5: 572–584.

Wyrtki, K., 1982. The Southern Oscillation, ocean–atmosphere interaction and El Niño. J. Mar. Tech. Soc., 10: 3–10.

CHAPTER 5

RESULTS FROM A MOIST EQUATORIAL ATMOSPHERE MODEL

M.K. DAVEY

ABSTRACT

Some model results are presented for the baroclinic response of a moist tropical atmosphere forced by fixed sea-surface temperature. In particular, the effect of varying the saturation moisture content of the lower atmosphere is investigated.

1. INTRODUCTION

Sensible heating of the atmosphere by some sea-surface temperature distribution drives a circulation that generally includes convergence and upward motion. In the tropics, particularly, this motion can lead to precipitation and sufficient latent heat release to enhance the large-scale circulation, substantially modifying the atmospheric response to the SST. Simple models that include dynamic moisture effects have been developed (see for example Gill, 1982; Lorenz, 1984), and this paper describes some results from the Gill model, forced by fixed SST.

As described in the next section, this model contains as a parameter the saturated moisture content q_s. Precipitation rate (and hence latent heating rate) depends on q_s, and we find that as q_s is increased more intense precipitation occurs on smaller scales. [This effect was also found by Held and Suarez (1978) in a two-level model.] For modest values of q_s steady solutions are obtained, but as q_s is raised oscillatory behaviour occurs. Zonally independent results are described in Section 3, showing the onset of a regular alternation of wet and dry periods at the equator.

For the preliminary two-dimensional experiments, idealised January and July Pacific SST patterns were used. July results with various values of q_s are presented in Section 4; steady January results are described elsewhere in this volume (Gill, 1985). As q_s is increased irregular oscillations appear, and convective regions shrink to the grid scale. Further work with finer resolution is in progress.

2. THE MODEL

A brief description of the model is given here, further details can be found in Gill (1982, 1985). A sinusoidal vertical structure for the first baroclinic mode in the atmosphere is assumed, for perturbations to a basic state with representative constant potential temperature gradient Θ_{oz}. For an equatorial beta-plane, the horizontal coordinates (x, y) are scaled by the equatorial Rossby radius $a = (c/2\beta)^{1/2}$ (about $10°$ of latitude). Horizontal velocity (u, v) is scaled by the gravity wave speed c of the baroclinic mode, and the time scale is a/c (about $1/4$ day). It is convenient to scale the mid-level potential temperature perturbation Θ by $\Theta_{oz}H$, where H is the mid-level to sea-surface height.

These variables are related by the linear shallow-water equations. For velocity we have:

$$u_t = \tfrac{1}{2}yv + \Theta_x - u/\tau + \nu\nabla^2 u \tag{1a}$$

$$v_t = -\tfrac{1}{2}yu + \Theta_y - v/\tau + \nu\nabla^2 v \tag{1b}$$

where τ and ν are constant coefficients for damping and dissipation. With mid-level vertical velocity scaled by cH/a, continuity gives:

$$w = -(u_x + v_y) = -\nabla u \tag{2}$$

For potential temperature:

$$\Theta_t = -w + Q + (\Theta_F - \Theta)/\tau + \nu\nabla^2\Theta \tag{3}$$

Thus mid-level potential temperature is forced toward Θ_F, which here is a temporally constant term with the same spatial pattern as the underlying SST. (Indication of a direct relation between mid-level heating and SST can be found in GCM results; e.g. Rowntree, 1976.) The time scale for this effect is τ, which is set to 10 for the results described below. (The dissipation ν is set to 0.03.) The term Q in eqn. (3) represents latent heating, and is dynamically determined via a moisture budget.

This moisture budget is best described in dimensional terms first. (Dimensional variables will be indicated by the superscript $\hat{\ }$.) Suppose $\hat{q}$ is the precipitable moisture in a column of air, measured in units of depth of liquid water/unit area, and let $\hat{q}_s$ be the saturated (maximum allowable) value. We assume all moisture is contained in the lower atmosphere, so there is no vertical transfer through the mid-level. Then in a region where $\hat{q} = \hat{q}_s$ and the lower-level flow is convergent, the moisture flux convergence must be balanced by a precipitation rate:

$$\hat{P} = -\nabla(\hat{u}\hat{q}) = \hat{q}_s\hat{w}/H \tag{4}$$

The corresponding rate of latent heat release is:

$$\rho_w\hat{P}L$$

where ρ_w is the density of liquid water and L is the latent heat of vaporisation. Suppose this heat is transferred to some depth H of air of density ρ and specific heat c_p: the rate of heating of the air is:

$$\hat{Q} = \frac{\rho_w L}{\rho H c_p}\hat{P} \tag{5}$$

Dimensionally, the term $-w + Q$ on the RHS of eqn. (3) gives a net heating rate:

$$\hat{w}\Theta_{oz}[\hat{q}_s/\hat{q}_{crit} - 1] \tag{6}$$

where:

$$\hat{q}_{crit} = H\Theta_{oz}\frac{\rho H c_p}{\rho_w L}$$

is a critical moisture content.

If $\hat{q}_s < \hat{q}_{crit}$ then cooling due to the upward motion of an air parcel exceeds the latent heating, and the system is locally stable. However, when $\hat{q}_s > \hat{q}_{crit}$ there is a net heating by these effects, and a tendency for amplification of the upward motion. This amplification can be limited by other terms on the RHS of eqn. (3), or by advection of surrounding dry air into the wet region.

When $\hat{q}$ is scaled by $\hat{q}_{\text{crit}}$, and $\hat{P}$ by $\hat{q}_{\text{crit}} a/c$, the non-dimensional moisture equation is:

$$q_t + \nabla(uq) = (q_s - q)/\tau + \nu \nabla^2 q - P \tag{7}$$

The first term on the RHS represents the tendency to restore q to saturation by evaporation in the absence of any other effects. The time scale τ and diffusivity ν are chosen to be the same as in eqns. (1) and (3) to keep the number of parameters to a minimum in this elementary model.

Precipitation is positive and equal to $-\nabla(uq)$ when $q = q_s$ and $\nabla(uq) < 0$, otherwise $P = 0$. The non-dimensional heating rate Q is equal to P.

3. ZONALLY INDEPENDENT FORCING

To force the model, simple analytic expressions for Θ_F were constructed to (crudely) represent west and east Pacific, January and July, meridional SST profiles for $|y| < 4$ (i.e. 40°S–40°N). For the west Pacific:

$$\Theta_{FW}(\text{January}) = 1 - y^2/16 \tag{8a}$$

and:

$$\Theta_{FW}(\text{July}) = 1 - (y - y_0)^2 (y_0^2 + 2yy_0 + 16)/(y_0^2 - 16)^2 \tag{8b}$$

with $y_0 = 1.5$; and for the east Pacific:

$$\Theta_{FE} = 0.6(1 - y^2/16) + A \exp - (y - 1)^2 \tag{8c}$$

with $A = 0.1$ (Jan.) or 0.3 (July).

Zonally independent results are given here for the January west case, for which Θ_F is symmetric about the equator, with a maximum at the equator. Figure 1 shows $q_s - q$ and w for $q_s = 1$ (i.e. when latent heating matches cooling by upward motion; locally neutral), at time $t = 40$. By this time a steady state has been established, with a region of strong

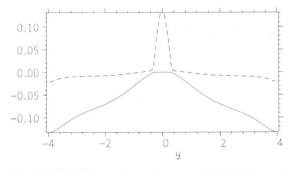

Fig. 1. Zonally independent case with $q_s = 1$, showing moisture $q_s - q$ (solid) and vertical velocity w (dashed) at time $t = 40$, for January forcing Θ_{FW}.

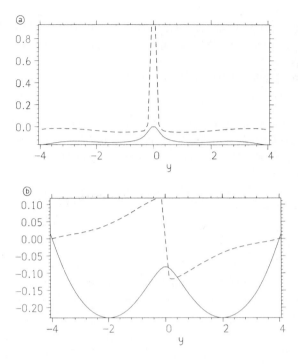

ⓐ

ⓑ

Fig. 2. Zonally independent case with $q_s = 1.2$, at time $t = 22$ for forcing Θ_{FW}. (a) $q_s - q$ (solid) and w (dashed); (b) zonal velocity u (solid) and meridional velocity v (dashed).

upward motion prominent at the equator, which coincides with the wet region where $q = q_s$. [Similar results for $q_s = 8/9$ given in Gill (1985) show a broader, weaker wet region.]

In Fig. 2 the same variables are shown for $q_s = 1.2$ (locally unstable), at time $t = 22$. The wet region is considerably narrower than for $q_s = 1$, with enhanced upward motion and stronger precipitation. Downward motion on either side is also increased, and there is relatively low moisture in these adjacent dry regions. The u and v (low-level) velocities given in Fig. 2b show easterly zonal flow with a double jet structure, and equatorward

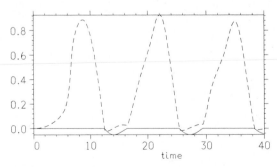

Fig. 3. Zonally independent case with $q_s = 1.2$, for forcing Θ_{FW}. Time series of $q_s - q$ (solid) and w (dashed) at $y = -0.06$, for $t = 0\text{–}40$.

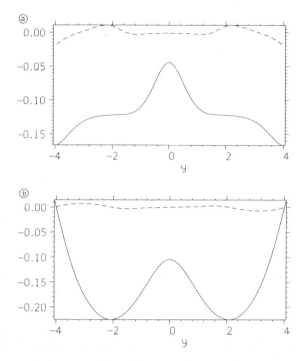

Fig. 4. Zonally independent case with $q_s = 1.2$, at time $t = 40$, for forcing Θ_{FW}. (a) $q_s - q$ (solid) and w (dashed); (b) low-level velocities u (solid) and v (dashed).

meridional flow. The sharp change in v on crossing the equator is an indication of the strong vertical motion there – for this x-independent model, $w = - v_y$.

The situation displayed in Fig. 2 is not stable. From the time series for w and $q_s - q$ in Fig. 3 (for a point almost on the equator), it can be seen that short-lived dry spells periodically replace the usual wet conditions. As a region of strong precipitation develops, the air in adjacent areas of sinking becomes correspondingly drier. Advection and diffusion of this dry air into the wet region then suppress the strong convection. Figure 4 shows $q_s - q$, w, and u and v, during a dry spell ($t = 40$): at this time $q < q_s$ everywhere, the meridional circulation is very weak, but the zonal velocity is little changed from that at $t = 22$. The SST maximum at the equator subsequently quickly re-establishes upward motion there, and precipitation returns.

The absence of an interactive ocean is clearly a limitation. We would expect the surface atmospheric flow to drive Ekman divergence in the oceanic upper layer, with associated upwelling tending to erase the SST maximum at the equator (see, for example, Pike, 1971).

4. ZONALLY VARYING FORCING

For the experiments with zonally varying forcing an east–west periodic channel was used, with $-8 \leqslant x \leqslant 8$ (e.g. 120°E–80°W) and $-4 \leqslant y \leqslant 4$ (40°S–40°N). The forcing

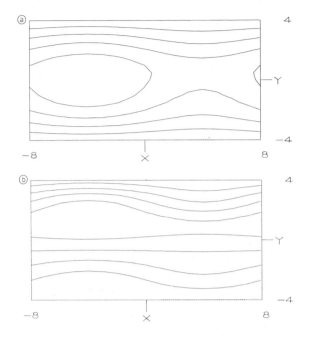

Fig. 5. Forcing patterns Θ_F based on Pacific sea-surface temperature patterns. (a) January; (b) July.

is a simple combination of the meridional profiles (8):

$$\Theta_F = \tfrac{1}{2}[\Theta_{FE} + \Theta_{FW} + (\Theta_{FE} - \Theta_{FW})\sin(\pi x/8)] \tag{9}$$

January and July Θ_F are illustrated in Fig. 5. The equations of Section 2 are integrated in time on a finite-difference C-grid with resolution of half a Rossby radius (much coarser than for the above x-independent integrations).

As q_s is increased behaviour qualitatively similar to the x-independent case is found. For $q_s = 1$, the vertical velocity map in Fig. 6a shows steady upward motion in a band north of the equator, corresponding to a model ITCZ. The wet region corresponds roughly to the area within the solid contours in Fig. 6a. This ITCZ has some zonal structure, being strongest in the western portion of the map. The low-level horizontal velocity field in Fig. 6b shows predominantly easterly flow, with cross equatorial flow leading into westerlies near the wet regions. [This case is further discussed in Gill (1985) along with January $q_s = 1$ results.]

When q_s is increased to 1.05 the wet regions shrink and intensify, as can be seen from the w map in Fig. 7. As for $q_s = 1$ a steady state is reached, but the zonal band of precipitation has now broken up into smaller (but larger than grid-scale) stable structures. (The retention of the non-linear advection terms in the moisture eqn. (7) is essential for this case: without them the model is disastrously unstable.)

Further increase in q_s gives rise to unsteady behaviour. For $q_s = 1.2$ Fig. 8 shows time series for $q - q_s$ and w at a point north of the equator in the western region ($x = -5.75$, $y = 1.75$). This point is usually dry, with fluctuating weak vertical motion, but there is a

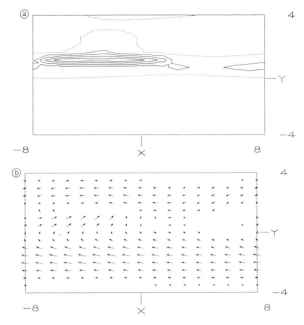

Fig. 6. The steady response to the "July" forcing with $q_s = 1$. (a) Mid-level vertical velocity w at $t = 80$, with contour interval 0.055. The zero contour is dotted, positive contours are solid. (b) Low-level horizontal velocity, maximum speed 0.29 (not shown when speed < max/4).

brief and intense wet event lasting from $t = 65$ to $t = 72$. In contrast to the x-indepen-dent model, there is no sign of a regular cycle of wet and dry episodes.

Low-level horizontal velocity maps for $t = 60$, 70 and 80 are provided in Fig. 9 for this case. During this period we see that the strong circulation associated with a convective region near the centre of the diagram weakens and decays away, while other convective

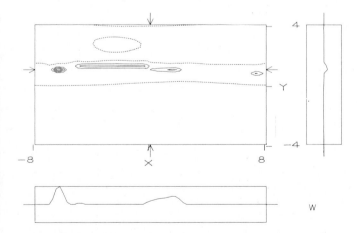

Fig. 7. Steady response to "July" forcing pattern with $q_s = 1.05$, showing vertical velocity w at time $t = 40$, with contour interval $ci = 0.26$. The zero contour is dotted, positive contours are solid. Arrows indicate positions of the cross-sections shown below and to the right of the contour map.

48

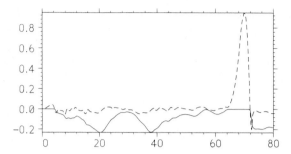

Fig. 8. Time series of $q_s - q$ (solid) and w (dashed) for "July" forcing with $q_s = 1.2$, at $x = -5.75$, $y = 1.75$, for $t = 0–80$.

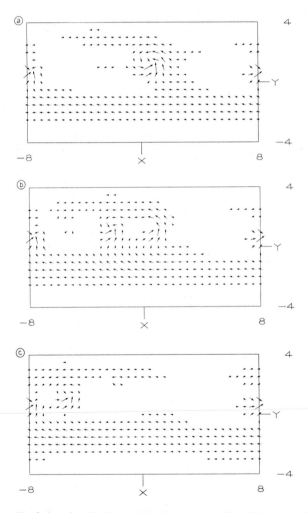

Fig. 9. Low-level horizontal velocity maps for "July" forcing with $q_s = 1.2$. (a) time $t = 60$, maximum speed 0.74 (not shown when speed < max/4); (b) $t = 70$, maximum speed 0.72; (c) $t = 80$, maximum speed 0.66.

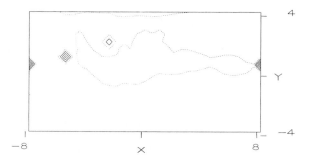

Fig. 10. Vertical velocity w for "July" forcing with $q_s = 1.2$ at time $t = 80$, with contour interval 0.69.

regions develop to the west. These convective regions do not propagate; rather, several preferred sites seem to grow and decay in turn. Convection amplifies at one point, driving and driven by strong local convergence. Inward advection of surrounding dry air then "shuts down" this positive feedback, but the strong local circulation apparently can meanwhile trigger convection at nearby locations.

Such conclusions must be regarded as speculative however, because these wet regions are actually single grid-point features when $q_s = 1.2$, as is evident in the w map in Fig. 10 for $t = 80$. Further experiments with higher resolution are in progress, with the aim of understanding the nature and mechanism of the unsteady behaviour.

REFERENCES

Gill, A.E., 1982. Studies of moisture effects in simple atmospheric models: the stable case. Geophys. Astrophys. Fluid Dyn., 19: 119–152.
Gill, A.E., 1985. Elements of coupled ocean atmosphere models for the tropics. In: J.C.J. Nihoul (Editor), Coupled Ocean–Atmosphere Models. (Elsevier Oceanography Series, 40) Elsevier, Amsterdam, pp. 303–327 (this volume).
Held, I.M. and Suarez, M.J., 1978. A two-level primitive equation atmospheric model designed for climatic sensitivity experiments. J. Atmos. Sci., 35: 206–229.
Lorenz, E.N., 1984. Formulation of a low-order model of a moist general circulation. J. Atmos. Sci., 41: 1933–1945.
Pike, A.C., 1971. Intertropical Convergence Zone studied with an interacting atmosphere and ocean model. Mon. Weather Rev., 99: 469–477.
Rowntree, P.R., 1976. Tropical forcing of atmospheric motions in a numerical model. Quart. J. R. Meteorol. Soc., 98: 290–321.

CHAPTER 6

LARGE-SCALE MOMENTUM EXCHANGE IN THE COUPLED ATMOSPHERE–OCEAN

JOHN A.T. BYE

ABSTRACT

It is shown that the exchange of momentum between the atmosphere and the ocean is essentially a two-way process. Thus the surface shearing stress, $\boldsymbol{\tau}_s = \boldsymbol{\tau}_w - \boldsymbol{\tau}_F$ consists of two independent parts – one due to the air (the wind stress, $\boldsymbol{\tau}_w$) and the other due to the water (which we will call the "understress", $\boldsymbol{\tau}_F$). This is in distinction to the single-term representation ($\boldsymbol{\tau}_w$) over a solid surface. The understress which retards the surface current, has the approximate form $\boldsymbol{\tau}_F \sim \rho_a K |u_a| u_o$ where ρ_a is the density of air, and u_a is the wind velocity with the associated drag coefficient K and u_o is the surface current. It is suggested that the primary dissipation process for the deep stratified World ocean may occur in the air–sea boundary layer rather than say near the bottom. This is demonstrated by considering the relative importance of bottom stress and understress for an individual oceanic eddy.

For a scale of order 100 km which averages over the mesoscale eddies in the ocean, but resolves the synoptic disturbances in the atmosphere, it is also shown that the understress acting on the mean current has the approximate form, $\bar{\boldsymbol{\tau}}_F \sim \rho_a K (1 + \xi^2) |\bar{u}_a| \bar{u}_o$ in which $\xi = (\overline{u_o'^2})^{1/2}/|\bar{u}_o|$ is the intensity of the surface two-dimensional turbulence in the ocean, and ' and ‾ denote respectively a mean and a fluctuation. On using $\rho_a \sim 1\,\mathrm{kg\,m^{-3}}$, $K \sim 2 \times 10^{-3}$, $\bar{u}_a \sim 10\,\mathrm{m\,s^{-1}}$, and $\xi \sim 2$ we obtain $\bar{\boldsymbol{\tau}}_F \sim 0.1\,\bar{u}_o\,\mathrm{N\,m^{-2}}$, and it follows that $\bar{\boldsymbol{\tau}}_F$ and $\bar{\boldsymbol{\tau}}_w$ are both of the same order of magnitude.

Finally, if dissipation in the air–sea boundary layer is the dominant process, it is deduced that the ratio of the dissipation rates for water and air should be $\sim 1/30$, and hence by a thermodynamic argument that the meridional energy fluxes between the two fluids should be partitioned in a similar manner.

1. INTRODUCTION

An understanding of the physics of the air–sea interface is very important in general circulation studies for the ocean and the atmosphere. The ocean receives by far the major part of its momentum through the interface. It will be shown below that to regard the momentum transfer as essentially a one-way process (with subsequent dissipation say by bottom friction) is a great injustice to the ocean, and that indeed important transfers of momentum occur in the opposite direction, from the ocean to the atmosphere.

The theory developed from a need to understand the energy dissipation processes for the large-scale ocean circulation following an editorial (Gill, 1980) in which it was pointed out that the traditional oceanic dissipation mechanisms of bottom and internal friction may not be adequate to explain the large-scale circulation patterns.

Accordingly an attempt was made to re-examine the frictional processes immediately below the ocean surface to determine whether a systematic dissipation mechanism may be operating therein which is significant for the large-scale dynamics. A brief note (Bye, 1980) suggested that this was a fruitful venture, and it is the purpose here to examine some of the consequences of this mechanism in detail.

It will be shown that the shearing stress at the air—sea interface consists of two independent parts — the wind stress, and a water stress, which we will call the "under-stress". This is in contrast to the single-term formulation for a solid surface. For the oceanic general circulation, the understress appears of comparable importance to the wind stress.

2. THE SURFACE SHEARING STRESS

The continuity of shearing stress (τ_s) at the ocean surface (cf. Phillips, 1966, p. 224) yields the pair of relations:

$$|\tau_s| = \rho_a u_*^2 = \rho_o w_*^2 \tag{1}$$

where ρ_a and u_* refer respectively to the density and friction velocity of air, and ρ_o and w_* are the corresponding properties for water. An alternative form of eqn. (1) using drag coefficients for air and water is the approximation:

$$
\begin{aligned}
\tau_s &= \rho_a K |u_a - u_s|(u_a - u_s) \\
&= \rho_o K' |u_s - u_o|(u_s - u_o)
\end{aligned} \tag{2}
$$

where u_a and u_o refer respectively to the surface wind and the surface current at the

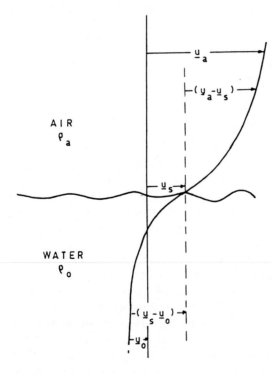

Fig. 1. The air—sea coupled boundary layer.

edge of the constant stress layers in each fluid, with the associated drag coefficients K and K', and u_s is the Lagrangian surface drift velocity (Fig. 1).

On solving eqn. (2) for u_s we obtain:

$$u_s = \frac{\epsilon u_a + u_o}{1 + \epsilon} \tag{3}$$

where $\epsilon = (K\rho_a/K'\rho_o)^{1/2}$. Thus the surface drift velocity comprises a wind drift velocity $[\epsilon u_a/(1 + \epsilon)]$ and a current drift velocity $[u_o/(1 + \epsilon)]$. Observations of the wind drift velocity (e.g. Kraus, 1972) usually show it to be about 3% of the surface wind (measured at 10 m) and suggest that $\epsilon \sim 1/30$, and predict that the surface current would be reduced by a factor (~ 0.97) in the current drift velocity.

On assuming that $\rho_o = 10^3 \, \text{kg m}^{-3}$ and $\rho_a = 1.2 \, \text{kg m}^{-3}$ we obtain:

$$(\rho_a/\rho_o)^{1/2} \sim 0.034$$

and hence on comparing this result with the observational estimate for ϵ it appears also that $K/K' \sim 1$.

The air boundary layer has been extensively examined (e.g. Sheppard et al., 1972), but the water boundary layer is not nearly as well understood (Csanady, 1984). A full treatment of the coupled Ekman layer dynamics however is not essential to the main ideas of the paper, for which it is sufficient that the free-stream velocities u_a and u_o are good approximations to the surface geostrophic velocities (Bye, 1985).

The wind drift velocity clearly arises through the action of the wind stress which is that part of the shearing stress due to the atmosphere. The retardation of the surface current in the current drift velocity by a parallel argument arises from that part of the shearing stress due to the ocean, which we will call the understress. Thus, on substituting eqn. (3) into (2) we obtain the expression:

$$\boldsymbol{\tau}_s = \boldsymbol{\tau}_w - \boldsymbol{\tau}_F \tag{4}$$

in which:

$$\boldsymbol{\tau}_w = C u_a \tag{5}$$

and:

$$\boldsymbol{\tau}_F = C u_o \tag{6}$$

are, respectively, the wind stress and the understress

$$C = \frac{\rho_a K |u_a - u_o|}{(1 + \epsilon)^2} \tag{7}$$

and since $\epsilon \ll 1$ and normally $|u_a| \gg |u_o|$:

$$C \sim \rho_a K |u_a|$$

Equation (5) is an approximate quadratic drag relation for u_a, whereas eqn. (6) is an approximate linear relation for u_o. It is emphasised that it is desirable to regard the shearing stress as consisting of two parts, since these arise from two independent general circulations, the one in the upper fluid (air) and the other in the lower fluid (water).

On assuming that the dissipation occurs in the constant stress layer, the net rate of dissipation/unit area by the ocean–atmosphere system:

$$L = \tau_s \cdot (u_a - u_o)$$
$$= \tau_s \cdot (u_a - u_s) + \tau_s \cdot (u_s - u_o) \qquad (8)$$

in which the first and the second terms represent respectively the dissipation rate/unit area in the air and water. On substituting for u_s from eqn. (3) we find that:

$$\frac{\tau_s \cdot (u_s - u_o)}{\tau_s \cdot (u_a - u_s)} = \epsilon \qquad (9)$$

and hence the ratio of the dissipation rates in the interfacial boundary layer in the ocean and the atmosphere is $\sim 1/30$. The thermodynamic consequences of this result are discussed in Section 5.

We examine now the energy fluxes associated with the interfacial dissipation process. Firstly, there is a flux of energy/unit area from the air to the water of magnitude, $\tau_s \cdot u_s$. Hence the *downward* rate of working across the lower boundary of the constant stress layer:

$$\tau_s \cdot u_o = \tau_s \cdot u_s - \tau_s \cdot (u_s - u_o) \qquad (10)$$

is the difference between the energy flux across the interface and the dissipation flux in the water boundary layer. An alternative, less familiar but more fundamental expression of this result, is:

$$(-\tau_s \cdot u_o) = -\tau_F \cdot (u_a - u_o) \qquad (11)$$

in which $(-\tau_s \cdot u_o)$ is the rate of working of the ocean on the constant stress layer.

Similarly the rate of working of the atmosphere on the constant stress layer, is:

$$\tau_s \cdot u_a = \tau_s \cdot (u_a - u_s) + \tau_s \cdot u_s$$

or:

$$\tau_s \cdot u_a = \tau_w \cdot (u_a - u_o) \qquad (12)$$

and since energy is conserved the sum of these two rates of working is equal to the dissipation rate of the coupled system, and on substituting eqns. (11) and (12) [or eqn. (4)] in eqn. (8) we obtain the expression:

$$L = \tau_w \cdot (u_a - u_o) - \tau_F \cdot (u_a - u_o) \qquad (13)$$

Note that each fluid contributes to the dissipation rate of the coupled system, and that in the rates of working expressed in terms of the relative velocity $(u_a - u_o)$, wind stress is the relevant stress in air, and understress in water.

3. HORIZONTAL AVERAGING

The results of Section 2 refer to mean horizontal velocities at a point in the air–sea boundary layer. Implicit in general circulation studies, however, is a horizontal spatial average. We now obtain the horizontal average dissipation rates and hence the shearing stress acting on the mean velocities using a simple model of two-dimensional turbulence. This analysis indicates the relative significance of the process for the general circulations of the ocean and atmosphere. From eqn. (13) using eqns. (5) and (6), the spatial average rate of dissipation/unit area by the ocean–atmosphere system:

$$\bar{L} = \overline{Cu_a^2} + \overline{Cu_o^2} - 2\overline{Cu_a \cdot u_o} \tag{14}$$

where the overbar denotes a horizontal spatial average. On now assuming that $|u_o| \ll |u_a|$ and using eqn. (7) we have that:

$$\overline{Cu_a^2} = \frac{K\rho_a}{(1+\epsilon)^2} \overline{|u_a|u_a^2}$$

and assuming that the horizontal turbulences in the ocean and the atmosphere are independent:

$$\overline{Cu_o^2} = \frac{K\rho_a}{(1+\epsilon)^2} \overline{|u_a|} \overline{u_o^2}$$

and:

$$\overline{Cu_a \cdot u_o} = \frac{K\rho_a}{(1+\epsilon)^2} \overline{|u_a|u_a \cdot u_o}$$

where in each expression $K\rho_a/(1+\epsilon)^2$ has been assumed to be a constant. These results may be evaluated approximately by assuming that the horizontal turbulences in the atmosphere and in the ocean are isotropic. Thus on defining $u_o = \bar{u}_o + u_o'$ and $u_a = \bar{u}_a + u_a'$ where u_o' and u_a' denote turbulent fluctuations, we have, using expressions given in Bye (1970):

$$\overline{Cu_a^2} = \bar{C}\bar{u}_a^2 (1 + \tfrac{9}{4}\xi_a^2) \qquad \xi_a \ll 1 \tag{15}$$

in which $\xi_a = |u_a'^2|^{1/2}/|\bar{u}_a|$ is the turbulent intensity of the atmospheric turbulence and $\bar{C} = K\rho_a|\bar{u}_a|/(1+\epsilon)^2$, and:

$$\overline{Cu_o^2} = \bar{C}\bar{u}_o^2 (1 + \tfrac{1}{4}\xi_a^2)(1 + \xi_o^2) \qquad \xi_a \ll 1 \tag{16}$$

where $\xi_o = |u_o'^2|^{1/2}/|\bar{u}_o|$ is the turbulent intensity of the oceanic turbulence, and:

$$\overline{Cu_a \cdot u_o} = \bar{C}\bar{u}_a \cdot \bar{u}_o(1 + \tfrac{3}{4}\xi_a^2) \qquad \xi_a \ll 1 \tag{17}$$

On substituting eqns. (15)–(17) in (14) we obtain:

$$\bar{L} = \bar{\tau}_w \cdot [(1 + \tfrac{9}{4}\xi_a^2)\bar{u}_a - (1 + \tfrac{3}{4}\xi_a^2)\bar{u}_o] - \bar{\tau}_F \cdot [(1 + \tfrac{3}{4}\xi_a^2)\bar{u}_a - (1 + \tfrac{1}{4}\xi_a^2)(1 + \xi_o^2)\bar{u}_o] \tag{18}$$

where $\bar{\tau}_w = \bar{C}\bar{u}_a$ and $\bar{\tau}_F = \bar{C}\bar{u}_o$. Note that the coefficients of $\bar{u}_a$ and $\bar{u}_o$ differ due to the properties of the two-dimensional turbulence in the two fluids. On using the identity:

$$\tau_F \cdot u_a = \tau_w \cdot u_o$$

equation (18) can be rearranged to yield:

$$\bar{L} = [\bar{\tau}_w(1 + \tfrac{9}{4}\xi_a^2) - \bar{\tau}_F(1 + \tfrac{3}{4}\xi_a^2)] \cdot \bar{u}_a - [\bar{\tau}_w(1 + \tfrac{3}{4}\xi_a^2) - \bar{\tau}_F(1 + \tfrac{1}{4}\xi_a^2)(1 + \xi_o^2)] \cdot \bar{u}_o$$

Hence the effective shearing stresses which act on the spatial mean surface current and on the spatial mean surface wind, respectively:

$$\tau_s^- = \bar{\tau}_w(1 + \tfrac{3}{4}\xi_a^2) - \bar{\tau}_F(1 + \tfrac{1}{4}\xi_a^2)(1 + \xi_o^2) \tag{19}$$

and:

$$\tau_s^+ = \bar{\tau}_w(1 + \tfrac{9}{4}\xi_a^2) - \bar{\tau}_F(1 + \tfrac{3}{4}\xi_a^2) \tag{20}$$

are not identical. This rather surprising result has important consequences, cf. Section 4.

4. CONSEQUENCES FOR THE GENERAL CIRCULATION DYNAMICS IN THE OCEAN AND THE ATMOSPHERE

(a) In the ocean

Firstly we compare the relative importance of understress and bottom stress for an eddy. The ratio of the dissipation rates in the surface and the bottom boundary layers:

$$r = \frac{\tau_F \cdot u_o}{\tau_B \cdot u_B} \sim \frac{\rho_a K |u_a| u_o^2}{\rho_o K_B |u_B| u_B^2}$$

where u_B is the bottom velocity and K_B is the bottom drag coefficient. On assuming for a typical eddy that $|u_a| \sim 10\,\mathrm{m\,s^{-1}}$, $|u_o| \sim 0.5\,\mathrm{m\,s^{-1}}$, $|u_B| \sim 0.1\,\mathrm{m\,s^{-1}}$, $K \sim 2 \times 10^{-3}$ and $K_B \sim 2 \times 10^{-3}$ we find that $r \sim 3$. Hence the dissipation rate by the understress is at least as important as that by the bottom stress.

It is instructive therefore to consider the nature of the oceanic general circulation in the limit in which surface dissipation is dominant. Several important consequences follow from the results of Section 3 applied over a scale of order 100 km which averages over the mesoscale eddies in the ocean but resolves the synoptic disturbances in the atmosphere. For this scale of averaging $\xi_a = 0$, and the effective shearing stress acting on the spatial mean current [eqn. (19)] reduces to the expression:

$$\tau_s^- = \bar{\tau}_w - \rho_o \gamma \bar{u}_o \tag{21}$$

in which $\gamma = (1/\rho_o)\bar{C}(1 + \xi_o^2)$ is a dynamic coefficient of understress. The fundamental reason for the enhancement of the coefficient for the spatial mean understress by the factor $(1 + \xi_o^2)$ is that the mean ocean circulation must generate an eddy field of sufficient intensity to attain a balance between vorticity imparted by the winds and lost by the currents. In the deep ocean in which bottom currents are small in comparison with surface currents, the major part of the dissipation occurs at the surface. Hence, since the surface-current speeds generally are smaller than the surface-wind speeds, a compensatory increase in the two-dimensional turbulent intensity in the ocean relative to that in the atmosphere on the same scale of averaging is required.

The energetics of this process are described by the spectral theory of two-dimensional turbulence which predicts an energy flux towards the low wavenumbers from an intermediate wavenumber at which energy is injected by large scale instability processes.

The key to understanding this energy flux is a Lagrangian frame of reference. Consider two water particles which travel from A to B — the first a real particle which follows a natural trajectory in the turbulent fluid, the second a hypothetical particle which follows a direct path at the mean fluid velocity. The real particle clearly travels at a greater speed and over a greater path length than the hypothetical particle, and consequently its rate of dissipation of energy is greater. On average the dissipation rate measured relative to the mean velocity is increased by the factor $(1 + \xi_0^2)$, cf. Bye (1980).

In wavenumber space, this process is equivalent to the high wavenumbers which convolute the path of the real particle passing the dissipative energy flux to the lowest wavenumber which is the path followed by the hypothetical particle with the enhanced dissipation rate. On again using the typical magnitudes for the atmospheric quantities, $|u_a| \sim 10 \, \mathrm{m\,s^{-1}}$ and $K \sim 2 \times 10^{-3}$ and assuming that the horizontal turbulent intensity in the ocean $\xi_0 \sim 2$ we find that $\gamma \sim 10^{-4} \, \mathrm{m\,s^{-1}}$. As suggested by the discussion on the eddy this value indicates that understress is highly significant for the oceanic general circulation. Thus for a mean current of $1 \, \mathrm{m\,s^{-1}}$, the spatial mean understress has a magnitude of $0.1 \, \mathrm{N\,m^{-2}}$ which is comparable with the wind stress. The corresponding dissipation rate:

$$\overline{\boldsymbol{\tau}_F \cdot \boldsymbol{u}_o} = \rho_o \gamma \bar{u}_o^2$$

would be:

$$0.1 \, \bar{u}_o^2 \, \mathrm{W\,m^{-2}}$$

Let us now consider briefly the vorticity balance for the large-scale circulation. The linearized and frictionless (except for the understress) transport vorticity equation for a constant depth ocean on a β-plane is:

$$\beta V = \frac{1}{\rho_o} \left(\frac{\partial \tau_{sy}}{\partial x} - \frac{\partial \tau_{sx}}{\partial y} \right)$$

where $\beta = df/dy$, in which $f = 2\Omega \sin \theta$ where Ω is the angular speed of rotation of the Earth and θ is latitude, and (U, V) are the components of volume transport, and we omit the overbars denoting mean quantities, and the superscripts denoting oceanic stresses. On substituting eqn. (21), we obtain:

$$\beta V = \frac{1}{\rho_o} \left(\frac{\partial \tau_{wy}}{\partial x} - \frac{\partial \tau_{wx}}{\partial y} \right) - \left(\frac{\partial}{\partial x} \gamma v_o - \frac{\partial}{\partial y} \gamma u_o \right) \tag{22}$$

This equation is the transport vorticity equation with understress. On assuming a momentum scale depth for the ocean, H, and that γ is a constant, we obtain:

$$\beta V = \frac{1}{\rho_o} \left(\frac{\partial \tau_{wy}}{\partial x} - \frac{\partial \tau_{wx}}{\partial y} \right) - R \left(\frac{\partial V}{\partial x} - \frac{\partial U}{\partial y} \right) \tag{23}$$

where $R = \gamma/H$. Equation (23) is exactly the Stommel model (Stommel, 1948) of the ocean circulation, and on substituting $\gamma \sim 10^{-4} \, \mathrm{m\,s^{-1}}$ and $H \sim 100 \, \mathrm{m}$, we obtain $R \sim 10^{-6} \, \mathrm{s^{-1}}$, which is the value normally used for the transport dissipation coefficient in large-scale dynamical studies.

The main feature of eqn. (22) is that the dissipation process is associated with the surface current rather than the bottom current. This is comforting for ocean modellers since in the deep thermohaline ocean, subsurface motions are less well known than the surface circulation. Indeed, the use of transport dissipation coefficients in many pioneering studies is justified at least as a good first approximation.

(b) In the atmosphere

The shearing stress acting on the mean wind is independent of the oceanic turbulence, cf. eqn. (20). Thus the understress would be proportional to the mean current, and usually of relatively small account.

On scales less than ~ 100 km, however, perturbation understresses of typical magnitude $0.01 \, \text{N m}^{-2}$ would exist which are localised geographically by the oceanic eddy fields. These stresses may constitute important sources of vorticity for the atmosphere in the early development of meteorological systems. There is a large literature on the transfer of momentum from the atmosphere to the ocean (cf. LeBlond and Mysak, 1978) in the sense of loss by the surface wind $(\boldsymbol{u}_a)$ and gain by the surface current $(\boldsymbol{u}_o)$, for which the downward rate of working is $(\boldsymbol{\tau}_w \cdot \boldsymbol{u}_o)$. We have drawn attention to the need for study of the back transfer which, by the principle of action and reaction, has an identical upward rate of working $(\boldsymbol{\tau}_F \cdot \boldsymbol{u}_a)$.

5. OCEANIC MERIDIONAL HEAT TRANSPORT

We consider briefly the thermodynamics of the coupled ocean–atmosphere system. For the coupled system Paltridge (1978) has shown that the global rate of dissipation (D) has the form:

$$D = - \int_{T_S}^{T_N} Y/T \, \mathrm{d}T > 0 \tag{24}$$

where Y is the meridional energy flux, T is the corresponding dissipation temperature, and T_S and T_N are the dissipation temperatures at the south and north poles, respectively. We may also express D as the sum:

$$D = \int^A L \, \mathrm{d}A + D'$$

where A is the surface area of the ocean, D' is the dissipation rate due to solid boundaries both on land and in the sea. On now partitioning between the ocean and the atmosphere using eqn. (9), and considering the ocean and the atmosphere separately we obtain the analogous pairs of expressions:

$$D_o = - \int_{T_{oS}}^{T_{oN}} Y_o/T_o dT_o \qquad (25)*$$

$$D_a = - \int_{T_{aS}}^{T_{aN}} Y_a/T_a dT_a$$

and:

$$D_o = \int^{A} \boldsymbol{\tau_s} \cdot (\boldsymbol{u_s} - \boldsymbol{u_o}) dA + D_o'$$

$$D_a = \int^{A} \boldsymbol{\tau_s} \cdot (\boldsymbol{u_a} - \boldsymbol{u_s}) dA + D_a' \qquad (26)$$

in which the suffixes a and o refer throughout to the atmosphere and the ocean, and:

$$D = D_o + D_a, D' = D_o' + D_a' \text{ and } Y = Y_o + Y_a \qquad (27)$$

In the limit in which the major part of the dissipation occurs in the air–sea boundary two important conclusions can be drawn.

(i) Equation (26) yields the explicit expression:

$$D_o/D_a = \epsilon \qquad (28)$$

where ϵ is the local ratio of the dissipation rates/unit area, cf. eqn. (9).

(ii) The dissipation temperatures in the two fluids are equal. Hence from eqn. (25) we have:

$$\frac{D_o}{D_a} = \frac{\int_{T_S}^{T_N} \frac{Y_o}{T} dT}{\int_{T_S}^{T_N} \frac{Y_a}{T} dT}$$

and using (i) it follows that:

$$Y_o/Y_a = \epsilon$$

Order of magnitude arguments suggest that this result may not be very sensitive to the inclusion of solid boundary dissipation. For example, if the atmospheric dissipation rates over land and sea are approximately in proportion to their respective areas we have

---

*The estimates of eqn. (25), by the argument of Paltridge that the temperature structure of the atmosphere–ocean system is specified by the overall constraint of maximum entropy, are stable to small perturbations in the meridional fluxes, since any transfer of dissipative energy flux from one fluid to the other is precluded by the principle of action and reaction, discussed in Section 4(b). This can be illustrated by considering the classical example of a block sliding down an inclined plane. If the temperatures of the block and the plane are supposed given, the dynamics of the sliding process are completely determined.

$D_a' \sim 0.3 D_a$ and if, as suggested by the discussion on the dissipation by an oceanic eddy, $D_o' \sim 0.2 D_o$, we would obtain $D_o/D_a \sim 0.9\epsilon$. It is reasonable therefore to replace eqn. (28) in (i) by:

$$D_o/D_a \sim \epsilon \tag{30}$$

as a measure of the global ratio of the dissipation rates, but to remember that this ratio would not be a uniform estimate for the local ratio of the dissipation rates/unit area. In (ii) the dissipation temperatures would be always similar.

We conclude that at most latitudes:

$$Y_o/Y_a \sim \epsilon \tag{31}$$

the exact ratio depending on geography and the dynamics of the general circulations of the fluids. Thus the oceanic meridional heat transport should have a value of about 1/30 the atmospheric heat transport. Recent experimental estimates (Hastenrath, 1982) indicate that near $30°N$ where the maximum northward fluxes occur $Y \sim 6EW$, and $Y_o \sim 2EW$, which yields a ratio, 0.5. Our conclusions suggest that this ratio is much too large, and argue for a reassessment downwards of Y_o.

6. CONCLUSIONS

In summary, air–sea momentum interaction involves two coupled fluids geostrophically sliding relative to one another, the sliding motions being accommodated by small frictional velocities within the planetary layers, matched by the surface drift velocity. This relative motion allows exchange of momentum across the fluid interface, which is enhanced by high wind speeds, and occurs on horizontal scales determined by the two-dimensional turbulence of the initiating fluid. In this sense the ocean surface is transparent to the passage of momentum, notwithstanding the great difference in the densities of air and water.

The significance of the two-way momentum exchange process for the ocean is amenable to testing through an eddy resolving oceanic general circulation model in which the only dissipation mechanism is through the understress. Such a model should produce eddy fields of similar intensity to observation if understress is the dominant dissipation process in the baroclinic ocean.

For the atmosphere, it would be interesting to compare the predictions of an atmospheric general circulation model with and without feedback on the surface shearing stress from an oceanic eddy field, especially with regard to the regions of cyclogenesis.

REFERENCES

Bye, J.A.T., 1970. Eddy friction in the ocean. J. Mar. Res., 28: 124–134.
Bye, J.A.T., 1980. Energy dissipation by the large scale circulation. Ocean Modelling, 31: 12–13.
Bye, J.A.T., 1985. Momentum exchange at the sea surface by wind stress and understress. Q.J. R. Meteorol. Soc. (submitted).
Csanady, G.T., 1984. The free surface turbulent shear layer. J. Phys. Oceanogr., 14: 402–411.

Gill, A.E., 1980. Editorial in Ocean Modelling, 29: p. 10.

Hastenrath, S., 1982. On meridional heat transports in the World ocean. J. Phys. Oceanogr., 12: 922–927.

Kraus, E.B., 1972. Atmosphere–Ocean Interaction. Oxford Univ. Press, Oxford, 284 pp.

LeBlond, P.H. and Mysak, L.A., 1978. Waves in the Ocean. Elsevier, Amsterdam.

Paltridge, G.W., 1978. The steady-state format of global climate. Q.J. R. Meteorol. Soc., 104: 927– 946.

Phillips, O.M., 1966. The Dynamics of the Upper Ocean. Cambridge Univ. Press, Cambridge, 261 pp.

Sheppard, P.A., Tribble, D.T. and Garratt, J.R., 1972. Studies of turbulence in the surface layer over water (Lough Neagh). Part I. Instrumentation, Programme profiles. Q.J. R. Meteorol. Soc., 98: 627–641.

Stommel, H., 1948. The westward intensification of wind-driven ocean currents. Trans. Am. Geophys. Union, 29: 202–206.

CHAPTER 7

CLIMATE VARIABILITY STUDIES WITH A PRIMITIVE EQUATION MODEL OF THE EQUATORIAL PACIFIC

M. LATIF, E. MAIER-REIMER and D.J. OLBERS

ABSTRACT

A primitive equation equatorial model has been developed to study climate variability due to wind and thermodynamic forcing in an equatorial region. The model basin extends from 30° S to 30° N and zonally over 140°, has a variable horizontal resolution (50–800 km) and 13 vertical levels. Experiments are performed with observed annual cycle as well as 32 years of observed bimonthly wind data. A preliminary analysis of these experiments shows that the model is capable of simulating the basic pattern of annual as well as interannual variability of the Pacific Ocean. In particular, the model response shows evidence of the major El Niños which occurred between 1947 and 1978.

1. INTRODUCTION

The equatorial Pacific Ocean exhibits the most prominent interannual signal of climate variability known as El Niño. During these events warm surface waters appear for several months over the entire equatorial zone. The relation of the phenomenon to the wind anomaly field has clearly been revealed (Wyrtki, 1975) but also the imbedding of the El Niños in global interannual climate variations has been demonstrated by the connections to the Southern Oscillation (e.g., Wright, 1977) and to teleconnections to North American weather indices (e.g., Horel and Wallace, 1981). Apparently, modelling the Pacific Ocean circulation is an essential prerequisite of understanding and eventually predicting this part of global climate variability. A major step in this task was the clear attribution of observed sealevel variations to equatorially trapped waves forced by realistic wind anomalies (e.g., Busalacchi and O'Brien, 1981; Cane, 1984), a piece of work which was the logical consequence of a series of equatorial wave process studies as those of Hurlburt et al. (1976), McCreary (1976) and Cane (1979a, b). However, these models do not include thermodynamic processes and these are necessary ingredients to predict sea-surface temperature which is the essential parameter in air–sea interaction studies. There is a range of models designed for this purpose where high-resolution, general circulation models represent the highest degree of complexity. In this paper we present an investigation of the response of such a model to realistic wind forcing.

High-resolution oceanic circulation models (OGCM) have traditionally been used to study the mean steady state of the ocean circulation. In the equatorial region, however, the wind forcing is highly variable on time scales comparable with the intrinsic oceanic time scales so that the oceanic adjustment cycles to changing winds appear as relevant to study as the asymptotic mean state of the circulation. Philander has investigated the response of an equatorial OGCM to abruptly and periodically changing winds in a series of papers (Philander and Pacanowski, 1980, 1981; Philander, 1981) and has revealed the importance of nonlinear advective effects in addition to the wave processes for the response of the upper ocean. Simulations of the seasonal cycle of the tropical Atlantic Ocean as well as the 1982/83 El Niño event have been performed (S.G.H. Philander, pers. commun., 1984) and have clearly demonstrated the ability of OGCM's to reproduce the observed variations of the thermal and velocity structure of the equatorial ocean in response to the actual wind forcing.

The numerical model used in the present study differs slightly from the model used by Philander. The dimensions of the model are those of the equatorial Pacific. We use variable resolution in the latitudinal and longitudinal directions. In particular in the central Pacific the resolution is considerably coarser than in Philander's model. There is no horizontal diffusive transport of heat and the remaining friction and diffusion parameters are slightly larger than the values used by Philander.

The wind fields are bimonthly data for 32 years resulting from an objective analysis of Barnett (1983) of the data set of Wyrtki and Meyers (1975a, b). It represents a heavily filtered state of the actual Pacific wind field but it was the only wind field available for the range of three decades. The experiment was performed as a sequence, spinning up the model from rest by mean winds, followed by forcing the model with the seasonal cycle and finally by the 32 years of realistic winds. In this paper we present a preliminary analysis of the resulting data set with respect to the mean state, the seasonal cycle and interannual variability of the model. Also, the simulation of El Niño is discussed.

2. THE MODEL

The numerical model used in this study is based on primitive equations in the Boussinesq and hydrostatic approximations on an equatorial β-plane:

$$u_t + (u \cdot \nabla)u + \beta y k \times u + \nabla p = A_h \nabla^2 u + (A_v u_z)_z \tag{2.1}$$

$$\nabla \cdot u = 0 \tag{2.2}$$

$$T_t + (u \cdot \nabla)T = (K_v T_z)_z \tag{2.3}$$

There is no horizontal diffusion of heat in the model because there is no physical or numerical reason to incorporate any horizontal mixing. Moreover, the values of diffusion coefficients are notoriously uncertain anyhow.

The pressure is given by:

$$p = g\zeta - g\alpha \int_z^0 T dz' \tag{2.4}$$

where ζ is the surface elevation which is not eliminated as in the GFDL-type models (e.g., Bryan, 1969; Philander and Pacanowski, 1980) but kept as a prognostic variable calculated from the vertically integrated continuity equation:

$$\zeta_t + \nabla \cdot \int_{-H}^0 u dz = 0 \tag{2.5}$$

This equation is numerically integrated by an implicit algorithm which damps the surface mode but is neutral with respect to the geostrophic mode. The baroclinic part (2.1) and (2.3) is integrated explicitly with a time step of two hours.

The model ocean is a rectangular box with the zonal extent of the Pacific (14,400 km) and a latitudinal extent of 6600 km and a constant depth of 4000 m. An Arakawa E-grid is used with variable grid distance: the meridional resolution decreases from 50 km at the equator to 420 km at the northern and southern boundaries, the longitudinal resolution is 50 km at the coasts and decreases to 800 km in the center. The 13 vertical levels are spaced 10 m near the surface and the spacing increases to 500 m below 500 m depth. The vertical velocity is computed between the levels. The eddy viscosities A_v and A_h and the diffusivity K_v are chosen constant with values $A_v = 15$, $A_h = 10^8$ and $K_v = 1$ in units $cm^2 s^{-1}$.

The model is forced at the surface by wind and heating:

$$A_v u_z = t_{-0} \qquad \text{at } z = 0 \tag{2.6}$$

$$K_v T_z = \gamma(T_0 - T) \tag{2.7}$$

where the wind stress t_{-0} and the forcing temperature T_0 are prescribed. The wind stress field is described in Section 3, the forcing temperature was taken constant as $T_0 = 26°C$. The constant γ determines the time scale on which the surface temperature would relax to T_0 in the absence of advection (Haney, 1971). With the thickness $d = 10$ m of our first layer this time scale d/γ is chosen as 28 days.

At the bottom (which is flat) and at lateral boundaries a no-slip condition is used. We apply a no-heat flux condition at the bottom. Since there is no explicit horizontal diffusion of temperature, no further boundary conditions are needed. Notice, however, that with these boundary conditions the model ocean is thermally isolated at the walls and the bottom. So starting with an initial temperature field $T < T_0$, the ocean will continually become warmer due to the surface heat flux (2.7).

3. THE WIND FIELD

The wind field was taken from an objective analysis of the Wyrtki and Meyers (1975a, b) data set by Barnett (1983). This analysis yielded bimonthly wind vectors for the period 1947–1978 in the equatorial strip 30°S–30°N and 150°E–80°W. The annual cycle of this data set is given on a $2° \times 10°$ grid, wind anomalies have been smoothed on a $4° \times 20°$ grid. Barnett's analysis used a filtering technique based on empirical orthogonal functions which eliminated small-scale noise. Figure 1 shows an example of wind anomalies before and after filtering for the particular bimonth. As pointed out by Barnett from a statistical point of view, only ten EOF's (which make about half of the observed variance) of the anomaly field can be classified as signal. The first five EOF's have been included in the analysed wind field. These describe about 45% of the variance. Apparently, the filtering is rather strong and the variability in the model can be expected to be too weak. However, though being not optimal for an equatorial variability study the wind field was the only one available to us, covering this extended range of three decades.

The wind vectors were interpolated onto the model grid and wind stress vectors were

66

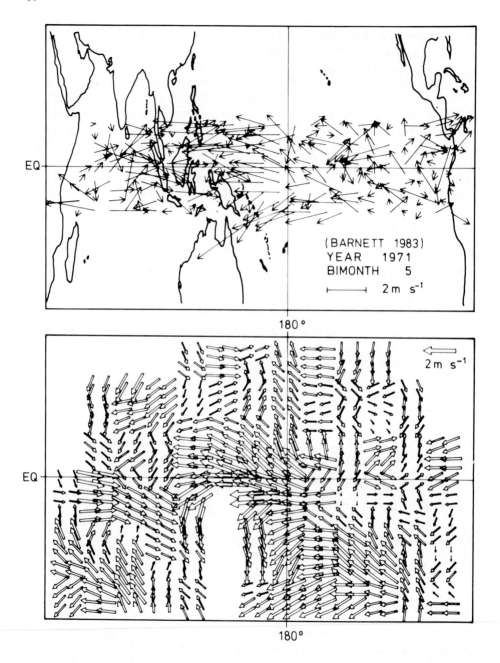

Fig. 1. Original wind anomalies (upper panel, from Barnett (1983)) and reconstructed anomaly field (lower panel) for bimonth 5 of 1971.

calculated with a constant drag coefficient of 1.5×10^{-3} and interpolated linearly in space and time. For the equatorial wave guide the zonal wind stress component is the most important forcing mechanism and essential in understanding the thermal response.

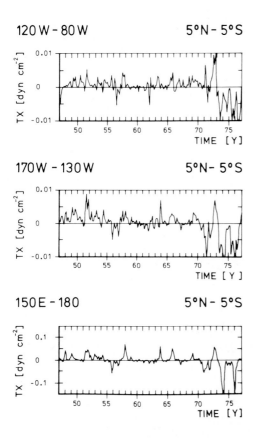

Fig. 2. Time series of zonal wind anomaly averaged over the indicated areas.

Figure 2 displays time series of the anomaly of the zonal wind stress, averaged over three different regions 5°S–5°N in the eastern, central and western Pacific. Most obvious is the predominance of westerly anomalies over the entire Pacific after the El Niño year 1972. Notice further that the variance in the eastern and central Pacific is much less than in the western Pacific.

4. THE EXPERIMENTS

The sequence of experiments is displayed schematically in Fig. 3a. Initially the ocean was at rest with a horizontally uniform stratification displayed in Fig. 3b. At $t = 0$ the mean annual wind was switched on. These mean winds were obtained by averaging the climatological seasonal cycle of Barnett's (1983) data. After seven months of integration the system was close to a steady state. Integration was then carried on using the seasonal cycle of the wind field. After three cycles the model reached an almost cyclostationary state. Then the wind anomalies for the period 1947–1978 were imposed upon the seasonal cycle in order to stimulate the interannual variability of the Pacific Ocean over these 32 years.

68

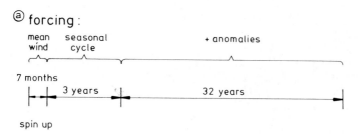

(a) forcing :

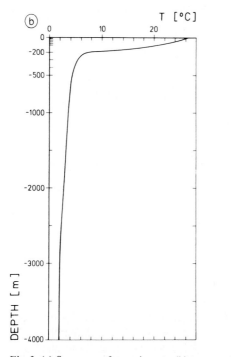

Fig. 3. (a) Sequence of experiments; (b) temperature profile at the start of the spin-up phase.

Results will be presented for the situations after the spin-up experiment, for the seasonal cycle, and during the 32-year run.

4.1. The spin-up experiment

After the seven months of the spin-up phase the model was not in complete equilibrium but the major features of the equatorial currents and the temperature field were clearly established. A meridional section of the zonal velocity and the temperature field in the upper 300 m is shown in Fig. 4a. There is a westward surface current with velocities up to 30 cm s^{-1}, on top of an eastward undercurrent with a speed of about 50 cm s^{-1} in its core at about 100 m depth. The width of the undercurrent, about 3°, is controlled both by horizontal friction and by inertial effects since the scales $(A_h/\beta)^{1/3} \cong 125$ km and also $(U/\beta)^{1/2} \cong 100$ km have comparable magnitude. At about 8°N the equatorial counter

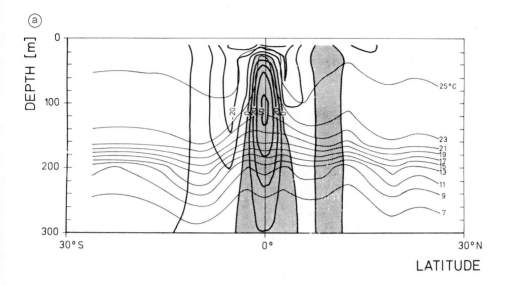

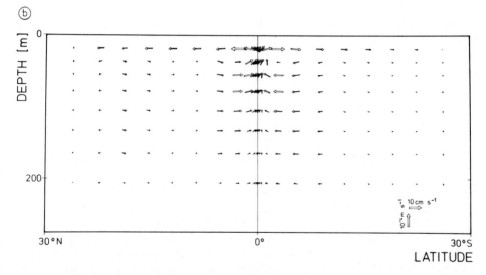

Fig. 4. (a) Meridional cross-section of zonal velocity and of temperature along the central meridian at the end of the spin-up phase; shaded area has eastward velocities; (b) meridional circulation along the central meridian at the end of the spin-up phase.

current appears with velocities up to $10 \, \mathrm{cm \, s^{-1}}$. The temperature field reveals the familiar spreading in the equatorial strip, associated with the geostrophic convergence and equatorial upwelling. This meridional circulation for the same section is displayed in Fig. 4b. Vertical velocities get up to $0.5 \times 10^{-2} \, \mathrm{cm \, s^{-1}}$ above the core of the undercurrent.

Figure 5 shows the SST at the end of the spin-up. The cold water tongue due to up-welling is clearly present with a pool of 21°C water located around 120°W and an increase to about 26°C at the western boundary. This zonal gradient is too weak: in the

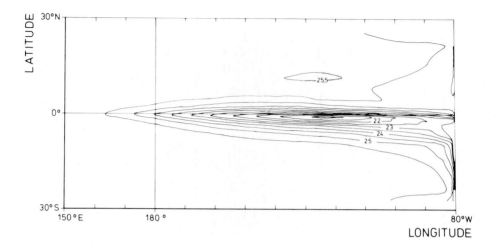

Fig. 5. Sea-surface temperature at the end of the spin-up phase.

atlas of Robinson (1976) we find a pool of 22°C water at 95°W and about 29°C in the western equatorial Pacific.

4.2. The seasonal cycle

After forcing the model three years with the seasonal cycle of the wind stress an almost cyclostationary state was reached, showing no significant changes in the model variables between the second and the third cycle. The data of the seasonal cycle presented in this section were, however, extracted from the following 32 years of integration with the full wind forcing.

In Fig. 6 we compare the resulting seasonal cycle of the temperature at the surface and at 100 m depth at some points along the equator with the corresponding data of the Robinson (1976) atlas. These figures again demonstrate the failure of the model to simulate the zonal temperature gradient correctly. While the mean temperatures agree quite well at the eastern coast they are about 4°C too cool in the western part. However, some features are reproduced by the model quite well as the decrease of the annual range of the SST and the decrease of the vertical temperature gradient from east to west. However, the annual range of SST is too small in the model, the maximum is about 1°C in the eastern Pacific as compared to about 3°C in the Robinson data. Notice that the model range may be a little underestimated due to the bimonthly averaging.

4.3. The 32-years experiment

In the last stage of the experiment the model was forced with the 32 years of the actual wind stress field taken from Barnett's (1983) analysis. Only bimonthly averages of the velocity vector and the temperature at all grid points could be saved on tape. A preliminary analysis of this averaged model response is presented in this section.

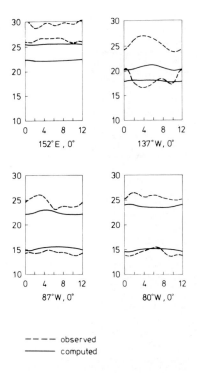

--- --- observed
——— computed

Fig. 6. Seasonal cycle of temperature on the equator at the surface and at 100 m for different longitudes. Dashed lines represent the observed cycles taken from Robinson (1976).

Figure 7 shows a longitude–time section of the model SST on the equator. The eastern and central response is obviously dominated by the seasonal cycle, most apparent in the cold-water pool centred at about 130°W. In the western response (emphasized by the 24.0°C isotherm) intra- and interannual variations appear to have larger amplitudes than the seasonal cycle. Since model SST anomalies have comparable variances along the equator (as shown below in the center panel of Fig. 8a, b) we can attribute the blurring of seasonal variations in the western part to the low amplitudes of the seasonal cycle discussed in the last section.

A further obvious feature in Fig. 7 is the slow increase of SST in the eastern and central part from the beginning of the experiment until 1971–72. The increase is about 1°C over these 25 years, the trend being most obvious in the temperature of the cold water pool and the slow shift of the 22.5°C isotherm to the west. The western SST appears not to participate in this trend. An explanation of the warming will be given later.

After 1972 the warming phase abruptly stops and the SST in the eastern part cools by about 1°–2°C. Also, the seasonal amplitude amplifies. This behaviour can be attributed to the sudden occurrence of the large wind stress anomalies over the entire equatorial area east of the date line, which we have pointed out in Fig. 2. Observations of SST do not indicate such a continuous decrease after the El Niño of 1972. Thus, either the strong

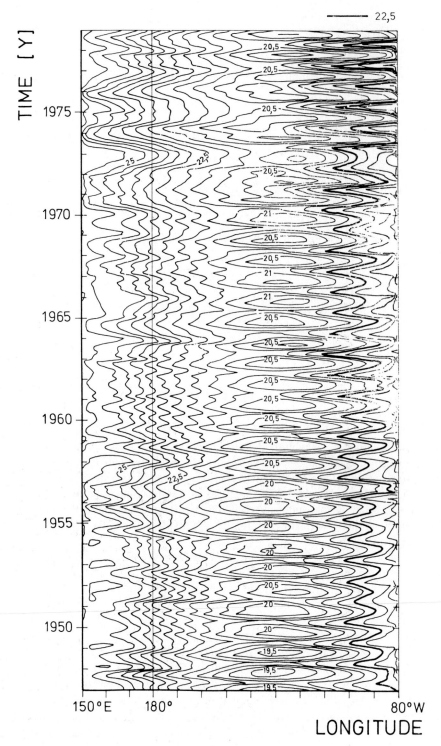

Fig. 7. Longitude–time cross-section of computed sea-surface temperature on the equator from 1947 to 1978.

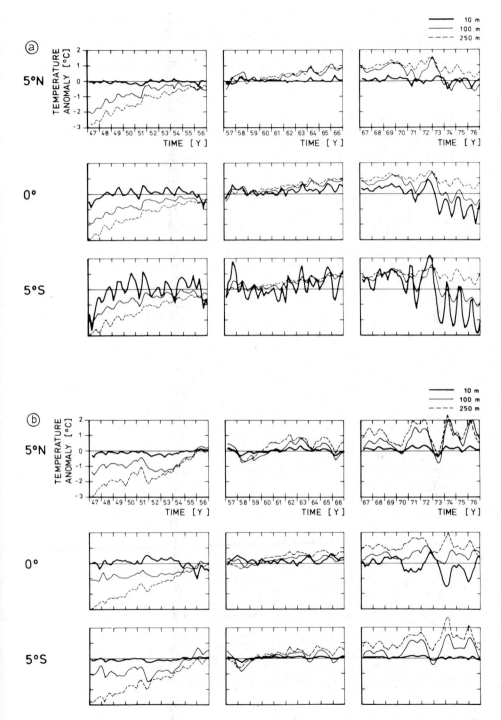

Fig. 8. (a) Time series of temperature anomaly at 80°W for 10, 100 and 250 m depth and for 5°N, 0° and 5°S; (b) same as (a), but for 165°E.

74

wind anomaly after 1972 is overemphasized in Barnett's analysis or the sensitivity of the model SST response to wind anomalies is too large. This latter hypothesis, however, is unlikely since the SST variability in the model is too low. This was revealed already in the seasonal amplitudes of Fig. 6 in the last section, but also SST anomalies exemplified

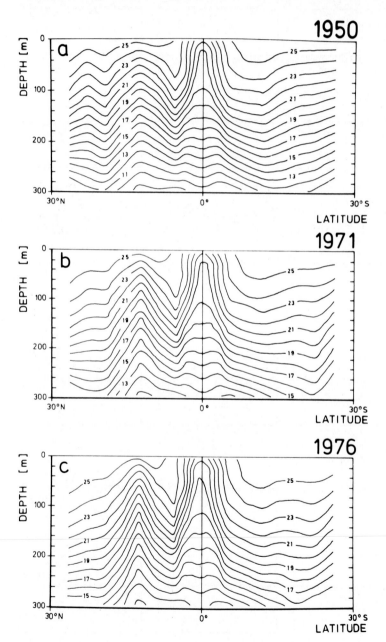

Fig. 9. Meridional cross-section of temperature along the central meridian for (a) Jan./Feb. 1950; (b) Jan./Feb. 1971; and (c) Jan./Feb. 1976.

in the following figures have too small variances.

Figures 8a and b display time series of temperature anomalies at 5°N, the equator and 5°S at the eastern coast (80°W) and in the western part at 165°E. Some features of the SST anomaly at the equator (middle panel) have been discussed guided by the full signal in Fig. 7: the slow increase at 80°W until 1972 and the drop with an increased annual amplitude thereafter. While the SST anomalies at 5°S, 80°W show these features in much pronounced form they are entirely absent at 5°N, 80°W and in the western part. Notice also the large regional differences in the SST variances: at 5°N and the equator they are small and of comparable magnitude in the eastern and western part, at 5°S we find SST anomaly amplitudes of 1°C with peak values over 2°C in the eastern part and only very low values in the western part.

The behaviour of the temperature anomalies at 100 m and 250 m depth is dominated by a much larger warming than the surface temperatures discussed so far. The temperature increases at all locations shown in Fig. 8 by about 2°C at 100 m and about 4°C at 250 m. At deeper levels (not shown) the increase diminishes again, the anomaly at 700 m is an almost straight line with a 1°C increase over 32 years and at 2000 m we find less than 0.5°C increase. Still, the basic stratification and structure of the velocity field is not destroyed in the course of the integration: Fig. 9 shows meridional sections of temperature in the center of the basin shortly after the beginning and at the end of the experiment.

The warming of the model can clearly be attributed to the continuous input of heat at the surface. The total input

$$Q = \rho c_p \gamma \int_{\text{surface}} d^2 x (T_0 - T) \qquad (4.1)$$

was fairly constant during the 32 years integration and amounts to 1.25×10^{15} W, about the amount that should leave the equatorial Pacific by meridional transport (e.g., Stommel, 1980). If the total heat received during the 32 years is distributed over 1000 m the temperature increase is about 3.5°C in agreement with our results in the deeper layers. The surface temperature itself does not follow this trend because of the relatively rigid boundary constraint implied by the surface heating parameterization (2.7). Vertical advection and surface heat input are of comparable size for the relaxation time of 28 days used in the boundary condition (2.7).

5. EL NIÑO EVENTS

The capability of reduced gravity models and models based on several baroclinic modes to simulate El Niño events by using a realistic wind as forcing has been demonstrated quite successfully in the recent years (e.g., Busalacchi and O'Brien, 1981; Busalacchi and Cane, 1984; Cane, 1984). The connection between pycnocline height and sealevel simulated in these models and SST, which is the essential parameter of air–sea interactions, is however not at all simple, as demonstrated by Schopf and Cane (1983). It appears thus a necessary prerequisite of coupled models to test the performance of

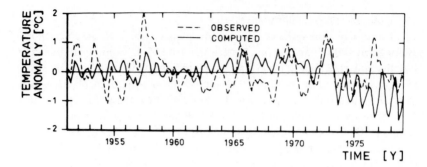

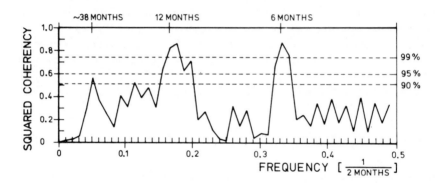

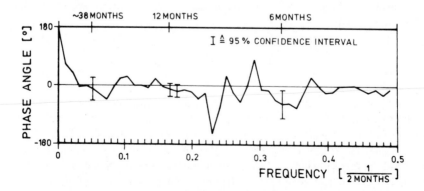

Fig. 10. Time series of computed (full line) and observed (dashed line) sea-surface tempera
anomaly at Galapagos (top), coherence spectrum (centre) and phase spectrum (bottom) of the
series.

models which include the thermodynamics. In fact, this was one of the aims of the project.

During the time span 1947–1978 of the 32 years integration there were three major El Niños in the years 1957/58, 1965 and 1972. The model was able to simulate the El

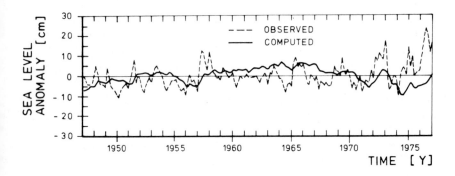

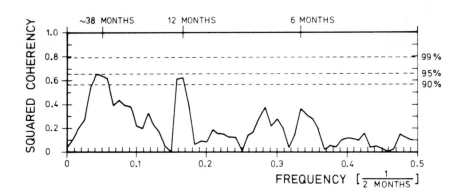

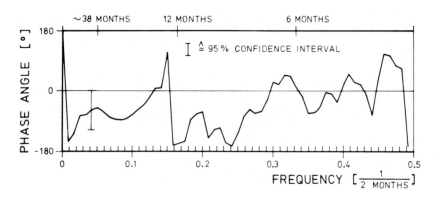

Fig. 11. Same as Fig. 10, but for sealevel anomaly at Talara (Peru).

Niño-type pattern in these years, however, as pointed out in the last sections, the SST anomalies in the model have generally too small amplitudes and this defect also applies to the model El Niños.

The occurrence of anomalous warm water at the eastern coast in the years 1957/58 and 1972 is already evident in the time series of temperature anomalies shown in Fig. 8a, particularly on the equator and at 5°S. A closer comparison of SST anomalies at Galapagos is given in Fig. 10. The simulated anomalies reproduce the warm phases in 1957/58, 1965 and 1972, generally with smaller amplitudes. The colder phases between El Niños do not occur in the model, except for the sudden cooling after 1972. The transition from the cool phase to the 1976 El Niño, clearly visible in the Galapagos SST is not accomplished at all.

Squared coherence and phase of the observed and simulated SST at Galapagos are shown in Figs. 10b and c, computed for the period 1951–1978. Significant peaks (at 99%) occur at the annual and semiannual period and there is a less significant peak (at 90%) at a period of about 38 months. The phase vanishes almost consistently over the whole frequency range. Though being only marginally significant the peak at 38 months was found in other cross-spectra for the eastern Pacific. A further example is given in Fig. 11 showing observed and simulated anomalies of sealevel at Talara as well

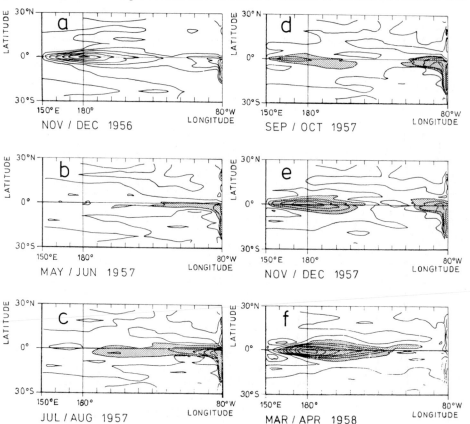

Fig. 12. Evolution of computed sea-surface temperature anomaly during the 1957 El Niño. Shaded areas indicate positive anomalies greater than 0.15°C. Contour interval is 0.15°C.

as their coherence and phase. Rasmusson and Carpenter (1982) have found pronounced peaks in the spectra of the Southern Oscillation index, in the SST at the eastern coast and the sealevel pressure in the eastern Pacific at periods between 36.6 and 42.7 months. The wind field has a peak in this frequency range and apparently the wind forces the model consistently to produce the coherence and correct phase with the corresponding signal in the observed SST and sealevel.

The succession of the SST-anomaly pattern for the El Niños 1957/58 and 1972 is presented in Figs. 12 and 13. By and large the development of the pattern is similar to the El Niño phases as e.g. described in the composite of Rasmusson and Carpenter (1982), but some details do not agree. In most cases these discrepancies can be traced back to differences in the wind anomaly field of Barnett's (1983) analysis and Rasmusson and Carpenter's composite. Consider the development of the model 1957/58 El Niño presented in Fig. 12. In November/December 1956 (Fig. 12a) we find a cold anomaly over the whole equatorial Pacific with a cold pool of water off the eastern coast and around the date line. In early 1957 this starts to disappear and in May/June 1957 (Fig. 12b) a positive anomaly has developed at the east coast which extends along the equator into the eastern Pacific and continues to grow and spread westward until July/August

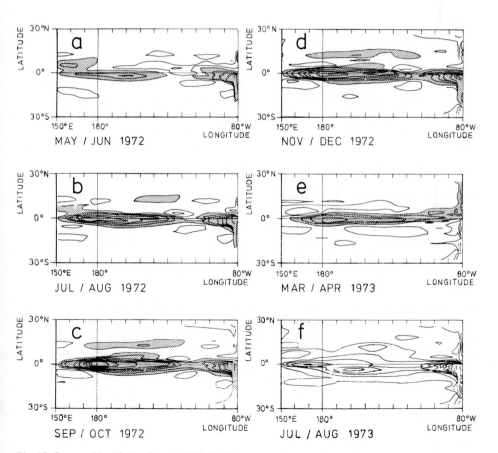

Fig. 13. Same as Fig. 12, but for the 1972 El Niño.

1957 (Fig. 12c). So far this whole sequence agrees quite well with the composite. In September/October 1957 (Fig. 12d), however, the warm-water tongue starts to separate due to the development of cold anomalies in the central Pacific east of about 140°W. This entire pattern still grows in amplitude (positive as well as negative anomalies) during November/December 1957 (Fig. 12e). In the following six months the eastern positive anomaly disappears and the western positive anomaly grows and propagates eastward (Fig. 12f) until in the middle of 1958 the whole equatorial region is anomalously warm and the second peak of the 1957/58 El Niño develops. The sequence of Fig. 13 shows a similar development for the 1972 El Niño, except that the warm phase in 1972 and early 1973 is followed by basinwide cold anomalies in the middle of 1973.

The splitting of the warm-water tongue during the warming phase is less pronounced in the 1972 El Niño. In both cases this phenomenon can be attributed to strong easterly wind anomalies between 120° and 140°W which are not present in Rasmusson and Carpenter's (1982) maps. These wind anomalies converge south of the equator and cause anomalous upwelling of cold water which leads to the splitting of the tongue. We should emphasize that the anomalous wind in this region may be real and that the resulting upwelling only succeeds to produce a cold SST anomaly because the simulated warm anomalies are too small.

6. DISCUSSION

The intent of the experiments reported here is to test the climate variability of a high-resolution equatorial ocean model forced by realistic winds. Of interest was mainly the response in the thermal structure of the upper ocean, in particular SST, since this is the key oceanic parameter for modelling the coupled ocean—atmosphere system. As briefly outlined in the introduction, the equatorial ocean plays a major role in such a project when dealing with climate variability with time scales up to a decade.

To summarize the results, the model performed reasonably well to reproduce the basic pattern of the thermal and velocity structures of the Pacific Ocean. Amplitudes of the seasonal cycle as well as of interannual variations like the El Niño events are generally too small. However, the overall pattern of these variations are simulated quite well. Besides this realistic variability the model exhibits an artificial warming trend due to continuous heating at the surface and isolated lateral walls at the northern and southern boundaries.

One obvious reason for any disagreement between the model variability and observations is definitely the poor quality of the wind data set. It is clear that the original wind data are severely undersampled, especially in the western Pacific and the southern hemisphere which are the crucial areas for predicting the wave and advective contributions to SST variability. Furthermore, a large part of the variance in the original wind data has been filtered by the objective analysis of Barnett (1983). For this reason work is now in preparation on running the model with other wind data for shorter periods.

ACKNOWLEDGEMENTS

We appreciate the support of Tim Barnett for making the wind data and SST data available to us. Many thanks also to Klaus Wyrtki for allowing us to use the sealevel data

We also like to thank Mrs. Marion Grunert for preparing the figures and Mrs. Ulla Kircher for typing the manuscript.

REFERENCES

Barnett, T.P., 1983. Interaction of the Monsoon and Pacific trade wind system at interannual time scales. Part I: The equatorial zone. Mon. Weather Rev., 111 (4): 756–773.
Bryan, K., 1969. A numerical method for the study of the circulation of the ocean. J. Comp. Phys., 4: 347–376.
Busalacchi, A.J. and O'Brien, J.J., 1981. Interannual variability of the equatorial Pacific in the 1960's. J. Geophys. Res., 86 (C11): 10901–10907.
Busalacchi, A.J. and Cane, M.A., 1984. Hindcasts of sea level variations during the 1982/1983 El Niño. Submitted to J. Phys. Oceanogr.
Cane, M.A., 1979a. The response of an equatorial ocean to simple wind stress patterns: I. Model formulation and analytic results. J. Mar. Res., 37: 232–252.
Cane, M.A., 1979b. The response of an equatorial ocean to simple wind stress patterns: II. Numerical results. J. Mar. Res., 37: 253–299.
Cane, M.A., 1984. Modelling sea level during El Niño. Submitted to J. Phys. Oceanogr.
Haney, R.L., 1971. Surface thermal boundary condition for ocean circulation models. J. Phys. Oceanogr., 1: 241–248.
Horel, J.D. and Wallace, J.M., 1981. Planetary-scale atmospheric phenomena associated with the Southern Oscillation. Mon. Weather Rev., 109: 813–829.
Hurlburt, H.E., Kindle, J.C. and O'Brien, J.J., 1976. A numerical simulation of the onset of El Niño. J. Phys. Oceanogr., 5: 621–631.
McCreary, J., 1976. Eastern tropical ocean response to changing wind systems: With application to El Niño. J. Phys. Oceanogr., 6: 632–645.
Philander, S.G.H., 1981. The response of equatorial oceans to a relaxation of the trade wind field. J. Phys. Oceanogr., 11: 176–189.
Philander, S.G.H. and Pacanowski, R.C., 1980. The generation of equatorial currents. J. Geophys. Res., 85: 1123–1136.
Philander, S.G.H. and Pacanowski, R.C., 1981. Response of equatorial oceans to periodic forcing. J. Geophys. Res., 86 (C3): 1903–1916.
Rasmusson, E.N. and Carpenter, T.H., 1982. Variations in tropical sea surface temperature and surface wind fields associated with the Southern Oscillation/El Niño. Mon. Weather Rev., 110: 354–384.
Robinson, M.K., 1976. Atlas of North Pacific Ocean, monthly mean temperatures and mean salinities of the surface layer. Naval Oceanographic Office, Ref. Publ. 2, Washington, D.C.
Schopf, P.S. and Cane, M.A., 1983. On equatorial dynamics, mixed layer physics and sea surface temperature. J. Phys. Oceanogr., 13: 917–935.
Stommel, J., 1980. Asymmetry of interoceanic fresh-water and heat fluxes. Proc. Natl. Acad. Sci., 77: 2377–2381.
Wright, P.B., 1977. The Southern Oscillation-patterns and mechanisms of the teleconnections and the persistence. HIG-77-13, Hawaii Inst. of Geophys., University of Hawaii, Honolulu, Hawaii.
Wyrtki, K., 1975. El Niño – The dynamic response of the equatorial Pacific Ocean to atmospheric forcing. J. Phys. Oceanogr., 5: 572–584.
Wyrtki, K. and Meyers, G., 1975a. The trade wind field over the Pacific Ocean. Part I: The mean field and the annual variation. Rep. HIG-75-1. Hawaii Inst. Geophys., University of Hawaii, Honolulu, Hawaii, 26 pp.
Wyrtki, K. and Meyers, G., 1975b. The trade wind field over the Pacific Ocean. Part II: Bimonthly fields of wind stress; 1950 to 1972. Rep. HIG-75-2. Hawaii Inst. Geophys., University of Hawaii, Honolulu, Hawaii, 16 pp.

CHAPTER 8

RESPONSE OF THE UK METEOROLOGICAL OFFICE GENERAL CIRCULATION
MODEL TO SEA-SURFACE TEMPERATURE ANOMALIES IN THE TROPICAL
PACIFIC OCEAN

T.N. PALMER

ABSTRACT

A number of tropical Pacific SST anomaly experiments, run on the UK Meteorological Office 11-level general circulation model in perpetual January mode, are described. It is found that the model's extratropical response can be statistically significant as far downstream as the European continent depending on the anomaly used, but is sensitive to the specification of orography in the model. A realistic response to a composite El Niño SST anomaly is obtained in the extratropics provided envelope orography replaces the standard orographic specification. A negative East Pacific anomaly run revealed that aspects of the tropical response are not readily explained by linear theory and some possible reasons for this are proposed. The extratropical response appears in some experiments to be qualitatively consistent with a downstream Rossby wavetrain and this has been tested using a barotropic model. In other experiments, and in the Southern Hemisphere, this is less clear. The extent to which the extratropical response is maintained by cyclogenesis in mid-latitudes remains to be firmly established though preliminary results suggest it is important. The extratropical response is sensitive to relatively weak SST anomalies in the tropical West Pacific.

1. INTRODUCTION

Some details are given in this paper of a number of tropical Pacific SST anomaly experiments run on the UK Meteorological Office high-resolution 11-level (grid point) GCM. The model has been described by Saker (1975), and the horizontal resolution used for the experiments described below is $2\frac{1}{2}° \times 3\frac{3}{4}°$. The GCM has a penetrative convection scheme (Lyne and Rowntree, 1976) and the version used here has zonally averaged non-interactive cloud amounts. SSTs and sea-ice are fixed throughout the integration.

The integrations are summarised in Table 1. They consist of experiments run with

TABLE 1

Integrations on the 11-level model

Experiment	SST anomaly	Orography
a	control	standard
b	+ Fig. 1a	standard
c	− Fig. 1a	standard
d	Fig. 1b	standard
e	control	2σ envelope
f	Fig. 1b	2σ envelope
g	Fig. 1c	2σ envelope

84

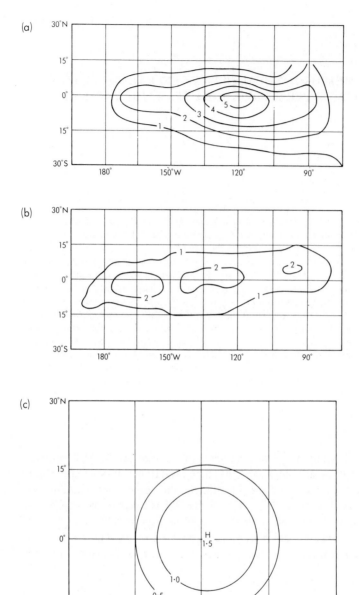

Fig. 1. (a) SST anomaly (K) for experiments *b* and *c* (December 1982); (b) SST anomaly (K) for experiments *d* and *f* (enhanced Rasmusson and Carpenter, 1980, composite, as used by Blackmon et al., 1983); and (c) SST anomaly (K) for experiment *g* (warm West Pacific).

both plus and minus the observed SST anomaly for December 1982 (Fig. 1a); experiments with an enhanced version of the composite Rasmusson and Carpenter (1982) El Niño SST anomaly (as used by Blackmon et al., 1983; Fig. 1b), both with standard and envelope orography; and an experiment with a small (idealised) warm SST anomaly in the tropical West Pacific (Fig. 1c). The integrations have all been initialised with data from 28 December 1972, and run for 540 days in a perpetual January mode, except for experiment g which has, at present, been run for 90 days.

2. EXPERIMENTS

2.1. Experiments a, b and c; a strong El Niño anomaly in the East Pacific

Addition of the observed SST anomaly for December 1982 (Fig. 1a) to climatological SST (Fig. 2) produced a strong absolute maximum in SST near 120°W. The associated anomalous rainfall is illustrated in Fig. 3a, with a maximum value of 12 mm per day, lying close to this SST maximum. To the west of this maximum, 850 mb wind anomalies are westerly with strengths up to about $10 \, \mathrm{m \, s^{-1}}$ (Fig. 3b) cancelling the normal easterly trade winds and converging into the region of maximum precipitation. The observed 850 mb wind anomaly and satellite-sensed outgoing longwave radiation anomalies (OLR) are shown in fig. 4 of Rasmusson and Wallace (1983). The agreement in the general strength and direction of the wind anomalies is good, although, for the period December–February, the time-averaged maximum OLR anomaly is positioned about 30° further west than the model's rainfall anomaly.

The 540-day mean 200 mb geopotential height anomalies (in dm) for the experiments with positive and negative SST anomalies are shown in Fig. 4a and b, respectively. In Fig. 4a there are positive anomaly centres to the north and south of the rainfall anomaly maximum, at 15°N and 15°S and about 120°W. In the Northern Hemisphere there is some indication of a wavetrain with centres over the western United States, eastern Canada, and the British Isles. The response in mid-latitudes is approximately equivalent barotropic. The response in the Southern Hemisphere is somewhat different, corresponding to a dominant signal in the anomalous zonal mean flow.

In Fig. 4b the anomaly field in the tropics is similar to the negative of Fig. 4a except

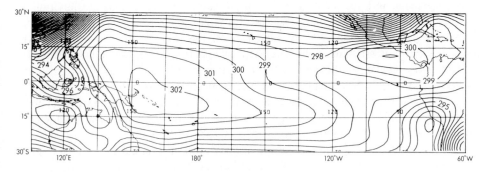

Fig. 2. Climatological winter SSTs (K) in the tropical Pacific.

(a)

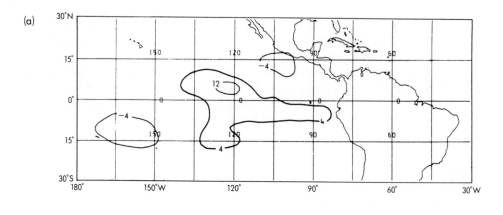

(b)

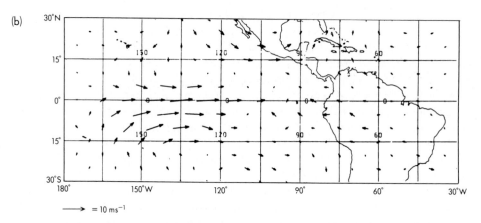

⟶ = 10 ms⁻¹

Fig. 3. (a) convective rain anomaly (mm/day); and (b) 850 mb wind anomaly for experiments *b–a* (December 1982).

that the cyclonic centres are nearly 15 degrees to the west of the corresponding anti-cyclonic centres, as is the maximum negative rainfall anomaly relative to the corresponding maximum for the positive SST anomaly experiment. It is interesting to note that, whereas the magnitude of the anomalous cyclonic centres is similar to that of the corresponding anticyclonic centres, the magnitude of the anomalous precipitation is not, with the negative rainfall anomaly (not illustrated) some two times smaller than the corresponding positive anomaly. In mid-latitudes, the response to the negative SST anomaly is quite uncorrelated with the pattern in Fig. 4a. In fact, the pattern in Fig. 4b is qualitatively similar to the Pacific/North American (PNA) teleconnection pattern, discussed by Horel and Wallace (1981), with anomaly centres over the Aleutian Islands, Hudson Bay and the southeastern states of America. Whilst the response in Fig. 4a does not show a strong PNA teleconnection pattern characteristic of the observed December–February 200 mb height anomaly composite (Quiroz, 1983), the December 200 mb height anomaly did not itself show a strong PNA pattern. Preliminary results from a recent experiment with the observed December–February 1982/83 composite SST anomaly (see Quiroz, 1983), and envelope orography (see Section 2.2) show a stronger

PNA pattern which agrees well with observations. It is interesting to note in this respect that the December–February composite SST anomaly was weaker, and positioned further west than the December only SST anomaly (see Quiroz, 1983).

In order to test that the difference fields illustrated in Fig. 4 are statistically significant, t-statistics of the six 90-day mean fields have been produced, treating them as independent samples. Of course this latter assumption cannot be completely justified. Recent tests (D.A. Mansfield, pers. commun., 1984) have shown that a GCM's evolution depends strongly on its initial conditions up to about day 15 of an integration. Hence, only about 2/3 of a given 90-day period can be thought of as essentially independent of its neighbouring periods.

A second problem in relating a model-generated t-statistic to the atmosphere, concerns the model's low-frequency variability. The model's low-frequency variability is smaller than corresponding atomospheric values so for a given size of anomaly, the model may show stronger significance than the atmosphere.

The t-statistics for the fields shown in Fig. 4a and b are illustrated in Fig. 5a and b respectively. With each experiment comprising six 90-day periods, there are ten degrees of freedom in the t-test. A two-sided significance of 1% with this number of degrees of freedom corresponds to a value of about three. (With half this number of degrees of freedom the corresponding value is about four.) In Fig. 5a, in the Northern Hemisphere, the low centres over the United States and the British Isles are strongly significant. The tropical high centres are also significant at 1%, though the high over the Canadian east coast is not. Much of the southern hemispheric response is significant at 1%. In Fig. 5b the tropical low centres and the Northern Hemisphere mid-Pacific high are strongly significant. The high/low dipole downstream of these anomaly centres is barely significant. Parts of the Southern Hemisphere response also appear to be significant at 1%.

Bearing in mind the caveats discussed above, it appears that with such large SST anomalies, the atmospheric response can be strongly significant, not only over the Pacific, but further downstream as far as the European continent.

In discussing reasons for the asymmetry between the positive and negative SST anomaly experiments, it is important to bear in mind that the anomaly fields do not represent differences from a zonally averaged basic state. In the control integration SSTs are highest in the tropical West Pacific. Associated with this there is large-scale ascent and a precipitation maximum over the West Pacific with general descent over the East Pacific, forming part of the so-called Walker circulation. A decrease in SST east of the dateline may not substantially change the rainfall field in the East Pacific, as it is already small. The largest rainfall differences are likely to occur near the western edge of such a negative SST anomaly where its addition may reverse the sense of large-scale vertical motion. On the other hand, a large increase in SST in the East Pacific may completely reverse the Walker Circulation in the East Pacific and give rise to substantial anomalous convection over the SST anomaly. Hence, the longitudinal position of the latent heating anomaly would not be expected to be identical for positive and negative SST anomaly fields. As we have noted above (see also Sections 2.2 and 2.3), the extratropical PNA pattern appears to be more readily excited the further west, over the Pacific, the latent heating anomaly occurs.

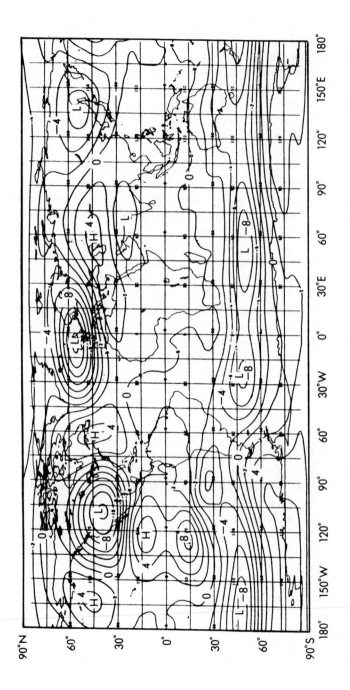

(a)

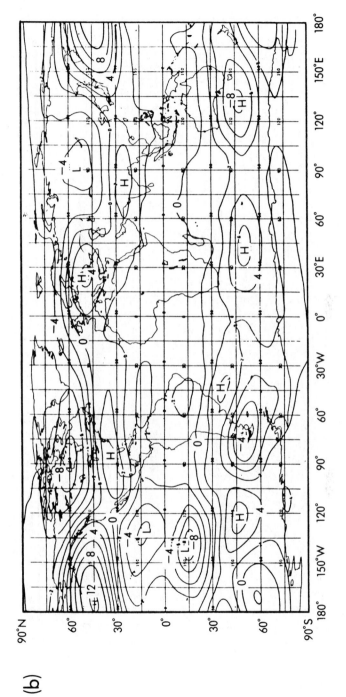

Fig. 4. 200 mb geopotential height anomaly (dm); (a) for experiments $b-a$ (+ December 1982); and (b) for experiments $c-a$ (− December 1982).

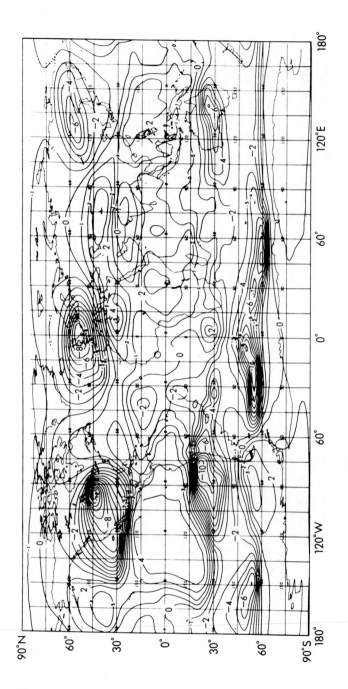

(a)

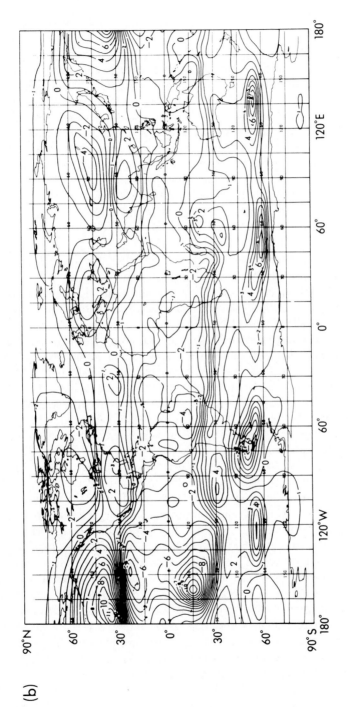

(b)

Fig. 5. (a) t-statistics of 200 mb height anomaly for experiments $b-a$; and (b) t-statistics of 200 mb height anomaly for experiments $c-a$.

2.2. Experiments d–f; Rasmusson and Carpenter composite El Niño anomaly

The convective precipitation associated with this SST anomaly (Fig. 1b) is illustrated in Fig. 6. Maximum values occur near the dateline, again consistent with the position of warmest waters in the anomaly experiment. The tropical 200 mb wind anomalies are illustrated in Fig. 7, for day 1 (Fig. 7a), for days 1–10 averaged (Fig. 7b), and for days 1–540 (Fig. 7c). It can be seen that the westerly anomalies over Africa and the Indian Ocean spread eastward from the SST anomaly. The easterlies that spread westwards, on the other hand, appear to be weak and possibly contained within the West Pacific. The anticyclone pairs are positioned to the east of the precipitation anomaly.

The 200 mb geopotential height anomaly, illustrated in Fig. 8a, shows in the PNA region a deep low over the Gulf of Alaska and a weak ridge over eastern North America. This pattern does not compare particularly well with the PNA pattern, or the response to a similar SST anomaly, reported by Blackmon et al. (1983), or Shukla and Wallace (1983).

In common with many other high-resolution GCMs, the 11-level model develops a winter-time climate drift with polar temperatures several degrees colder than observed. Associated with this, zonal winds are too strong, and baroclinic waves are not sufficiently damped over land. These systematic errors affect the results of our sensitivity experiments. However, Wallace et al. (1983) have shown that a respecification of orography in a GCM can substantially alleviate these errors (although other effects due to sub-grid scale orographically induced gravity wave drag in the upper troposphere and lower stratosphere may also help alleviate these errors; see Palmer and Shutts, 1984).

The 200 mb geopotential height anomaly for the experiment with the enhanced Rasmusson and Carpenter SST anomaly, and twice standard deviation (2σ) envelope orography (minus control with 2σ envelope orography) is shown in Fig. 8b. (With 2σ envelope orography, the model orographic heights are enhanced by twice the subgridscale standard deviation of orographic height, as determined by a ten-minute resolution specification of orography; Wallace et al., 1983.) Figures 8a and b show substantial differences in the extratropical Northern Hemisphere. With envelope orography there is a low at about 45°N, 160°W, a high over central Canada, and a low to the east of the U.S.A. This pattern conforms more closely with the PNA teleconnection pattern, and Blackmon et al.'s response. The notion that the correct positioning of storm-track activity is important in obtaining a realistic time-mean response to an SST anomaly is suggested by Fig. 9 which shows the high-pass filtered anomalous eddy streamfunction forcing $-\nabla^{-2}[\nabla \cdot (v'\zeta')]$ (using a poor man's filter (Lorenz, 1979) with three-day variances) for two 90-day periods of the composite SST anomaly runs. Figure 9a, for the standard orography runs, shows anomalous cyclonic eddy forcing in excess of $15\,\mathrm{m^2\,s^{-2}}$ (shaded) extending across the Pacific into the U.S.A. (an eddy forcing of $-10\,\mathrm{m^2\,s^{-2}}$ would spin-up a time-mean cyclonic streamfunction of $-1 \times 10^6\,\mathrm{m^2\,s^{-1}}$ in about one day). For the envelope orography experiments, there is strong cyclonic forcing in the mid-Pacific near 165°W, though in the East Pacific, the eddy forcing is anticyclonic. The differences between Figs. 9a and b appear to be consistent with the difference in the positions of the negative 200 mb height anomalies over the North Pacific, in Figs. 8a and b.

The sensitivity of the response to different orographic specifications highlights the

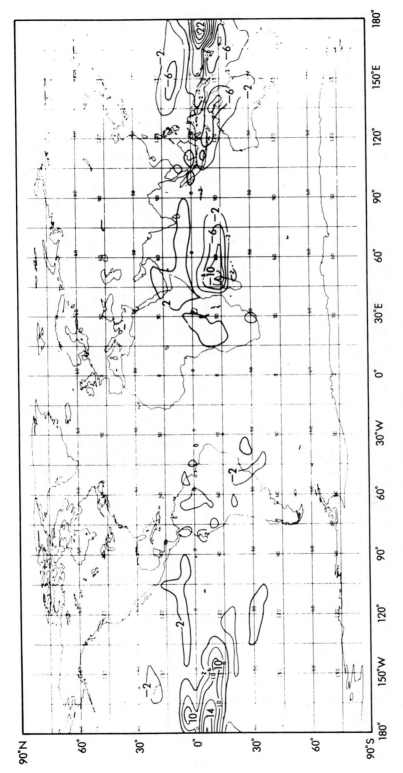

Fig. 6. Convective rain anomaly (mm/day) for experiment *d–a* (RC composite).

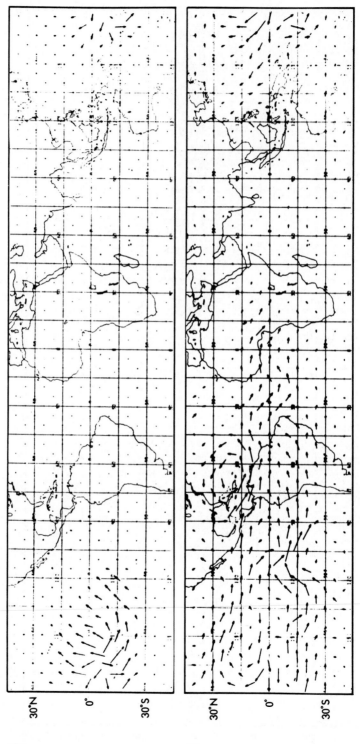

(a)

(b)

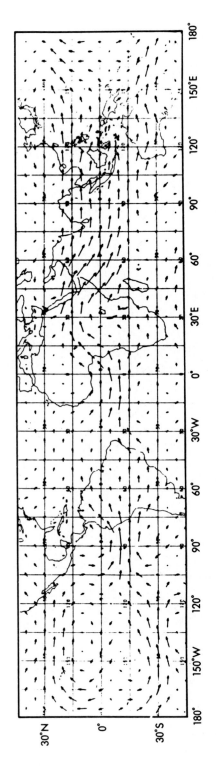

(c)

Fig. 7. 200 mb wind anomaly for experiments $d-a$ (RC composite); (a) day 1; (b) days 1−10; and (c) days 1−540.

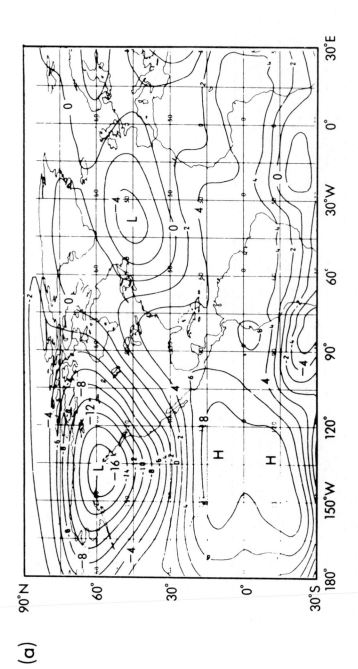

(a)

97

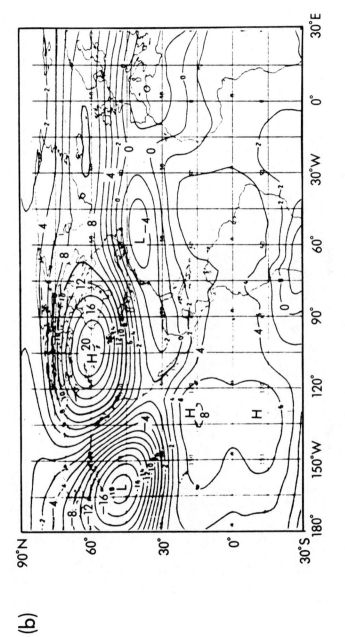

Fig. 8. 200 mb geopotential height anomaly (dm); (a) experiments $d-a$ (RC composite, standard orography); and (b) experiments $f-e$ (RC composite, envelope orography).

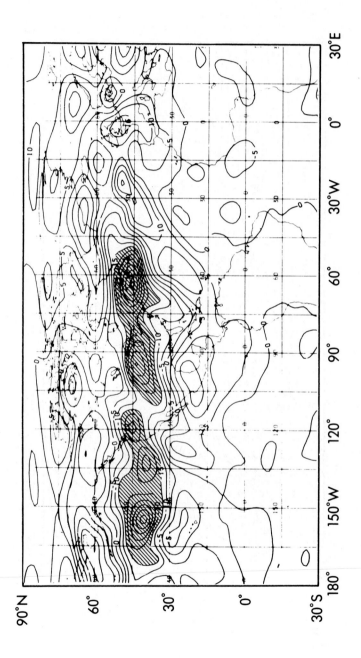

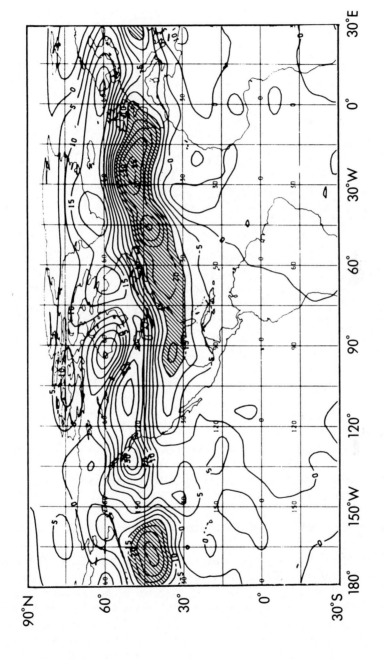

Fig. 9. Anomalous high-pass eddy streamfunction forcing at 250 mb (m² s⁻²) over a 90-day period (values less than − 15 m² s⁻² are shaded); (a) experiments $d-a$ (RC composite, standard orography); and (b) experiments $f-e$ (RC composite, envelope orography).

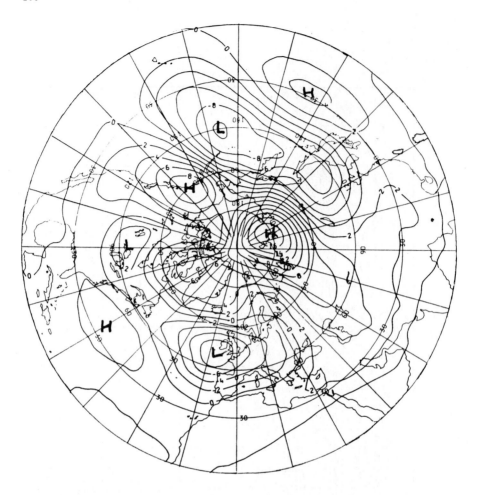

Fig. 10. 500 mb geopotential height anomaly (dm) for experiments *g−e* (warm West Pacific).

importance of the model's basic climatology in the determination of the impact in the extratropics of tropical SST forcing.

2.3. Experiments g and e; the response to a warm tropical West Pacific SST anomaly

The precipitation anomaly associated with the warm West Pacific SST anomaly has positive values near the SST anomaly with a maximum of 10 mm per day at about 5°N, 155°E. Figure 10 illustrates the 500 mb geopotential height anomaly. There is a high centre at 30°N, 155°E, to the north of the SST anomaly, and a downstream response with centres just east of the dateline, and over Alaska. Note also the anomalous easterly flow in high latitudes. This pattern is remarkably similar to that occurring in the severe El Niño winter of 1976/77. Palmer and Mansfield (1984) have argued that the difference between the extratropical response to the 1976/77 El Niño event, and the response to the 1972/73 or 1982/83 El Niño events may be due to small differences in SST in the tropical West Pacific. (An analysis of SST in the tropical West Pacific in January 1977, from Meteorological Office historical SST archives, showed a pool of anomalously warm water in excess of 1 K; see Palmer and Mansfield, 1984.) The response to a warm West Pacific anomaly reported in Palmer and Mansfield is similar to that shown here except that the distance between anomaly centres over the Pacific is less here. This is consistent with the reduction in stationary Rossby wavelength brought about by the influence of envelope orography in weakening the zonal mean flow. Integration g is continuing to 540 days.

3. COMPARISON OF RESULTS WITH LINEAR THEORY

3.1. Tropical response

Gill's (1980) model (see also Heckley and Gill, 1984) describes the linear response of the tropical atmosphere to a diabatic heating anomaly. This type of model has been used to provide the atmospheric component of tropical coupled atmosphere/ocean models, so it is clearly important to understand the extent to which it can describe the atmospheric response to a fixed SST anomaly.

The time evolution of the tropical 200 mb wind anomaly associated with the Rasmusson and Carpenter SST anomaly was illustrated in Fig. 7. The propagation of a westerly mode with small meridional wind component, to the east of the precipitation anomaly, is consistent with the Kelvin mode component of Gill's solution. At 850 mb this mode is much weaker, however (see, e.g., Fig. 3b); whereas the low-level linear solution is equal and opposite to the upper-level solution.

The time evolution shown at 200 mb also shows a weaker, and less extensive mode to the west of the precipitation anomaly. The easterly wind anomalies do not appear to extend into the Indian Ocean. Furthermore, the anticyclonic anomalies are located to the northeast and southeast of the precipitation anomalies (cf. Figs. 6 and 8).

All these facts suggest that the linear tropical Rossby mode response may not be simulated in the GCM. In the linear model the anticyclonic doublet is part of the Rossby mode propagating to the west of the diabatic heating.

A related problem of accounting for the tropical GCM response in terms of linear theory is posed by results from the positive and negative 1982 El Niño anomaly experiments. As discussed, the magnitude of the anomalous 200 mb cyclonic doublet associated with the negative SST anomaly experiment is equal to the corresponding anticyclonic doublet associated with the positive SST anomaly experiment, despite the fact that the magnitude of the anomalous precipitation was two to three times smaller for the negative experiment.

It is possible, therefore, that nonlinear modifications to Gill's model may be essential. For example, the forcing in his model is provided (with standard notation) by the linearised form, $-f\nabla \cdot v$, of the divergence forcing $-(f + \zeta)\nabla \cdot v$. In the tropics where f is small, $\zeta \sim f$, providing a basic asymmetry between positive and negative anomalous diabatic heating. For negative forcing, ζ will reinforce f to enhance the linearised effect; conversely for positive forcing ζ will cancel f to reduce the linearised effect (see also the discussion in Section 3.2). The importance of nonlinearity in the vorticity budget of the real atmosphere during the 1982/83 winter has been stressed recently by Sardeshmukh and Hoskins (1984).

A second effect which may be relevant in explaining the apparent asymmetry between Kelvin and Rossby modes is the zonal inhomogeneity of the basic state flow. In particular it is found that the upper level absolute vorticity is small over the tropical West Pacific in the control run with very weak meridional gradient north of the equator. It is possible that propagation of the equatorial Rossby mode over the West Pacific was inhibited by the presence of this weak gradient in the control integration.

3.2. Extratropical response

The notion of equivalent barotropic tropically forced Rossby wavetrains propagating into the extratropics has been developed by Hoskins and Karoly (1981) and Webster (1981) and extended by Simmons (1982) and Simmons et al. (1983), to explain the extratropical response to anomalous diabatic heating. Linearised about a zonally symmetric superrotational basic state, the theory predicts that these wavetrains should propagate on great circle paths.

It is clear that the GCM experiments by no means provide unequivocal support for the notion that the wavetrain theory accounts for the major extratropical response. Whilst results from the West Pacific anomaly experiment (Section 2.3) and the positive December 1982 El Niño anomaly experiment (Section 2.1) suggest a wavetrain-like response into the Northern Hemisphere, results from the other experiments (and the Southern Hemisphere responses) are less clear.

One component of the general circulation which is likely to be essential in accounting for the extratropical response is the (nonlinear) interaction of the transient eddies (extra-tropical cyclones for example) with the stationary flow, as discussed by Opsteegh and Vernekar (1981), and Kok and Opsteegh (1984). For example, one important systematic error in the climatology of the model with standard orography is the inability to fill baroclinic lows over land. This effect is largely removed by the introduction of envelope orography. It was commented in Section 2.2 that the introduction of envelope orography into the Rasmusson and Carpenter SST anomaly experiments tended to position the low anomaly centre further west, moving it from the west coast of Canada to the Aleutian Islands. It is quite possible that this resulted from eddy mean-flow interaction with the correct positioning of storm tracks in runs which included envelope orography (see Fig. 9). The remainder of the PNA pattern over the North American continent may correspond to an enhanced ridge and downstream trough forced by the strengthened mean-flow upstream over the Pacific flowing over (the envelope of) the Rockies.

On the other hand, results using the barotropic model described by Simmons (1982) suggest that the forced Rossby mode response can reproduce the qualitative extratropical pattern of the GCM response to at least one SST experiment. Figure 11a shows the GCM's anomalous 200 mb streamfunction response to the positive December 1982 El Niño anomaly (Fig. 1a). Figure 11b shows the barotropic model's response to (an ideali-sation of) the GCM's anomalous 200 mb divergence over the tropical East Pacific. The barotropic model's basic state was provided by the (540-day mean) 200 mb stream-function of the control run (*a*). A five-day damping was applied so that the forced solution reached a steady state, and no unstable modes were allowed to grow. The

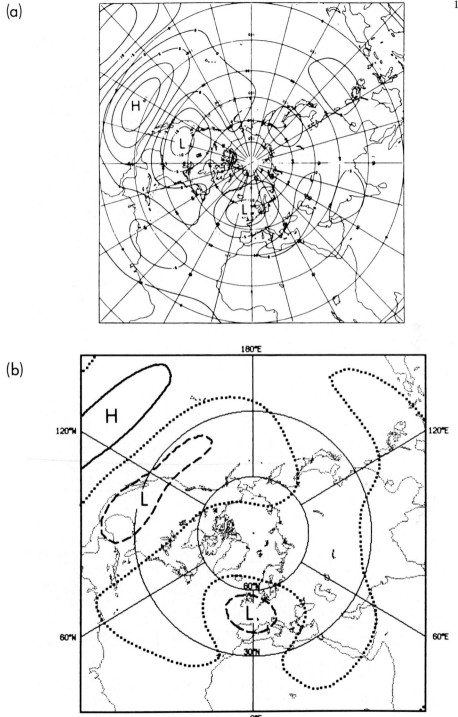

Fig. 11. (a) 200 mb streamfunction anomaly for experiments $b-a$ (+ December 1982); and (b) streamfunction response from barotropic model forced by the 200 mb divergence anomaly in experiments $b-a$ over the tropical East Pacific.

104

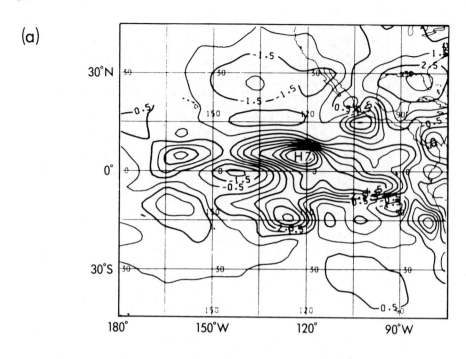

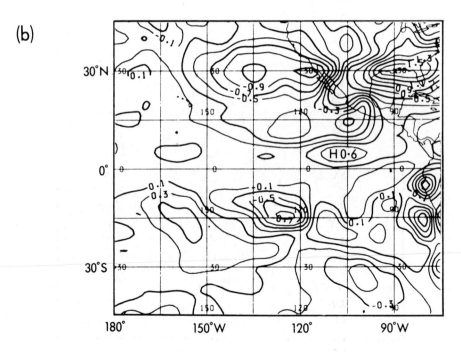

Fig. 12. (a) anomaly of $\nabla \cdot v$ at 200 mb (contour interval $0.5 \times 10^{-6} \text{ s}^{-1}$) for experiments $b-a$; and (b) anomaly of $(f + \zeta)\nabla \cdot v/2\Omega$ at 200 mb (contour interval $0.2 \times 10^{-6} \text{ s}^{-1}$) for experiments $b-a$.

qualitative correspondence between Figs. 11a and 11b is good. Note, for example, the low centres over the United States and the British Isles in both experiments.

One should treat this result with caution, however. As we have discussed above, the anomalous nonlinear divergence forcing $(f + \zeta)\nabla \cdot v$ may be considerably weaker than the linear counterpart used to force the barotropic model. For example, Fig. 12 shows the anomalous values of $\nabla \cdot v$ and $(f + \zeta)\nabla \cdot v/2\Omega$ for the positive December 1982 anomaly experiment at 200 mb. Whereas the anomalous divergence clearly shows the effect of the tropical SST anomaly, contours of $(f + \zeta)\nabla \cdot v/2\Omega$ do not. It is possible that the tropical height anomaly is in a (nonlinear modon-like) balanced state, with diabatic heating acting to offset radiative and other nonconservative effects, but not acting as a direct forcing for the nondivergent flow. As before, anomalous synoptic-scale eddies to the north may provide a substantial forcing to the time-averaged mid-latitude downstream response. In this respect, it is interesting to speculate that growth of mid-latitude cylones over the East Pacific may be *directly* affected by the enhanced moisture field in the tropical Pacific as a result of the SST anomaly. Work is now in progress analysing the present model runs to establish the interplay between these different components of the general circulation.

6. CONCLUSIONS

In this paper we have discussed results from experiments designed to test the sensitivity of the wintertime climatology of the UK Meteorological Office general circulation model to sea-surface temperature anomalies in the tropical Pacific Ocean. The following summarises the conclusions of this paper.

(1) The GCM extratropical response is more sensitive to SST anomalies in the tropical West Pacific than in the tropical East Pacific. Small differences (~ 1 K) in SST in the tropical West Pacific may explain the variability of the response from one El Niño event to another (cf. Palmer and Mansfield, 1984).

(2) The response in the model to a large SST anomaly in the East Pacific can be statistically significant, not only over the Pacific Ocean, but also downstream as far as the European continent.

(3) The model's response to a tropical SST anomaly does depend strongly on the model's basic climatology. With standard orographic specification, the Northern Hemisphere wintertime flow is too strong and baroclinic waves are not damped sufficiently over land. By altering the orographic specification in such a way as to reduce the strength of the flow and position more correctly the extratropical storm tracks, the response to a composite El Niño anomaly corresponds more closely with observed atmospheric anomaly patterns.

(4) Certain aspects of the model's tropical response to an SST anomaly cannot be readily explained by linear theory on a zonally homogeneous basic state. Possible modifications of such theory to account for the model results may include the relative vorticity contribution, ζ, to the divergence forcing $(f + \zeta)\nabla \cdot v$, and the imposition of weak basic state absolute vorticity gradients over the tropical West Pacific.

(5) The anomalous tropical divergence forcing from one experiment has been used to generate a downstream wavetrain response in a linear barotropic model. The results

compare favourably with the GCM response, both in the subtropics and in mid-latitudes. However, the forcing $f\nabla \cdot v$ used in the linear model appears to be substantially larger than the nonlinear forcing $(f + \zeta)\nabla \cdot v$. It is possible that enhanced cyclogenesis to the north of the SST anomaly may act to maintain the mid-latitude time-averaged response. A more detailed account of these, and other experiments is in preparation.

ACKNOWLEDGEMENTS

My thanks to Dr. A.J. Simmons for help in carrying out the integration described in Section 3, to Dr. D.A. Mansfield for considerable assistance (in producing the eddy diagnostics in particular) and to Dr. G.J. Shutts for helpful discussions. This research was partially supported by the EEC contract, Number CLI-034-81UK(H).

REFERENCES

Blackmon, M.L., Geisler, J.E. and Pitcher, E.J., 1983. A general circulation model study of January climate anomaly patterns associated with interannual variations of equatorial Pacific sea surface temperatures. J. Atmos. Sci., 40: 1410–1425.

Gill, A.E., 1980. Some simple solutions for heat-induced tropical circulation. Q. J.R. Meteorol. Soc., 106: 447–462.

Heckley, W.A. and Gill, A.E., 1984. Some simple analytical solutions to the problem of forced equatorial long waves. Q. J.R. Meteorol. Soc., 110: 203–218.

Horel, J.D. and Wallace, J.M., 1981. Planetary-scale atmospheric phenomena associated with the Southern Oscillation. Mon. Weather Rev., 109: 813–829.

Hoskins, B.J. and Karoly, D.J., 1981. The steady linear response of a spherical atmosphere to thermal and orographic forcing. J. Atmos. Sci., 38: 1179–1196.

Kok, K. and Opsteegh, J.D., 1985. On the possible causes of anomalies in seasonal mean circulation patterns during the 1982/83 El Niño event. J. Atmos. Sci. (in press).

Lorenz, E.N., 1979. Forced and free variations of weather and climate. J. Atmos. Sci., 36: 1367–1376.

Lyne, W.H. and Rowntree, P.R., 1976. Development of a convective parameterization using GATE data. Meteorol. Off. 20 Technical Note II/70. UK Meteorological Office, Bracknell.

Opsteegh, J.D. and Vernekar, A.D., 1982. A simulation of the January standing wave pattern including the effects of transient eddies. J. Atmos. Sci., 39: 734–744.

Palmer, T.N. and Mansfield, D.A., 1984. Response of two atmospheric general circulation models to sea-surface temperature anomalies in the tropical East and West Pacific. Nature, 310: 483–485.

Palmer, T.N. and Shutts, G.J., 1984. Preliminary results of the effect of a parameterization of gravity wave drag in the Meteorological Office 11-layer operational model. Meteorol. Off. 13 Branch Memorandum 147, UK Meteorological Office, Bracknell.

Quiroz, R.S., 1983. The climate of the "El Niño" Winter of 1982–83. A season of extraordinary climate anomalies. Mon. Weather Rev., 111: 1685–1706.

Rasmusson, E. and Carpenter, T., 1982. Variations in tropical sea surface temperature and surface wind fields associated with the Southern Oscillation/El Niño. Mon. Weather Rev., 110: 354–384.

Rasmusson, E. and Wallace, J.M., 1983. Meteorological aspects of the El Niño/Southern Oscillation. Science, 222: 1195–1202.

Rowntree, P.R., 1972. The influence of tropical east Pacific Ocean temperatures on the atmosphere. Q, J.R. Meteorol. Soc., 98: 290–321.

Saker, N.J., 1975. An 11-layer general circulation model. Meteorol. Off. 20 Technical Note II/30, UK Meteorological Office, Bracknell.

Sardeshmukh, P.D. and Hoskins, B.J., 1985. Vorticity Balances in the Tropics during the 1982–3 El-Niño Southern Oscillation event. Q. J. R. Meteorol. Soc. (in press).

Shukla, J. and Wallace, J.M., 1983. Numerical simulation of the atmospheric response to equatorial Pacific sea surface temperature anomalies. J. Atmos. Sci., 40: 1613–1630.

Simmons, A.J., 1982. The forcing of stationary wave motion by tropical diabatic heating. Q. J. R. Meteorol. Soc., 108: 503–534.

Simmons, A.J., Wallace, J.M. and Branstator, G., 1983. Barotropic wave propagation and instability and atmospheric teleconnection patterns. J. Atmos. Sci., 40: 1363–1392.

Wallace, J.M., Tibaldi, S. and Simmons, A.J., 1983. Reduction of systematic forecast errors in the ECMWF model through the introduction of an envelope orography. Q. J.R. Meteorol. Soc., 109: 683–718.

Webster, P.J., 1981. Mechanisms determining the atmospheric response to sea surface temperature anomalies. J. Atmos. Sci., 38: 554–571.

CHAPTER 9

TRANSIENT EFFECTS DUE TO OCEANIC THERMAL INERTIA IN AN ATMOSPHERIC MODEL COUPLED TO TWO OCEANS

A.H. TAYLOR, D. MUIR and N. MURDOCH

ABSTRACT

A model of the mixed layers of the North Pacific and North Atlantic Oceans has been coupled to a hemispherical, quasigeostrophic, beta-plane model of the atmosphere. The two oceans are represented by the sectors $90°W-90°E$ and $0°-45°W$, respectively, with the North American and Eurasian land masses by $45°-90°W$ and $0°-90°E$. The mixed-layer model is a simplification of that of Kraus and Turner (1967). The meridional heat flux in each ocean is prescribed during each run of the model and several experiments were carried out between which these transports were varied. This paper considers the role of oceanic thermal inertia in these experiments. Each experiment began with a mixed layer which was everywhere in dynamic equilibrium with the unperturbed heat transport. The results show how the mixed layer affects the dynamics of the model. There are teleconnections between the mixed layers of the two oceans with time scales of months. The model shows an east–west oscillation reminiscent of the seesaw in northern hemisphere temperatures described by Van Loon and Rogers (1978), which involves the northernmost mixed layer of the oceans.

INTRODUCTION

Manabe (1983) highlighted three ways in which the oceans affect the climate, via: (1) the hydrological cycle; (2) their thermal inertia; and (3) the transport of heat. The first of these is not considered in the present model. Murdoch and Taylor (in prep.) have investigated the influence on the climate of the meridional oceanic heat flux using a hemispherical, quasigeostrophic, beta-plane model of the atmosphere which is coupled to two land masses and two oceans, each ocean having an interactive mixed layer. The transport of heat within each ocean was prescribed throughout any experiment so that its impact on the atmospheric circulation could be determined. The experiments showed that: changes in the surface temperature spanning the whole northern hemisphere accompanied perturbations of the oceanic heat fluxes; zonal wind strengths were inversely related to the strength of the heat transport, the wind changes being generally such as to oppose any change in the heat flux but with effects in both oceans; and the regions of deep convection in the northern Atlantic were important during the readjustment between climatic phases. These results were obtained by analysing the average conditions during the last 150 days of each 450-day run and assumed that the averages provided an indication of the equilibrium towards which the model was moving. The present paper examines the role of the thermal inertia of the oceans in the dynamics of the climatic system by studying the behaviour of the mixed layer during the evolution with time of this model.

The model uses an unchanging annual-average heating function throughout the calculations so that the heat content of the mixed layer is approximately constant. The real mixed layer will adjust to seasonal changes and so the response times of the thermal and potential energy contents of the layer to external forcing must be less than six months and may be similar to the decay time of sea-surface temperature anomalies (about two months, Frankignoul, 1979; Wells, 1982). The atmospheric index cycle has a period of about 20 days so that some response of the mixed layer to these atmospheric fluctuations can be anticipated. Therefore, the mixed layer may be expected to affect the dynamics of the atmosphere over monthly periods, and this is reflected in the large-scale transients to be described. The transients that occur exhibit zonal patterns which are similar to features observed in climatic data.

MODEL DESCRIPTION

The complete model is described in detail by Murdoch and Taylor (in prep.). The atmospheric part, which is that of Gordon and Davies (1977), is a two-level, quasi-geostrophic, beta-plane system, incorporating a model-dependent surface energy balance equation and, at the 500 mb level, a time-dependent heating function with associated feedback relationships due to radiative exchanges with the surface. The surface conditions are interactive; the albedo, which represents snow and ice cover, being a function of temperature. When the temperature at any point is below 272K the local albedo is changed to that of ice and if the area is sea it is treated as land. Radiation transmission parameters, sensible heat loss, latent heat loss and condensation heating were from Smagorinsky (1963); sensible and latent heat fluxes were allowed to vary linearly with surface temperatures about these mean values.

Surface temperatures over land, T_*, were calculated by assuming thermal equilibrium. Over the sea this is not permissible because heat can be stored in, or released from, the water column. The temperature, T_*, and depth, h, of each ocean's surface mixed layer were calculated by the method of Horgan, Davies and Gadian (1983) which is a simplification of that of Kraus and Turner (1967). Water temperature beneath the mixed layer is assumed to be a constant function of latitude (T_h). T_* and h were determined by ensuring that the thermal and potential energies of the layer were conserved. The atmospheric thermal and potential energy forcing functions are written as $S + B - M$ and $G - D + S/\gamma$, respectively, in which: S represents the contribution from solar heating; B the sum of all black-body, latent heat and sensible heat losses from the sea surface; G the rate of generation of mixing energy by the surface winds (deduced from the atmospheric model fields); S/γ the generation of buoyancy by the penetrative solar radiation (γ is the coefficient of decay of S with depth) and D the rate of dissipation of energy within the layer. Following Stevenson (1979), D was taken proportional to the depth of the layer, thereby ensuring no irreversible accumulation of potential energy. The term M is the oceanic heat-flux divergence which represents the meridional transport of heat within each ocean. Zonally averaged values of M (from Smagorinsky, 1963) were used and variations in the oceanic heat flux were implemented by changing M.

EXPERIMENTS

The model (Fig. 1a) partitions the Northern Hemisphere into four sectors: Pacific (90°E–90°W), North America (90°–45°W), Atlantic (45°W–0°) and Eurasia

a

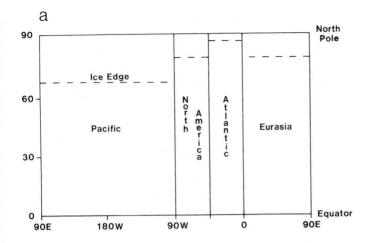

b

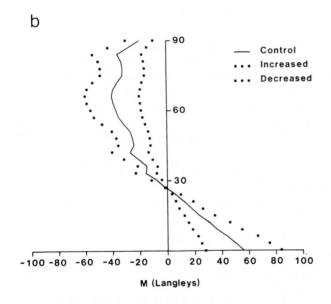

M (Langleys)

Fig. 1. a. The geometry used in the model. The extent of sea- and land-ice for the control run are shown. b. The latitudinal distribution of heat-flux divergence used in each ocean during the control run and during the runs with increased M and decreased M. Each unit of M represents 1 Langley in the Pacific or 4 Langleys in the Atlantic.

$(0°–90°E)$. The grid-spacing is $7.5°$ of longitude and $2.9°$ of latitude. There is no topography.

The experiments were designed to investigate the sensitivity of the system to changes in the oceanic heat-flux divergence (M) occurring in either ocean. In this model, M is prescribed in any run. There is considerable uncertainty about the relative strengths of the zonally averaged oceanic heat flux divergence in the two oceans. Meridional heat transports in the North Pacific and North Atlantic Oceans are similar, despite the greater area of the Pacific (Bryan, 1982); partly because there is a larger return flow of heat in the Pacific and partly because of the wide connection between the Atlantic and Arctic Oceans. Therefore, in the model, M was set up to be four times as great in the Atlantic sector as in the Pacific sector so that the two model oceans transported the same amount of heat northwards. The divergence was distributed uniformly with longitude. This arrangement was used for the control run.

The initial values of the stream functions were obtained by running the model without a mixed layer until the atmosphere came into equilibrium with the specified climatological heating (the zonally averaged heating functions of Smagorinsky, 1963). Subsequent initialisation of the mixed layer was carried out using the surface temperature and surface wind fields, averaged over several energy index cycles. By requiring that thermal and potential-energy forcing functions be simultaneously zero, that is:

$$S + B - M = 0 \tag{1}$$

and:

$$G + D - S/\gamma = 0 \tag{2}$$

these average fields could be used to provide initial oceanic-surface temperatures and mixed-layer depths. The first control run was obtained from the succeeding 450 days after the mixed layer became active. The calculation without a mixed layer was continued as a second control run.

Four experiments with differing M-values were carried out (Fig. 1b): (A1) M was increased by 50% in the Atlantic sector; (A2) M was decreased by 50% in the Atlantic sector; (P1) M was increased by 50% in the Pacific sector; and (P2) M was decreased by 50% in the Pacific sector.

Each of these experiments started with the same initial mixed layer and atmosphere as did the first control run and ran for at least 450 days. Comparing the pair of opposite runs A1 and A2 (or P1 and P2) to see which responses reverse sign provides an indication of signal to noise ratio.

RESULTS

Figure 2 shows the temporal variation of the atmospheric eddy kinetic energy for the control run with and without a mixed layer. When the oceans have a mixed layer the kinetic energy varies more smoothly with time, and this is consistent with analyses of models under stochastic forcing (e.g., Frankignoul, 1979) in which the thermal inertia of the ocean transforms a white noise spectrum of sea-surface temperature variations into a spectrum with greater energy in the low-frequency regime. However, Fig. 2 shows this reddening to be more complex, for peaks at 30 days and 75–100 days are smeared into a broad band centred on a period of 50 days.

The presence of oceanic mixed layers in the model leads to a generally warmer climate (Murdoch and Taylor, in prep.). For instance, although ice cover on land does not change appreciably, the extent of sea-ice is reduced by about six degrees of latitude. This warming can be understood if the development in time of the model is examined in the

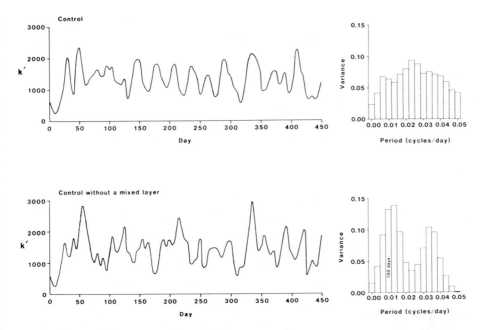

Fig. 2. Time series of mean eddy kinetic energy for the control run with and without a mixed layer. The low-frequency spectrum is shown for each series.

light of the atmospheric dynamics and thermodynamics. The initial stages of Fig. 2 illustrate the way this difference in climate arises. Both time series show a trough at about day 10, after which each graph climbs to a peak. The series from the experiment with a mixed layer has a much higher first peak. This occurs because the presence of a heat-absorbing mixed layer requires that more heat be transported northwards by the atmospheric eddies to produce a given temperature change. So, more vigorous eddies can build up before the temperature gradient at the 500 mb level is forced back towards baroclinic stability. During this period, heat that would otherwise be lost by radiation to space is stored in the mixed layer. Such a storage of heat occurs during each of the atmosphere's energetic phases and this heat can be released at less energetic times to impede the formation of ice. Thus, the mixed layer affects the climate of the model by influencing the atmospheric thermodynamics via the 500-mb temperatures and hence the dynamics of the atmosphere through the meridional temperature gradient. The mixed layer is therefore seen to have a significant effect on the system even in the absence of seasonal forcing.

The fluctuations shown in Fig. 2 are dominated by the atmospheric energy cycle. However, the experiments A1, A2, P1 and P2 having modified oceanic heat fluxes each began with mixed layers which were not in "equilibrium" since, due to the altered heat transports, their thermal and potential-energy forcing functions were not zero. As a result the system gradually adjusts during each run towards conditions which satisfy eqns. (1) and (2) and fluctuations with longer time scales are also observed. These fluctuations can be seen in Fig. 3 which displays for two typical cases, A1 and A2, successive 50-day averages of the heat stored in the mixed layer (after subtracting that of the control run) at a southern latitude and at a latitude near the northernmost exposed region of each ocean. For the whole of the Pacific and the southern part of the Atlantic the changes are slightly oscillatory and it is possible that an equilibrium is being approached. In the northern Atlantic, however, a clear trend is apparent. Mixed-layer depths in the two oceans (Fig. 3) also show less trend at low latitudes but, at high latitudes, the depths show a trend which has the same sign in both oceans. All of these trends change sign when the perturbation in oceanic heat flux is reversed. The mixed layers at high latitudes will take longer to achieve a dynamic equilibrium than the layers at low latitudes because they are much deeper and so require the transfer of a considerably larger quantity of heat to cause a given temperature change.

At low latitudes, the trend in the heat content of the mixed layer is the same in both oceans and corresponds to the underlying trend in the mixed-layer depth (Figs. 3 and 4). The winds in the model respond rapidly (i.e. within 50 days) to the changing oceanic heat flux and the trend in the layer depth represents a slower adjustment to these winds. At high latitudes this is also true of the Pacific in runs A1, A2 (Fig. 3) and the Atlantic in P1, P2 (e.g., Fig. 4). Thus, there is a teleconnection between the two oceans affecting most latitudes. The northern mixed layer in the ocean whose heat flux is perturbed shows a trend in its heat content which is the direct result of the changed heat transport and is in the opposite direction to that of the other ocean.

Superimposed on this overall trend, the mixed-layer depths in the northern region of the ocean show an oscillation of period about 400 days which is out of phase between the two oceans, thereby having the appearance of an east—west seesaw. The seesaw has its greatest amplitude in the high-latitude area of each ocean. As a result, the response in the Atlantic decreases towards middle latitudes and at 57°N in this ocean the oscillation is less clear. This transient response of the model is examined further in Fig. 4 which contains the same graphs for experiment P1 together with the corresponding time series of surface temperatures and zonal-wind components. They are typical of the four runs. At low latitudes where the mixed layers are shallow the fluctuations in surface temperature are relatively large. The temperatures and zonal winds appear inversely related to the layer depths indicating the importance of variations in wind mixing. In this "trade-wind" region, the winds are predominantly zonal easterlies, thus a positive zonal wind difference means a reduced wind strength. There is a similar set of relationships in high latitudes but, because of the trends in the layer depth and storage, the temperatures, and especially the winds, are less clearly related to the layer depths. The interpretation is also complicated because 57°N is in a band of westerlies, while 75°N is in a band of easterlies.

The mixed layers of the two oceans are coupled by means of the zonal winds which transmit changes over one ocean downstream to the other. Equations describing this ocean—ocean interaction can be obtained from those describing the thermal and potential-energy balances of the mixed layer, viz:

$$\frac{\partial}{\partial t} \left[(T_* - T_h)h \right] = S + B - M \tag{2}$$

where T_h is the temperature below the surface layer, and:

115

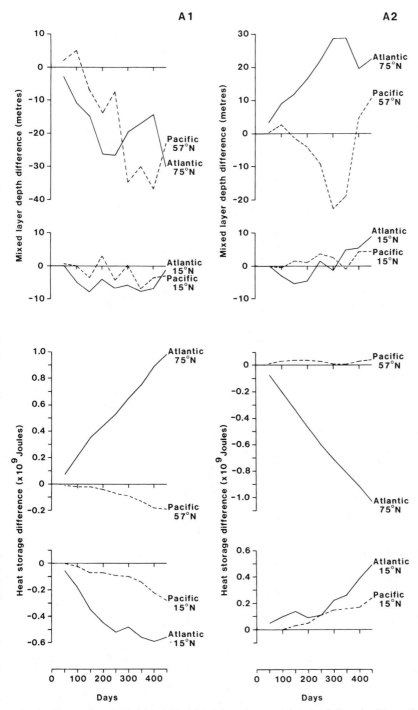

Fig. 3. Time series of 50-day averages from experiments A1 and A2 for mixed-layer depth and heat content of the mixed layer at latitudes in the northern and southern regions of each ocean. The series are averaged zonally across each ocean and expressed as deviations from the corresponding series of the control run.

116

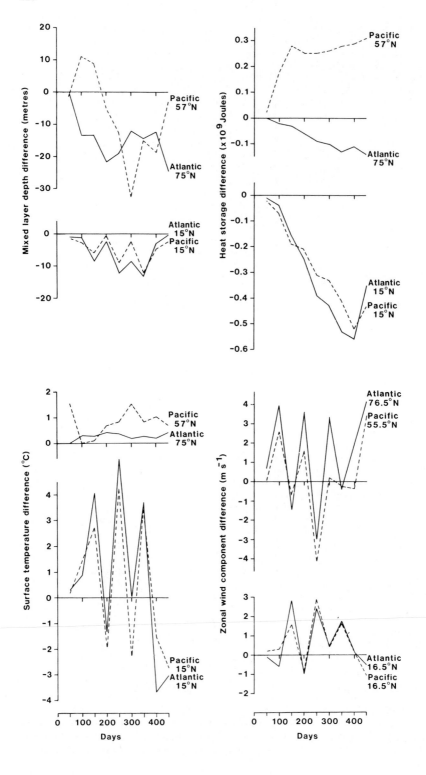

$$\frac{\partial}{\partial t}[(T_* - T_h)h^2] = G - D + S/\gamma \tag{3}$$

Figures 3 and 4 indicate that the total heat content of the mixed layer does not oscillate but shows merely an almost linear variation with time. This implies that on

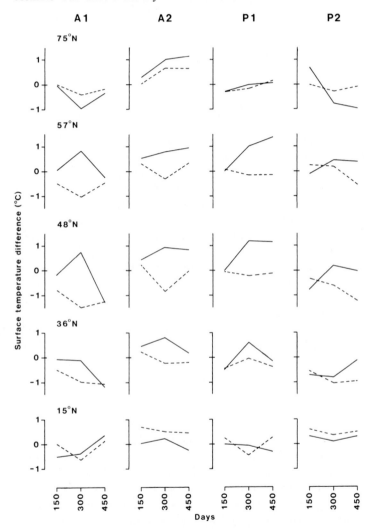

Fig. 5. Time series of 150-day averages from experiments A1, A2, P1 and P2 for surface temperatures at 15°, 36°, 48°, 57° and 75° N. The series are averaged zonally across each continent and expressed as deviations from the corresponding series for the control run. The variation of 50-day means about these values is ± 0.75° at mid-latitudes, lower at the highest latitudes and ± 0.25° at low latitudes.

Fig. 4. Time series of 50-day averages from experiment P1 for mixed-layer depth, heat content of the mixed layer, surface temperature and component of zonal wind at latitudes in the northern and southern regions of each ocean. The series are averaged zonally across each ocean and expressed as deviations from the corresponding series for the control run.

118

occasions the mixed layer can shallow without changing its heat content, which is an unrealistic feature of the model. However, even with a more accurate treatment of potential-energy losses, it is likely that fluctuations analagous to those shown here will occur. Thus, the thermal forcing is approximated as constant in time. In the model D was written as $D'h$ where D' was a decreasing function of latitude. Applying eqns. (2) and (3) to each ocean, a pair of equations describing the transient behaviour of the oceanic mixed layers in the model can be derived:

$$(S + B_A - M_A)h_A + (S + B_A - M_A)t \frac{\partial h_A}{\partial t} + H_A \frac{\partial h_A}{\partial t} = G_A - D'h_A + S/\gamma \tag{4}$$

$$(S + B_P - M_P)h_P + (S + B_P - M_P)t \frac{\partial h_P}{\partial t} + H_P \frac{\partial h_P}{\partial t} = G_P - D'h_P + S/\gamma \tag{5}$$

where suffix "A" refers to the Atlantic and "P" to the Pacific, and H_A and H_P are the initial heat contents of the two oceans' mixed layers. Equations (4) and (5) are coupled by the rates of wind mixing, G_A and G_P; each of these will be a complicated function of the temperature and mixed-layer depth distributions in the two oceans. It is these functions which will determine the periods of any oscillations. The dissipation terms $D'h_A$ and $D'h_P$ will tend to dampen any transients, and this effect will decline northwards with D'. As a result both the trend and the oscillation are most pronounced at high latitudes (Figs. 3 and 4).

An oscillatory teleconnection is also present in the continental-surface temperatures. These temperatures are much more variable in time than oceanic temperatures, so that the effect is most clearly illustrated by using 150-day averages (Fig. 5). While temperature fluctuations at the highest and lowest latitudes are similar over the two continents, in middle latitudes (45°–60°N) America and Eurasia show variations that are out of phase. In general, the temperature over each continent tends to be similar to that of the ocean upstream of it (i.e. to the west in the westerly wind belt).

DISCUSSION

Changing the oceanic heat flux in the model leads to a transient response with a distinct structure which shows coupling between the oceans. Although at low latitudes little or no longitudinal variation is shown, in middle and high latitudes there is a pronounced zonal asymmetry, with an east–west seesaw of period about 400 days occurring in each of the four runs. The mixed layers of the northern regions of the ocean and their response to varying wind-mixing and varying heat storage form an important element of this oscillation. The oscillation is also reflected in continental-surface temperatures. Having only one cycle per experiment, we cannot conclude that this mode is truly periodic; it may be a temporary feature of the adjustment process. Van Loon and Rogers (1978) have described a seesaw between winter temperatures at Greenland and those in northern Europe. This pattern is related (Rogers and Van Loon, 1979) to the strength of the zonal winds and is accompanied by large anomalies in the atmosphere–ocean–ice system, some of which persist through the subsequent spring and summer. The idealised calculations presented here suggest that such a seesaw effect could be a transient mode of the climatic system, perhaps triggered by an alteration of the oceanic heat transport.

ACKNOWLEDGEMENTS

This work forms part of the Physical Processes programme of the Institute for Marine Environmental Research, a component of the Natural Environment Research Council, and is funded in part by the Ministry of Agriculture, Fisheries and Food. N. Murdoch was supported by a grant from the E.E.C. Climate Research Progam. Preliminary calculations for this study were carried out by C. Reeve. We are grateful to J.A. Stephens and T. Woodrow who prepared the figures and typed the text.

REFERENCES

Bryan, K., 1982. Poleward heat transports by the ocean: observations and models. Annu. Rev. Earth Planet. Sci., 10: 15−38.

Frankignoul, C., 1979. Large-scale air−sea interaction and climate predictability. In: J.C.J. Nihoul (Editor), Marine Forecasting: Predictability and Modelling in Ocean Hydrodynamics. Elsevier, Amsterdam, pp. 35−37.

Gordon, H.B. and Davies, D.R., 1977. The sensitivity of climatic characteristics in a two level general circulation model to small changes in solar radiation. Tellus, 29: 484−501.

Horgan, M.J., Davies, D.R. and Gadian, A.M., 1983. Variable depth oceanic mixed layer interaction in a general circulation model. In: A.L. Berger and C. Nicolis (Editors), New Perspectives in Climate Modelling. Elsevier, Amsterdam, pp. 249−268.

Kraus, E.B. and Turner, J.S., 1967. A one dimensional model of the seasonal thermocline. Tellus, 19: 88−97.

Manabe, S., 1983. Oceanic influence in climate, studies with mathematical models of the joint ocean− atmosphere system. In: Large Scale Oceanographic Experiments in the World Climate Research Program. Vol 2, W.C.R.P. Publ. Ser. No. 1, WMO, pp. 1−27.

Murdoch, N. and Taylor, A.H., in prep. Effect of variation of meridional oceanic heat transport in an atmospheric model with a mixed layer. Submitted to J. Climatol.

Rogers, J.C. and Van Loon, H., 1979. The seesaw in winter temperatures between Greenland and Northern Europe. Part II: Sea ice, sea surface temperatures and winds. Mon. Weather Rev., 107: 509−519.

Smagorinsky, J., 1963. General circulation experiments with the primitive equations. Mon. Weather Rev., 91: 99−164.

Stevenson, J.W., 1979. On the effect of dissipation on seasonal thermocline models. J. Phys. Oceanogr., 9: 57−64.

Van Loon, H. and Rogers, J.C., 1978. The seesaw in winter temperatures between Greenland and Northern Europe. Part I: General description. Mon. Weather Rev., 106: 296−310.

Wells, N.C., 1982. Ocean−atmosphere interaction − a current perspective. Weather, 37: 162−164.

CHAPTER 10

GCM SENSITIVITY TO 1982–83 EQUATORIAL PACIFIC SEA-SURFACE TEMPERATURE ANOMALIES

M.J. FENNESSY, L. MARX and J. SHUKLA

ABSTRACT

Three control and anomaly simulation pairs run with the Goddard Laboratory for Atmospheric Sciences (GLAS) climate model have been analyzed in order to investigate the atmospheric response to the 1982–83 tropical sea-surface temperature anomalies. The observed 1982–83 SST anomalies obtained from the Climate Analysis Center were applied to two separate 75-day control simulations, starting on 16 Dec. 1982 and 16 Dec. 1979, respectively, and a third 60-day control simulation starting on 1 Jan. 1975.

In each experiment the equatorial Pacific precipitation increased significantly in a wide band stretching from just east of the dateline to the South American coast, in agreement with observed outgoing longwave radiation (OLR) anomalies. West of this region the precipitation was reduced in the anomaly simulations. As in previous GCM experiments, the major contributor to the tropical precipitation changes was the low-level moisture convergence. The largest evaporation differences were around 4 mm per day and occurred over the regions of highest SST in the anomaly simulations. The tropical sea-level pressure field showed a marked Southern Oscillation pattern, with a magnitude of roughly 2 mb and a node at the dateline. There was a strong ($\sim 10 \, \mathrm{m \, s^{-1}}$) increase in the equatorial eastern Pacific 850 mb westerlies as well as a large ($\sim -20 \, \mathrm{m \, s^{-1}}$) easterly wind anomaly at 200 mb. The latter anomaly was flanked by strong ($\sim 20 \, \mathrm{m \, s^{-1}}$) westerly anomalies at roughly 30°S and 30°N.

In agreement with earlier simulations with composite SST anomalies, the tropical precipitation anomalies for 1982–83 were also closely related to the extent of very warm ($\geqslant 29°$C) sea-surface waters.

Each experiment had anomalous anticyclonic circulations aloft straddling the equator in the eastern Pacific, although they were weaker and more eastward than those observed. The extra-tropical response varied between the three experiments, as well as between months of a given experiment. Over North America the ensemble average anomaly minus control 300 mb geopotential height difference field resembled the observed February or March anomaly field more than the typical PNA-like pattern. Other extra-tropical responses were difficult to interpret, although they were clearly equivalent barotropic in structure and showed a much stronger dependence on initial conditions than was noted for the tropics.

INTRODUCTION

Several general circulation model (GCM) studies (Rowntree, 1972; Julian and Chervin, 1978; Keshavamurty, 1982; Blackmon et al., 1983; Shukla and Wallace, 1983) simulated many of the basic features of the observed atmospheric response to SST anomalies representative of the mature stage of previous El Niño events. The tropical features included an eastward shift of the region of maximum convective activity, increased low-level equatorial Pacific westerlies, and anomalous anticyclonic couplets straddling the equator aloft. The most pronounced extra-tropical anomaly observed was also simulated, that being the Pacific–North American (PNA)-like pattern (Wallace and Gutzler, 1981)

122

in the upper-level geopotential height field over the northeast Pacific and North America. Although the general features of the "composite El Niño event" were correctly simulated in most of these experiments, significant variations between experiments utilizing different initial conditions were noted (Shukla and Wallace, 1983).

The present study examines the response of the GLAS climate model to the much larger 1982–83 SST anomalies as obtained from CAC.

MODEL AND INTEGRATIONS

The model used is an improved version of the GLAS climate model used by Shukla and Wallace (1983) and documented by Shukla et al. (1981). A thorough description of the model and its climatology is given by Randall (1982). It is global in extent with a 4° lat. × 5° long. grid in the horizontal. Its nine sigma layers of equal thickness are centered at approximately 65, 175, up to 945 mb. A Matsuno scheme is used for time integration, and a Shapiro filter is applied at each model half hour to the pressure, potential temperature, and wind fields. To maintain stability near the poles, Fourier filtering of the zonal wind flux and pressure gradient terms is performed at each time step (7.5 min.). This is preferable to the earlier used split-grid which generated spurious flow components. The planetary boundary layer (PBL) parameterization was also changed to that of Deardorff (1972) as modified by Randall (1976). The model generates supersaturation clouds at all nine levels and cumulus clouds (Arakawa, 1969) at the lowest six levels. In the current version only the supersaturation clouds are allowed to interact with radiation. Shortwave radiation and other physical processes are determined at each model simulated half hour. Longwave radiation is calculated at 5-h intervals and applied each half hour. Dry convective adjustment is performed at each time step. Also differing from the earlier version is the ground hydrology, which carries two temperatures, the ground temperature and saturated ground temperature (Mintz and Serafini, 1981), and the prognostic temperature variable is θ instead of T.

The climatological boundary condition datasets for control integrations are unchanged from those used in the Shukla and Wallace (1983) and the Shukla et al. (1981) version. Surface albedos are prescribed for land, ocean and desert grid points. Climatological monthly mean sea-surface temperatures are interpolated to their daily values.

Important improvements in the current GLAS climate model simulations are the removal of the climate drift towards unrealistically high temperatures in the tropics, which was noted by Shukla and Wallace (1983), as well as a more realistic sea-level pressure pattern. Further details of the models' climatology can be found in Randall (1982).

Three separate experiments (control and anomaly simulation pairs) designed to simulate the mature phase of the 82–83 event were performed with the following initial conditions: (1) observed initial conditions on 16 Dec. 1982; (2) initial conditions on 16 Dec. 1979 taken from a 2-year model control run after one year of simulation; and (3) observed initial conditions on 1 Jan. 1975. The first two experiments covered 75 days,

Fig. 1. January 1983 sea-surface temperature fields for (a) anomaly minus control differences, (b) control simulation, and (c) anomaly simulation. (Anomaly obtained from CAC.) Units are °C. Dashed contours are negative.

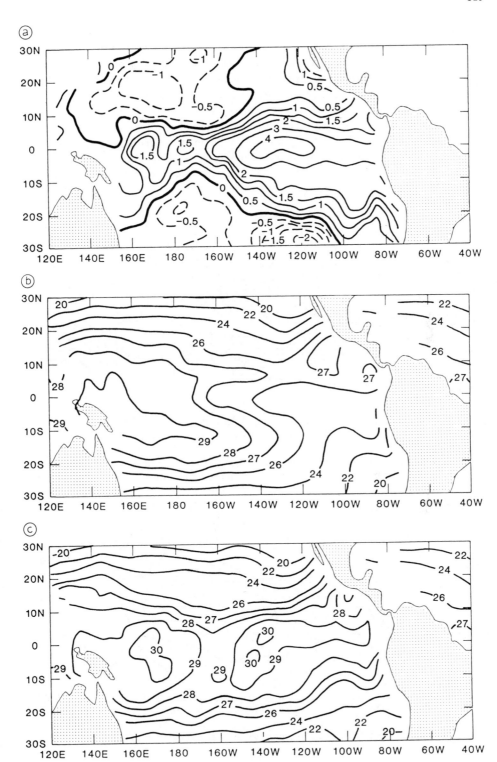

the third one 60 days. The Dec. 1982, Jan. 1983 and Feb. 1983 observed monthly SST anomalies were obtained from the Climate Analysis Center on magnetic tape. The SST fields used in the anomaly simulations were obtained by adding these monthly anomalies to the monthly climatological SST fields for the region of the Pacific between 40°S and 60°N. For the actual integration the monthly boundary condition datasets are linearly interpolated to daily values. The January control and anomaly SST fields used in each experiment are shown in Figs. 1b and c, respectively. Notable is the greatly extended region of very warm ($\geqslant 29°C$) SST water in the anomaly simulation. The January SST anomaly field is representative of the other months of experiments, all of which had a much larger region of very warm SST in the anomaly simulation than in the previously noted GCM studies.

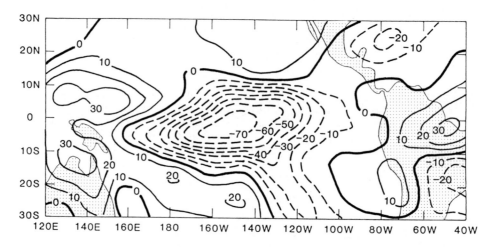

Fig. 2. January 1983 observed outgoing longwave radiation anomaly. Units are W m^{-2}. Obtained from Lau and Chan (from NOAA Polar Orbiters). Dashed contours are negative.

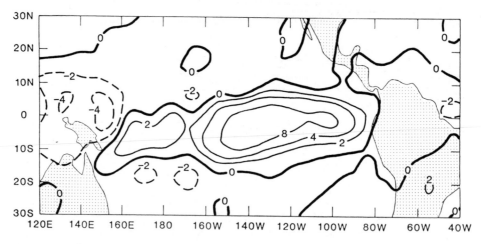

Fig. 3. 11–60-day ensemble average anomaly minus control precipitation difference field. Units are mm per day.

RESULTS

Most of the differences (anomaly minus control) were quite similar in the three experiments, thus we will concentrate on averages of all three and refer to this as the ensemble average. We will note any appreciable individual departures from the ensemble averages when they occur. For comparisons to the observed fields the reader is referred to Quiroz (1983) and the NMC CAC Special Climate Diagnostics Bulletins (1982–83).

The ensemble average 11–60-day precipitation difference field (hereafter differences referenced are anomaly minus control) shown in Fig. 3 exhibits a wide region of enhanced precipitation from 160°E to 80°W in the anomaly simulation, as well as a region of decreased precipitation to the west. These differences, as well as the positive differences over SW Brazil and the Gulf of Mexico agree well with the CAC January 1983

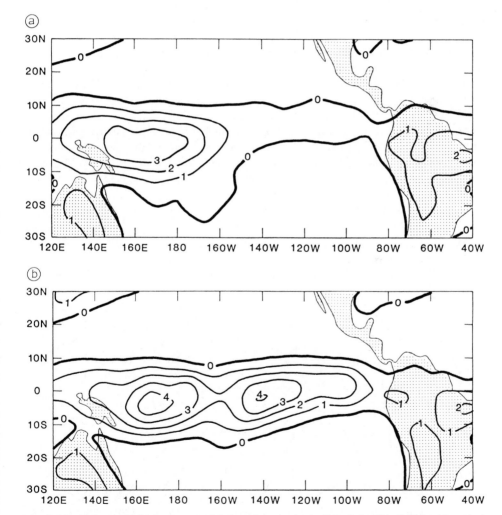

Fig. 4. 11–60-day ensemble average modeled total atmospheric diabatic heating field for (a) control, and (b) anomaly. Units are °C per day.

OLR anomalies (Fig. 2). The exceptions are the eastward extension of the simulated anomaly and the failure to capture the observed anomalous precipitation to the south-east, associated with a shift in the South Pacific Convergence Zone. However, the overall similarity between the simulated precipitation anomalies and the observed OLR anomalies is striking. There is an obvious direct relation between the positive precipitation diffe-rences and the extent of the very warm ($\sim 29°C$) SST in the anomaly simulation (Fig. 1c). These precipitation differences developed quickly (within 2 weeks) and persisted throughout the course of the experiment. The precipitation differences are reflected in the 11–60-day ensemble average total atmospheric diabatic heating fields for the control (Fig. 4a), and anomaly (Fig. 4b) simulations. Notable in the anomaly simulation diabatic heating field are two distinct maxima which occur approximately in the same position as the 30°C SST maxima (Fig. 1c) at roughly 160°E and 140°W. Although similar maxima are not found in the precipitation difference field (Fig. 3), they do appear in the total anomaly simulation precipitation field (not shown). The westward of the two maxima is due more to the heavy precipitation already present in the control simulation, rather than to the precipitation anomaly, which is relatively small in the region. The diabatic heating differences were greatest in the mid troposphere (not shown).

The 11–60-day ensemble average evaporation differences (Fig. 5) reached around 4 mm per day over the regions of warmest SST in the anomaly simulation. Thus, the main contributor to the greatly enhanced precipitation was the highly anomalous moisture convergence (not shown) which had greatest differences in the 800–1000 mb layer. This agrees with the results of Shukla and Wallace (1983), and it is mainly due to the enhanced low-level equatorial westerlies to the west of the SST anomaly (Fig. 6). The large evaporation differences were due to the combined effects of warmer SST and stronger surface winds.

The sea-level pressure (SLP) differences showed a strong (~ 2 mb) Southern Oscillation signal, as seen in the 11–60-day ensemble average difference (Fig. 7). This suggests that the SST anomalies are capable of producing the sea-level pressure pattern of the Southern Oscillation as observed in the mature stage of the El Niño event.

All three experiments had anomalous anticyclonic circulations aloft straddling the

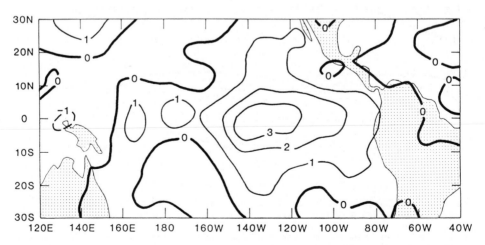

Fig. 5. 11–60-day ensemble average anomaly minus control evaporation difference. Units are mm per day. Dashed contours are negative.

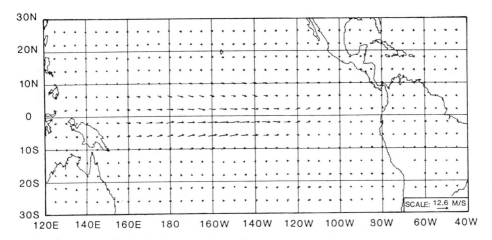

Fig. 6. 11–60-day ensemble average anomaly minus control 850 mb vector wind difference. Maximum = 12.6 m s^{-1}.

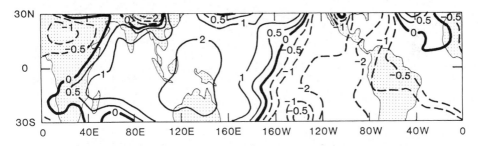

Fig. 7. 11–60-day ensemble average anomaly minus control sea-level pressure difference. Contours are ± 0.5, 1, 2, 4 mb. Dashed contours are negative.

equator in the eastern Pacific, although they were much weaker and shifted eastward compared to those observed. The $\sim 10°$ eastward shift of the anticyclonic couplet observed from January to February was approximately simulated in each experiment. A PNA-like pattern is evident in the observed 200-mb northern hemisphere height anomaly field for January 1983, although by February, the anomalous pattern, though still very strong, undergoes large changes.

The 11–60-day ensemble average 300 mb geopotential height difference field (Fig. 8) shows a pattern over North America which resembles more the observed February or March anomaly pattern rather than the PNA or January pattern. This could perhaps be explained by the eastward extension of the anomalous precipitation in the simulations, which did not occur in the observations until a couple of months later. None of the patterns over North America were significant at 95% on a univariate t-test (Chervin and Schneider, 1976). However, the increased heights throughout the tropical belt, the pattern off the southern tip of South America, and the large positive region over NW Eurasia were all significant at 99% on this same test.

The observed 200-mb zonal wind field for winter 1983 (not shown) had a large ($\sim$ 20 m s^{-1}) negative anomaly over the central and eastern equatorial Pacific, flanked by large (~ 20 m s^{-1}) positive anomalies at roughly 30°S and 30°N. The 11–60-day

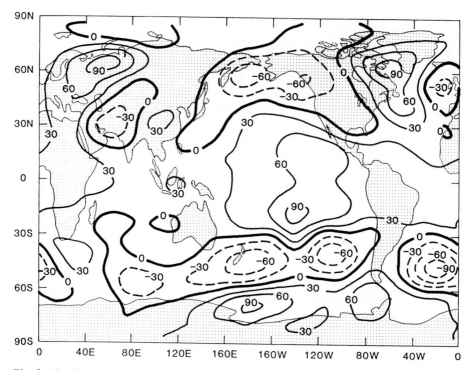

Fig. 8. 11–60-day ensemble average 1982–83 anomaly minus control 300 mb geopotential height difference. Units are geopotential meters. Dashed contours are negative.

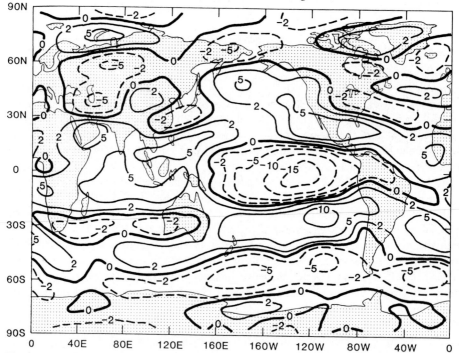

Fig. 9. 11–60-day ensemble anomaly minus control 300 mb zonal wind difference. Contours are ± 2, 5, 10, 15, 20 m s⁻¹. Dashed contours are negative.

ensemble average 300 mb zonal wind difference field (Fig. 9) shows that the negative anomaly was correctly simulated, although the positive anomalies flanking it are both too weak and shifted eastward as compared to the observed anomalies. However, it does seem that the model, at least qualitatively, correctly simulated the local strengthening of the subtropical jet of either hemisphere.

In conclusion, the results in the tropics seem clear. The SST anomalies force anomalous precipitation and heating over the areas of very warm ($\geqslant 29°C$) SST. In addition, they result in anomalous low-level westerlies and upper level easterlies in the equatorial Pacific. The upper-level easterlies are related to a forced anomalous upper-level anticyclone couplet straddling the equator. The SST anomalies produce the SO signal of sea-level pressure correctly.

The extra-tropical results are not nearly so clear. There is a strong extra-tropical response, although it varies greatly with both time and initial conditions. The first couple of weeks of the 1982–83 experiment did show a PNA like pattern in the 300 mb geopotential height difference field. These results suggest the need for further studies with simpler models and observations to understand the mechanisms which determine the influence of tropical heating anomalies on mid-latitude circulation.

ACKNOWLEDGEMENTS

The authors would like to thank Dr. Richard Reynolds of the Climate Analysis Center for providing the sea-surface temperature data used in this study. We would also like to thank Ms. Lora Wright for her careful typing of the manuscript, and Ms. Laura Rumburg for drafting the figures.

REFERENCES

Arakawa, A., 1969. Parameterization of cumulus convection. Proc. WMO/IUGG Symposium on Numerical Prediction, Tokyo, IV8, pp. 1–6.

Blackmon, M.L., Geisler, J.E. and Pitcher, E.J., 1983. A general circulation model study of January climate anomaly patterns associated with interannual variation of equatorial Pacific sea surface temperatures. J. Atmos. Sci., 40: 1410–1425.

Chervin, R.M. and Schneider, S.H., 1976. On determining the statistical significance of climate experiments with general circulation models. J. Atmos. Sci., 33: 405–412.

Deardorff, J.W., 1972. Parameterization of the planetary boundary layer for use in general circulation models. Mon. Weather Rev., 100: 93–106.

Julian, P.R. and Chervin, R.M., 1978. A study of the Southern Oscillation and Walker circulation phenomenon. Mon. Weather Rev., 106: 1433–1451.

Keshavamurty, R.N., 1982. Response of the atmosphere to sea surface temperature anomalies over the equatorial Pacific and the teleconnections of the Southern Oscillation. J. Atmos. Sci., 39: 1241–1259.

Mintz, Y. and Serafini, V., 1981. Monthly normal global fields of soil moisture and land-surface evapotranspiration. Int. Symposium on Variations in the Global Water Budget, Oxford, August 10–15, 1981.

National Meteorological Center Climate Analysis Center, 1982–83. Special Climate Diagnostics Bulls., Nos. 1–6.

Quiroz, R.S., 1983. The climate of the "El-Niño" winter of 1982–83, a season of extraordinary climatic anomalies. Mon. Weather Rev., 11: 1685–1706.

Randall, D.A., 1976. The interaction of the planetary layer with large-scale circulations. Ph.D. thesis, Univ. of Calif. at Los Angeles, Calif., 247 pp.

Randall, D.A., 1982. Monthly and seasonal simulations with the GLAS climate model. Proc. Workshop on Intercomparison of Large-Scale Models Used for Extended Range Forecasts of the European Centre for Medium-Range Weather Forecasts, Reading.

Rowntree, P.R., 1972. The influence of tropical east Pacific Ocean temperatures on the atmosphere. Q. J.R. Meteorol. Soc., 98: 290–321.

Shukla, J. and Wallace, J.M., 1983. Numerical simulation of the atmospheric response to equatorial Pacific sea surface temperature anomalies. J. Atmos. Sci., 40: 1613–1630.

Shukla, J., Straus, D., Randall, D., Sud, Y. and Marx, L., 1981. Winter and summer simulations with the GLAS climate model. NASA Tech. Memo. 83866, 282 pp. [NTIS N8218807].

Wallace, J.M. and Gutzler, D.S., 1981. Teleconnections in the geopotential height field during the Northern Hemisphere winter. Mon. Weather Rev., 109: 784–812.

CHAPTER 11

THE GLOBAL CLIMATE SIMULATED BY A COUPLED ATMOSPHERE–OCEAN GENERAL CIRCULATION MODEL: PRELIMINARY RESULTS

W. LAWRENCE GATES, YOUNG-JUNE HAN and MICHAEL E. SCHLESINGER

ABSTRACT

The results of a synchronously coupled ocean–atmosphere GCM are presented in comparison with observation. The model consists of a six-layer oceanic GCM and a two-layer atmospheric GCM, each with a global horizontal resolution of four degrees latitude and five degrees longitude. Momentum, heat and moisture are exchanged at the ocean surface every hour as a function of the model's evolving sea-surface temperature and surface wind, with full diurnal and seasonal variations of solar radiation retained.

In an extended interannual integration the coupled model simulates a climate which is similar in most respects to that produced by the uncoupled atmospheric GCM (with climatological sea-surface temperatures); a major exception is the coupled model's failure to simulate a realistic seasonal distribution of surface wind, sea-surface temperature and precipitation over the tropical Pacific Ocean. In January the coupled GCM simulates an El Niño-like collapse of the trades and the appearance of warm water across the equatorial Pacific; however, this occurs every year as part of the model's seasonal cycle, rather than as a sporadic interannual event. The major errors in the simulated sea-surface temperature itself are excessively cold water off the east coasts of Asia and North America, and excessively warm water off the western coasts of the tropical continents; these errors are evidently due to systematic errors in the simulated surface heat balance. These errors also cause a progressive loss of sea ice in the southern ocean, while the Arctic sea ice is realistically simulated. Further analysis of the results is underway in order to more fully understand the reasons for the model's behavior, and thereby to improve its ability to simulate large-scale ocean–atmosphere interaction.

INTRODUCTION

A comprehensive model of the global ocean–atmosphere system is a major and long-standing goal of climate dynamics research. Progress toward this goal during the past decade has been slow and difficult, although reasonably steady; it has been slow because of the small number of global general circulation models (GCMs) available and because of the need for the simulation and analysis of large volumes of data; it has been difficult because of the inadequacies of both atmospheric and oceanic GCMs and the general lack of sufficient observations in the ocean. The proper depiction and/or parameterization of processes at the air–sea interface in particular is critical for the development of a satisfactory coupled model, and may require higher resolution than has been customarily employed in GCMs. On the other hand, the satisfactory portrayal of at least some aspects of the large-scale climate may not require the resolution of meso-scale oceanic structure and motion, provided their effects can be satisfactorily parameterized.

The present research represents our exploration of the possibility that significant information on the dynamics of the ocean–atmosphere system may be obtained with a coupled model of relatively coarse oceanic resolution. The first such inquiry was made by Manabe and Bryan (1969), who showed that the inclusion of a dynamic ocean could have a substantial effect on the middle- and high-latitude climate (see also Manabe, 1969). These early studies, however, considered idealized geography, did not include seasonal variations, and used a coupling scheme in which the ocean's response was accelerated about 100 times relative to that of the atmosphere. The important climatic role of the oceans was subsequently confirmed in studies by Manabe et al. (1975, 1979), Bryan et al. (1975) and Washington et al. (1980), who considered realistic global geography but still retained asynchronous coupling schemes. The first synchronous interaction between the atmosphere and ocean was permitted in a GCM simulation by Bryan et al. (1982), who found that the transient climate response to increasing atmospheric carbon dioxide was substantially controlled by the ocean. In this case, however, Bryan and his colleagues returned to the idealized geography and annual average insolation used in their earlier studies.

Here we report what we believe is the first extended simulation of a GCM with realistic global geography in which the atmosphere and ocean are permitted to interact on a synchronous basis. Like previous studies, however, this model has a resolution of several hundred km in both ocean and atmosphere and therefore addresses only the larger scales of the coupled system. The exploration of the successes and failures of such a relatively coarse-grid model in the simulation of global climate is the primary goal of our present research, and is a logical extension of the development of separate GCMs for the atmosphere and ocean which has taken place at Oregon State University over the past several years (Gates and Schlesinger, 1977; Schlesinger and Gates, 1980; Han, 1984a, b). In comparison to the simpler coupled models in which prescribed or highly parameterized air–sea interaction is used to study idealized regional time-dependent processes (e.g., Anderson and McCreary, 1984; or Philander et al., 1984), the present model is fully coupled, both dynamically and thermally, in a realistic global domain. Although the present integration has not been carried to equilibrium in the oceanic domain (nor, as it turns out, in terms of the distribution of sea ice), nevertheless its time-dependent evolution illustrates important features of the coupled ocean–atmosphere system.

With the sea-surface temperature now an internally determined variable, the primary physical constraints on the coupled ocean–atmosphere model are the prescriptions of the radiative properties of clouds and the parameterizations of the surface fluxes of heat, momentum and moisture in the context of the two-layer atmosphere. Along with the imposed seasonal (and diurnal) variation of solar radiation at the top of the model atmosphere and the distribution of properties at the earth's surface, these constraints have served to determine a global model climate which is different in several respects from that found when only the uncoupled atmospheric portion of the model is used. The analysis of these differences provides important insight on the ocean's role in the determination of the modeled climate (Gates, 1979), and is of particular value when the present coupled model is used as a control for experiments with increased atmospheric carbon dioxide. The results presented here are preliminary in the sense that they are drawn from the first portion of a continuing integration, and have not been given extensive physical or statistical diagnosis. Further details of the simulated oceanic and atmospheric climate, however, are given in the companion papers in this volume by Han et al. (1985) and Schlesinger et al. (1985), with the latter paper also showing preliminary results of the coupled model's response to increased CO_2.

OVERVIEW OF MODEL

Atmospheric and oceanic GCMs

The atmospheric portion of the coupled model is basically the same as that described by Schlesinger and Gates (1980, 1981) and documented by Ghan et al. (1982). This is a two-layer primitive-equation GCM formulated in σ-coordinates with a top at 200 mb, and includes surface orography as resolved by a horizontal spherical grid of four degrees latitude and five degrees longitude. The model contains comprehensive physics, including internally predicted cloudiness and both diurnal and seasonal variations of insolation. Although there is not an explicit planetary boundary layer, the model simulates a surface temperature and surface wetness in concert with the maintenance of surface heat and moisture budgets.

A number of relatively minor changes have been made in the version of the OSU atmospheric GCM used in the present coupled run which were not present in the uncoupled atmospheric control run. These include a slight reduction in cloud albedo (in order to reduce the erroneously high planetary albedo), and relaxation of the conditions for the formation of low-level clouds under stable statification (in order to lower the excessive surface flux of solar radiation over the eastern equatorial oceans). Another change made in an effort to reduce what appeared to be excessively large surface winds was the restriction that the surface wind (found by extrapolation from the model's two tropospheric levels at approximately 400 and 800 mb) not exceed the wind at the lower model level. While these changes have in some cases changed the local accuracy of the coupled model, they have not significantly changed the model's overall performance.

The oceanic portion of the coupled model is basically the same as that described by Han (1984a, b). This is a six-layer primitive-equation model of the world ocean with realistic lateral and bottom topography as resolved by the four degrees latitude and five degrees longitude grid. In distinction from the oceanic GCM described by Han (1984a, b), the model version used in the present coupled simulation has been extended to include the Arctic Ocean and includes a bulk sea-ice model. The GCM simulates the changes of salinity (with a constraint of a prescribed surface salinity) and the formation and melting of sea ice by means of a simple bulk-ice parameterization as described in Han et al. (1985). Although a version of the model exists with an imbedded upper mixed layer, that version was not used in the present coupled run.

Domain and solution procedure

The vertical structure of the coupled model is shown in Fig. 1 along with the primary dependent variables simulated during the course of integration. Here u and v are the horizontal wind and current, T the temperature, q the atmospheric water-vapor mixing ratio, s the oceanic salinity, and $\dot{\sigma}$ and w the atmospheric and oceanic vertical velocity, respectively. The surface boundary condition for the atmosphere is: $\dot{\sigma} = 0$ (at $\sigma = 1$), while that for the ocean is: $w = 0$ (at $z = 0$); at the ocean bottom the condition $w = V \cdot \nabla h$ is imposed (where h is the ocean depth), while at the top of the model atmosphere (at 200 mb) the condition $\dot{\sigma} = 0$ is used. The three-dimensional oceanic grid is illustrated in Fig. 2, which also serves to show the horizontal resolution of the model and

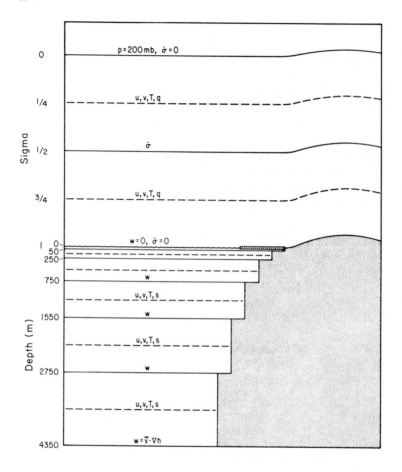

Fig. 1. The vertical structure of the coupled model, together with the primary dependent variables and boundary conditions in the atmosphere and ocean. Here the σ-coordinate atmospheric GCM determines the horizontal velocity components u and v, the temperature T and the water-vapor mixing ratio q at two tropospheric levels near 400 and 800 mb, and the vertical velocity $\dot\sigma$ at an intermediate level; the z-coordinate oceanic GCM determines the horizontal current components u and v, the temperature T and the salinity s at six levels intermediate to those at which the oceanic vertical velocity w is determined. The boundary condition $\dot\sigma = 0$ is imposed at the model top at 200 mb and at the surface, while the conditions $w = 0$ and $w = \mathbf{V} \cdot \nabla h$ are imposed at the ocean surface and ocean bottom, respectively, where h is the ocean depth.

the continental surface orography. The coincidence of the horizontal grid in both the oceanic and atmospheric portions of the model is computationally convenient; the ocean's grid, however, prevents the effective resolution of boundary currents.

The coupling of the atmospheric and oceanic portions of the model is straightforward: the net surface heat flux over the oceans (calculated as a function of the evolving sea-surface and surface air temperatures, the surface wind, the surface evaporation, and the net surface radiation) serves as forcing for the determination of the ocean's surface and interior temperature, while the surface wind stress is similarly used for the calculation of the ocean currents. As noted earlier, in the present model the oceanic salinity is found

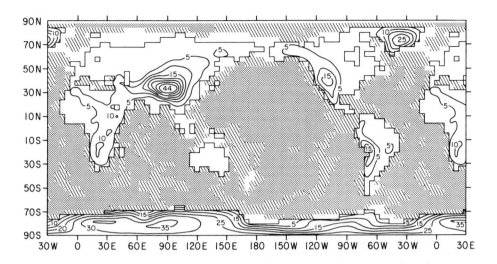

Fig. 2. The global domain of the coupled model, showing the continental outline and orography (in 10^2 m), and the oceanic depth resolved by the model's four degrees latitude and five degrees longitude grid. Here the unshaded oceanic area is less than 750 m depth, the hatched area is between 750 and 2750 m depth, and the shaded area is between 2750 and 4350 m depth.

from a climatologically prescribed surface (skin) salinity, rather than from the simulated evaporation–precipitation difference (see Han et al., 1985). Atmospheric advection takes place every 10 min., and the air temperature is updated once each hour as a result of diabatic heating; the surface (and interior) winds are updated hourly as a result of friction, which coincides with the basic one-hour time step of the ocean model. By such a synchronous exchange of heat and momentum at the sea surface both the ocean and atmosphere model are subject to full diurnal and seasonal forcing, and no time smoothing or averaging is employed during the course of their simultaneous integration. The coupled model requires approximately 4.5 h on a CRAY-1 computer for each year's simulation, approximately one-fourth of which is used for the ocean.

UNCOUPLED MODELS' PERFORMANCE

As background to the results from the coupled model, it is useful to describe briefly the performance of the uncoupled atmospheric and oceanic GCMs. In these independent integrations the necessary boundary conditions at the surface were taken from observed climatological data, in contrast to their automatic provision as an internal part of the solution in the coupled model. For the atmospheric GCM these conditions were the monthly climatological distributions of sea-surface temperature and sea ice, while for the oceanic GCM the monthly climatological distributions of wind stress, net insolation, air temperature and mixing ratio at the ocean surface were used (Han, 1984b).

The performance of the uncoupled atmospheric GCM is illustrated by the average distribution of sealevel pressure simulated for January during the last five years of a ten-year atmospheric control run shown in Fig. 3. In comparison with the observed average

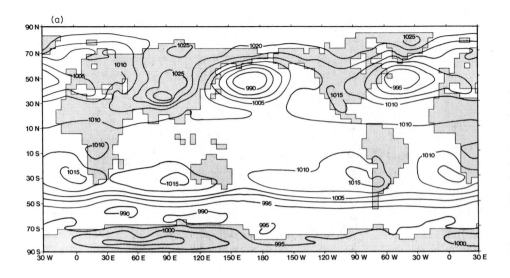

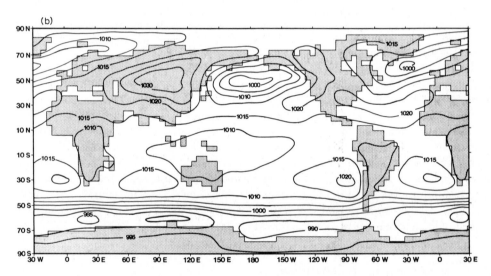

Fig. 3. The average January sealevel pressure (in mb) simulated by the uncoupled atmospheric GCM with climatological sea-surface temperature (above), and the observed average January distribution of sealevel pressure (below; from Schutz and Gates, 1971).

January sealevel pressure also shown in Fig. 3, the atmospheric GCM is seen to reproduce the observed large-scale features reasonably well. The most obvious shortcoming of the pressure simulation is the model's underestimation of the strength of the subtropical highs and its overestimation of the strength of the Icelandic and Aleutian lows. This latter error is evidence of the model's overemphasis of the amplitude of the planetary-scale waves in the winter hemisphere, and is also present in earlier versions of the GCM (Schlesinger and Gates, 1980).

Further evidence of the atmospheric GCM's performance in an uncoupled mode is given in Fig. 4, where the average precipitation rate simulated in January during the last

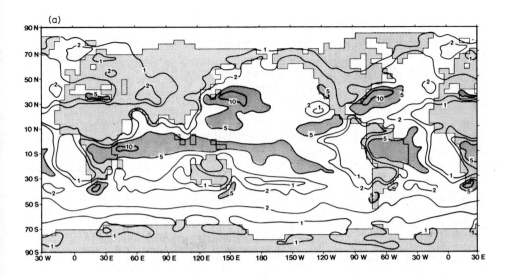

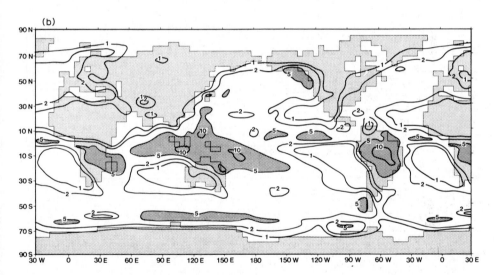

Fig. 4. The average January precipitation rate (in mm per day) simulated by the uncoupled atmospheric GCM with climatological sea-surface temperature (above), and the observed average January distribution of precipitation (below; from Jaeger, 1976). Here areas with precipitation rates greater than 5 mm per day are shaded.

five years of a ten-year atmospheric control integration is shown along with the observed climatological January precipitation. Using the climatological sea-surface temperature, the model is reasonably successful in depicting the large-scale rainfall distribution, especially that over the Indian Ocean, Africa, South America, and the South Atlantic Ocean. Over the central equatorial Pacific and Indonesia, however, the model evidently simulates too little rainfall, although in this region there is considerable simulated interannual precipitation variability (Schlesinger and Gates, 1981). The model's most prominent precipitation error occurs over the western North Pacific and Atlantic Oceans, off the east coasts of Asia and North America. This feature is related to the simulation of strong surface westerlies on the south side of the semi-permanent low-pressure centers in the North Pacific and Atlantic oceans, and consequent vigorous evaporation, convection and rainfall over the (fixed) warm water offshore. Heretofore these and other errors related to the surface heat flux in the atmospheric GCM have not been given much attention, since in an uncoupled simulation the maintenance of a realistic surface heat balance is dominated by the prescription of sea-surface temperature itself.

The performance of the uncoupled oceanic GCM is best illustrated by the sea-surface temperature, since this is the most important output of an ocean model insofar as the atmosphere is concerned. The average sea-surface temperature simulated in January in an

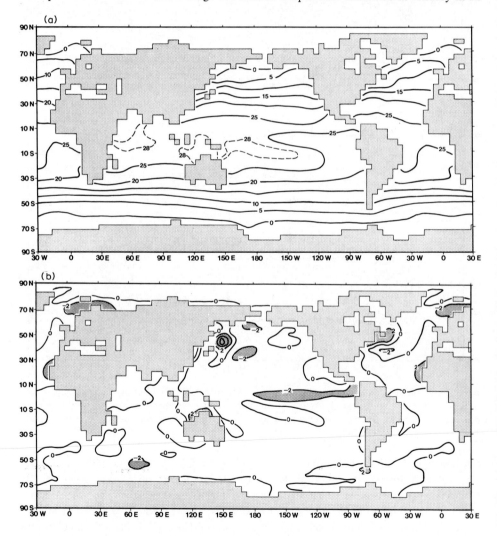

Fig. 5. The average January sea-surface temperature (in °C) simulated by the uncoupled oceanic GCM with climatological surface wind stress, net insolation, air temperature and mixing ratio (above), and the departure of the simulated January sea-surface temperature from the observed climatological January mean (below; from data of Alexander and Mobley, 1976).

extended interannual integration of the oceanic GCM with monthly climatological surface
forcing is shown in Fig. 5. In comparison with the observed average January sea-surface
temperature, this is considered to be a satisfactory simulation: both the large-scale
pattern and magnitude of the sea-surface temperature are given with good accuracy in
almost all regions of the world ocean. This is shown in Fig. 5 by the difference between
the oceanic GCM's January simulation and the observed climatological January distribution of sea-surface temperature; here the discrepancies are less than 2°C nearly
everywhere, and there is little evidence of systematic or large-scale error. The effective
prescription of the surface air temperature in this model, however, may account for much
of this agreement (see Han, 1984b). The larger errors which do occur appear to be related
to the model's inability to accurately depict the intensity of the boundary currents with
its relatively coarse horizontal mesh. In spite of this, it is interesting to note the model's
successful simulation (oversimulation, in fact) of the cold (upwelled) water in the eastern
equatorial Pacific. This and other aspects of the oceanic GCM's performance discussed
by Han et al. (1985) show that the present model is comparable to those developed
earlier by Bryan (1969), Bryan and Lewis (1979) and Meehl et al. (1982).

RESULTS FROM AN EXTENDED COUPLED INTEGRATION

The performance of the uncoupled models discussed above indicates that they are
separately capable of simulating most of the observed large-scale features of the atmospheric and oceanic climate, provided they are given appropriate climatological boundary
conditions at the (ocean) surface. This testimony to the models' overall internal fidelity,
however, says little about their behavior when coupled in a fully interactive manner.
Since it cannot be expected that the ocean model will be able to produce sea-surface
temperatures which are as accurate as those observed, nor that the atmospheric model
will be able to simulate fully accurate surface fluxes over the ocean, it is inevitable that
the sea-surface temperature simulated by the coupled model will be less accurate than
that in the uncoupled models. This may in turn cause other aspects of the simulated
climate to deteriorate relative to that given by the uncoupled models, while the
simulation of other aspects may be improved by the presence of large-scale air—sea
interactions.

The initial conditions for the integration of the coupled model were taken as those
simulated for 1 November by the separate integrations of the uncoupled GCMs. For the
atmosphere these initial data were taken from the second year of a ten-year control
integration (with monthly climatological sea-surface temperature) of a version of the
atmospheric GCM nearly identical with that used later in the coupled run. For the ocean,
the coupled run's initial conditions were taken from the ninth year of an extended
control integration of the oceanic GCM (with monthly climatological surface forcing).
These separately generated fields were then placed in juxtaposition and progressively
updated as a result of the evolution of the joint or coupled model.

Time-dependent evolution

Before discussing the model's simulation of the seasonal cycle or its characteristic
portrayal of the geographical distribution of global climate, it is useful to consider

first the interannual variation of the global or hemispheric integrals of selected variables.

One of the more basic quantities of the coupled atmosphere—ocean climate system is the net radiation at the top of the (model) atmosphere. This is shown in Fig. 6 for the first sixteen years of integration of the coupled GCM, along with the corresponding average observed variation. The model is seen to successfully simulate the maximum of (incoming) net radiation in January, and to underestimate the observed minimum of (outgoing) net radiation in June; this latter shortcoming may be related to the model's underestimation of cloud cover in the southern winter and northern summer seasons. The annual average of the model's simulated net radiation is between 4 and $5\,\mathrm{W\,m^{-2}}$, and no attempt has been made to reduce the solar constant to achieve a zero radiation balance. In fact, the model should show a net downward (positive) flux of radiation at its top, since there is known to be an effective computational energy sink in the model of about $2\,\mathrm{W\,m^{-2}}$, and since the model does not consider the heat equivalent of the energy dissipated by friction, which is about 2 or $3\,\mathrm{W\,m^{-2}}$.

The corresponding variation of the simulated globally averaged net surface heat flux is shown in Fig. 7, along with the average seasonal variation of the observed flux. Since over monthly and annual time scales there is little net heating of the ground, these data essentially reflect the heat flux at the ocean surface. As was the case for the net radiation at the top of the atmosphere, the model fails to simulate the observed magnitude of the June minimum of negative (upward) net heat flux from the oceans to the atmosphere. The model also, however, fails to simulate the maximum positive (downward) net heat flux which is observed in February. These features appear to be associated with the sensible and latent heat flux from the mid- and high-latitude oceans during winter, which the model systematically overestimates. Figure 7 also shows evidence of a gradual rise in the net downward heat flux simulated from September to February in successive years of the integration, although the average annual surface heat flux has not changed by more than $1\,\mathrm{W\,m^{-2}}$.

The variation of the globally averaged sea-surface temperature is shown in Fig. 8, along

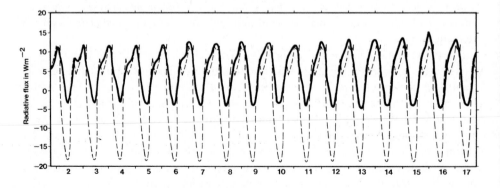

Fig. 6. The variation of the globally averaged net radiation at the top of the atmosphere simulated by the coupled model (full line) and the observed globally averaged seasonal cycle (dashed line; from Ellis et al., 1978). The integration of the coupled GCM was begun on 1 November of year 1 in the chronology of an earlier atmospheric GCM integration.

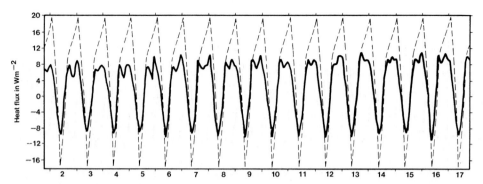

Fig. 7. The variation of the globally averaged net surface heat flux simulated by the coupled model (full line), and the observed globally averaged seasonal cycle (dashed line; from Esbensen and Kushnir, 1981).

with the observed average seasonal cycle. Although this statistic undergoes relatively small seasonal changes due to its inherent conservatism and the opposing effects of the seasons in the northern and southern hemispheres, it is clear that the coupled model has not reproduced the observed variation. In fact, from almost the beginning of the integration the seasonal variation of the simulated average sea-surface temperature is out of phase with that observed, especially after about ten years' simulation. This behavior is evidently the result of the model's progressive simulation of reduced Antarctic sea ice (discussed below), which permits excessive warming (cooling) of the high-latitude southern ocean during the southern summer (winter) while the observed variation is dominated by seasonal changes in the northern hemisphere oceans. If the ocean south of 70°S is excluded from the global average, the simulated mean sea-surface temperature is in reasonably good agreement with observation. The relatively large interannual variations of the simulated average sea-surface temperature also reflect the response of the tropical oceans to varying surface winds.

The simulated variation of the globally and mass-averaged (tropospheric) air temperature is shown in Fig. 9, together with the average seasonal variation of the observed globally averaged mean temperature. Aside from a systematic model

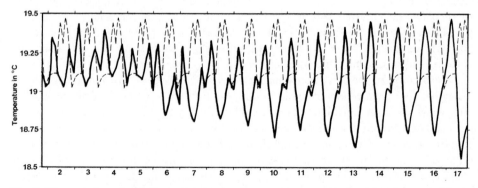

Fig. 8. The variation of globally averaged sea-surface temperature simulated by the coupled model (full line), and the observed globally averaged seasonal cycle (dashed line: from data of Alexander and Mobley, 1976).

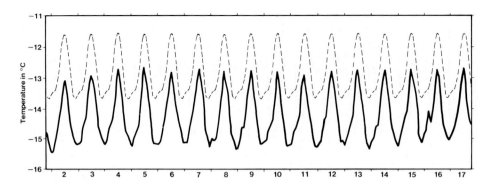

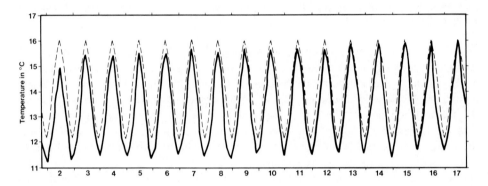

Fig. 9. The variation of globally averaged surface air temperature (above) and mass-averaged tropospheric temperature (below), as simulated by the coupled model (full line) and as observed (dashed line; from Jenne, 1975).

underestimate of the observed mean by about 1.5°C (due partly to the model's coarse vertical resolution), the observed seasonal cycle is reasonably well simulated. The coupled model, moreover, shows no sign of a long-term drift in mean temperature after an apparent spin-up during the first few years. Also shown in Fig. 9 is the variation of the simulated and observed average surface air temperature. Here the coupled model is seen to simulate the observed seasonal variation rather well, although there is evidently a gradual increase in the model's maximum average air temperature during successive northern summers. This is consistent with the slight increase of the net downward radiation and surface heat flux noted earlier.

The variation of the total sea-ice mass given in Fig. 10 shows that the coupled model undergoes a progressive loss of sea ice during the first ten years or so of the integration, after which an approximate interannual equilibrium is achieved. The average mass of sea ice simulated toward the end of the run, however, is about half that given by the un-coupled oceanic GCM in near-equilibrium with observed surface forcing, a difference principally due to ice in the southern ocean. Further insight into the behavior of the coupled model's sea ice is given in Fig. 11, which shows the sea-ice area in both the northern and southern hemispheres. In the Arctic the model provides a reasonably accurate simulation of the observed mean seasonal variation of sea-ice area; in the

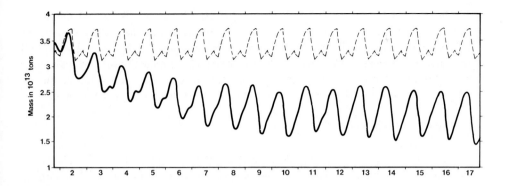

Fig. 10. The variation of the total global sea-ice mass simulated by the coupled model (full line), and that given by the uncoupled ocean GCM in near-equilibrium with observed surface forcing (dashed line; from Han, 1984b).

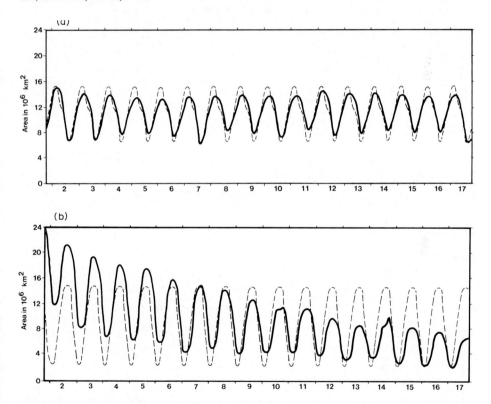

Fig. 11. The variation of sea-ice area in the northern hemisphere (above) and southern hemisphere (below), as simulated by the coupled model (full line) and as observed (dashed line; from Zwally et al., 1983).

Antarctic, however, there is a progressive reduction of the simulated mean sea-ice area, together with a reduction in the simulated amplitude of the seasonal variation. Although the model was started with approximately twice the observed sea-ice area in the southern hemisphere (as given by the uncoupled ocean model), it appears likely that the model would simulate the Antarctic to be free of sea ice during at least part of the year if the simulation were continued. This behavior must be considered a defect of the present model, and is responsible for the anomalous variation of the model's averaged sea-surface temperature noted earlier; it does not, however, appear to have prevented the simulation of an approximate seasonal climatic equilibrium in at least the northern hemisphere.

Global climate distribution

In order to examine the geographical structure of the coupled model's simulation of global climate, we here consider the average global distribution of selected climate

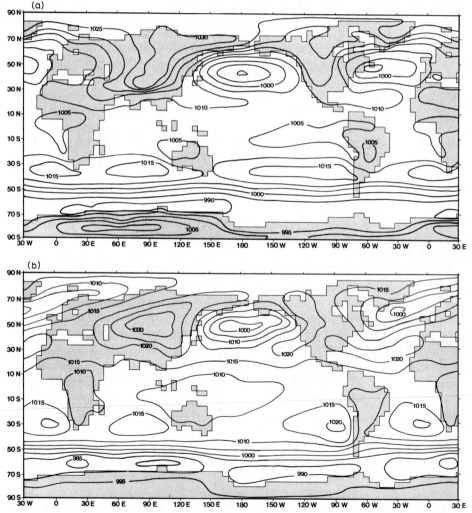

Fig. 12. The average January sealevel pressure (in mb) simulated by the coupled model (above), and the observed average January distribution (below; from Schutz and Gates, 1971).

variables for January; this will also permit comparison of the model's performance with that of the uncoupled atmospheric and oceanic models shown earlier.

The model's simulation of January sealevel pressure, averaged over the last four years of the coupled integration, is shown in Fig. 12, along with the observed climatological distribution. A comparison of Figs. 3 and 12 shows that while the coupled model has made a small improvement relative to the uncoupled atmospheric GCM in the simulation of the principal features of sealevel pressure, it resembles the atmospheric GCM's result more than it resembles observation. The coupled model's simulated sealevel pressure is,

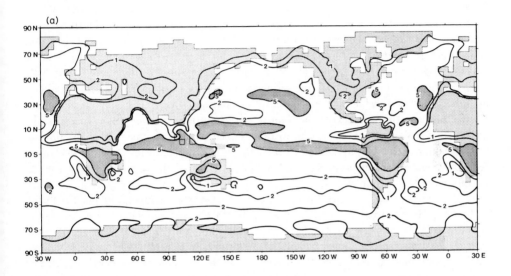

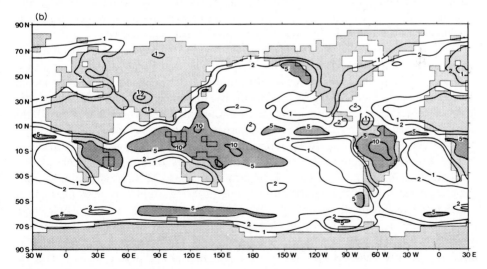

Fig. 13. The average January precipitation rate (in mm per day) simulated by the coupled model (above), and the observed average January distribution (below; from Jaeger, 1976). Here areas with precipitation rates greater than 5 mm per day are shaded.

however, noticeably lower than that of the uncoupled atmospheric GCM off the west coasts of the tropical continents; this difference is likely a result of the coupled GCM's simulation of warmer-than-observed sea-surface temperatures in these areas.

The distribution of the coupled model's simulation of the average January precipitation rate is shown in Fig. 13, together with the observed climatological distribution. Over Indonesia, over the Indian Ocean, and off the east coasts of North America and Asia the simulation may be considered an improvement over the uncoupled model's result (shown in Fig. 4). Over the South Pacific Ocean the coupled model has failed to reproduce the precipitation associated with the South Pacific convergence zone, and has displaced the zone of precipitation observed near $5°N$ in the eastern Pacific about 10^3 km to the south. Furthermore, the dry zones observed to the west of Africa, South America and Australia are systematically underestimated. These changes in tropical precipitation collectively represent a deterioration of the results found with the uncoupled atmospheric model (and climatological sea-surface temperatures); they are principally reflections of systematic errors in the simulated tropical sea-surface temperature, which is in turn closely coupled to the atmospheric circulation. Overall, the coupled model has slightly lowered the simulated total January precipitation relative to that found in the uncoupled GCM; most of this is due to the coupled model's suppression of the precipitation maxima previously simulated in the eastern North Pacific and Atlantic oceans. In July the coupled model produces similar modifications of the precipitation relative to that found in the uncoupled case.

The coupled model's simulation of the January sea-surface temperature itself is shown in Fig. 14. Here the most notable feature is the $28°C$ water simulated across the entire equatorial Pacific, along with a similar warm-water pool in the Atlantic. This distribution, in fact, resembles that observed during the mature phase of an El Niño, and is simulated in the model only during the northern winter months. In the western North Pacific and Atlantic the coupled model has failed to simulate the relatively warm water associated with the Kuroshio and Gulf Stream, with the result that the offshore distribution of simulated sea-surface temperature is somewhat more zonal than observed.

These and other aspects of the coupled model's simulated sea-surface temperatures are clearly seen in the lower portion of Fig. 14, which shows the simulation's errors relative to the observed January climatological distribution. Here the model's failure to portray the relatively cold water off the west coasts of the tropical continents is evident, resulting in local sea-surface temperature errors of $4°-6°C$; as discussed by Han et al. (1985), this error is believed to be partly caused by excessive solar radiation at the surface (due to deficient cloudiness), and partly by deficient coastal currents and upwelling. Also clearly seen is the model's failure to simulate sufficiently warm water off the east coasts of the mid-latitude continents, which yields local errors as large as $10°C$; as already noted, this error is related to both the simulation of strong surface westerlies in these areas and to a deficient strength of the simulated western boundary currents. A third characteristic error apparent in Fig. 14 is the simulation of a broad band of water in the southern ocean between about $50°-70°S$ which is between $2°-4°C$ warmer than observed. This error is evidently caused by excessive solar radiation at the surface and by deficient simulated upwelling around Antarctica, and is closely related to the progressive loss of sea ice in the southern ocean. This general pattern of sea-surface temperature error is similar to that found in previous (asynchronous) simulations with coupled

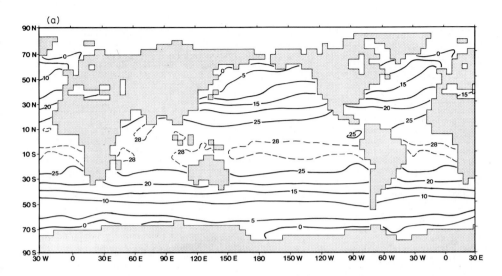

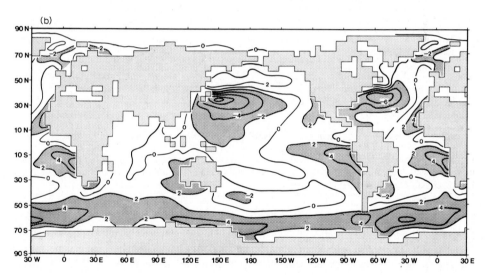

Fig. 14. The January sea-surface temperature (in °C) simulated by the coupled model (above), and the departure of the simulated sea-surface temperature from the observed (below; from data of Alexander and Mobley, 1976). Here simulation errors greater than 2°C are shaded.

ocean—atmosphere GCMs (Manabe et al., 1979; Washington et al., 1980), and may be characteristic of coarse-resolution coupled models.

Since the surface wind is an important factor in air—sea interaction, especially over the equatorial ocean, the average January surface wind simulated in the tropical Pacific is shown in Fig. 15. Here the southeasterly winds which are observed near the South American coast and which cross the equator near 120°W are not well simulated by the coupled model, although the trades near 10°N and 20°S are reasonably well portrayed. Of most consequence to the equatorial sea-surface temperature, however, is the model's

(a)

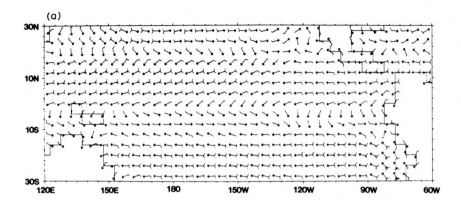

(b)

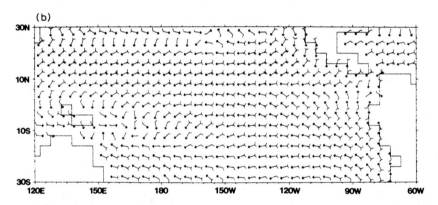

Fig. 15. The average January surface wind simulated over the tropical Pacific Ocean by the coupled model (above), and as observed (below; from National Climatic Center's TDF-11 surface marine deck). Here each half barb denotes 1 m s^{-1} speed.

simulation of weak surface easterlies or even westerlies across most of the Pacific near and just south of the equator. This circulation has served to effectively remove the upwelling of relatively cold water which is usually found in this area in January, and has permitted the appearance of warm surface water across the entire Pacific as seen in Fig. 14; this in turn is related to the anomalous precipitation in this area seen in Fig. 13 and to the tropical sea-level pressures shown in Fig. 12. This wind distribution bears a striking resemblance to an El Niño, although instead of occurring sporadically as in nature it is simulated during January of every year of the simulation. The causes or controlling factors of this behavior have not yet been determined, but it is known that the easterlies found across the equatorial Pacific in other seasons are reasonably well simulated.

CONCLUDING REMARKS

The integration of the synchronously coupled ocean–atmosphere GCM reported here may be considered a success from the viewpoint of the solution's overall configuration and stability, and demonstrates that full dynamical and thermal coupling over a global domain is a feasible undertaking for periods of a decade or two. During this time the ocean's deeper waters will not in general be in equilibrium unless the model's initial conditions are fortuitously chosen, although the atmosphere and upper ocean will have achieved effective adjustment on seasonal and annual time scales. From the viewpoint of the simulation of interannual ocean–atmosphere variability, however, the present integration cannot be considered completely successful since it has not simulated significant interannual variations of tropical sea-surface temperature and has not simulated a realistic El Niño-Southern Oscillation. Although the model has indeed simulated an annual El Niño-like event in the equatorial Pacific, this must be considered part of the model's regional season cycle rather than a true El Niño; its presence does show, however, that the model is at least capable of such a response. The model's failure in this respect is apparently due to shortcomings in the GCM's portrayal of the surface heat exchange.

Further study of the coupled model's simulation is underway, including analysis of the simulated seasonal cycle and a diagnosis of the components of the ocean-surface heat budget in both low and high latitudes. Of particular interest is the search for systematic relationships among the time and space distributions of the surface wind, sea-surface temperature, surface heating and precipitation in low latitudes. The present coupled GCM provides an opportunity to examine the possible mechanics of air–sea interaction on both synoptic and seasonal time scales. Such analysis should prove useful in the improvement of the present model and in the design of more realistic simplified models of the coupled ocean–atmosphere system, which in turn may be used in more extensive sensitivity studies. The present model may also be used in experiments designed to show the transient response of the coupled ocean–atmosphere system to a variety of initial conditions or external forcing, in addition to the response to increased CO_2 discussed by Schlesinger et al. (1985).

ACKNOWLEDGEMENTS

We would like to thank Robert L. Mobley and William McKie for their management of the integration of the coupled GCM and their assistance in its analysis, Larry Holcomb for drafting the figures, and Elizabeth Webb for typing the manuscript. We also thank NCAR for the provision of the necessary time on their CRAY-1 computer which was accessed via a remote terminal at the OSU Climatic Research Institute. This research was supported by the National Science Foundation and the U.S. Department of Energy under grant ATM 8205992.

REFERENCES

Alexander, R.C. and Mobley, R.L., 1976. Monthly averaged sea surface temperatures and ice-pack limits on a 1° global grid. Mon. Weather Rev., 104: 143–148.
Anderson, D.L.T. and McCreary, J.P., 1984. Slowly propagating disturbances in coupled ocean–atmosphere model. (unpubl. manuscr.).
Bryan, K., 1969. Climate and the ocean circulation: III. The ocean model. Mon. Weather Rev., 97: 806–827.

150

Bryan, K. and Lewis, L.J., 1979. A water mass model of the world ocean. J. Geophys. Res., 84: 2503–2517.

Bryan, K., Manabe, S. and Pacanowski, R.C., 1975. A global ocean–atmosphere climate model. Part II. The oceanic circulation. J. Phys. Oceanogr., 5: 30–46.

Bryan, K., Komro, F.G., Manabe, S. and Spelman, M.J., 1982. Transient climate response to increasing atmospheric carbon dioxide. Science, 215: 56–58.

Ellis, J.S., Vonder Haar, T.H., Levitus, S. and Oort, A.H., 1978. The annual variation in the global heat balance of the Earth. J. Geophys. Res., 83: 1958–1962.

Esbensen, S.K. and Kushnir, Y., 1981. The heat budget of the global ocean: An atlas based on estimates from surface marine observations. Report No. 29, Climatic Research Institute, Oregon State University, Corvallis, Oreg., 27 pp. + 188 charts.

Ghan, S.J., Lingaas, J.W., Schlesinger, M.E., Mobley, R.L. and Gates, W.L., 1982. A documentation of the OSU two-level atmospheric general circulation model. Report No. 35, Climatic Research Institute, Oregon State University, Corvallis, Oreg., 395 pp.

Gates, W.L., 1979. The effect of the ocean on the atmospheric general circulation. Dyn. Atmos. Oceans, 3: 95–109.

Gates, W.L. and Schlesinger, M.E., 1977. Numerical simulation of the January and July global climate with a two-level atmospheric model. J. Atmos. Sci., 34: 36–76.

Han, Y.-J., 1984a. A numerical world ocean general circulation model, Part I. Basic design and barotropic experiment. Dyn. Atmos. Oceans, 8: 107–140.

Han, Y.-J., 1984b. A numerical world ocean general circulation model, Part II. A baroclinic experiment. Dyn. Atmos. Oceans, 8: 141–172.

Han, Y.-J., Schlesinger, M.E. and Gates, W.L., 1985. An analysis of the air–sea–ice interaction simulated by the OSU-coupled atmosphere–ocean general circulation model. In: J.C.J. Nihoul (Editor), Coupled Ocean–Atmosphere Models. (Elsevier Oceanography Series, 40) Elsevier, Amsterdam, pp. 167–182 (this volume).

Jaeger, L., 1976. Monatskarten des Niederschlags für die ganze Erde. Ber. Dsch. Wetterdienstes, No. 139, Offenbach, 38 pp.

Jenne, R.L., 1975. Data sets for meteorological research. NCAR Tech. Note TN/1A-III, Natl. Cent. Atmos. Res., Boulder, Colo., 194 pp.

Manabe, S., 1969. Climate and the ocean circulation: II. The atmospheric circulation and the effect of heat transfer by ocean currents. Mon. Weather Rev., 97: 775–805.

Manabe, S. and Bryan, K., 1969. Climate calculations with a combined ocean–atmosphere model. J. Atmos. Sci., 26: 786–789.

Manabe, S., Bryan, K. and Spelman, M.J., 1975. A global ocean–atmosphere climate model. Part I. The atmospheric circulation. J. Phys. Oceanogr., 5: 3–29.

Manabe, S., Bryan, K. and Spelman, M.J., 1979. A global ocean–atmosphere climate model with seasonal variation for future studies of climate sensitivity. Dyn. Atmos. Oceans, 3: 393–426.

Meehl, G.A., Washington, W.M. and Semtner Jr., A.J., 1982. Experiments with a global ocean model driven by observed atmospheric forcing. J. Phys. Oceanogr., 12: 301–312.

Philander, S.G.H., Yamagata, T. and Pacanowski, R.C., 1984. Unstable air-sea interactions in the tropics. J. Atmos. Sci., 41: 604–613.

Schlesinger, M.E. and Gates, W.L., 1980. The January and July performance of the OSU two-level atmospheric general circulation model. J. Atmos. Sci., 37: 1914–1943.

Schlesinger, M.E. and Gates, W.L., 1981. Preliminary analysis of the mean annual cycle and interannual variability simulated by the OSU two-level atmospheric general circulation model. Report No. 23, Climatic Research Institute, Oregon State University, Corvallis, Oreg., 47 pp.

Schlesinger, M.E., Gates, W.L. and Han, Y.-J. 1985. The role of the ocean in CO_2-induced climate change: Preliminary results from the OSU-coupled atmosphere–ocean general circulation model. In: J.C.J. Nihoul (Editor), Coupled Ocean–Atmosphere Models. (Elsevier Oceanography Series, 40) Elsevier, Amsterdam, pp. 447–478 (this volume).

Schutz, C. and Gates, W.L., 1971. Global Climatic Data for Surface, 800 mb, 400 mb: January, R-915-ARPA, The Rand Corporation, Santa Monica, Calif., 173 pp.

Washington, W.M., Semtner Jr., A.J., Meehl, G.A., Knight, D.J. and Mayer, T.A., 1980. A general circulation experiment with a coupled atmosphere, ocean, and sea ice model. J. Phys. Oceanogr., 10: 1887–1908.

Zwally, H.J., Comiso, J.C., Parkinson, C.L., Campbell, W.J., Carsey, F.D. and Gloersen, P., 1983. Antarctic sea ice cover 1973–1976 from satellite passive microwave observations. NASA SP-459, National Aeronautics and Space Administration, Washington, D.C., 170 pp.

CHAPTER 12

FREE EQUATORIAL INSTABILITIES IN SIMPLE COUPLED ATMOSPHERE–OCEAN MODELS

ANTHONY C. HIRST

ABSTRACT

The stability of free linear equatorial waves in two versions of a simple coupled atmosphere–ocean model is investigated here. In case I, the simple traditional assumption is made that sea-surface temperature (SST) anomalies are proportional to anomalous ocean mixed-layer depth. In case II, SST anomalies result from anomalous advection of a mean westward temperature gradient. The coupling is by windstress on the ocean and heating of the atmosphere proportional to the underlying SST anomaly. Both the atmosphere and ocean are represented by one-layer models on the equatorial β-plane. The matrix method with variable resolution is used to find eigenvalues and eigenvectors of the coupled system.

In case I, the oceanic Kelvin, Yanai and gravest inertia-gravity waves are all destabilized while the oceanic Rossby waves are damped by the coupling. In contrast, the gravest oceanic Rossby wave is destabilized while the Kelvin wave is damped in case II. This difference between the two cases is easily explained in terms of the configuration of induced atmospheric motion relative to the oceanic velocities. Only long wavelength ($\gtrsim 8000 \, \mathrm{km}$) Kelvin (case I) or gravest Rossby (case II) waves are unstable at most probable values of the coupling coefficients. These results demonstrate that details of the model physics are crucial in determining the stability of ocean waves in coupled atmosphere–ocean models.

INTRODUCTION

The ability of idealized ocean–atmosphere coupling to destabilize equatorial ocean waves is suggested by the growth of perturbations in the ocean basin model of Philander et al. (1984). The extent to which instabilities in such simply coupled models relate to the development of real climatic disturbances such as ENSO events is unclear; it is possible that adoption of more realistic ocean thermodynamics may completely change the stability properties of the model. In this study, the stability, phase speed and structure of free equatorial waves in two versions of a simple coupled atmosphere–ocean model are compared, in order to more fully understand the nature of instabilities in such models and to determine the effects of a simple change in the method used to determine the sea-surface temperature (SST) anomaly on the stability of the model waves. The first version of the model (case I) is essentially identical to that of Philander et al. (1984), in which the SST anomaly is assumed proportional to the perturbation in the ocean mixed-layer depth. In the second version (case II), the SST anomaly is determined by anomalous advection in the presence of a background westward temperature gradient. Observations indicate that SST anomalies are well-correlated with mixed-layer depth only in the far eastern Pacific (Gill, 1982), while anomalous advection seems to be a principal cause of SST anomalies over most of the equatorial Pacific (Gill, 1983; Schopf and Harrison, 1983).

A stability analysis for "Kelvin" waves in a coupled atmosphere–ocean model similar to that in case I was performed by Lau (1981); however he assumed that meridional velocities and Coriolis accelerations are zero. Such assumptions cannot be made a priori, since heating by an ocean Kelvin wave will continuously excite atmospheric Rossby and inertia-gravity waves (Matsuno, 1966) which in turn excite higher meridional mode ocean waves. The resulting meridional velocities may be significant. In this study, oceanic and atmospheric motions are governed by the complete linear shallow-water equations on the equatorial β-plane.

MODEL AND METHOD

The atmosphere

Atmospheric motion in response to a heat source $Q(x,y,t)$ is given by:

$$U_t - \beta y V + \phi_x + AU = 0 \tag{1a}$$

$$V_t + \beta y U + \phi_y + AV = 0 \tag{1b}$$

$$\phi_t + c_a^2(U_x + V_y) + B\phi = Q \tag{1c}$$

Equations (1a–c) represent the baroclinic mode of a two-level model if U and V are the difference between the upper- and lower-level wind components and ϕ is the thickness between the levels (Matsuno, 1966). This interpretation is adopted here, since latent heating in the tropics excites mainly baroclinic mode motions (Gill, 1980; Lim and Chang, 1983). Values of the gravity wave speed c_a quoted in the literature for the baroclinic mode range from $15\,\mathrm{m\,s^{-1}}$ (Lau, 1981) to $66\,\mathrm{m\,s^{-1}}$ (Philander et al., 1984). Here two values of c_a are tried, $c_a = 25\,\mathrm{m\,s^{-1}}$ and $c_a = 66\,\mathrm{m\,s^{-1}}$. Values of the damping coefficients A and B are listed in Table 1.

TABLE 1

Values of the coefficients and parameters for case I and case II. Values given here for K_S, K_H^I and K_H^{II} are the "most probable" values

	Parameter	Value
Both cases	A	$2.3 \times 10^{-6}\,\mathrm{s^{-1}}$
	B	$2.3 \times 10^{-6}\,\mathrm{s^{-1}}$
	a	$1.16 \times 10^{-7}\,\mathrm{s^{-1}}$
	b	$1.16 \times 10^{-7}\,\mathrm{s^{-1}}$
	c	$1.16 \times 10^{-7}\,\mathrm{s^{-1}}$
	c_0	$1.4\,\mathrm{m\,s^{-1}}$
	Ca	alternatively 25 and 66 $\mathrm{m\,s^{-1}}$
	K_S	$8 \times 10^{-8}\,\mathrm{s^{-1}}$
Case I	K_H^I	$3.5 \times 10^{-3}\,\mathrm{s^{-1}}$
Case II	K_H^{II}	$2.5 \times 10^{-2}\,\mathrm{s^{-1}}$
	$\bar{T}_x$	$5.0 \times 10^{-7}\,\mathrm{°C\,m^{-1}}$

The ocean

Case I

The ocean model consists of a mixed layer of mean depth H overlying a deep cold quasi-motionless layer. Anomalous windstress (τ^x, τ^y) acts as a body force on the mixed layer. Motions in the mixed layer are given by:

$$u_t - \beta yv + g^* h_x + au = \tau^x \tag{2a}$$

$$v_t + \beta yu + g^* h_y + av = \tau^y \tag{2b}$$

$$h_t + H(u_x + v_y) + bh = 0 \tag{2c}$$

$$T = \kappa h \tag{2d}$$

The zonal (u) and meridional (v) velocity components are associated with variations in h, the mixed-layer depth perturbation. g^* is the reduced gravity. T is the SST anomaly and is assumed to be proportional to h. A crude estimate of the proportionality constant was found here to be $\kappa = +0.03°\text{C m}^{-1}$, from a regression of SST on mixed-layer depth using bi-monthly mean data for the far eastern equatorial Pacific from Gill (1982). a and b are damping coefficients; their values are given in Table 1.

The results presented in this report for both cases I and II are obtained using a value of $c_0 = (g^* H)^{1/2} = 1.4 \text{ m s}^{-1}$ (following Philander et al., 1984), corresponding to a mixed-layer depth $H = 70$ m. This value of c_0 may be appropriate for conditions in the far eastern equatorial Pacific (where $H \lesssim 70$ m; Gill, 1982).

Solutions have also been obtained for $c_0 = 2.5 \text{ m s}^{-1}$ (i.e. $H = 200$ m), this value being appropriate for the first baroclinic mode in the central and west equatorial Pacific. These solutions are qualitatively similar to those for $c_0 = 1.4 \text{ m s}^{-1}$, and are not presented here.

Case II

The ocean model is the same as in case I except that SST anomalies result from anomalous advection of a background westward temperature gradient, $\bar{T}_x$, present in the mixed layer. Thus motions are given by:

$$u_t - \beta yv + g^* h_x + \alpha g H T_x + au = \tau^x \tag{3a}$$

$$v_t + \beta yu + g^* h_y + \alpha g H T_y + av = \tau^y \tag{3b}$$

$$h_t + H(u_x + v_y) + bh = 0 \tag{3c}$$

$$T_t + \bar{T}_x u + cT = 0 \tag{3d}$$

where α and g are the coefficients of termal expansion and gravity. The westward temperature gradient is $5°\text{C}$ per 10,000 km.

Solutions were also computed using an ocean model as in eqns. (3a–d), except that the effects of T on the buoyancy are ignored by deleting the terms $\alpha g H T_x$ and $\alpha g H T_y$ from eqns. (3a–d) (cf. Rennick, 1983). The solutions were not significantly different from those determined using the full eqn. (3), and are not further discussed.

Coupling parameterisation

Anomalous windstress and latent heating are parameterised according to the traditional simple assumptions:

$$(\tau_x, \tau_y) = -K_S(U, V) \qquad Q = K_H T \tag{4}$$

Solutions are determined for various values of K_S and K_H, nevertheless most probable values are estimated as follows: A value for K_S is estimated using the bulk aerodynamic formula:

$$K_S = \frac{\rho_a C_D |\bar{V}|}{\rho_0 H} = 8 \times 10^{-8} \, \text{s}^{-1}$$

Symbols have their usual meanings and standard values, the background surface wind $|\bar{V}| = 4 \, \text{m s}^{-1}$, and $H = 70 \, \text{m}$. Note that a larger value of H, such as in the central Pacific, results in a smaller K_S and thus inhibits the effects of atmosphere–ocean coupling.

A crude estimate of K_H is determined via a regression analysis using SST and outward longwave radiation (OLR) anomalies for the equatorial Pacific obtained from Arkin et al. (1983, maps 1 and 10) and rainfall observations obtained from NOAA (1981–1983). The resulting value is $K_H = 3.5 \times 10^{-3} \, \text{m}^2 \, \text{s}^{-3} \, \text{K}^{-1}$.

Thus for case I: $Q = K_H^I(g^* h)$ where $K_H^I = 3.5 \times 10^{-3} \, \text{s}^{-1}$

and for case II: $Q = K_H^{II}(\alpha g H T)$ where $K_H^{II} = 2.5 \times 10^{-2} \, \text{s}^{-1}$

It is noted that a correlation between OLR and SST anomalies in the equatorial Pacific for 1981–1983 performed here yielded a correlation coefficient of -0.64. Thus, for this period, relationship (4) had some validity in the equatorial Pacific.

The estimated values of the coupling coefficients for case I (K_H^I and K_S) are much smaller than those used by Philander et al. (1984).

Free mode analysis

An eigenvector analysis is performed for the model eqns. (1) and (2) in case I or (1) and (3) in case II. The method of analysis is outlined in this section for case I; that for case II is essentially identical.

Equations (1) and (2) are non-dimensionalised, using a length scale of $L = (c_0/\beta)^{1/2} = 2.5 \times 10^5 \, \text{m}$ and a time scale of $T = (c_0\beta)^{-1/2} = 1.8 \times 10^5 \, \text{s}$. Solutions are then sought of the form:

$$U(x, y, t) = U(y) e^{i(kx - \sigma t)}$$

where k is given and σ is an eigenvalue to be determined. It is convenient to make the transform $y = 1/2 \sinh (y')$, so that a finite difference scheme will have higher resolution near the equator ($y = 0$). This higher resolution is necessary in order to properly resolve the oceanic equatorial waves. The boundary conditions are:

$$(U, V, \phi, u, v, h, T) = 0 \text{ at } y' = \pm L \tag{5}$$

where L is sufficiently large.

The derivatives are approximated by centered finite differences, and, on application of the boundary condition (5) and symmetry conditions, eqns. (1) and (2) become the linear algebraic system $A\xi = i\sigma\xi$ where $i\sigma$ and ξ are the eigenvalues and eigenvectors to be found and A is a $(6N-2)\times(6N-2)$ or $(6N-4)\times(6N-4)$ matrix for the symmetric and antisymmetric eigenvectors respectively; N is the number of internal grid points. The QR method was used to find the eigenvalues and eigenvectors of A.

Tests using various N and/or L indicate that $N = 12$ and $L = 4.2$ gave results close to convergence for the lower order meridional modes, in both cases I and II. The results in the following two sections are in general obtained using the values $N = 12$ and $L = 4.2$.

The correct convergence of the finite difference scheme was checked by comparing the eigenvalues and eigenvectors for the lower order meridional modes obtained in the absence of coupling with the known theoretical values (Matsuno, 1966). The scheme was further verified by a comparison of the computed results when $c_a = c_o$ with the results of a parabolic cylinder function (D_n) analysis (Philander et al., 1984).

STABILITY AND DISPERSION

In the following sections, attention is restricted to the effects of coupling on oceanic Kelvin, Yanai (i.e. mixed Rossby-gravity), $n = 1$ Rossby (R1), $n = 2$ Rossby (R2) and $n = 0$ inertia-gravity (IGEO) waves. The oceanic inertia-gravity waves of higher meridional mode and atmospheric waves are not significantly affected by the coupling in either case I or II. The oceanic Rossby waves of higher meridional mode are not well resolved by the finite difference scheme.

Experiments demonstrate that the eigenvalues of the coupled system in both cases I and II depend on the product of the coupling coefficients.

Case I

The effect of case I atmosphere—ocean coupling on the frequency and stability of oceanic equatorial waves with a wavelength of 10,000 km ($k = 0.16$) is shown in Fig. 1. Here $c_0 = 1.4\,\mathrm{m\,s^{-1}}$ and $c_a = 25\,\mathrm{m\,s^{-1}}$. At very small values of the coupling coefficients ($K_H^I K_S$), the damping rates of all the waves are very close to the uncoupled oceanic damping rate determined by a and b. As $K_H^I K_S$ is increased, the Kelvin, Yanai and IGEO waves become less damped and eventually unstable. Meanwhile, the Rossby waves become more damped and eventually are replaced by highly damped unphysical eigenvectors. The frequency, hence phase speed, of all waves decreases as $K_H^I K_S$ is increased, however at high values of $K_H^I K_S$, the Kelvin wave frequency begins a gradual increase.

Lau (1981) found that his "Kelvin" waves are unstable and stationary at sufficiently large $K_H^I K_S$. The results here show that the unstable Kelvin wave in fact has an eastward propagation, albeit slower than in the absence of coupling.

The growth rates and frequencies of the waves as a function of wavenumber k are shown in Fig. 2 for the "most probable" values of the coupling coefficients. Kelvin waves of wavelength greater than 6000 km ($k < 0.26$) are unstable, the maximum growth rates occur for wavelengths around 16,000 km ($k = 0.1$). The stabilities of the Yanai and IGEO waves (not shown) are not greatly affected by the coupling at any wavelength.

158

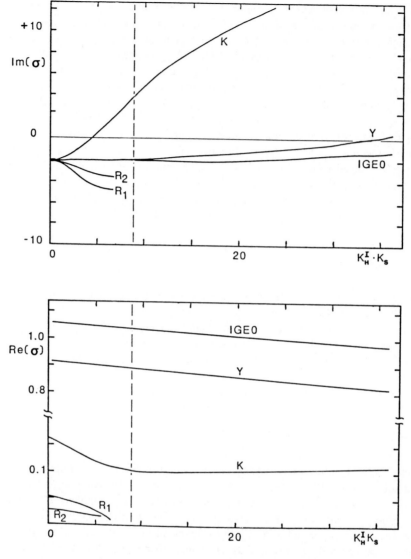

Fig. 1. Growth rate $[Im(\sigma)]$ and frequency $[Re(\sigma)]$ of ocean waves as a function of the product of coupling coefficients, $K_H^I K_S$, for case I with $c_a = 25$ m s^{-1}. Symbols refer to Kelvin (K), $n = 1$ Rossby $(R1)$, $n = 2$ Rossby $(R2)$, Yanai (Y) and $n = 0$ inertia-gravity $(IGE0)$ waves. $Im(\sigma)$ is in 10^{-2} non-dimensional units, $K_H^I K_S$ and $Re(\sigma)$ are in nondimensional units. A value of 10×10^{-2} for $Im(\sigma)$ corresponds to an e-folding growth time of 20 days. Dashed line indicates "most probable" $K_H^I K_S$.

The Rossby waves are highly damped, especially at shorter wavelengths. The frequencies of all waves at all wavelengths are decreased by the coupling. In particular, the Kelvin wave is now dispersive. The phase speed and group velocity are about $0.6c_0$ at long wavelengths, where the wave is unstable; at shorter wavelengths the phase speed increases towards c_0 and the group velocity is slightly above c_0.

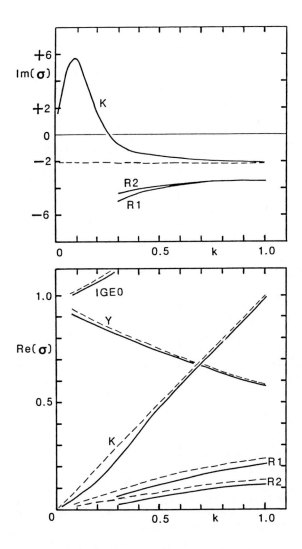

Fig. 2. Growth rate $[Im(\sigma)]$ and frequency $[Re(\sigma)]$ of oceanic waves as a function of the wavenumber (k) for most probable values of the coupling coefficients, in case I with $c_a = 25$ m s^{-1}. Dashed lines indicate computed values in the absence of coupling. Symbols are as in Fig. 1. $Im(\sigma)$ is in 10^{-2} non-dimensional units, $Re(\sigma)$ and k are in nondimensional units. A wavelength of 10,000 km corresponds to $k = 0.16$.

The values of $K_H^I K_S$ at which $Im(\sigma) = 0$ for the Kelvin and Yanai waves are shown as a function of wavenumber in Fig. 3. The curves computed with $c_a = 25$ m s^{-1} are compared to that for the Kelvin wave computed with $c_a = 66$ m s^{-1}. As $K_H^I K_S$ is increased the Kelvin wave first becomes unstable at ultra long wavelengths of 20,000–80,000 km. The Yanai wave first becomes unstable at the shorter wavelength of 8000 km $(k = 0.2)$ when $c_a = 25$ m s^{-1}. The effect of the larger value of c_a is to inhibit the occurrence of instability, this tendency being more pronounced at shorter wavelengths.

160

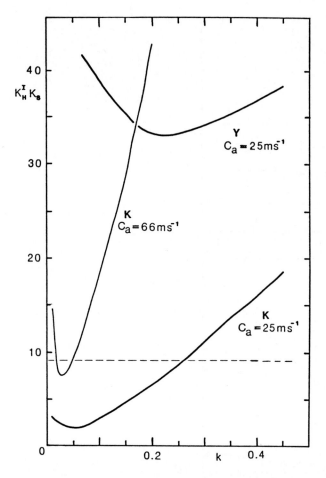

Fig. 3. Value of $K_H^I K_S$ at which $Im(\sigma) = 0$ for oceanic Kelvin (K) and Yanai (Y) waves as a function of wavenumber k, in case I with $c_a = 25$ m s^{-1} and with $c_a = 66$ m s^{-1}. $K_H^I K_S$ and k are in nondimensional units. Dashed line indicates most probable value of $K_H^I K_S$.

A detailed analysis for $c_a = 66$ m s^{-1} not shown here indicates that the effects of coupling on ocean waves are generally less than when $c_a = 25$ m s^{-1}.

Philander et al. (1984) used $c_a = 66$ m s^{-1} in their model; however, the value of $K_H^I K_S$ in their model was a factor of 20 larger than that considered "most probable" here. Kelvin waves with wavelength over 4000 km ($k < 0.4$) were found to be unstable in these conditions. Thus the instability observed in their model is probably a destabilized oceanic Kelvin wave.

Case II

The effect of case II coupling on the frequency and stability of oceanic equatorial waves is displayed in Figs. 4–6. Figure 4 shows that the gravest Rossby (R1) wave is destabilized by case II coupling. Meanwhile the coupling damps Kelvin and $n = 2$ Rossby

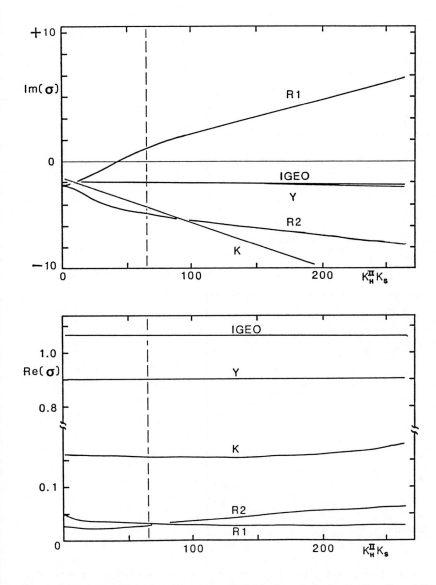

Fig. 4. Growth rate $[Im(\sigma)]$ and frequency $[Re(\sigma)]$ of ocean waves as a function of the product of coupling coefficients $K_H^{II} K_S$, for case II with $c_a = 25$ m s^{-1}, otherwise as for Fig. 1.

(R2) waves and does not significantly affect the frequency and stability of the Yanai and IGEO waves. All this is in marked contrast to case I (Fig. 1).

The growth rates and frequencies of the waves as a function of wavenumber are displayed in Fig. 5 for the "most probable" value of $K_H^{II} K_S$. As in case I, the coupling more strongly affects the long waves. Unlike in case I, the long R1 wave is unstable while the long Kelvin and R2 waves are highly damped. The dispersion diagram suggests that the frequencies of the Kelvin and Rossby modes do not tend to zero as $k \to 0$, but rather approach small nonzero values.

162

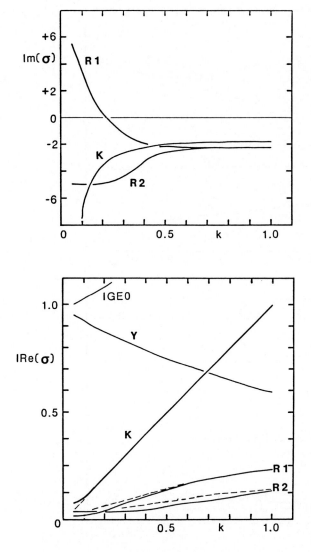

Fig. 5. Growth rate [$Im(\sigma)$] and frequency [$Re(\sigma)$] of oceanic waves as a function of the wave number (k) for most probable values of the coupling coefficients, for case II with $c_a = 25 \text{ m s}^{-1}$; otherwise as for Fig. 2.

Figure 6 shows that adoption of a larger value of c_a acts to suppress instability, especially at shorter wavelengths, as in case I.

The inability of case II-coupling to affect the Yanai and IGEO waves, evident in Figs. 4 and 5, has a simple explanation. The frequencies of these waves at small k are large; thus there is not sufficient time for the associated zonal currents to build up large SST anomalies. Hence the response of the atmosphere to the oceanic Yanai or IGEO waves is very weak.

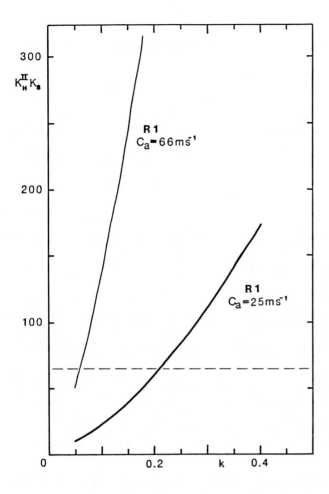

Fig. 6. Value of $K_H^{II} K_S$ at which $Im(\sigma) = 0$ for gravest ocean Rossby (R1) waves as a function of wavenumber k, in case II with $c_a = 25$ m s^{-1} and with $c_a = 66$ m s^{-1}; otherwise as for Fig. 3.

STRUCTURE OF FREE MODES

The marked contrast between the effects of case I and case II coupling on the stability of oceanic waves is easily explained with reference to the structure of the free modes illustrated in Figs. 7A–D. We first refer to Figs. 7A and C for the Kelvin wave in case I and case II, respectively. In either case, the response of the atmosphere to heating is as expected (Matsuno, 1966), with equatorial surface westerlies at and to the west of the heat source (i.e. positive SST anomaly). Now in case I, the positive SST anomaly coincides with the ocean Kelvin wave crest, so atmospheric westerlies overlie oceanic eastward motion and energy is transferred from the atmosphere to the ocean; the oceanic Kelvin wave gains energy and grows. In case II, advection produces a positive SST anomaly centered a quarter cycle behind the oceanic Kelvin wave crest, so the

164

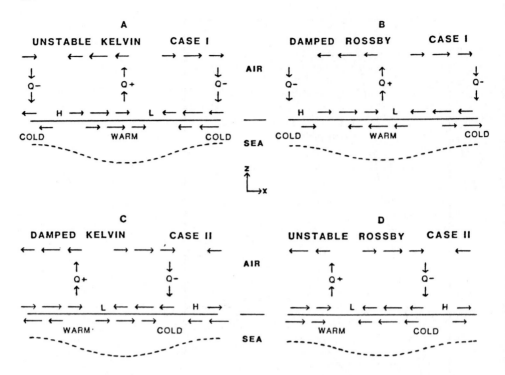

Fig. 7. Schematic illustration of equatorial motion associated with the oceanic Kelvin and gravest Rossby modes in case I (A and B) and case II (C and D). The solid horizontal line represents the ocean surface, the dashed line represents the bottom of the mixed layer. Arrows indicate anomalous ocean currents and atmospheric motions. Maximum positive and negative anomalies of SST are indicated by "warm" and "cold", of atmospheric heating by $Q+$ and $Q-$, of surface atmospheric pressure by H and L, respectively.

atmospheric response is shifted west and easterlies overlie the oceanic eastward motion; the oceanic Kelvin wave loses energy and decays. A similar phase shift of the SST anomaly relative to the ocean flow explains the difference in the stability of the gravest Rossby wave between cases I and II (Figs. 7B and D).

The magnitude of the atmospheric motion relative to that of the ocean motion is strongly dependent on the ratio of the coupling coefficients, K_H^I/K_S or K_H^{II}/K_S. Larger values of this ratio are associated with stronger atmospheric motion. The most probable values of the coupling coefficients give $|U|_{max}/|u|_{max} = 10.5$ for the unstable Kelvin wave in case I (Fig. 7A) and $|U|_{max}/|u|_{max} = 5.3$ for the unstable R1 wave in case II (Fig. 7D). The surprisingly low value of $|U|_{max}/|u|_{max} = 2$ obtained by Philander et al. (1984) may partly result from their very large value for K_S.

CONCLUSION

Certain equatorial ocean waves are destabilized by idealized atmosphere—ocean coupling in models that allow a dynamic response in both the atmosphere and the ocean. The effect of atmosphere—ocean coupling on the stability of ocean waves depends on the distribution of atmospheric heating in relation to the ocean velocities. In case I, maximum heating coincides with the crest of the ocean wave and Kelvin waves are destabilized while Rossby waves are damped. In case II, maximum heating is a quarter cycle out of phase with the crest of the ocean wave and gravest Rossby waves are destabilized while Kelvin waves are damped. In either case, long and low-frequency waves (i.e. long oceanic Rossby and Kelvin waves) are the most affected by the coupling.

The results demonstrate that the behaviour of waves in coupled atmosphere–ocean models is crucially dependent on the method employed to compute the SST anomaly. In particular, the results explain the occurrence of amplifying eastward propagating perturbations in the ocean basin coupled model of Philander et al. (1984), in which SST anomalies are set proportional to h (as in case I), and the dominance of westward propagating perturbations in the ocean basin coupled models of Rennick (1983) and Gill (1985), in which SST results from anomalous advection (as in Case II). Proper determination of the SST anomaly is thus essential for successful modelling of ENSO events.

ACKNOWLEDGEMENTS

Hearty thanks to Prof. John A. Young for many useful discussions and to Prof. Rolland Stull for help with preparation of the manuscript. This study was supported by NSF grant ATM-144-S482.

REFERENCES

Arkin, P.A., Kopman, J.D. and Reynolds, R.W., 1983. 1982–1983 El Niño/Southern Oscillation event quick look atlas. NOAA/National Weather Service, Climate Analysis Center, Washington, D.C.
Gill, A.E., 1980. Some simple solutions for heat-induced tropical circulation. Q.J.R. Meteorol. Soc., 106: 447–462.
Gill, A.E., 1982. Changes in thermal structure of the Equatorial Pacific during the 1972 El Niño as revealed by bathythermograph observations. J. Phys. Oceanogr., 12: 1373–1387.
Gill, A.E., 1983. An estimation of sea-level and sea-current anomalies during the 1972 El Niño and consequent thermal effects. J. Phys. Oceanogr., 13: 586–605.
Gill, A.E., 1985. Elements of coupled ocean–atmosphere models for the tropics. In: J.C.J. Nihoul (Editor), Coupled Ocean–Atmosphere Models. (Elsevier Oceanography Series, 40) Elsevier, Amsterdam, pp. 303–327 (this volume).
Lau, K.M., 1981. Oscillations in a simple equatorial climate system. J. Atmos. Sci., 38: 248–261.
Lim, H. and Chang, C.P., 1983. Dynamics of teleconnections and Walker circulations forced by equatorial heating. J. Atmos. Sci., 40: 1897–1915.
Matsuno, T., 1966. Quasi-geostrophic motions in the equatorial area. J. Meteorol. Soc. Jpn., 44: 25–42.
NOAA, 1981–1983. Monthly climatic data for the world. Vols. 34–36. Natl. Clim. Cent., Asheville N.C.
Philander, G.H., Yamagata, T. and Pacanowski, R.C., 1984. Unstable air–sea interactions in th tropics. J. Atmos. Sci., 41: 604–613.
Rennick, M.A., 1983. A model of atmosphere–ocean coupling in El Niño. Trop. Ocean–Atmo: Newsl., 15: 2–4.
Schopf, D.S. and Harrison, D.E., 1983. On equatorial waves and El Niño: Influence of initial states o wave-induced currents and warming. J. Phys. Oceanogr., 13: 936–948.

CHAPTER 13

AN ANALYSIS OF THE AIR–SEA–ICE INTERACTION SIMULATED BY THE OSU-COUPLED ATMOSPHERE–OCEAN GENERAL CIRCULATION MODEL

YOUNG-JUNE HAN, MICHAEL E. SCHLESINGER and W. LAWRENCE GATES

ABSTRACT

Preliminary results from a 16-year simulation of the OSU coupled atmosphere–ocean general circulation model (GCM) are evaluated by comparison with observations and other contemporary coupled GCM calculations. The results are also compared with those from control runs of the individual atmosphere and ocean GCMs to show some of the unique characteristics of the air–sea–ice coupling.

The CGCM simulation was started with initial conditions obtained from the individual ocean and atmosphere model runs which were made with prescribed boundary conditions appropriate for each model. After a few months of the coupled run several conspicuous sea-surface temperature (SST) errors developed in the eastern tropical oceans and in the western oceans of the Northern Hemisphere. There is also an early indication of a steady SST warming and sea ice melting in the southern oceans at high latitudes. It took only a few months of the integration for the tropical SST errors to reach maximum values of $5°–6°C$, while the middle and high latitude errors grew steadily during the entire period of the 16-year integration. Both the annual mean and seasonal variation of the simulated Antarctic sea-ice area are much smaller than the observed. Interestingly, all existing coupled model results, regardless of the degree of model sophistication, exhibit more or less the same error characteristics as found in the present simulation.

In an effort to isolate the causes of such model errors a heat budget analysis of the upper ocean has been made using the simulated data. The results indicate that the relatively rapid initial warming in the eastern tropical oceans is due to excessively large downward insolation during summer. The less-than-observed cloud amount simulated in these regions is the probable cause of this insolation error. The SST errors in the western oceans are mainly due to the excessively large latent and sensible heat fluxes simulated there during winter. The fact that the regional patterns of these particular SST errors coincide with those of the simulated Aleutian and Icelandic lows indicates a significant contribution to the error by the erroneous atmospheric surface wind.

Because of the complicated feedback nature of the air–sea–ice interaction, the SST and sea-ice errors in the southern high latitude oceans are difficult to diagnose. Nevertheless, at least the initial phase of the error development is found to be due to the excessively large insolation simulated during summer. In addition, the cooling by the Ekman transport simulated in these oceans is much weaker than in the ocean control simulation, and also contributed to the gradual SST increase and sea-ice melting around Antarctica.

1. INTRODUCTION

The present paper is a sequel to the paper by Gates et al. (1985) in this volume and is intended to evaluate several specific details of the air–sea–ice interactions simulated by the OSU coupled general circulation model (GCM). Our particular interest in the present study is the evaluation of those physical processes which were responsible for certain large-scale features of the simulated sea-surface temperature (SST) and sea ice.

In Gates et al. (1985) it is shown that the January mean SST simulated by the GCM, when compared with observation, displays several conspicuous discrepancies, particularly in the eastern tropical oceans and the western oceans of the Northern Hemisphere. Also shown in that study are the less-than-observed sea ice and the excessively warm SSTs simulated in the vicinity of Antarctica. Interestingly, similar discrepancies have also been found in previous coupled-model studies reported by Manabe et al. (1979) and Washington et al. (1980), and thus appear to be characteristic errors of all existing coupled GCMs. The SST and sea ice simulated over these oceanic regions, in the light of their implications for the global climate as suggested in many recent studies (e.g., Rasmusson and Carpenter, 1982), are of particular importance in coupled studies.

The equilibrium of the SST and sea ice in the real ocean is the result of a surface energy balance that involves the air, sea and ice. But the energy balance also depends on the state of the SST, sea-ice and atmospheric variables such as the surface air temperature and cloud amount. Therefore, to trace cause and effect, it is essential to monitor the time evolution of the surface energy balance, along with the key variables involved. While it is difficult to perform such an analysis using actual observational data because of their usual incompleteness in space and time, this does not present a problem using the temporally and spatially regular coupled GCM simulation data. The analysis of that data allows us to closely examine the time evolution of both the state of the coupled system and the associated energy balance.

The present coupled model was assembled from existing component models of the atmosphere, ocean and sea ice. The component models, when tested independently using sets of prescribed boundary conditions, reproduced many of the observed features of climate but not without systematic biases (see Schlesinger and Gates, 1980; Han, 1984). It was anticipated that these biases would likely influence the climate simulated by the coupled GCM and this has been substantiated by Gates et al. (1985) wherein errors closely related to the component-model biases were clearly discernible. On the other hand, other discrepancies such as the warming of the southern ocean SST and the sea-ice decrease developed rather slowly, showing a unique characteristic of the air–sea–ice coupling. Although cause and effect are hard to trace for this type of error, the initial development of the errors was triggered by known model biases as will be discussed later.

In the following we first describe the coupled model and its simulation in Section 2. The major characteristics of the simulated SST and sea-ice fields are then presented and compared with the corresponding observations in Section 3. In Section 4 we present results from a local heat budget analysis for the upper ocean with the intention of identifying the major factors that result in the regional SST and sea-ice error in the coupled-model simulation. A discussion of the findings of this study and suggestions for further research are given in Section 5.

2. DESIGN OF THE COUPLED MODEL EXPERIMENT

2.1. Basic model

Since a description of the basic coupled model is given by Gates et al. (1985), we only recapitulate here the schematics of the ocean and sea-ice models. As shown in Fig. 1 the

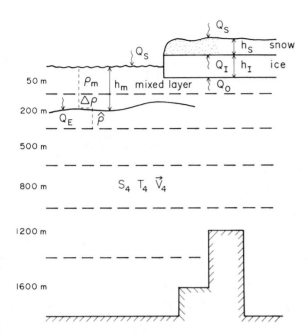

Fig. 1. Schematic of the ocean general circulation model. The horizontal velocity V, temperature T and salinity S are carried at the middle of each layer. The ice and mixed-layer depth (h_I, h_m) and the mixed-layer density ρ_m are predicted by the ice and mixed-layer models, respectively. The snow depth h_s is prescribed from climatology in the ocean control run, but is calculated based on the predicted precipitation in the coupled run. The net interface fluxes Q_S, Q_I, Q_0 and Q_E are computed appropriately in the model. The density jump $\Delta\rho$ is computed as the difference between the density of the mixed-layer (ρ_m) and that of the layer immediately below the mixed layer ($\hat{\rho}$).

ocean model has six layers of unequal mass in the vertical, sufficient to resolve the deepest part of the world ocean. The actual number of vertical layers was determined so that it best fits the local ocean depth averaged over 4° latitude x 5° longitude in the horizontal. The uppermost layer of 50 m represents the ocean surface or mixed layer, and the temperature predicted at the center of this layer is used to compute the surface heat flux (Q_s). If the predicted SST is below the freezing point of sea water $(-1.9°C)$, sea ice is immediately generated such that the release of latent heat restores the SST back to the freezing point. Once sea ice is present, snow is allowed to accumulate on top of it. In this case the sea-ice depth is thermodynamically predicted taking into account the fluxes of heat at the top of the snow, at the interface between snow and sea ice, and at the bottom of the sea ice. Details of the sea-ice model will not be presented here, since the present model closely follows the basic formulation of Semtner (1976) and Parkinson and Washington (1979).

The mixed-layer model depicted in Fig. 1 is also normally an integral part of the present ocean model. This model is basically of the Kraus and Turner (1967) type, but is imbedded in the ocean GCM such that it can interact with the sea water below the mixed layer. At the time the present coupled integration was made, the mixed-layer model had not been tested enough to be confidently implemented; we therefore omitted it from the present experiment.

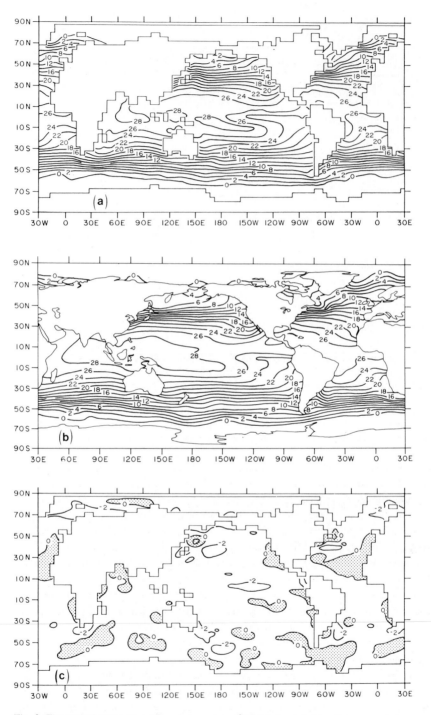

Fig. 2. The annual-mean sea-surface temperature (°C) (a) simulated by the uncoupled ocean GCM with climatological atmospheric conditions, (b) observed from Alexander and Mobley (1976), and (c) the departure of the simulated SST from the observations.

2.2. Initial and boundary conditions

The initial states of the ocean and sea ice were taken from the results of an extended control run (40 years) of the ocean GCM in which the upper boundary conditions were prescribed from observed atmospheric climatology. The initial condition for the atmospheric component was taken similarly from a ten-year atmospheric control run in which the SST was prescribed from monthly climatology. That these initial SST and sea-ice distributions are fairly realistic may be seen in Figs. 2 and 5, respectively. The boundary conditions are the no-slip condition at the bottom and the free-slip condition at the lateral boundaries, with all but the surface assumed to be impermeable to salt and heat. In coupling, the ocean was allowed to exchange momentum and heat with the atmosphere as computed by bulk flux formulas at one hour intervals, which is also the time step for the ocean's integration. In principle, the surface salinity flux can also be calculated in terms of the net surface water flux (precipitation minus evaporation) and the net fresh-water flux from the major rivers. In the present experiment, however, we prescribed the surface skin salinity from climatology (Levitus, 1982) in order to minimize the uncertainties of the surface salt flux.

3. SIMULATED SST AND SEA ICE

Since the initial state of the ocean for the coupled-model simulation was obtained from the ocean control run, it was not necessarily compatible with the atmospheric initial condition that was taken from the atmospheric control run. It was therefore necessary to let the coupled model run over an extended period of time (16 years) so that both the ocean and atmosphere may mutually adjust. Considering the long oceanic thermal adjustment time scale (100–1000 years), the present 16-year coupled integration is hardly long enough to yield a true oceanic equilibrium. A close inspection of the time behavior of the coupled system, however, reveals a relatively rapid initial adjustment of the upper three ocean layers, accompanied by very small changes in the deeper ocean. By year 16 of the coupled integration, the ocean surface layer has already reached near equilibrium everywhere except at high latitudes in the southern oceans. In that area the SST and sea ice are still changing and show little sign that they are near their equilibrium. This is undoubtedly due to the nonlinear air–sea–ice interaction which is unique in this part of the world ocean. It is estimated that it would take an additional two decades of integration to obtain an approximate equilibrium (from a projection based on fig. 11 in Gates et al., 1985). Nevertheless, the first 16-year integration has provided sufficient data to address many important questions concerning air–sea–ice modelling.

3.1. Uncoupled ocean control simulation

Since the results of the ocean control run can serve as a useful reference for interpreting the coupled-model results, we first show the annual-mean SST simulated by the ocean control (Fig. 2a), the observed SST (Fig. 2b), and the differences between the two (Fig. 2c). The large-scale patterns of the simulated SST are seen to agree with the observations reasonably well. There are, however, several significant regional differences as also noted by Gates et al. (1985). For example, in the western tropical Pacific, the simulated 28°C isotherm occupies a relatively narrower region compared to the observations, while in the eastern tropical Pacific the simulated cold tongue of the 26°C isotherm extends too far to the west. The difference map shows that the simulated SSTs are approximatley 1°–2°C colder than the observed SSTs in the tropical Pacific. A heat-budget analysis of the (uncoupled) ocean model (Han, 1984) indicates that the model simulates an overly strong equatorial upwelling which keeps the tropical SST too cold. The other regional SST differences occur in the western boundary regions where the horizontal grid is too coarse to resolve the observed narrow boundary currents.

It is likely that the overall success of the control run in simulating the actual SST is a direct result of the prescribed upper surface boundary conditions, especially the prescribed surface air temperature and specific humidity which already embody many important effects of the actual SST. However, even with such strong constraints, the simulated SST can be regionally erroneous if the oceanic heat transport is erroneous and comparable with the surface fluxes. This point will be further elaborated as we discuss the SST simulated by the coupled model.

3.2. Coupled simulation

The annual-mean SST simulated by the coupled model and its difference from the climatological distribution are shown in Figs. 3a and 4a, respectively. Again, on the global scale the simulated SST pattern is generally similar to climatology (Fig. 3b). On the regional scale, however, there are several large discrepancies. Especially notable is the excessively cold SST in the western part of the northern hemisphere oceans. There is also a conspicuous absence of cold water in the eastern tropical oceans, and there is excessively warm water in the vicinity of Antarctica. In comparison with previous studies (see Fig. 4b, c), it can be seen that these regional SST patterns are similar in all coupled-model studies, although the magnitude of the errors varies from one model to another. As to the Antarctic SST error, Manabe et al. (1979) found that the major causes were the excessively large surface insolation and weak equatorward Ekman heat transport simulated by their model. On the other hand, Washington et al. (1980) speculated that the same type of error in their model was probably due to the use of the large horizontal heat diffusion coefficient ($10^4 \, m^2 \, s^{-1}$). As for the other SST errors mentioned above, no concrete explanation has been given in previous studies. Nevertheless, a critical test of coupled models may be whether they can be made realistic enough to simulate these climatically important regional SSTs. Therefore, we shall investigate this important question in detail in Section 4.

To illustrate the performance of the sea-ice model in the present coupled experiment we present in Fig. 5 the simulated sea-ice distribution for February, along with that observed and that simulated in the (uncoupled) ocean control run. The areal extent of the northern sea ice in both the coupled and control cases is in good agreement with observations. The southern sea ice in the coupled run, however, is significantly underpredicted while that of the control run is comparable with observation. A similar comparison made for the month of August (not shown) reveals that the coupled model again

errors are clearly related to the much colder-than-observed SST simulated in the same regions.

In the high-latitude oceans of both hemispheres the simulated heat fluxes are given only over the ice-free areas and are difficult to compare with the observed values; one can see, however, that the simulated values are generally less than observed estimates almost everywhere in theses latitudes. Interestingly, the SST simulated over these regions (see Fig. 4) is generally warmer than observation. This is in contrast to the case of the western oceans where the SST is much colder than observation. A detailed calculation of the heat flux components shows that in the western oceans the relatively small upward heat flux is due to the lack of sufficient latent and sensible heat fluxes, which are in turn due to the excessively cold SST. In the high latitudes, on the other hand, excessively large insolation turned out to be the major factor that reduced the net upward heat flux and thus maintained the relatively warmer SST.

Because of the nonlinear nature of the surface energy exchanges, the simulated net surface heat flux over the eastern tropical oceans is not much different from the observed estimate, while the simulated SST over the same oceanic regions is much higher than observations. An immediate implication of this result is that the steady state of the coupled system is not always uniquely related to a particular energy balance at the surface. In the next section we therefore examine the time history of the solutions and its relation to the history of the surface energy balance.

4. LOCAL UPPER-OCEAN HEAT BUDGET ANALYSIS

The time behavior of the SST, sea ice and other key variables were monitored during the entire period of the 16-year coupled run. To our surprise, during the first year of the

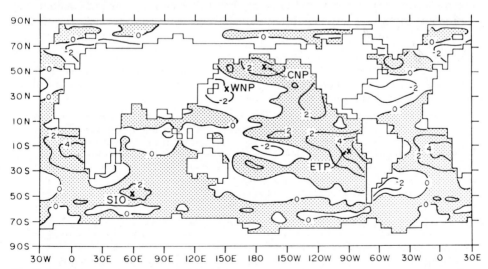

Fig. 7. The sea-surface temperature changes during the first year of the 16-year coupled run (°C). The areas of net warming during this period are stippled. The regions where the local ocean-surface heat budget was computed are indicated by the crosses in the southern Indian Ocean (*SIO*), eastern tropical Pacific Ocean (*ETP*), central North Pacific Ocean (*CNP*), and western North Pacific Ocean (*WNP*).

integration most of the SST errors described in the previous sections were already evident. In Fig. 7 we show the SST changes which occurred during the first year of integration. Over the eastern parts of the tropical oceans the sea surface warmed by $\sim 2°-6°C$, while the SST dropped by $\sim 2°-4°C$ over the western oceans in the Northern Hemisphere. There were also definite warming trends in the high-latitude oceans of both hemispheres. Clearly such rapid SST changes should be expected if there were excessive heat-flux imbalances during the initial phase of the coupled-model integration. We therefore examined all of the heat-flux components simulated over those areas where the largest SST changes occurred.

The local areas for which the heat-balance calculations were made are indicated by crosses in Fig. 7, and the results are summarized in Table 1. In the same table we also show the results obtained for the case of the uncoupled ocean control run. The second column of Table 1 shows the net SST changes which occurred over the first year of the coupled run, along with those obtained for the control case. Obviously the net SST changes in the control case (40-year integration) are very small everywhere, indicating an almost perfect steady state in the solution. The net surface heat fluxes are the sum of the surface sensible, latent and radiative fluxes, and are shown in the third column of Table 1. The fourth column shows the upper-ocean (50 m) heat transports by both advection and diffusion, while the last column shows the net heat flux into the ocean surface layer from below by convective overturning (which is modelled to occur whenever the density of sea water becomes unstably stratified, as happens mostly in the middle and high-latitude oceans during the winter season).

From the table, the net SST change of $3°C\,yr^{-1}$ over the southern Indian Ocean (SIO) in the coupled case is seen to balance the net downward surface heat flux, while the diffusive-advective cooling of $5°C\,yr^{-1}$ balances the convective warming. A separate calculation shows that the simulated surface insolation of $200\,W\,m^{-2}$ in this part of the ocean is greater than the observed value by almost a factor of two, and this undoubtedly contributed to the SST warming. From the table one may also argue, however, that the

TABLE 1

The local ocean heat budget (in units of $°C\,yr^{-1}$) for the cross-marked points shown in Fig. 7. The numbers in the upper left of each box are for the first year of the coupled run and those in the lower right are for the 40-year uncoupled ocean control run

Location	Net Temp. Change	Net Surface Heat Flux	Diffusion and Advection	Convective Adjustment
60E, 50S SIO	+3 / ~0	+3 / ~0	−5 / −8	+5 / +8
90W, 14S ETP	+4.5 / ~0	+4 / ~0	+0.5 / ~0	~0 / ~0
175W, 54N CNP	+3.5 / ~0	~0 / −6.5	+4 / +2	~0 / +4.5
155E, 34N WNP	−2.0 / ~0	−15 / −12	−2.5 / −1.5	+15.5 / +13.5

heat transport by the diffusive-advective processes was perhaps too small, thereby contributing to the same SST warming. In comparison with the uncoupled control run, this transport was indeed smaller, a finding which cannot be directly verified by observation.

In the eastern tropical Pacific (ETP) the net SST increase of $4.5°C\,yr^{-1}$ is accounted for almost entirely by the net downward surface heat flux ($4°C\,yr^{-1}$). In comparison with the control case, we found the net surface insolation to be the major factor causing the surface heat flux imbalance at the surface in this location. The table also indicates that the advective-diffusive heat transport was probably underestimated in both the coupled and control cases. The model grid resolution of $4°$ latitude $\times$ $5°$ longitude is obviously too coarse to resolve the known tropical circulations. It is therefore quite possible that the simulated heat transport may also be erroneous. Unfortunately, at present a more precise error estimate cannot be made due to the lack of reliable observations.

At the high latitudes of the central North Pacific (CNP) the upward surface heat flux is much less in the coupled case ($\sim 0°C\,yr^{-1}$) than in the control case ($\sim 6.5°C\,yr^{-1}$). A more detailed calculation revealed that the simulated surface insolation in the coupled case is about twice the observed estimate, and this balances the upward sensible plus latent-heat fluxes. In the real ocean, however, the latter is larger than the surface insolation, and usually causes convective overturning of the surface and subsurface water (as simulated in the control run). This is rather convincing evidence that the surface insolation at this location was overestimated and caused the SST warming in the high-latitude ocean.

Finally, the SST drop of $2°C\,yr^{-1}$ in the western part of the North Pacific is an early indication of the excessively cold SST obtained in the final stage of the coupled run (see Fig. 4a). This initial tendency was found to be due to the excessively large latent and sensible heat fluxes which are reflected in the net surface heat flux shown in Table 1. As discussed in Gates et al. (1985), the southward shift of the simulated Aleutian low and the associated strong cyclonic winds greatly enhanced both the latent and sensible heat fluxes from the ocean surface. The enhancement of these fluxes in turn increased the convective heat flux (see Table 1), but the net effect including the diffusive advective flux was a drop in SST. As the integration proceeded, the reduced SST in turn reduced both the sensible and latent heat fluxes as is evident in Fig. 6.

5. DISCUSSION AND CONCLUDING REMARKS

The present evaluation of the OSU atmosphere—ocean coupled GCM is considered preliminary in that only a few specific simulated features (e.g., SST and sea ice) have been evaluated. A more comprehensive analysis of the simulation will be useful for the further clarification of some important issues raised in connection with the upper-ocean heat budget analysis. Here we shall briefly recapitulate these issues and offer a few remarks on possible additional analyses and model improvements appropriate for future experiments.

An important finding from the local heat budget analysis was the apparent overestimation of the simulated surface insolation in the high-latitude oceans. The surface insolation is basically controlled by the effective atmospheric transmissivity which depends critically on the amount, type and optical properties of the simulated clouds.

Manabe et al. (1979) attributed their large simulated insolation in the southern high-latitude oceans to the lack of sufficient cloud cover which they specified in the model. (They used the observed northern hemisphere climatology.) In the present coupled GCM clouds are predicted by the model and only their optical properties are specified. In comparison with observation, the total cloud cover over Antarctica simulated by the coupled model was found to be realistic. We intend to examine in more detail the simulated radiative processes over the southern high-latitudes to determine the cause of the excessive insolation simulated there.

Another issue associated with the heat budget in the southern high-latitude oceans involves the relative importance of the oceanic heat transport compared to the surface fluxes. We found that the simulated poleward heat transport by diffusion is much smaller (by an order of magnitude) than the surface fluxes, while the equatorward heat flux by the simulated Ekman transport is comparable in magnitude to the surface fluxes. Since the Ekman transport is basically driven by the surface wind stress, it was useful to compare the simulated surface winds with observations. We found that the simulated westerly winds over the southern high-latitude oceans were somewhat too weak. The significance of this discrepancy in terms of the Ekman heat transport is hard to judge in the real ocean. But in comparison with the ocean control run, the discrepancy was found to be significant enough to affect the overall heat budget and thus the SSTs and sea ice in these regions.

In the analysis of the SST and associated heat budget over the eastern part of the tropical Pacific, the simulated surface insolation was found to be overestimated because of the underestimated cloud cover. In actuality the marine-layer stratus clouds are rather outstanding features in this part of the Pacific. The present two-layer atmospheric GCM does not adequately reproduce this cloud, and thus allows solar radiation to reach the ocean surface without significant attenuation.

From the oceanic point of view, on the other hand, the conclusion drawn above may be an oversimplification. In the eastern parts of the tropical Pacific, the local upwelled cold water and the southeasterly current associated with the subtropical gyre are known to be important factors in the local ocean heat balance. Our heat budget calculation using the simulated data, however, shows no such effects in either the coupled run or in the control run (see Table 1). In the coupled case, Gates et al. (1985) found significant discrepancies in the simulated tropical surface winds (fig. 15 in their paper), and thus attributed the relatively weak simulated upwelling to these wind errors. In the uncoupled ocean the surface wind stresses were prescribed from the observed estimates (Han and Lee, 1983). Yet the net contribution from the oceanic processes was almost zero in the local heat budget. In effect, the simulated SST there was determined entirely by the local surface heat-flux balance. It is then logical to conclude that even in the control case the seemingly realistic SST in the eastern tropical Pacific is fortuitous, and simply reflects the imposed atmospheric surface climatology. Considering the importance of the tropical SSTs suggested in many El Niño–Southern Oscillation studies, the tropical performance of the ocean model is disappointing and deserves special attention for improvement.

Turning now to the western North Pacific, the simulated SSTs and the associated surface heat fluxes are evidently related to the simulated atmospheric surface circulations, which were found to be stronger and shifted southward by about ten degrees of latitude compared to their observed positions. More importantly, these atmospheric biases are evident even in the case of the uncoupled atmospheric control run in which the observed SSTs were prescribed as a lower boundary condition. An important question then is: "Why are there such biases in the atmospheric GCM?" To answer this question one may need to fully diagnose the dynamics of the simulated atmospheric motions. In fact, the simulated surface circulations are directly related to momentum sources and sinks, and may need to be analyzed within the context of the total angular momentum balance of the simulated atmosphere.

Aside from such diagnostic analysis, several improvements of the ocean model are under consideration. In an attempt to improve the model resolution for the tropical oceans, a meridionally stretched coordinate is being tested. This new coordinate is an attractive alternative to our present uniform grid, because it better resolves the tropical features while allowing about the same computational efficiency as the present model. Another area of ocean model improvement which draws our attention is the coupling of the ocean GCM with the bulk models of the mixed layer and sea ice. The latter submodels have been tested mostly in one-dimensional cases without seriously considering their coupling (Niiler and Kraus, 1977). We found, however, that the simulated SST and sea ice are fairly sensitive to the particular coupling methods actually used in the model. This aspect of coupled modelling deserves more attention in future research.

ACKNOWLEDGEMENTS

We would like to thank William McKie and Robert L. Mobley for their execution of the model integration and their assistance in its analysis, Cindy Romo for drafting the figures, and Leah Riley for typing the manuscript. The computations were performed at the National Center for Atmospheric Research. This research was supported by the National Science Foundation and the U.S. Department of Energy under grant ATM 8205992.

REFERENCES

Alexander, R.C. and Mobley, R.L., 1976. Monthly averaged sea surface temperatures and ice-pack limits on a 1° global grid. Mon. Weather Rev., 104: 143–148.
Gates, W.L., Han, Y.-J. and Schlesinger, M.E., 1985. The global climate simulated by a coupled atmosphere–ocean general circulation model: Preliminary results. In: J.C.J. Nihoul (Editor), Coupled Ocean-Atmosphere Models. (Elsevier Oceanography Series, 40) Elsevier, Amsterdam, pp. 131–151 (this volume).
Han, Y.-J., 1984. A numerical world ocean general circulation model, Part II. A baroclinic experiment. Dyn. Atmos. Oceans, 8: 141–172.
Han, Y.-J. and Lee, S.-W., 1983. An analysis of monthly mean wind stress over the global ocean. Mon. Weather Rev., 111: 1554–1566.
Kraus, E.B. and Turner, J.S., 1967. A one-dimensional model of the seasonal thermocline: II. The general theory and its consequences. Tellus, 19: 98–106.
Levitus, S., 1982. Climatological Atlas of the World Oceans. NOAA Professional Paper No. 13, U.S. Gov. Printing Office, Washington, D.C., 173 pp.
Manabe, S., Bryan, K. and Spelman, M.J., 1979. A global ocean–atmosphere climate model with seasonal variation for future studies of climate sensitivity. Dyn. Atmos. Oceans, 3: 393–426.

Niiler, P.P. and Kraus, E.B., 1977. One-dimensional models of the upper ocean. In: E.B. Kraus (Editor), Modelling and Prediction of the Upper Layers of the Ocean. Pergamon, New York, N.Y., pp. 143—173.

Parkinson, C.L. and Washington, W.M., 1979. A large-scale numerical model of sea ice. J. Geophys. Res., 84: 311—337.

Rasmusson, E.M. and Carpenter, T.H., 1982. Variations in tropical sea surface wind fields associated with the Southern Oscillation/El Niño. Mon. Weather Rev., 110: 354—384.

Schlesinger, M.E. and Gates, W.L., 1980. The January and July performance of the OSU two-level atmospheric general circulation model. J. Atmos. Sci., 37: 1914—1943.

Semtner, A.J., 1976. A model for the thermodynamic growth of sea ice in numerical investigations of climate. J. Phys. Oceanogr., 6: 379—389.

Washington, W.M., Semtner Jr., A.J., Meehl, G.A., Knight, D.J. and Mayer, T.A., 1980. A general circulation experiment with a coupled atmosphere, ocean, and sea ice model. J. Phys. Oceanogr., 10: 1887—1908.

CHAPTER 14

OBSERVED LONG-TERM VARIABILITY IN THE GLOBAL SURFACE
TEMPERATURES OF THE ATMOSPHERE AND OCEANS

ABRAHAM H. OORT and MARY ANN C. MAHER

ABSTRACT

Based on historical surface observations available at the U.S. National Climatic Center at Asheville, N.C., the monthly global fields of air–surface temperature, sea-surface temperature and sea–air temperature difference were analyzed for a 45-year period since May 1920. All available data were used. A clear long-term trend was found in the monthly time series of sea–air temperature difference for the tropics, being about 0.5°C less during the 1920's and 1930's than during the 1960's and 1970's and with a gradual change in the years between. Probably this trend is not real, but due to the gradual change in observing the sea-surface temperature from using buckets to engine-intake data. Our analyses suggest a reasonable way to correct the historical sea-surface temperatures for changes in observing techniques, making them more useful for climate research. Clear evidence of El Niño/Southern Oscillation episodes are found throughout the 45-year period, although they may have been less frequent during the 1920's and 1930's than during the later years.

1. INTRODUCTION

The long-term variations of the surface temperatures over the Earth were analyzed on a global scale independently by Mitchell (1968) and Budyko (1969). Both concluded that a global heating of 0.5°–0.7°C had taken place since the 1880's, and that the peak in temperature was reached around the 1940's with a consequent slow decrease in temperature until the 1960's. Recently, Jones et al. (1982) have updated the earlier curves and have shown a rise in temperature since the 1970's. All these studies tend to be land-biased since land stations with only a few island data were used.

Recently the plentiful ship records have been utilized, e.g., Fieux (1980), Paltridge and Woodruff (1981), Barnett (1984), Wright and Wallace (1984) and Folland et al. (1984), to determine more accurately the oceanic contributions to the global temperature changes. Several biases in the ship records have been identified connected with differences in observing practices and in instrumentation. These biases make the interpretation of the historical records very difficult; a further discussion will be given in Section 2.1. Nevertheless, the general trend in the surface-air temperature over the oceans was shown to have some resemblance with the trend over land with a heating from the early 1900's until the 1940's of about 0.6°C. However, the cooling observed over land since the 1940's was not clear in the ship records and the trend was even shown as a slight heating during the 1950's and 1960's by Paltridge and Woodruff (1981) and Barnett (1984).

Fortunately, a long-term and very careful research effort at the British Meteorological Office reported by Folland et al. (1984) has now provided a more reliable and probably definitive estimate of the global temperature trend over the oceans. It shows a temperature

184

variation in close agreement with the temperature trend over land derived by Jones et al. (1982) with a historical maximum temperature in the 1940's, a weak minimum in the late 1960's and a gradual rise in temperature since the early 1970's. There is some evidence in their curves of a time lag of several years in the sense that the temperature change over the oceans lags that over land. This is what one might expect in view of the large heat capacity of the oceans. It also is in agreement with preliminary results by Spelman and Manabe (1984, fig. 15) who used a general circulation model to study the atmospheric response over land and ocean to an increase in atmospheric CO_2. Hansen et al. (1981) and others have, in fact, speculated that the rise in surface-air temperature since the 1970's is associated with the observed supposedly man-made increase in atmospheric CO_2 that finally may have overcome the threshold of meteorological "noise" in the temperature records.

In Section 3 of the present study we give a detailed account of the month-to-month changes in global sea-surface temperature (T_s) and air temperature (T_A) for the 1920– 1973 period. No corrections were applied to the reported data to provide a "pure" picture of the data. The main point of discussion will be the clear, historical trend in the sea–air temperature difference ($T_s - T_A$). This trend appears too large to be physically real and understandable since it would imply a large change in global air–sea exchange of sensible and latent heat. Therefore we interpret the trend in $T_s - T_A$ to be due to a change in observational practice and/or instruments for measuring the sea surface temperature, namely the transition from bucket to engine-intake temperatures. The $T_s - T_A$ curves will enable us to provide a reasonable correction to the T_s data.

2. DATA

2.1. Data source and data distribution

The surface marine observations (Tape deck TDF-11) as assembled by the National Climatic Data Center (NCDC) in Asheville, N.C., formed the source of our data over the oceans. An idea of the temporal evolution of the data can be obtained from Fig. 1. It shows a monthly plot of the number of $2° \times 2°$ ocean squares with three or more ship reports during a month for the period May 1920–April 1973. A period of eight years (May 1939–April 1947) around World War II was not analyzed because of sparse data coverage. The number of station grid points was fairly uniform, on the order of 1500, during the period 1921–1937; it reached a value of 2000–2500 grid points during the early 1950's and a maximum value of about 4000 grid points during the late 1960's. The decrease since 1969 is associated with the time lag between taking the ship observations and archiving them into the final data set at NCDC. (We acquired this data set from NCDC in early 1978.)

Some maps of the spatial distribution of the ship data for selected Januaries through-out the 53-year period are shown in Fig. 2. They show, in agreement with Fig. 1, the best spatial coverage during January 1968. Because of the general lack of data south of about 40°S throughout the period all analyses south of 40°S cannot be trusted. Except for the Indian Ocean, the ocean areas north of 40°S are reasonably well covered with data during the 45 years considered.

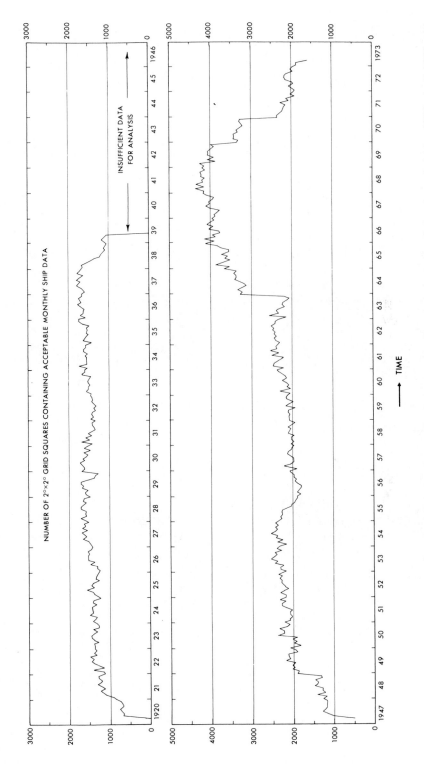

Fig. 1. Time plot of the number of $2° \times 2°$ ocean squares with three or more ship reports during a month for the period May 1920–April 1973 with an 8-year break around World War II.

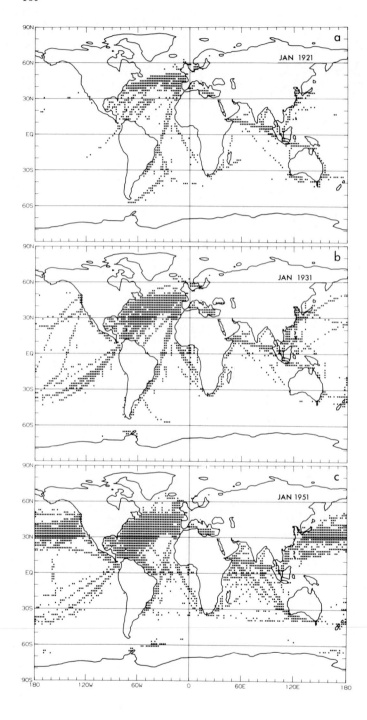

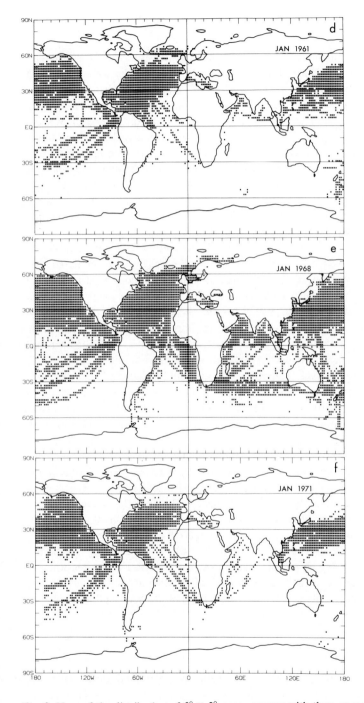

Fig. 2. Maps of the distribution of $2° \times 2°$ ocean squares with three or more ship reports during the month of January for the years 1921 (a), 1931 (b), 1951 (c), 1961 (d), 1968 (e), and 1971 (f).

2.2. Data reduction and analysis

As described more completely in Oort (1983, pp. 23–24) the original ship reports were first converted to a uniform format and gross error checks were performed. Atmospheric temperatures above 40°C or below − 20°C, and sea-surface temperatures above 35°C or below − 4°C were discarded. In a second round through the data, the long-term 6-month winter and summer means and standard deviations for each Marsden Square (10° latitude × 10° longitude) were computed using all validated data from the 20 years, 1950–1969. These means and standard deviations were then used in a third round through the data to discard those original reports that deviated more than three standard deviations from the normal seasonal mean for the particular Marsden Square. If a ship did not report an acceptable wind speed, wind direction and atmospheric temperature at the same time, it was not used. For each month of the 45-year period the validated data were accumulated in 2° latitude by 2° longitude squares.

Before proceeding with the objective analysis of the monthly data, the anomalies of the 2° × 2° grid point data from a 10-year (1963–1973) climatology were computed. This was done before the analysis, because the departures from climatology tend to be much smaller than the climatological mean values and therefore are easier to analyze. We found that the analyzed fields of the total quantities (means plus departures) were too sensitive to the particular structure of the mean field and the input data. Slight changes in the distribution of reporting stations, or in their values, would lead to very large spurious anomalies in the extrapolated values over data-sparse regions, especially in regions of strong gradients in the mean field. Therefore, the monthly *anomaly* 2° × 2° grid point data were used as input to the analysis scheme. Since the anomalies tend to be small and tend to change sign between different regions, the anomaly analyses do not give excessive year-to-year variations as would be the case in the analyses of the mean plus anomaly fields.

The objective analysis scheme used was one that generated fields that resembled as closely as possible the fields obtained by a careful hand analysis. We used a technique called CRAM (Conditional Relaxation Analysis Method) as described by Harris et al. (1966) with zonal averages of the data as a first-guess field. The field of the departures from the first guess field was extended to those grid points which had no nearby reporting ship data by solving Poisson's equation at the unknown grid points while the grid points with ship data were kept as fixed boundary values. The Laplacian of the first guess field was used as the forcing term in Poisson's equation. After some smoothing of the resulting field, the entire cycle was repeated leading after a few iterations to the final analysis. For further details of the data reduction and analyses methods, see Oort (1983).

2.3. Data problems

Some of the problems connected with the spatial and temporal gaps in the ship archives were mentioned before in connection with Figs. 1 and 2. However, there are other possible biases in the data that are more directly related to the measuring techniques used in taking the basic observations and to the changes in these techniques. We may mention the artificial increase in air temperature resulting from daytime heating of the ship (e.g. Ramage, 1984), and the increase in elevation of the thermometer above sea level as ships increased in size affecting both day- and nighttime readings of the temperature (Folland et al., 1984). In case of the sea-surface temperature there has been a change from uninsulated to insulated buckets (Folland et al., 1984). However, the most dramatic change was probably the gradual shift from bucket to engine-intake measurements beginning around 1940 (e.g. Saur, 1963; Barnett, 1984). An excellent summary of the various difficulties in the observations can be found in Folland et al. (1984) and in several unpublished recent memoranda from the British Meteorological Office.

In the present project we decided to use all available data without distinguishing either between night- and daytime air temperatures, or between bucket and engine-intake sea temperatures. The reason for this procedure was that it was often not known what techniques were actually used in taking the observations, and if the technique was known, it was quite debatable what the sign and magnitude of the corrections should have been.

3. RESULTS

3.1. Sea–air temperature differences

Since the $T_s - T_A$ differences and their temporal evolution will play an important role in our further discussions we will first show in Figs. 3a and b the climatological $T_s - T_A$ values for the months of January and July. These maps were based on the surface ship data from the 10-year standard period, May 1963–April 1973. In the tropics between about 30°S and 20°N we find generally positive $T_s - T_A$ values, ranging between 0° and + 1°C. North of 20°N there is a strong seasonal cycle, especially along the east coasts of North America and Asia. In winter, the air streaming off the northern continents at midlatitudes is, of course, much colder than the ocean with temperature differences reaching values of + 10°C near the coast. On the other hand during summer, the continental air masses tend to be warmer than the oceans, and differences down to − 3°C are found in Fig. 3b along the east coasts.

3.2. Temperatures averaged over the tropical belts

The monthly departures from the climatological mean values for the 1963–1973 period are shown in Figs. 4a and b for the northern and southern tropics. Each figure shows the unsmoothed curves for T_A, T_s and $T_s - T_A$ during the period May 1920–April 1973 averaged over a tropical belt (0°–30° latitude) that covers one fourth of the globe. In these figures there is some sign of a rising trend in the atmospheric temperature from the 1920's until the 1940's, as mentioned earlier in this paper. Superimposed on this trend are interesting "surges" of warming and cooling presumably associated with El Niño/Southern Oscillation (ENSO) events. Figures 4a and b show a remarkable, high simultaneous correlation between what happens in the northern and in the southern tropics.

Regarding the sea-surface temperature curves, we find strong similarities with the corresponding curves for the atmosphere except for a much stronger long-term warming

(a)

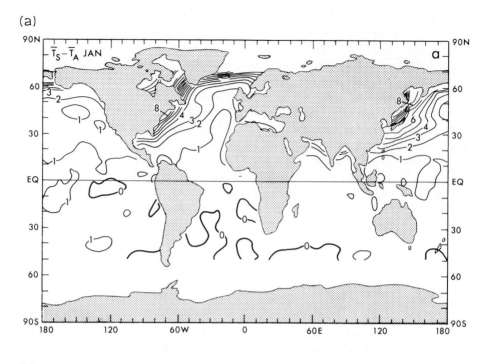

(b)

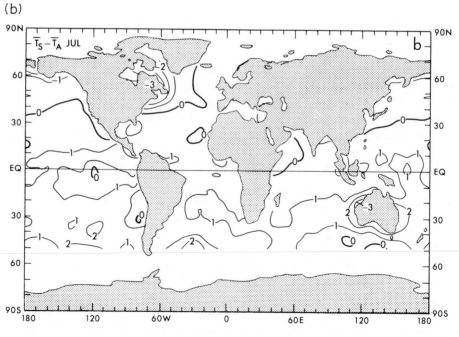

Fig. 3. Maps of the sea–air temperature difference, $T_s - T_A$, for mean January (a) and July (b) conditions in °C. North of 70°N and south of 50°S no analyses were attempted because of sparse ship data.

trend in the T_s curves. The $T_s - T_A$ difference curves at the bottom of Figs. 4a and b give a clue as to what may be happening. Before the 1940's we find that the $T_s - T_A$ differences are steady with a magnitude of about $-0.6°C$, but that since about 1938 the differences tend to decrease to zero values in the late 1950's. Based on Haney's (1971) work, it seems plausible to assume that such a large change in air–sea temperature difference cannot be real. If true, it would imply a change in the net air–sea exchange of energy on the order of $20\,W\,m^{-2}$, too large to be acceptable. The most obvious conclusion is that the change from bucket to engine-intake measurements is the principal reason for the general trend in $T_s - T_A$. The magnitude of the change of about $0.5°C$ is of the same sign and order of magnitude as suggested by Saur (1963) based on direct comparisons.

3.3. Geographical distribution of trend in $T_s - T_A$

The geographical distribution of the trend in the $T_s - T_A$ difference between the recent 20-year period, 1953–1973, and the earlier 20-year period, 1920–1939, is shown in Fig. 5 for annual-mean conditions. There is a clear signal of a rather uniform trend, on the order of $0.5°C$, between about $40°S$ and $40°N$. For still unknown reasons, the trend decreases in magnitude and possibly reverses sign at middle and high latitudes.

The same trend but zonally averaged over all oceans is shown in Fig. 6 for the year and the four seasons. In the tropics the seasonal variations are found to be quite small. The apparent similarity of the trend for the four independent seasonal samples, and the spatial homogeneity shown before in Fig. 5 tend to support our arguments for an instrumental cause of the trend and for its artificial nature.

3.4. Possible implications for the variations in global mean atmospheric temperature

It was shown by Horel and Wallace (1981) and Pan and Oort (1983) that the variations in the tropical sea-surface temperatures (SST) have important effects on the temperature of the tropical atmosphere. When the SST are high in the equatorial eastern Pacific, i.e. during ENSO events, the atmospheric temperatures in the entire tropical troposphere were found to be warmer than usual, and vice versa for low SST. This is illustrated in Fig. 7 taken from Pan and Oort (1983). It shows that the temperature integrated over the entire Northern Hemisphere atmosphere between the surface and 50 mb is highly correlated with the SST record. The strongest correlations occur when the atmosphere lags the ocean by $+6$ months (see Fig. 8). Locally in the tropics the correlations are largest at shorter lags of a few months.

We will now speculate that the relation between SST and mean hemispheric temperature found for the 15-year period, 1958–1973, also will hold up during the longer 45-year SST record presented in this paper. In other words, we suggest that our 45-year SST record, May 1920–April 1973, in the equatorial eastern Pacific may actually indicate the history of what has happened with the temperature in the entire tropical atmosphere. The uncorrected SST record is shown in Fig. 9a, and the corrected curve, after a uniform correction of $0.5°C$ was applied before the 1940's, in Fig. 9b. This simple correction for the instrumental change from bucket to engine-intake measurements is probably the best correction we can apply at this time to the data set as discussed in the previous section on the $T_s - T_A$ differences. In Fig. 9 we have used as

192

(a)

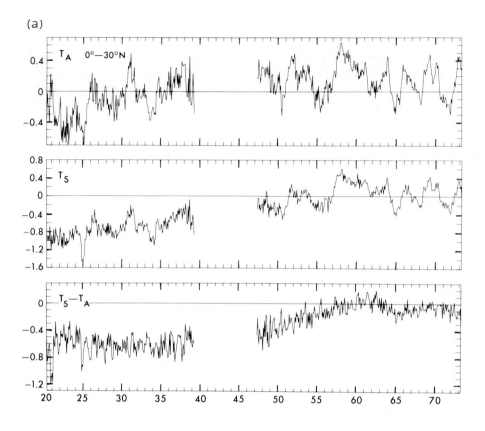

an index the mean temperature averaged over a tropical strip in the eastern Pacific instead of the local temperature in the key region at 130°W (see Fig. 7). The reason we preferred to use here the area averages rather than the local values was that the results would be less influenced by possible variations in ship data coverage (see Fig. 2) during the 45-year period of record. Nevertheless, the two time series are almost identical in shape for the 1958–1973 period, as can be seen by comparing Figs. 7a and 9c.

Assuming that the SST record in Fig. 9 also represents the mean atmospheric temperature record, we find a signal that is dominated by ENSO events and we find very little long-term trend. Possibly there is a weak tendency for the 1920's to have been somewhat cooler than the 1950's and 1960's. Worth noting is the less frequent occurrence of ENSO events during the 1920's and 1930's compared to the later decades. This fact may be significant, but needs further confirmation from other data studies as to its reality. If confirmed, it may help to clarify the general climatic conditions under which ENSO events tend to occur and when they do not.

4. SUMMARY AND CONCLUDING REMARKS

The surface marine observations archived at the National Climatic Data Center in Asheville, N.C., were used to analyze the monthly fields of T_A, T_s and $T_s - T_A$ over the

(b)

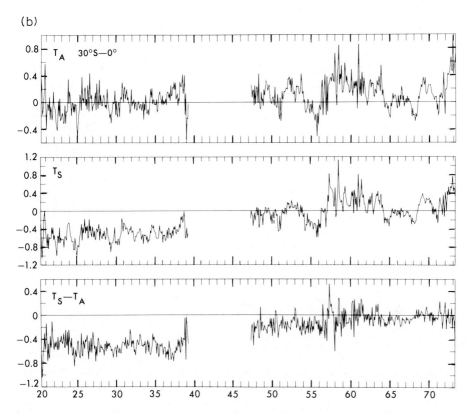

Fig. 4. a. Time plot of the surface-air temperature (top) the sea-surface temperature (middle) and the sea—air temperature difference (bottom) averaged over the ocean areas of the Northern Hemisphere tropics (0°–30° N) in units of °C. The tit-marks along the abscissa indicate January of the corresponding year. b. Same as (a) except for the Southern Hemisphere tropics (30° S–0°).

globe during the periods May 1920–April 1939 and May 1947–April 1973. No corrections were applied to the reported data during the data reduction process. The oceans between about 50°S and 70°N were reasonably well covered by ship reports throughout the period except in the southeastern Pacific Ocean, and in parts of the Indian and South Atlantic Oceans. The best data coverage was found during the late 1960's.

Time series of the mean monthly values of T_A and T_s, averaged over the tropics, showed clear evidence of important ENSO events, and also showed a weak tendency for general warming from the early 1920's to the 1940's. A careful analysis of the $T_s - T_A$ difference as a function of time and also with respect to its geographical distribution led us to conclude that there is a spurious trend in $T_s - T_A$, probably related to the gradual change in bucket to engine-intake observations of T_s during the 1940's and early 1950's.

If we assume that the close direct relationship between the T_s in the eastern equatorial Pacific Ocean and T_A averaged over the entire Northern Hemisphere mass, which was found by Pan and Oort (1983) during the 1958–1973 period, also holds for the present

194

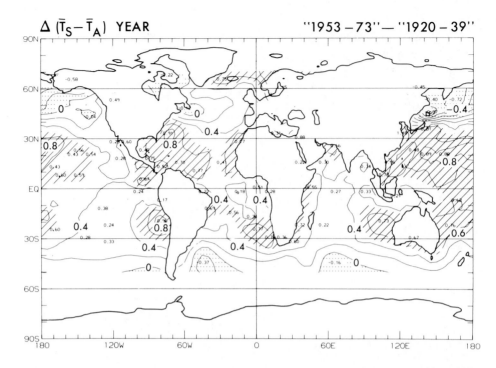

Fig. 5. Map of the change in the $T_s - T_A$ difference between the 1953–1973 and the 1920–1939 periods for annual-mean conditions in units of °C. The shaded areas indicate regions where the change is greater than $+0.5$°C.

45-year long record, the corrected T_s curve in Fig. 9 must also represent a likely scenario of what has happened in the global atmosphere during those 45 years. Based on this (admittedly crude) evidence we find that there has not been any substantial long-term trend in the mean atmospheric temperature, and that the predominant interannual signal has been associated with ENSO events.

With the recently completed Comprehensive Ocean–Atmosphere Data Set (COADS) developed by Fletcher et al. (1983) in cooperation with the National Climatic Data Center and the National Center for Atmospheric Research it will be possible to improve the present analyses and extend this study possibly back to the 1880's and beyond 1973.

ACKNOWLEDGEMENTS

We would like to thank Kirk Bryan for his review of our manuscript and for helpful comments. One of us (M.A.M.) was supported by the Federal Junior Fellowship Program during her stay at GFDL.

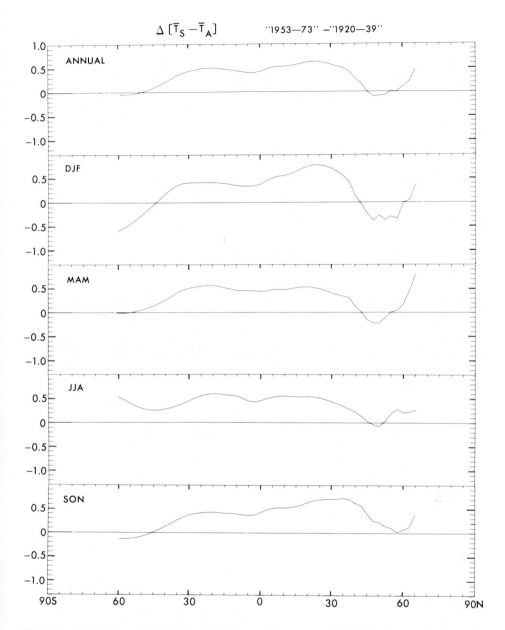

Fig. 6. Profiles of the zonally averaged change in the $T_s - T_A$ difference between the 1953–1973 and the 1920–1939 periods for the annual mean and the four seasons in units of °C (*DJF* = December–February, *MAM* = March–May, etc.).

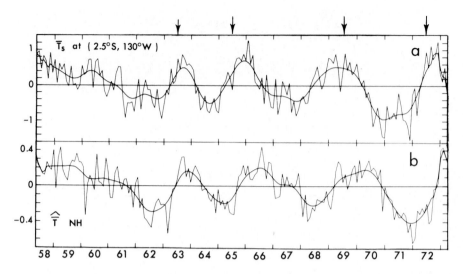

Fig. 7. Time series of (a) the monthly-mean sea-surface temperature in a key region (2.5°S, 130°W) representative of the general conditions in the eastern equatorial Pacific Ocean, and (b) the air temperature averaged vertically and horizontally over the entire mass of the Northern Hemisphere for a 15-year period, after Pan and Oort (1983). (The mean annual cycle has been removed.) The smoothed lines were obtained by applying a Gaussian filter in time. The arrows at the top of the figure indicate the El Niño events during the 15-year period, according to Rasmusson and Carpenter (1982). The tit-marks along the abscissa indicate January of the corresponding year.

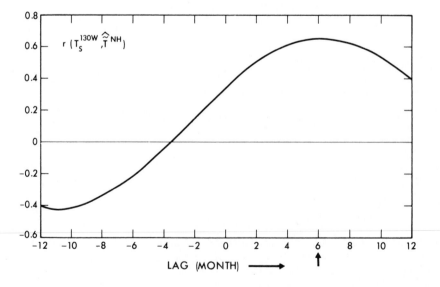

Fig. 8. Correlation coefficient between the two time series shown in Fig. 7 at various lags. A positive lag indicates that anomalies in the Northern Hemisphere air temperature tend to follow anomalies in the sea-surface temperature from Pan and Oort (1983).

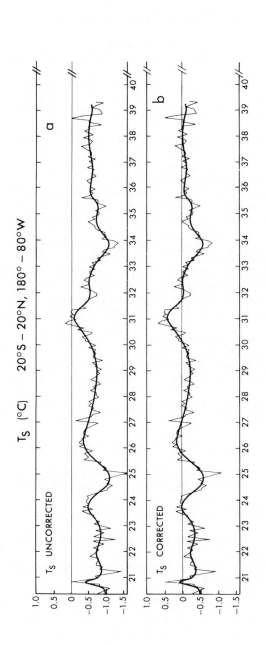

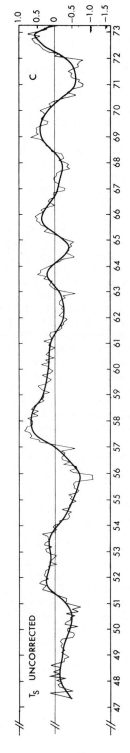

Fig. 9. Time series of the uncorrected, (a) and (c), monthly-mean sea-surface temperature anomalies in the eastern equatorial Pacific Ocean (20° S–20° N, 180° –80° W) in units of ° C. The anomalies were taken from the 1963–1973 mean conditions for each calendar month. A corrected series for the 1921–1939 period is shown in (b); the correction applied was a uniform addition of + 0.5° C to all temperatures before 1940 in connection with the change in reporting practice from mainly bucket to mainly engine-intake temperature measurements. The tit-marks along the abscissa indicate January of the corresponding year.

REFERENCES

Barnett, T.P., 1984. Long-term trends in surface temperature over the oceans. Mon. Weather Rev., 112: 303–312.

Budyko, M.I., 1969. The effect of solar radiation variations on the climate of the earth. Tellus, 21: 611–619.

Fieux, M., 1981. Evolution à long terme des températures de surface de la mer. In: Proc. Int. Conf. on the Evolution of Planetary Atmospheres and Climatology of the Earth. Nice, 16–20 October 1978, Centre National d'Etudes Spatiales, 7 pp.

Fletcher, J.O., Slutz, R.J. and Woodruff, S.D., 1983. Towards a comprehensive ocean–atmospheric data set. Trop. Ocean–Atmos. Newslett., 20: 13–14.

Folland, C.K., Parker, D.E. and Kates, F.E., 1984. Worldwide marine temperature fluctuations 1856–1981. Nature, 310: 670–673.

Haney, R.L., 1971. Surface thermal boundary conditions for ocean circulation models. J. Phys. Oceanogr., 1: 241–248.

Hansen, J., Johnson, D., Lacis, A., Lebedeff, S., Lee, P., Rind, D. and Russel, G., 1981. Climate impact of increasing atmospheric carbon dioxide. Science, 213 (4511): 957–966.

Harris, R.G., Thomasell Jr., A. and Welsh, J.G., 1966. Studies of techniques for the analysis and prediction of temperature in the ocean, Part III: Automated analysis and prediction. Interim Report, prepared by Travelers Research Center, Inc. for U.S. Naval Oceanographic Office, Contract No. N62306-1675, 97 pp.

Horel, J.D. and Wallace, J.M., 1981. Planetary-scale atmospheric phenomena associated with the southern oscillation. Mon. Weather Rev., 109: 813–829.

Jones, P.D., Wigley, T.M.L. and Kelly, P.M., 1982. Variations in surface air temperatures: Part I. Northern Hemisphere, 1881–1980. Mon. Weather Rev., 110: 59–70.

Mitchell Jr., J.M., 1968. A preliminary investigation of atmospheric pollution as a cause of the global temperature fluctuation of the past century. In: A. Singer (Editor), Global Effects of Environmental Pollution. Reidel, Dordrecht, pp. 139–155.

Oort, A.H., 1983. Global Atmospheric Circulation Statistics, 1958–1973. NOAA Prof. Pap. No. 14, U.S. Govt. Printing Office, Washington, D.C., 180 pp.

Paltridge, G. and Woodruff, S., 1981. Changes in global surface temperature from 1880 to 1977 derived from historical records of sea surface temperature. Mon. Weather Rev., 109: 2427–2434.

Pan, Y.H. and Oort, A.H., 1983. Global climate variations connected with sea surface temperature anomalies in the eastern equatorial Pacific Ocean for the 1963–1973 period. Mon. Weather Rev., 111: 1244–1258.

Ramage, C.S., 1984. Can shipboard measurements reveal secular changes in tropical air–sea heat flux? J. Climate Appl. Meteorol., 23: 187–193.

Rasmusson, E.M. and Carpenter, T.H., 1982. Variations in tropical sea surface temperature and surface wind fields associated with the Southern Oscillation/El Niño. Mon. Weather Rev., 110: 354–384.

Saur, J.F.T., 1963. A study of the quality of sea water temperatures reported in logs of ships' weather observations. J. Appl. Meteorol., 2: 417–425.

Spelman, M.J. and Manabe, S., 1984. Influence of oceanic heat transport upon the sensitivity of a model climate. J. Geophys. Res., 89: 571–586.

Wright, P.B. and Wallace, J.M., 1984. Problems in the use of ship observations for the study of interdecadal climate changes. Mon. Weather Rev. (in press).

CHAPTER 15

MULTIVARIATE ANALYSIS OF SENSITIVITY STUDIES WITH ATMOSPHERIC GCM'S

CLAUDE FRANKIGNOUL

ABSTRACT

Because the sample size in sensitivity studies with atmospheric GCMs is much smaller than the dimension of the GCM fields, the multivariate signal-to-noise analysis of the experiments requires a drastic reduction in dimensionality. This can be done by applying a hypothesis-testing strategy, where the anticipated GCM response to a prescribed change is represented by an a priori sequence of guessed patterns characterized by only a few parameters. The method is discussed and briefly illustrated by a sea-surface temperature anomaly experiment made with two versions of the GISS GCM. The sea-surface temperature anomaly is in the North Pacific, and is shown to have a significant influence at the planetary scale on the wintertime circulation of the two models. The response patterns are, however, very different.

1. INTRODUCTION

General circulation model (GCM) fields, like their atmospheric counterparts, have a large natural variability which reflects the instability of the atmospheric circulation. Since the dominant scale of the fluctuations is large, different grid points and variables in GCMs do not provide independent information, and the statistical significance in sensitivity studies must be treated as a multivariate problem. The main difficulty with this approach is that the sample size (roughly the number of independent realizations) is always much smaller than the dimension of the GCM fields, so that a multivariate signal-to-noise analysis cannot be done straightforwardly. Thus, GCM users have mostly relied on simpler univariate tests of significance, even though this strategy is very unsatisfactory, and often leads to unsubstantiated claims.

It has been suggested recently that the multivariate method could be applied to GCM experiments, although only a limited amount of details can be considered in the assessment of statistical significance. This was first shown by Hasselmann (1979), who suggested using a hypothesis-testing strategy where the anticipated GCM response to a prescribed change is represented by an a priori sequence of guessed patterns. The guesses, which can be derived from prior knowledge or simpler dynamical models, are characterized by only a few parameters. Thus, the dimensionality can be smaller than the sample size, and multivariate tests of significance become applicable. The method was established for the asymptotic case where the sample size is large, and it has been extended to the more realistic case where the sample size is limited by Hannoschöck and Frankignoul (1984), who applied it to a sea-surface temperature (SST) anomaly experiment with the GISS (Goddard Institute for Space Studies) general circulation model.

In this paper, we want to briefly present the new multivariate approach and illustrate it by comparing the response of two versions of the GISS GCM (Models I and II) to the same mid-latitude SST anomaly. More details can be found in Hannoschöck and Frankignoul (1984) for the Model I experiment, and in Molin and Frankignoul (in prep.) for the Model II experiment. The results presented here are based on the use of a sequence of "non-dynamical" guesses, and they demonstrate that the SST anomaly has a significant large-scale influence on the climate of the two models. This is of interest since only local effects have been unambiguously detected in most mid-latitude experiments, but it does not elucidate the mechanisms of the model response (e.g., linear or nonlinear behavior?). The new method is well-adapted to this purpose, but suitable "dynamical" guesses have first to be constructed.

The duality of the hypothesis-testing strategy should be stressed in the present context of atmosphere and ocean modelling, because it provides a link between general circulation models and simpler dynamical models. Indeed, the simpler models (e.g., a linear wave model) can be used to formulate the guessed response for the GCMs, i.e., to help interpreting the noisy GCM data. However, the method also allows the use of GCM simulations to test the validity of simpler models, or to verify parameterizations. Although no example will be given here of the interplay between GCMs and simple models, some simple parameterizations of the diabatic heating associated with the SST anomaly will be briefly discussed.

2. THE SST ANOMALY EXPERIMENTS

To investigate the influence of a mid-latitude SST anomaly on the wintertime circulation, we have used two versions of the GISS coarse resolution ($8° \times 10°$ grid) GCM described by Hansen et al. (1983). The first version (Model I), which has seven σ-levels, is rather realistic but has a somewhat deficient eddy kinetic energy, and the mean zonal wind reaches excessive velocity in the stratosphere. These deficiencies have been remedied in Model II, which has nine σ-levels and simulates quite successfully the atmospheric general circulation and its seasonal variations. In both models, cloud cover, snow depth, ground temperature and moisture are computed, while sea-surface temperature and sea ice are prescribed.

The SST anomaly used in both experiments is rather similar to the "classical" one observed in the North Pacific during the winter of 1976–1977, but it is nearly twice as large (Fig. 1). The Model I experiment consists of three long runs in a statistically steady state in the perpetual January mode: one 8-month control run, one 8-month anomaly run and one additional 15-month control run. The Model II experiment consists of five February control runs taken from the basic 5-year run described in Hansen et al. (1983), and five February anomaly runs obtained by introducing the SST anomaly on each preceding November 1. Additional data on Model II interannual variability was obtained from five other February anomaly runs obtained by introducing similarly an SST anomaly of comparable magnitude in the subtropical North Pacific.

In the following, we shall focus on the methodology and illustrate it by an analysis of the mean tropospheric response to the SST anomaly. In both versions of the GCM, the response seems primarily equivalent barotropic. Hence, although we shall, for practical reasons, consider different variables in each model (sea-level pressure has a larger sample

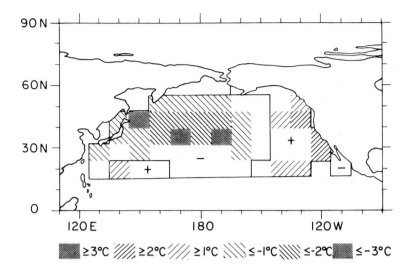

Fig. 1. The prescribed SST anomaly, in °C.

size than geopotential height in Model I, and the response is more significant in the upper troposphere in Model II), the results of the two experiments are directly comparable.

3. UNIVARIATE ANALYSIS

Since multivariate tests of significance are a generalization of univariate tests, it is convenient to first discuss the more familiar one-dimensional problem. Let us consider one variable at one grid point, and let $x^c(n)$ $n = 1, N$ and $x^a(m)$ $m = 1, M$ denote its mean value in N independent control runs and M independent anomaly runs, respectively. If the runs are long enough, $x^c(n)$ and $x^a(m)$ are normal variables, and the null hypothesis that the control and anomaly runs have the same true mean can be tested since the statistic:

$$t = \frac{\bar{x}^a - \bar{x}^c}{[(1/N + 1/M)s]^{1/2}} \tag{1}$$

should then be distributed as a t-variable with $N + M - 2$ degrees of freedom. In eqn. (1), the overbar denotes the sample mean and s is an unbiased pooled estimate of the variance, given by:

$$s = \frac{\sum\limits_{n=1}^{N} [x^c(n) - \bar{x}^c]^2 + \sum\limits_{m=1}^{M} [x^a(m) - \bar{x}^a]^2}{N + M - 2} \tag{2}$$

The result of the two-sample t-test is illustrated for the Model I experiment in Fig. 2 which shows the mean change in sea-level pressure. Slashed areas indicate grid points where the null hypothesis was rejected at the 95% level of confidence, based on independent 6-month pieces. No smoothing was applied. Although the areas where the

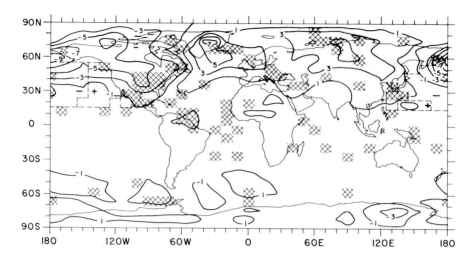

Fig. 2. Mean sea-level pressure difference (in mb) between anomaly and control runs in the Model I experiment. The slashes indicate grid points where the null hypothesis is rejected at the 95% significance level using the two-sample t-test (after Hannoschöck and Frankignoul, 1984).

null hypothesis was rejected are rather scattered, there is a large rate of rejection in the northern hemisphere (15%). This is suggestive of a significant response, and is also found for the other variables.

Because of the finite number of grid points, *global* significance requires that more than 5% of the individual tests yield rejection of the null hypothesis at the 95% level of confidence, in the absence of an a priori hypothesis. As discussed by Livezey and Chen (1983), the critical rejection rate can be inferred from the binomial distribution for a finite sample of independent variables. It is found that, if the grid points in Fig. 1 were independent, the change in the mean sea-level pressure would be significant north of $30°S$. However, taking into account the interdependence between grid points should drastically reduce the number of independent tests, hence increase the effective threshold. Unfortunately, the increase cannot be quantified. Instead, the significance must be tested with a multivariate method.

In the Model II experiment, the univariate tests suggested no significant response at all, except possibly for the air temperature just above the SST anomaly.

4. MULTIVARIATE ANALYSIS

Since GCM variables and grid points are not independent, we now describe each run by an l-dimensional vector which may represent all variables and grid points. Using the same notations in vector form, we can in principle test the null hypothesis by considering the statistic:

$$T^2 = (1/N + 1/M)^{-1}(\bar{x}^a - \bar{x}^c)'S^{-1}(\bar{x}^a - \bar{x}^c) \tag{3}$$

where the prime denotes the transposed vector and S is an unbiased estimate of full rank

l of the true error covariance matrix, and is given by:

$$S = \frac{\sum\limits_{n=1}^{N} [x^c(n) - \bar{x}^c] [x^c(n) - \bar{x}^c]' + \sum\limits_{m=1}^{M} [x^a(m) - \bar{x}^a] [x^a(m) - \bar{x}^a]'}{N + M - 2} \tag{4}$$

The two-sample Hotelling T^2 statistic (3) is the direct analogue of the univariate t^2, and the null hypothesis is rejected at the 95% level of confidence if:

$$T^2 > \frac{(N + M - 2)l}{N + M - l - 1} \qquad F_{\alpha;\, l,\, M + N - l - 1} \tag{5}$$

where F is Fisher's distribution with l and $N + M - l - 1$ degrees of freedom.

In practice, the test (5) cannot be applied to the GCM case because the number of independent runs, typically of order 10, is much smaller than the dimensionality of the GCM fields, typically of order 10^4. Thus, the error covariance matrix S is of much reduced rank, and the conditions of validity for (5) are not met.

To use the multivariate test, we must strongly reduce the dimensionality. Since the significance conditions encountered in multidimensional tests are very stringent when the sample size does not largely exceed the dimensionality of the fields, the GCM data must be projected onto a truncated set of say p new basis vectors, where p is smaller than the number of degrees of freedom. In practice, this severely limits the amount of detail of the GCM experiment that can be evaluated. Furthermore, one must implicitly make a priori assumptions on the general structure of the GCM response to the prescribed change, since one has to choose a highly truncated representation. It is therefore equivalent to the use of the hypothesis-testing strategy of Hasselmann (1979), where the anticipated structure of the GCM response is specified a priori by a sequence of guessed patterns.

Suppose that we can formulate a first guess $g_1 = (g_{1,1}, g_{1,2} \ldots, g_{1,n})$ of the expected GCM response on the basis of some prior empirical or theoretical knowledge (g_1 specifies a pattern and its amplitude). Often, we shall also be able to estimate the likely structure of the deviation of the first guess from the true response, hence an improved guess will be a linear combination of two guess vectors g_1 and g_2. In general, we shall define a sequence of p guesses g_α, $\alpha = 1, 2, \ldots p$, where p does not exceed the number of available number of degrees of freedom. The guesses have been ordered a priori according to their anticipated contribution to the total response, and they could be "non-dynamical", for instance a sequence of large-scale spherical harmonics as considered below, or they could be "dynamical", for instance the prediction of a simpler linearized model.

The hypothesis that the true GCM response can be represented by a linear combination of the p guesses, g_α is written in the l-dimensional space as:

$$\bar{x}^a - \bar{x}^c = \sum_{\alpha=1}^{p} \gamma_\alpha g_\alpha + r \tag{6}$$

where γ_α are scalar parameters and r a residual noise. Clearly, the relevance of the guesses is indicated by the (unknown) true value of γ_α. An estimate of the true value of the scalar parameters is obtained by minimizing the residual error $|r|^2$, which yields:

$$\tilde{\gamma}_\alpha = \sum_{\beta=1}^{p} G_{\alpha\beta}^{-1} (\bar{x}^a - \bar{x}^c)' g_\beta \tag{7}$$

with $G_{\alpha\beta} = g_\alpha' g_\beta$. Since $(\bar{x}^a - \bar{x}^c)$ has a multivariate normal distribution, the parameters $\tilde{\gamma}_\alpha$ are also normally distributed. The null hypothesis that there is no response to the prescribed change is that the true parameters $\gamma_1, \gamma_2, \ldots, \gamma_p$ are all zero. The dimension has been reduced and we can consider the test statistic:

$$T^2 = (1/N + 1/M)^{-1} \tilde{\gamma}' \mathbf{\Gamma}^{-1} \tilde{\gamma} \tag{8}$$

where the $p \times p$ error covariance matrix $\mathbf{\Gamma}$ has full rank p and is derived from the $l \times l$ matrix S by:

$$\Gamma_{\alpha\beta} = d_\alpha' S d_\beta \tag{9}$$

with

$$d_\alpha = \sum_{\beta=1}^{p} G_{\alpha\beta}^{-1} g_\beta \tag{10}$$

The null hypothesis is rejected at the $\alpha\%$ level of confidence if:

$$T^2 > \frac{(N+M-2)p}{N+M-p-1} \qquad F_{\alpha; p, N+M-p-1} \tag{11}$$

If $p > 1$, we can test the model successively for increasing values of p. As discussed by Barnett et al. (1981), the choice of the model selection criteria (at which order to stop the hierarchy) should depend on how strongly one feels about the order in the sequence of guesses, and it must also be made a priori.

Hasselmann (1979) has shown in the asymptotic case when the sample size is very large how to use the properties of the noise field (the GCM natural variability) to optimize the signal-to-noise ratio. The finite sample case has been discussed by Hannoschöck and Frankignoul (1984), who concluded that the procedure in its present form was not advantageous if the number of available GCM runs was small. The optimization procedure will not be considered here.

5. THE ATMOSPHERIC RESPONSE TO THE SST ANOMALY

In this paper we shall use non-dynamical guesses to describe the atmospheric response to the SST anomaly, namely as an a priori sequence of large-scale spherical harmonics. Attempts to use simple linear wave models to predict the Model I and Model II response are described in Hannoschöck and Frankignoul (1984) and Molin and Frankignoul (to be published), respectively.

Since the three runs with Model I were unusually long, as many as 24 degrees of freedom could be obtained for the estimate of the error covariance matrix. Actually S was not calculated from independent runs as in eqn. (4), but it was obtained from the zero frequency estimates of the corresponding cross-spectral matrix, as discussed by Hannoschöck and Frankignoul. In the Model II experiment, S was obtained directly from the three sets of five Februaries described in Section 2, so that there are 12 degrees of freedom.

In both experiments, we have restricted our attention to one variable at a time (which avoids specifying relationships between variables), and considered the northern hemisphere only. Many theoretical studies suggest that, in the absence of resonance, the transfer function for the atmospheric response to diabatic heating is largest at the largest scales. Thus, we decided a priori to use as guess vectors a sequence of spherical harmonics of decreasing scale.

(1) *Model I.* We have considered sea-level pressure and have first eliminated some grid scale noise by expanding the field into even spherical harmonics with a triangular truncation at $n = 18$. The zonally symmetric function ($m = 0$) is also excluded, which leads to a lighter representation in terms of 179 even spherical harmonics, with practically no alteration of the sea-level pressure field. The mean sea-level pressure change between anomaly and control runs is represented in Fig. 3 (upper panel), after transformation back onto the original grid.

The guesses are given by the following (rather arbitrary) sequence of spherical harmonic functions y_n^m, where m is the zonal wavenumber and n the total wavenumber: $Y_1^1, Y_3^1, Y_2^2, Y_4^2, Y_3^3, Y_5^1, Y_5^3, Y_4^4, Y_6^2, Y_6^4, Y_5^5, Y_6^6$. Each function contains the cosine and sine modes, which are considered together in the analysis. To establish the validity of the a priori sequence of guesses, the Hotelling test (eqn. 11) was applied to estimate whether the amplitudes of the spherical harmonics are significantly different from zero, including successively more functions in the model testing. Since we have no strong confidence in the precise ordering of the sequence, we use a mild selection criterion and choose as optimal model the maximum-order model satisfying the significance test. Fig. 4 shows that the first four spherical harmonics can be retained at the 99% level of confidence, and the first seven at the 95% level. Thus, the North Pacific SST anomaly has a significant influence on the atmospheric circulation. The significant part of the mean sea-level pressure change (according to our sequence of guesses) is

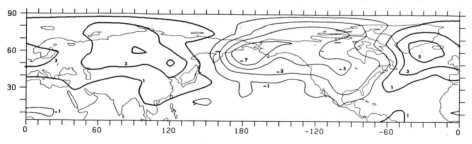

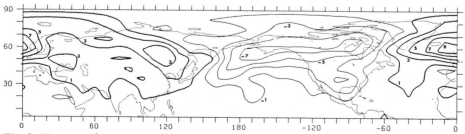

Fig. 3. *Upper panel*: Mean change in sea-level pressure (in mb) for the Model I experiment. *Lower panel*: Significant part of the mean sea-level pressure signal (in mb) at the 99% level (see text). (From Hannoschöck and Frankignoul, 1984.)

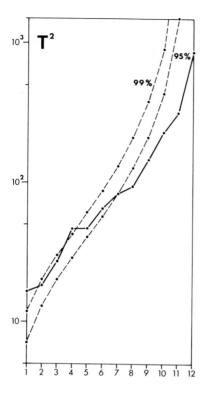

Fig. 4. Test statistic T^2 as a function of the number of guesses for the sea-level pressure changes in Model I experiment. The 95 and 99% significance bounds are given for the null hypothesis (after Hannoschöck and Frankignoul, 1984).

represented in Fig. 3 (lower panel) which suggests a planetary scale, predominantly zonal wavenumber one response. At higher levels, the response has the same character and seems to be equivalent barotropic.

The sensitivity of Model I at the largest scales is remarkable, and seems to reflect the dominance of very large scales in the natural variability of the model at low frequencies, which was illustrated in Hannoschöck and Frankignoul (1984) by an empirical orthogonal functions analysis. This model feature is not realistic, however, and therefore the representativeness of the Model I response to the SST anomaly may be questionable.

(2) *Model II.* The sea-level pressure and the geopotential height data in Model II were each treated similarly, with a stronger truncation at $n = 12$. The mean change in the 200 mb geopotential height between the five anomaly and the five control runs is shown in Fig. 5 (upper panel). It is typical of the upper tropospheric changes, and has smaller characteristic scales than the corresponding Model I changes.

Since only 12 degrees of freedom were available, we could not use the same sequence of spherical harmonics for the a priori guesses, as the response would then be restricted to the largest scales. Instead, we decided to order the sequence in terms of increasing

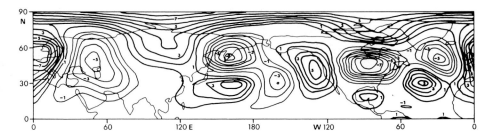

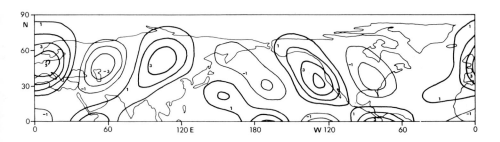

Fig. 5. *Upper panel*: Mean change in 200 mb geopotential height (in tens of meters) for the Model II experiment. *Lower panel*: "Significant part" of the signal (in tens of meters) at the 95% level (see text).

zonal wavenumber m and to choose for corresponding meridional structure that which dominated the February climatology in the control runs, thereby assuming that the SST anomaly will predominantly force the same spherical harmonics as orography and mean diabatic heating. The results of the Hotelling test are illustrated for the 200 mb level in Fig. 6. At the 95% level, there is a significant signal and the optimal model contains the first five spherical harmonics. As shown in Fig. 5, the significant part of the signal according to this hierarchy of guesses suggests a predominantly zonal wavenumber three to five-response for the geopotential height. The amplitude is about two times smaller than in the sample difference between anomaly and control runs, and thus only a fraction of the "observed" change seems significant. The response is clearly not a local one, and it extends throughout the northern hemisphere extra-tropical regions. At lower levels, the results are similar, but the signal-to-noise ratio deteriorates rapidly with height. For example, at 500 mb the five-spherical harmonics model is only significant at the 90% level, and at the ground (sea-level pressure) there is no significant response.

Thus, although the standard univariate tests of significance suggested no response in the Model II experiment, there is a small but significant response on the planetary scale. However, the characteristic scales are smaller than in Model I. This seems to correspond to the smaller (and more realistic) scales of the Model II natural variability. Both responses also seem to strongly differ from the response of the NCAR Community Climate Model to a rather similar SST anomaly (Pitcher et al., 1985).

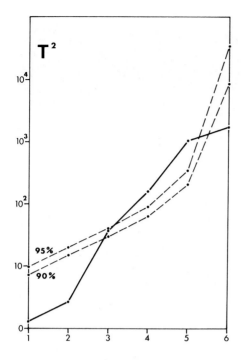

Fig. 6. Test statistic as a function of the number of spherical harmonics for the 200 mb geopotential height changes in the Model II experiment. The 90 and 95% significance bounds are given for the null hypothesis.

6. DISCUSSION

In this brief presentation, we have reviewed how a hypothesis-testing strategy can be used to evaluate the statistical significance of prescribed change GCM experiments. To stress the simplicity of the multivariate approach, we have purposely omitted describing how the statistical significance of the guessed response could be enhanced by applying pattern filtering methods. This is discussed by Hasselmann (1979) and Hannoschöck and Frankignoul (1984).

To illustrate the power of the new method, we have discussed some aspects of the response of two versions of the GISS GCM to the same SST anomaly in the North Pacific. Using a hierarchy of non-dynamical guesses, it was shown that in both cases there was a significant response on the planetary scale. However, the responses differ widely, having shorter scales and a smaller amplitude in the more realistic Model II. It is puzzling that the response of GCM's to mid-latitude SST anomalies seems so strongly model-dependent, and that its characteristic scales are similar to that of the model natural variability. The causes might be elucidated when using dynamical guesses. Interestingly, the GCM experiments with equatorial SST anomalies discussed elsewhere in this volume also suggest very different extratropical response in different models, while the tropical responses are generally similar.

Although the multivariate method is well-suited to the use of simpler dynamical models to provide guesses for the GCM response, or conversely to the use of GCMs to test simpler models, we have not discussed this aspect here. A difficulty of such analysis is that the SST anomaly is prescribed in GCM experiments, while the simpler models generally respond to changes in the diabatic heating. Since the relation between SST and diabatic heating anomalies is very poorly known, particularly in the middle latitudes, there is a need to first establish relevant parameterizations. For this purpose, we have used the multivariate method to test two very simple relations in the Model II experiment (North Pacific region only). First we have tested whether the diabatic heating anomaly (integrated over the whole troposphere) was linearly related to the SST anomaly. The hypothesis was rejected, and it was found instead that the anomaly heating was proportional to the anomaly in air–sea temperature difference. On the other hand, it was found for a subtropical SST anomaly that the changes in diabatic heating were significantly correlated with the SST anomaly, and not with the air–sea temperature difference. This illustrates the well-known differences in the dynamics of the two regions, and the larger influence of convection and moisture convergence in the tropical regions. The consequences of these findings are being investigated.

ACKNOWLEDGEMENTS

We would like to thank Drs. J. Hansen and G. Russel who generously provided the GCM data, and Mr. A. Molin who let us use some of his unpublished results. This research was supported in Paris by a research grant from the CNEXO and at MIT by the National Science Foundation, Climate Dynamics Program, under grant ATM-8116047.

REFERENCES

Barnett, T.P., Preisendorfer, R.W., Goldstein, L.M. and Hasselmann, K., 1981. Significance tests for regression model hierarchies. J. Phys. Oceanogr., 11: 1150–1154.
Hannoschöck, G. and Frankignoul, C., 1984. Multivariate statistical analysis of a sea surface temperature anomaly experiment with the GISS general circulation model. J. Atmos. Sci. (submitted).
Hansen, J., Russel, G., Rind, D., Stone, P., Lacis, A., Lebedeff, S., Ruedy, R. and Traves, L., 1983. Efficient three-dimensional global models for climate studies: Models I and II. Mon. Weather Rev., 111: 609–662.
Hasselmann, K., 1979. On the signal-to-noise problem in atmospheric response studies. In: D.B. Shaw (Editor), Meteorology of the Tropical Oceans. R. Meteorol. Soc., London, pp. 251–259.
Livezey, R.E. and Chen, W.Y., 1983. Statistical field significance and its determination by Monte Carlo techniques. Mon. Weather Rev., 111: 46–59.
Pitcher, E.J., Blackmon, M.L., Bates, G.T. and Munoz, S., 1984. The effect of midlatitude Pacific sea-surface-temperature anomalies on the January climate of a general circulation model. J. Atmos. Sci. (in press).

CHAPTER 16

INTERANNUAL AND SEASONAL VARIABILITY OF THE TROPICAL ATLANTIC OCEAN DEPICTED BY SIXTEEN YEARS OF SEA-SURFACE TEMPERATURE AND WIND STRESS

JACQUES SERVAIN, JOËL PICAUT and ANTONIO J. BUSALACCHI

ABSTRACT

Monthly fields of sea-surface temperature and wind stress for the tropical Atlantic are carefully established for January 1964 to December 1979. The availability of these gridded data sets permits a few simple measures of the seasonal and interannual variability of the tropical Atlantic. Throughout most of the basin the amplitude of seasonal SST variability is greater than the amplitude of monthly SST anomalies. Interannual fluctuations of SST are largest in areas of large seasonal SST variability such as coastal and equatorial upwelling zones. In these regions monthly SST anomalies may reach $1° - 2°C$. As opposed to SST, the largest year-to-year changes in wind stress do not necessarily coincide with the locations of the largest seasonal cycles. The amplitude of monthly wind stress anomalies is greater than the amplitude of the seasonal cycle except in the vicinity of the ITCZ. The range of monthly wind stress anomalies along the equator and eastern boundary, expressed in units of wind speed squared, is $10-20 \, m^2 \, s^{-2}$. Of particular note are the differing phase relationships with the seasonal cycle for the interannual fluctuations of SST and wind stress within the upwelling zones of the Northern and Southern Hemispheres. Moreover, sample time series of SST and wind stress are characterized, at times, by anomalous events persisting for more than one year. This evidence is suggestive of anomalous conditions in the tropical Atlantic that are not the result of a simple phase shift or change in amplitude of the seasonal cycle.

INTRODUCTION

The role played by the tropical oceans in affecting climate appears to be very important although not well understood. There is evidence to suggest that the dramatic El Niño/Southern Oscillation phenomena of the Pacific Ocean may have an influence on the climate of North America (Horel and Wallace, 1981). The attention directed toward El Niño has enhanced our knowledge of the interannual variability of the tropical Pacific primarily through historical data analyses. Based on the analyses of sea-surface temperature (SST) in the Gulf of Guinea by Merle et al. (1980) and theoretical consideration of the adjustment time scale of the basin (Philander, 1979), it is commonly asserted that the tropical Atlantic Ocean variability is dominated by a large seasonal cycle. A French–U.S. experiment (FOCAL–SEQUAL) is designed to increase our understanding of this seasonal variability. However, large SST anomalies can develop in the tropical Atlantic Ocean (Hisard, 1980; Merle, 1980a) which in turn may correlate with droughts in Brazil and the Sahel (Hastenrath, 1976; Markham and McLain, 1977; Lamb, 1978; Moura and Shukla, 1981). Most of these studies focus on specific abnormal events, primarily from 1967 to 1968. Apart from the EOF analysis of SST by Weare (1977), little is known about the long-term time history of SST or the importance of interannual

212

SST variability in the tropical Atlantic Ocean. Likewise, the nonseasonal wind stress variability of the tropical Atlantic is only known for specific periods of time such as during the GATE and FGGE experiments.

As a prelude to a possible Tropical Ocean Global Atmosphere (TOGA) experiment in the tropical Atlantic long-term historical data analyses are needed. Therefore we have constructed a complete set of monthly SST and wind stress fields for 16 years based on nearly two million merchant ship reports. This unique data set provides a basis for jointly analyzing both the seasonal and interannual fluctuations of SST and wind stress of the tropical Atlantic Ocean. Furthermore, these data will serve as historical points of reference for coupled ocean—atmosphere models and forcing functions for independent atmosphere and ocean models. An example of the utilization of the wind stress data for ocean modelling is presented.

The purpose of this paper is to serve as an introduction to the combined SST/wind stress data set and provide specific examples of the seasonal and interannual variabilities of these fields. First the techniques used to transform individual merchant ship observations into monthly data on $2° \times 2°$ grids are summarized. Then maps of the amplitude of seasonal and interannual SST and wind stress fluctuations are presented in addition to selected time series. Finally, the SST and wind stress variability are briefly compared with one another in certain regions where the variability is notable.

DATA PROCESSING

Prior to the start of the FOCAL—SEQUAL experiment a historical data analysis group obtained all the available individual ship observations in the tropical Atlantic through December, 1979 (TDF-11) from the National Climate Center, Asheville, N.C. The low number of observations in the Southern Hemisphere led us to choose a study area extending from 20°S to 30°N and 60°W to the African continent (Fig. 1a). Due in part to the relatively small size of the tropical Atlantic Ocean, the data density is greater and more evenly distributed than in the tropical Pacific Ocean, especially within the equatorial belt. As a result of a greater international emphasis on securing ship reports, the number of observations since World War II increased suddenly after 1963. Hence the present analysis covers the period 1964—1979 and is based on nearly two million wind observations and a slightly lower number of SST observations. A histogram of the total number of observations per month within the study area is presented in Fig. 1b. Except for the first seven months of 1971 and three months in mid-1974 the number of observations is sufficient (>5000 per month) to provide reasonable monthly estimates of both parameters for 1964—1979. Estimates of the standard error of wind stresses computed from the TDF-11 file are given in Hellerman and Rosenstein (1983).

The sequence of processing steps for the SST and wind stress analyses is outlined in Fig. 2. The initial processing step was to compute the average monthly SST and wind stress for each 2° latitude by 5° longitude box. For comparison, the size of each quadrangle is half that used in analyses of the tropical Pacific wind field by Wyrtki and Meyers (1976) and Goldenberg and O'Brien (1981). In the present study the term "wind stress" represents the two components $\hat{\tau}^x = 1/N \sum_{i=1}^{N} W_i^x |W_i|$ and $\tau^y = 1/N \sum_{i=1}^{N} W_i^y |W_i|$

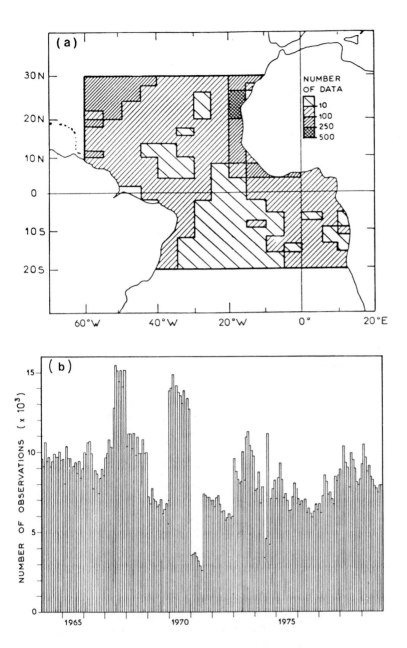

Fig. 1(a). Area of study and mean monthly number of observations per 5° longitude × 2° latitude box. (b) Total number of observations per month in the area of study for 1964–1979.

where W_i is the wind velocity of the ith observation of a total of N in the box being considered, and W_i^x, W_i^y the zonal and meridional wind components, respectively. To obtain a dimensionally correct value of the wind stress, τ^x and τ^y would be multiplied by an air density and drag coefficient which may or may not be constant according to various empirical formulae. The data within each grid were subjected to a series of tests

214

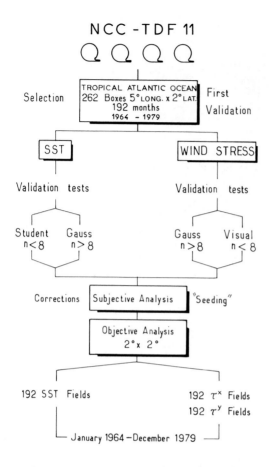

Fig. 2. Flow chart of data processing steps for SST and wind stress.

when computing the monthly averages. Each temperature measurement outside the range $10°–31°C$ or each wind measurement greater than $25\ \mathrm{m\,s^{-1}}$, such as that representing a tropical cyclone, was rejected. The data within any $5° \times 2°$ box were completely ignored if there were at most two SST observations or only one wind observation. For less than eight SST observations per grid square, the mean value was tested against the climatic mean and the adjacent data using a Student's t-test of homogeneity. An iterative approach was then used to correct the 10% of the grid boxes in this category (less than 3% of all possible boxes) that failed the test. For less than eight wind observations this automatic test was replaced by a subjective analysis requiring that the direction and magnitude at each point in question be visually inspected. Twenty percent of the grid points containing less than eight wind observations were corrected. When the number of SST or wind observations was greater than or equal to eight (75% of all possible boxes) the range of the data within each box was tested against a normal distribution. Depending on the scatter of the data, 2–10% of the data in such a grid square were rejected. The last in this series of tests was the inspection of each monthly SST and wind stress map for any remaining spurious anomalies. Two percent of all boxes required subjective analysis that took into account data from surrounding grid points in time and space.

Having obtained monthly mean data sets on a 2° latitude by 5° longitude grid, an objective analysis method based on Cressman (1959) was used to obtain a gap-free data base of monthly averaged SST and wind stress components on a 2° × 2° grid. The convergence of this method can give spurious extrapolations when there are large areas of missing data, e.g. in the south-center of the study area. This was prevented by "seeding" such regions with a few subjectively analyzed data. A total of 2% of all possible grid points were entered in this manner. The choices of influence radii and iteration count for the objective analysis scheme were based on visual comparisons with contour fields of the

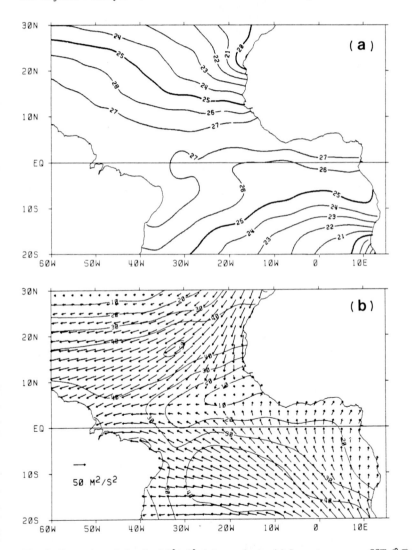

Fig. 3. Examples of the final 2° × 2° data products: (a) Long-term mean SST (°C) 1964–1979; (b) long-term mean wind stress vectors (m² s⁻²) 1964–1979.

216

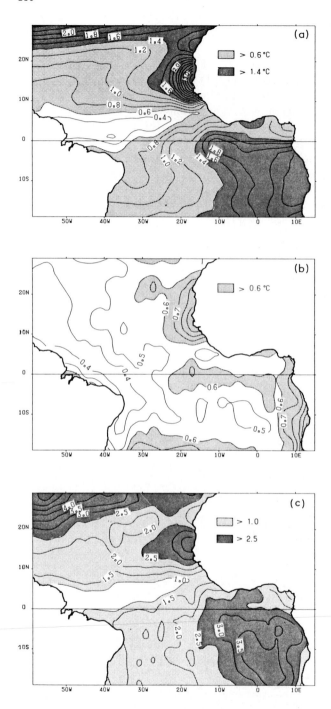

Fig. 4(a). Standard deviation, σ_s, of the mean seasonal cycle of SST; (b) standard deviation, σ_i, of the anomalies from the mean seasonal cycle of SST; and (c) ratio σ_s/σ_i.

5° x 2° data, correlation analyses, and advice from meteorologists experienced with this technique. The radii of influence, in descending order, were: 1400, 1000 and three iterations at 600 km for SST, and 1600, 1200 and four iterations at 800 km for the wind stress components. The 16-year means of SST and wind stress are presented as examples of the final 2° x 2° products (Fig. 3).

VARIABILITY OF SST

In this and the following section, measures of the seasonal and interannual variability will be provided by standard deviations of the variable of interest. The standard deviation of the 12 months for the mean seasonal cycle, σ_s, is used as an estimate of the seasonal variability. The standard deviation of the 192 monthly departures from that seasonal cycle, σ_i, is used as a measure of the interannual variability. This estimate of the inter-annual variability will take into account phase shifts of the mean seasonal cycle, amplitude changes in the seasonal cycle, and events disassociated from this cycle. Further-more, to determine what months of the seasonal cycle have the largest interannual variability, the standard deviation about the mean of a particular month over 16 years, σ_{im}, is calculated. In what follows the distribution of the data density (Fig. 1a) should always be kept in mind. Discussion of the variability in regions where the number of observations is small has been kept to a minimum.

The subsurface thermal structure of the tropical Atlantic Ocean is characterized by a sloping thermocline which rises from west to east (Merle, 1980b). This zonal distribution is also reflected in the mean SST field (Fig. 3a) where the coolest waters are found in the eastern tropical Atlantic. The spatial structure of the amplitude of seasonal SST variability (Fig. 4a) resembles the mean state and is similar to the annual and semi-annual decompositions performed by Merle and Le Floch (1978). Regions with the greatest seasonal variability coincide with regions of low SST. The shallowing thermocline in the eastern tropical Atlantic together with vertical mixing induces maximum seasonal SST variability ($\sigma_s = 1.4° - 3.4°C$) in the seasonal upwelling zones along the coasts of Mauritania and Senegal (NW Africa), the northern and southern coasts of the Gulf of Guinea, and along the equator near 10°W. The location of the minimum SST variability ($\sigma_s < 0.4°C$), coincident with the thermal equator and hence the mean Intertropical Convergence Zone (ITCZ), is obvious between 0° and 7°N. Poleward of 25°N mid-latitude seasonal changes begin to appear.

The interannual variability of SST, as depicted in Fig. 4b, is maximum ($\sigma_i = 0.6° - 1.0°C$) in the coastal and equatorial upwelling regions. One conspicuous difference between the location of maximum seasonal and interannual variability is the relatively narrow offshore extent of interannual SST changes along the southern coast of the Gulf of Guinea. The minimum interannual SST fluctuations ($\sigma_i < 0.4°C$) are offshore the northern coast of Brazil and Guiana; areas where the seasonal signal is also small. Due to the paucity of observations in the south-center limit of the study area (less than four observations per month) it is difficult to assert that the corresponding interannual variability is real and influenced by mid-latitude variability.

An estimate of the amplitude of the seasonal signal with respect to interannual changes is obtained by forming the ratio of the standard deviation of the seasonal variability to

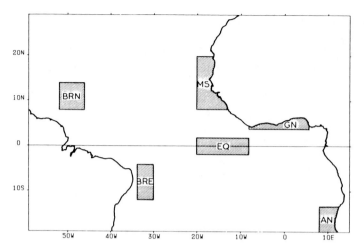

Fig. 5. Geographical location of sample time series of SST.

the standard deviation of the interannual variability, σ_s/σ_i (Fig. 4c). Obviously, high values of this ratio ($\sigma_s/\sigma_i > 2.5$) are found in the regions where the annual signal is important, i.e. north of $20°$N, in the vicinity of seasonal coastal upwelling regions, and inside the Gulf of Guinea. A similar ratio, σ_s/σ_{im}, has been calculated relative to each of the 12 months of the mean year to determine how the ratio of seasonal to interannual variability changes as a function of season. In the Gulf of Guinea σ_s/σ_{im} reaches 3–4 from February to May and July to September, and decreases to 2–3 from October to January and during June. These results complete those of Merle et al. (1980) who used the Dakar–Cape Town shipping transect across the Gulf of Guinea to study the inter-annual variability for the months of February and August. In the present work February and August are months in which the interannual signal is relatively small. For the western and central equatorial Atlantic Ocean σ_s/σ_{im} is less than two for all months and is always less than one in the vicinity of the thermal equator where seasonal changes are the smallest.

Having presented a brief overview of the amplitudes of the seasonal and interannual SST changes across the tropical Atlantic basin, we will now focus on the time histories of SST for a few specific areas. Geographic locations of interest are shown in Fig. 5. For convenience, each individual zone is denoted by a geographical abbreviation: MS for offshore Mauritania–Senegal (NW Africa), BRN and BRE for north and east of Brazil, GN for the northern Coast of the Guinea Gulf, EQ for an equatorial zone, and AN for offshore Angola (SW Africa). The choice of these six areas is based on the information provided by Fig. 4 and limited to regions with a sufficient number of observations. Four of these areas correspond to the upwelling zones of the eastern side of the basin where the seasonal and interannual SST variabilities are found to be the largest. The two locations off Brazil, symmetric about the thermal equator, are chosen as specific examples of the SST variability, albeit small, in the western tropical Atlantic.

The seasonal cycles of these areas are presented in Fig. 6. The interannual variability about each month of the mean seasonal cycle, σ_{im}, is also given. The four areas south of the thermal equator (GN, EQ, AN, BRE), as expected, have similar phase with maximum

SST appearing during the austral summer. The opposite phase applies for the two locations in the northern tropical Atlantic (MS, BRN). Year to year perturbations to the seasonal cycle are small or do not change much as a function of the season at BRN, BRE, and GN. The interannual variability at MS, EQ, and AN is, on average, greatest at specific times of the year. Interannual fluctuations for the coastal upwelling region off Mauritania and Senegal (MS) are out of phase with the seasonal cycle. The largest SST anomalies occur when SST is lowest. Analysis of the individual monthly anomalies presented in Fig. 7 indicates that most of the variability is associated with changes in the amplitude, timing, and duration of the cold season. The range of SST maxima for 1964–1979 was $27.3°–27.8°C$ while the range of SST minima was $20.3°–22.8°C$. Year-to-year changes in the seasonal cycle within the equatorial band (EQ) are primarily at the beginning of the cold season. Unlike MS and EQ, interannual SST anomalies for the coastal upwelling region in the Southern Hemisphere (AN) are not linked to a cold season. Monthly SST anomalies at AN tend to be in phase with the seasonal cycle, i.e. the largest SST anomalies nearly coincide with the highest SST.

The range of monthly SST anomalies is $± 2°C$ in the eastern part of the basin and $± 1°C$ in the west (Fig. 7). The most significant anomalous events persist for several months to more than one year as indicated by the low-pass filtered time series of the SST anomalies. For example, along Mauritania and Senegal (MS) the period of most persistent anomalies occurred during 1968–1970 when SST was lower than normal in 1968 ($- 1°C$) and the cold seasons of the following two years had SST higher than usual ($+ 2°C$, 1969; $+ 1°C$, 1970). The largest contribution to this nonseasonal variability is approximately at a period of 38 months with an amplitude of $0.4°C$. For comparison purposes, the annual and semi-annual amplitudes are $2.9°$ and $0.5°C$, respectively. The extreme temperature anomalies within the central equatorial zone (EQ) occurred in 1967 ($- 1.5°C$) and 1968 (for discussion of this particular event in which SST was $2°C$ higher than usual see Lamb, 1978; Hisard, 1980; Merle, 1980a; and Servain, 1984). The maximum Fourier amplitude of this time series is $0.4°C$ with an approximate period of 27 months. The dominant amplitudes of the seasonal cycle are $2.0°C$ at 12 months and $0.7°C$ at 6 months.

Having presented the temporal characteristics within these regions, an estimate of the spatial structure of the anomalies is of interest. Towards that end, the correlation coefficients from the cross correlation of the mid-point of regions MS, AN, EQ, and GN with all surrounding grid points are mapped in Fig. 8. The degrees of freedom used to determine the 95% significance levels were obtained by dividing the total record length by the integral time period needed to obtain two independent realizations (Davis, 1976). The SST anomalies of the Mauritania–Senegal region are significantly correlated with SST anomalies at distances of up to 1100 km offshore. Lagged cross correlations indicate the anomalous SST variability of MS leads the SST variability further west by one month. This is contrasted with the Angola upwelling region where the offshore extent of significantly correlated SST anomalies is limited to approximately 700 km. These results remain the same if the correlations are made relative to other points within the MS and AN regions. The spatial structure for the SST anomalies of EQ and GN are similar to each other. The off-equator and offshore length scales of significantly correlated anomalies are 700–800 km with along-equator and alongshore scales somewhat larger.

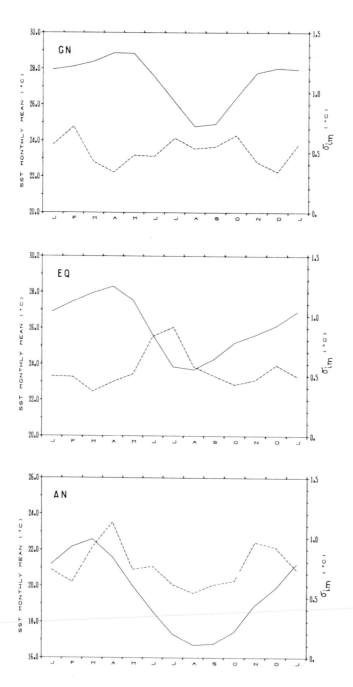

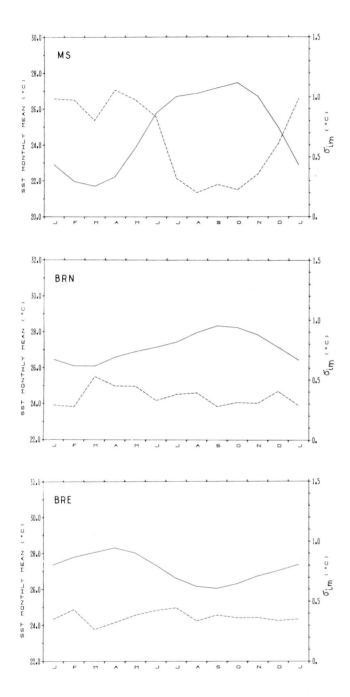

Fig. 6. Monthly mean seasonal cycle of SST (solid) and the corresponding standard deviations, σ_{im} (dashed), of the 16 monthly anomalies from each monthly mean for the regions of Fig. 5.

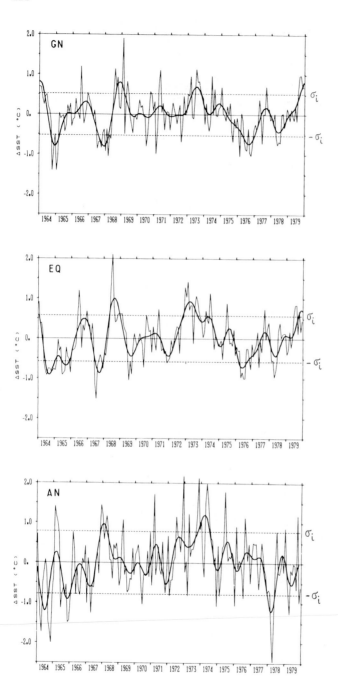

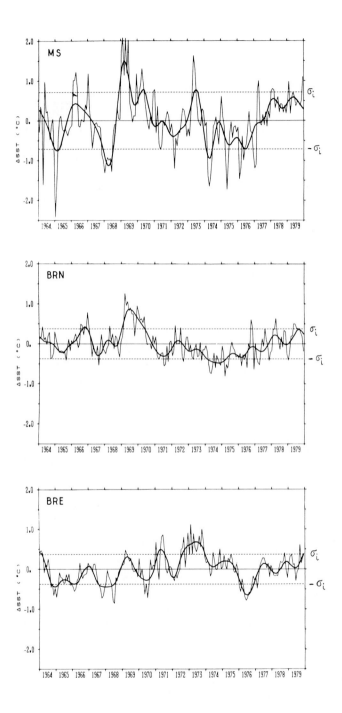

Fig. 7. Monthly SST anomalies (thin line) from the mean seasonal cycle for each region of Fig. 5, 1964–1979. Low-pass (> 12 months) filtered time series (thick line) of the SST anomalies. The dashed line indicates the standard deviation, σ_i, of the monthly anomalies.

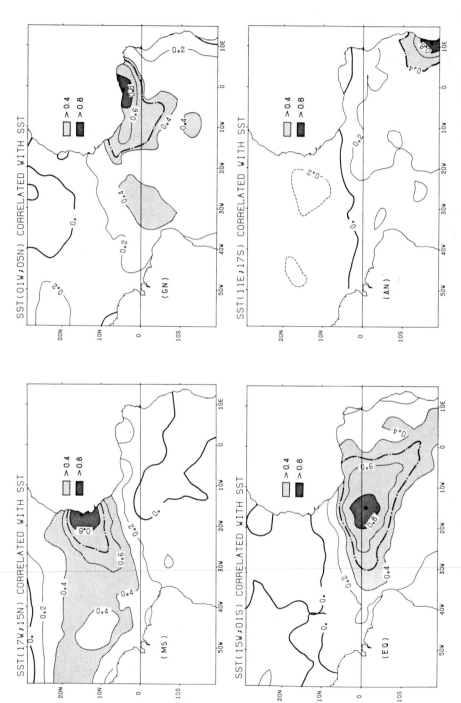

Fig. 8. Correlation coefficients at zero lag for the cross correlation of the SST anomalies at the mid-point of regions MS, AN, EQ and GN with the SST anomalies of all surrounding grid points. The dashed-dotted line indicates the 95% significance level.

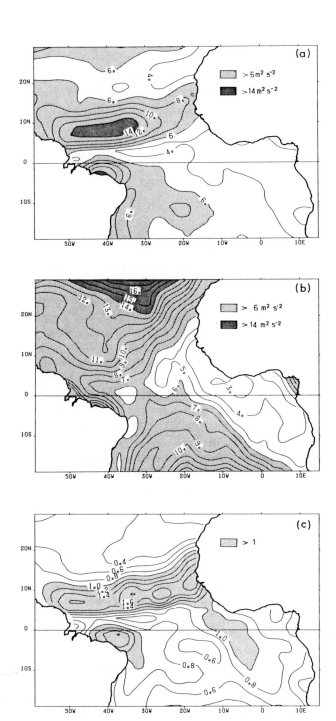

Fig. 9(a). Standard deviation, σ_s, of the mean seasonal cycle of zonal wind stress; (b) standard deviation, σ_i, of the anomalies from the mean seasonal cycle of the zonal wind stress; and (c) ratio σ_s/σ_i.

VARIABILITY OF WIND STRESS

Regions of the maximum seasonal variability of the zonal (Fig. 9a) and meridional (Fig. 10a) wind stress are contained within an envelope defined by the seasonal excursion of the ITCZ. The maxima of the zonal wind stress fluctuations ($\sigma_s = 10-15\,\mathrm{m^2\,s^{-2}}$) straddle the mean position of the ITCZ whereas the largest deviations of the meridional wind stress ($\sigma_s = 15-20\,\mathrm{m^2\,s^{-2}}$) occur along this line. These maps closely resemble the corresponding charts of Hellerman (1980) when a drag coefficient of 2.5×10^{-3} is used to compute the wind stress. (The analysis performed by Hellerman was based on the wind stress calculated by Bunker (1976), in which individual wind observations were converted to wind stress using a drag coefficient that was a function of atmospheric stability and wind speed.)

Unlike SST, regions of significant interannual wind stress fluctuations do not necessarily coincide with the regions of maximum seasonal variability. The largest year-to-year changes in the zonal wind stress ($\sigma_i = 12-18\,\mathrm{m^2\,s^{-2}}$) are at the poleward extremes of the study area (Fig. 9b). This interannual signal is likely associated with multiple low pressure patterns propagating in each winter hemisphere. The amplitude of the interannual variability at lower latitudes is less than $4\,\mathrm{m^2\,s^{-2}}$ in the Gulf of Guinea and between 4 and $8\,\mathrm{m^2\,s^{-2}}$ in the remainder of the equatorial Atlantic. The amplitude of the interannual variability of the meridional wind stress is nearly constant ($\sigma_i = 6-8\,\mathrm{m^2\,s^{-2}}$) over broad regions of the tropical basin (Fig. 10b) contrary to that for the zonal wind stress. Minima are noted inside the Gulf of Guinea and offshore the coasts of Guiana and Brazil.

The ratios of the amplitude of the seasonal cycle to the amplitude of interannual perturbations, σ_s/σ_i, for both wind stress components are presented in Figs. 9c and 10c. The distribution of these ratios is similar to that of the seasonal cycle. Monthly anomalies of the wind stress are greater than the amplitude of the seasonal variability ($\sigma_s/\sigma_i < 1$) for large regions of the tropical Atlantic including the equatorial zone in contrast to that of SST. The seasonal wind stress changes are dominant only in the vicinity of the ITCZ.

As examples of wind stress time series obtainable from this 16-year data set we choose to present the wind stress variability of selected equatorial and coastal zones (Fig. 11). The choice of these areas is restricted to regions with a significant number of observations which coincide with regions where the wind stress is believed to play an important role in the local or remote forcing of the ocean. The meridional wind stress along Mauritania and Senegal (MS), the meridional wind stress along Angola (AN), and the zonal wind stress along the northern coast of the Gulf of Guinea (GN) are presented as indications of the alongshore wind variability of three coastal upwelling regions. The equatorial strip is arbitrarily divided at 20°W to distinguish the zonal and meridional wind stress of the eastern equatorial Atlantic (EQE) from the zonal wind stress of the western equatorial Atlantic (EQW).

The seasonal cycle and the seasonal dependence of the interannual variability, σ_{im}, are depicted in Fig. 12. The amplitude of wind stress anomalies is small and not closely coupled to the seasonal cycle in areas where the mean wind stress is small, i.e. EQE and GN. The amplitude of meridional wind stress anomalies in the two main upwelling zones of western Africa, MS and AN, is largest during the time of year in which the seasonal mean is the largest. This is not the situation, however, for the zonal wind stress in the western equatorial Atlantic (EQW). The largest anomalies in this region occur during June

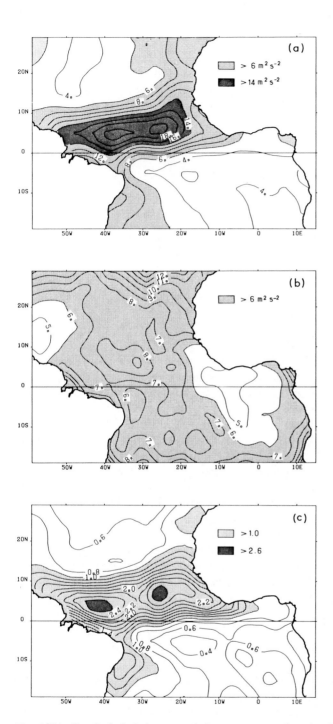

Fig. 10(a). Standard deviation, σ_s, of the mean seasonal cycle of the meridional wind stress; (b) standard deviation, σ_i, of the anomalies from the mean seasonal cycle of the meridional wind stress; and (c) ratio σ_s/σ_i.

228

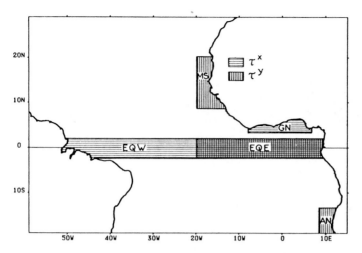

Fig. 11. Geographical location of sample time series of wind stress components.

which is the mid-point of the seasonal intensification of the equatorial easterlies.

The amplitude of individual monthly anomalies (Fig. 13), as alluded to above, is small in the eastern equatorial Atlantic (GN and EQE). One aspect worth noting is the evidence of a slight long-term trend to both components of the wind stress along the equator in the east. The range of monthly anomalies of the meridional wind stress at Mauritania–Senegal (MS) and Angola (AN) is 15 and 25 $m^2 s^{-2}$, respectively. Of the two locations, Angola has the most significant sustained anomalies. The largest Fourier amplitudes for this nonseasonal variability of the meridional wind stress are 2.2 $m^2 s^{-2}$ at approximately 32 months for MS and 3.2 $m^2 s^{-2}$ at 64 months for AN. The amplitudes of the seasonal cycles within these coastal upwelling zones are 14.2 $m^2 s^{-2}$ at 12 months and 4.5 $m^2 s^{-2}$ at 6 months for MS and 4.0 $m^2 s^{-2}$ at 12 months and 4.3 $m^2 s^{-2}$ at 6 months for AN. Maximum anomalies of the zonal equatorial wind stress west of 20°W (EQW) are greater than 10 $m^2 s^{-2}$. The equatorial easterlies were persistently weaker than normal for 1964–1968 and stronger than normal for 1976–1979. A similar change in amplitude between the 1960's and 1970's has been detected in an EOF analysis of tropical Pacific wind data (Legler, 1983). The largest contribution to the low-frequency variability is 2.2 $m^2 s^{-2}$ at a period of approximately 27 months compared with 8.2 $m^2 s^{-2}$ at 12 months and 0.6 $m^2 s^{-2}$ at 6 months.

DISCUSSION

The 16 years of gridded SST and wind stress resulting from the objective analysis provide an opportunity to jointly analyze the two fields for a few of the regions discussed earlier. Previous studies have addressed the relationship between the local wind field and SST at seasonal time scales for the two major coastal upwelling regions of western Africa. Wooster et al. (1976) and Speth and Kohn (1983) indicate that within the coastal upwelling zone along Mauritania and Senegal the seasonal SST variability is positively correlated with the meridional wind stress. Comparison of the predominantly annual signals depicted in Figs. 6 and 12 bears this out. Berrit (1976) suggests that along the southern coast of the Gulf of Guinea the seasonal variation of SST is in agreement with the alongshore wind stress between $15°-20°$S but no agreement exists further north. We note, however, a mismatch of periodicities in the SST and overlying meridional wind stress off Angola (AN) (Figs. 6 and 12). The seasonal SST variability within AN is predominantly annual whereas the meridional wind stress is comprised of equal contributions at the annual and semi-annual periods.

Yet, little is known about the relation, if any, between perturbations to the seasonal cycles of SST and local wind stress. Cross correlation of the Mauritania–Senegal (MS) time series of monthly anomalies of SST (Fig. 7) and meridional wind stress (Fig. 13) indicate the two time series are positively correlated with a maximum correlation coefficient of 0.36 at zero lag. The same does not hold true of the upwelling zones along the southern coast of the Guinea Gulf where SST anomalies are not significantly correlated with meridional wind stress anomalies. For example, the zero lag correlation between SST and τ^y off Angola is -0.04.

Servain et al. (1982) compared monthly anomalies of SST inside the Gulf of Guinea for 1923–1938 with the interannual variability of the wind stress inside and remote from the Gulf of Guinea. The maximum correlation was between SST anomalies in the Gulf of Guinea and anomalies of the zonal equatorial wind stress off Brazil. When similar correlations are calculated for the time series presented here we find that the equatorial SST anomalies for $20°-10°$W (EQ, Fig. 7) are positively correlated with the equatorial zonal wind stress anomalies west of $20°$W (EQW, Fig. 13) at a correlation coefficient of $r = 0.30$ about zero lag. Anomalies of SST along the northern coast of the Gulf of Guinea (GN) lag the zonal wind stress fluctuations of region EQW by one month with $r = 0.29$. These correlation coefficients increase when the wind stress averaging area is reduced and centered further west. Correlations with wind stress anomalies within the Gulf of Guinea are, as in Servain et al. (1982), considerably less. The maximum correlation between SST anomalies in region EQ and zonal equatorial wind stress anomalies east of $20°$W (EQE) is -0.17 at zero lag and the correlation with the meridional wind stress anomalies within EQE is 0.04 at zero lag. The seasonal dependencies of the EQ-SST and EQW-τ^x monthly anomalies suggest that the positive correlations with the western equatorial easterlies are mainly associated with anomalous conditions during the upwelling season. The time series of σ_{im} (Figs. 6 and 12) indicate that the largest wind stress anomalies within EQW precede by one month the largest SST anomalies within EQ that occur in July.

It was suggested in Servain et al. (1982) that, based on their findings and following Moore et al. (1978), impulsive wind stress fluctuations in the western equatorial Atlantic would excite equatorially trapped Kelvin waves that in turn would induce a remotely forced response in the Gulf of Guinea. Busalacchi and Picaut (1983) used a single baroclinic-mode linear model to study the dynamic response of the tropical Atlantic to periodic climatological forcing. The results of their study indicated that the annual wind-driven response within an idealized Gulf of Guinea was due to equatorial wind stress fluctuations remote from the gulf. The 16 years of wind data presented here have recently been used to force the same model mentioned above. For the purposes of the present discussion we will restrict our attention to the model results pertaining to the influence of the zonal wind stress anomalies along the equator.

ZONAL WIND STRESS

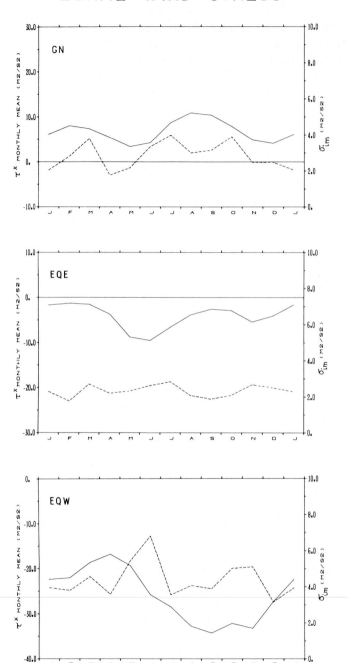

MERIDIONAL WIND STRESS

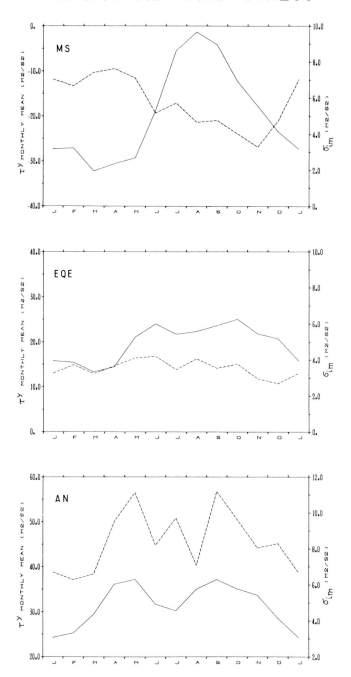

Fig. 12. Monthly mean seasonal cycle of wind stress components (solid) and the corresponding standard deviations, σ_{im} (dashed), of the 16 monthly anomalies from each monthly mean for the regions of Fig. 11.

ZONAL WIND STRESS

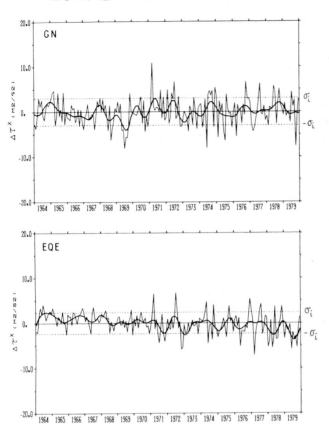

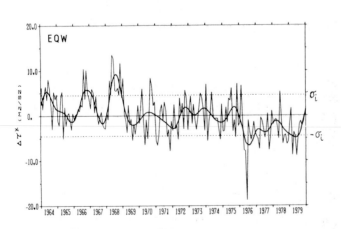

MERIDIONAL WIND STRESS

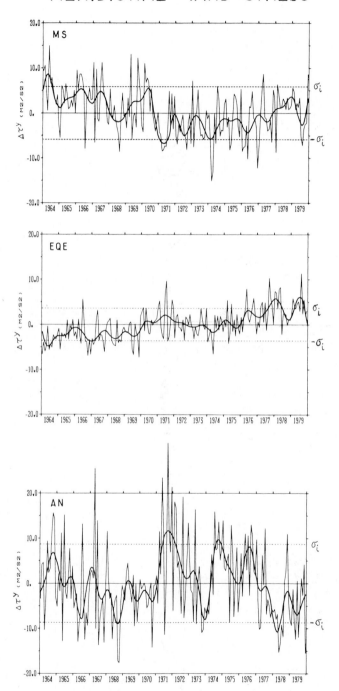

Fig. 13. Monthly wind stress component anomalies (thin line) from the mean seasonal cycle for each region of Fig. 11, 1964–1979. Low-pass (> 12 months) filtered time series (thick line) of these anomalies. The dashed line indicates the standard deviation, σ_i, of the monthly anomalies.

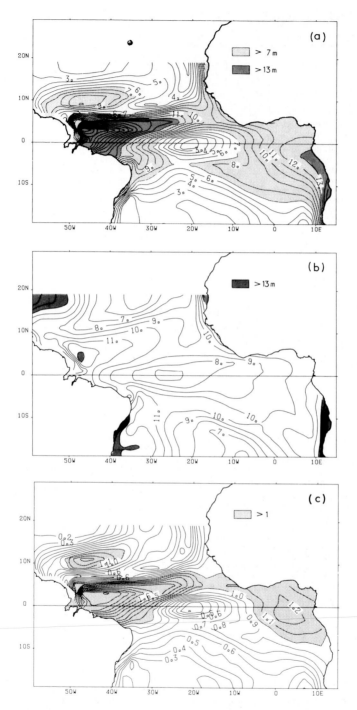

Fig. 14(a). Standard deviation, σ_s, of the mean seasonal cycle of the model pycnocline depth; (b) standard deviation, σ_i, of the anomalies from the mean seasonal cycle of the model pycnocline depth; and (c) ratio σ_s/σ_i.

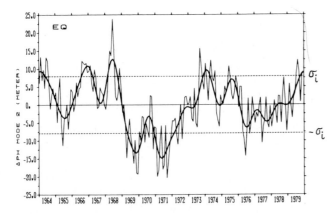

Fig. 15. Monthly model pycnocline anomalies from the mean seasonal cycle, 1964–1979, for the equatorial region of Fig. 5.

Standard deviations of seasonal and interannual changes in depth of the model pycnocline (Fig. 14) are presented in the same manner as previous plots of SST and wind stress. The distribution of the standard deviation of the seasonal variability is similar to the distribution of the amplitude of the annual variability of model pycnocline depth presented in Busalacchi and Picaut (1983). Within the Gulf of Guinea the standard deviation of the monthly anomalies in pycnocline depth is on the same order as the standard deviation of the seasonal cycle (Fig. 14c). A time series of monthly anomalies of pycnocline depth (Fig. 15) is presented for the same equatorial area as in Fig. 5. In response to the monthly anomalies of wind stress the largest and most prolonged anomalies of the model pycnocline depth are the deep period of mid-1966 through 1968 followed by the shallow period of 1969 through 1972. Cross correlation of this time series with the zonal wind stress anomalies to the west (EQW, Fig. 13) is maximum at zero lag with $r = 0.55$. The interannual pycnocline variability at the equatorial eastern boundary lags the zonal wind stress of EQW by one month with a correlation coefficient of 0.59. The maximum correlation between the anomalies of the pycnocline variability at the equatorial eastern boundary and anomalies of the zonal equatorial wind stress east of 20°W (EQE) is 0.15 with the wind stress leading by two months. Hence this simple modelling example suggests that the dynamic response of the Gulf of Guinea to monthly anomalies of the wind stress may be significant when compared to the seasonal cycle, and like the response to seasonal forcing may be related to the wind stress variability west of the gulf.

SUMMARY AND CONCLUSIONS

The availability of carefully established fields of SST and wind stress encompassing a period of 16 years has permitted the calculation of a few simple measures of the seasonal and interannual variability for the tropical Atlantic. Standard deviations of the mean seasonal cycle have been compared with the standard deviation of the monthly anomalies. Although the amplitude of specific monthly anomalies of SST and wind stress are greater than the amplitude of the seasonal cycle, in terms of a periodic signal the amplitude of the annual period remains dominant over the amplitude of lower frequencies.

Year-to-year changes in SST are largest in regions where the seasonal SST signal is large. The regions of largest seasonal SST changes are located where the mean SST is low, such as zones of coastal and equatorial upwelling. Throughout a large percentage of the basin the amplitude of the seasonal cycle of SST is greater than the amplitude of monthly SST anomalies. However, time series of monthly anomalies of SST indicate several events in which SST differs from the monthly mean by $1°-2°C$ for periods of several months to more than one year. The monthly anomalies for the coastal upwelling zones of Mauritania–Senegal and Angola are largest during the same months of the year but different phases of the seasonal cycle. Maximum anomalies at Mauritania–Senegal occur when SST is low while maximum anomalies along Angola occur when SST is high. The offshore extent of significantly correlated SST anomalies off Mauritania–Senegal tends to be larger than that of the corresponding upwelling region along Angola. The largest SST anomalies along the equator near $10°W$ occur at the beginning of the cold season. The spatial structure of these anomalies is similar to those along the zonally oriented coast of the Gulf of Guinea.

Locations of significant interannual fluctuations of wind stress, unlike SST, do not coincide with the locations of the largest seasonal cycles. The standard deviation of monthly wind stress anomalies is greater than the standard deviation of the seasonal cycle throughout the basin except in the vicinity of the ITCZ where the seasonal signal is the largest. The largest monthly anomalies of the meridional wind stress along Mauritania–Senegal and Angola occur when the seasonal mean wind stress is the strongest. The largest perturbations to the seasonal cycle of the zonal wind stress in the western equatorial Atlantic occur in June, which is the mid-point of the seasonal intensification. Each of these areas is characterized by anomalous events that can persist for more than one year. Anomalies of such duration cannot be the result of a simple phase shift or change in amplitude of the seasonal cycle. Of particular importance are the anomalies of the western equatorial Atlantic that may be capable of inducing a dynamic response in the Gulf of Guinea greater than or equal to the response to the mean seasonal forcing.

This presentation of 16 years of gridded SST and wind stress has indicated several aspects of the temporal and spatial structure of the tropical Atlantic worthy of further study. A scalar EOF analysis of SST and a vector EOF analysis of the wind stress are underway. Questions pertaining to the difference in the ratios of seasonal to interannual variability of SST and wind stress, the different relationships between local winds and SST for anomalous and seasonal fluctuations within upwelling areas, the relationship between SST anomalies of different parts of the basin, and the causal nature of specific SST events need to be addressed by studies with coupled atmosphere–ocean models.

ACKNOWLEDGEMENTS

This work was made possible by the support of CNEXO under Contract 84–3149, CNRS (ERA-766 and ASP-040175), NOAA cooperative agreement NA80RAH00002 to the Joint Institute for Marine and Atmospheric Research of the University of Hawaii, and NASA RTOP-161-20-31. The assistance rendered by Marc Seva, Vincent Verbeque, Sharon Lukas, Pascal Lecomte, and Jerry Herwehe is gratefully acknowledged.

REFERENCES

Berrit, G.R., 1976. Les eaux froides côtieres du Gabon à l'Angola sont-elles dues à un upwelling d'Ekman? Cah. ORSTOM, Sér. Oceanogr., 14: 273–278.

Bunker, A.F., 1976. Computations of surface energy flux and annual air–sea interaction cycles of the North Atlantic Ocean. Mon. Weather Rev., 104: 1122–1140.

Busalacchi, A.J. and Picaut, J., 1983. Seasonal variability from a model of the tropical Atlantic Ocean. J. Phys. Oceanogr., 13: 1564–1588.

Cressman, G.P., 1959. An operational objective analysis system. Mon. Weather Rev., 87: 367–374.

Davis, R.E., 1976. Predictability of sea surface temperature and sea level pressure anomalies over the North Pacific Ocean. J. Phys. Oceanogr., 6: 249–266.

Goldenberg, S.B. and O'Brien, J.J., 1981. Time and space variability of tropical Pacific wind stress. Mon. Weather Rev., 109: 1190–1207.

Hastenrath, S., 1976. Variations in low-latitude circulation and extreme climate events in the tropical Americas. J. Atmos. Sci., 33: 202–215.

Hellerman, S., 1980. Charts of the variability of the wind stress over the tropical Atlantic. GATE Sup. II. Deep-Sea Res., 26: 63–75.

Hellerman, S. and Rosenstein, M., 1983. Normal monthly wind stress over the world ocean with error estimates. J. Phys. Oceanogr., 13: 1093–1104.

Hisard, P., 1980. Observation de réponse de type "El Niño" dans l'Atlantique tropical oriental Golfe de Guinée. Oceanol. Acta, 3: 69–78.

Horel, J.D. and Wallace, J.M., 1981. Planetary scale atmospheric phenomena associated with the Southern Oscillation. Mon. Weather Rev., 109: 813–829.

Lamb, P.J., 1978. Case studies of tropical Atlantic surface circulation pattern during recent sub-Sahara weather anomalies, 1967–1968. Mon. Weather Rev., 106: 282–291.

Legler, D.M., 1983. Empirical orthogonal function analysis of wind vectors over the tropical Pacific. Bull. Am. Meteorol. Soc., 64: 234–241.

Markham, C.G. and McLain, D.R., 1977. Sea surface temperature related to rain in Ceara, northeastern Brazil. Nature, 265: 320–323.

Merle, J., 1980a. Variabilité thermique annuelle et interannuelle de l'océan Atlantique équatorial Est. L'hypothèse d'un "El Niño" Atlantique. Oceanol. Acta, 3: 209–220.

Merle, J., 1980b. Seasonal heat budget in the equatorial Atlantic. J. Phys. Oceanogr., 10: 464–469.

Merle, J. and Le Floch, J.F., 1978. Cycle annuel moyen de la température dans les couches supérieures de l'océan Atlantique intertropical. Oceanol. Acta, 1: 271–276.

Merle, J., Fieux, M. and Hisard, P., 1980. Annual signal and interannual anomalies of sea surface temperature in the eastern equatorial Atlantic. Gate Sup. II, Deep-Sea Res., 26: 77–101.

Moore, D.W., Hisard, P., McCreary, J.P., Merle, J., O'Brien, J.J., Picaut, J., Verstraete, J.M. and Wunsch, C., 1978. Equatorial adjustment in the eastern Atlantic. Geophys. Res. Lett., 5: 637–640.

Moura, A.D. and Shukla, J., 1981. On the dynamics of droughts in Northeast Brazil: Observations, theory and numerical experiments with a general circulation model. J. Atmos. Sci., 38: 2653–2675.

Philander, S.G.H., 1979. Variability of the tropical oceans. Dyn. Atmos. Oceans, 3: 191–208.

Servain, J., 1984. Réponse océanique à des actions éloignées du vent dans le Golfe de Guinée en 1967–1968. Oceanol. Acta, 7: 297–307.

Servain, J., Picaut, J. and Merle, J., 1982. Evidence of remote forcing in the equatorial Atlantic Ocean. J. Phys. Oceanogr., 12: 457–463.

Speth, P. and Kohne, A., 1983. The relationship between sea surface temperatures and winds off northwest Africa and Portugal. Oceanogr. Trop., 18: 69–80.

Weare, B.C., 1977. Empirical orthogonal analysis of Atlantic Ocean surface temperature. Q. J. R. Meteorol. Soc., 103: 467–478.

Wooster, W.S., Bakun, A. and McLain, D.R., 1976. The seasonal upwelling cycle along the eastern boundary of the North Atlantic. J. Mar. Res., 34: 131–141.

Wyrtki, K. and Meyers, G., 1976. The trade wind field over the Pacific Ocean. J. Appl. Meteorol., 15: 698–704.

CHAPTER 17

AN ATMOSPHERE–OCEAN COUPLED MODEL FOR LONG-RANGE NUMERICAL FORECASTS

GUO YU-FU, WANG XIAO-XI, CHEN YING-YI and CHAO JIH-PING

THE PRINCIPLE OF THE MODEL

An atmosphere–ocean coupled model for long-range forecasts (LRF) with the time-range from a month to a season has been developed (Chao et al., 1982). There are two basic ideas in this model which are different from that of the usual general circulation model (GCM). The first one is that, since we are only interested in the temporal evolution of the anomalous component rather than the climatological one, the climatological component can thus be removed from the total field equations by dividing all the system variables into their climatic and anomalous components and only the time-dependent anomalous system is left. However, the observational climate values are still utilized. This

TABLE 1

Correlation coefficients of the anomalous fields between the prediction and observation over the Northern Hemisphere

Cases		T_s'			ϕ'		
		prediction		persistence	prediction		persistence
		A^a	B^b		A^a	B^b	
Jan.–Feb.	1976	0.40	0.07	−0.01	0.11	−0.16	−0.13
	1977	0.36	0.18	0.25	0.42	0.05	0.34
	1978	0.50	0.29	0.58	0.48	0.22	0.30
April–May	1976	0.27	0.20	0.26	0.44	0.30	0.25
	1977	0.40	0.33	0.38	0.33	0.21	0.07
	1978	0.40	0.25	0.32	0.05	0.05	0.19
July–Aug.	1976	0.15	0.15	0.27	0.33	0.29	0.47
	1977	0.47	0.26	0.46	0.43	0.17	0.42
	1978	0.43	0.28	0.40	0.20	0.16	0.28
Oct.–Nov.	1976	0.57	0.32	0.50	0.55	0.16	0.21
	1977	−0.09	−0.01	−0.05	−0.18	−0.29	−0.07
	1978	0.35	−0.01	0.30	0.34	−0.06	0.32
Average		0.35	0.19	0.31	0.29	0.09	0.22

[a] Two time-steps.
[b] One time-step.

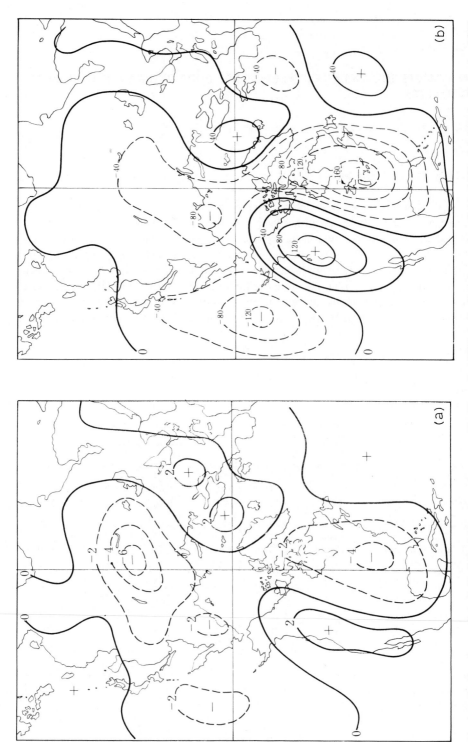

Fig. 1. a. Predicted anomalous field of Earth's surface temperature for November 1976. b. Predicted anomalous field of 500 mb height for November 1976.

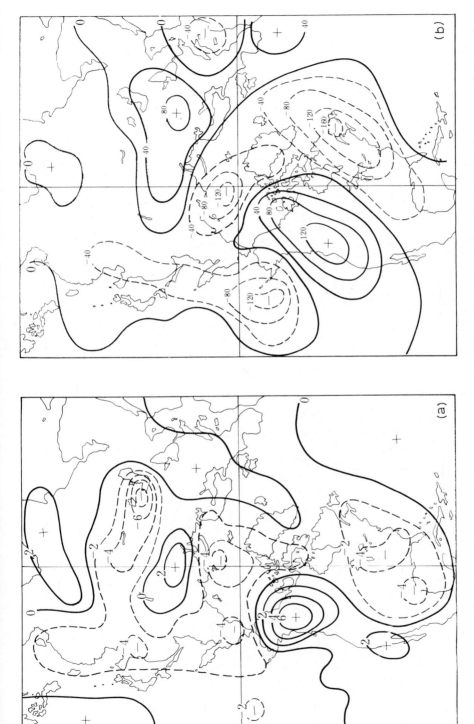

Fig. 2. a. Observed anomalous field of Earth's surface temperature for November 1976. b. Observed anomalous field of 500 mb height for November 1976.

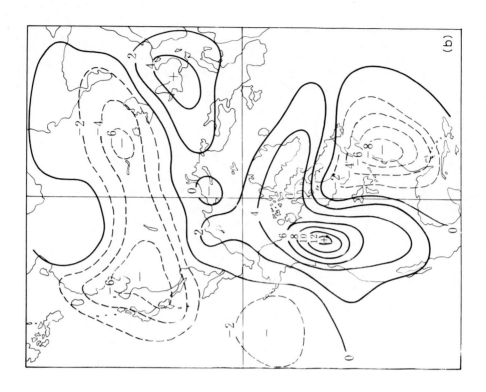

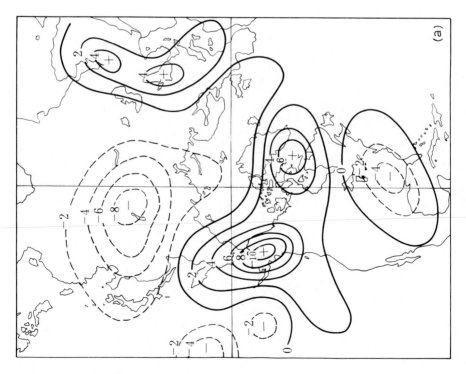

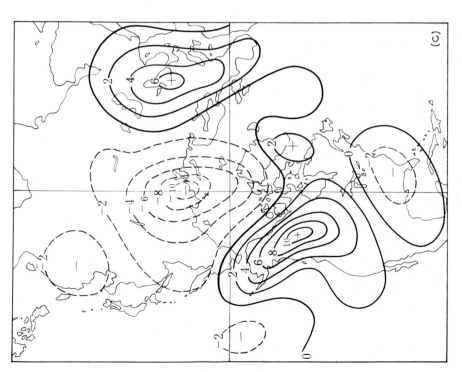

243

Fig. 3. a. Predicted anomalous field of Earth's surface temperature for February 1977 (three-level). b. Predicted anomalous field of Earth's surface temperature for February 1977 (one-level). c. Observed anomalous field of Earth's surface temperature for February 1977.

244

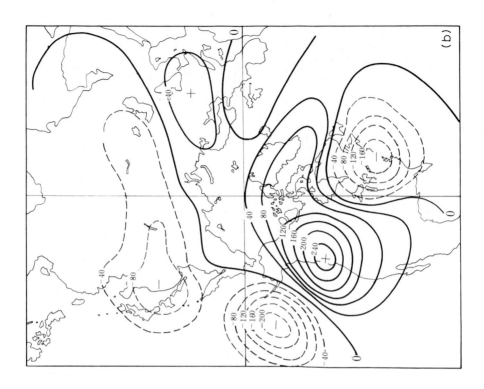

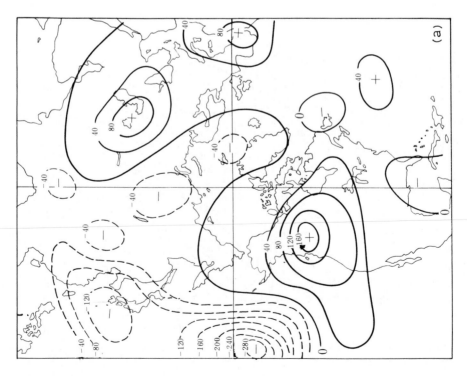

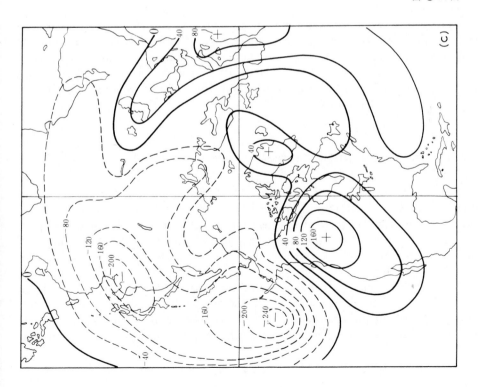

Fig. 4. a. Predicted anomalous field of 500 mb height for February 1977 (three-level). b. Predicted anomalous field of 500 mb height for February 1977 (one-level). c. Observed anomalous field of 500 mb height for February 1977.

(c)

procedure is the same one used by Reynolds to remove the turbulent component from the time mean flow. Unlike Reynolds, we are only interested in the time evolution of the anomaly rather than the mean or climatological flow which is supposedly well known. Since the climatic components are assumed to be time independent for short period, say 1–3 months, this equation system can be ignored leaving only the time-dependent anomalous system. Usually, the equation for the anomaly includes the Reynolds term, but these terms are not considered in the present model. The current model uses the geostrophic approximation. The system consists of two major equations. One is the non-adiabatic vorticity equation for atmosphere, and the other is the thermal equation for the underlying ocean and land. The two equations are connected at the Earth's surface through the interface conditions which are the heat balance and where the vertical velocity is zero.

The second important idea of this model is as follows. It can be shown by analysing the linear case of this atmosphere–ocean system that there are two basic types of dynamical processes corresponding to different time scales. The fast one, transient Rossby wave, has a period of the order of one week. The other is slow with a period of the order of several months. The slow one is produced by the heating of the ocean, by the interaction of atmosphere and ocean. Since the growth rate of the shorter time-scale waves is about one order of magnitude larger than that of the long time-scale fluctuation, this leads to one of the difficulties for the long-range numerical forecasts, because the evolution of the long-range process of smaller amplitude will be distorted by the short-range process with large amplitude. One way to overcome this difficulty is to filter out these high-frequency dynamic events from the numerical model of long-range or short-term climate forecasting. According to this idea, the transient Rossby wave can be filtered by omitting the partial derivative with respect to time in the atmospheric vorticity equation. However, the time derivative term is still kept in the thermal equation for the underlying ocean and land. With this assumption, the vorticity equation becomes time-independent, i.e. it is only a balance relationship between the anomaly geopotential height field and the Earth's surface heating field. This model may therefore be called a filtered anomaly model (FAM).

EXAMPLES OF A ONE-MONTH PREDICTION

Using this method, the predicted results of 12 cases were described in the paper of Miyakoda and Chao (1982) using the correlation coefficients between the prediction and observation over the Northern Hemisphere. These results are also described in Column B in Table 1. The predicted skill is lower than that of persistence. All these examples of one month prediction are obtained by a one-level model for the atmosphere, with a time-step of one month.

Recently, this model has been improved. The time-step has been reduced to half a month. The 12 examples are re-calculated and the results are shown in column A in Table 1. The results for anomalous Earth's surface temperature T_s' and anomalous 500-mb geopotential height are much better than the early ones and the average correlation coefficients for predicted T_s' and ϕ' are 0.35 and 0.29 which are slightly higher than that of the persistence 0.31 and 0.22, respectively.

In order to test the ability to predict the blocking events, another case is considered. A spectacular blocking event over the west coast of America lasted for four months, from November 1976 to February 1977. The best developing month is January 1977 which had been predicted by a GCM of GFDL as well as by the FAM (Chao and Caverly, 1982; Miyakoda and Chao, 1982). An interesting and important point is that the GCM and the FAM are quite different, yet the predictions of both models are similar to each other in the map of monthly anomaly 500-mb geopotential height. Figures 1a, b and 2a, b are the predicted charts as well as the corresponding observations in November 1976. The comparison shows that in general the predictions are in agreement with the observations in the blocking region. The correlation coefficients of the predicted anomaly fields for T_s' and ϕ' are 0.57 and 0.55, respectively, and those of persistance are 0.50 and 0.21, respectively.

Another way to improve the predictive skill of the model is to increase the number of layers of the model atmosphere. A three-level filtered anomaly model has been developed, and predictions of monthly averaged anomalous fields of Earth's surface temperature and the geopotential height at 300, 500, and 700 mb, respectively, have been carried out. The preliminary experiments with the three-level model show that it could improve the forecast results of the one-level model and has the potential even for making seasonal forecasts. Figures 3 and 4 show the predicted results of February 1977 for both of these models. The correlation coefficients for predicted T_s' and ϕ' by the three-level model are 0.77 and 0.51, respectively, which are much better than those by the one-level model (0.36 and 0.42, respectively).

CONCLUSION

According to the results obtained above, we believe that the FAM has the potential ability for long-range forecasts. The FAM also requires much less computer time than a GCM. This model is only suitable for middle latitudes, because the geostrophic approximation has been used. The dynamical effects of ocean currents also become important in the tropics. Another disadvantage of this model is that the solution depends to some extent on the value of physical parameters and in general, the predicted intensity of the anomaly centres is weaker than the observed one. These problems can be solved in the future.

REFERENCES

Chao, J.P., Guo, Y.F. and Xin, R.N., 1982. A theory and method of long-range numerical weather forecasts. J. Meteorol. Soc. Jpn., 60: 282–291.
Chao, J.P. and Caverly, R., 1982. Anomaly model and its application to long-range forecasts. Proceedings of the Sixth Annual Climate Diagnostic Workshop, Lamont-Doherty Geological Observatory, Columbia University, Palisades, N.Y., October 14–16, 1981, pp. 316–319.
Miyakoda, K. and Chao., J.P., 1982. Essay on dynamical long-range forecasts of atmospheric circulation. J. Meteorol. Soc. Jpn., 60: 292–308.

CHAPTER 18

ON THE SPECIFICATION OF SURFACE FLUXES IN COUPLED
ATMOSPHERE–OCEAN GENERAL CIRCULATION MODELS

J.F.B. MITCHELL, C.A. WILSON and C. PRICE

ABSTRACT

The turbulent surface fluxes of heat and momentum from a high-resolution atmospheric model are presented and assessed. The errors in computing the fluxes from monthly mean atmospheric model data are calculated, and the consequences for coupled ocean models are discussed.

1. INTRODUCTION

The coupling between the ocean and the atmosphere is effected entirely by the fluxes of momentum and heat (latent and sensible) at the air–sea interface. The atmospheric circulation is dependent on the temperature, and to a lesser extent, the roughness of the ocean surface. The ocean-surface temperature in turn depends on the ocean circulation which is forced by the transfer of momentum and heat from the atmosphere. Errors in coupled ocean–atmosphere simulations may arise from errors in the atmospheric model, the oceanic model, or both. In this paper, we assess a high-resolution atmospheric model by comparing the simulated stress and turbulent heat flux with climatological estimates, and discuss the likely implications for coupling to an ocean model.

The fluxes of heat and momentum from both the model and climatological data are calculated using the bulk aerodynamic formulae, involving the product of a drag coefficient, surface wind and vertical gradient. In the model, the fluxes of heat and momentum from the atmosphere to the ocean are calculated timestep by timestep, and so fully take into account the correlations in time between drag coefficients, windspeed and vertical gradients. In regions where such correlations are large, the surface fluxes of heat and momentum will differ from those obtained from drag coefficients, wind speed and vertical gradients derived from time-averaged atmospheric variables. Hence an ocean driven by atmospheric fluxes accumulated timestep by timestep will be forced differently than one driven by fluxes derived from time-averaged data.

This is analogous to the differences obtained in climatological estimates of surface fluxes derived from daily rather than monthly mean atmospheric variables (Kraus and Morrison, 1966; Esbensen and Reynolds, 1981). Here we calculate the errors in model fluxes derived from monthly averaged atmospheric model data, and compare them with the errors found in parallel studies made with observational data.

A similar error may occur in bringing coupled ocean–atmosphere models to equilibrium. The thermal relaxation time of the ocean is several orders of magnitude larger than that of the atmosphere. On the other hand, oceanic general circulation models currently used in climate studies are generally computationally less expensive than atmospheric models. In order to bring a coupled ocean–atmosphere model to equilibrium, it has been the practice to run the atmospheric model for a short period (typically one month) with fixed ocean temperatures and then use the mean data to force the ocean model over a much longer period of time (e.g., Manabe et al., 1979a; Washington et al., 1980), a process often referred to as "asynchronous coupling". During the asynchronous period, the ocean is driven by the fluxes of momentum and latent heat accumulated timestep by timestep during the synchronous period. The sensible heat flux is derived from the mean low-level winds, temperatures and humidities to ensure convergence to equilibrium, and so short-term correlations between atmospheric variables are ignored.

2. THE MODEL

The atmospheric model is global with 11 layers in the vertical; a limited area version was used with data from the GARP Atlantic tropical experiment (GATE; Lyne et al., 1976). It is a primitive equation model using σ (pressure/surface pressure) as a vertical coordinate, and a regular $2.5° \times 3.75°$ latitude–longitude grid. The seasonal and diurnal variations of solar radiation are represented, and the radiative fluxes are a function of temperature, water vapour, carbon dioxide and ozone concentrations, and prescribed zonally averaged cloudiness. Sea-surface temperatures and sea-ice extents are prescribed from climatology, and updated every five days.

The surface exchanges of momentum, heat and moisture are determined by:

$$F_\chi = -C_\chi V(Z_1) \Delta\chi(Z_1) \tag{1}$$

where F_χ is the mean vertical flux of χ; C_χ is the bulk transfer coefficient at height Z; $V(Z_1)$ is the mean wind speed at a specified height Z_1 above the surface but within the atmospheric boundary layer; and $\Delta\chi(Z_1) = \chi(Z_1) - \chi_0$ is the difference between the value of χ at Z_1 and its surface value χ_0.

The bulk transfer coefficients are chosen using Method I of Clarke (1970), which is based on the Monin-Obukhov similarity hypothesis for the fully turbulent boundary layer. In practice, the value of C_χ over the ocean is tabulated against the bulk Richardson number R_{iB} assuming a roughness length of 10^{-4} m (Fig. 1).

$$R_{iB} = \frac{gZ_1 [\Delta\theta(Z_1) + 0.61 T\Delta q(Z_1)]}{T V^2(Z_1)} \tag{2}$$

where g is the acceleration due to gravity; Z_1 is the height of the centre of the lowest model layer; and T is a representative temperature for the bottom layer. θ, q and V are respectively the potential temperature, specific humidity and wind speed in the lowest model layer. Note that this takes into account the effect of stability, as indicated by the sharp increase in C_χ as one moves from stable ($R_{iB} > 0$) to unstable ($R_{iB} < 0$) regions in Fig. 1. No allowance is made for an increase in surface roughness with windspeed when calculating the bulk transfer coefficient for momentum. In deriving quantities from monthly mean data, the monthly mean windspeed (average of values diagnosed each time-step) is used for $V(Z_1)$ in eqn. (1), and the monthly mean values of temperature, humidity and surface velocity are used in calculating C_χ and $\Delta\theta(Z_1)$.

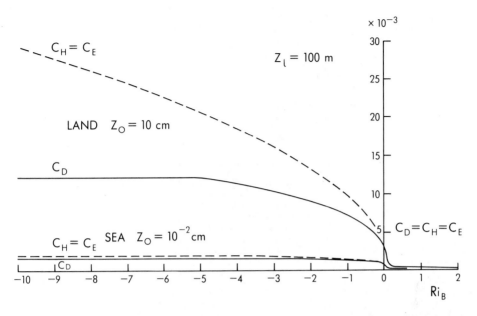

Fig. 1. Surface-layer bulk transfer coefficients used in the Meteorological Office's 5-layer model. R_{iB} is the bulk Richardson number (see text) and C_D, C_H, C_E are the bulk transfer coefficients for momentum, sensible and latent heat, respectively.

3. RESULTS AND DISCUSSION

3.1. Diagnosed fluxes

3.1.1. Momentum

The model's diagnosed monthly mean wind stress field (Fig. 2), averaged over three years, is qualitatively similar to that derived from climatological data (e.g., Han and Lee, 1981; Hellerman and Rosenstein, 1983). In January (Fig. 2a), the maximum magnitude of westerly stress (about 2.5 dynes cm^{-2}) in the north of the major ocean basins is similar to that found in the climatological estimates. However, the easterly stress in the northern tropics exceeds 0.5 dynes cm^{-2} in only a few places whereas the climatologies suggest that it should exceed 1.0 dynes cm^{-2} over much of the tropical Pacific and Atlantic. The easterly stress in the southern hemisphere tropics also appears to be substantially underestimated. A belt of maximum westerly stress is found near 45°S in accordance with observations, though weaker. In July (Fig. 2b) the model field is qualitatively similar to the climatological estimates, but generally weaker. The strength of the westerly stress around Antarctica is much closer to the observed data than in January, but there is a considerable underestimation of the easterly stress over the subtropical Atlantic and Pacific oceans, and also over the northeastern Indian Ocean associated with the monsoon.

The underestimation of the strength of the surface stress in an atmospheric model which is to be coupled to a dynamical model of the ocean is a serious shortcoming. For example Bryan et al. (1975), and Washington et al. (1980) attribute the anomalous

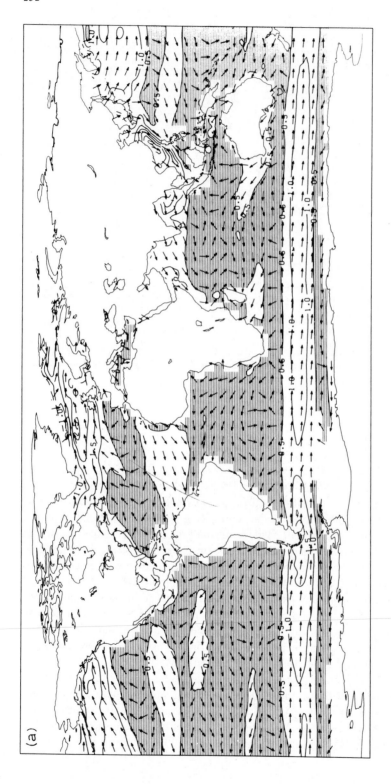

(a)

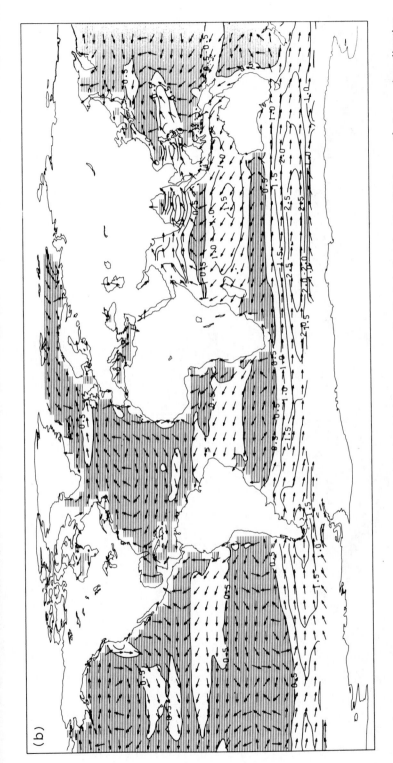

Fig. 2. Model surface stress (three-year mean). Contours every 0.5 dyne cm^{-2} (0.05 N m^{-2}). Shaded where less than 0.5 dyne cm^{-2}. Arrows show direction only. (a) January; (b) July.

254

(a)

255

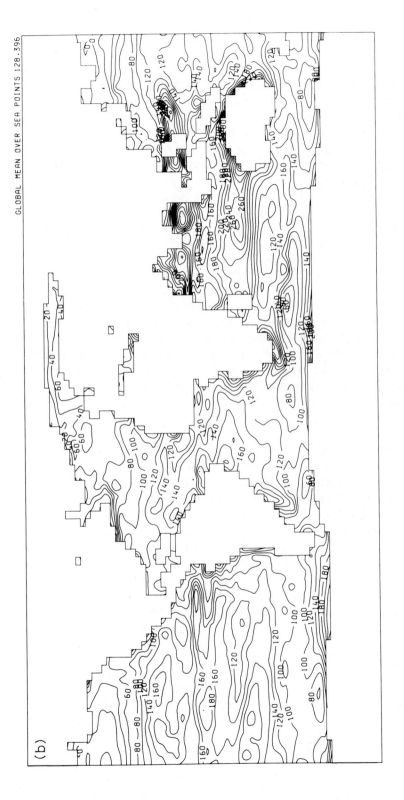

Fig. 3. Model turbulent (sensible + latent) heat flux (three-year mean). Contours every $20\,\mathrm{W\,m^{-2}}$. (a) December, January and February; (b) June, July and August.

oceanic warming and the consequent underestimate of sea-ice cover round Antarctica in their coupled ocean—atmosphere models in part to the weakness of the westerly circumpolar flow, and the associated equatorward Ekman surface drift. Similarly, a coupled model will fail to produce sufficient equatorial upwelling, and the pronounced east—west surface-temperature contrast which characterizes the equatorial Pacific if the atmospheric model does not produce sufficient easterly stress. In higher northern latitudes, a weakening of the surface stress will contribute to a weakening of the western boundary currents.

One reason for the underestimate of the surface stress by atmospheric models may be lack of horizontal resolution. In general, the surface flow becomes stronger as resolution is increased, though in the Northern Hemisphere in winter, the mid-latitude depression belt becomes excessively deep (see for example Manabe et al., 1979b). Similar trends are found in the present model. If the horizontal diffusivity, included to remove computational instability, is also reduced as the resolution is enhanced, the energy in the transient flow is increased, and is in closer agreement with observations.

The Meteorological Office model produces stronger than observed westerly flow in northern mid-latitudes in January, yet apparently still simulates the magnitude of the surface stress correctly (Fig. 2a). The drag coefficient fixed for momentum does not include a dependence on wind strength [except in the wrong sense through the bulk Richardson number in eqn. (2)]. There is a growing body of evidence that the momentum drag coefficient should increase with wind speed, to allow for the increase in surface roughness [Wu (1982) has attempted to produce a simple empirical relationship between wind strength and the momentum drag coefficient]. Increasing the drag coefficient for high wind speeds would undoubtedly increase the surface stress in the model. However, the atmospheric flow will be decelerated as a result, reducing the net increase.

3.1.2. Turbulent heat fluxes

The turbulent heat fluxes (sensible and latent heat, Fig. 3) are generally similar in both magnitude and geographic distribution to available climatological estimates (Budyko, 1963; Bunker, 1976; Esbensen and Kushnir, 1981). Peak values of over 200 W m^{-2} occur in winter along the sea-ice margin, and, in the Northern Hemisphere, off the eastern seaboard of continents. Large values also occur in the tropics, though there is a minimum along the equator, particularly in the East Pacific, due to the local surface-temperature minimum associated with oceanic upwelling. In the summer hemisphere, the fluxes in middle and high latitudes are small. Note the large cooling of the ocean off India associated with the summer monsoon. Further discussion of the simulation of sea-surface temperatures using the model's turbulent heat flux is given by Gordon and Bottomley (1985).

3.2. Errors in sampling due to the use of monthly mean data

The surface fluxes of momentum and the turbulent (sensible plus latent) heat fluxes were: (a) accumulated timestep by timestep (*diagnosed*) as presented in Section 3.1; and (b) calculated in retrospect from monthly time means of low-level temperature, wind and humidity using the model's boundary layer algorithm (*derived*). The difference (derived— diagnosed) indicates the likely error introduced into climatological estimates by using monthly mean data, or in the surface forcing of an ocean model coupled asynchronously to an atmospheric model.

3.2.1. Surface stress

A comparison of Fig. 2 (diagnosed stress) and Fig. 4 (derived–diagnosed stress, with the contour interval reduced by a factor of ten) indicates that *magnitudes* of diagnosed and derived stresses differ by about 10%, though locally the differences exceed 20%. In general, the difference in stresses is in the opposite *direction* to the diagnosed fields, indicating a consistent underestimation of the stress due to using monthly mean data. This is in agreement with the findings of Esbensen and Reynolds (1981) using observational data, who found that the monthly mean stress at various weather ships was larger when estimated from data when divided into 16 direction categories, each with a mean wind speed, as opposed to a single monthly mean wind speed and direction. (Note that here, the "derived" surface stress is calculated from:

$$\tau = \rho C_D |V| V$$

where ρ is the density; C_D is the bulk transfer coefficient for momentum; and $|V|$ is wind speed averaged over each model timestep, but V is the monthly mean vector wind, whereas Esbensen and Reynolds appear to have used:

$$\tau = \rho C_D |V|^2 \hat{V}$$

where $\hat{V}$ is the unit vector in the direction of the mean vector wind. Since $|V| \geqslant |V|$, our "derived" wind stress will tend to be smaller than using their "direction only" wind rose method.)

3.2.2. Turbulent fluxes

Differences (Figs. 5a, b, 6a, b) are generally less than $10\,\mathrm{W\,m^{-2}}$, though they exceed $20\,\mathrm{W\,m^{-2}}$ locally. There are few points where the difference exceeds 10%, in agreement with Esbensen and Reynolds findings using observational data. The sign of the difference varies, and so cannot be minimised by a uniform fractional change in the magnitude of the fluxes. Note the general underestimation of the surface cooling in the vicinity of the sea-ice margins. This is probably due to the correlation between large (unstable) drag coefficients and large air–sea temperature differences which occur when equatorward winds advect cold air from sea-ice over the warm ocean. The surface cooling may be underestimated in a similar fashion in an ocean model forced by prescribed low-level atmospheric data (wind, temperature and humidity), leading to an underestimation of the extent of sea-ice and intensity of meridional circulation.

There is also a consistent over-estimation of the heat flux in lower latitudes ($30°$N and $30°$S in December, January, and February; $20°$N and $35°$S in June, July, and August).

4. CONCLUDING REMARKS

The UK Meteorological Office's 11-layer model appears to underestimate the surface stress over the ocean. This may be due to insufficient horizontal resolution and a lack of a wind-speed dependence in the calculation of the momentum drag coefficient. The simulated turbulent heat fluxes are broadly similar to climatological estimates. As found

258

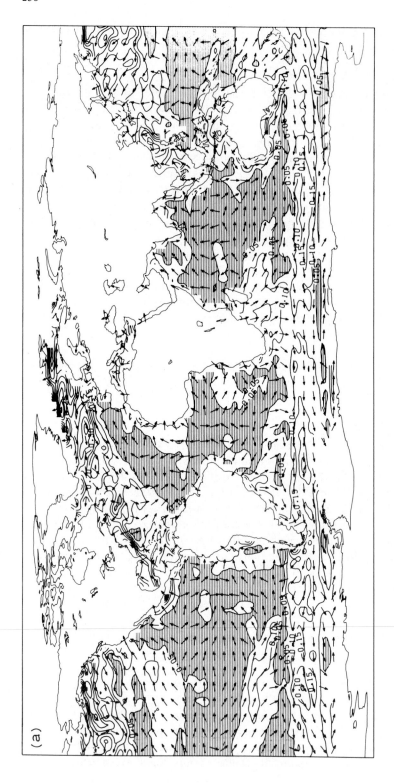

(a)

259

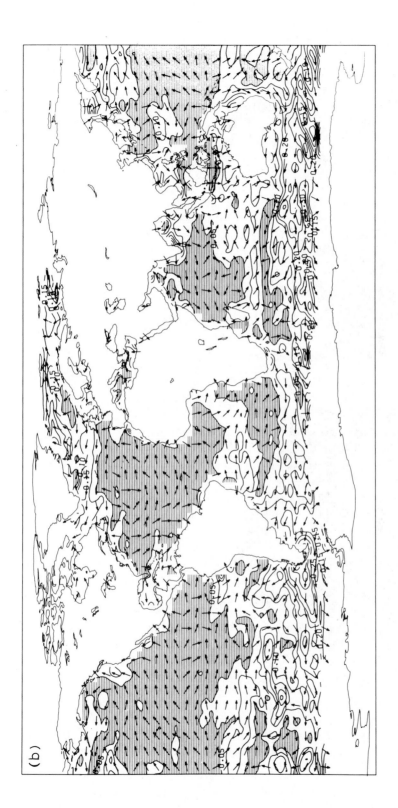

Fig. 4. Error in model surface stress due to using monthly mean data (three-year mean). Contours every 0.05 dyne cm^{-2}. Shaded where less than 0.05 dyne cm^{-2}. Arrows show directions only. (a) January: (b) July.

(a)

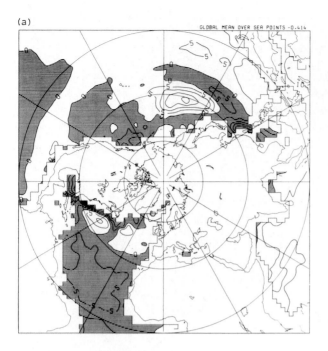

(b)

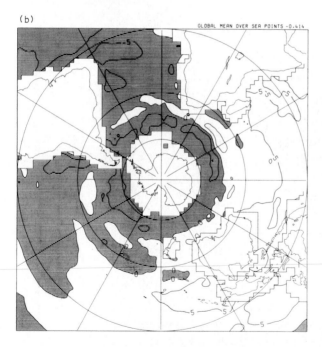

Fig. 5. Error (derived–diagnosed) in model turbulent heat flux due to using mean data (three-year mean). Contours every 5 W m^{-2}; shaded where underestimated. December, January and February. (a) Northern Hemisphere; (b) Southern Hemisphere.

(a)

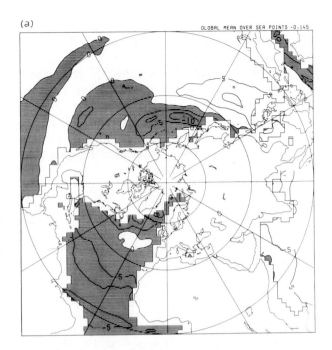

(b)

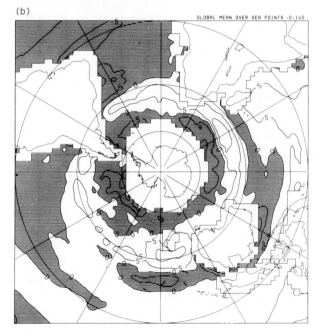

Fig. 6. Error (derived–diagnosed) in model turbulent heat flux due to using mean data (three-year mean). Contours every 5 W m^{-2}; shaded where underestimated. June, July and August. (a) Northern Hemisphere; (b) Southern Hemisphere.

in observational studies, the use of monthly mean data can lead to an underestimation of the surface stress (by up to 20%), and a geographically dependent bias in the surface fluxes (up to 5 or $10\,W\,m^{-2}$). If more accuracy is required for coupled simulations, the wind stress should be accumulated from the atmospheric model timestep by timestep, as should be the turbulent heat flux in the case of synchronous coupling. An asynchronously coupled ocean—atmosphere model will converge to a slightly different equilibrium to a synchronously coupled model, due to the inevitable bias in the estimate of the surface fluxes. Although this bias is generally less than 10%, it is particularly pronounced along the sea-ice margins, where small differences in simulations could be amplified by the strong feedback between temperature and albedo.

REFERENCES

Bryan, K., Manabe, S. and Pacanowski, R.L., 1975. A global ocean—atmosphere climate model, part II. The oceanic circulation. J. Phys. Oceanogr., 5: 30—46.

Budyko, M.I., 1963. Atlas of the Heat Balance of the Earth. Akad. Nauk SSSR, Prezidium. Mezhduvedomstvennyi Geofiz, Komitet.

Bunker, A.F., 1976. Computation of surface energy flux and annual air—sea interaction cycles of the North Atlantic Ocean. Mon. Weather Rev., 104: 1122—1140.

Clarke, R.H., 1970. Recommended methods for the treatment of the boundary layer in numerical models. Aust. Meteorol. Mag., 18: 51—73.

Esbensen, S.K. and Kushnir, Y., 1981. The heat budget of the global ocean. An atlas based on estimates from surface marine observations. Climate Research Institute Report 29, Oregon State University, Corvallis, Oreg.

Esbensen, S.K. and Reynolds, R.W., 1981. Estimating monthly averaged air—sea transfer of heat and momentum using the bulk aerodynamic method. J. Phys. Oceanogr., 11: 457—465.

Gordon, C. and Bottomley, M., 1985. The parameterization of the upper ocean mixed layer in coupled ocean—atmosphere models. In: J.C.J. Nihoul (Editor), Coupled Ocean—Atmosphere Models. (Elsevier Oceanography Series, 40) Elsevier, Amsterdam, pp. 613—635 (this volume).

Han, Y.J. and Lee, S.W., 1981. A new analysis of monthly mean wind stress over the global ocean. Climate Research Institute Report No. 26, Oregon State University, Corvallis, Oreg.

Hellerman, S. and Rosenstein, M., 1983. Normal monthly wind stress over the world ocean with error estimates. J. Phys. Oceanogr., 13: 1093—1104.

Kraus, E.B. and Morrison, R.E., 1966. Local interactions between the sea and the air at monthly and annual timescales. Q. J. R. Meteorol. Soc., 92: 114—127.

Lyne, W.H., Rowntree, P.R., Temperton, C. and Walker, J., 1976. Numerical modelling using GATE data. Meteorol. Mag., 105: 261—271.

Manabe, S., Bryan, K. and Spelman, M.J., 1979a. A global ocean—atmosphere climate model with seasonal variation for future studies of climate sensitivity. Dyn. Atmos. Oceans, 3: 393—426.

Manabe, S., Hahn, D.G. and Holloway Jr., J.R., 1979b. Climate simulations with GFDL spectral models of the atmosphere: Effect of spectral truncation report of JOC study conference on climate models: Performance, intercomparison and sensitivity studies. GARP Publ. Ser., 22: 41—94.

Washington, W.M., Semtner Jr., A.J., Meehl, G.A., Knight, D.J. and Mayer, T.A., 1980. A general circulation experiment with a coupled atmosphere, ocean and sea-ice model. J. Phys. Oceanogr., 10: 1887—1908.

Wu, J., 1982. Wind stress coefficients over sea surface from breeze to hurricane. J. Phys. Oceanogr., 12: 9704—9706.

CHAPTER 19

IMPORTANCE OF COUPLING BETWEEN DYNAMIC AND THERMODYNAMIC PROCESSES AT THE SEA SURFACE: THE LARGE-SCALE, OCEANIC POINT OF VIEW

BENOIT CUSHMAN-ROISIN

ABSTRACT

Combination of dynamic and thermodynamic processes (shear stress and buoyancy flux) at the sea surface can lead to important consequences that either process alone could not possibly explain. The present work focuses on the large-scale, mean ocean circulation at mid-latitudes. After new developments are brought to the classical thermocline theory, it is shown how the wind-driven and buoyancy-induced currents can interact and combine in such a way as to lead to frontal formation. The Subtropical Front of the North Pacific Ocean is then explained as a manifestation of this interaction rather than Ekman convergence.

Finally, it is stressed that model sensitivity through frontogenesis requires accurate parameterizations of sea-surface processes and fine spatial resolution in global climate models.

1. INTRODUCTION

Since air—sea interactions rely sensitively on both dynamic and thermodynamic sea-surface processes (shear stress and buoyancy flux), detailed considerations on the interactions between these two different processes either in the atmosphere or the ocean are in order. In particular, one should address the question as how the oceanic response to both dynamic and thermodynamic forcings differs from the superposition of the particular responses to each forcing alone. Indeed, wind-driven and buoyancy-induced currents add up in a very nonlinear way. Wind-driven currents advect the density field which in turn affects the buoyancy currents. On the other side, geostrophic buoyancy currents do not contribute to horizontal advection but produce convergence and divergence on a beta-plane. The focus of this presentation is to demonstrate one particular and somewhat unexpected consequence of this combination, namely the capability of an ocean basin to form large-scale fronts.

Another motivation for such study arises from a look at the classical air—sea interaction cycle (dark arrows on Fig. 1). In this cycle, the ultimate source of all energy, the sun, affects the atmospheric temperature around the globe which in turn drives the winds. Surface winds induce oceanic currents through shear at the air—sea interface. These oceanic currents affect the distribution of heat in the ocean and, in particular, the sea-surface temperature (SST). Finally, by means of surface heat flux, the SST can modify the atmospheric temperature distribution. Although this cycle demonstrates the existence of an air—sea feedback, it is incomplete, for two other equally important processes are at work (open arrows on Fig. 1). These are the effects of the atmospheric temperature on the SST and the existence of oceanic thermal currents. Of course, all air—sea interaction models of reasonable sophistication incorporate these last processes, but it is

264

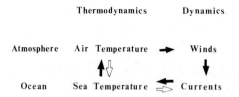

Fig. 1. The air—sea interactions. Dark arrows indicate processes that lead to a feedback cycle, while open arrows are other important processes.

worth isolating their particular impact on the large-scale ocean circulation. The large-scale, long-time ocean circulation is of particular interest since it involves length and time scales also found in the atmosphere, and such matching of scales is a prerequisite to effective air—sea interactions.

The cost of global air—sea and air—sea—land interaction models usually requires a compromise on the spatial resolution, and the typical resolution is generally insufficient for the ocean. As it will be shown here, dynamic and thermodynamic forcings can be responsible for setting up large-scale ocean fronts, which, to be adequately incorporated in such interaction models, ought to be well resolved. At this stage, the North Pacific Subtropical Front (see other chapters in this volume) is not resolved. It is however anticipated that the persistence of such a feature can largely affect the climate.

In order to demonstrate the spontaneous emergence of large-scale oceanic fronts in an ocean submitted to both dynamic and thermodynamic forcings, the presentation is organized in the following manner. As a prerequisite, Section 2 summarizes a few observations of the Subtropical Front and its associated Countercurrent as they are found in the mid-latitudes of the North Pacific Ocean. It is also argued that Ekman convergence cannot be the sole frontogenetical mechanism responsible for the front. Section 3 develops a generalization of the classical thermocline theory and demonstrates the possibility of temperature discontinuities in the open ocean. Section 4 is an application of the preceding theory to the mid-latitude North Pacific Ocean. The theory is able to reproduce the Subtropical Front and its associated Countercurrent. The frontogenetical mechanism is then investigated and traced back to the convergence of northward-flowing thermal currents on a beta plane. Section 5 briefly recapitulates this mechanism, warns climate modellers of the unexpected oceanic response at a short length scale, and stresses the importance of the feedback that such phenomenon can have on atmospheric dynamics.

2. THE SUBTROPICAL FRONT

Much of the credit for the discovery and documentation of the Subtropical Front and Subtropical Countercurrent in the North Pacific Ocean must go to Japanese investigators. From calculations of the Sverdrup transport from wind-stress data during springtime, Yoshida and Kidokoro (1967a, b) predicted the existence of an eastward current at a latitude of the Subtropical Gyre where one would otherwise expect a westward flow feeding the western boundary current. For this reason, they coined the name Subtropical Countercurrent. Later, Uda and Hasunuma (1969) confirmed such eastward flow from direct current-meter data as well as from geostrophic calculations based on hydrographic sections. The Subtropical Countercurrent is seen as a shallow, density-driven geostrophic flow ("thermal wind") associated with a Subtropical Front. The two features are two facets of a single phenomenon. Detailed charts of current directions in the western North Pacific prepared by Uda and Hasunuma (1969) prove clearly, in spite of some seasonal variations, the existence of an eastward current, significantly separated from the Kuroshio, and that originates around 20°N and tilts slightly northward as it flows eastward.

A more recent update of observations in the western North Pacific is found in a paper by Hasunuma and Yoshida (1978). There, they presented a long-term mean geopotential anomaly which testifies to the existence of the Subtropical Countercurrent as a steady feature. Vertical density sections show that the shallow eastward current is associated with a density front, located around 18°–20°N in the western North Pacific.

Observations of the Subtropical Front in the central and eastern North Pacific are reported by Roden (1972, 1975, 1980a and b). In particular, large-scale satellite coverage (Roden, 1980b) shows that the Subtropical Front can be traced across almost the entire zonal extent of the North Pacific, tilting slightly northward toward the east. Near the American coast, the Subtropical Front joins the California Current front which is, however, of a different, coastal origin. Although satellite data are the only present source of ocean-wide, detailed coverage, the sea-surface temperature that they provide may not accurately reflect subsurface conditions, and caution must be in order. In addition, Schroeder (1965) presented meridional sections of monthly temperatures in the North Atlantic Ocean. No essential difference is found between the structure of Subtropical Front in the North Atlantic and that in the North Pacific. The phenomenon has, therefore, a very general character. For the sake of simplicity, however, the present work focuses on the North Pacific Ocean only.

A first model leading to an eastward current across the middle of the Subtropical Gyre was presented by Haney (1974). A crucial feature of that model is the combination of a dynamical forcing (surface wind stress) and thermodynamical forcing (surface heat flux with warming near the equator and cooling at high latitudes). The results show an eastward countercurrent, whose shallowness suggests that it is of thermal origin. Haney's model demonstrates the importance of thermodynamically forced currrents insofar as being comparable to wind-driven currents. However, as a result of the wide span covered by the model (50°S–50°N), a lack of resolution precluded the representation of a zonal temperature front such as the Subtropical Front. More recently, Takeuchi (1980) modelled a smaller region (equator to 45°N) with essentially the same physics but with much finer horizontal and vertical resolution. His numerical results clearly show a Subtropical Front-Countercurrent structure where the current pattern basically consists of the classic wind-driven circulation superimposed on which is an eastward thermal flow along the Subtropical Front. This feature is found in the upper hundred meters of the water column only. Below, the current pattern consists of the same wind-driven circulation but with an opposite thermal return flow. In summary, the combination of surface wind stress (dynamical forcing) and surface heat flux (thermodynamical forcing) sets up a hybrid flow pattern consisting of a barotropic wind-driven circulation and a baroclinic thermally driven cell. But a question remains: Why do these two flow fields combine in such a way as to produce a large temperature gradient?

Based on wind data and hydrographic sections, Roden (1975, 1976) pointed out that there is a strong relation between the position of the Subtropical Front and the convergence of the surface Ekman transports. Westerlies to the north generate southward Ekman transports while the Trades to the south generate northward Ekman transports; convergence takes place at the latitude of zero zonal wind, i.e., around 28°–30°N.

But, there are two major objections to the likelihood of Ekman convergence as the only frontogenetical mechanism. First, in the western North Pacific, the Subtropical Front is found year-round near 20°N with, at times, a branch as south as 17°N, while the zonal wind component vanishes around 28°N (Kutsuwada, 1982). The second objection is more fundamental. In a numerical experiment designed to evaluate the role of Ekman convergence in the Subtropical Gyre, Takeuchi (1980) ran his model with meridional instead of zonal winds, but characterized by the same spatial curl distribution. Aside from minor differences at the uppermost level, his solution was identical to the one obtained with the zonal winds. This experiment leads to believe that the subsurface dynamics, that include the Subtropical Front and Countercurrent, are not dependent on the actual wind stress but only on its curl distribution. This important result demonstrates that convergence of Ekman transports can hardly be the sole mechanism responsible for the maintenance of the Subtropical Front-Countercurrent structure.

3. THERMOCLINE THEORY

Large-scale, long-time oceanic motions lead to the so-called Sverdrup dynamics, which will be used here to demonstrate the role of coupling between wind-driven and density-driven currents. The governing equations are (Veronis, 1969):

$$-fv = -p_x \tag{1}$$

$$fu = -p_y \tag{2}$$

$$p_z = \alpha g T \tag{3}$$

$$u_x + v_y + w_z = 0 \tag{4}$$

$$uT_x + vT_y + wT_z = (KT_z)_z \tag{5}$$

where x, y and z are the eastward, northward and upward coordinates along which the velocity components are u, v and w, respectively; f is the variable Coriolis parameter ($\beta = df/dy$), p the dynamical pressure, g the gravitational acceleration, T a temperature variable which is meant to include all buoyancy effects, α the corresponding expansion coefficient, and K a space- and time-dependent diffusivity. For the sake of mathematical feasibility, K is first set to zero (this section) and then reintroduced (next section). The accompanying upper and lower boundary conditions are:

$$w = \text{curl}_z(\tau/f) \qquad \text{at } z = 0 \tag{6}$$

$$-KT_z = k(T - T_a) \qquad \text{at } z = 0 \tag{7}$$

$$w = 0 \qquad \text{at } z = -H \tag{8}$$

i.e. the wind influence is through vertical pumping at the base of the Ekman layer ($z = 0$) while the buoyancy flux at the same level is proportional to the buoyancy difference between the uppermost layer of fluid in the model and that above (T_a can be interpreted as an equivalent atmospheric temperature if one assumes that the equivalent heat flux at the surface is transmitted to the substratum). The above parameterizations do not intend to be accurate but only to be representative enough to illustrate quite simply the dynamics at play. Finally, the model requires lateral boundary conditions which will be mentioned later.

Following the methodology of Welander (1959), one defines a quantity M by $p = M_z$ in terms of which all other quantities are expressed [using eqns. (1) to (4)]:

$$u = -\frac{1}{f}M_{yz}, v = \frac{1}{f}M_{xz}, w = \frac{\beta}{f^2}M_x, p = M_z, T = \frac{1}{\alpha g}M_{zz} \qquad (9)$$

Equation (5) can then be written in terms of the function M alone:

$$M_{xz}M_{yzz} - M_{yz}M_{xzz} + \frac{\beta}{f}M_x M_{zzz} = 0 \qquad (10)$$

Welander (1959) showed that the Sverdrup dynamics yield three conserved quantities, which expressed in terms of M are M_{zz} (buoyancy), fM_{zzz} (potential vorticity), and $M_z - zM_{zz}$ (Bernoulli function). Since a streamline is a line along which all three quantities are constant, there must exist in the most general case a functional relationship between these three quantities. The particular structure of this latter function depends ultimately on the vertical and lateral boundaries of the problem. A functional relationship between the three quantities ensures that their isosurfaces have lines of intersection rather than points. That these lines be the streamlines is an additional requirement provided by eqn. (10).

The a-priori assumption is made here that the potential vorticity is a function of buoyancy alone:

$$fM_{zzz} = F(M_{zz}) \qquad (11)$$

where F is an arbitrary, non-negative function of its argument. The sign restriction is required for static stability ($T_z \geqslant 0$). Within the context of the two restrictions stated above (no diffusivity and no dependence on the Bernoulli function), the arbitrariness of F ensures generality of the solution. Equation (11) can be written in the differential form $dM_{zz}/F(M_{zz}) = dz/f$ which has the integral:

$$L(M_{zz}) = \frac{z}{f} + C \qquad (12)$$

where L is the integral of $1/F$ and is thus an arbitrary, monotonic function of its argument, and C is a constant of integration, a function of x and y. Since the function L is monotonic, it can be inverted to yield:

$$M_{zz} = G\left(\frac{z}{f} + C\right) \qquad (13)$$

where G is an arbitrary monotonic function of its argument. Two further integrations yield:

268

$$M = A + Bz + f^2h\left(\frac{z}{f} + C\right) \qquad (14)$$

where h is the integral of the integral of G, and A and B are two additional constants of integration that are functions of x and y in general. Since the starting function F is arbitrary so is h, and expression (14) is interpreted as the most general type of solution for (10) within the framework of (11). The implementation of this expression (14) in the governing equation (10) permits the elimination of the vertical dependence and leaves a constraint between the three functions A, B and C:

$$B_x C_y - B_y C_x + \frac{\beta}{f^2} A_x = 0 \qquad (15)$$

Two complementing relationships for A, B and C are obtained from the upper and lower boundary conditions (6) and (8) (no buoyancy flux can be imposed in the absence of diffusivity):

$$A + f^2h(C) = HB_0 \qquad (16)$$

$$A + f^2h\left(C - \frac{H}{f}\right) = HB \qquad (17)$$

where $B_0(x, y) = (f^2/\beta H)\int^x \mathrm{curl}_z(\tau/f)\mathrm{d}x$ can be evaluated from the knowledge of the surface wind stress. Elimination of A and B followed by a further integration yields an algebraic implicit relationship between C and B_0:

$$Q\left(C - \frac{H}{f}\right) - \frac{f^2}{H}\left[h\left(C - \frac{H}{f}\right) - h(C)\right] - fh'\left(C - \frac{H}{f}\right) = B_0(x, y) \qquad (18)$$

where Q, like h, is an arbitrary function of its argument. Assuming now that lateral boundary conditions imposed with the appropriate degrees of freedom have led to the specification of the two functions Q and h, one can in principle find C as a function of

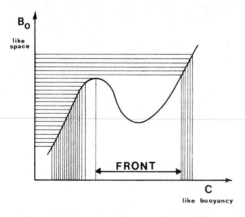

Fig. 2. Sketch of a possible relationship between C and B_0. Since C is related to temperature and B_0 to space, a discontinuity of C for continuously varying B_0 corresponds to a front.

x and y. However, the implicit character of eqn. (18) may lead to multiple solutions, which, physically, correspond to the existence of fronts. Figure 2 schematically illustrates a possible situation where the continuity of space requires discontinuity of C. Since the buoyancy is proportional to $h''(C)$ and since $h'' = G$ is a monotonic function, there is a direct relationship between C and buoyancy, and the discontinuity in C is interpreted as a front.

In summary, the analysis of this section shows that large-scale, stationary fronts can exist in the presence of wind-driven and buoyancy-driven currents. There, in the absence of vertical diffusion, the thermal currents if any are forced by the lateral conditions (presence of a heterogeneous western boundary current, for instance), but the conclusion remains valid if these currents are forced by a surface buoyancy flux transmitted downward by diffusion, as the next section shows, or if these currents are forced by a specified distribution of buoyancy in the surface Ekman layer (Killworth, 1985).

4. APPLICATION TO THE NORTH PACIFIC OCEAN

The above generalized theory anticipates the existence of discontinuous solutions that correspond to fronts. The present section demonstrates that, indeed, this can be the case in the North Pacific Ocean for which such discontinuity is predicted, and a new explanation for the Subtropical Front is proposed.

Firstly, in order to incorporate thermal currents forced by surface buoyancy flux conditions, the vertical diffusion term of eqn. (5) is reinserted, and the upper boundary condition (7) is used. This can be done provided that the vertical dependence of the diffusivity K is proportional to h''/h'''. Secondly, a choice of the function h is made; a particular form of the solution (14) for the vertical structure of the thermocline that has been used previously is the exponential function (Needler, 1967, 1971; Anderson and Killworth, 1979; Clarke, 1982) with:

$$f^2 h\left(\frac{z}{f} + C\right) = \frac{\alpha g f^2}{a^2} \, T_s \exp{(az/f)} \tag{19}$$

where a is a dimensional constant that one can take equal to $5 \times 10^{-7} \, \mathrm{m^{-1} \, s^{-1}}$ so that the e-folding depth of variation in the vertical, f/a, is about 150 m. The function $T_s(x, y)$ that replaces $C(x,y)$ through the relation $agT_s = a^2 \exp{(aC)}$ corresponds to the subsurface temperature at the base of the Ekman layer ($z = 0$). Thirdly, a simple but representative wind-stress field is selected consisting only of a zonal component with meridional structure, $\tau^x(y)$.

With all above particularizations, eqns. (15)–(17) yield a single, first-order, hyperbolic equation for T_s (Cushman-Roisin, 1984):

$$\frac{x}{\beta H} \tau_{yy}^x \, T_{sx} - \frac{1}{\beta H} \tau_y^x T_{sy} + \frac{\tau^x}{fH} \, T_{sy} - \frac{a}{f}\left(\frac{\tau^x}{f}\right)_y T_s - \frac{\alpha g \beta}{af}\left(1 - \frac{2f}{aH}\right) T_s T_{sx}$$

$$= -\frac{ak}{f}(T_s - T_a) \tag{20}$$

This governing equation for the surface temperature, $T_s(x, y)$, is the last and key equation that remains to be solved. One recognizes in the first two terms zonal and meridional advection by the barotropic, non-divergent, Sverdrup flow as would be obtained from a vertically homogeneous, wind-driven model. The third term represents meridional advection by a divergent Ekman return flow (in reaction to the surface Ekman drift, not included). The fourth and fifth terms correspond to vertical heat exchanges due to Ekman pumping and divergence of geostrophic isothermal flow on a beta-plane, respectively. The latter of the two is the only non-linear term in the equation. The term on the right-hand side represents the vertical exchange of heat with the Ekman layer and atmosphere. Finally, it should be noted that boundary conditions on T_s must still be imposed. Once T_s has been determined from eqn. (20), all other variables can be easily reconstructed:

$$u = \frac{x}{\beta H} \tau_{yy}^x - \frac{\alpha g f}{a^2} F' T_{sy} - \frac{\alpha g \beta}{af} \left[\left(1 - \frac{az}{f} \right) \exp \left(\frac{az}{f} \right) - \frac{2f}{aH} \right] T_s \tag{21}$$

$$v = - \frac{1}{\beta H} \tau_y^x + \frac{\tau^x}{fH} + \frac{\alpha g f}{a^2} F' T_{sx} \tag{22}$$

$$w = - \left(1 + \frac{z}{H} \right) \left(\frac{\tau^x}{f} \right)_y + \frac{\alpha g \beta}{a^2} F T_{sx} \tag{23}$$

$$p = \frac{B_0}{H} - \frac{x}{\beta H} (f \tau_y^x - \beta \tau^x) + \frac{\alpha g f^2}{a^2} F' T_s \tag{24}$$

$$T = T_s \exp \left(\frac{az}{f} \right) \tag{25}$$

where $F(z, y) = \exp(az/f) - (1 + z/H)$ and $F' = \partial F / \partial z$. Among the various terms, one recognizes the non-divergent wind-driven Sverdrup flow, the Ekman pumping and its associated meridional Ekman return flow and the baroclinic isothermal flow that is divergent on a beta-plane.

The flow along isotherms does not contribute to horizontal advection, but, on a beta-plane, does contribute to vertical advection through the convergence of its meridional component. It is this component of the vertical velocity [second term of eqn. (23)] which yields the only nonlinearity in the governing equation (20) for the surface temperature. Quite surprisingly, because of the x derivative, this term resembles a contribution to zonal advection and will be treated as such in what follows.

Equation (20) is a first-order, hyperbolic equation, which is best solved by the method of characteristics. Because of the nonlinear term, the characteristics are not uniquely defined as curves in the (x, y) plane regardless of boundary conditions. On the contrary, their slope depends on the evolution of T_s along them, starting from their origin on the boundary. It is thus possible for characteristics to intersect and for a discontinuity in T_s (a thermal front) to be present.

From the inspection of eqn. (20), one notes that the basin is covered by two families of characteristics: one originates from the northern portion of the western boundary, while the other originates from the eastern wall. The front is expected to be situated along the separation line between these two families. To determine the location and shape of this frontal line, one needs to integrate along the characteristics, a task for which one first ought to set boundary conditions on T_s. For the family of characteristics originating from the upper portion of the western boundary, one can take T_s equal to the value of T_a at the middle of the basin (20°C), arguing that particles leaving the western boundary current are warmer than their environment. For the family of characteristics originating from the eastern wall, it is chosen to impose T_s to be uniform with latitude (also 20°C). This is analogous to requiring no depth-averaged zonal currents through the boundary. Although this condition may seem too weak, a better choice requires to include eastern-boundary dynamics (Pedlosky, 1983) which are not yet completely understood and are too irrelevant to the open-ocean, minimal model developed here. As a consequence, the following results will reproduce the open-ocean phenomena but will miss the California-current dynamics.

For these lateral boundary conditions, the solution of eqn. (20) was numerically evaluated by the method of characteristics (see Cushman-Roisin, 1984, for additional information). The resulting temperature distribution at the base of the Ekman layer ($z = 0$) is presented on Fig. 3. On that figure, a thermal front is clearly apparent. Its position agrees with the one of the Subtropical Front in the North Pacific Ocean (see Section 2), except in the eastern portion of the basin. There, the absence of California-current dynamics do not force the front to meet with the coastal cold-water front generated by this latter current. Agreement with the numerical results of Takeuchi (1980) suggests that the present simple model retains the essential frontogenetical physics, which now should be traced back through the mathematics in order to identify the frontogenetical mechanism that can be responsible for large-scale, open-ocean fronts such as the Subtropical Front.

But, before tackling this task, it is worth discussing the currents associated with the temperature distribution of Fig. 3. From the knowledge of the subsurface temperature field, T_s, all three-dimensional fields can be computed from eqns. (21) to (25). In particular, horizontal currents (Fig. 4) consist of the wind-driven, clockwise, subtropical gyre superimposed on which is an eastward countercurrent of thermal origin and flowing along the Subtropical Front. With depth, this latter thermal current reverses yielding no depth-averaged flow while the wind-driven current is of barotropic nature. The vertical

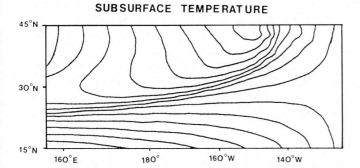

Fig. 3. Temperature distribution at the base of the surface Ekman layer as obtained numerically by the method of characteristics. Note the Subtropical Front running northeastward across the basin.

SUBSURFACE CURRENTS

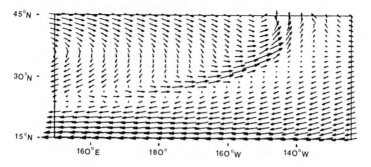

Fig. 4. Total currents (wind-driven and thermally forced) at the base of the surface Ekman layer. Note the eastward-flowing Subtropical Countercurrent, as a thermal current weakened by the wind-driven circulation.

velocity field (not displayed here — see Cushman-Roisin, 1984) shows that the frontal region is one of downwelling, such that the convergence necessary for that downwelling maintains the front.

5. DISCUSSION

The previous two sections led to the evidence that the combination of dynamical and thermodynamical forcings can be responsible for large-scale oceanic fronts, and it now remains to elucidate the particular frontogenetical mechanism at work. This task is accomplished by tracing back through the mathematics the term that leads to a discontinuity.

The temperature discontinuity is explained by intersecting characteristics, a phenomenon attributable to the nonlinear term in the governing equation. In turn, this term is traced to the vertical velocity induced by the convergence of geostrophic thermal flow on a beta-plane. Since $w \sim \beta z T_x / f^2$, positive (negative) divergence occurs if the thermal flow has a southward (northward) component. Due to the presence of an eastern obstacle, the thermal flow must have a meridional component. If this latter is directed southward ($T_x < 0$), it brings divergence, smooth temperature variations by spreading of isotherms, slow thermal flow, adjustment to the atmospheric conditions, and a resulting zonal temperature distribution ($T_x = 0$), i.e. a contradiction. It results that the meridional component ought to be directed at least slightly northward bringing downwelling, convergence, large temperature gradients, fast thermal flow, the possibility of finding anomalously warm water and thus a front. The convergence zone acts as a barrier on each side of which water masses flow and sink while remaining identifiable as of different origin.

The mechanism, based on the beta-plane convergence of thermal currents, leads to a situation similar to the one in the North Pacific Ocean. However, it is not claimed that this mechanism is the sole process responsible for the Subtropical Front, since Ekman dynamics can also certainly contribute to maintaining and strengthening the front. The point to be made here is a warning for air—sea interaction models. Since combination of dynamical (surface stress) and thermodynamical (buoyancy flux) forcings at the air—sea interface can lead to the spontaneous formation of large-scale oceanic fronts, and since such oceanic barrier can affect atmospheric conditions on a long-time scale (few months to years), it is imperative to recognize this phenomenon and to include it in the models by providing adequate spatial resolution, at least in the oceanic part of the models. The resolution needed to incorporate the appropriate dynamics (100 km or less) is quite high.

Moreover, since small-scale phenomena such as fronts are sensitive to other dynamics at the same scale, it is anticipated that the precise position and strength of a front maintained by beta-plane convergence of thermal currents can be highly sensitive to the parameterization chosen to represent processes at that scale, and thus the present work also stresses the need to refine and improve further the various parameterizations on which air—sea interaction models rely at scales around 100 km.

ACKNOWLEDGEMENTS

The author is indebted to Drs. Peter D. Killworth, James J. O'Brien and John Woods for stimulating remarks, to the Committee for Climate Changes and the Ocean for the invitation to the 16th Liège Colloquium, and to the Office of Naval Research for continuous support. This work is contribution No. 211 of the Geophysical Fluid Dynamics Institute at the Florida State University.

REFERENCES

Anderson, D.L.T. and Killworth, P.D., 1979. Non-linear propagation of long Rossby waves. Deep-Sea Res., 26: 1033–1049.
Clarke, A.J., 1982. The dynamics of large-scale, wind-driven variations in the Antarctic circumpolar current. J. Phys. Oceanogr., 12: 1092–1105.
Cushman-Roisin, B., 1984. On the maintenance of the Subtropical Front and its associated Counter-current. J. Phys. Oceanogr., 14: 1179–1190.
Haney, R.L., 1974. A numerical study of the response of an idealized ocean to large-scale surface heat and momentum flux. J. Phys. Oceanogr., 4: 145–167.
Hasunuma, K. and Yoshida, K., 1978. Splitting of the Subtropical Gyre in the western North Pacific. J. Oceanogr. Soc. Jpn., 34: 160–171.
Killworth, P.D., 1985. A simple wind- and buoyancy-driven thermocline model (Extended Abstract). In: J.C.J. Nihoul (Editor), Coupled Ocean—Atmosphere Models. (Elsevier Oceanography Series, 40) Elsevier, Amsterdam, pp. 513–516 (this volume).
Kutsuwada, K., 1982. New computation of the wind stress over the North Pacific Ocean. J. Oceanogr. Soc. Jpn., 38: 159–171.
Needler, G.T., 1967. A model for thermocline circulation in an ocean of finite depth. J. Mar. Res., 25: 329–342.
Needler, G.T., 1971. Thermocline models with arbitrary barotropic flow. Deep-Sea Res., 18: 895–903.
Pedlosky, J., 1983. Eastern boundary ventilation and the structure of the thermocline. J. Phys. Oceanogr., 13: 2038–2044.
Roden, G.I., 1972. Temperature and salinity fronts at the boundaries of the Subarctic—Subtropical transition zone in the western Pacific. J. Geophys. Res., 77: 7175–7187.
Roden, G.I., 1975. On North-Pacific temperature, salinity, sound velocity and density fronts and their relation to the wind and energy flux fields. J. Phys. Oceanogr., 5: 557–571.

Roden, G.I., 1976. On the structure and prediction of oceanic fronts. Naval Res. Rev., 29(3): 18–35.

Roden, G.I., 1980a. On the subtropical frontal zone north of Hawaii during winter. J. Phys. Oceanogr., 10: 342–362.

Roden, G.I., 1980b. On the variability of surface temperature fronts in the western Pacific, as detected by satellite. J. Geophys. Res., 85C: 2704–2710.

Schroeder, E.H., 1965. Average monthly temperatures in the North Atlantic Ocean. Deep-Sea Res., 12: 323–343.

Takeuchi, K., 1980. Numerical study of the Subtropical Front and the Subtropical Countercurrent. Doct. Diss., Ocean Research Institute, University of Tokyo, Tokyo, 45 pp.

Uda, M. and Hasunuma, K., 1969. The eastward Subtropical Countercurrent in the western North Pacific Ocean. J. Oceanogr. Soc. Jpn., 25: 201–210.

Veronis, G., 1969. On theoretical models of the thermocline circulation. Deep-Sea Res., 16 (suppl.): 301–323.

Welander, P., 1959. An advective model of the ocean thermocline. Tellus, 11: 309–318.

Yoshida, K. and Kidokoro, T., 1967a. A subtropical countercurrent in the North Pacific — An eastward flow near the Subtropical Convergence. J. Oceanogr. Soc. Jpn., 23: 88–91.

Yoshida, K. and Kidokoro, T., 1967b. A subtropical countercurrent (II) — A prediction of eastward flows at lower subtropical latitudes. J. Oceanogr. Soc. Jpn., 23: 231–246.

CHAPTER 20

THE SIGNIFICANT TROPOSPHERIC MIDLATITUDINAL EL NIÑO RESPONSE PATTERNS OBSERVED IN JANUARY 1983 AND SIMULATED BY A GCM

HANS VON STORCH and HARALD A. KRUSE

ABSTRACT

(1) Does the mean January 1983 500-mb height field differ significantly from the January fields observed in non-El Niño Januaries?
    yes: – Intense depression north of the anomaly;
           – mid-latitudinal high at Greenwich; and
           – intensified pattern with an irregularly distributed series of three maxima and minima.
(2) Does the ECMWF T21L15 GCM respond to an El Niño SST anomaly (similar to the winter 1982/83 SST anomaly) significantly?
    yes: – Intense depression north of the anomaly; and
           – three stably located mid-latitudinal highs at $90°$ W, $10°$ W and $70°$ E.
(3) Are the 1983 observed and the GCM generated response patterns coherent?
    yes: – The two patterns are significantly correlated; and
           – most of the synoptic scale details of the patterns coincide with respect to location and sign.
(4) Is it possible to simulate the response with sufficient accuracy by linearized models?
    no: – GCM experiments with negative anomalies point to a nonlinear relation "SST anomaly– circulation anomaly";
           – linear experiments show that the relation "tropospheric heating anomaly–circulation anomaly" is nonlinear; and
           – the link "SST anomaly–heating anomaly" is nonlinear, because the release of latent heat by condensation of moisture anomalies generated by evaporation at the sea surface depends essentially nonlinear on the large-scale flow.
(5) Conclusion: El Niño-type SST anomalies up to $+4$ K generate a unique response pattern in the 500-mb north-hemispheric mid-latitudinal height field. Cold anomalies ("La Niña") yield a varying and less significant mid-latitudinal response.

1. INTRODUCTION

There is no doubt about the existence of an effect of an El Niño SST anomaly on the tropical circulation, and about the appearance of the response patterns of, say, air pressure and precipitation. This uniqueness is due to the fact that the signals are relatively strong as compared with the weak natural variability of the tropical atmosphere.

In contrast, the remote effect on the mid-latitudes cannot be expected to be as large, and to detect a signal is much more difficult in the presence of the enormous natural variability of the extra-tropical atmosphere. Many efforts have been made to find a mid-latitudinal reaction in the observations as well as in numerical simulations, but the particular appearance of the response patterns remained uncertain since it varied from experiment to experiment. (A good review is given by Shukla and Wallace, 1983.) A good part of this uncertainty is due to the use of inadequate statistical assessment methods which reduces unfortunately to a discussion of insignificant patterns.

Our central ingredients of statistical methods for a signal recognition in the presence of a large natural variability are multi-variate test procedures for the assessment of whole vectors of, say, grid-point values, and an a-priori reduction of the number of parameters used to characterize the signal patterns. In this paper, we shall apply these methods both to observed and numerically simulated pressure distributions.

We shall first show that the north-hemispheric atmospheric circulation in terms of the 500-mb height in the winter 1982/83 was exceptional, i.e. significantly different from the normal variety of atmospheric states. Then the hypothesis is tested that this exceptional deviation from the long-term average is caused by the anomalous sea-surface temperature in the Pacific (Fig. 1a); for that we shall use the results of a ten-year simulation with the T21L15 GCM of the ECMWF. We show that the model simulates a significant and stable response to an El Niño-type SST anomaly (Fig. 1b), and that this response is similar to the observed one of 1982/83. From this analysis we conclude that a unique mid-latitudinal response pattern to El Niño anomalies exists.

(a)

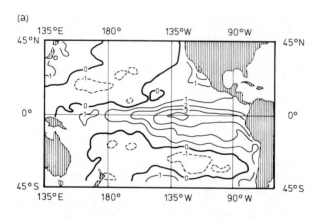

(b)

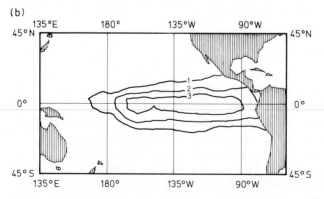

Fig. 1. SST anomaly in the Pacific: (a) Observed in winter 1982/83 (average December–February; from Arkin et al., 1983); contour interval: 1 K; (b) used in the ECMWF simulation [Twice standard Rasmusson/Carpenter anomaly, from Cubasch (1983)]; contour interval: 1 K.

281

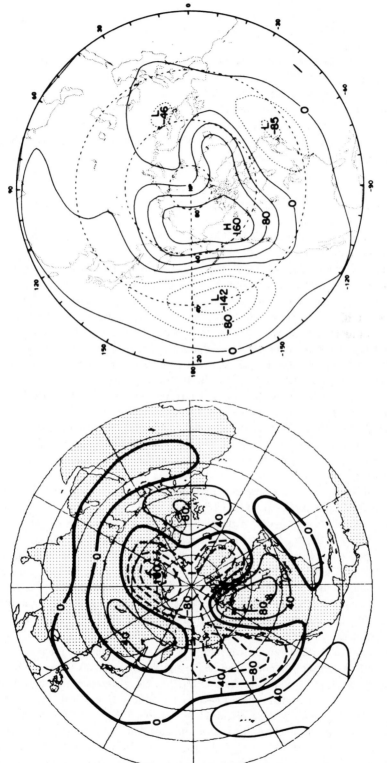

Fig. 5. Simulated response to El Niño SST anomalies given by Shukla and Wallace (1983; left) and Blackmon et al. (1983; right).

congruent remote responses. However, one of the model circulations exhibits Features 2 and 3 as well.

We hypothesize that the particulars of the mean January 1983 tropospheric circulation are related to the El Niño event. To test this hypothesis, we utilize the result of a numerical experiment with the T21L15 GCM performed at the European Centre for Medium Range Weather Forecast (ECMWF; Cubasch, 1983). These experiments consisted in the simulation of a total of nine winter seasons with a climatological SST distribution and of three winter seasons with a superimposed SST anomaly essentially equal to the January 1983 El Niño SST anomaly.

4. SIGNIFICANT EL NIÑO RESPONSE SIMULATION WITH THE ECMWF GCM

We investigated the ECMWF GCM's response to the prescribed El Niño SST anomaly with two approaches.

As first approach, we studied the complete north-hemispheric response in three anomaly experiments which differ by their initial conditions only. The data compression was done by a spectral expansion into a series of spherical surface harmonics (SSHs). A hierarchy of a few, large-scale SSHs, expected to represent the global character of the response, was established. A triangular truncation at total wavenumber seven was performed. Each member of this hierarchy was tested by a chi-square statistic. As the final significant response we selected that member of the hierarchy with maximum skill at the fixed significance level of 95% (Barnett et al., 1981).

These three tests resulted in a significant response in all three cases, which are displayed in Fig. 6. The selected members of the hierarchy differ with respect to the spectral resolution: the patterns of cases (b) and (c) consist of a superposition of 30 SSHs and are hard to distinguish from the respective untruncated patterns; pattern (a) shows up some differences in the longitudinal sector $0°-140°E$ due to a truncation to 18 SSH modes.

In all three cases, the response patterns are similar, insofar as the sequence of mid- and high-latitudinal relative extrema is stable within the $270°$ sector covering the central Pacific, North America, and parts of Eurasia. If we adopt the convention to term the sequence of extrema a "wave train" (e.g., Hoskins and Karoly, 1981), we may say that these wave trains emanate westward starting from a negative center near the date line.

As second approach we used the data compression from Section 2, namely an EOF expansion of the $30°-60°N$ meridional average. As test we chose the generalized randomized Mann-Whitney procedure, which is based on permutation arguments. Again we found that the difference of the mean "control" and the mean "anomaly" state is highly significant (99%).

To analyze the differences, we return as in Section 2 back to the grid-point space and plot the 95% bands of the control ensemble and the three curves simulated by the anomaly experiments (Fig. 7). Since we are interested in the differences, we subtracted the mean state of all 12 experiments mentioned so far and of three further experiments with a cold anomaly, which will be discussed in Section 6. Thus, in Fig. 7 the details of the climatological patterns as e.g. the Pacific and East-American troughs are missing.

As can be deduced from Fig. 7, the pattern entitled Feature 1 in Section 3 is found in all three experiments. At two locations this pattern has an exceptional amplitude of 60 and 85 m, respectively, the third curve has at that longitude a maximal deviation at the lower bound of the normal, namely about 20 m.

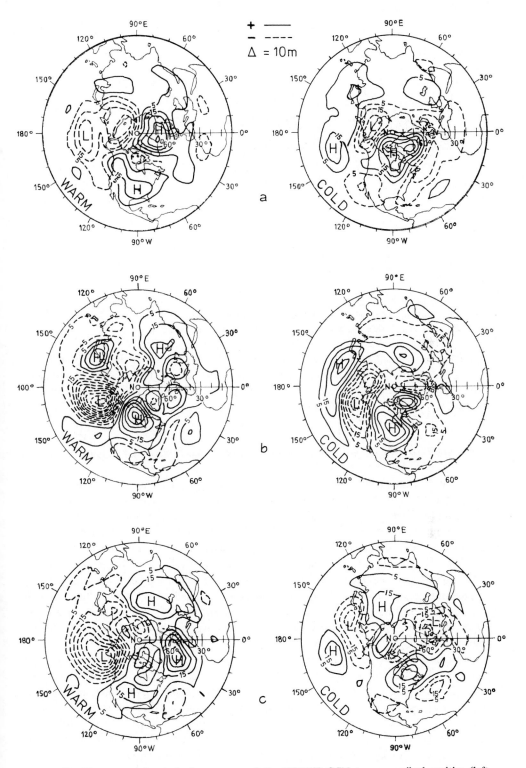

Fig. 6. Significant north-hemispheric response of the ECMWF GCM to a prescribed positive (left column) and negative (right column) El Niño SST anomaly. The negative anomaly experiment is discussed in Section 6. Contour interval: 10 m.

284

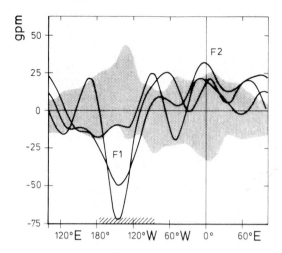

Fig. 7. The simulated SST anomaly response: 95% band of the 30°–60°N average of the mean 500-mb height built up of nine control experiments (stippled), and the corresponding states from three anomaly experiments (thin lines). Compare Features 1 and 2 with Fig. 6.

Furthermore, three maxima with stable location (namely about 90°W, 10°W and 70°E) are simulated, which are often found outside the 95% tube. The 10°W maximum coincides reasonably well with Feature 2 and the distribution of maxima and minima resembles Feature 3.

5. THE COINCIDENCE OF THE SIMULATED AND OBSERVED RESPONSE PATTERNS

In the foregoing section we found that the simulated response exhibits prominent details similar to those of the observed circulation anomaly of January 1983 discussed in Section 3. Now we shall ensure the similarity of these patterns as a whole in an objective way. For that, we define the "observed SST anomaly response pattern" as the difference of the normal Januaries (1967–82) average minus the El Niño January 1983 (see Fig. 2). We project the nine simulated "control fields" and the three "anomaly fields" onto this "observed response"; this procedure yields a total of 12 numbers, the scalar products, namely nine "controls" and three "anomalies". With the ordinary Mann-Whitney statistic, we tested the one-sided alternative that the "anomalies" tend to be larger than the "controls". We found this alternative acceptable with a risk smaller than 5%.

This means that the simulated (significant) response pattern is correlated with the (significant) observed one. A side-by-side comparison of the observed (Fig. 2) and the simulated response (Fig. 8) shows that both fields are really very similar in structure. Besides Features 1–3, the North-American and Central Siberian ridges and the East-European and Greenland/North-Atlantic troughs are common properties. The difference in the amplitude values results from the under-estimated spatial variance of the GCM and the longer averaging period (three months) for the simulated patterns.

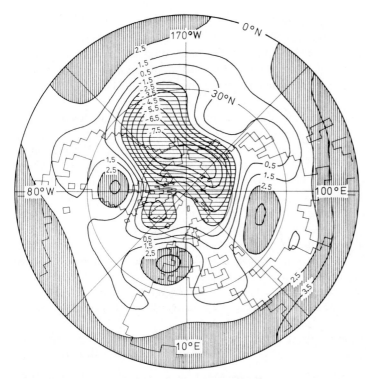

Fig. 8. Mean simulated response pattern: Average of anomaly runs minus average of control runs. Contour interval: 10 m.

6. DISCUSSION: THE LINEARITY HYPOTHESIS

A linear relationship between the hemispheric response and the SST anomaly would facilitate the understanding of the involved mechanisms and simplify the prediction of seasonal averages of atmospheric states from ocean surface states. We tested the existence of such a linear relationship by evaluating a series of three further GCM simulations with an SST anomaly distributed as the one studied above but with reversed sign (Cubasch, 1983). Following a proposal of Philander we denote such an anomaly as a "La Niña" event. A linear relationship would imply similar response shapes and amplitudes both for the warm El Niño and cold La Niña SST anomaly, but opposite sign.

In Section 4, we found a stably located negative center near the date line as the most pronounced response to the positive El Niño SST anomaly. At the same location, La Niña induces apparently a positive center, with half the amplitude. From this "principal center", wave trains emanate with wavelengths shorter than those in the warm anomaly experiments. However, the wave trains are not stable but vary with respect to strength and path. In fact, the application of the chi-square test sketched in Section 4 yields for La Niña significance levels of about 84% in two cases and below 50% in one case, whereas El Niño gave in all three cases responses significant at levels of more than 95%.

Thus, we conclude that the remote response of the 500-mb height field to the equatorial SST anomaly on the whole is nonlinear. A possible exception might be the subtropical center at the date line, which alludes to the possibility of a regional linearity.

The beauty of linear relationships gives the motivation to search for linear subsystems in the chain that leads from the ocean surface temperature anomaly to the global atmospheric response. The first link is the influence of the SST anomaly on the various diabatic heating processes, and the second link is the effect of these sources on the global atmospheric circulation.

Whereas diabatic heating anomalies due to sensible and radiative heat flux anomalies may be regarded as more or less linear in the SST anomaly, the latent heating is the essentially non-linear part of the game, even in simple parameterizations. From a simple two-layer model (Webster, 1981) the latent heat release has been estimated to be of the same order of magnitude as the sensible heat input, if the SST anomaly is placed in the tropics. Thus, the non-linear part of the heating processes plays an important part.

The second step to be considered is the dependence of the flow field on the heat sources. Here the question is whether the anomaly variables are small enough to allow for a linearization of the advective processes. To that end, the response of a GCM to a mid-latitudinal Pacific SST anomaly was compared to the response of a relatively simple linearized model to the forcing by the total heating induced by the SST anomaly as computed by the GCM (Hannoschöck, 1984). It turned out that only a very small fraction of the GCM response can be explained by a model with the linearized advection (Fig. 9).

Unfortunately, we cannot repeat Hannoschöck's study with our El Niño anomaly, because the heating anomaly generated by the ECMWF GCM is not available. Therefore,

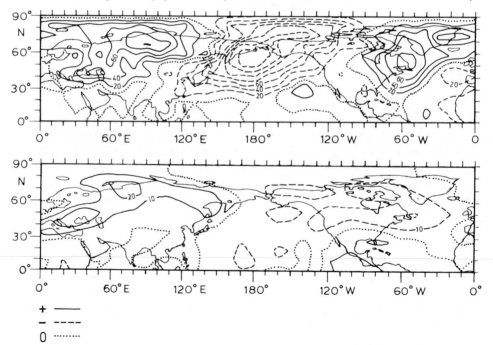

Fig. 9. Response to a mid-latitudinal SST anomaly of a GCM (top) and of a linearized model (bottom; from Hannoschöck, 1984).

we have to speculate whether Hannoschöck's result may be transferred to tropical conditions. However, even if the regional response to the tropical heating anomaly would be
linear, this response would in turn induce mid-latitudinal heating anomalies, again causing
non-linear reactions.

To summarize, the tropical SST anomaly produces a latent heating contribution that
is an essentially non-linear function of the flow field, at least in the tropics. Furthermore,
the global flow field depends non-linearly on these sources due to advection, at least in
the mid-latitudes.

REFERENCES

Arkin, P.A., Kopman, J.D. and Reynolds, R.W., 1983. Event quick look atlas: 1982–1983 El Niño/
   Southern Oscillation. – Nov. 1983. NOAA/National Weather Service, Numer. Model Centre,
   Washington, D.C. 20 233.
Barber, R.T. and Chavez, F.P., 1983. Biological consequences of El Niño. Science, 222: 1203–1210.
Barnett, T.P., Preisendorfer, R.W., Goldstein, L.M. and Hasselmann, K., 1981. Significance tests for
   regression model hierarchies. J. Phys. Oceanogr., 11: 1150–1154.
Bjerknes, J., 1966. A possible response of the atmospheric Hadley circulation to equatorial anomalies
   of ocean temperature. Tellus, 18: 820–829.
Blackmon, M.L., Geisler, J.E. and Pitcher, E.J., 1983. A general circulation model study of January
   climate anomaly patterns associated with interannual variations of equatorial Pacific sea surface
   temperatures. J. Atmos. Sci., 40: 1410–1425.
Cane, M.A., 1983. Oceanographic events during El Niño. Science, 222: 1189–1194.
Chen, W.Y., 1983. The climate of Spring 1983 – a season with persistent global anomalies associated
   with El Niño. Mon. Weather Rev., 111: 2371–2384.
Chiu, W.C., Lo, A., Weider Jr., D.H. and Fulker, D., 1981. A study of the possible relationship
   between the tropical Pacific sea surface temperature and atmospheric circulation. Mon. Weather
   Rev., 109: 1013–1020.
Cubasch, U., 1983. The response of the ECMWF global model to the El Niño anomaly in extended
   range prediction experiments. European Centre for Medium Range Weather Forecast, Tech.
   Rep., 38.
Hannoschöck, G., 1984. A multi-variate signal-to-noise analysis of the response of an atmospheric
   circulation model to sea surface temperature anomalies. Hamburger Geophysikalische Einzel-
   schriften, 67.
Hasselmann, K., 1979a. Linear statistical models. Dyn. Atmos. Ocean, 3: 501–521.
Hasselmann, K., 1979b. On the signal-to-noise problem in atmospheric response studies. In:
   Meteorology of Tropical Oceans. R. Meteorol. Soc., London, pp. 251–259.
Hoskins, B. and Karoly, D.J., 1981. The steady linear response of a spherical atmosphere to thermal
   and orographic forcing. J. Atmos. Sci., 38: 1179–1196.
Krueger, A.F., 1983. The climate of Autumn 1982, with a discussion of the major tropical Pacific
   anomaly. Mon. Weather Rev., 111: 1103–1118.
Livezey, R.E. and Chen, W.Y., 1983. Statistical field significance and its determination by Monte
   Carlo techniques. Mon. Weather Rev., 111: 46–59.
O'Brien, J.J., 1978. El Niño – an example of ocean–atmosphere interactions. Oceanus, 21(4): 40–46.
Preisendorfer, R.W. and Barnett, T.P., 1983. Numerical model–reality intercomparison tests using
   small sample statistics. J. Atmos. Sci., 40: 1884–1896.
Quiroz, R.S., 1983. The climate of the "El Niño" Winter of 1982–83, a season of extraordinary
   climate anomalies. Mon. Weather Rev., 111: 1685–1706.
Rasmusson, E.M. and Wallace, J.M., 1983. Meteorological aspects of the El Niño/Southern Oscillation.
   Science, 222: 1195–1202.
Shukla, J. and Wallace, J.M., 1983. Numerical simulation of the atmospheric response to equatorial
   Pacific sea surface temperature anomalies. J. Atmos. Sci., 40: 1613–1630.

Von Storch, H., 1982. A remark on Chervin/Schneider's algorithm to test significance of climate experiments with GCMs. J. Atmos. Sci., 39: 187–189.

Von Storch, H., 1984. An accidental result: The mean 1983 January 500 mb height field significantly different from its 1967–81 predecessors. Beitr. Phys. Atmos., 57: 440–444.

Von Storch, H. and Roeckner, E., 1983. On the verification of January GCM simulations, Proc. II International Meeting on Statistical Climatology, September 26–30, 1983, Lisboa.

Wagner, A.J., 1983. The climate of Summer 1982. A season with increasingly anomalous circulation over the equatorial Pacific. Mon. Weather Rev., 111: 590–601.

Webster, P.J., 1981. Mechanisms determining the atmospheric response to sea surface temperature anomalies. J. Atmos. Sci., 38: 554–571.

CHAPTER 21

RESPONSE OF A GFDL GENERAL CIRCULATION MODEL TO SST FLUCTUATIONS OBSERVED IN THE TROPICAL PACIFIC OCEAN DURING THE PERIOD 1962–1976

NGAR-CHEUNG LAU and ABRAHAM H. OORT

ABSTRACT

The results of a special 15-year integration of an atmospheric general circulation model are compared with observations. In the $30°S-30°N$ strip over the Pacific, the lower boundary of the model is forced by sea-surface temperatures which vary continuously according to actual observations during the period January 1962–December 1976. Everywhere else, the sea-surface temperatures follow a normal annual cycle without year-to-year variations. Using global teleconnection maps and time series of certain tropical circulation indices, such as the low-level and upper-level zonal winds, the surface pressure and the 200-mb height, it is shown that the dominant spatial modes of atmospheric variability in the tropics are very well simulated. Some of these modes are conspicuously absent from an earlier integration in which the surface temperatures everywhere were prescribed to follow the normal seasonal cycle.

1. INTRODUCTION

A 15-year integration of an atmospheric general circulation model (GCM) was performed at the Geophysical Fluid Dynamics Laboratory (GFDL) with continuously varying sea-surface temperatures (SST) specified at the lower boundary of the model. In a tropical strip of the Pacific ($30°S-30°N$, $120°E-80°W$), observed SST during the period January 1962–December 1976 were prescribed, whereas the SST everywhere else were constrained to evolve through fifteen identical annual cycles. The main purpose of this integration was to find out how realistic the response of the atmospheric model was and, by inference, how much of the variability in the real atmosphere might be accounted for by the SST anomalies in the tropical Pacific Ocean.

In a previous 15-year integration, Manabe and Hahn (1981) and Lau (1981) showed that a GCM with climatological SST conditions prescribed everywhere can reproduce both the amplitude and characteristic spatial patterns of the observed atmospheric variability in middle latitudes. However, the simulated variability in the tropics was much below the observed level, and no signal of the El Niño/Southern Oscillation (ENSO) was detected. These two papers raised the important issue of whether SST anomalies in the tropics are an essential element for a successful simulation of various atmospheric phenomena associated with ENSO. The importance of the ENSO phenomena in the year-to-year variations of the tropical climate has, of course, been realized for many years (Walker and Bliss, 1932, 1937; Berlage, 1966; Bjerknes, 1969). Recent observational studies (e.g. Rasmusson and Carpenter, 1982) have brought the significance of these tropical phenomena to the attention of an even broader sector of the scientific community.

In the present paper we will present some results from a 15-year simulation with actual, temporally and spatially varying SST anomalies prescribed in the tropical Pacific Ocean, and compare them with the corresponding statistics for the observed atmosphere. We will show that most of the deficiencies of the earlier integration with fixed SST have been removed, and that the principal modes of variability in the tropical atmosphere are now realistically simulated.

2. MODEL CHARACTERISTICS AND EXPERIMENTAL DESIGN

The GCM used in this study is a global model with nine levels in the vertical. Horizontal variations are represented spectrally by spherical harmonics with a rhomboidal truncation at fifteen wavenumbers, which is equivalent to a grid resolution of approximately 7.5° in longitude and 4.5° in latitude. Seasonal changes in the insolation are imposed at the top of the model atmosphere. A comprehensive hydrologic cycle is incorporated, with water vapor being one of the predicted variables. In the radiative transfer calculations, the surface albedo is allowed to change only when snow cover or sea ice are present, and the cloud cover is assumed to be a function of height and latitude only. Other details of the model formulation have been given in Gordon and Stern (1982).

The 15-year integration was started on the first day of the calendar year. The initial atmospheric conditions are furnished by the data for January 1 of an arbitrary year from another long-term model run conducted at GFDL (see Manabe and Hahn, 1981, and Lau, 1981, for details). Except for the tropical Pacific, the surface temperatures over the oceans are constrained to evolve through fifteen identical seasonal cycles, as defined using the climatological data published by the U.S. Naval Oceanographic Office. Within the tropical belt over the Pacific (30°S–30°N, 120°E–80°W), the lower boundary condition at each model grid point is prescribed to follow the local SST fluctuations observed during the period January 1962–December 1976. This particular 15-year period was selected primarily because it encompasses four rather well-defined El Niño events. The SST data fields were kindly provided by E.M. Rasmusson and T.H. Carpenter of the Climate Analysis Center. The SST in the model are updated every day. The daily values are obtained by linear interpolation from the monthly mean data, with the assumption that the monthly mean values apply at the middle of each month.

3. COMPARISON OF MODEL RESULTS WITH OBSERVATIONS

3.1. Global teleconnection maps

In this subsection we shall present the temporal correlation patterns between the SST fluctuations in the equatorial eastern Pacific and the corresponding fluctuations of selected atmospheric variables at all model grid points. These so-called teleconnection charts offer a global view of the relationship between SST anomalies in the region of forcing and large-scale atmospheric circulation features located elsewhere. Pan and Oort

(1983) have demonstrated that the SST fluctuations in the region located between 10°S and 5°N at 130°W are most indicative of the behavior of various warm- and cold-water episodes. We shall henceforth correlate various atmospheric parameters with the SST data averaged over this "key region". The SST time series thus obtained is very similar to that shown in Fig. 3a. All computations are performed with zero lag, and are based on monthly mean values from which the normal seasonal cycle has been removed. For each model result shown, the corresponding pattern based on observational data for the 10-year period from May 1963 to April 1973 is also displayed. Most of these latter maps have been discussed in detail by Pan and Oort (1983).

3.1.1. Near-surface and 200-mb zonal winds

In Fig. 1 are shown the distributions of correlation coefficients between SST fluctuations averaged over the key region defined above and the zonal wind component at (b) 950 mb and (d) 200 mb, as computed using model data. The corresponding observed patterns are presented in panels (a) for surface wind and (c) for 200-mb wind. For both the observed and simulated atmospheres, the strongest correlations are found in the proximity of the key region. In the lower troposphere (Figs. 1a and b), warmer than normal SST in the eastern tropical Pacific are accompanied by eastward wind anomalies west of the key region, and by westward wind anomalies further east. The polarity of these correlations is reversed in the upper troposphere (Figs. 1c and d), where warm-water episodes are associated with westward anomalous flow immediately to the west of the key region, and eastward flow along the remainder of the equatorial belt. This three-dimensional configuration of zonal wind anomalies implies the occurrence of low-level convergence, upper-level divergence, and therefore rising motion over the key region when the local SST are warmer than average. Analogously, it is seen that the same warm water episodes are associated with enhanced subsidence over the western tropical Pacific and Indonesia. The anomalous features during El Niño periods hence act to weaken or even reverse the Walker Circulation, which normally prevails in the zonal plane over the equatorial Pacific. During cold episodes the opposite conditions occur, with stronger rising motions over the western tropical Pacific, and sinking over the eastern Pacific.

Another feature worth noting in the 200-mb charts (Figs. 1c and d) is the tendency for stronger westerlies near 30°N and 30°S over the eastern Pacific during warm-water episodes, and weaker westerlies during cold-water episodes. The correlations at locations much beyond the subtropical oceans are generally rather weak.

Over much of the tropics and subtropics, the model and observational results presented in Fig. 1 are in good agreement with each other. Notable discrepancies are discernible in the 200-mb charts over the North Atlantic and Europe, perhaps as a consequence of the absence of interannual variability in the prescribed SST in the Atlantic Ocean.

3.1.2. Surface pressure or 1000-mb height

The correlation coefficients between SST fluctuations in the key region and the global distributions of 1000-mb height are displayed in Fig. 2, for (a) observations and (b) model. The global-scale east—west seesaw in the surface pressure pattern associated with

(a)

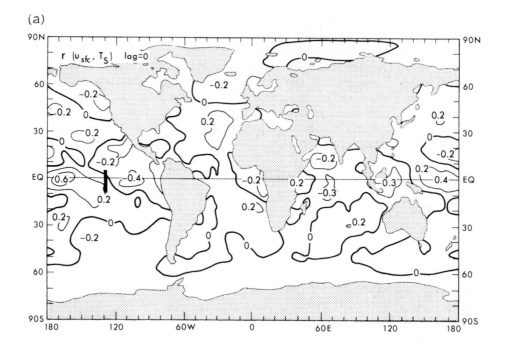

(b)

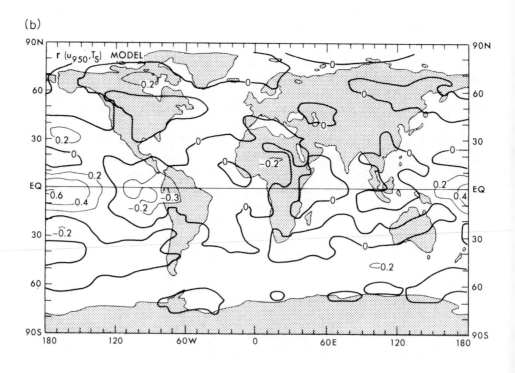

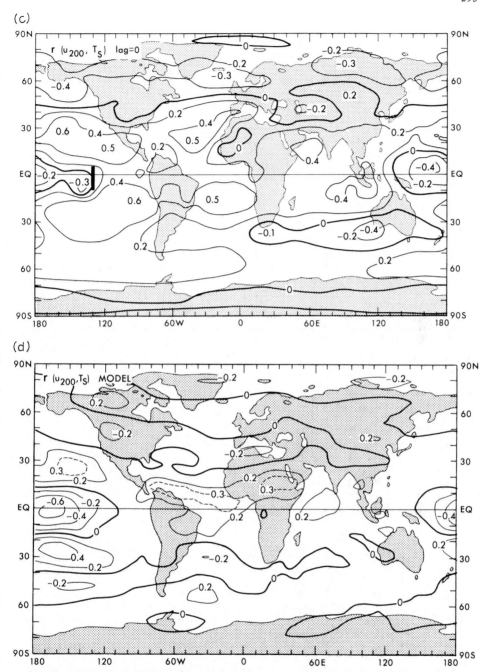

Fig. 1. Maps of the correlation coefficient (r) between the monthly sea-surface temperature anomalies in the eastern equatorial Pacific Ocean (T_s) and the anomalies in the eastward component of the wind (u) near the Earth's surface and in the upper troposphere. (a) The observed $r(u_{sfc}, T_s)$; (b) the model-derived $r(u_{950}, T_s)$; (c) the observed $r(u_{200}, T_s)$; (d) the model-derived $r(u_{200}, T_s)$. The observed maps are from Pan and Oort (1983). T_s was averaged over the key region from $10°$S to $5°$N at $130°$W (indicated by a solid bar in panels a and c). Values of $|r| \geqslant 0.3$ are considered to be significant at the 95% confidence level for a 120-month sample (see Pan and Oort, 1983).

(a)

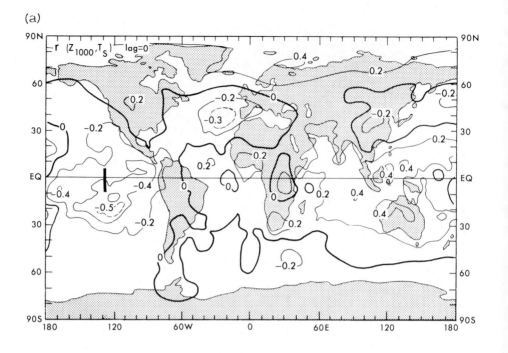

(b)

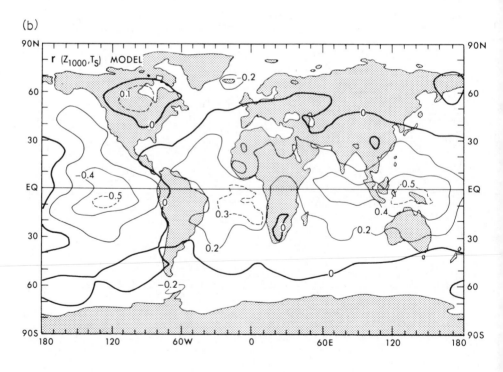

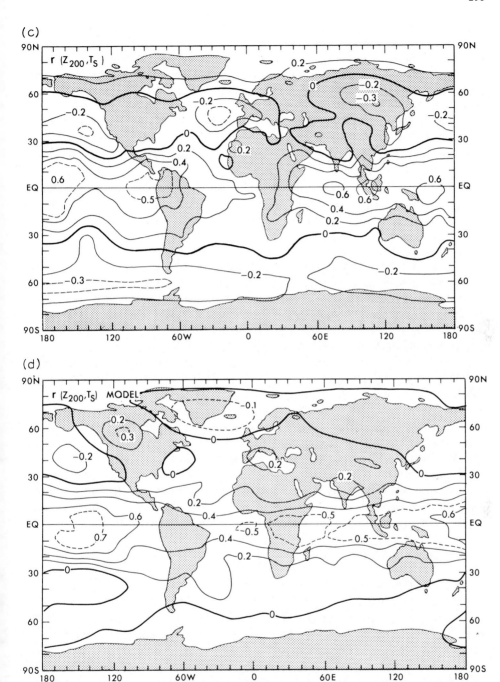

Fig. 2. Maps of the correlation coefficient between the monthly sea-surface temperature anomalies in the eastern equatorial Pacific Ocean and the anomalies in the geopotential height field (z). (a) The observed $r(z_{1000}, T_s)$; (b) the model-derived $r(z_{1000}, T_s)$; (c) the observed $r(z_{200}, T_s)$; (d) the model-derived $r(z_{200}, T_s)$. The observed maps were constructed using Oort's (1983) data for the May 1963–April 1973 period and Pan and Oort's (1983) method of analysis (also see legend of Fig. 1).

SST anomalies in the key region is clearly portrayed in these charts. Warm-water episodes are seen to be accompanied by below-normal surface pressure over the tropical Pacific east of the date line, and by above-normal pressure over the western tropical Pacific, Indonesia, large parts of the Indian Ocean and tropical Atlantic. The polarities of these anomalies are reversed during cold-water episodes. This global-scale phenomenon, often referred to as the Southern Oscillation (SO), is known to be the principal mode of variability in the tropical circulation. The evidence presented here demonstrates the close relationship between the SO and SST fluctuations in both the observed and model atmospheres. With the exception of the pattern over the North Atlantic, the western half of the Indian Ocean, and Europe, there exists good agreement between the observations and the model simulation. The discrepancies over the tropical Atlantic and Indian Oceans may partly be a result of the climatological SST conditions prescribed in those regions (see also discussion in Section 4).

3.1.3. 200-mb height

The observational studies by Horel and Wallace (1981) and Pan and Oort (1983) suggest that ENSO episodes tend to be associated with warming of the troposphere along the entire equatorial belt. The fidelity of the model simulation in this regard may be tested by examining the correlations between SST fluctuations and 200-mb height, shown in Fig. 2 for (c) observations and (d) model. The 200-mb height field is a good indicator of the thickness of the atmospheric column between the surface and the tropopause, and hence provides a representative measure of the vertical mean temperature of the entire troposphere. The model pattern (Fig. 2d) is characterized by uniformly high positive values along the whole equatorial belt, thus indicating above-normal (below normal) tropospheric temperatures over most of the tropics during warm (cold) water events. The degree of zonal symmetry is not as strong in the observed pattern (Fig. 2d), which yields relatively weaker correlations over central Africa and the eastern equatorial Atlantic. Notable differences are also present in the midlatitude belt between 30° and 60°N. The observations indicate an almost continuous belt of negative correlations along this belt, whereas the model pattern exhibits positive maxima over North America and the Mediterranean region. Again the absence of interannual variability in the SST prescribed in the Atlantic may be one of the important reasons for these discrepancies.

3.2. Time series of tropical circulation indices

We proceed to present here several monthly time series depicting important aspects of the tropical circulation during the 15 years of model integration. In order to highlight the low-frequency fluctuations associated with ENSO, a smoothed version of each time series is also displayed. The smoothing is achieved by processing the monthly-mean values through a 15-point Gaussian-type filter with weights 0.012, 0.025, 0.040, 0.061, 0.083, 0.101, 0.117 and 0.122, the last weight being applied at the center point. The same filter was used in Pan and Oort (1983). The climatological seasonal cycle was subtracted from the SST and all model data to obtain anomalies. These anomalies were then normalized by their respective standard deviations, which were calculated separately for each calendar month. The model results are compared with corresponding observed time series for the 1962–76 period, as reported by various investigators. Since the latter

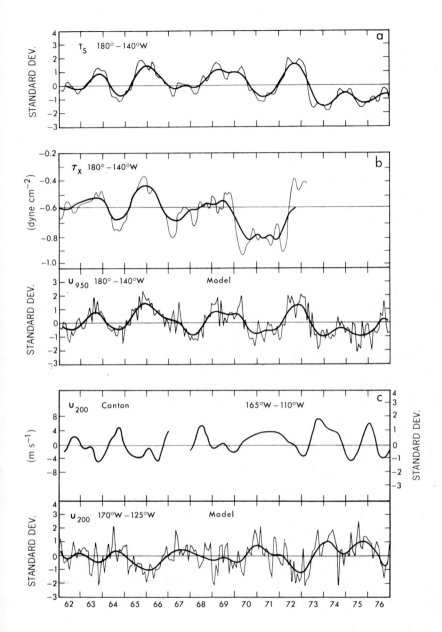

Fig. 3. Time series of monthly-mean anomalies during the 15-year period, January 1962–December 1976. (a) Observed T_s averaged over the equatorial strip 5°S–5°N, 180°–140°W; (b) observed surface-wind stress averaged over strip 4°S–4°N, 180°–140°W from Philander and Rasmusson (1985) (dyne cm^{-2}), and model-derived u_{950} averaged over strip 5°S–5°N, 180°–140°W; (c) observed u_{200}: for 1962–1967 period, low-pass filtered data at Canton Island from Julian and Chervin (1978), and for 1968–1976 period, 5-month running mean curve for the strip 5°S–5°N, 165°–110°W from Climate Analysis Center (1984), and model-derived u_{200} averaged over the strip 5°S–5°N, 170°–125°W. The smoothed curves in the model results were obtained by applying a 15-point Gaussian filter to the monthly data. All model results are normalized by the standard deviations (see text).

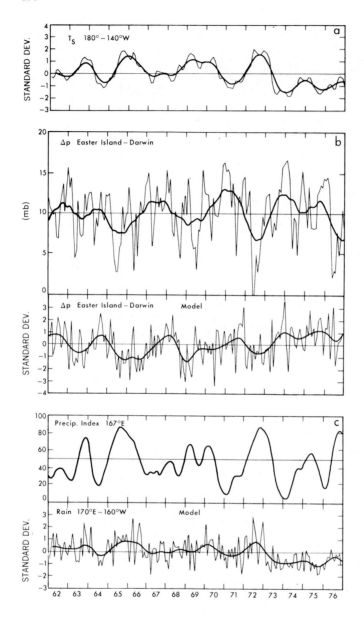

Fig. 4. Time series of monthly-mean anomalies during the 15-year period January 1962–December 1976. (a) Observed T_s in equatorial strip $5°S-5°N$, $180°-140°W$; (b) observed and model-derived differences in surface pressure, Easter Island minus Darwin; the observed series is from Wyrtki (1982) with a 12-month running mean smoothing; (c) an observed precipitation index (in %) at $5°S$, $167°E$ from Philander and Rasmusson (1985), and the model-derived rainfall averaged for the zone $5°S-5°N$, $170°E-160°W$ (see also legend of Fig. 3).

results were computed using somewhat different analysis procedures, only qualitative comparisons between observations and model can be made here.

Fluctuations of various atmospheric parameters will be discussed in relation to the SST conditions in the eastern equatorial Pacific. In the top panels of both Figs. 3 and 4 are shown the time series of observed SST data averaged over the equatorial Pacific between 180° and 140°W. As mentioned in Section 2, this data set was provided by Rasmusson and Carpenter, and has been used to specify the lower boundary condition of the GCM in this experiment. The time series in Figs. 3a and 4a are seen to be dominated by the four warm-water episodes in 1963, 1965, 1969 and 1972, with cold-water episodes in between.

3.2.1. Surface and 200-mb zonal wind

The surface-wind conditions over the equator east of the date line are shown in Fig. 3b. The observed values are for the zonal wind stress in dyne cm^{-2} from Philander and Rasmusson (1985). In agreement with the results in Figs. 1a and b, we find a high positive correlation near zero-lag between warm SST episodes and westerly winds over the equator just east of the date line. Model and observed results agree very well.

In Fig. 3c we compare the model results for the 200-mb zonal wind with the observed time series over the eastern equatorial Pacific. Again fairly good agreement is found. Comparison with the SST curve in Fig. 3a reveals a high negative correlation as was discussed before in connection with Figs. 1c and d. The curves in Fig. 3c indicate that the Walker Circulation tends to be weaker or even reversed during warm-water episodes, and vice versa during cold-water episodes.

3.2.2. Sea-level pressure and precipitation

In Fig. 4 are shown the variations of (b) the pressure difference between Easter Island and Darwin, one of the classical indices of the Southern Oscillation, and (c) the precipitation at the Equator near the date line. As is well known, abnormally low pressures over the eastern equatorial Pacific and high pressures over Indonesia tend to accompany warm ENSO episodes, thus leading to a negative correlation between the SST anomalies and the pressure differences, as shown clearly in Fig. 4b. At the same time, abnormally high amounts of precipitation occur over the equator near and to the east of the date line,

TABLE 1

Contemporary correlation coefficients between the time series of the (Northern Hemisphere) winter mean SST in the equatorial eastern Pacific Ocean and various winter-mean atmospheric indices. See Horel and Wallace (1981) for definitions of the 200-mb height and PNA indices

	Observed[a]	Model
Sea-level pressure difference:		
Tahiti–Darwin	−0.83	−0.84
200-mb height index	0.80	0.77
700-mb PNA index	0.46	0.53
Fanning rainfall	0.79	0.71
Christmas Island rainfall	0.64	0.71
Canton rainfall	0.82	0.74

[a]From Horel and Wallace (1981).

as shown in both the observed and model curves. The reversed conditions prevail during cold SST episodes.

In summary, we find good agreement during both warm- and cold-water events between observed and model data for the tropical Pacific Ocean. This same excellent agreement is also shown in more quantitative terms in Table 1, in which the model results are compared with various observed station statistics published by Horel and Wallace (1981). The 200-mb height index shown in this table is a weighted average of the 200-mb height anomalies at five stations spread more or less evenly around the tropical belt (see Horel and Wallace, 1981, eqn. 1). The strong positive correlation between this index and the SST fluctuations is indicative of the tropospheric warming during ENSO episodes, as was pointed out in Section 3.1.3.

3.3. Extratropical response to SST anomalies

Up to this point we have focussed the discussion on the tropics. However, another important question concerns the atmospheric response in midlatitudes to abnormal SST in the equatorial Pacific. Horel and Wallace (1981) have discerned a definite relationship between SST variability and the Pacific–North American (PNA) circulation pattern. This relationship is illustrated in Fig. 5a, which shows a simultaneous (zero-lag) correlation map of the winter 700-mb geopotential height field with the SST anomalies in the eastern equatorial Pacific based on a sample of 28 winters. A similar map was

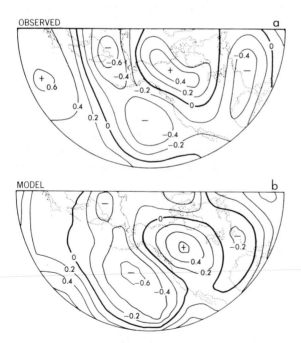

Fig. 5. Maps of the correlation coefficient between the monthly sea-surface temperature anomalies in the eastern equatorial Pacific Ocean and the anomalies in the 700-mb geopotential height field for the Pacific–North American region north of 20°N. (a) Observed $r(z_{700}, T_s)$ from Horel and Wallace (1981); (b) model-derived $r(z_{700}, T_s)$.

computed using data for 15 model winters and is presented in Fig. 5b. It shows a PNA correlation pattern that closely resembles the observed one.

Thus, in spite of the high level of meteorological noise in middle latitudes associated with synoptic weather systems, a significant signal is detectable during Northern Hemisphere winter. However, as is demonstrated in Table 1, the correlation between SST and the PNA index (see definition in Horel and Wallace, 1981) is substantially weaker than correlations involving tropical indices. It is also worth noting that, in agreement with observations, the PNA signal in the model atmosphere is much weaker during the summer months.

4. SUMMARY AND CONCLUSIONS

The results of a 15-year integration with an atmospheric general circulation model forced by seasonally varying insolation and realistic surface boundary conditions were discussed and compared with observations. At the lower boundary of the model the SST pattern in the tropical Pacific Ocean was specified to vary from month-to-month and year-to-year according to actual ship observations as analyzed by Rasmusson and Carpenter (1982) for the period January 1962–December 1976. Outside the tropical Pacific, the SST followed the observed normal seasonal cycle and were not subject to interannual variations.

Manabe and Hahn (1981) and Lau (1981) have previously analyzed an experiment with the lower boundary being forced by a normal seasonal cycle everywhere, and detected no significant modes of variability over the tropics. On the contrary, it is demonstrated that the present experiment produces a realistic simulation of the year-to-year variations associated with the ENSO events in the tropical atmosphere. Thus we can conclude that a considerable fraction of the interannual variability in the tropical atmosphere may be attributed to interannual variability in the tropical SST, especially in the Pacific.

Through correlation maps between the zonal wind component and the SST in the eastern equatorial Pacific we showed a global picture of the atmospheric response in the near-surface and 200-mb wind fields. In excellent agreement with observations, a reversed Walker Circulation in the equatorial zonal plane, with westerly anomalies at the surface and easterly anomalies aloft, was found to prevail during warm SST conditions, and vice versa for cold SST conditions. Correlation maps for the geopotential height field showed east–west shifts of atmospheric mass over the equatorial Pacific and general heating in the tropical atmosphere during warm-water episodes, also in conformity with the observed features. However, certain differences between model and observations became apparent over the midlatitudes and over the Atlantic and Indian Oceans. These discrepancies may be partially accounted for by the absence in this experiment of any interannual variability of SST in the Atlantic and Indian Oceans, whereas Pan and Oort (1983, fig. 8b) have noted that SST in all oceans tend to be positively correlated along most of the equator. It is hence likely that another model integration incorporating SST variability at all longitudes might yield an even more realistic simulation.

The low-frequency component of the fluctuations of selected tropical indices observed during the 1962–1976 period also showed good correspondence with the model results.

Outside the tropics, the Pacific—North American circulation pattern was well simulated during the Northern Hemisphere winter season.

ACKNOWLEDGMENTS

The authors would like to express their thanks to J. Shukla for helpful suggestions concerning the design of this experiment, to Syukuro Manabe, Douglas Hahn and Donahue Daniel for the use of their General Circulation Model results, to the members of the Scientific Illustration Group at GFDL for drafting the figures, and to Joyce Kennedy for typing the manuscript.

REFERENCES

Berlage, H.P., 1966. The Southern Oscillation and world weather. R. Neth. Meteorol. Inst., Meded. Verh., No. 88, 152 pp.
Bjerknes, J., 1969. Atmospheric teleconnections from the Equatorial Pacific. Mon. Weather Rev., 97: 163—172.
Climate Analysis Center, 1984. Climate Diagnostics Bulletin, May 1984. NOAA/National Weather Service, Washington, D.C.
Gordon, C.T. and Stern, W.F., 1982. A description of the GFDL global spectral model. Mon. Weather Rev., 110: 625—644.
Horel, J.D. and Wallace, J.M., 1981. Planetary-scale atmospheric phenomena associated with the Southern Oscillation. Mon. Weather Rev., 109: 813—829.
Julian, P.R. and Chervin, R.M., 1978. A study of the Southern Oscillation and Walker Circulation phenomena. Mon. Weather Rev., 106: 1433—1451.
Lau, N.-C., 1981. A diagnostic study of recurrent meteorological anomalies appearing in a 15-year simulation with a GFDL general circulation model. Mon. Weather Rev., 109: 2287—2311.
Manabe, S. and Hahn, D.G., 1981. Simulation of atmospheric variability. Mon. Weather Rev., 109: 2260—2286.
Oort, A.H., 1983. Global Atmospheric Circulation Statistics, 1958—1973. NOAA Prof. Pap. No. 14, U.S. Gov. Printing Office, Washington, D.C., 180 pp.
Pan, Y.H. and Oort, A.H., 1983. Global climate variations connected with sea surface temperature anomalies in the eastern equatorial Pacific Ocean for the 1963—1973 period. Mon. Weather Rev., 111: 1244—1258.
Philander, S.G. and Rasmusson, E.M., 1985. The Southern Oscillation and El Niño. Issues in Atmospheric and Oceanic Modeling. Adv. Geophys. Ser., Academic Press.
Rasmusson, E.M. and Carpenter, T.H., 1982. Variations in tropical sea surface temperature and surface wind fields associated with the Southern Oscillation/El Niño. Mon. Weather Rev., 110: 354—384.
Walker, G.T. and Bliss, E.W., 1932. World Weather V. Mem. R. Meteorol. Soc., 4: 53—84.
Walker, G.T. and Bliss, E.W., 1937. World Weather VI. Mem. R. Meteorol. Soc., 4: 119—139.
Wallace, J.M. and Gutzler, D.S., 1981. Teleconnections in the geopotential height field during the Northern Hemisphere winter. Mon. Weather Rev., 109: 784—812.
Wyrtki, K., 1982. The Southern Oscillation, ocean—atmosphere interaction and El Niño. Mar. Tech. Soc. J., 16: 3—10.

CHAPTER 22

ELEMENTS OF COUPLED OCEAN–ATMOSPHERE MODELS FOR THE TROPICS

A.E. GILL

ABSTRACT

Experiments with a simple coupled model involving integrations in longitude and time only gave some interesting results. First, an initial-value run gave a warm event with many features similar to that observed. Secondly, a self-excited but weakly growing oscillation with a period of about one year could be produced. However, some features of the physics of the model need improvement, one being the relationship between heating rate in the atmosphere and the sea-surface temperature. A moist model has been designed to give this relationship and to produce corresponding fields of low-level wind, position of the ITCZ, etc. Results for "January" and "July" simulations show quite realistic features, and suggest that the most important elements of a coupled model may be capable of being given a simple mathematical form.

1. INTRODUCTION

The major climatic anomaly of 1982/83 drew attention to the need to understand the way in which the tropical ocean and atmosphere can work together to produce such dramatic and damaging events. Much can be learnt from studies of individual parts of the system, but eventually the parts have to be put together to interact so that it can be seen whether they can reproduce the observed type of behaviour. At present, such studies are rather experimental, so one of the best approaches is to try to build a rather simple model but at the same time trying to capture the main features of the physics. This article describes one such attempt.

The study of the 1982/83 event by Gill and Rasmusson (1983) highlights some of the features which a coupled model needs to contain. In particular, the following four processes need to be modelled: (1) The production of anomalous precipitation zones in the tropical Pacific in response to anomalous sea-surface temperature patterns; (2) the production of anomalous wind patterns in the tropical zone due to the anomalous latent heat release associated with the precipitation anomalies; (3) the anomalous water movements in the equatorial Pacific ocean in response to the anomalous winds; and (4) the changes in sea-surface temperature in the tropical Pacific which are a consequence of the anomalous water movements. It can be seen that these four elements combine to give a self-interacting coupled system. Although this system is unlikely to contain *all* the processes necessary to reproduce what is observed, it is hard to envisage a realistic model that ignores any of the above four elements. It therefore seems a good starting point for exploring possibilities.

Having decided to combine the four elements listed above, the next step is to look for the simplest way to model each element in a way which captures the essence of the process but does not lead to complicated mathematics. Gill and Rasmusson (1983) found that the most important features of the 1982/83 event in the tropical Pacific

304

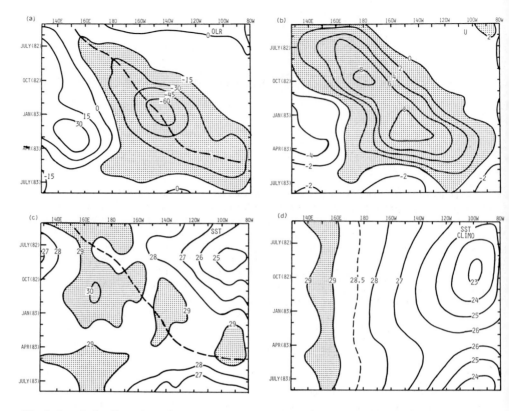

Fig. 1. Longitude–time plots along the equator of monthly mean values of (a) outgoing longwave radiation anomaly during 1982/83 in $W m^{-2}$ (representative of atmospheric heating anomaly); (b) westerly wind anomaly during 1982/83 in $m s^{-1}$; (c) SST during 1982/83 in °C; and (d) SST climatological mean values (from Gill and Rasmusson, 1983).

could be represented by longitude-time plots of properties averaged over a strip about $10°$ wide centred on the equator. Figure 1 shows the changes observed in the position of the precipitation zone, in the zonal-wind anomaly and in the sea-surface temperature. This leads to the idea of constructing a mathematical model in which the essential variations are only in longitude (or eastward distance x) and time t. Variations in the vertical coordinate z or in northwards distance y from the equator are assumed to have simple forms.

The model that was constructed on these lines is described below. It should perhaps be mentioned that the work was actually done in 1982 before most of the observations relevant to the 1982/83 event were available, as this may explain some of the choices made.

2. THE OCEAN MODEL

An ocean model was constructed to determine changes in currents in the ocean due to changes in the wind stress. The model is the same as that described in Gill and Rasmusson

(1983). The simplifications are that (a) the model is linear; (b) there is only one vertical mode that is regarded as significant; and (c) the long-wave (east–west scale ≫ north–south scale) approximation can be made. The non-dimensional equations satisfied by the surface current anomaly (u_w, v_w) and surface elevation anomaly η are:

$$\frac{\partial u_w}{\partial t} - \tfrac{1}{2} y v_w = -\frac{\partial \eta}{\partial x} + \tau \tag{2.1}$$

$$\tfrac{1}{2} y u_w = -\frac{\partial \eta}{\partial y} \tag{2.2}$$

$$\frac{\partial \eta}{\partial t} + \frac{\partial u_w}{\partial x} + \frac{\partial v_w}{\partial y} = 0 \tag{2.3}$$

where τ is the non-dimensional zonal-wind stress anomaly. The suffix "w" for water is used to denote ocean variables and the suffix "a" will be used for air. The scales for the east–west and north–south directions are *not* the same. The east–west scale used is based on the wind-stress scale taken to be the atmospheric Rossby radius:

$$a_a = (c_a/2\beta)^{1/2} \tag{2.4}$$

where c_a is the phase speed of long internal waves in a non-rotating atmosphere, and β is the gradient of the Coriolis parameter. The value of a_a is taken to be equivalent to $10°$ of longitude. The non-dimensional coordinate x therefore ranges from zero at the western boundary of the Pacific to 14 at the eastern boundary. The north–south scale is taken as the oceanic Rossby radius:

$$a_w = (c_w/2\beta)^{1/2} \tag{2.5}$$

with $c_w = 2.8 \text{ m s}^{-1}$. The scale for u_w is c_w and that for v_w is $(a_w/a_a)\,c_w \approx \tfrac{2}{9}\,c_w$. The scale for t is a_a/c_w (4.6 days) and that for η is the equivalent depth $c_w^2/g = 78$ cm. The wind stress scale is $\rho_w c_w^2/B a_a$ where ρ_w is the density of water and B^{-1} is the "equivalent forcing depth" (Gill, 1982a). The value chosen for B is 3.5×10^{-3} as in Wunsch and Gill (1976), making the wind-stress scale 2.0 N m^{-2}. (This seems a very unrealistic value and perhaps the basin width would be a better east-west scale to use than a_a. However, in dealing with the atmospheric model, there are some advantages in using a_a and this was the choice that was made!)

The numerical method of solving the equations is described by Blundell and Gill (1983). The wind stress τ and the variable:

$$q^w = \eta + u_w \tag{2.6}$$

are expanded in the parabolic cylinder functions $D_n(y)$:

$$(\tau, q^w) = \sum_{n=0}^{\infty} (X_n^w, q_n^w) D_n(y) \tag{2.7}$$

where X_n^w and q_n^w are functions of x and t, only satisfying:

$$(2n+1)\frac{\partial q_{n+1}^w}{\partial t} - \frac{\partial q_{n+1}^w}{\partial x} = nX_{n+1} - X_{n-1} \qquad n = -1,1,3,\ldots \tag{2.8}$$

with the convention that X_{-2} is zero. (For $n = -1$ it is the Kelvin wave equation, while

$n = 1, 3, \ldots$ correspond to planetary waves.) In practice the system is truncated at a particular value of n, namely $n = 19$. The boundary condition at the eastern boundary is one of no normal velocity giving:

$$(n + 1)q_{n+1}^{w} = q_{n-1}^{w} \qquad n = 1,3,5,\ldots \tag{2.9}$$

This is used to determine the amplitudes of all the reflected planetary waves in terms of the incident Kelvin wave. At the western boundary, the condition is the single condition of no normal mass flux $\int u \, dy$, which gives:

$$q_0^{w} = q_2^{w} + 1 \cdot (q_4^{w} + 3(q_6^{w} + 5(q_8^{w} + \ldots \tag{2.10}$$

This gives the Kelvin wave amplitude in terms of the amplitudes of the incident planetary waves.

In summary, the above equations show how part 3 of the four elements of the model is put in mathematical form. The wind stress is expanded in the form (2.7) to give the forcing terms on the right-hand sides of the first-order eqns. (2.8). For the planetary waves ($n = 1,3,\ldots$), the time integration proceeds in time steps of three units by tracing back along characteristics from each grid point ($x = 0,1,2,\ldots, 13$), and interpolating in space. Equation (2.9) is used to give the value at the eastern boundary. The same method is used for the Kelvin wave ($n = -1$), but proceeding in time steps of one unit (three steps to each planetary wave step) and integrating from the western boundary where eqn. (2.10) is used.

3. THE SEA-SURFACE TEMPERATURE ANOMALY

Now consider the fourth element or link in the chain, i.e. the problem of calculating changes in sea-surface temperature given the anomalous water movements as calculated in the previous section. The philosophy adopted here was one of opting for mathematical simplicity by focusing on a particular process which previous studies (Gill, 1983; Gill and Rasmusson, 1983) led me to believe were essential to providing the energy for driving the large events. This process is the advection eastwards along the equator of the warm pool normally confined to the western Pacific. This provides a large anomalous source of heat in the central Pacific which drives the atmospheric perturbations.

For modelling purposes, it is supposed that it is sufficient to consider an equatorial surface-temperature perturbation $\theta(x, t)$ only, which in practice will be an average over some finite range of latitudes (say $5°S–5°N$). It is assumed that this equatorial surface temperature is changed by advection due to a mean zonal current $\bar{u}_w$ only (taken as the average over $|y| < 2$). The perturbation is one on a mean zonal gradient, with a temperature scale chosen to make this gradient unity where its gradient is uniform. This gives a temperature scale of:

$$\Delta T = 0.58 \text{ K} \tag{3.1}$$

assuming the gradient is 0.58 degrees per $10°$ of longitude. In practice, the gradient G goes to zero near the western boundary, so the non-dimensional gradient was given the form:

$$G = \tanh(0.4x) \tag{3.2}$$

which is close to unity over most of the range.

The other process included was a Newtonian cooling with non-dimensional decay time s^{-1} so the equation for θ used was:

$$\frac{\partial \theta}{\partial t} + s\theta = -\bar{u}_w G(x) \tag{3.3}$$

The heat budget studies in Gill (1983) suggest a decay time of about 1.4 years which corresponds to a value:

$$s = 0.1 \tag{3.4}$$

In practice, values ranging from 0.01 to 0.09 were tried in addition to $s = 0$.

The formula for $\bar{u}_w$ in terms of the q's works out to be:

$$\bar{u}_w = 0.373q_0 - 0.741q_2 - 0.373q_4 + 3.66q_6 - 174q_8 + 399q_{10} + \ldots \tag{3.5}$$

which is calculated at each value of x and eqn. (3.3) is used to advance θ in time steps of three units (i.e. about a fortnight).

4. THE ATMOSPHERE MODEL

Next consider element 2 in the list given in Section 1 of links in the chain of interactions, namely the atmospheric model. Here the model used is that of Gill (1980), which proved remarkably useful for relating wind anomalies to heating anomalies in the 1982/83 event (Gill and Rasmusson, 1983). The model is closely related to that for the ocean, but because the response time of the atmosphere is so much shorter, the atmosphere is assumed to be in equilibrium with the forcing all the time. The non-dimensional form of the equations is:

$$\epsilon u_a - \tfrac{1}{2} Y v_a = -\frac{\partial p}{\partial x} \tag{4.1}$$

$$\tfrac{1}{2} Y u_a = -\frac{\partial p}{\partial Y} \tag{4.2}$$

$$\epsilon p + \frac{\partial u_a}{\partial x} + \frac{\partial v_a}{\partial Y} = -Q \tag{4.3}$$

where (x, Y) are the horizontal coordinates, but both scaled on the atmospheric Rossby radius a_a and (u_a, v_a) are the non-dimensional low-level velocity components. For consistency with the surface stress scale $\rho_w c_w^2/Ba_a$, the velocity scale adopted was:

$$\frac{\rho_w c_w^2}{\rho_a c_D U B a_a} \tag{4.4}$$

where c_D is the surface drag coefficient [the value of $(\rho_a/\rho_w)c_D$ is taken to be 2×10^{-6}], and U is a typical mean wind speed taken to be $6\,\mathrm{m\,s^{-1}}$. This makes the velocity scale $170\,\mathrm{m\,s^{-1}}$. This mean wind needs to be taken into consideration when using a linear formula to convert the low-level wind u_a to a surface stress τ, and account needs also to be taken of the fact that the mean wind is stronger at the central longitudes than near the boundaries. The non-dimensional formula used here was:

$$\tau = u_a \cos [0.15 (x - 7)] \tag{4.5}$$

To continue with the scales appropriate to the variables in eqns. (4.1), (4.2) and (4.3), the one for surface pressure anomaly p is $\rho_a c_a$ times the velocity scale, i.e. about 11 mb. The decay distance ϵ^{-1} was taken to be 5 in units of Rossby radii following Zebiak (1982). The unit for heating rate Q is $\delta\theta/a_a$ times the velocity unit, where $\delta\theta$ is the potential temperature difference between the surface and a middle level, say 500 mb. For $\delta\theta = 20$ K, the scale for heating rate is 260 K/day. The equivalent precipitation rate is 580 mm/day obtained by multiplying by $\rho_a c_p H/\rho_w L_v$ where c_p is the specific heat of air and L_v the latent heat of vaporisation of water. Of course, these scales are much larger than observed values, but that is simply a consequence of the scales adopted for the non-dimensionalisation.

5. RELATION BETWEEN THE SURFACE TEMPERATURE AND HEATING ANOMALIES

The most difficult part of the four elements to construct was the relation between surface temperature and heating rate. The approach adopted was to suppose the heating rate anomaly Q at any longitude depended only on the surface temperature anomaly $\theta(x)$ at that longitude and had a fixed meridional structure that was the same at all longitudes. Further, this relationship would be determined empirically using the composites of Rasmusson and Carpenter (1982), which give fields during five stages of a warm episode in their figs. 17–21. These fields do not include Q, but do give the low-level divergence which is, according to the atmospheric model, close to being in proportion with Q (for ϵ small). Thus the relationship adopted was:

$$\text{low-level divergence} = -\lambda^* \theta S(Y) \tag{5.1}$$

where λ^* is a constant and $S(Y)$ is a shape factor with value 1 at the equator. The composites give divergences of around 2×10^{-6} s^{-1} for each degree Kelvin of temperature anomaly, so it was assumed that a typical value of λ^* is around:

$$\lambda^* = 2 \times 10^{-6} \text{ K}^{-1} \text{ s}^{-1} \tag{5.2}$$

although it would be legitimate to experiment with a range of values with similar orders of magnitude. To obtain the non-dimensional value λ of λ^*, note that the divergence scale is a_a^{-1} times the velocity scale, i.e. 153×10^{-6} s^{-1} and the temperature scale is 0.58 K. Thus:

$$\lambda = \frac{\lambda^* \rho_a c_p U B a_a^2 \Delta T}{\rho_w c_w^2} = \frac{2 \times 0.58}{153} = 0.008 \tag{5.3}$$

For the shape factor, it was decided to take the first two terms in the expansion of a delta function (cf. Heckley and Gill, 1984), namely:

$$S(Y) = 2/3 D_0(Y) - 1/3 D_2(Y) \tag{5.4}$$

This curve is shown in Fig. 2. Versions with more terms were also tried.

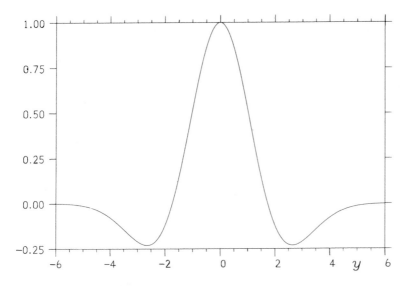

Fig. 2. The shape factor $S(Y)$.

The advantage of this form is that there are only three terms in the expansion of:

$$q^a \equiv u_a + p = q_0^a D_0(Y) + q_2^a D_2(Y) + q_4^a D_4(Y) \tag{5.5}$$

and these satisfy the simple equations (Gill, 1980):

$$dq_0^a/dx + \epsilon q_0^a = -2/3\lambda\theta$$
$$dq_2^a/dx - 3\epsilon q_2^a = 1/3\lambda\theta \tag{5.6}$$
$$dq_4^a/dx - 7\epsilon q_4^a = -1/3\lambda\theta$$

The function q_0^a is obtained by integrating from west to east from zero at the western boundary $x = 0$. The integration proceeded in steps of one unit of distance ($10°$) using the value of θ at the central point and the analytic relation between $q(m+1)$ and $q(m)$ that would apply if $\theta(m+1/2)$ were constant. The functions q_2^a and q_4^a were obtained by integrating westwards from the eastern boundary $x = 14$ using the same technique. The boundary condition used was:

$$q_2^a = 1/2q_0^a \qquad q_4^a = 1/4q_2^a \tag{5.7}$$

corresponding to zero zonal velocity due to the presence of the Andes. This seemed to produce slightly more realistic fields than making the q's zero, although the differences were not usually large.

Having calculated the q's, the low-level zonal wind is calculated from:

$$2u_a = q_0^a D_0(Y) + q_2^a [D_2(Y) - 2D_0(Y)] + q_4^a [D_4(Y)] \tag{5.8}$$

and this can be inserted in eqn. (4.5) to give the surface stress. This has to be expanded by eqn. (2.7) in cylinder functions based on the *oceanic* Rossby radius, so $D_m(Y)$ needs to be expanded in terms of $D_n(Y)$. The method of doing this is described in the Appendix.

310

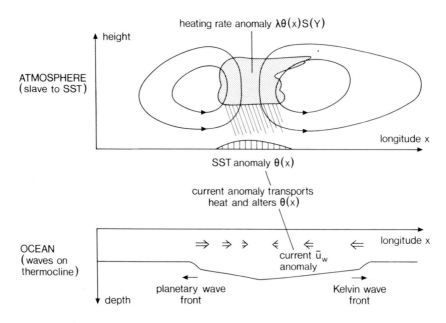

Fig. 3. Schematic of the coupled model as shown by vertical sections in the equatorial plane. The upper panel represents the atmosphere in equilibrium with an SST anomaly, while the lower panel indicates the transient nature of the ocean response.

Now all the elements of the coupled model have been assembled and are summarised in Fig. 3 by means of zonal sections along the equator. The upper panel shows the convection (heating) anomaly zone above the surface temperature anomaly, and the winds associated with it. Not shown is the full three-dimensional structure which is determined by the shape factor $S(Y)$. The atmosphere is always in equilibrium with the surface temperature, so is a functional of $\theta(x)$ and so can be described as a "slave atmosphere" having no independent behaviour of its own.

The ocean on the other hand, does have its own transient behaviour, so if a localised wind field were suddenly switched on, wave fronts would move out as depicted in the lower panel. These are shown to indicate that the model includes such transients. The connection with the atmosphere is that the response involves zonal currents along the equator which create temperature anomalies by advection down the mean zonal gradient.

6. BEHAVIOUR IN THE UNCOUPLED MODEL

Now consider the properties of the system which has been formulated in the previous sections. The atmosphere is always in equilibrium with the ocean so that the state of the atmosphere is completely determined by the function $\theta(x)$. In fact θ is given at 14 grid points only, so these 14 values determine the wind anomalies and the stress on the ocean. The magnitude of the stress is, by eqns. (5.6), (5.8) and (4.5), proportional to the coupling coefficient λ, which is one of the two parameters of the system. As λ tends to zero, the stress tends to zero and the ocean becomes decoupled from the atmosphere.

Unlike the atmosphere, the ocean can exhibit all sorts of transient fluctuations, and its properties are described by a whole set of functions q_n^w and θ. It is helpful to consider the behaviour of these transients in the absence of coupling ($\lambda = 0$). So take the case where the long-term behaviour is exponential in time, i.e.:

$$q_n^w \propto \exp(\sigma t) \tag{6.1}$$

The free solution of eqn. (2.8) satisfying the eastern boundary condition [eqn. (2.9)] is then:

$$q_{n+1}^w = \frac{\exp[(2n+1)\sigma(x-14) + \sigma t]}{(n+1)(n-1)\ldots 1} \tag{6.2}$$

Substituting in the western boundary condition eqn. (2.10) at $x = 0$ then gives:

$$1 = 1/2z - 1/2(-1/2)z^2/2! + 1/2(-1/2)(-3/2)z^3/3! - \ldots \tag{6.3}$$

where the terms are written in the same order as the terms in eqn. (2.10) which give rise to them, and z is given by:

$$z = \exp(-56\sigma) \tag{6.4}$$

The number 56 is the product of 14, the width of the basin, and 4, the sum of the slownesses (inverse wave speeds) of the Kelvin wave and first planetary wave.

The expression (6.3) can be recognised as a binomial expansion corresponding to:

$$(1-z)^{1/2} = 0 \tag{6.5}$$

So if all the terms q_{n+1}^w are included, eqn. (6.5) implies $z = 1$ and eqn. (6.4) gives $\sigma = 0$, so the transients do not decay with time and no energy is lost. If, on the other hand, the expansion (2.11) in cylinder functions is truncated, the binomial expansion (6.3) is truncated to the same number of terms and σ is no longer zero. For instance, if only the first planetary wave is included, only the first term appears on the right-hand side of eqn. (6.3) and so $z = 2$. By eqn. (6.4) this corresponds to $\sigma = -0.012$, a decay time of about one year. If two planetary waves are included, the decay time is 1.9 years and for three planetary waves the decay time is 2.7 years. Thus truncation leads to weak decay. With the 19 waves normally used, this decay is negligible.

The remaining feature of the uncoupled solution is the solution θ of the temperature eqn. (3.3). This has forced solutions which will exhibit the same weak decay rate as the zonal velocity $\bar{u}_w$, i.e. as the q's. Ignoring this decay, it is possible to have steady solutions for the ocean with non-decaying currents and a corresponding fixed temperature anomaly. In addition, eqn. (3.3) has free solutions which decay exponentially at the rate s.

7. RESULTS FOR THE COUPLED MODEL

The coupled model gave some interesting results, but also had some unsatisfactory features which caused me to move on to examine them and suspend work on the coupled model for the time being. It seemed appropriate for a symposium on coupled models to discuss the set-up of the model and what has been done so far, even though there is scope for many improvements.

One type of experiment that was carried out was an initial value experiment with initial values contrived to favour a warm episode. This was done by setting a sea-level distribution favourable for producing eastward currents on the equator, and by starting with positive temperature anomalies. The initial sea level was set up by prescribing q_0^w, q_2^w and q_4^w to give subsequent Kelvin wave and planetary wave behaviour of a suitable kind. Actual values used are given by:

$$
\left.
\begin{aligned}
q_0^w &= 0.216 \cos (\pi x/28) \\
q_2^w &= -0.144 \sin (\pi x/14) \\
q_4^w &= \begin{cases} 0.432 \cos (\pi(x-3)/9) & \text{for } x < 7.5 \\ 0 & \text{for } x > 7.5 \end{cases}
\end{aligned}
\right\}
\qquad (7.1)
$$

The idea was to prescribe planetary waves which on reflection from the western boundary would produce a Kelvin wave such that a strong eastward current would result, thus creating a positive temperature anomaly. It was found that only low-order planetary waves could produce significant effects.

The initial values of $\theta(x)$ can be seen in Fig. 4 which shows longitude−time plots of the solution. These show the first 16 months, this duration being chosen as being the same as that for Fig. 1 and for the plots in Gill and Rasmusson (1983). The values of the parameters used are:

$$\lambda = 0.012 \qquad s = 0.03 \qquad\qquad\qquad (7.2)$$

both a bit larger than the estimates made from observations. The value of s does not matter too much, but λ was increased by 50% from that given by eqn. (5.3) as the winds produced did not seem strong enough.

It can be seen that a large eastward current surge is produced and this causes a positive temperature anomaly of 2.5 degrees to occur in the central Pacific. High sea levels were produced at the eastern boundary. The wind system that the model produces consists of westerly anomalies west of the temperature anomaly and easterly anomalies to the east.

Comparing with the Rasmusson and Carpenter composites, the temperature anomaly in the central Pacific looks quite reasonable in terms of its duration and its location. However, the warm anomaly in the east Pacific is missing because the mechanism for producing that anomaly is not included in the model. This mechanism involves upwelling of anomalously warm water and advection (see discussion in Gill, 1982d, 1983 and in Gill and Rasmusson, 1983).

The westerly wind anomaly to the west of the temperature anomaly is quite realistic, but the equally strong easterly anomaly to its east is not. One important reason for this is the absence of the warm anomaly in the east Pacific. Another could be non-linear effects − these tend to enhance the westerlies (Gill and Philips, in prep.). Also the westerlies are not as strong as they should be. This could result from the effects mentioned above, or perhaps the coupling coefficient should be even larger than the value given by eqn. (7.2).

Comparing with the 1982/83 event, the initial conditions produce a good temperature

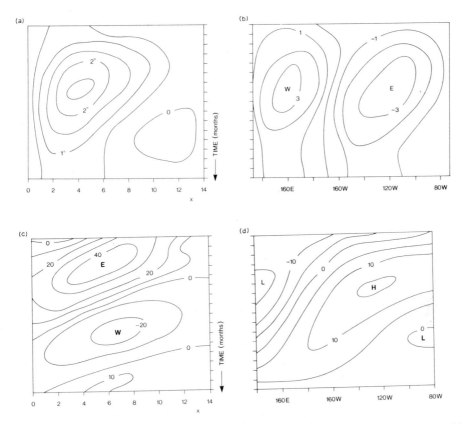

Fig. 4. Longitude–time plots along the equator of the results of an initial-value calculation with the coupled model. Fields shown are (a) SST anomaly θ in °C; (b) zonal-wind anomaly in m s^{-1} ($W =$ westerlies, $E =$ easterlies); (c) zonal-current anomaly u_w in cm s^{-1} ($E =$ eastward, $W =$ westward); and (d) sea level in cm ($H =$ high, $L =$ low).

anomaly, but the feedback processes do not result in winds which cause the anomaly to migrate further east. Again this can be due to the effects mentioned in the previous paragraph.

If the model is run with the same parameters for a long time (more than a decade), a self-excited oscillation eventually emerges. Figure 5 shows longitude–time plots for a 16-month period at the end of such a run. Weak amplification occurs, but the amplitudes increase only a few percent each year. The main feature is that the growing disturbance is oscillatory in character with a period close to a year. The wave is not stationary but shows a tendency for propagation towards the west. Temperature and current pertur-bations are greatest near the central longitude, whereas the largest wind fluctuations are somewhat to the west. Sea-level fluctuations tend to be largest near the boundaries.

The period and growth rate change a little with changes in parameters, but the increase in amplitude per year is not large for values near the ones quoted [λ and s between zero and the values given by eqn. (7.2)]. Negative as well as positive growth rates were found, and values could change with altering the shape factor $S(Y)$. The period is determined basically by the time for a Kelvin wave and first planetary wave to go back and forth

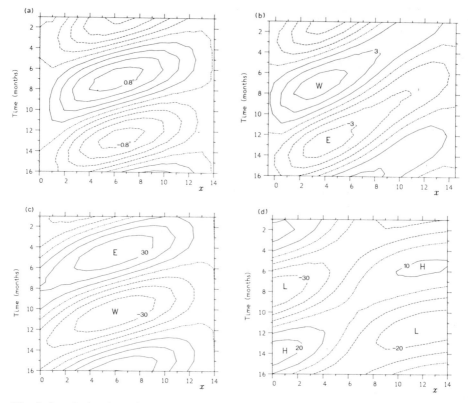

Fig. 5. Longitude–time plots along the equator showing the properties of a weakly growing self-excited oscillation. Values are shown for the last 16 months of a long run for the same fields as in Fig. 4. Units as in Fig. 4. Note that the contour interval for θ is smaller.

across the basin. This is about nine months in the absence of atmospheric feedback, but the period is increased by the feedback effects.

The self-excited mode of oscillation obtained in the model does not seem particularly relevant to the El Niño/Southern Oscillation because the period is wrong and the growth rate is so small. The initial value approach can produce effects that look more like a warm event, but completely dominate the self-excited mode. Therefore it was felt that the model as currently formulated does not have the necessary physics to explain why warm events should occur quasi-periodically with a spacing of 2–5 years. Also, the events of 1982/83 diverted my attention from the coupled model and made me more interested in re-examining some of the component parts, particularly the atmospheric part. The work on non-linear effects has already been mentioned. Another feature examined was the relationship between the sea-surface temperature anomaly and the latent heat release in the atmosphere. The remainder of the article is devoted to this topic.

8. RELATING PRECIPITATION TO SST

In the version of the coupled model described above, the anomalous heating (in practice related to anomalous precipitation) was placed at the same longitude as the SST anomaly and made proportional to SST. This means, in effect, that the relationship between precipitation and SST is controlled by a linear process. Experience with convection suggests this might be a rather poor approximation because convection is a strongly non-linear process where the rising motion tends to be concentrated over hot spots while the compensating descent is much more widely distributed in space. The observations of the 1982/83 event (Gill and Rasmusson, 1983) tend to confirm the idea that convection is strongly related to the actual SST, whereas correlations between precipitation anomalies and SST *anomalies* are less direct. If that is the case, there is a need for a simple model which relates the precipitation field to the SST field in the tropics. The model must therefore have some means of representing moist processes.

The simplest way I have found to do this is to use the same equations as in Gill (1980) for heat and momentum, but to add a moisture equation of the same form as used in two previous papers (Gill, 1982b, c). As in Gill (1980), the dependent variables apart from moisture, can be taken to vary sinusoidally with height (see schematic in Fig. 6), so that in non-dimensional form the horizontal velocity is $(u, v) \cos z$, the vertical velocity is $w \sin z$, the perturbation pressure is $p \cos z$ and the perturbation potential temperature is $\theta \sin z$. Then the hydrostatic equation gives:

$$p = -\theta \tag{8.1}$$

the incompressibility relation gives:

$$w + \frac{\partial u}{\partial x} + \frac{\partial v}{\partial Y} = 0 \tag{8.2}$$

and the non-dimensional form of the momentum equation is:

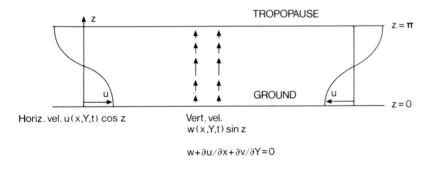

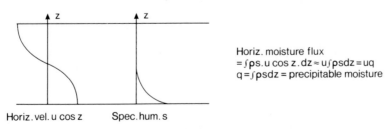

Fig. 6. Schematic of the vertical structure of the fields as assumed in the moist model.

$$\frac{\partial u}{\partial t} - \tfrac{1}{2} Yv = \frac{\partial \theta}{\partial x} - \epsilon u \qquad (8.3)$$

$$\frac{\partial v}{\partial t} + \tfrac{1}{2} Yu = \frac{\partial \theta}{\partial Y} - \epsilon v \qquad (8.4)$$

with ϵ being a non-dimensional Rayleigh friction. These are the same as eqns. (4.1) and (4.2) with the same non-dimensionalisation, but the suffix "a" has been dropped, the time-dependent terms are included, the long-wave approximation has not been made and the hydrostatic relation [eqn. (8.1)] has been utilised. The heat equation [cf. eqn. (4.3)] is:

$$\frac{\partial \theta}{\partial t} + w = Q - \epsilon\theta \equiv P - \epsilon(\theta - \theta_s) \qquad (8.5)$$

and has a particular form for the heating rate Q which is explained below. In this equation ϵ represents a Newtonian cooling effect.

The moisture equation utilizes the fact that the moisture is concentrated near the surface and nearly all in the lower half of the troposphere (see Fig. 6). Thus if q is the total precipitable moisture, the horizontal moisture flux is close to (uq, vq), and the moisture equation can be written as:

$$\frac{\partial q}{\partial t} + \frac{\partial}{\partial x}(uq) + \frac{\partial}{\partial Y}(vq) = -P - \epsilon(q - \bar{q}) \qquad (8.6)$$

P represents the precipitation rate and $-\epsilon(q - \bar{q})$ the evaporation rate. The coefficient ϵ is given the same value as in the other equations for convenience. Since precipitation is associated with condensation and the release of latent heat, P appears as a source term in the non-dimensional heat eqn. (8.5). $\bar{q}$ is a saturation value for q.

The term θ_s in eqn. (8.5) represents the sea-surface temperature (relative to some prescribed value), the idea being that θ, which is representative of, say, the 900–300 mb thickness, is forced by thermal processes towards the sea-surface temperature. In other words, the troposphere tends to be warmer when the sea is warmer. Thus, in the absence of latent heat effects, the heating rate Q varies in proportion to the sea-surface temperature perturbation, as in the model of Zebiak (1982). Rather similar features to those assumed in the above model are found in the general circulation model results of Rowntree (1976) in that the 900 and 300 mb height varied in opposite phase (as they would with a cos z variation with height) and the 900–300 mb thickness varied with longitude in a similar way to surface temperature.

Equation (8.6) can be linearised by reference to a saturation value $\bar{q}$ which will be taken as spatially uniform. Thus if:

$$q = \bar{q} + q' \qquad (8.7)$$

the linearised form of eqn. (8.6) is, after use of eqn. (8.2):

$$\frac{\partial q'}{\partial t} - \bar{q}w = -P - \epsilon q' \qquad (8.8)$$

This equation can have two different forms, namely:

(a) If $q' = 0$ and $w > 0$, then:

$$P = \bar{q}w \tag{8.9}$$

and q' remains zero. In other words, when the atmosphere is saturated and there is low-level convergence, it rains and the moisture budget [eqn. (8.9)] determines the rainfall rate.

(b) Otherwise $P = 0$ and eqn. (8.8) reduces to:

$$\frac{\partial q'}{\partial t} + \epsilon q' = \bar{q}w \tag{8.10}$$

Thus, when the atmosphere is below saturation or is just starting to become dry through subsidence, eqn. (8.10) applies and the moisture equation is decoupled from the other equations. This is in contrast to the case when it is raining, and substitution from eqn. (8.9) in (8.5) gives:

$$\frac{\partial \theta}{\partial t} + (1 - \bar{q})w = -\epsilon(\theta - \theta_s) \tag{8.11}$$

For small $\bar{q}$, the term involving w shows the buoyancy effects are reduced when it is raining because the latent-heat release partially compensates for the adiabatic cooling in the ascending air. If $\bar{q}$ is greater than unity, the sign of the buoyancy term is *reversed* because latent heat is greater than adiabatic cooling. This corresponds to a conditionally unstable atmosphere. Various solutions of these "moist shallow-water equations" can be found in the two papers already mentioned for the non-rotating case and for an f-plane. The above form of the equations, however, is for an equatorial beta-plane.

9. SOLUTION FOR THE ONE-DIMENSIONAL NON-ROTATING CASE

An appreciation for the structure of solutions of these equations can be obtained by first considering the one-dimensional case, i.e. where variations are with Y only and conditions are uniform in x. As a further simplification, take the non-rotating case where the Coriolis terms are omitted from eqn. (8.4); u is identically zero; there is no friction; and the sea-surface temperature varies sinusoidally with Y. The "steady-state" solution is depicted schematically in Fig. 7a. Air rises over the warmed part of the ocean, so leading to moisture convergence and rain over the hot spot in a region $|Y| < \xi$ say. Since the motion is steady, eqn. (8.4) yields:

$$\frac{\partial \theta}{\partial Y} = 0 \tag{9.1}$$

and so θ is uniform in space. Although velocities do not vary, the temperature has to increase with time because latent heat is being released continually and Newtonian cooling is being neglected. Because of eqn. (9.1), however, this rate of increase is a constant A in space, i.e.:

$$\frac{\partial \theta}{\partial t} = A \tag{9.2}$$

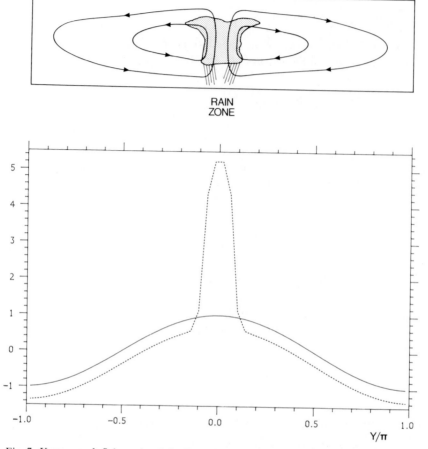

Fig. 7. *Upper panel*: Schematic of the flow. *Lower panel*: The heating rate function $\epsilon\theta_s$ which is also the SST and the solution for w in the dry case (solid line). The vertical velocity w found numerically for the moist case with $\bar{q} = 8/9$, $\epsilon = 0.1$ and $\nu = 0.01$ (broken line).

For the non-rotating case, the length scale can be chosen at will, so suppose the forcing $\epsilon\theta_s$ is equal to $\cos Y$ so that eqn. (8.5) gives:

$$A + w = \cos Y \qquad Y > \xi \tag{9.3}$$

in the non-precipitating zone and eqn. (8.11) gives:

$$A + (1 - \bar{q})w = \cos Y \qquad Y < \xi \tag{9.4}$$

in the precipitating zone. Two conditions now have to be satisfied. The first is that $w = 0$ at the boundary of the precipitating zone, so:

$$A = \cos \xi \tag{9.5}$$

Secondly, the integral of the vertical velocity has to vanish, i.e.:

$$\int\limits_{0}^{\pi} w\, dY = 0 \qquad (9.6)$$

and so:

$$\tan \xi = \xi + (1 - 1/\bar{q})\pi \qquad (9.7)$$

For example, when $\bar{q} = 8/9$, $\xi = 0.92$, $A = 0.605$, $w_{max} = 3.55$ and $w_{min} = -1.61$. A numerical solution obtained for this value of $\bar{q}$ but with a small value of friction included, is shown in Fig. 7.

It is interesting to consider the changes as $\bar{q}$ increases from zero towards unity. For low moisture levels (small $\bar{q}$), there is not much latent-heat release and nearly half the region (i.e. where there is ascent in the dry case) becomes a precipitation zone. The rate of heating A which is a measure of the total heat released is given approximately by:

$$A \approx (\bar{q}/\pi)(1 + \bar{q}/2 + \ldots)$$

As $\bar{q}$ increases, the precipitating zone narrows but the mass and moisture fluxes into the zone increase as well and so the rate of heating increases. There is a very rapid rise as $\bar{q}$ gets very close to unity, an approximate formula for $(1 - \bar{q})$ small being:

$$A \approx 1 - 1/2 \, [3\pi(1 - \bar{q})]^{2/3}$$

For instance, when $\bar{q}$ increases from 8/9 to 1, i.e. by 12%, A increases from 0.605 to unity, i.e. by 65%.

10. SOLUTIONS FOR THE ONE-DIMENSIONAL ROTATING CASE: AN ITCZ

The mathematics is a little harder where rotation is included, but analytic solutions can still be found. Using the experience of the non-rotating case, it can be seen that the simplest case is one without friction (to simplify the mathematics) and when $\bar{q} = 1$, so the precipitation is confined to a strip of infinitesimal width. This latter condition is an idealisation of the intertropical convergence zone (ITCZ). The effect of not including friction is that, if there is net heat put in, the atmosphere must keep warming up at a constant rate. Also, the heating function will generally vary with latitude so the pressure gradient will change uniformly with time and so there will be a constant zonal acceleration. This is reminiscent of the "Yoshida jet" in oceanography which is being forced by a constant wind stress. The flow in the meridional plane is quite steady, however, and this flow is not altered very much if weak friction and Newtonian cooling are added to limit the warming and the zonal flow.

The equations in the absence of friction are:

$$w = -\frac{\partial v}{\partial Y} \qquad \frac{\partial u}{\partial t} = 1/2 Yv \qquad 1/2 Yu = \frac{\partial \theta}{\partial Y} \qquad (10.1)$$

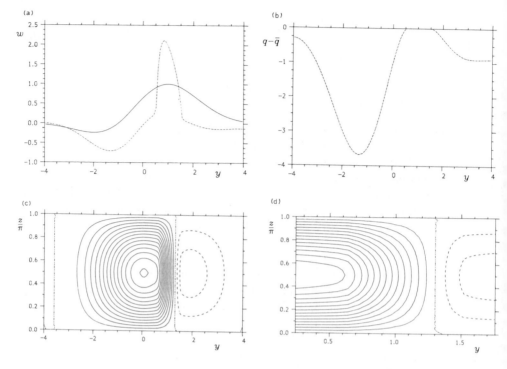

Fig. 8. Results for the moist case on the equatorial beta-plane with $\bar{q} = 8/9$, $\epsilon = 0.1$ and $\nu = 0.03$, (a) vertical velocity w (dotted line) and surface temperature θ_s (solid line); (b) perturbation moisture q'; (c) circulation in the meridional plane over the whole equatorial zone $|Y| < 4$; and (d) expanded view of the circulation in the neighbourhood of the ITCZ.

from eqns. (8.2), (8.3) and (8.4), while (8.5) gives:

$$\frac{\partial \theta}{\partial t} + w = Q \tag{10.2}$$

Differentiating eqn. (10.2) with respect to Y and substituting from eqn. (10.1) gives:

$$\frac{\partial^2 v}{\partial Y^2} - 1/4 Y^2 v = -\frac{\partial Q}{\partial Y} \tag{10.3}$$

Usually, the ITCZ is off the equator, so for Q take the asymmetric function used by Gill (1980) namely:

$$\epsilon \theta_s = Q = (1 + Y) \exp(-1/4 Y^2) \tag{10.4}$$

which has a maximum value of $2 \exp(-\tfrac{1}{4}) = 1.56$ at $Y = 1$. For the dry case, i.e. no moist effects, the solution of eqn. (10.3) is simply:

$$
\begin{aligned}
v &= V(Y) \equiv (6/5 - 1/3 Y - 1/5 Y^2) \exp(-1/4 Y^2) \\
w &= \mathring{W}(Y) \equiv (1/3 + Y - 1/6 Y^2 - 1/10 Y^3) \exp(-1/4 Y^2) \\
\partial \theta / \partial t &= (2/3 + 1/6 Y^2 + 1/10 Y^3) \exp(-1/4 Y^2) \\
\partial u / \partial t &= (3/5 Y - 1/6 Y^2 - 1/10 Y^3) \exp(-1/4 Y^2)
\end{aligned}
\tag{10.5}
$$

There is a continued heating even in the dry case, so eqn. (10.2) gives part of the Q creating a meridional flow by balancing w and the remainder giving net heating by balancing $\partial\theta/\partial t$. The way Q is partitioned depends on the dynamics as is seen in eqn. (10.3) and the equatorial Rossby radius is an important scale determined by the dynamics, as the left-hand side of eqn. (10.3) demonstrates. (The scale does not appear explicitly because it is the one used for the non-dimensionalisation.)

The effect of moisture is very easy to calculate in the case $\bar{q} = 1$ because the latent-heat release is confined to a single point $Y = 1$ where Q is a maximum, so there is a delta-function singularity here. The solution for such a singularity is given in section 6 of Gill (1980) and displayed in fig. 11.18 of Gill (1982a). It involves the parabolic cylinder function $U(y)$ of order $1/2$ which is the free solution of eqn. (10.3), i.e. the solution when $Q = 0$. For decay at infinity, it has the form (Abramowitz and Stegun, 1965):

$$U(Y) = \left(1 - Y + 1/2 \cdot \frac{Y^2}{2!} - 3/2 \cdot \frac{Y^3}{3!} + 1/2 \cdot 5/2 \cdot \frac{Y^4}{4!} - 3/2 \cdot 7/2 \cdot \frac{Y^5}{5!} + \ldots\right) \exp(-1/4Y^2)$$

(10.6)

The strength of the delta-function is determined by the condition that $w = 0$ at $Y = 1$ and hence:

$$w = \begin{cases} W(Y) - W(1)U'(Y)/U'(1) & Y > 1 \\ \\ W(Y) - W(1)U'(-Y)/U'(-1) & Y < 1 \end{cases}$$

(10.7)

The pressure, which is proportional to θ, is automatically continuous by eqn. (10.2). The solution for v is:

$$v = \begin{cases} V(Y) + W(1)U(Y)/U'(1) & Y > 1 \\ \\ V(Y) - W(1)U(-Y)/U'(-1) & Y < 1 \end{cases}$$

(10.8)

by continuity [first of eqn. (10.1)].

The amount of precipitation can be obtained from the flux of moisture into the singularity, namely:

$$\int_{1-}^{1+} P\,dY = -[\bar{q}v]_{1-}^{1+} = -W(1)[U(1)/U'(1) + U(-1)/U'(-1)] = 3.163$$

Numerical solutions for the equatorial beta-plane case were calculated by integrating in time until a steady state was reached. In these cases a small friction was included — ϵ was 0.1 and there was a lateral viscosity ν and diffusion ν with non-dimensional value of 0.03. The results are shown in Fig. 8 and show the same character as the analytic solution obtained above. There is strong rising motion in the ITCZ producing a strong Hadley circulation. Most of the exchange of air is on the equatorial side of the convergence zone, as is found in practice. For instance, Fig. 9 shows the meridional circulation as found by Frank (1983) from a composite of GATE data, the flow being relative to the ITCZ in all the cases used for the composite. The model calculation is

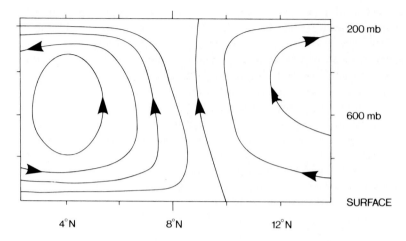

Fig. 9. Observed meridional flow relative to the ITCZ in the Atlantic during GATE. The data are taken from the composites given by Frank (1983) with the mean northward flow removed. The centre of the cloud band (marked 8°N in the diagram) was used as the reference latitude in each case. The contour interval is 1.8 ton m^{-1} s^{-1} (180 mb m s^{-1}).

too strongly biased toward the equatorial side, but that may be due to the temperature gradients being just as steep there whereas in practice they would be smaller than on the poleward side.

11. TWO-DIMENSIONAL SOLUTIONS FOR THE TROPICS

The results for the one-dimensional model are promising and the following features are noteworthy: (1) They exhibit the strong asymmetry between the regions of ascending and descending air; (2) the ascent is over the warmest part of the ocean and gives a structure like the ITCZ; (3) the descent is spread over the tropical zone on a scale determined by the equatorial Rossby radius; and (4) most of the descent is on the equatorial side, between the ITCZ and say 20° latitude in the other hemisphere; a small part of the descent is poleward of the ITCZ.

However, to use the model for coupled simulations, it is necessary to do two-dimensional calculations and here numerical methods need to be used. Work of this nature is being done in collaboration with Dr. M.K. Davey and some results are shown in Figs. 10 and 11.

The computational domain chosen was an equatorial beta-plane channel defined by $|Y| < 4$ and with period 16 in the x-direction. The unit for both x and Y is $a_a = 10°$ so the channel goes from 40°S to 40°N and its longitudinal extent is commensurate with the Pacific Ocean (120°E–80°W). The two cases shown in the figures are based on January (Fig. 10) and July (Fig. 11) SST distributions, but grossly simplified as follows. For each case, analytical expressions were constructed to resemble meridional profiles of SST for (a) the west Pacific; and (b) the east Pacific. Profile (a) applies at $x = -4$ and profile (b) at $x = +4$ with an overall structure of period 16. The formula used for profile (a) was:

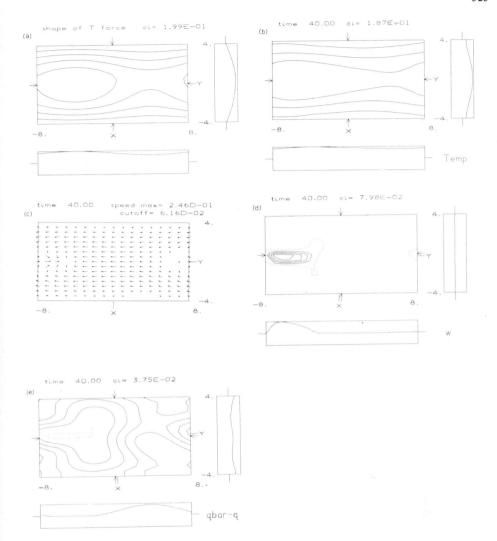

Fig. 10. Results from the "January" simulation for an equatorial beta-plane channel $|Y| < 4$ with $\bar{q} = 1$, $\epsilon = 0.1$ and $\nu = 0.03$. The channel is periodic in the zonal direction with period 16 units. (a) The prescribed surface temperature distribution θ_s, which forces the flow; (b) mid-level temperature (or thickness) θ (surface pressure is proportional to $-\theta$ and upper-level height anomaly is proportional to θ); (c) low-level wind (the upper-level wind is equal and opposite); (d) vertical velocity w; and (e) moisture perturbation (dryness), $-q'$. In each case, positive contours are solid, the dotted contour is zero and broken contours are negative. Side panels show profiles on the sections between the arrows.

$$\theta_{sw} = 1 - (Y - Y_0)^2 (Y_0^2 + 2YY_0 + 16)/(Y_0^2 - 16)^2$$

with $Y_0 = 0$ for January and 1.5 for July. This curve has the property of being zero at $Y = \pm 4$ and having a maximum value of unity at $Y = Y_0$. The formula for profile (b) was:

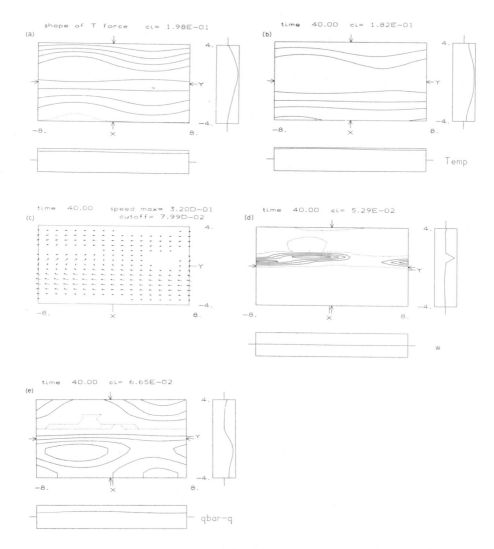

Fig. 11. As for Fig. 10, but the "July" simulation.

$$\theta_{SE} = 0.6(1 - Y^2/16) + A \exp[-(Y-1)^2]$$

with $A = 0.1$ for January and 0.3 for July. The maximum value of this function is 0.67 for January and 0.87 for July. If we take the unit as $15°$ and the base value as $15°$, then this corresponds to maxima of $25°$ for January and $28°$ for July in the east as opposed to a maximum of $30°$ in the west for both January and July. The results shown are for $\bar{q} = 1$, $\epsilon = 0.1$ and $\nu = 0.03$.

Figure 10 shows results for January. The vertical velocity field picks out the region of strong convection which, like the SST field, has its centre on the equator and represents the monsoon rainfall over Indonesia. The strongest convection is to the west of the SST maximum. The q'-field shows the driest area is in the east and south of the equator, with a secondary dry in the Northern Hemisphere. Note also the moist areas extend both westwards and to the northeast and southeast of the precipitating zone rather like some observed cloud patterns. The low-level winds are predominantly easterly, except for the strong westerly flow in the monsoon region. The monsoon area also shows as a low in surface pressure (proportional to $-\theta$). The interesting result for this season is that the strong east–west contrast in SST produces strong east–west contrasts in other fields as well and so has a strong Walker-cell component. Comparison with July simulation is illuminating.

Figure 11 shows results for July when the east–west SST contrast was much less. Now the vertical velocity field is much more zonal showing an ITCZ at most longitudes. There is significant cross-equatorial flow in the west feeding westerly winds in the narrow zone between the equator and the ITCZ. The surface pressure has a broad zone of low values between, say, 5°S and 15°N. The driest area is now in the Southern Hemisphere at the longitudes where the precipitation zone is found in the north, so there is now a strong Hadley-cell component of the circulation. All-in-all, the results of the two-dimensional simulations are very encouraging and indicate the moist model will indeed be useful for simple coupled model calculations.

12. CONCLUSIONS

The two-dimensional version of the shallow-water equations with moisture effects gives some realistic-looking precipitation patterns and so seems a useful step in the quest for a simple model which determines the heating rate and corresponding low-level wind distribution for a given sea-surface temperature pattern. Since the model is non-linear, however, there is a question of how to use it in a coupled model. If the coupled model is to deal with the actual state of the ocean–atmosphere system, then the problem immediately becomes harder because the system has to be capable of reproducing a realistic mean state. The alternative is to reconstruct the full SST field from the anomaly, calculate the winds and hence construct an anomaly wind field.

The other part of the coupled system which most needs improvement is the calculation of SST anomaly. One deficiency was the neglect of the anomalies near the eastern boundary, and there are devices for introducing this. For instance, an anomaly of 1 K for each rise in surface elevation of 5 cm could be created on the boundary and advected westward into the interior by a prescribed mean flow. If attempts are made to be more sophisticated, one soon gets to the stage of requiring a reasonable ocean model with thermodynamics, and again there is the problem of the model being capable of producing a realistic mean state.

There is much to be learned using simple coupled models at many levels of sophistication. The experiments with the system reported here show that events with duration and characteristics somewhat like those that occur can be reproduced, but the model did not seem capable of reproducing a natural tendency for warm events to repeat at intervals of a few years.

The study of the physics of individual components of the coupled system is, of course, very important if a realistic model is to eventuate. Moist processes are an essential part of the physics, so a special effort to understand how these act in large-scale tropical systems is well worthwhile, and it does seem possible to obtain realistic results from models which are conceptually of the utmost simplicity.

ACKNOWLEDGEMENT

I would like to thank Jeffrey Blundell for doing much of the numerical work reported.

APPENDIX: EXPANSIONS OF CYLINDER FUNCTIONS

To calculate the coefficients a_{mn} in the series:

$$D_m(\alpha y) = \sum_{n=0}^{\infty} a_{mn} D_n(y)$$

the orthogonality relation can be used to give:

$$\sqrt{2\pi}\, n!\, a_{mn} = \int_{-\infty}^{\infty} D_m(\alpha y) D_n(y)\, dy$$

Because the integrand is odd when $m + n$ is odd, the coefficients are zero then. Next a recurrence relation can be found using:

$$2\frac{d}{dy}[D_m(\alpha y)D_n(y)] = \alpha [mD_{m-1}(\alpha y)D_n(y) - D_{m+1}(\alpha y)D_n(y)]$$

$$+ n [nD_m(\alpha y)D_{n-1}(y) - D_m(\alpha y)D_{n+1}(y)]$$

which comes from applying the formulae for D'_m, etc., and integrating this over y-space. The result is:

$$\alpha(ma_{m-1,n} - a_{m+1,n}) + a_{m,n-1} - (n+1)a_{m,n+1} = 0 \qquad (A1)$$

This formula will give all the coefficients provided those with $m = 0$ are known. The coefficient a_{00} is easily found by direct integration:

$$a_{00} = [2/(1 + \alpha^2)]^{1/2} \qquad (A2)$$

Similarly, a_{02}, a_{04}, etc. can be calculated using the polynomial expressions for D_2, D_4, etc. to give:

$$n(1 + \alpha^2)a_{0,n} = (1 - \alpha^2)a_{0,n-2} \qquad (A3)$$

The formulae (A1), (A2) and (A3) are sufficient to determine all the non-zero coefficients a_{mn}.

REFERENCES

Abramowitz, M. and Stegun, I.A., 1965. Handbook of Mathematical Functions. Dover.
Blundell, J.R. and Gill, A.E., 1983. Equatorial ocean response to wind forcing. Ocean Modell., 53.
Frank, W.M., 1983. The structure and energetics of the East Atlantic Intertropical Convergence Zone. J. Atmos. Sci., 40: 1916–1929.

Gill, A.E., 1980. Some simple solutions for heat-induced tropical circulation. Q. J. R. Meteorol. Soc., 106: 447–462.

Gill, A.E., 1982a. Atmosphere–Ocean Dynamics. Academic Press, New York, N.Y., 662 pp.

Gill, A. E. 1982b. Studies of moisture effects in simple atmospheric models: the stable case. Geophys. Astrophys. Fluid Dyn., 19: 119–152.

Gill, A.E., 1982c. Spontaneously growing hurricane-like disturbances in a simple baroclinic model with latent heat release. In: L. Bengtsson and M.J. Lighthill (Editors), Intense Atmospheric Vortices. Springer, Berlin, pp. 111–129.

Gill, A.E., 1982d. Changes in thermal structure of the equatorial Pacific during the 1972 El Niño as revealed by bathythermography observations. J. Phys. Oceanogr., 12: 1373–1387.

Gill, A.E., 1983. An estimation of sea-level and surface-current anomalies during the 1972 El Niño and consequent thermal effects. J. Phys. Oceanogr., 13: 586–606.

Gill, A.E. and Rasmusson, E.M., 1983. The 1982/83 climate anomaly in the equatorial Pacific. Nature, 306: 229–234.

Heckley, W.A. and Gill, A.E., 1984. Some simple analytical solutions to the problem of forced equatorial long waves. Q. J. R. Meteorol. Soc., 110: 203–217.

Rasmusson, E.M. and Carpenter, T.H., 1982. Variations in tropical sea-surface temperature and surface wind fields associated with the Southern Oscillation/El Niño. Mon. Weather Rev., 110: 354–384.

Rowntree, P.R., 1976. Tropical forcing of atmospheric motions in a numerical model. Q. J. R. Meteorol. Soc., 98: 290–321.

Wunsch, C. and Gill, A.E., 1976. Observations of equatorially-trapped waves in Pacific sea-level variations. Deep-Sea Res., 23: 371–390.

Zebiak, S.E., 1982. A simple atmospheric model of relevance to El Niño. J. Atmos. Sci., 39: 2017–2027.

CHAPTER 23

THE MEAN RESPONSE OF THE ECMWF GLOBAL MODEL TO THE COMPOSITE EL NIÑO ANOMALY IN EXTENDED RANGE PREDICTION EXPERIMENTS

U. CUBASCH

INTRODUCTION

Numerical experiments have been carried out, using a low-resolution version of the ECMWF general circulation model, to investigate the response of the atmosphere to the El Niño sea-surface temperature anomaly. The experiments described here have been carried out as part of an international project for investigating the effect of SST anomalies. The appearance of a strongly developed "El Niño" SST anomaly in the winter of 1982/83 gives these experiments additional importance and makes comparisons with observations, which have never been as comprehensive as for this event, easier. First results of these experiments have been published in Cubasch (1983).

THE EXPERIMENT

The model

The model used in the experiments was a low-resolution version of ECMWF's first spectral model (Baede et al., 1979), using the comprehensive operational parameterisation package for the simulation of the physical aspects (Tiedtke et al., 1978). It has 15 levels in the vertical using sigma surfaces and resolves horizontally up to a total zonal wave-number 21 (i.e., it is a T21L15 model). Diabatic processes, which include the parameterisation of radiation, large-scale condensation, turbulent vertical diffusion and cumulus convection, were calculated on a Gaussian grid with a spacing of about $5.6°$ of latitude.

The solar angle was adjusted every 12 hours to its climatological value, but there was no diurnal cycle; the climatological soil moisture content, the SST and deep-soil temperature were modified every fourth day (for more details see Fig. 1).

The data

The initial data were taken from a 10-year integration of the T21L15 model. This integration has been described in Rilat and Thepaut (1982) and Deque (1984). This run serves as the control experiment.

The data for the SST, the soil wetness and the deep soil temperature have been derived from the operational ECMWF climatology (Louis, 1980). This climatology, which consists of monthly averages on a $1.875°$ grid, has been interpolated onto the Gaussian grid

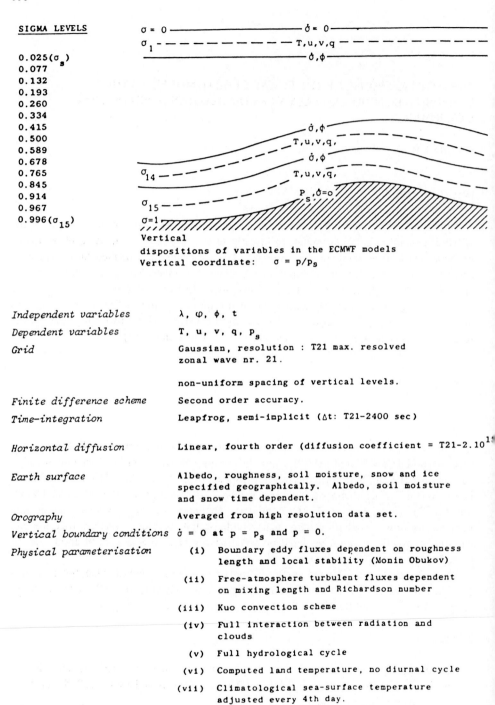

Vertical
dispositions of variables in the ECMWF models
Vertical coordinate: σ = p/p$_S$

Independent variables	λ, φ, φ, t
Dependent variables	T, u, v, q, P$_s$
Grid	Gaussian, resolution : T21 max. resolved zonal wave nr. 21. non-uniform spacing of vertical levels.
Finite difference scheme	Second order accuracy.
Time-integration	Leapfrog, semi-implicit (Δt: T21-2400 sec)
Horizontal diffusion	Linear, fourth order (diffusion coefficient = T21-2.10$^{1\cdot}$
Earth surface	Albedo, roughness, soil moisture, snow and ice specified geographically. Albedo, soil moisture and snow time dependent.
Orography	Averaged from high resolution data set.
Vertical boundary conditions	σ̇ = 0 at p = p$_s$ and p = 0.
Physical parameterisation	(i) Boundary eddy fluxes dependent on roughness length and local stability (Monin Obukov) (ii) Free-atmosphere turbulent fluxes dependent on mixing length and Richardson number (iii) Kuo convection scheme (iv) Full interaction between radiation and clouds (v) Full hydrological cycle (vi) Computed land temperature, no diurnal cycle (vii) Climatological sea-surface temperature adjusted every 4th day.

Fig. 1. Characteristics of the ECMWF model.

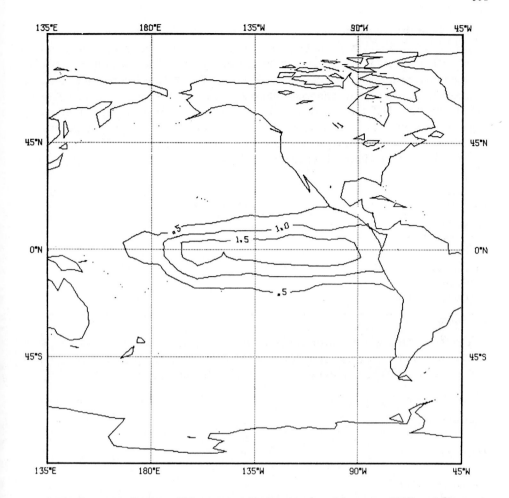

Fig. 2. The "composite" El Niño SST anomaly (after Rasmusson and Carpenter, 1982); unit: K.

of the T21 model. The average value for every month was taken as representative on the 15th of that month; for the other days values have been linearly interpolated between those. This climatology was adjusted every 4th day.

In all experiments the "composite" El Niño SST anomaly due to Rasmusson and Carpenter (1982) has been used (Fig. 2). This SST represents an average of six El Niño events between 1949 and 1976. In the experiments discussed here the SST anomaly in its mature state, as it exists during December, January and February, has been super-imposed on the climatological SST.

The set up of the experiments

The model was restarted from model-generated data from the 30 October in different years of the ten-year forecast and integrated for 150 days. The only difference between ten-year run and the new experiments was the imposed SST anomaly in the central

Pacific. Since the composite El Niño anomaly represents the average over many El Niño events of varying strength, it can be assumed that it is weaker than many anomalies that occur; it has therefore been multiplied by a factor of 2. It is still slightly weaker than the one observed during winter 1982/83 (Arkin et al., 1983). As well as positive anomalies it is possible to find negative anomalies which are similar in shape to positive ones (Weare, 1982); therefore half the experiments have been carried out with the corresponding negative anomalies. The experiments with the negative anomaly make it possible to estimate to what extent the response to an SST anomaly is linear.

The following experiments have been carried out:

Initial date	El Niño anomaly
30 October year 1	* 2
30 October year 5	* 2
30 October year 9	* 2
30 October year 1	* −2
30 October year 5	* −2
30 October year 9	* −2

The initial dates where chosen so that every 150-day forecast fell in the winter season of the Northern Hemisphere, centered around January. The initial years have been chosen in this way in order to compensate for a trend in the 10-year integration.

RESULTS

The tropics

The precipitation

An increase in the SST leads to a labilisation of the adjacent air masses and leads in conditionally unstable environments (as commonly found in the tropics) to an increased convective activity and increased precipitation. This process can also be found in these simulations (Fig. 3). The precipitation increases over the positive anomaly and decreases over the negative one. West of the region with increased (decreased) rainfall one finds an area of decreased (increased) precipitation for the positive (negative) anomaly. This means that the rainfall has not only been enhanced or diminished by the altered SST, but also suffers a lateral displacement. This effect, which has also been observed, leads to a severe drought over Indonesia in the case of an El Niño event (Rasmusson and Wallace, 1983). The fact that the rainfall has altered particularly in the western part of the anomaly can be explained by the climatological distribution of the SST. The eastern part of the Pacific ocean is colder than the western part. This is the cause of the "Walker" circulation with ascending motion over the west Pacific and descending motion over the east Pacific. Since the saturation mixing ratio depends non-linearly on the absolute temperature, the possibility of an increase in precipitation is relatively higher in the west Pacific. The same non-linearity can be assumed to be the cause of the stronger increase of precipitation in case of the positive anomaly than decrease in case of the negative anomaly.

333

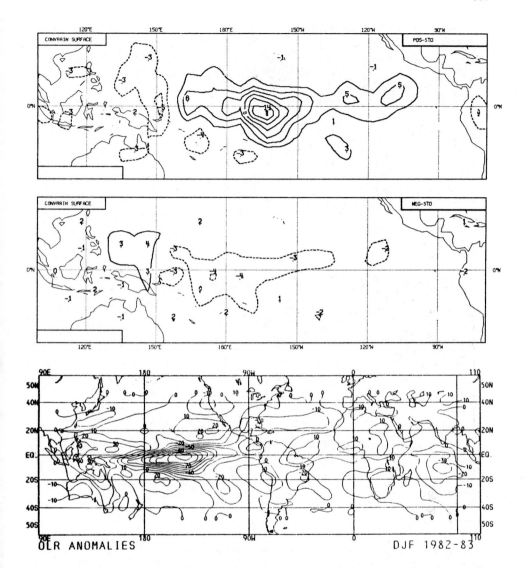

Fig. 3. The change in the precipitation by the El Niño anomaly. *Top*: Positive anomaly; *middle*: Negative anomaly; unit: mm/day; *bottom*: The anomaly of the outgoing longwave radiation observed during December, January and February 1982/83 (after Arkin et al., 1983); unit: W/m⁻².

It is interesting to compare these model results with observations. The current most comprehensive dataset has been collected during the El Niño year 1982/83. Since the precipitation over sea areas is hardly known, it has been determined indirectly from satellite data via the outgoing longwave radiation. Clouds emit and absorb infrared radiation like a black body. The amount of outgoing longwave radition is therefore dependent on the temperature at the cloud tops. The higher the cloud, the lower is the

334

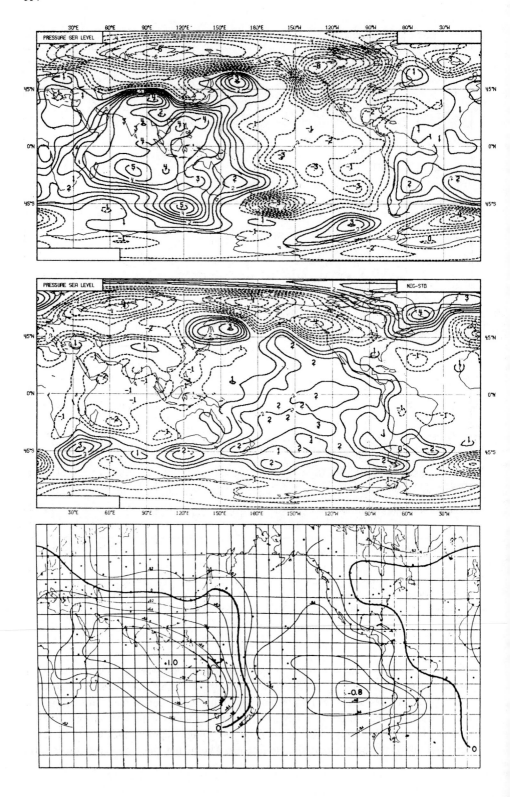

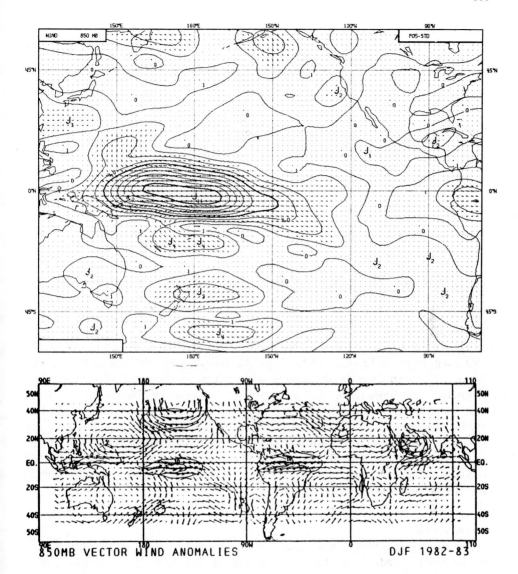

Fig. 5. The change of the wind field at 850 mb for the positive anomaly (top; unit: m s^{-1}); winter 1982/83 observed wind anomaly (bottom; after Arkin et al., 1983).

radiation and the larger is (probably) the precipitation. As one can see in Fig. 3 the anomaly in the outgoing longwave radiation observed during winter 1982/83 agrees quite well with the pattern of the precipitation increase in the case of the positive anomaly.

Fig. 4. The change in the surface pressure. *Top*: Positive anomaly; *middle*: Negative anomaly; unit: mb; *bottom*: The change of the surface pressure by the "Southern Oscillation" (after Berlage, 1957; correlation pattern).

m/s

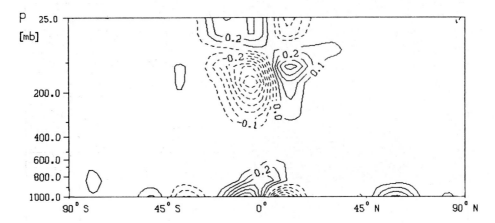

Fig. 6. The change of the meridional wind (positive–negative anomaly case), zonally averaged; unit: m s^{-1}.

The surface pressure

The surface pressure (Fig. 4) is lowered over the warm anomaly and raised over the cold anomaly as can be expected for dynamical reasons. The amplitude of this pressure change (maximum 3 mb for the positive anomaly) agrees quite well with observations (Rasmusson and Wallace, 1983). It is noteworthy that in the tropics this pressure change happens in a form which is known in the literature as "Southern Oscillation" (Berlage, 1957): A pressure change in the central and eastern Pacific is accompanied by a pressure change in the remaining tropical atmosphere of the opposite sign. These experiments show clearly that the state of the "Southern Oscillation" can be controlled by the SST anomaly and is correlated with the El Niño. This is consistent with the observations of Rasmusson and Carpenter (1982).

The windfield

For the positive anomaly one finds at 850 mb a decrease of the easterlies to the west of the anomaly and east of it an increase (Fig. 5). This leads to a convergence of air masses which is compensated by enhanced vertical movement and divergence near the tropopause. Together with an enhanced meridional wind in the lower layers and near the tropopause (Fig. 6) it becomes clear that the increased SST leads to an enhanced Hadley circulation as demanded by Bjerknes (1966). In the case of the negative anomaly this effect reverses, but it is again not as strongly developed.

The change of the wind field in the case of the positive anomaly can be compared with observations from the winter 1982/83. The convergence into the region of the heaviest precipitation agrees in its shape and strength quite well with the observed pattern.

The mid-latitudes of the Northern Hemisphere

The heightfield

The mean state. Hoskins and Karoly (1981) assumed in their experiments that the effect of the anomaly in the mid-latitudes is essentially linear to a first approximation. This implies that a positive anomaly creates a wavetrain with the reverse phase to that of a negative anomaly. This could not be verified in any of the experiments (Cubasch, 1983). Every anomaly seems to create the same wavetrain (Fig. 7) whose phase depends only on the initial state. The phase of the anomaly experiments seems to be the reversal of the phase difference between the control run and the model climate so that the average of both comes close to the model climate. Therefore, following the theory of Charney and DeVore (1979) that the atmosphere knows (greatly simplified) only two stable states, it might be possible that the anomaly just forces the model atmosphere from one state to the other. Another explanation might be given by the results of T. Palmer (pers. commun., 1983). He moved the SST anomaly along the equator, but in mid-latitudes the position of the wavetrain stayed unchanged. It might be that the negative anomaly which creates a sideward displacement of the area of maximum rainfall is interpreted by the model atmosphere as a sidewards shifted warm anomaly; this then creates a similar wavetrain to the positive anomaly. The invariance of the wavetrain has been also found by other research institutes M. Blackmon (pers. commun., 1984) and could be found in the simulation with a barotropic model by Simmons (1982). It has still not yet become clear whether it is a resonant mode in the model atmosphere or the effect of barotropic instability in preferred regions.

A strong signal as function of the sign of the anomaly can be found in the pressure change of the Aleutian low. As suggested by Bjerknes (1966, 1969), the positive anomaly deepens the Aleutian low. This deepening is more than a factor of 2 larger than for the negative anomaly. Over Newfoundland both anomalies create a positive pressure deviation, while over Europe the positive anomaly increases the pressure, but the negative anomaly decreases it. Both anomalies generate a pressure increase over Asia, but the magnitude is smaller for the negative anomaly than for the positive anomaly. The mean pressure gradient between pole and equator is increased in the case of the positive anomaly.

If one takes as verification of the effects of the El Niño anomaly the results of Van Loon and Rogers (1981) which described the pressure change in mid-latitudes as a function of the state of the "Southern Oscillation", it is found that the response for single years from the positive anomaly resembles these observations. In the mean over all experiments, however, the similarity is small. The wavetrains which appear in the model simulations can rather be compared with the first EOF, i.e. the major part of the variance, of the height field in the Northern Hemisphere as described by Wallace and Gutzler (1981).

Statistical significance. The ambiguous relation between the sign of the anomaly and the phase of the wavetrain demands a statistical significance test. The aim is to assess if those wavetrains have really been caused by the anomalies or if they are the product of the internal variability within the model atmosphere. Lau (1981) and Volmer et al. (1984) pointed out in their analysis of a 15-year integration that the models are quite

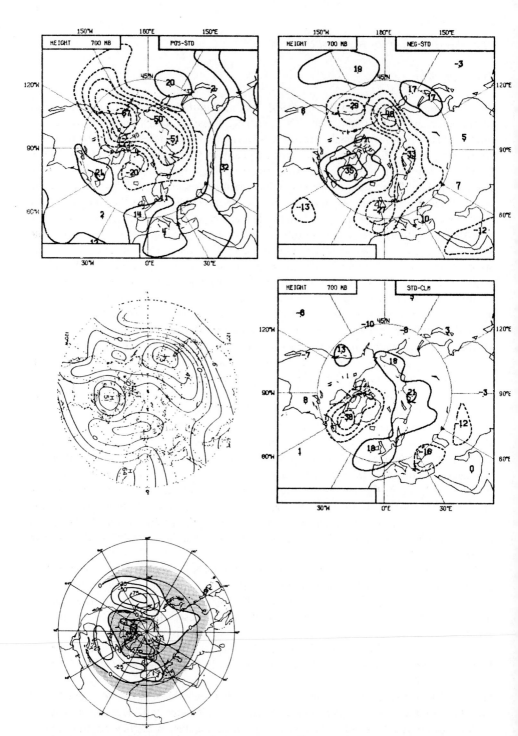

Fig. 7. The change in the 700-mb height field. *Top left*: Positive anomaly; *top right*: Negative anomaly; unit: m; *middle left*: First EOF of the observed winter circulation (after Wallace and Gutzler, 1981); *middle right*: Difference of the control experiments to the model climate; *bottom left*: Observed height-field difference for winter with the state of the "Southern Oscillation" (after Van Loon and Rogers, 1981).

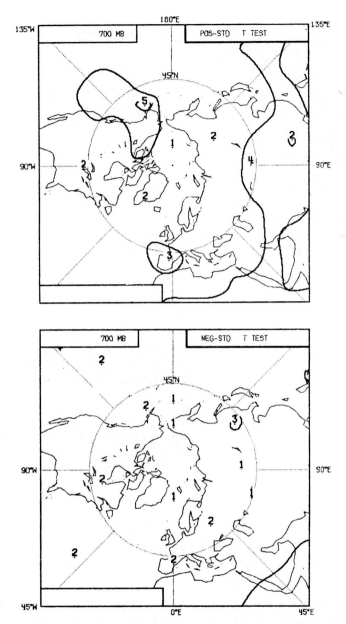

Fig. 8. The statistical significance of the change in the 700-mb height field (*t*-test). The contour includes areas with a significance of more than 95%. Top: Positive anomaly; bottom: Negative anomaly.

capable of generating long-lasting circulation anomalies, even without changes in the boundary forcing.

The significance test used here was a modified version of the Student's *t*-test proposed by Chervin and Schneider (1976). It assumes that each point is independent of each other (univariate). In order to take the internal variability into account, the anomaly runs

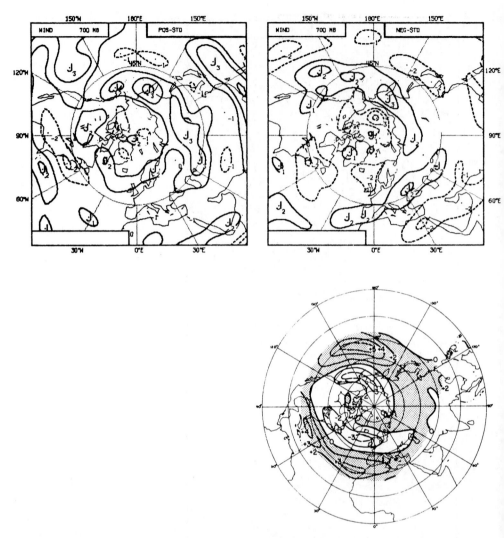

Fig. 9. The change of the zonal wind in 700 mb. *Top left*: Positive anomaly; *top right*: Negative anomaly; unit: m s⁻¹; *bottom right*: Observed wind field difference as function of the state of the "Southern Oscillation" (after Van Loon and Rogers, 1981).

are compared against the model climate. The difference of the height-field related to the model climate is not much different to the difference of the height field to the control-experiments. Due to the larger sample it is to be expected that the "*t*-test" will be more stable if it is related to the model climate due to the greater number of degrees of freedom.

The null-hypothesis is that the mean state of the anomaly experiment is not significantly different from the winter of the ten-year integration. This hypothesis is rejected at the 95% level if the value of *t* exceeds 2.228 (Sachs, 1979). This threshold value can be obtained from tables for the appropriate number of degrees of freedom, in this case ten, under the assumption that the variance is not significantly altered in the anomaly experiment.

This significance test (Fig. 8) shows clearly that the deepening of the Aleutian low in the case of the positive anomaly is a significant change in the model's circulation. Other significant changes can be found over Europe and large areas of Asia. This means that the El Niño SST anomaly does not only influence the Aleutian low as suggested by Bjerknes (1966, 1969), but also the Monsoon circulation over the Asian continent probably in connection with the change of the state of the "Southern Oscillation". However the variability over the Asian land areas is quite small so that a comparatively small change in the height field is indicated as significant by this statistical test.

The negative anomaly on the other hand generates only a few areas with a significant change in the height field. Since by chance 5% of the area are indicated as significant it can be assumed that the change in the height field by the negative anomaly is generally insignificant and therefore the mid-latitudes hardly respond to the altered SST. The wavetrain in the case of the negative anomaly seems, therefore, rather to be some mode in the model atmosphere which might have been excited by the anomaly or has been caused by the internal variability in the model.

A statistically comprehensive evaluation of these experiments can be found in the contribution of Von Storch and Kruse in this volume.

The wind field

The zonal wind in the subtropics (Fig. 9) shows an unequivocal dependence on the sign of the anomaly. The positive anomaly enhances, as suggested by Bjerknes (1966, 1969), the subtropical jet while the negative anomaly weakens it. The structure of the wind change bears only little resemblance with observations for both the positive and negative anomalies. The change in the polar jet is less clear; it is stronger over the north Pacific for both the positive and the negative anomaly, while over the north Atlantic and Europe it increases for the positive anomaly and decreases for the negative one so that the net effect appears to be small in the zonal mean.

CONCLUSIONS

The effect of the positive El Niño SST anomaly on the general circulation of the atmosphere during winter in the Northern Hemisphere can be described according to the model results in the following way:

(1) The SST anomaly leads to increased precipitation (the amount of which depends non-linearly on the size of the total SST) over the western part of the anomaly. The air masses are destabilised over the warm water which leads to an increased upward motion over the anomaly which generates stronger convective activity. The increased upward mass transport over the anomaly region is compensated by increased sinking motion in the subtropical high-pressure belt, as well as in the areas west of the anomaly. In summary the anomaly leads, as already suggested by Bjerknes (1966, 1969), to an intensification of the Hadley circulation. At the same time the surface pressure decreases in the area of the anomaly but increases in the remainder of the tropical belt. Therefore it seems to be likely that the SST anomaly is closely correlated with the state of the "Southern

Oscillation". The anomaly generates a state of the oscillation which has been described in the literature as "low/wet".

(2) The intensification of the Hadley circulation leads to an intensification of the subtropical jet; the Aleutian low deepens significantly. Any other impact of the anomaly on the circulation in the mid-latitudes can not be proven unequivocally since the fluctuations in this region are quite strong and hide any possible signal.

(3) The results obtained by the integrations with a negative SST support the findings of the simulation with the positive anomaly in the tropics. In mid-latitudes they contradict them in some points. However, the statistical significance of the runs with the negative anomaly is small.

(4) It is noteworthy that the results of this simulation with the positive anomaly compare quite well with the events of the El Niño winter in 1982/83 in the tropics as well as in mid-latitudes, while agreement with the older observations appears to be poor.

REFERENCES

Arkin, P.A., Kopman, J.D. and Reynolds, R.W., 1983. 1982/83 El Niño/Southern Oscillation event quick look atlas. NOAA/National Weather Service, NMC/CAC, Washington, D.C.
Baede, A., Jarraud, M. and Cubasch, U., 1979. Adiabatic formulation and organisation of ECMWF's spectral model. Tech. Rep. No. 15, ECMWF.
Berlage, H.P., 1957. Fluctuations of the general circulation of more than one year, their nature and their prognostic value. Kon. Ned. Meteorol. Inst., Meded. Verh., No. 69, 152 pp.
Bjerknes, J., 1966. A possible response of the atmospheric Hadley circulation to the equatorial anomalies of the ocean surface temperature. Tellus, 18: 820–829.
Bjerknes, J., 1969. Atmospheric teleconnections from the equatorial Pacific. Mon. Weather Rev., 97: 163–172.
Charney, J.G. and DeVore, J.G., 1979. Multiple flow equilibria in the atmosphere and blocking. J. Atmos. Sci., 36: 1205–1216.
Chervin, R.M. and Schneider, S.H., 1976. On determining the statistical significance of climate experiments with general circulation models. J. Atmos. Sci, 33: 405–412.
Cubasch, U., 1983. The response of the ECMWF global model to the El Niño anomaly in extended range prediction experiments. Tech. Rep. No. 38, ECMWF.
Deque, M., 1984. Analyse en composantes principales d'une simulation de 10 ans de circulation générale. Note de Travail de l'établissement d'études et de recherches météorologiques no. 80, Météorologie Nationale.
Horel, J.D. and Wallace, M.J., 1981. Planetary scale atmospheric phenomena associated with the interannual variability of sea-surface temperature in the equatorial Pacific. Mon. Weather Rev., 109: 813–829.
Hoskins, B. and Karoly, D., 1981. The steady linear response of an atmosphere to thermal and orographic forcing. J. Atmos. Sci., 38: 1179–1196.
Lau, N.C., 1981. A diagnostic study of recurrent meteorological anomalies appearing in a 15 year simulation with the GFDL general circulation model. Mon. Weather Rev., 109: 2287–2286.
Louis, J.F., 1980. ECMWF forecast model documentation. ECMWF.
Rasmusson, E.M. and Carpenter, T., 1982. Variations in the tropical sea-surface-temperature and surface wind fields associated with the Southern Oscillation/El Niño. Mon. Weather Rev., 110: 354–384.
Rasmusson, E.M. and Wallace, J.M., 1983. Meteorological Aspects of the El Niño/Southern Oscillation. Science, 222: 1195–1202.
Rilat, M.F. and Thepaut, M.A., 1982. Analyse d'un cycle hydrologique dans un expérience de dix ans de simulation du climat. Note de Travail de l'Ecole Nationale de la Météorologie No. 20, Métérologie Nationale.

343

Sachs, L., 1979. Statistische Methoden. Springer, Berlin, 105 pp.

Simmons, A.J., 1982. The forcing of stationary wave motion by tropical diabatic forcing. Q.J.R. Meteorol. Soc., 108: 507–534.

Simmons, A.J., Wallace, M.J. and Branstator, G.W., 1982. Barotropic wave propagation and instability, and atmospheric teleconnection patterns. J. Atmos. Sci., 40: 1363–1392.

Tiedtke, M., Geleyn, J.F., Hollingsworth, A. and Louis, J.F., 1978. ECMWF-model parameterisation of subgrid processes. Tech. Rep. No. 10, ECMWF.

Van Loon, H. and Rogers, J.C., 1981. The southern oscillation. Part II: Associations with changes in the mid-latitudinal troposphere in the northern winter. Mon. Weather Rev., 109: 1163–1168.

Volmer, J.P., Deque, M. and Rousselet, D., 1984. EOF analysis of 500 mb geopotential: A comparison between simulation and reality. Tellus, 36: 36:336–347.

Wallace, J.M. and Gutzler, D.S., 1981. Teleconnections in the geopotential height field during the northern hemisphere winter. Mon. Weather Rev., 109: 784–812.

Weare, B.C., 1982. El Niño and tropical Pacific Ocean temperatures. J. Phys. Ocean, 12: 17–27.

CHAPTER 24

SIMPLE MODELS OF EL NIÑO AND THE SOUTHERN OSCILLATION

JULIAN P. McCREARY, Jr. and DAVID L.T. ANDERSON

ABSTRACT

Two simple coupled ocean–atmosphere models of El Niño and the Southern Oscillation (ENSO) are discussed. In both cases the ocean is a reduced-gravity system, following the response of the surface layer of the ocean; however, the two models differ considerably in their treatment of ocean thermodynamics and the atmosphere. Solutions to both have features that compare favorably with observations. In particular, they oscillate at the long time scales associated with ENSO.

The first model is dynamically simple in that both ocean thermodynamics and the model atmosphere are highly parameterized. Sea-surface temperature (SST) is assigned one of two values, warm or cool, according to whether the layer is deeper or shallower than an externally specified depth. The model atmosphere consists of two patches of zonal wind stress, τ_s and τ_w, that are idealizations of the annual cycle of the equatorial tradewinds and of Bjerknes' Walker circulation, respectively. τ_w is switched on only when the eastern ocean is in a cool state. An important property of the interaction of τ_w with the ocean is that the coupled system has two equilibrium states for the same values of model parameters: one state with τ_w switched on and the other with it off. The annual cycle, τ_s, is a "trigger" that enables the system to switch from one state to another.

The second model includes active ocean thermodynamics and a dynamical, rather than empirical, atmosphere. The atmosphere is a single-baroclinic-mode system driven by convection, Q, which depends solely on SST. For a broad range of model parameters it develops instabilities that grow to a finite amplitude and propagate slowly eastward. A major limitation of this second model is that it does not simulate the rapid onset of El Niño. This failure may be due to the fact that the system apparently cannot develop two equilibrium states.

1. INTRODUCTION

The phenomenon of El Niño and the Southern Oscillation (ENSO) exhibits a time scale of the order of 2–9 years. Atmospheric models suggest that the atmosphere itself does not have such low-frequency variability. An intriguing hypothesis is that the atmosphere and ocean form a strongly coupled system in the tropics with an inherent periodicity of 2–9 years. Several studies using simple, coupled ocean–atmosphere models have begun to explore this idea. These systems use ocean and atmosphere models that are dynamically considerably less sophisticated than general circulation models. Their advantage is that, due to their simplicity, it is easy to isolate and to understand the various processes at work in them. Their limitation, of course, is that they necessarily involve many simplifying assumptions.

This paper describes two coupled models that we have developed during the past several years. Both models oscillate at the long time scales associated with ENSO, but for very different reasons. In the first model (McCreary and Anderson, 1984), the ocean and atmosphere interact to allow two equilibrium states to exist for the same values of model parameters: a normal state with strong equatorial trades, and an El Niño state with weak trades. An additional process, either the annual cycle of the trades or a random wind event, acts as a "trigger" to switch the system from being near one equilibrium state to being near the other. In the second model (Anderson and McCreary, 1985), instabilities grow to a finite amplitude, and then propagate slowly eastward. The slow propagation of the disturbance across the ocean basin sets the time scale of the oscillation. Physical processes at work in these models are also involved in other models. Before proceeding further, it is useful to review briefly a few studies that are most relevant to our work.

The model of McCreary (1983) produced oscillatory solutions, and is similar in its formulation to that of McCreary and Anderson. The model ocean consisted of the single baroclinic mode of a two-layer ocean [as in eqns. (1) and (2)], and sea-surface temperature (SST) was either warm or cold according to whether the depth of the model interface was less than or greater than a specified depth [as in eqn. (3)]. The model atmosphere consisted of two patches of zonal wind stress that were assumed to interact with the ocean in a way suggested by the ideas of Bjerknes (1966, 1969); when the eastern ocean was cool the equatorial easterlies were assumed to strengthen in the central ocean [similar to eqn. (6)], thereby simulating Bjerknes' Walker circulation, and when the eastern ocean was warm the extra-equatorial easterlies were assumed to strengthen in the eastern ocean, thereby simulating an enhanced Hadley circulation (HC) there. Unlike the McCreary and Anderson model, however, the system did not have two equilibrium states. Instead, the HC provided a restoring force that prevented solutions from ever reaching any equilibrium state. Wind curl associated with the HC deepened the model interface to form a ridge in a region centered off the equator. This ridge propagated westward slowly as a Rossby wave, reflected from the western boundary as an equatorially trapped Kelvin wave, eventually affecting SST in the eastern ocean. It was the long time it took the ridge to propagate across the ocean basin that governed the period of the oscillation.

Hughes (1979, 1984) first developed a coupled system that possessed two equilibrium states. His model ocean consisted simply of an equatorial pycnocline which tilted in response to zonal wind stress, and all other properties of the tropical ocean were ignored. The model atmosphere was an externally specified equatorial easterly wind stress (a Walker circulation), that responded to the depth of the pycnocline in the eastern ocean. One equilibrium state had a large pycnocline tilt and strong easterlies, and the other had a small tilt and weak easterlies. In the latter study Hughes included the effect of the annual cycle of the trades, and suggested that El Niño events which are initiated during the warm part of the annual cycle tend to be large. Lau (1985) presented evidence for a bimodal climatic state associated with ENSO, and compared observations to the response of a stochastically forced, non-linear oscillator.

Lau (1982) and Philander et al. (1984) have developed coupled models that produced unstable disturbances. The model ocean in both studies is a linear, reduced-gravity model [as in eqns. (1) and (2)]. The model atmosphere is a linear, single-baroclinic-mode model that is forced by a heating function, Q [as in Gill, 1980, and eqns. (15)], where Q is taken to be directly proportional to the pycnocline-depth anomaly, $h - \bar{h}$. Lau (1982) greatly simplified his system by considering only coupling between equatorial Kelvin waves in the ocean and atmosphere, and found a simple dispersion relation for the coupled waves. One of the roots always had a phase speed very close to that of an atmospheric Kelvin wave. The other was an oceanic Kelvin wave for weak coupling. As coupling increased, the phase speed of the wave slowed to zero and then became positive imaginary, indicating that it was now a stationary, growing instability. Philander et al. (1984) solved their system numerically without resorting to Lau's simplification, and also found a rapidly growing instability. They discussed the relevance of this unstable growth to ENSO. Recently, Hirst (1985) and Yamagata (1985) have studied unstable solutions in similar coupled systems in greater detail, and found that eastward propagation at slow speeds is characteristic of them.

2. A COUPLED MODEL WITH TWO EQUILIBRIUM STATES

2.a. The model ocean

In a state of no motion the model ocean consists of an upper layer of thickness H and density ρ overlying a deep lower layer of density $\rho + \Delta\rho$. Vertical mixing is strong enough so that wind stress acts like a depth-independent body force in the upper layer. Linear equations of motion describing the response of the layer are:

$$u_t - \beta y v + p_x = \tau^x/H + \nu_h \nabla^2 u - \nu u$$
$$v_t + \beta y u + p_y = \tau^y/H + \nu_h \nabla^2 v - \nu v \tag{1}$$
$$p_t/c^2 + u_x + v_y = -\kappa p/c^2$$

where all quantities have their usual definitions. The pressure field is closely related to the thickness of the upper layer, h, by the relation:

$$h = H + p/g' \tag{2}$$

where $g' = g(\Delta\rho/\rho)$ and g is the acceleration of gravity. Values of model parameters are $\beta = 2 \times 10^{-11}\,\mathrm{m^{-1}\,s^{-1}}$, $\nu_h = 10^4\,\mathrm{m^2\,s^{-1}}$, $g' = 0.02\,\mathrm{m\,s^{-2}}$, $H = 100\,\mathrm{m}$, $\kappa = 5 \times 10^{-9}\,\mathrm{s^{-1}}$, $\nu = \kappa$, and $c = 1.25\,\mathrm{m\,s^{-1}}$. The horizontal mixing coefficient is large, but horizontal mixing does not play a significant role in the dynamics of the model. ν is set equal to κ throughout the study, but test runs not reported here show that solutions are virtually independent of its value.

Equations (1) are essentially the same equations that describe the response of one of the baroclinic modes of a continuously stratified ocean, and several parameter values are set by exploiting this similarity. Recent studies show that in the tropics, higher-order baroclinic modes affect the near-surface current and density fields of the ocean more than the first mode does (Philander and Pacanowski, 1980; McCreary, 1981; McCreary et al., 1984). For this reason, g', H and c take values that are typical for the second baroclinic mode of the tropical ocean.

Thermodynamics in the model are parameterized as in McCreary (1983). SST is assumed to be related to h according to:

$$\text{SST is} \begin{cases} \text{warm, } h \geqslant h_c \\ \\ \text{cool, } h < h_c \end{cases} \tag{3}$$

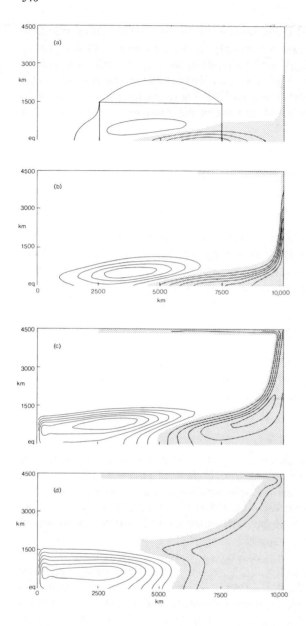

Fig. 1. The thickness of the model interface in response to τ_w alone at various times: (a) 1 month, (b) 3 months, (c) 13 months, and (d) 5 years. The location and structure of τ_w are indicated in panel (a), and are described precisely in eqns. (5). The contour interval is 10 m, there is no 100 m contour, and the shaded region indicates thickness less than 100 m. To avoid displaying irrelevant, weak fluctuations about 100 m, the shaded region is not shown everywhere in the basin. The figure illustrates the propagation of Kelvin and Rossby waves about the ocean basin. The initial rise of h_e in the figure is caused by the arrival of equatorial Kelvin waves generated directly by the wind. The gradual relaxation of h_e toward $\bar{h}_w$ is caused by the arrival of a second set of equatorial waves that are generated when Rossby waves reflect from the western ocean boundary.

where h_c is an, as yet, unspecified depth. See McCreary (1983) for a discussion of this parameterization.

The ocean basin is confined to the region $-L < y < L$ and $0 < x < D$, where $L = 4500$ km and $D = 10,000$ km. No-slip conditions are applied at the basin boundaries, and solutions to eqns. (1) are found numerically. In order to inhibit the propagation of mid-latitude coastal Kelvin waves, the value of κ is increased to $10\,\kappa$ within four grid points of the northern and southern boundaries. Details of the numerical scheme are outlined in McCreary (1983).

2.b. The model atmosphere

The model atmosphere is composed of two patches of zonal wind stress: τ_w and τ_s. Their forms are:

$$\tau_w = \tau_{0w} X(x) Y(y)$$
$$\tau_s = \tau_{0s} X(x) Y(y) \cos \sigma t \tag{4}$$

where:

$$X(x) = \cos \frac{2\pi}{D}(x - D/2), \quad D/4 < x < 3D/4$$

$$Y(y) = \tfrac{1}{2}\left(1 + \cos \frac{2\pi}{\lambda}y\right), \quad |y| < \lambda \tag{5}$$

$$X = Y = 0, \quad \text{otherwise}$$

In every case $\tau_{0w} = -0.05\,\mathrm{N\,m^{-2}}$, $\sigma = 2\pi\,\mathrm{yr^{-1}}$ and $\lambda = 3000$ km. The value of τ_{0s} is usually $0.015\,\mathrm{N\,m^{-2}}$. The location and horizontal structures of τ_w and τ_s are indicated in Fig. 1a. τ_w is an idealization of the increased equatorial wind field in the central Pacific associated with cool eastern Pacific temperatures (Pazan and Meyers, 1982), and τ_s simulates the annual cycle of the equatorial trade-wind field. The structure of the observed annual cycle of the Pacific equatorial trades is considerably different from that assumed here (Meyers, 1979), but the dynamics of oscillatory solutions do not involve details of this structure.

The wind stress acting on the ocean at any time is given by:

$$\tau^x = \begin{cases} \tau_w + \tau_s, & h_e < h_c \\ \\ \tau_s, & h_e \geqslant h_c \end{cases} \tag{6}$$

where h_e is the value of h at the eastern equatorial boundary, that is, at $x = 10,000$ km and $y = 0$. According to eqns. (3) and (6), SST in the eastern ocean acts as a switch that turns the central Pacific easterly wind patch on or off. When $h_e < h_c$ (SST in the eastern equatorial ocean is cool) the patch is present, and when $h_e \geqslant h_c$ (SST in the eastern equatorial ocean is warm) the patch is absent. Thus, τ_w interacts with the ocean in a manner like the Walker circulation hypothesized by Bjerknes (1966, 1969). τ_s is unaffected by the state of the ocean.

It is also possible to include an additional wind field in eqn. (6) that represents the steady background component of the trades τ_b (as in the McCreary, 1983, model). Since τ_b is steady and does not interact with the ocean, its presence is not essential to the dynamics of the coupled model. For this reason, we have ignored the steady part of τ_b throughout this paper [but not a slowly varying part; see the discussion of eqn. (8)]. The only advantage of including this component is that the values of h_c that allow the model to oscillate can be smaller (see the discussion at the end of Section 2c).

2.c. Oscillation conditions

The present model has two elements of ocean–atmosphere interaction that are essential for solutions to oscillate at long time scales. The first involves the particular nature of the response of the ocean to τ_w. This interaction provides a positive feedback mechanism for the system and allows the system to adjust to one or the other of two equilibrium states: one with τ_w switched on, the other with τ_w off. The second requires τ_s, and prevents the model from ever reaching either equilibrium state. As we shall see below, it is possible to describe the conditions under which oscillations will occur in terms of a single inequality.

Figures 1 and 2 illustrate the model response to switching on τ_w alone. Figure 1 shows the horizontal structure of the model interface at 1, 3 and 13 months and at 5 years. Since the solution is symmetric about the equator, it is shown only in the northern half-plane. It is possible to follow the progress of Kelvin and Rossby waves about the basin in the figure. Figure 2 shows the time development of h_e during the same period of time. τ_w directly generates equatorial Kelvin waves and Rossby waves. After one month the Kelvin waves have propagated into the eastern ocean and are just beginning

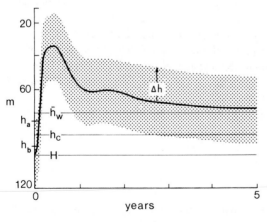

Fig. 2. The time development of the model interface in the eastern ocean, h_e, in response to τ_w alone. The curve first overshoots and then gradually relaxes towards its equilibrium value $\bar{h}_w$ (thin line). The amplitude of the annual cycle of h_e is Δh, and the shaded region indicates the range of values of h_e that are possible in response to τ_w and τ_s. With τ_w switched on the deepest possible value of h_e is h_b; with τ_w switched off the shallowest possible value is h_a. Line h_c is the level that determines whether the ocean is warm or cool. Provided that h_c intersects the shaded region both initially and at large times, eqn. (7) holds and solutions must oscillate.

to raise the interface there, but the slower Rossby waves have not yet progressed very far into the western ocean. After three months the equatorial Kelvin waves have reflected from the eastern boundary, and have caused a rapid decrease of h_e to 34 m. Coastal Kelvin waves are carrying the reflected upwelling signal poleward along the boundary, and Rossby waves are beginning to carry it westward. The wind-driven Rossby waves have just reached the western boundary, and reflect from the western boundary as equatorial Kelvin waves. After thirteen months coastal Kelvin waves are evident partly along the northern boundary where they are strongly damped by the increase of κ. In addition, h_e is no longer at its maximum value due to the arrival of equatorial Kelvin waves from the western boundary. Subsequently, h_e gradually approaches its equilibrium value, $\bar{h}_w = 74$ m, after several reflections of waves from both the eastern and western boundaries have occurred. After 5 years the ocean is very nearly in equilibrium with the wind. Only poleward of about 15°N in the eastern and central ocean is there an indication of mid-latitude Rossby waves propagating very slowly across the basin.

Positive feedback is present in the model in the following way. Suppose initially that h_e rises until $h_e < h_c$ so that SST in the eastern ocean becomes cool and τ_w switches on. According to Figs. 1 and 2, the ocean will subsequently adjust to a state where $h_e = \bar{h}_w < h_c$. Thus, this response acts to keep SST in the eastern ocean cool and τ_w switched on. Similarly, if h_e deepens until $h_e \geqslant h_c$ and τ_w switches off, the response of the ocean acts to keep SST warm and τ_w off. So, in the absence of τ_s, the system has two equilibrium states.

The seasonal wind field, τ_s, prevents the system from ever reaching either equilibrium state. Figure 2 also indicates the response of the model to $\tau_w + \tau_s$. Let Δh be the amplitude of h_e due to τ_s alone. The shaded region of the figure is formed by shifting the solid curve up and down a distance Δh, and so indicates the range of values of h_e that are possible due to $\tau_w + \tau_s$. It is apparent in the figure that sometime after τ_w switches on, the deepest possible value of h_e approaches $h_b \equiv \bar{h}_w + \Delta h$. Now suppose that $h_c < h_b$ so that line h_c must intersect the shaded region. In that case h_e will eventually be deeper than h_c, and shortly thereafter the eastern ocean must become warm and τ_w must switch off. Similarly, when τ_w is off, the shallowest possible value of h_e is $h_a \equiv H - \Delta h$. If $h_c > h_a$, then h_e eventually will be shallower than h_c and τ_w must switch on. Thus, a sufficient condition (it is also necessary) for the model to oscillate is that the inequality:

$$h_a < h_c < h_b \tag{7}$$

holds. Essentially this inequality means that, because of the presence of τ_s, the eastern ocean cannot remain forever either warm or cool.

It is now possible to understand why solutions can oscillate at long time scales. Because h_e overshoots its equilibrium value, the shaded region does not necessarily intersect the line h_c until several years have passed (in the figure this intersection occurs for $t > 2\ 3/4$ years), and so the eastern ocean cannot become warm until that time. Therefore, once τ_w switches on, it cannot switch off until h_e relaxes sufficiently close to $\bar{h}_w$.

Finally, the values of h_c that satisfy eqn. (7) are quite large (typically 87.3 m). It is difficult to imagine that fluctuations in thermocline depth centered about such deep values of h_c actually do influence SST as assumed in eqn. (3). The introduction of a steady τ_b eliminates this problem. Steady τ_b acts to raise the background value of h_e

from H to a considerably shallower level, and in that case the values of h_c that satisfy eqn. (7) are likewise much shallower. As stated above, however, the inclusion of a steady τ_b does not affect the dynamics of the model, and so is not included.

2.d. Results

2.d.i. Symmetric oscillations in time

Figure 3 shows the variation of h_e as a function of time (thick curve) when the amplitude of the annual cycle is 0.015, 0.013, and 0.0115 N m^{-2} in the upper, middle and lower panels, respectively, and when the value of h_c is $(\bar{h}_w + H)/2 = 87.3$ m (straight line). In order to emphasize the response due to τ_w alone, the figure also shows the variation of h_e minus the annual cycle (thin curve). As in Fig. 2, the shaded region indicates the range of possible values of h_e. Important features of the solutions are the

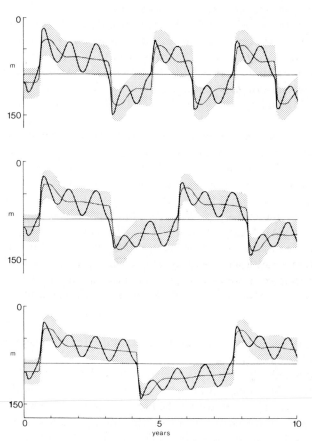

Fig. 3. The time development of h_e due to the wind stress given in eqn. (6) when $\tau_{os} = 0.015, 0.013$ and 0.0015 N m^{-2} in the upper, middle and lower panels, respectively. In all three panels $h_c = (\bar{h}_w + H)/2$, and as a result solutions are symmetric about the line h_c. The onset of warm (cool) periods are phase-locked to the warm (cool) part of the annual cycle. Phase-locking and symmetry imply that the possible oscillations periods are $2n + 1$ years ($n = 0, 1, 2, \ldots$); for realistic choices of τ_{os} the minimum period is 3 years.

following. First, they are all symmetric in time, that is, τ_w is switched on and off and SST in the eastern ocean is cool and warm for equal amounts of time. Symmetry is ensured by the choice of h_c to be the average of $\bar{h}_w$ and H. Due to initial transients, however, symmetry can take a few years to establish; for example, in the upper panel the cool period lasts for 2.5 years initially as opposed to 1.5 years subsequently. Second, all the solutions are phase-locked to the annual cycle: the onset of El Niño events always occurs near the warmest part of the annual cycle (that is, near times when h_e first becomes greater than h_c), whereas the onset of anti-El Niño events occurs near the coolest part. Finally, the oscillation period depends on the amplitude of τ_s, a reduction in amplitude lengthening the oscillation period.

A corollary of the symmetry and phase-locking properties is that oscillation periods for symmetric oscillations are necessarily an odd number of years. A period of one year, however, cannot realistically occur in the Pacific Ocean. To illustrate this point, it is again useful to refer to Fig. 2. The only way a period of one year can occur is if τ_w switches off after just 6 months. At 6 months, however, the curve is near its peak value (that is, h_e is near its minimum value). If the shaded region is to intersect line h_c at this time, it is necessary nearly to triple Δh, which requires that τ_{0s} must be greater than about $0.035 \, \text{N m}^{-2}$. This value is much larger than the observed amplitude of the annual cycle of the Pacific equatorial trades (Meyers, 1979), and consequently oscillation periods less than 3 years are not possible for realistic choices of τ_{0s}.

2.d.ii. Asymmetric oscillations in time

Observations indicate that the Southern Oscillation does not occur symmetrically, as do the solutions of Fig. 3. Weak central Pacific easterlies generally remain for a period of only 1–1.5 years. The model can produce asymmetric solutions provided that h_c is not fixed to be the average of $\bar{h}_w$ and H. Figure 4 shows solutions when h_c is 79.3 m and 95.3 m in the upper and lower panels, respectively, and with $\tau_{0s} = 0.015 \, \text{N m}^{-2}$. Thus, the two solutions are comparable to that in the upper panel of Fig. 3 except that h_c is shifted up and down a distance of 8 m. In the upper panel τ_w remains on only for 1.5 years, whereas in the lower panel τ_w remains off for 1.5 years in agreement with the observations. In both panels the oscillation period is 5 years, two years longer than that of the symmetric solution in Fig. 3. As for the symmetric oscillations, the minimum oscillation period is again 3 years when τ_{0s} has a realistic value.

2.d.iii. Aperiodic solutions

An important aspect of the Southern Oscillation is that it is not an exactly periodic phenomenon. Here we consider how it is possible for the model to produce solutions that oscillate with a range of periods typical of the Southern Oscillation. One way is to allow secular variability of the background trade winds, τ_b. Another is to allow the amplitude of τ_s to vary randomly.

Barnett (1977) has shown that there was a significant increase in the strength of the background tropical Pacific trade-wind field from 1950 to 1970. A similar result was shown by Reiter (1978). More recently, J.O. Fletcher (pers. commun., 1983) noted that this increase was part of a long-period fluctuation of the trades that has taken place during the past 120 years; the Pacific trades tended to weaken from 1860 until about 1930, and have strengthened since that time. Models of tropical ocean circulation

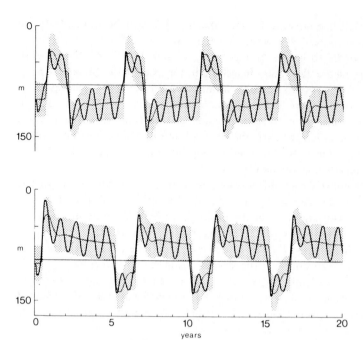

Fig. 4. As in Fig. 3, except that $\tau_{os} = 0.015\,\mathrm{N\,m^{-2}}$, and $h_c = 79.3$ and $95.3\,\mathrm{m}$ in the upper and lower panels, respectively. Without the symmetry property phase-locking implies only that the possible oscillation periods are n years ($n = 1, 2, 3, \ldots$); again, for realistic choices of τ_{os} the minimum period is 3 years. The response in the lower panel compares better with the observations that warm spells typically last only $1-1.5$ years.

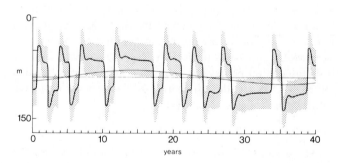

Fig. 5. As in Fig. 4, except that h_e is modified according to eqn. (8). This modification simulates the effect at the eastern ocean of a fluctuation in the background trade-wind field with a period of 40 years. Other parameter values are $\tau_{os} = 0.015\,\mathrm{N\,m^{-2}}$ and $h_c = 87.3\,\mathrm{m}$. For purposes of display, only the response with the annual cycle removed is shown. This very low-frequency variability acts to broaden the spectrum of the model oscillation, in better agreement with the observations.

(including the present one) suggest that such a fluctuation will affect the depth of the pycnocline in the eastern ocean, with an increase of easterly winds raising the pycnocline there. How does a change in h_e induced by low-frequency variability of τ_b affect the oscillation period of the model?

One way to answer this question is to include explicitly a low-frequency component of τ_b as part of the model atmosphere in eqn. (6). An equivalent way simply adds a varying amount to the value of h_e produced by the model. This latter approach has the advantage that it is not necessary to specify the horizontal structure of τ_b, and so it has greater generality, since a given low-frequency in h_e can arise from a wide variety of choices for τ_b. We therefore modify h_e according to:

$$h_e \rightarrow h_e + A \sin(\omega t + \phi) \tag{8}$$

Now h_e represents the depth of the pycnocline in the eastern ocean due to τ_w, τ_s and also the secular variability of τ_b. It is this modified value of h_e that enters eqn. (6) and controls when τ_w switches on and off.

Figure 5 shows h_e, as given in eqn. (8), for 40 years when $\omega = 2\pi\ 40\ \text{yr}^{-1}$, $\phi = 3\pi/4$, $A = 10\,\text{m}$, and when $h_c = 87.3\,\text{m}$ and $\tau_{0s} = 0.015\,\text{N}\,\text{m}^{-2}$. The period of 40 years was chosen to be longer than the time scales associated with the Southern Oscillation, and has no special significance beyond that. The thin curves show h_c (straight line) and $h_c + A \sin(\omega t + \phi)$, and the shaded region indicates the range of the annual cycle as in Fig. 1. Initially both cold and warm periods last for 1.5 years. As the pycnocline becomes shallower than normal $[A \sin(\omega t + \phi) < 0]$ the duration of cold spells increases, but the duration of warm spells does not. From year 15 on the pycnocline deepens, eventually becoming deeper than normal $[A \sin(\omega t + \phi) > 0]$ after year 25. As time passes, it becomes increasingly difficult for warm spells to last only 1.5 years. This tendency is particularly noticeable about year 35 when the pycnocline is most deep.

It is important to realize that Fig. 5 is only one realization of a 40-year time series. It is possible to produce plots which look considerably different in detail by adjusting slightly the amplitude A or phase ϕ. (For example, the long cool period centered around year 15 can easily be made to last longer than 5.5 years.) All realizations, however, exhibit the following properties. There is a lengthening of the cool period when the pycnocline is abnormally shallow, a tendency for the oscillation period to decrease as the pycnocline returns to its normal level, and a lengthening of the warm period when the pycnocline becomes abnormally deep.

As is evident in Fig. 3, the periods of low-frequency oscillations are quite sensitive to the amplitude of τ_s. It is likely, then, that random variability of τ_s will also act to broaden the spectrum of the model Southern Oscillation. Many different causes could be envisioned as generating this random variability, including causes external to the Pacific: coupling between the Tibetian plateau and the Pacific (Fu and Fletcher, 1982) or events propagating eastward from the Indian Ocean (Barnett, 1983) are two possibilities. To simulate this variability we add directly to the modified h_e in eqn. (8) an additional random amount, so that now:

$$h_e \rightarrow h_e + A \sin(\omega t + \phi) + B \tag{9}$$

where B is a random positive number that is reset every 6 months. Solutions not reported here (see McCreary and Anderson, 1984) show that, as expected, the random variability does broaden the spectrum of the model response. In addition, it also weakens the phase-locking of the model to the annual cycle. Whether phase-locking remains an important property of the model depends on the strength of the random variability relative to the annual signal, that is, on the ratio $B/\Delta h$.

3. A COUPLED MODEL THAT DEVELOPS SLOWLY PROPAGATING DISTURBANCES

3.a. The model ocean

The model ocean is an extension of the usual reduced-gravity system that allows the temperature of the layer to vary. The equations are:

$$(hu)_t + (huu)_x + (huv)_y - \beta yhv = -(\tfrac{1}{2}\alpha gh^2 T)_x + \tau^x + \nu_h\nabla^2(hu)$$

$$(hv)_t + (huv)_x + (hvv)_y + \beta yhu = -(\tfrac{1}{2}\alpha gh^2 T)_y + \tau^y + \nu_h\nabla^2(hv)$$

$$h_t + (hu)_x + (hv)_y = \frac{2\delta}{hT} - w + \gamma\left(\frac{T-T^*}{T}\right) \tag{10}$$

$$T_t + uT_x + vT_y = \frac{2}{h}\left[-\gamma(T-T^*) - \frac{\delta}{h}\right] + \nu_h\nabla^2 T$$

where T is the temperature excess of the surface layer over the deep inert layer, ρ_0, α and c_p are the density, thermal expansion coefficient and specific heat of water, respectively, and other quantities have their usual definitions. The values of α, β and ν_h are fixed throughout to be $0.0003°\mathrm{C}^{-1}$, $2.28 \times 10^{-11}\,\mathrm{m}^{-1}\,\mathrm{s}^{-1}$, and $2 \times 10^3\,\mathrm{m}^2\,\mathrm{s}^{-1}$, and the value of c_p is never needed. The dynamical form of these equations is discussed more fully in Anderson (1984), while the derivation of the forcing terms for the mixed layer follows closely that of Hughes (1980).

The surface stress is linearly related to the atmospheric velocity field according to:

$$\tau^x = \rho_a C_D' U, \qquad \tau^y = \rho_a C_D' V \tag{11}$$

where U and V are defined in eqns. (15), ρ_a is the density of the air taken to be $1.3\,\mathrm{kg}$ m^{-3}, and C_D' is a drag coefficient with the value $0.008\,\mathrm{m\,s}^{-1}$. This choice for C_D' ensures that the usual drag coefficient, $C_D = C_D'/U$, has a value of 0.0016 when the wind speed is $5\,\mathrm{m\,s}^{-1}$.

Equations (10) involve three thermodynamic processes: Q_s, δ, and w. Q_s is the heat flux into the layer. In reality, Q_s depends on many atmospheric parameters. Since the objective of this paper is to explore simple coupling mechanisms, the simple parameterization:

$$Q_s = -\rho_0 c_p \gamma(T-T^*) \tag{12}$$

is used, a form first suggested by Haney (1973). In this model T is always less than T^*, so Q_s always acts as a heat source for the layer. Turbulent mixing, $\rho_0\alpha g\delta$, represents the tendency for wind stirring to entrain fluid into the layer from below, thereby deepening the mixed layer and increasing its potential energy. Again for simplicity, we take $\delta = \delta_0/h$. w is a slow upwelling velocity in the deep ocean, presumably driven by a background thermohaline circulation.

The three thermodynamic processes involve four parameters: w, δ_0, γ and T^*. The values of w, δ_0 and γ are generally $4 \times 10^{-7}\,\mathrm{m\,s}^{-1}$, $4 \times 10^{-2}\,\mathrm{m}^3\,°\mathrm{C\,s}^{-1}$ and 3×10^{-6} $\mathrm{m\,s}^{-1}$, respectively. T^* varies with latitude according to:

$$T^* = 4 + (11.33 - 4)(1 + \cos 2\pi y/y_N)/2 \tag{13}$$

where $y_N = 9000\,\text{km}$. T^* has a maximum value of $11.33°C$ at the equator and decreases to $4°C$ at a distance of $4500\,\text{km}$ from the equator.

The choices of parameters in the preceding paragraph ensure that various aspects of the ocean response are physically realistic. For example, in the absence of dynamical processes (so that $u = v = 0$) and when T is uniform, the latter two equations of (10) have the steady solution:

$$\bar{T} = \frac{\gamma T^*}{(\gamma + w)}, \qquad \bar{h} = \left[\frac{\delta_0(\gamma + w)}{\gamma w T^*}\right]^{1/2} \tag{14}$$

Thus, the values of $\bar{T}$ and $\bar{h}$ at the equator are $10°C$ and $100\,\text{m}$, respectively. Note that if γ, δ_0 or w is set to zero, so that any one of the thermodynamic processes is neglected, the model cannot attain a sensible state of rest.

The model ocean consists of the finite-difference versions of eqns. (10) evaluated on an Arakawa C-grid, with T values stored at h points. The domain extends from the equator to $4500\,\text{km}$ north and may be either cyclic in x or a bounded basin. The horizontal resolution of the grid is $150\,\text{km}$ for all the solutions shown here. Kelvin waves along the northern boundary are artificially damped by including a damping term $-\kappa(h - \bar{h})$ in the third of eqns. (10), where κ decreases linearly from a value of $2.5 \times 10^{-7}\,\text{s}^{-1}$ at the boundary to zero $450\,\text{km}$ away. The equations are integrated forward in time with a time step of $1/8$ day.

3.b. The model atmosphere

The equations of motion of the model atmosphere are:

$$-\beta y V = -P_x - rU$$
$$\beta y U = -P_y - rV \tag{15}$$
$$U_x + V_y = \frac{Q}{c^2} - r\frac{P}{c^2}$$

where Q is the forcing by latent heat release, c is the speed Kelvin waves have in the undamped system, and r is a coefficient of Newtonian cooling. These equations describe the response of the first baroclinic mode of the atmosphere, a model that has already been successfully used in several studies of the tropical wind field associated with ENSO (Gill, 1980; Zebiak, 1982). The two free parameters in this model, r and c, have the values $3 \times 10^{-5}\,\text{s}^{-1}$ and $60\,\text{m}\,\text{s}^{-1}$, respectively, the same values used by Gill and by Zebiak. For these choices an equatorial Kelvin wave will be damped on a length scale of $c/\beta \sim 2000\,\text{km}$, while the lowest-order equatorial Rossby wave will decay on the scale $1/3\,(c/\beta) \sim 700\,\text{km}$.

Latent heat release over the ocean is assumed to be related to T according to:

$$Q = Q_0 \frac{T - T_c}{\bar{T}(0) - T_c} \theta(T - T_c) \tag{16}$$

where θ is the Heaviside step function, $\bar{T}(0)$ is the temperature calculated at the equator using eqn. (14), and Q_0 is an amplitude factor. Thus the strength of the forcing is taken to be proportional to the amount T, is above some critical value T_c, and is zero when T

is below T_c; unless stated otherwise, the value of T_c is always $\overline{T}(0) - 1.5$. Q_0 is chosen such that the strength of the surface wind stress as calculated from eqn. (15) is consistent with the typical observed equatorial values, and usually has the value of $0.05 \, \text{m}^2 \, \text{s}^{-3}$.

The response of the atmosphere over land is also found for the solutions in Section 3d. Latent heat release over land is parameterized according to:

$$Q = Q_1 X(x) \frac{\overline{T} - T_c}{\overline{T}(0) - T_c} \theta (\overline{T} - T_c) \tag{17}$$

where Q_1 is an amplitude factor usually taken equal to Q_0, and $X(x)$ is a modulating factor defined below.

A number of assumptions are inherent in this model. First, it is assumed that the atmosphere adjustment times are short compared with those of the ocean, and therefore that the atmosphere will be always in equilibrium with any evolving ocean surface temperature field. Second, the main forcing of the tropical atmosphere is considered to be the result of convection which occurs mainly in the middle troposphere (500–400 mb), thereby exciting a first-mode type of structure (Gill, 1980). Third, the atmospheric response to convection is assumed linear; this assumption is probably valid where convection is weak, but not where convection is strong. Fourth, it is assumed that convection depends only on local SST; this dependence again is unlikely to be strictly true, since convection also depends on surface fluxes and advection of moisture.

Equations (15) are solved numerically using the same technique as Zebiak (1982). Solutions are found by Fourier transforming the equations of motion in x and then solving for an equation in $\tilde{V}$ (the transform of V) alone. This equation is an ordinary differential equation in y, and in finite-difference form it reduces to a tridiagonal matrix equation that can be quickly solved for $\tilde{V}$. The quantities $\tilde{U}$ and $\tilde{P}$ are easily expressed in terms of $\tilde{V}$, and the solution is calculated by finding their discrete, inverse transforms. The domain extends from the equator to 4500 km north, and is cyclic in x with a horizontal resolution of 150 km. The atmosphere is updated only every five days, a procedure which is valid because the time scale of the solution is so long.

3.c. Results for a cyclic ocean

This section discusses solutions when both the atmosphere and ocean are cyclic in x with a circumference of 15,000 km. In each run, the model is initially assumed to be in a state of rest with values of h and T given by eqn. (14), and it is then forced by an imposed wind stress for 100 days. At the end of this time the imposed stress is switched off, and the solution is allowed to develop without further interference. For a wide range of model parameters, a large-scale disturbance subsequently grows to a finite amplitude and begins to propagate slowly eastward. The zonal structure of the disturbances (labelled here mode 1, 2, etc.) is determined by the zonal structure of the initial wind state. The disturbances are stable in the sense that once the model is in a given mode it remains in that mode.

Figure 6 shows time sections for 16 years of equatorial temperature, layer thickness, and wind stress. For the first 100 days, while the ocean is being forced by the imposed wind stress, the temperature remains quite uniform at the initial value of 10°C and there is little east—west slope to the thermocline. Starting about year 1 after the external forcing is removed, an east—west temperature gradient develops with the formation of warm and cool patches (Fig. 6a). The warm patches are associated with convection in the atmosphere (wherever $T > T_c = 8.5°C$ in Fig. 6a) and with a deep pycnocline (Fig. 6b). Regions of easterly and westerly winds exist east and west of the patch, respectively (Fig. 6c). The zonal scale of the disturbance is mode 1 so that there is one cool region and one warm region around the equatorial belt. The disturbance propagates eastward slowly, taking 5.5 years to circumvent the globe. This propagation speed is much slower than any uncoupled atmospheric or oceanic wave in the model (the oceanic Kelvin wave speed is 1.7 m s^{-1}), and is good evidence of the highly coupled nature of the disturbance.

The studies of Lau (1982) and of Philander et al. (1984) provide some insight into the cause of these disturbances. The instability in their models develops in the following way. Suppose initially that $h - \bar{h}$ is weakly positive in a patch on the equator, so that a Q exists to force the atmosphere. This Q drives westerlies to the west of the patch and easterlies to the east of it. Provided the amplitude of Q is sufficiently strong, the effect of this convergent wind field on the ocean is to pile up more water in the patch, thereby increasing $h - \bar{h}$ and Q once again. There is no limit to the size of h in their models, and so the instability grows indefinitely.

The growth of the disturbance in Fig. 6 is caused by a similar mechanism. The reason for this similarity is that, although Q is related to T rather than h, T is generally a monotonically increasing function of h. The disturbance does not grow indefinitely because ocean thermodynamics prevents T from ever increasing much beyond $\bar{T}(0)$.

A number of further calculations were performed, in order to determine the sensitivity of the coupled solutions to model parameters. The following is a brief summary of our conclusions (see Anderson and McCreary, 1984, for a more detailed discussion). The propagation speed of the disturbance is very nearly proportional to its wavelength. There is an inverse relationship between the oscillation period and Q_0. Solutions are quite sensitive to the value of T_c. When T_c is increased to $\bar{T}(0) - 1°$, so that the criterion for convection is more stringent, the horizontal scale of convecting regions decreases and the period increases to 11 years. Increasing the thermodynamic parameters, δ_0, γ and w led to an increase in propagation speed, whereas decreasing them sufficiently leads to the disturbance stalling. Finally, the solution is essentially unchanged when the advection terms were dropped from the temperature equation in (10), indicating that these terms do not play an important role in the dynamics of the oscillation.

3.d. Results for a bounded ocean

This subsection discusses solutions when the atmosphere is cyclic with a circumference of 30,000 km, but the ocean is confined to a bounded basin with a zonal extent of 15,000 km. Land comprises the rest of the globe and is also taken to be 15,000 km wide (see Fig. 7). Over the ocean, convection is parameterized as in eqn. (16) and so Q changes as T evolves, but over land it is usually fixed in time according to eqn. (17). Thus, there is one ocean basin representing the Pacific Ocean with land extending from "Indonesia" in the west round to "South America" in the east. To represent the strong convection over Indonesia and the weaker convection further east, the heating over land is modulated

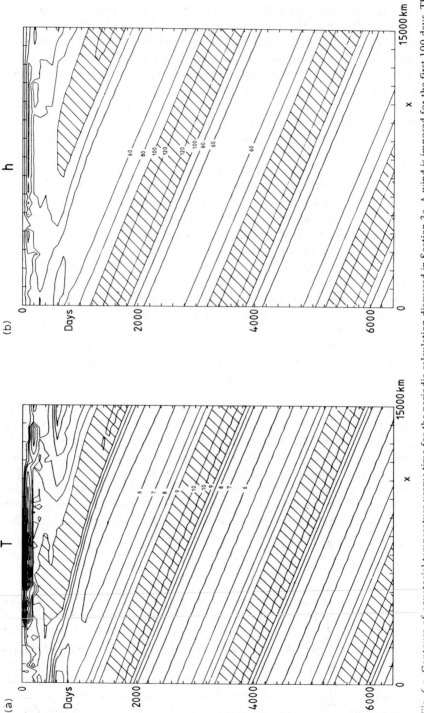

Fig. 6.a. Contours of equatorial temperature versus time for the periodic calculation discussed in Section 3c. A wind is imposed for the first 100 days. Thereafter the coupled model develops freely. The temperature is initially uniform at 10°C, but eventually an instability grows to finite amplitude and propagates eastward. The region where $T > 9°$ is crossed. Convection occurs whenever $T > \bar{T}(0) - 1.5 = 8.5°C$, and the 8.5°C contour is also included. b. As in (a), except showing contours of h. The region where $h > 100$ m is crossed.

361

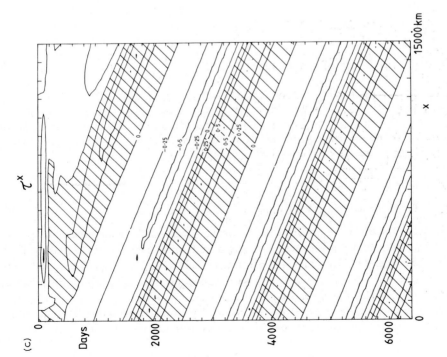

Fig. 6 (continued). c. As in (a), except showing contours of surface stress. Regions of positive τ^x are crossed.

362

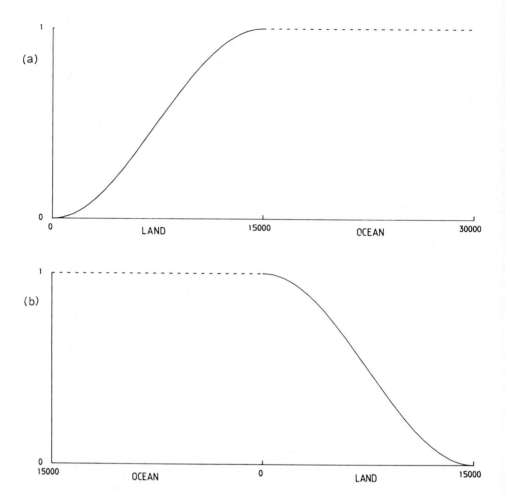

Fig. 7.a. Profile of Q along the equator used for the "Pacific Ocean" configuration discussed in Section 3c. The value of Q over land is given by the solid curve and does not change with time. Q over land is defined precisely in eqns. (17) and (18). The initial value of Q over the ocean is given by the dashed line, and its value subsequently changes as SST evolves. b. As in (a), except for the "Indian Ocean" configuration where $X(x)$ in eqn. (17) is given by eqn. (20).

by the function:

$$X(x) = 0.5\,[1 + \cos 2\pi(x - x_m)/x_w]\,, \qquad 0 < x < x_m \tag{18}$$

where $x_m = 15{,}000$ km, $x_w = 30{,}000$ km. Figure 7a shows the equatorial profile of Q over land (solid line) together with the initial value of Q over the ocean (dashed line). As for the cyclic-ocean solutions, the ocean is initially at rest and values of h and T are given by eqn. (14). Again, for a wide range of model parameters large-scale disturbances develop and propagate slowly eastward.

Figure 8 shows time sections for 16 years of equatorial temperature, layer thickness and wind stress. After an initial period of adjustment the model settles down into an

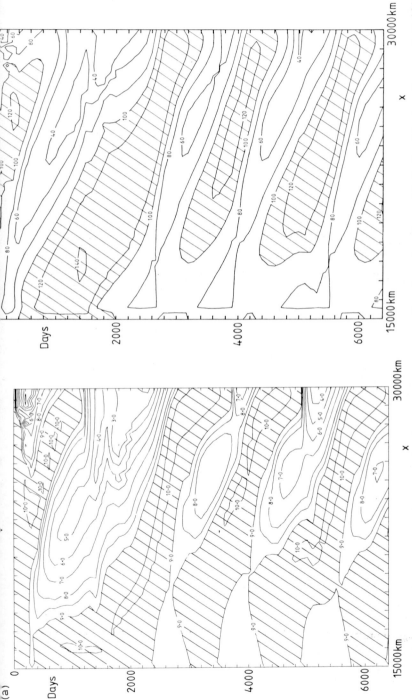

Fig. 8.a. Contours of equatorial temperature versus time for the "Pacific Ocean" configuration discussed in Section 3d, where the ocean is 15,000 km wide bounded by a single land mass of width 15,000 km, as in Fig. 7a. Regions of warm SST form in the west and propagate eastward. The ocean sometimes has a large temperature gradient between west and east, as after 2000 days. At other times the gradient is weak, resembling El Niño conditions, as after 2400 days. Regions where T is warmer than 9°C are crossed. Convection occurs wherever $T > 8.5°C$, and the 8.5°C contour is included.

b. As in (a), except showing contours of thermocline depth. The thermocline tilts most when the zonal temperature gradient is strong and least when the gradient is weak. Regions where h is greater than 100 m are crossed.

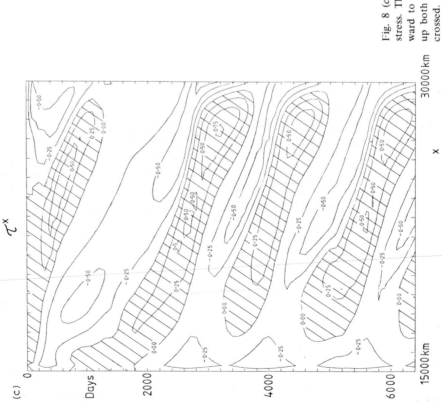

Fig. 8 (*continued*). c. As in (a), except showing contours of the zonal wind stress. The wind is eastward to the west of the pool of warm SST and westward to the east of it. In response to this wind field, the thermocline slopes up both west and east of the warm pool. Regions of positive wind stress are crossed.

oscillatory mode with a period of about 3.5 years. Instabilities develop in the western ocean, propagate eastward, and dissipate at the eastern boundary. The speed of eastward movement of the disturbances is variable, being slowest as the warm patch approaches the eastern boundary. In mid-ocean it travels at ~ 15 cm s^{-1}.

Disturbances are generated by the same mechanism as for the cyclic solutions. Essentially they can develop in any region where T remains sufficiently far above T_c. To illustrate, it is useful to follow the time development of a disturbance throughout its life cycle, from day 2800 to 4800 for example. At day 2800 the ocean begins to warm up in a region centered about 4000 km from the western boundary. This warm pool intensifies and begins to move eastward. By day 3600 it has moved to the central ocean 8000 km from the western boundary. At this time it is apparent in Figs. 8b and c that the warm pool is associated with a deepening of the pycnocline and a convergent wind field, just as for the cyclic solutions. After day 3600, the disturbance no longer increases in amplitude. Near day 4500 the anomaly has reached the eastern boundary and temperatures are a maximum there. The warm anomaly vanishes abruptly about day 4800, its disappearance clearly associated with the growth of another warm pool in the central and western ocean.

Some features of this solution are reminiscent of those observed in the 1982/83 El Niño event. Gill and Rasmusson (1984) discuss the time development of convection, winds and SST anomalies during the event. There was an eastward movement of warm SST, convection and wind anomalies during the event. The occurrence of westerly wind anomalies west of the SST anomaly and the north–south scale of the warm pool were both very similar to those produced in the model.

Other experiments suggest the following conclusions (see Anderson and McCreary, 1985, for a more detailed discussion). If the strength of convection over land, Q_1, is increased then the period lengthens; reducing Q_1 has the opposite effect. If Q_0 is increased by a factor of 1.5 then, as in the cyclic case, the period decreases; when Q_0 is reduced by a factor of 0.5 the warm patch which develops in the west never penetrates into the eastern ocean but stalls in mid-ocean. When T_c is changed to $\bar{T}(0) - 1°$ the oscillation period increases only slightly, in contrast to the cyclic case, even though the regions of convection were predictably smaller. Increasing δ_0, γ, and w increases the speed of propagation of the disturbances (and reduces their period), in agreement with the cyclic case. The atmospheric parameters, c and r, influence the ocean through their effect on the winds. If c is doubled to 120 m s^{-1}, then the model does not oscillate; the winds are weaker and the model locks into a steady state. If, however, Q_0, Q_1 and c are all doubled, then the winds are comparable with the control experiment and the behavior of the coupled system is very similar to the control run. Reducing r results in stronger winds, a very cold eastern Pacific and a shorter oscillation period, again a change similar to that caused by an increase in Q_0 and Q_1.

A seasonal cycle can be imposed on the model by modulating Q_1 according to:

$$Q_1 = Q_0(1 + a \sin \omega t) \tag{19}$$

where $\omega = 2\pi$ yr^{-1}. For values of a of order one, this change ensures that warm patches develop at the same time of year, so that the oscillation period is an integral number of years, but does not otherwise influence the instability.

Another solution investigates how the location of the convection over land influences

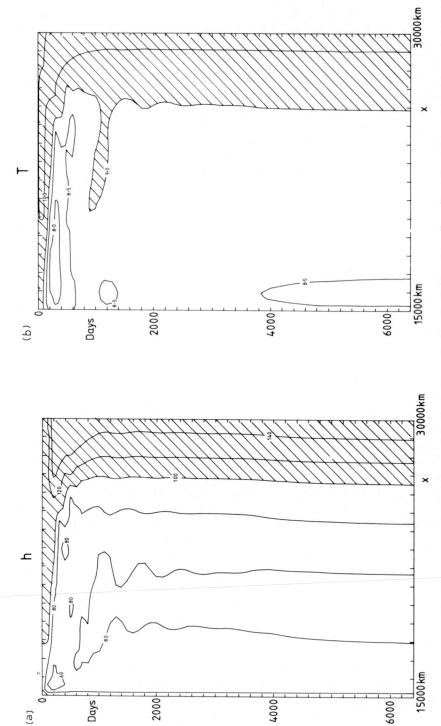

Fig. 9. Contours of equatorial temperature (a) and thermocline depth (b) vs. time for the "Indian Ocean" configuration, where convection over land is as in Fig. 7b. No instabilities develop in this case in contrast to the "Pacific Ocean" configuration of Fig. 8.

the disturbances. The region of strong convection was shifted to the eastern end of the ocean by replacing $X(x)$ in (17) with:

$$X(x) = 0.5(1 + \cos 2\pi x/x_{\mathrm{w}}), \qquad 0 < x < x_{\mathrm{m}} \tag{20}$$

Figure 7b shows the equatorial profile of Q that results from eqn. (20). With this configuration it is useful to regard the ocean as being the Indian Ocean, where the strong convection over Indonesia occurs at the eastern boundary. Figure 9 shows the resulting time development of equatorial temperature and layer thickness. In marked contrast to the solution of Fig. 8, the model locks into a state with westerlies over the ocean and with the eastern ocean warm. No disturbances develop or propagate. Evidently the presence of convection in the west is necessary for the model to be able to oscillate. The reason is likely that only when there are easterly winds over the ocean does the western ocean remain sufficiently warm for an instability to be able to develop.

4. SUMMARY AND DISCUSSION

This paper discusses two examples of simple, coupled ocean–atmosphere models that can oscillate at the long time scales associated with ENSO. The dynamics of the oscillations involve two, very different, mechanisms of ocean–atmosphere interaction. In the model discussed in Section 2 (Model 1), the interaction allows the coupled system to have two equilibrium states for the same values of model parameters. In the model discussed in Section 3 (Model 2), the interaction produces unstable disturbances that grow to a finite amplitude and propagate slowly eastward. Solutions compare favorably with observations of ENSO in some ways but not so well in others.

The ocean of Model 1 consists of the single baroclinic mode of a two-layer ocean, with several parameters (c, H, g' and κ) having values which correspond to the second baroclinic mode of a continuously stratified ocean. SST is very simply related to the depth of the model interface; if the interface is shallower (deeper) than a specified value, h_{c}, SST is cool (warm). The atmosphere of Model 1 consists of two patches of equatorial zonal wind stress. The patch denoted τ_{w} is switched on or off according to whether SST is cool or warm, and is analogous to Bjerknes' Walker circulation. The other, τ_{s}, is a highly idealized seasonally varying wind field.

Model 1 involves two elements of ocean–atmosphere interaction that are necessary for oscillations to occur. First, the interaction of τ_{w} with the ocean introduces positive feedback into the system, and allows for the existence of two equilibrium states: one with τ_{w} switched on, and the other with it off. Second, the annual wind, τ_{s}, prevents the system from every reaching either equilibrium state.

When τ_{w} switches on, equatorial Kelvin waves propagate into the eastern ocean, reflect from the eastern boundary as Rossby waves, and cause a rapid rise of h_{e}. Rossby waves, also directly forced by τ_{w}, eventually reflect from the western boundary as a second set of equatorial Kelvin waves. Their arrival in the eastern ocean begins a gradual deepening of h_{e} toward its equilibrium value. It is this overshoot and gradual relaxation of h_{e} that prevents oscillation periods from being less than three years; only when the ocean is sufficiently relaxed toward equilibrium can τ_{s} act to switch τ_{w} on or off. Thus, as in the McCreary (1983) model, long time scales are closely related to the slow propagation of Rossby waves across the ocean basin, but here they are generated in the tropics rather than the extra-tropics.

For reasonable choices of parameters, solutions compare favorably with the observations in several ways. For example, they oscillate at time scales typical of the Southern Oscillation with a minimum period of three years, the onset of the El Niño events is phase-locked to the annual cycle, and El Niño events last for only 1.5 years. In addition, the onset (and termination) of El Niño events is rapid, with an increase of h_e to its maximum value occurring only a few months after the collapse of τ_w. Finally, it is possible to broaden the spectrum of the model response by introducing secular changes in the background trades or random variability of the annual wind. Random variability also tends to weaken the phase-locking property of the model, so that sufficiently strong random events can cause the switching on or off of τ_w at any time of the year.

An unsatisfying aspect of Model 1 is that both the model atmosphere and ocean thermodynamics are specified externally in eqns. (6) and (3), respectively. It would be much more satisfying if these relations developed internally, as the properties of a system of dynamical equations. We developed Model 2 specifically to overcome these limitations.

The ocean of Model 2 is an extension of the reduced-gravity equations that does include active thermodynamics. The model atmosphere is a linear, single-baroclinic-mode model that is always in equilibrium with a heat source, Q. The wind stress that drives the ocean is directly proportional to the atmospheric velocity field, and Q is a simple function of ocean temperature. Model parameters are chosen to ensure that the uncoupled models respond as realistically as possible to various prescribed forcings. For all the solutions the atmosphere is cyclic in x. One set of solutions is found when the ocean is also cyclic in x. Another set is found in a domain that consists of a bounded ocean and a land mass, each 15,000 km wide. Convection over land can be located either to the east or west of the ocean, as in Fig. 7.

When both the atmosphere and the ocean are cyclic, the model is spun-up for 100 days with a prescribed wind stress. This wind is then switched off and the model allowed to develop without external forcing. Instabilities develop and grow to a finite amplitude. The process underlying their growth is the following one. A localized heating, Q, is associated with westerly winds to the west and easterly winds to the east of the heating. Provided the amplitude of Q is sufficiently large, this convergence in the surface winds causes h and T to increase. Increasing T leads to a stronger Q and a rapid growth of the instability. Ocean thermodynamics prevents T from increasing much beyond $T(0)$, and so the instability does not continue to grow indefinitely. Large-amplitude disturbances always propagate eastward. Their rate of propagation depends on their zonal scale (faster for longer disturbances), on the strength of convection (faster for larger Q_0), and on the mixing parameter (faster for larger values of δ_0, γ and w), but is not sensitive to the advection of heat.

In the bounded-ocean case and with convection over land to the west (Indonesia), instabilities develop in the western or central ocean and propagate slowly eastward. When they reach the eastern boundary they weaken and eventually vanish. Their disappearance is related to the growth of another disturbance in the west. The dependence of solutions on parameters is similar to that in the cyclic case (except for the dependence on T_c). The strength of convection over land also influences the solution. An increase in Q_1

lengthens the period between the development of warm patches, and vice versa. If convection is strongest over land to the east of the ocean (as in the Indian Ocean), no instabilities ever develop.

The model compares favorably with observations in several ways. The production of oscillatory solutions with long time scales is a very robust feature of the model. Disturbances always develop in the western or central ocean and propagate eastward (Fig. 8). Finally, ENSO-type disturbances are not observed to form in the Indian Ocean, a property consistent with the model response in Fig. 9.

Solutions differ from observations in other ways. Observations do not show easterlies to the east of the heating whereas this is a strong feature of this model. Nobre (1983) has suggested that easterlies can occur when convection is weak but that they are suppressed when it is strong; this implies that a nonlinear correction to the linear atmosphere model used here may be necessary to represent adequately the tropical atmosphere. The propagation speed of disturbances is too slow. (Several parameters can affect this speed, and it is possible to tune the model to reproduce more faithfully the observed speed.) Finally, a serious deficiency is that the model never produces abrupt ENSO features; for example, there is no rapid onset phase.

The two simple models discussed in the body of the paper, as well as those mentioned in the introduction, illustrate nicely various possible mechanisms of tropical ocean–atmosphere interaction. Their agreement with aspects of the observations is encouraging. Their limitations are evident, but should not be regarded as discouraging; rather, they are valuable because they suggest directions of future research. For example, the fact that Model 2 fails to simulate the rapid onset and intermittancy of El Niño indicates that the system lacks, or misrepresents, important processes involved in the oscillation dynamics. A number of processes that could be important immediately come to mind. We are currently improving Model 2 in two ways: by including two ocean basins (representing the Indian and the Pacific Oceans), and by developing a model atmosphere that includes a humidity equation, thereby allowing a better parameterization of convection. Our goal is to develop a coupled system that will allow the rapid onset of El Niño, perhaps by a mechanism similar to that present in Model 1.

ACKNOWLEDGEMENTS

This research was sponsored by the National Science Foundation under grant No. OCE 79-19698 through NORPAX and grant No. ATM 82-05491, and by the Natural Environment Research Council under grant No. GRS/4658. The dynamical framework of the model ocean (eqn. 10) is similar to one formulated in 1978 with A.E. Gill, but never published. We are indebted to Robert Wells; without his programming assistance this paper would not have been possible.

REFERENCES

Anderson, D.L.T., 1984. An advective mixed layer model with applications to the diurnal cycle of the low-level East African Jet. Tellus, 36A: 278–291.

Anderson, D.L.T. and McCreary, J.P., 1985. Slowly propagating disturbances in a coupled ocean–atmosphere model. J. Atmos. Sci. (in press).

Barnett, T.P., 1977. The principal time and space scales of the Pacific trade wind fields. Mon. Weather Rev., 34(2): 221–236.

Barnett, T.P., 1983. Interaction of the monsoon and Pacific trade wind system at interannual time scales. Part I: The equatorial zone. Mon. Weather Rev., 111: 756–773.

Bjerknes, J., 1966. A possible response of the atmospheric Hadley circulation to equatorial anomalies of ocean temperature. Tellus, 18: 820–829.

Bjerknes, J., 1969. Atmospheric teleconnections from the equatorial Pacific. Mon. Weather Rev., 97: 163–172.

Fu, C. and Fletcher, J.O., 1982. The role of the surface heat source over Tibet in interannual variability of the Indian summer monsoon. Trop. Ocean–Atmos. Newslett., 14:6–7.

Gill, A.E., 1980. Some simple solutions for heat-induced tropical circulation. Q. J. R. Meteorol. Soc., 106: 447–462.

Gill, A.E. and Rasmusson, E.M., 1984. The 1982/83 climate anomaly in the Equatorial Pacific. Nature, 306: 229–232.

Haney, R.L., 1971. Surface thermal boundary conditions for ocean circulation models. J. Phys. Oceanogr., 1: 241–248.

Hirst, A.C., 1985. Free equatorial instabilities in simple coupled atmosphere–ocean models. In: J.C.J. Nihoul (Editor), Coupled Ocean–Atmosphere Models. (Elsevier Oceanography Series, 40) Elsevier, Amsterdam, pp. 153–165 (this volume).

Hughes, R.L., 1979. A highly simplified El Niño model. Ocean Modell., 22.

Hughes, R.L., 1980. On the equatorial mixed layer. Deep-Sea Res., 27A: 1067–1078.

Hughes, R.L., 1984. Developments on a highly simplified El Niño model. Ocean Modell., 54.

Lau, K.-M., 1982. Oscillations in a simple equatorial climate system. J. Atmos. Sci., 38: 248–261.

Lau, K.-M., 1985. Subseasonal scale oscillation, biomodal climatic state and the El Niño/Southern Oscillation. In: J.C.J. Nihoul (Editor), Coupled Ocean–Atmosphere Models. (Elsevier Oceanography Series, 40) Elsevier, Amsterdam, pp. 29–40 (this volume).

McCreary, J.P., 1981. A linear stratified model of the equatorial undercurrent. Philos. Trans. R. Soc. London, Ser. A, 298: 603–635.

McCreary, J.P., 1983. A model of tropical ocean–atmosphere interaction. Mon. Weather Rev., 111(2): 370–389.

McCreary, J.P. and Anderson, D.L.T., 1984. A simple model of El Niño and the Southern Oscillation. Mon. Weather Rev., 112: 934–946.

McCreary, J.P., Picaut, J. and Moore, D.W., 1984. Effects of remote annual forcing in the eastern tropical Atlantic. J. Mar. Res., 42: 45–81.

Meyers, G., 1979. Annual variation in the slope of the 14°C isotherm along the equator in the Pacific Ocean. J. Phys. Oceanogr., 965: 885–891.

Nobre, C.A., 1983. Tropical heat sources and their associated large-scale atmospheric circulation. Ph.D. Thesis, M.I.T.

Pazan, S.E. and Meyers, G., 1982. Interannual fluctuations of the tropical Pacific wind field and the Southern Oscillation. Mon. Weather Rev., 110: 587–600.

Philander, S.G.H. and Pacanowski, R.C., 1980. The generation of equatorial currents. J. Geophys. Res., 85: 1123–1136.

Philander, S.G.H., Yamagata, T. and Pacanowski, R.C., 1984. Unstable air–sea interactions in the tropics. J. Atmos. Sci., 41: 604–613.

Reiter, E.R., 1978. The interannual variability of the ocean–atmosphere system. J. Atmos. Sci., 35(3): 349–370.

Yamagata, T., 1985. Stability of a simple air–sea coupled model in the tropics. In: J.C.J. Nihoul (Editor), Coupled Ocean–Atmosphere Models. (Elsevier Oceanography Series, 40) Elsevier, Amsterdam, pp. 637–657 (this volume).

Zebiak, S.E., 1982. A simple atmospheric model of relevance to El Niño. J. Atmos. Sci., 39: 3017–2027.

CHAPTER 25

THE RESPONSE OF THE OSU TWO-LEVEL ATMOSPHERIC GENERAL CIRCULATION MODEL TO A WARM SEA-SURFACE TEMPERATURE ANOMALY OVER THE EASTERN EQUATORIAL PACIFIC OCEAN

STEVEN K. ESBENSEN

ABSTRACT

The OSU two-level atmospheric general circulation model (OSU-AGCM) has been used to simulate the most probable response of the atmosphere to warm equatorial sea-surface temperatures which are characteristic of the mature phase of El Niño–Southern Oscillation events. The last sixty days of three pairs of ninety-day control and experimental integrations were composited to define anomalous flow features.

The simulated local response in the tropics is characterized by an increase in precipitation (~ 2 mm per day) and the total heating (~ 50 W m^{-2}) over the central and western equatorial Pacific. Anomalous precipitation rates are in excess of the anomalous evaporation rates. In agreement with observations, the anomalous westerly flow appears on the western side of the sea-surface temperature anomaly, on and just to the south of the equator. The upper-level flow over the central equatorial Pacific becomes more easterly, but most of this increase can be attributed to the nondivergent part of the flow.

The statistically significant part of the remote response in the geopotential height field resembles the teleconnections observed by Horel and Wallace (1981) and the observed anomaly patterns presented by Van Loon and Rogers (1981) and Chen (1982). Large variability was found to exist between pairs of control and experimental integrations which differ only in their initial conditions. It is suggested that this variability is due to low-frequency planetary-scale oscillations which would occur in the absence of anomalous equatorial sea-surface temperatures.

1. INTRODUCTION

The documentation of the empirical relationship between an index of the Southern Oscillation and pressure and temperature measurements over North America by Walker and Bliss (1932)[*] posed the intriguing problem of determining what dynamical mechanism could be responsible for such remote planetary-scale correlation patterns. A breakthrough of a conceptual and empirical nature occurred when Bjerknes (1969) established that there exist year-to-year fluctuations in tropical sea-surface temperature and precipitation which are clearly related to the Southern Oscillation and to the year-to-year fluctuations of sea-level pressure over the North Pacific. Bjerknes hypothesized that an increase in tropical precipitation would increase the strength of the Hadley circulation which could then strengthen the middle-latitude westerlies by transferring larger amounts of absolute angular momentum to middle latitudes.

While Bjerknes focused on the north–south teleconnections with middle latitudes through the Hadley circulation, Namias (1969) emphasized the teleconnections between

---

[*]See the review article by Montgomery (1940) for a review of Sir Gilbert Walker's work on this subject.

the North Pacific region and the climatic patterns over North America. Namias hypothesized that the anomalous fluxes of sensible and latent heat might warm (or cool) the atmosphere over large "pools" of anomalously warm (or cool) water over the North Pacific, resulting in a standing Rossby-wave pattern extending over North America.

Recently Horel and Wallace (1981) showed that the year-to-year fluctuations of the 700 mb height field over the entire North Pacific/North American sector contain a pattern which is clearly related to tropical events associated with the Southern Oscillation. In particular, when the December–February sea-surface temperatures are anomalously warm in an equatorial belt extending from the South American coast to the central Pacific, the 700 mb height anomalies are negative in a broad belt over the North Pacific, positive over northwestern Canada, and negative over the eastern United States. These observed correlations were found to have a strong resemblance to the steady-state solutions of the linearized primitive equations on a sphere forced by a tropical heat source which were presented by Hoskins and Karoly (1981) and Webster (1981). Horel and Wallace suggested that the middle latitude teleconnections with the equatorial Pacific may therefore be interpreted as a planetary wave pattern forced by the anomalously large release of latent heat which is associated with the warm equatorial sea-surface temperatures.

Simmons (1982) summarized and extended the studies of Hoskins and Karoly (1981) and Webster (1981). In some respects, Simmons' work supports Horel and Wallace's (1981) suggestion. It also shows that the teleconnection patterns do not seem to be sensitive to the vertical profiles of diabatic heating in the tropics, and a non-linear time-dependent barotropic model appears to give results which are consistent with linear, steady-state, primitive equation models. However, the results of these idealized models cannot be taken as proof of the stationary wave hypothesis for teleconnections. The magnitude of the extratropical response in the North Pacific–North American sector is very sensitive to the longitudinal and latitudinal placement of the tropical forcing with respect to the western Pacific westerly wind maximum in the upper troposphere. Also, as Simmons (1982) points out, it may be unrealistic to assume that the zonally varying basic flow is maintained by a constant forcing rather than a forcing which changes as the tropical circulation anomalies evolve.

A feedback mechanism which has been ignored in most idealized model calculations is the sensible and latent heat flux from the sea surface which plays a crucial role in the hypothesis of Namias (1969). Of particular interest is the result of Hendon and Hartmann (1982) which shows that the heat-flux feedback from the sea surface tends to reduce the remote response to a tropical diabatic heating anomaly in a linear, steady-state primitive equation model for which the zonal-wind component does not vary in the zonal direction. However, they find that surface heat-flux feedback strongly amplifies the local response to a diabatic heat source in middle latitudes and can significantly alter the remote response. Hendon and Hartmann also found that the middle latitude teleconnections with tropical heat sources are dramatically reduced by inclusion of dissipation in the form of Ekman pumping from the boundary layer.

Clearly, theoretical studies have not yet produced a completely satisfactory explanation of the Pacific–North American teleconnection pattern documented by Horel and Wallace (1981). More theoretical work is needed. However, these studies have provided us with a wealth of analytical and qualitative tools for exploring observational data and for interpreting the results of general circulation model experiments in which realistic orography and most, if not all, important feedback mechanisms are explicitly represented.

The purpose of this study is to document the response of the OSU Atmospheric General Circulation Model (AGCM) to a tropical sea-surface temperature anomaly which is characteristic of the mature phase of an El Niño event. We will address the degree of similarity between the model results and the teleconnection patterns documented by Horel and Wallace (1981) and others. We will also attempt to identify the mechanisms involved in the remote response using recent theoretical studies as a guide.

This study is not the first of its kind and is, in fact, part of a cooperative effort of the World Group for Numerical Experimentation to establish the relative importance of sea-surface temperature anomalies on atmospheric flow patterns averaged over months or seasons.

Two early AGCM experiments were performed with the expressed purpose of testing Bjerknes' (1969) hypothesis for the North Pacific teleconnection pattern. The first integration of this type was performed by Rowntree (1972) who attempted to verify Bjerknes' (1969) hypothesis by introducing an equatorial sea-surface temperature anomaly in the central and eastern Pacific into a nine-level hemispheric AGCM developed at the Geophysical Fluid Dynamics Laboratory (GFDL) with January insolation and surface conditions. The second attempt was by Julian and Chervin (1978) using a global six-layer AGCM developed at the National Center for Atmospheric Research (NCAR), Boulder, Colorado. Although the sea-surface temperature anomalies in both of these studies produced changes in the general circulation which gave support to Bjerknes' hypothesis, they were performed without the benefit of recent observational and theoretical studies which provide a vastly improved framework for interpreting the results of the AGCM integrations. Also, the hemispheric model used by Rowntree as well as the NCAR model had large systematic errors in their simulated planetary-scale wave patterns which we now realize can have important effects on the simulated teleconnection patterns. Furthermore, the averaging period used by Rowntree (1972) was too short to separate the teleconnection signal from the meteorological "noise".

Two recent AGCM experiments by Blackmon et al. (1983) and Shukla and Wallace (1983) have overcome some of the earlier difficulties faced by Rowntree (1972) and Julian and Chervin (1978). Both recent studies use advanced, multilevel, global AGCMs with reasonably realistic simulated climatologies. Both use realistic sea-surface temperature anomalies in the central and eastern equatorial Pacific. Both integrate for a sufficiently long period to separate the teleconnection signal from meteorological noise. And, both obtain teleconnection patterns which are reasonably similar to the observations of Horel and Wallace (1981).

The previous AGCM experiments provide a benchmark for measuring the results presented here and a point of departure for the interpretations. The two-level structure of the OSU AGCM is both an advantage and a disadvantage for this type of study. The OSU model is a logical extension of linearized, two-level primitive equation or quasi-geostrophic models on which much of our theoretical understanding of large-scale atmospheric dynamics is based. This simplifies the interpretation of simulated dynamical features. On the other hand, features which are sensitive to the details of the vertical distribution of diabatic heating or which have complex vertical structures cannot be simulated. Also, the lack of vertical propagation of wave energy into the stratosphere may introduce significant errors into the planetary-wave structure of the model. However, the author believes that the OSU AGCM is one of a hierarchy of models which can give insight into the response of the global atmosphere to tropical circulation anomalies.

2. EXPERIMENTAL DESIGN

2.1. The two-level OSU AGCM

The OSU AGCM is a two-level primitive equation model whose horizontal structure is represented by a $4° \times 5°$ latitude–longitude lattice of grid points. The model reproduces many important features of the general circulation. Its performance has been documented recently by Schlesinger and Gates (1980) while the details of the methods of numerical integration and the parametrization of physical processes are described fully by Ghan et al. (1982).

The model contains parametrizations for both convective and layer clouds which affect both the solar and longwave radiation budgets. The wind stress and the sensible and latent heat fluxes at the Earth's surface are parameterized by the bulk aerodynamic method.

2.2. The control integration

The control integrations for this experiment consist of 90-day integrations of the OSU AGCM with sea-surface temperatures held fixed at the average January values given by Alexander and Mobley (1976) and the solar insolation values for the 15th day of January. The initial conditions for the control integrations are taken from the 15th day of January of the fifth, eighth and eleventh year of a model integration in which solar radiation and sea-surface temperatures were allowed to pass through their annual cycles.

2.3. The experimental integration

The experimental integrations are identical to the control integrations except that a sea-surface temperature anomaly which is characteristic of a mature El Niño event is added to the average January sea-surface temperatures given by Alexander and Mobley (1976). The SST anomaly field was digitized directly from fig. 21a of Rasmusson and Carpenter (1982) onto the OSU $4° \times 5°$ latitude–longitude grid. The anomaly was set to zero on the boundary and outside of a spatial domain over the Pacific Ocean which is bounded by 30°S, 30°N, 120°E and the North and South American continents. In addition all anomaly values along latitudes 26°S and 26°N and along longitude 125°E were multiplied by one-half.

The SST anomaly field used in the experimental integration is shown in Fig. 1. Warm sea-surface temperatures exist in an equatorial belt extending eastward from 170°W to near the South American coast. Weak negative anomalies exist poleward and to the west of the warm anomaly. Weak warm anomalies can be seen near the Asian and Australian continents.

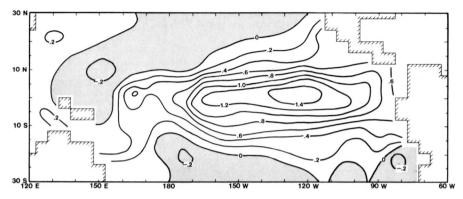

Fig. 1. Observed sea-surface temperature (°C) anomaly representative of the mature phase of a composite El Niño event (from Rasmusson and Carpenter, 1982).

The Rasmusson and Carpenter (1982) "mature phase" anomaly in Fig. 1 is a composite of the SST anomalies found during December–February in the year following a maximum El Niño anomaly along the South American coast. During the year preceding the mature phase, the warm temperatures spread from the South American coastal region to the central Pacific. The SST anomalies are past their peak, but occupy a very large portion of the tropical Pacific region. The mature phase anomaly also has a qualitative resemblance to the leading eigenvector of the principal component analysis performed on Pacific sea-surface temperatures by Weare (1982).

3. APPROACH TOWARDS EQUILIBRIUM

Neither the control nor the experimental integrations are in statistical equilibrium at the beginning of our 90-day integration period. For the control integrations, the lack of equilibrium is primarily due to the thermal "mass" and "dissipation" of the atmosphere which result in the phase lag and damping of the atmospheric response to the annual cycles of solar insolation and sea-surface temperature. When these external conditions are suddenly held constant at their January values, it takes a finite amount of time for the model atmosphere to come into equilibrium with the new steady-state conditions. For the experimental integrations, additional transients are introduced by the impulsive change in the sea-surface temperatures over the tropical Pacific Ocean.

As a measure of the approach toward statistical equilibrium we have chosen the global mass-averaged temperature and kinetic energy. Ideally, one would like to check other model variables, other statistical moments, and their spatial distributions. However, this is beyond our present resources and beyond the exploratory nature of the present research.

In Fig. 2 we see that the mass-averaged temperature of the composite control integration decreases by $\sim 1/2°C$ during the first fifteen days and then increases slowly with time. The time history of the composite experimental integration is similar. It is only during the 30–90 day period that it becomes apparent that the experimental temperature is systematically warmer than the control temperature by $\sim 0.2°C$.

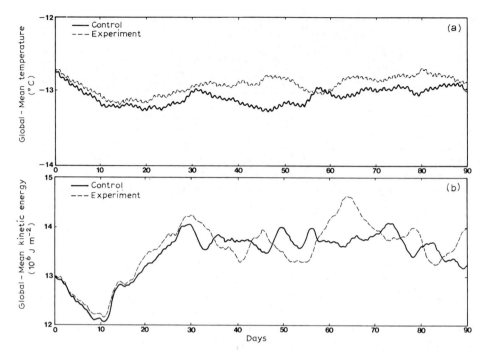

Fig. 2. Time series of a (a) global mass-averaged temperature (°C) and (b) global kinetic energy (10^6 J m^{-2}) for composite of three El Niño experiments (solid line) and control (dash line) integrations.

The small decrease in temperature at the beginning of the integrations can be explained if we note that the annual cycle of the mass-averaged temperature is dominated by the annual cycle in the Northern Hemisphere. Since the phase of the air-temperature cycle will lag behind the solar insolation we expect that the Northern Hemisphere air temperature will be warmer at the beginning of the integration.

The overall increase in mass-averaged kinetic energy during the first month can be explained in a similar manner. The kinetic energy undergoes a larger annual cycle in the Northern Hemisphere and lags the annual cycle of the insolation. We therefore expect larger values of the kinetic energy to develop with time in the perpetual January integration. With the presence of an anomaly of sea-surface temperature over a large portion of the tropical oceans, we might expect to see anomalously large values of the globally mass-averaged kinetic energy. But after the first 30 days, the differences between the two integrations are not obvious, and both undergo rather large fluctuations with periods in the order of two weeks.

In summary, we find that the transients of the global mass-averaged temperature and kinetic energy which are related to the change from seasonally varying solar insolation and sea-surface temperatures to fixed external conditions appear to be as large as the transients due to the impulsive increase in the sea-surface temperature over much of the tropical Pacific Ocean. The differences of temperature between the control and the experiment become obvious after about 30 days of integration while the kinetic energy differences are not as systematic. The results suggest that we may make meaningful comparisons of the mean states of the control and experiment during the last 60 days of the 90-day integrations.

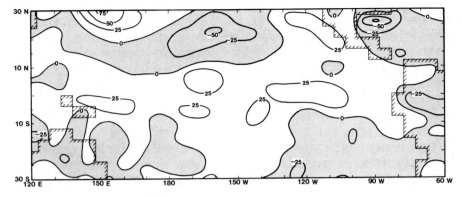

Fig. 3. Composite difference of the total heat flux (W m^{-2}) from the ocean to the atmosphere in the tropical Pacific region between El Niño experimental and control integrations.

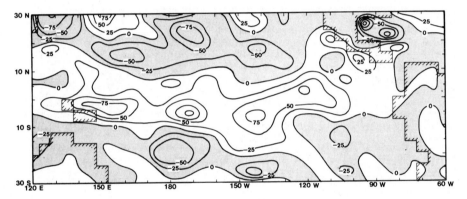

Fig. 4. Composite difference of atmospheric heating (W m^{-2}) between El Niño experimental and control integrations in the tropical Pacific region.

4. THE LOCAL EQUATORIAL RESPONSE

The simulated region of anomalous upward-directed heat flux from the surface of the eastern equatorial Pacific Ocean in Fig. 3 coincides roughly with the region of anomalously warm sea-surface temperatures. Intuition would suggest that this anomalous flux of heat would converge in the vicinity of the sea-surface temperature anomaly thereby generating the available potential energy needed to strengthen the circulation over the tropical Pacific. In a general sense the results are consistent with intuition. However, the simulated heat-flux anomaly field more closely resembles the vertically integrated heating anomaly field shown in Fig. 4 than it does the sea-surface temperature field. This is evidence that local radiative–convective adjustment to a change in sea-surface temperature can explain only a small fraction of the simulated atmospheric response. Most of the response is probably due to interactions between the dynamics and the parameterized physical processes of the model.

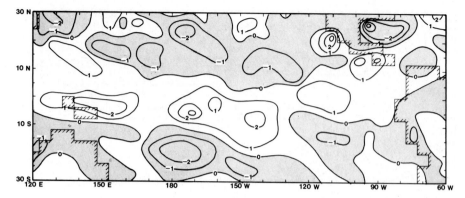

Fig. 5. Composite difference of precipitation rate (mm per day) between El Niño experiment and control integrations in the tropical Pacific region.

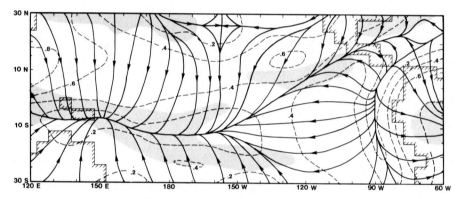

Fig. 6. Composite of horizontally divergent flow (10^4 kg m^{-1} s^{-1}) in the layer from the surface to 600 mb for the control integrations. Precipitation rates greater than 4 mm per day are indicated by shading.

We note here that the precipitation anomaly field shown in Fig. 5 closely resembles the vertically integrated heating field in Fig. 4. Using the fact that 1 mm per day of precipitation is equivalent to a flux of heat of approximately 30 W m^{-2}, we see that the latent-heat release explains a large fraction of the total heating. The parameterized convergences of the radiative and sensible heat fluxes may play an important role in setting up the anomalous pattern of total heating but their magnitudes are smaller than the latent-heat release.

The divergent part of the large-scale flow provides the essential link between the heating and the rotational part of the circulation. Figure 6 shows the divergent flow in the layer from the surface to 600 mb for the control integrations. Precipitation rates greater than 4 mm per day are indicated by shading. The OSU AGCM characteristically produces a very strong intertropical convergence zone (ITCZ) in the western and central Pacific Ocean near 6°S. As we move our attention eastward, the ITCZ in the Southern Hemisphere becomes less distinct near 150°W and shifts into the Northern Hemisphere along a line which runs northeastward towards Central America and southern Mexico.

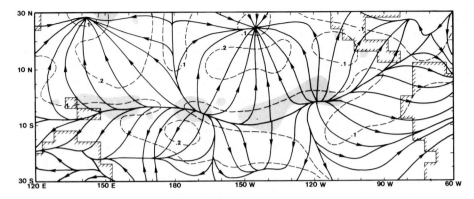

Fig. 7. Composite difference of the horizontally divergent flow (10^4 kg m^{-1} s^{-1}) in the layer from 600 to 200 mb between the El Niño experimental and control integrations. The anomalous atmospheric heating values greater than 50 W m^{-2} are indicated by shading.

The precipitation rates increase along this section of the ITCZ from a minimum near 140°W to a maximum just off the Central American coast. The model also produces a band of precipitation in the South Pacific which is oriented from west-northwest to east-southeast from about 170°E to 120°W which bears a qualitative resemblance to the cloud cover associated with the South Pacific convergence zone.

When the anomalous sea-surface temperature field is introduced, important changes occur in the divergent component of the circulation. Figure 7 shows the anomalous divergent flow in the layer from 600 mb to 200 mb with the regions of large anomalous vertically integrated heating indicated by shading. The surface to 600 mb flow pattern is nearly identical except for a 180° difference in direction.

The most obvious change in the divergent flow is the strengthening of the Hadley circulation in the Northern Hemisphere over the Pacific Ocean. This anomalous circulation feature appears to be closely related to the heating. In general, the flow at upper levels diverges away from the regions of anomalous heating and converges toward regions of anomalous cooling. From theoretical considerations, this type of behavior is expected for low-frequency disturbances in tropical regions with deep heating.

The total horizontal mass flux of the control integration is shown in Fig. 8. In the lower layer, relatively strong easterly flow appears in a band near 8°N in the western Pacific region. Slightly stronger easterly flow appears also in the upper layer. The upper-level easterlies extend into the Southern Hemisphere over the simulated ITCZ to near 10°S. However, the lower-level flow in this region is weak and has a complex latitudinal structure.

The flow patterns in the eastern Pacific are dominated by the flow around the sub-tropical high-pressure zones at low levels and by upper-level westerlies everywhere except for a region near 165°W in the south-central Pacific where there is northward cross-equatorial flow, and just off the equatorial South American coast where the flow is easterly. Lower-level easterlies and upper-level westerlies are found near 125°W.

The structure of the upper-level flow anomalies over the central Pacific in Fig. 9 is consistent with the pattern observed during the 1982–83 El Niño/Southern Oscillation event (Arkin et al., 1983) with the flow becoming more anticyclonic north and south of the equator. The flow pattern appears to be more consistent with the hypothesis that the anomalies are part of a stationary Rossby-wave pattern than with the classical hypothesis involving a strengthened Hadley circulation. Figure 9 shows that in a band from 10° to 30°N, the flow becomes more easterly in both layers near 165°W. By comparing Fig. 7 with Fig. 9 we cannot rule out the possibility that the easterly anomalies in the lower layer of this region are being maintained by the strengthened Hadley circulation through Coriolis accelerations. However, in the upper layer, the dynamics producing the anomalous easterlies are overcoming the effects of the Coriolis accelerations acting on the divergent part of the flow. This constitutes evidence against the classical hypothesis that warm sea-surface temperatures are associated with a strengthening of the westerlies in the upper branch of the Hadley circulation due to increased angular momentum transport by the Hadley circulation, at least in a regional sense.

The anomalous upper-level flow over the central equatorial Pacific is easterly, as one might expect with a strengthening and extension of the Walker circulation toward the SST anomaly. However, it is clear from a comparison of Fig. 9 with Fig. 7 that nearly all the increase is in the nondivergent part of the flow. Thus, interpretations of the mass flow using two-dimensional streamfunctions in an equatorial plane may be misleading. Regional changes in the Walker circulation should be described using the divergent part of the three-dimensional circulation.

In the lower layer, the flow within 15° latitude of the equator is in qualitative agreement with the surface flow presented by Rasmusson and Carpenter (1982). Figure 9 shows that the anomalous flow on the western side of the SST is westerly near and just to the south of the equator; cyclonic curvature of the streamlines exists to the north and south of the equator.

5. THE REMOTE RESPONSE

Although the sea-surface temperature anomaly introduced in the experimental integration was confined to the equatorial Pacific Ocean, large changes in simulated physical and dynamical quantities can be seen in regions that are remote from the equatorial Pacific. Some of the features of the remote response are probably due to the natural and possibly unpredictable variability of the OSU-AGCM (see Section 6). However, a part of the remote response is related to the Southern Oscillation. In this section we will document the regional characteristics of the global, remote response.

5.1. Geopotential height

The simulated general circulation at middle and high latitudes in the Northern Hemisphere is significantly changed by the presence of the equatorial SST anomaly. Figure 10 shows an alternating series of large low and high geopotential height anomalies beginning with a broad region of low heights extending northwest—southeast from northern Asia to the Aleutian area, and ending near 35°W and 40°N over the Atlantic Ocean.

The simulated 800 mb height anomaly patterns qualitatively resemble the spatial structure of the observed correlation between the wintertime Pacific sea-surface

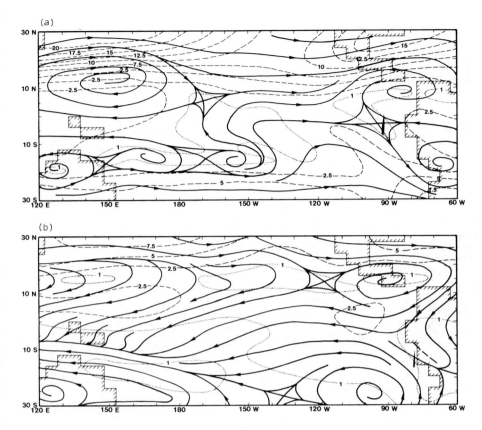

Fig. 8. Composite horizontal mass flux (10^4 kg m^{-1} s^{-1}) in (a) the upper layer from 600 to 200 mb and (b) the lower layer from the surface to 600 mb for the control integrations.

temperature index and the wintertime 700 mb height field presented by Horel and Wallace (1981). Both have the broad low region over northern Asia and the North Pacific; both have anomalous ridging over Central Canada; and both show anomalously low heights downstream from the Canadian ridging. However, there are discrepancies in the phases of the patterns, particularly over the model Atlantic Ocean where the simulated pattern appears to be significantly different from observations.

The 800 mb height anomaly pattern in Fig. 10b may also be compared with the wintertime 700 mb height anomaly composites of Van Loon and Rogers (1981) and Chen (1982). For compositing purposes these studies used conventional indices of the Southern Oscillation which have been shown to have a very high correlation with equatorial sea-surface temperature anomalies (e.g., Horel and Wallace, 1981). The simulated 800 mb anomalies again have a pattern similar to the composite 700 mb anomaly pattern for December, January and February, although the amplitudes are larger by a factor of 2 or more. Note that the Van Loon and Rogers (1981) pattern is the difference between extreme phases of the Southern Oscillation, not the deviation from a climatological mean state.

In magnitude and pattern the simulated anomalies also resemble the observed height

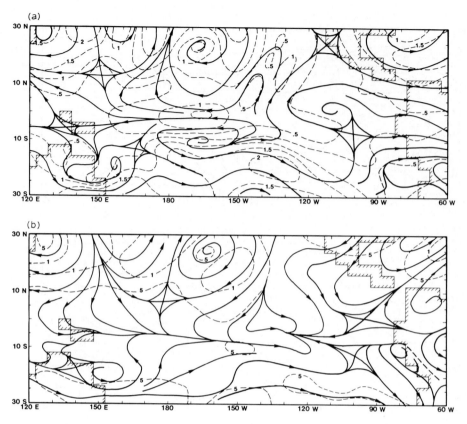

(a)

(b)

Fig. 9. Composite difference of horizontal mass flux (10^4 kg m^{-1} s^{-1}) in (a) the upper layer from 600 to 200 mb and (b) the lower layer from the surface of 600 mb between El Niño experimental and control integrations.

anomaly patterns of the 500 mb height field documented by Wallace and Gutzler (1981) for the Pacific–North American (PNA) region. However, the observed composite anomalies are based on the ten highest and ten lowest values of a monthly index which measures the strength of the monthly averaged height anomalies in the PNA pattern. Therefore, the composite anomaly pattern is not a direct measure of the remote response of the atmosphere to the slowly evolving equatorial SST anomalies nor does it represent a deviation of the atmosphere from the climatological mean conditions.

5.2. Evaporation and precipitation

A noteworthy feature of the remote response of the OSU-AGCM shown in Fig. 11 is a tendency for larger evaporation on the western sides and smaller evaporation on the eastern sides of the anomalously low upper-level heights over the North Pacific and near the eastern coast of North America. This is consistent with the simulated boundary-layer temperature anomaly field (not shown) which shows cooler temperatures on the western sides of the low-height anomalies and warmer temperatures on the eastern sides.

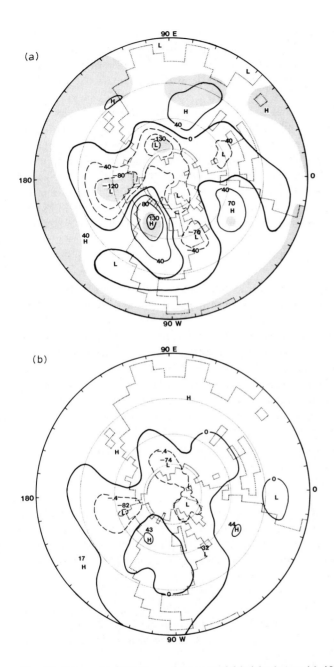

Fig. 10. Composite difference of geopotential height (m) at (a) 400 mb and (b) 800 mb between El Niño experimental and control integrations. The areas containing values which are significantly different from zero at above the 95% level are indicated by shading. The significance test of Katz (1982) was used.

Small-scale features of the evaporation anomaly fields appear to be associated with orography such as the very large deficit along the coastline in the Gulf of Alaska. The magnitude of the oceanic anomalies in the middle and high latitudes of the Southern Hemisphere are generally smaller than the Northern Hemisphere anomalies by a factor of 2 or 3.

Although the evaporation anomalies have a well-organized and large-scale character, the precipitation anomalies shown in Fig. 12 exhibit large variability in space. In terms of the vertically integrated water vapor budget, this suggests that the anomalous supply of moisture from the surface is relatively uniform while the loss in the form of precipitation is a relatively sporadic event. Since the vertically integrated storage of moisture over a 60-day period is less than the sources and sinks represented by the evaporation and precipitation, it is likely that synoptic-scale horizontal advection plays an important role in the water budget of precipitation anomaly patterns on the time scale of a season.

5.3. Tropospheric heating

The remote response of the tropospheric heating field to the SST anomaly is the same order of magnitude as the local tropical response. Figure 13 shows the global pattern of vertically integrated heating.

The largest middle latitude anomalies of vertically integrated heating occur over the oceans of the Northern Hemisphere. A large positive anomaly covers much of the North Pacific and is undoubtedly related to the increased precipitation and sea-surface evaporation to the west of the anomalously low geopotential heights at upper levels as noted in the previous subsection. Decreased heating is found near Japan and the North American coast with large negative values in the Gulf of Alaska. In the North Atlantic, an extensive region of heating is found just off the North American coast with cooling in the Gulf of Mexico.

5.4. Wind

Since winds averaged over a 60-day period are quasi-geostrophic, the geopotential height anomalies in Fig. 10 give a good indication of changes in the horizontal flow of air at middle and high latitudes. However, it is useful to inspect the wind field itself to give a qualitative description of the remote response in terms of the climatological features of the model.

Figure 14 shows the vertically integrated horizontal mass flux in the 600–200 mb layer for the control and experimental integrations. In the Northern Hemisphere the largest differences between the control and experimental integrations are found in the Pacific–North American sector. Other smaller changes, can be seen in the area of north-central Asia and in the Southern Hemisphere.

The changes in the Pacific–North American sector are related to the eastward shift and amplification of the ridge near the northwestern coast of the North American continent. In the control integration the flow over Alaska and the Gulf of Alaska is weak and from the west to the west-northwest. In the experiment, the ridge strengthens, and the flow over the Gulf of Alaska strengthens and shifts to the west-southwest. Strong

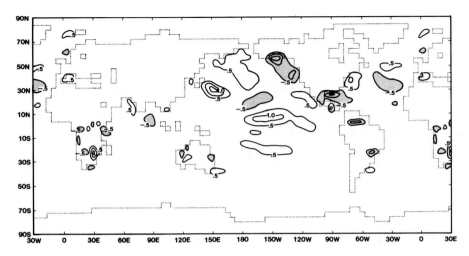

Fig. 11. Composite difference of surface evaporation rate (mm per day) between El Niño experimental and control integrations.

northwesterly flow on the eastern side of the ridge in central Canada in the experiment replaces the westerlies associated with the broad trough in the control.

In terms of magnitude, the western Pacific jet stream changes very little when the SST anomaly is introduced — increasing by only $\sim 0.7 \times 10^4 \, \mathrm{kg \, m^{-1} \, s^{-1}}$ which is equivalent to a $\sim 2 \, \mathrm{m \, s^{-1}}$ average wind increase in the upper layer. But it undergoes a change in its areal coverage due to an increase of the longitudinal extent of the westerlies along the jet axis into the eastern Pacific. The $15 \times 10^4 \, \mathrm{kg \, m^{-1} \, s^{-1}}$ isoline moves $10°-15°$ of longitude to the east and the region of strong westerlies expands to the northeast. The winds to the south of the jet maximum weaken while strengthening to the north. This results in an apparent northward shift of the position of the maximum.

In the control integration, the flow over northern Asia is weak and disorganized. The flow around the northern edge of the Tibetan Plateau is mainly from the east. In the experiment, the flow over north-central Asia strengthens and becomes north-westerly.

The response of the flow in the Southern Hemisphere is remarkably large considering the fact that easterlies dominate the near-equatorial region of the Southern Hemisphere except in the Eastern Pacific and the fact that there are no large-scale mountains or planetary-scale masses at middle latitudes. The upper level westerlies with values $\gtrsim 10 \times 10^4 \, \mathrm{kg \, m^{-1} \, s^{-1}}$ (or $\gtrsim 25 \, \mathrm{m \, s^{-1}}$) near $45°\mathrm{S}$ in the control integration are confined to a narrow latitude belt of $\sim 15°$ latitude. The magnitudes change very little with longitude. In the experimental integration, the belt of westerlies broadens, shifts toward the equator, and develops more zonal variation. The westerly anomaly field has three maxima just off the east coasts of the Australian, South American and African continents plus a maximum in the southeastern Pacific near $125°\mathrm{W}$.

386

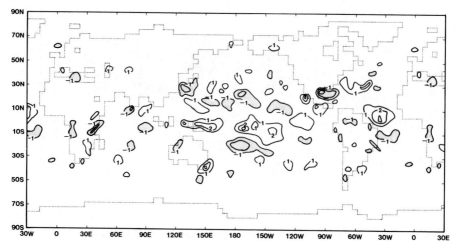

Fig. 12. Composite difference of precipitation rate (mm per day) between El Niño experimental and control integration.

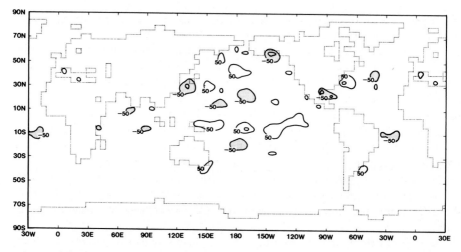

Fig. 13. Composite difference of atmospheric heating (W m^{-2}) between El Niño experimental and control integrations.

6. YEAR-TO-YEAR VARIABILITY

In a non-linear model we must admit that there exists a possibility that the difference between the 60-day averaged anomaly fields from any two control or experimental integrations which differ only in the choice of different initial conditions may be as large as the anomaly fields themselves. Figure 15 shows the average 400 mb geopotential height anomaly fields for the last 60 days of each experiment-control pair. The differences between realizations are very large. Although the pattern of anomalous heights is qualitatively consistent over the Pacific–North American sector, the magnitudes vary by as much as 250 m.

(a)

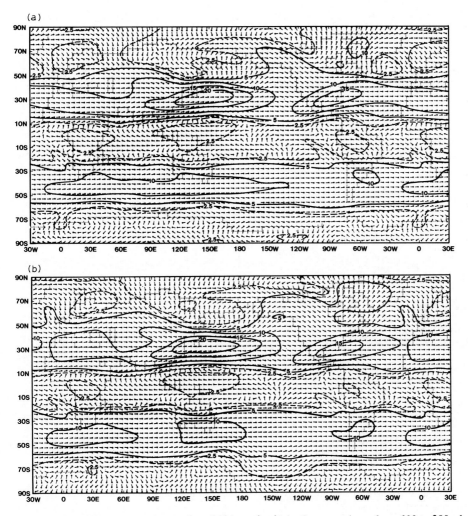

Fig. 14. Composite horizontal mass flux (10^4 kg m^{-1} s^{-1}) in the upper layer from 600 to 200 mb for (a) El Niño experiment and (b) control integrations.

The most optimistic interpretation is that deterministic forecasts of El Niño–Southern Oscillation events using coupled ocean–atmosphere models will require very careful initialization of the atmospheric portion of the model, even though the output required may be monthly to seasonal averages. Recently, we have shown (Qiu and Esbensen, 1985) that the OSU AGCM with a fixed annual cycle of sea-surface temperatures contains a pattern of month-to-month variability of geopotential height in the Pacific–North American sector which resembles the remote response presented in Fig. 10. Unless these large month-to-month fluctuations can be predicted in a deterministic sense, they may lead to large uncertainty in the predicted flow pattern for a given winter season.

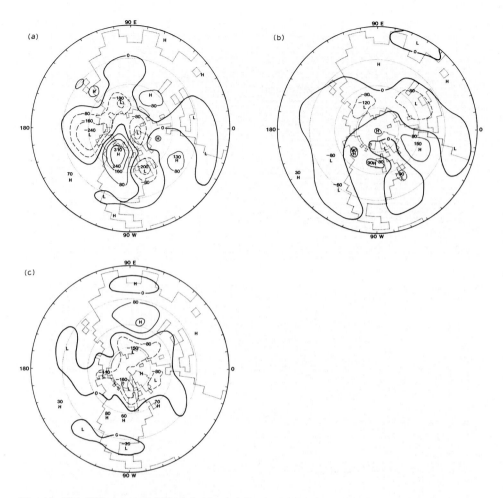

Fig. 15. The differences of 400 mb height (m) between experimental and control integrations for three experiment-control pairs.

7. DISCUSSION

In this study we have documented the remote response of the OSU two-level AGCM to an equatorial sea-surface temperature anomaly in the Eastern Pacific. Over the Pacific– North American sector, the model produces a large negative anomaly of the geopotential height which extends northwestward from the Aleutians to northern Asia, and a large positive anomaly over northwestern Canada. Other downstream regions of alternating low and high heights were of marginal statistical significance. The remote response qualitatively resembles the teleconnections observed by Horel and Wallace (1981) and the observed anomaly patterns presented by Van Loon and Rogers (1981) and Chen (1982). In these studies, the patterns were isolated using indices of the Southern Oscillation. In addition, the simulated pattern also resembles the Pacific–North American

(PNA) teleconnection pattern identified in monthly averaged geopotential height data by Wallace and Gutzler (1981).

One of the most striking results is the large variability between pairs of integrations which differ only in their initial conditions. Similar results have been found by Shukla and Wallace (1983). Our interpretation is that the remote response to anomalous equatorial sea-surface temperatures in any given month can be obscured by low-frequency planetary oscillations which can occur in the absence of any variation in the climatological sea surface from its normal annual cycle. Using the same reasoning we can expect that since the PNA pattern is similar to the remote response pattern associated with the Southern Oscillation, extreme events may occur when the two phenomena are in phase.

The next step in our research is to interpret the results of our control and experimental integrations using diagnostic methods and simple theoretical models. Preliminary calculations with the linear baroclinic model of Kang (1984) containing parameterized effects of cumulus convection and air–sea interaction have given a pattern which qualitatively resembles the results presented here. These results will be reported on at a later date.

ACKNOWLEDGEMENTS

The atmospheric general circulation model used in this study was conceived by Y. Mintz, designed by A. Arakawa, improved by M. E. Schlesinger and run by R. M. Mobley. Without these individuals, this paper could not have been written. The author is very grateful for the assistance provided by Mr. Guoqing Qiu in the preparation of the manuscript and would also like to thank Professors Y.-J. Han, W. L. Gates, H.-L. Pan and M. E. Schlesinger for helpful discussions. S. Swartz assisted with the digitizing of the Carpenter and Rasmusson data, L. Holcomb drafted the figures, and N. Zielinski typed the manuscript.

This material is based upon work supported by the National Science Foundation under Grant No. ATM-8205992.

REFERENCES

Alexander, R.C. and Mobley, R.L. 1976. Monthly average sea-surface temperatures and ice-pack limits on a 1° global grid. Mon. Weather Rev., 104: 143–148.
Arkin, P.A., Kopman, J.D. and Reynolds, R.W., 1983. 1982–1983 El Niño/Southern Oscillation event quick look atlas. NOAA/National Weather Service, National Meteorological Center and Climate Analysis Center, Washington, D.C., 82 pp.
Bjerknes, J., 1969. Atmospheric teleconnections from the equatorial Pacific. Mon. Weather Rev., 97: 162–172.
Blackmon, M.L., Geisler, J.E., Hohn, E. and Pitcher, E.J., 1983. A general circulation model study of January climate anomaly patterns associated with interannual variation of equatorial Pacific sea surface temperatures. J. Atmos. Sci., 40: 1410–1425.
Chen, W.Y., 1982. Fluctuations in Northern Hemisphere 700 mb height field associated with the Southern Oscillation. Mon. Weather Rev., 110: 808–823.
Ghan, S.J., Lingaas, J.W., Schlesinger, M.E., Mobley, R.L. and Gates, W.L., 1982. A documentation of the OSU two-level atmospheric general circulation model. Report No. 35, Climatic Research Institute, Oregon State University, Corvallis, Oreg. 395 pp.

Hendon, H.H. and Hartmann, D.L., 1982. Stationary waves on a sphere: sensitivity to thermal feedback. J. Atmos. Sci., 39: 1906–1920.

Horel, J.D. and Wallace, J.M., 1981. Planetary scale atmospheric phenomena associated with the Southern Oscillation. Mon. Weather Rev., 109: 813–829.

Hoskins, B.J. and Karoly, D., 1981. The steady, linear response of a spherical atmosphere to thermal and orographic forcing. J. Atmos. Sci., 38: 1179–1196.

Julian, P.R. and Chervin, R.M., 1978. A study of the Southern Oscillation and Walker circulation phenomenon. Mon. Weather Rev., 106: 1433–1451.

Kang, I.-S., 1984. Quasi-stationary atmospheric responses to various types of large-scale forcing. Ph.D. Thesis, Oregon State University, Corvallis, Oreg., 152 pp.

Katz, R.W., 1982. Statistical evaluation of climate experiments with general circulation models: A parametric time series modeling approach. J. Atmos. Sci., 39: 1446–1455.

Montgomery, R.B., 1940. Report on the work of G.T. Walker. Mon. Weather Rev., 39 (Suppl.): 1–22.

Namias, J., 1969. Seasonal interactions between the North Pacific and the atmosphere during the 1960's. Mon. Weather Rev., 97: 173–192.

Qiu, G. and Esbensen, S.K., 1985. Teleconnections simulated by the OSU AGCM. Report No. 51, Climatic Research Institute, Oregon State University, Corvallis, Oreg., 42 pp.

Rasmusson, E. and Carpenter, T., 1982. Variations in tropical sea surface temperature and surface wind fields associated with the Southern Oscillation/El Niño. Mon. Weather Rev., 110: 354–384.

Rowntree, P.R., 1972. The influence of tropical east Pacific Ocean temperatures on the atmosphere. Q. J. R. Meteorol. Soc., 98: 290–321.

Schlesinger, M.E. and Gates, W.L., 1980. The January and July performance of the OSU two-level atmospheric general circulation model. J. Atmos. Sci., 37: 1914–1943.

Shukla, J. and Wallace, J.M., 1983. Numerical simulation of the atmospheric response to equatorial Pacific sea surface temperature anomalies. J. Atmos. Sci., 40: 1613–1630.

Simmons, A.J., 1982. The forcing of stationary wave motion by tropical diabatic heating. Q. J. R. Meteorol. Soc., 108: 503–534.

Van Loon, H. and Rogers, J.C., 1981. The Southern Oscillation. Part II: Associations with changes in the middle troposphere in the northern winter. Mon. Weather Rev., 109: 1163–1168.

Walker, G.T. and Bliss, E.W., 1932. World Weather V. Mem. R. Meteorol. Soc., 4: 119–139.

Wallace, J.M. and Gutzler, D.S., 1981. Teleconnections in the geopotential height field during the Northern Hemisphere winter. Mon. Weather Rev., 109: 784–812.

Weare, B.C., 1982. El Niño and tropical Pacific Ocean surface temperatures. J. Phys. Oceanogr., 12: 17–27.

Weare, B.C. and Nasstrom, J.S., 1982. Examples of extended empirical orthogonal function analyses. Mon. Weather Rev., 110: 481–485.

Webster, P.J., 1981. Mechanisms determining the atmospheric response to sea surface temperature anomalies. J. Atmos. Sci., 38: 554–571.

CHAPTER 26

A PRELIMINARY STUDY OF INTERMONTHLY CHANGES IN SEA-SURFACE TEMPERATURE ANOMALIES THROUGHOUT THE WORLD OCEAN

MARTIN R. NEWMAN and ANN M. STOREY

ABSTRACT

Before deciding whether it would be desirable for ocean models to be coupled with atmospheric models to improve atmospheric forecasts 1–3 months ahead, we need to know the magnitude of sea-surface temperature anomaly changes on appropriate space and time scales. Using the Meteorological Office Historical Sea-Surface Temperature data set for the period 1951–80, global maps of the standard deviation of monthly mean sea-surface temperature anomaly changes between successive months have been produced on the 5° latitude/longitude space scale. Statistics from five ocean areas whose sea-surface temperature anomalies are thought to influence the Northern Hemisphere atmospheric circulation have been calculated in more detail for four three-month seasons, including calculations of the return periods for 0.5°, 0.75° and 1.0°C anomaly changes over one, two and three months. For 1.0°C changes occurring over a period of one month starting in a specific month, the return period is 19 or more years. For similarly defined changes over two or three months in middle latitudes, except in winter, the return period is generally between 10 and 15 years and, exceptionally, as little as six years. In middle latitudes in winter, and in the tropics in all seasons, changes of 1°C starting in a specified calendar month are expected only once every 19 or more years.

INTRODUCTION

There have been many studies of the behaviour of sea-surface temperature anomalies through time in specific geographical regions such as the tropical east Pacific (Rasmusson and Carpenter, 1982) and at ocean weather ships (Garwood and Adamec, 1982) particularly in the context of simple ocean-model verification. There has not been a similar study on the global scale because of the lack of suitable data. The Meteorological Office Historical Sea-Surface Temperature data set is now sufficiently comprehensive to enable such a project to be undertaken. A knowledge of the statistics of sea-surface temperature anomaly variation through time is important as it provides constraints on physical theories of large-scale air—sea interaction. Such statistics also provide a concise, although imperfect, description of the temporal behaviour of sea-surface temperature anomalies (SSTA) which is essential for the development and evaluation of monthly to seasonal atmospheric forecast models for which sea-surface temperature anomalies may be used as input parameters. This paper tries to answer two especially important questions: (1) what is the average size of intermonthly anomaly changes; and (2) what are the extreme values of anomaly changes which can occur?

The need for this study is illustrated by Fig. 1 which shows a SSTA time series in a 10° latitude by 10° longitude area in the tropical Northeast Atlantic just off the African coast, that lies within the area which Rowntree (1976) has highlighted as of importance

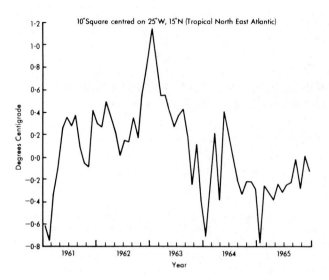

Fig. 1. Monthly SSTA time series from the 1951–1980 mean.

to the West European atmospheric circulation. The very rapid rise in SSTA between February 1961 and May 1961 and the equally large rise between October 1962 and January 1963 (perhaps associated with the severe winter of 1962–63 in northern Europe) illustrate the ability of sea-surface temperature anomalies to vary considerably from month to month.

The Meteorological Office Historical Sea-Surface Temperature data set contains monthly mean data averaged over 5° latitude by 5° longitude areas for most of the oceans between 80°N and 60°S. The data were derived from ship reports using a complex analysis procedure (Minhinick and Folland, 1984). In this study data from 1951–80 were used. We were particularly interested in the larger intermonthly changes in sea-surface temperature anomaly which occur rather infrequently. It was therefore important that the data were of high quality and so before the intermonthly anomaly changes were evaluated additional quality control procedures were used to remove doubtful values. These additional procedures involved comparison of the data with surrounding data both in time and space.

INTERMONTHLY LOCAL AREA SEA-SURFACE TEMPERATURE ANOMALY CHANGES – A GLOBAL VIEW

Figure 2 shows the mean lifetime of monthly mean anomalies with a magnitude in excess of 1°C for data averaged over 10° latitude by 10° longitude areas i.e., the average number of consecutive months with an anomaly in excess of 1°C between January 1951 and December 1980. In mid-latitudes the longest mean lifetimes are four months, although two months is more typical. The longer lifetimes (3–5 months) in the tropical East and Central Pacific mainly reflect the El Niño phenomenon. The mean lifetime of anomalies is comparable with the timescales (one to three months) on which extended range forecasts are currently produced (WMO, 1980) and so the fact that appreciable variations of SSTA are likely during some extended range forecasts merits further consideration.

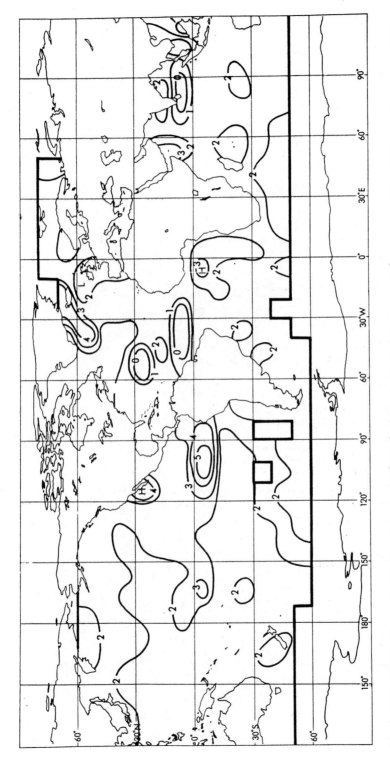

Fig. 2. Mean lifetime (in months) of a monthly mean $10° \times 10°$ SSTA of magnitude in excess of 1°C, based on 1951–1980 data. Note: A zero lifetime implies that no 1°C anomaly was found in the area between 1951 and 1980.

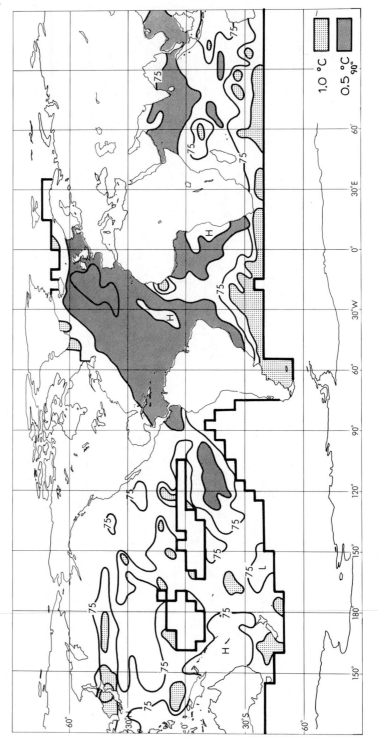

Fig. 3. Standard deviation of the change in monthly mean SSTA from January to February, based on 1951–1980 data; contour interval 0.25°C.

395

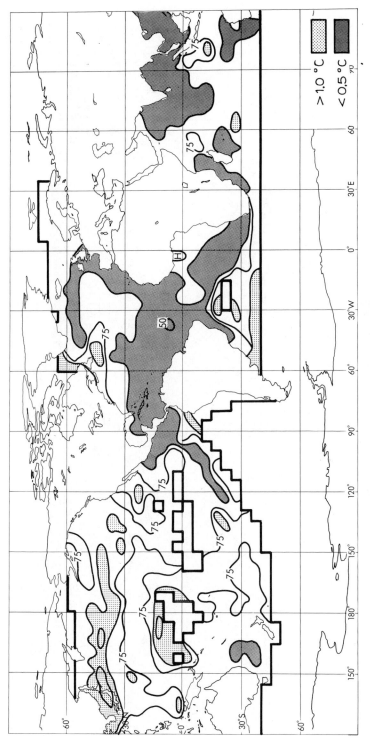

Fig. 4. Standard deviation of the change in monthly mean SSTA from July to August, based on 1951–1980; contour interval 0.25°C.

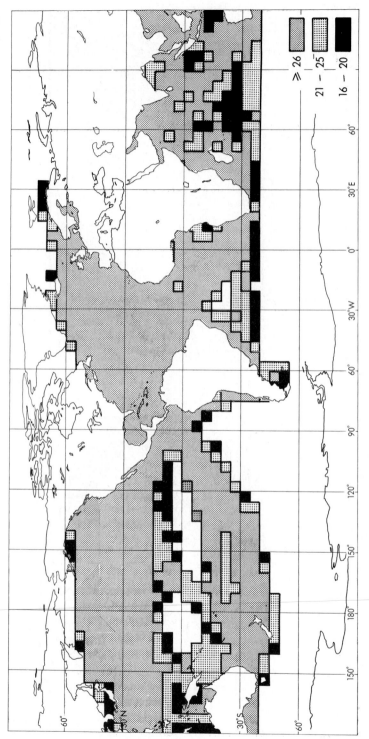

Fig. 5. Number of anomaly pairs used in the calculation of January–February SSTA changes.

The standard deviation of SSTA changes between one set of periods and another is a measure of the typical magnitude of these anomaly changes. The standard deviation of the SSTA changes is simply related to the standard deviation of the anomalies in each set and the lag correlation between anomalies in one set and those in the other. Let T_i be the SSTA in calendar month i and $\rho_{i,i-n}$ be the correlation of the SSTA in calendar month i with that n months before. Then the variance of SSTA-change from month $i-n$ to month i is (angle brackets represent ensemble means):

$$\langle (T_i - T_{i-n})^2 \rangle$$

$$= \langle T_i^2 \rangle + \langle T_{i-n}^2 \rangle - 2 \langle T_i T_{i-n} \rangle$$

$$= \langle T_i^2 \rangle + \langle T_{i-n}^2 \rangle - 2\rho_{i,i-n} \sqrt{[\langle T_i^2 \rangle \langle T_{i-n}^2 \rangle]}$$

So it can be seen that high standard deviations may result either from a large magnitude of the monthly anomalies or low intermonthly correlations.

Figures 3 and 4 show the standard deviations of anomaly changes from January to February and from July to August for $5° \times 5°$ areas. (Charts were also produced for April–May and October–November.) Only areas in which at least 15 of the possible 30 pairs of anomaly values were available are shown. Figure 5 shows the geographical distribution of sample sizes for the January to February anomaly changes and is very similar to the charts (not shown) appropriate to the other intermonthly changes which we have studied. In the Pacific Ocean north of 15°N, in the Atlantic north of 20°N, and in the northern Indian Ocean sample sizes were generally larger than 25 (maximum possible was 30). Data were available on between 20 and 25 occasions in most of the remaining areas leaving only a very small proportion of $5° \times 5°$ areas with 15–19 values. To aid the interpretation of these figures Table 1 shows 5% confidence limits on standard deviations as a function of estimated standard deviation and sample size.

TABLE 1

5% confidence limits on standard deviations as a function of sample size

Estimated standard deviation	Sample size	Upper	Lower
0.5	30	0.67	0.40
	25	0.64	0.39
	20	0.73	0.38
	15	0.79	0.37
0.75	30	1.00	0.60
	25	1.04	0.58
	20	1.09	0.57
	15	1.18	0.55
1.0	30	1.34	0.80
	25	1.39	0.78
	20	1.46	0.76
	15	1.58	0.73
1.25	30	1.68	0.99
	25	1.73	0.98
	20	1.82	0.95
	15	1.97	0.91

398

TABLE 2

Oceanic areas reported to be associated with atmospheric circulation anomalies

Area	Author
Central North Pacific	Namias (1969)
North Atlantic off Newfoundland	Ratcliffe and Murray (1970)
Tropical West Pacific	Palmer and Mansfield (1984)
Tropical East Pacific	Many (e.g. Rowntree, 1972; Pan and Oort, 1983)
Tropical Northeast Atlantic	Rowntree (1976)

Similar charts, not shown, were also produced illustrating standard deviations of changes in monthly mean $5° \times 5°$ SSTA from January, April, July and October to: (1) the month two months later; and (2) the month three months later. The following general features were noted:

(1) Higher standard deviations of anomaly changes occur in the vicinity of western boundary currents.

(2) A strong annual cycle of standard deviation is observed in middle latitudes, with lower standard deviations in local winter than in local summer.

(3) In general larger standard deviations occur in the middle latitudes in the Pacific than in the Atlantic Ocean.

(4) Large intermonthly standard deviations are observed in the tropical East Pacific despite the long anomaly lifetimes noted above.

The standard deviations of the one-month changes vary from $0.2°C$ in mid-ocean areas of deep mixed layer in winter to more than $1.3°C$ in the extension of the Kuroshio current into the North Pacific current in summer. Standard deviations of the three-month anomaly changes vary from $0.3°C$ (in the Northeast Atlantic from January–April) up to a maximum of $1.8°C$ in the Northwest Pacific in the first half of the year. The geographical distribution of the three-month changes is broadly similar to that of the one-month changes.

INTERMONTHLY REGIONAL SCALE SEA-SURFACE TEMPERATURE ANOMALY CHANGES

Various authors (listed in Table 2) have presented evidence that anomalies in the Northern Hemisphere atmospheric general circulation may be associated with SSTA in particular areas on a space scale rather larger than $5°$ of latitude by $5°$ of longitude. These "key" oceanic areas are of particular interest to UK long-range (monthly to seasonal) forecasters (Folland, 1984) and five of these regions were studied (see Fig. 6 and Table 3). The precise boundaries of the areas were chosen subjectively, guided by the work of the authors in Table 2 and helped by an analysis of the spatial coherence of the sea-surface temperature anomaly fields. This was done by correlating the monthly mean of the anomalies in a given $5° \times 5°$ area with all other available $5° \times 5°$ areas in the same month.

Time series of changes in monthly mean SSTA in the regions over one, two and three months beginning in each calendar month were prepared. The results are presented in

399

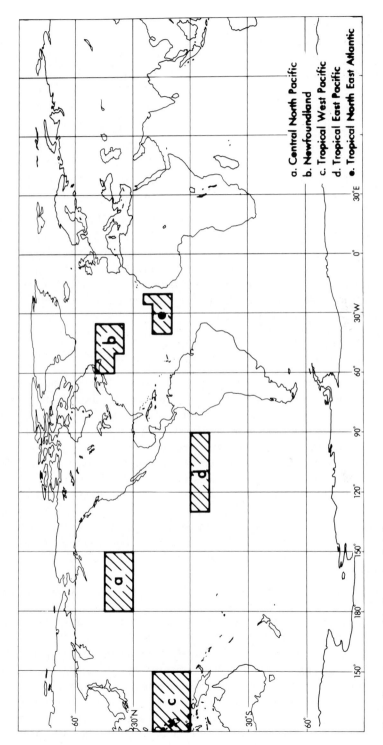

a. Central North Pacific
b. Newfoundland
c. Tropical West Pacific
d. Tropical East Pacific
e. Tropical North East Atlantic

Fig. 6. Ocean areas used in the regional study.

TABLE 3

Ocean areas used in the regional study

Area	Location	
Newfoundland	$35° - 60°$W	$40° - 50°$N
	$35° - 50°$W	$35° - 40°$N
Central North Pacific	$150° - 180°$W	$30° - 45°$N
Tropical East Pacific	$90° - 130°$W	$10°$S–Eq
Tropical Northeast Atlantic	$20° - 40°$W	$10° - 20°$N
	$20° - 25°$W	$20° - 25°$N
Tropical West Pacific	$120° - 150°$E	Eq–$20°$N

Tables 4 and 5 for changes starting in January, April, July and October. Table 4 shows the standard deviation of the area-mean SSTA changes and Table 5 shows the return period, R, calculated separately for each pair of months shown for a change in anomaly of magnitude at least $0.5°$, $0.75°$ and $1°$C (Tables 5a, b and c, respectively), from a method given by Jenkinson (1969). In this method the changes from the N years sampled are ranked in descending order. A change associated with the mth rank is expected to be reached or exceeded once every:

$$R = \frac{N + 0.38}{m - 0.31} \text{ yrs}$$

Confidence limits for the return periods have not been obtained. However, the return periods for a given start-month region vary smoothly through the year suggesting that the results are at least qualitatively correct. In addition to the general features noted in the previous section, several further points are of interest:

(1) In the central North Pacific there is evidence that large anomaly changes often occur in spring, across the winter to summer transition (associated with the rapid shallowing of the mixed layer, noted by Garwood and Adamec, 1982). In contrast the area off Newfoundland seems most prone to large changes in summer.

(2) In the tropical East Pacific the standard deviations of changes starting in April and January are larger than those in other seasons, presumably reflecting the rapid rise and sharp decline respectively in SSTA during an El Niño event (Rasmusson and Carpenter, 1982). The return periods for a given anomaly change are smallest for changes starting in April and January.

(3) Return periods for changes of $1°$C between specified months are generally ten years or more. For changes of $0.5°$C in mid-latitudes the return periods are generally less than ten years. There is considerable variability with location and season in the return period for changes of $0.5°$C in the tropics.

(4) There is not a direct relationship between the standard deviation and return periods. This implies that the distribution of anomaly changes is not the same at all locations. No statistical assessment has been made to confirm this.

(5) The calculations carried out above are for return periods of changes starting in specific calendar months. Those for return periods of changes of SSTA anomaly starting in *any* month are considerably less and will be reported in another paper.

TABLE 4

Standard deviations of intermonthly SSTA changes (in °C)

Area	Changes from											
	January			April			July			October		
	J/F	J/M	J/A	A/M	A/J	A/J	J/A	J/S	J/O	O/N	O/D	O/J
Newfoundland	0.25	0.28	0.43	0.28	0.58	0.81	0.63	0.69	0.70	0.44	0.56	0.53
Central North Pacific	0.32	0.38	0.55	0.41	0.57	0.62	0.34	0.49	0.60	0.41	0.64	0.60
Tropical East Pacific	0.32	0.46	0.68	0.31	0.40	0.54	0.24	0.30	0.35	0.27	0.31	0.34
Tropical North-east Atlantic	0.35	0.53	0.51	0.28	0.38	0.43	0.21	0.27	0.35	0.26	0.39	0.46
Tropical West Pacific	0.18	0.25	0.26	0.24	0.29	0.35	0.22	0.31	0.30	0.24	0.21	0.22

TABLE 5a

Return periods (years) of SSTA changes of magnitude $\geq 0.5°C$

Area	Changes from											
	January			April			July			October		
	J/F	J/M	J/A	A/M	A/J	A/J	J/A	J/S	J/O	O/N	O/D	O/J
Newfoundland	12	12	6	9	3	2	2	2	2	3	2	2
Central North Pacific	9	5	4	6	3	3	7	4	2	4	2	3
Tropical East Pacific	12	4	2	12	3	3	19	12	6	+[a]	7	3
Tropical North-east Atlantic	7	3	3	9	5	4	19	12	5	12	6	3
Tropical West Pacific	+[a]	19	9	19	12	6	19	9	12	19	19	+[a]

[a]An entry of "+" in the table indicates that no change of anomaly of 0.5°C was observed during the appropriate months between January 1951 and December 1980 and so a return period could not be calculated.

A 1°C SSTA, or sometimes rather less, in a "key" area often appears to be associated with significant atmospheric anomalies. Therefore an anomaly change of 1°C might be expected to be significant for the long-range forecast problem. So the return periods for changes of anomaly of 1°C give an indication of how often the extended range forecaster need to be concerned by SSTA changes during the period of the forecast. Table 5 shows return periods for anomaly changes of 0.5°, 0.75° and 1.0°C. The results show that changes of 1°C from a given month to the next, at those times of year studied, are

402

TABLE 5b

Return periods (years) of SSTA changes of magnitude $\geqslant 0.75°C$

| Area | Changes from | | | | | | | | | | | |
| | January | | | April | | | July | | | October | | |
	J/F	J/M	J/A	A/M	A/J	A/J	J/A	J/S	J/O	O/N	O/D	O/J
Newfoundland	+	+	12	+	7	3	4	3	3	12	6	5
Central North Pacific	12	+	9	12	5	9	19	7	6	12	5	4
Tropical East Pacific	+	6	4	19	+	5	+	19	+	+	19	+
Tropical North-east Atlantic	12	5	6	+	+	12	+	19	9	+	10	6
Tropical West Pacific	+	+	+	+	+	19	+	19	+	+	+	+

TABLE 5c

Return periods (years) of SSTA changes of magnitude $\geqslant 1.0°C$

| Area | Changes from | | | | | | | | | | | |
| | January | | | April | | | July | | | October | | |
	J/F	J/M	J/A	A/M	A/J	A/J	J/A	J/S	J/O	O/N	O/D	O/J
Newfoundland	+	+	+	+	9	6	12	7	6	19	12	18
Central North Pacific	+	+	19	19	19	19	+	19	12	+	7	11
Tropical East Pacific	+	+	+	+	+	19	+	+	+	+	+	+
Tropical North-east Atlantic	+	19	19	+	+	19	+	19	+	+	19	18
Tropical West Pacific	+	+	+	+	+	+	+	+	+	+	+	+

expected to occur once every 19 or more years. The frequency with which changes of $1°C$ are expected to occur from a given calendar month to the next calendar month, two or three months hence, varies considerably with location and season. The return period of a $1°C$ change is lowest in the mid-latitude areas (except in winter) where it may be as low as six years from July to October (in the Newfoundland area). It is emphasized that these statistics refer to changes starting in the four *given* months. The atmosphere is not affected by sea-surface temperature anomalies equally at all times of year and $1°C$ may not be an appropriate benchmark in all cases. For instance, the work of Palmer and Mansfield (1984) suggests that the atmosphere is sensitive to changes of only $0.5°C$ in the tropical West Pacific. The return periods for $0.5°C$ changes in the tropical West

Pacific are considerably shorter than for 1°C changes; for example nine years between July and September and six years between April and July. So a proper evaluation of the importance of SSTA changes to extended-range forecasting requires parallel information on the sensitivity of the atmosphere to the anomalies.

CONCLUSION

Examples have been given of quasi-global fields of the standard deviation of inter-monthly sea-surface temperature anomaly changes on the 5° × 5° space scale. As we are chiefly interested in the effect of anomalies on the large-scale atmospheric flow for long-range forecasting, we concentrated most of our work on five ocean areas believed to affect the general circulation of the Northern Hemisphere. At a given starting time of year, SSTA changes between calendar months (one, two or three months apart) of magnitude 1°C or larger are estimated to typically occur once every ten years. In some areas the estimated return period is much greater; indeed in some regions there was no such SSTA change between 1951 and 1980 of magnitude 1°C or greater. Because of the way they are constructed some current statistical long-range forecasting techniques may implicitly take into account changes of SSTA through the forecast period. If dynamical models are to take SSTA changes into account they will need to do so explicitly. Further studies of the effects of SSTA and of changes in SSTA will be necessary to decide whether the additional complexity of SSTA modelling need be seriously considered.

ACKNOWLEDGEMENTS

Thanks are due to Chris Gordon for several helpful discussions.

REFERENCES

Folland, C.K., 1984. Recent Developments in Monthly Long-Range Forecasting for the United Kingdom. Synoptic Climatology Branch Memorandum No. 145.
Garwood, R.W. and Adamec, D., 1982. Model simulations of seventeen years of mixed layer evolution at ocean station Papa. Naval Postgraduate School, Monterey, Report Number NP 568-82-006.
Jenkinson, A.F., 1969. Statistics of extremes; Estimation of maximum floods. Weather Meteorol. Office, Tech. Note 98, ch. 5.
Minhinick, J. and Folland, C.K., 1984. The Meteorological Office Historical Sea Surface Temperature Data Set. (MOHSST) Synoptic Climatology Branch Memorandum No. 137.
Namias, J., 1969. Seasonal interactions between the North Pacific Ocean and the atmosphere during the 1960's. Mon. Weather Rev., 97(3): 173–192.
Palmer, T. and Mansfield, D., 1984. Response of two atmospheric general circulation models to sea surface temperature anomalies in the tropical east and west Pacific. Nature, 310: 483–485.
Pan, Y.H. and Oort, A.H., 1983. Global climate variations connected with sea surface temperature anomalies in the eastern equatorial Pacific Ocean for the 1958–73 period. Mon. Weather Rev., 111: 1244–1258.
Rasmusson, E. and Carpenter, T., 1982. Variations in tropical sea surface temperature and surface wind fields associated with the Southern Oscillation/El Niño. Mon. Weather Rev., 110: 354–384.

Ratcliffe, R.A.S. and Murray, R., 1970. New lag associations between North Atlantic sea temperature and European pressure applied to long range weather forecasting. Q. J. R. Meteorol. Soc., 96: 226–246.

Rowntree, P.R., 1972. The influence of tropical east Pacific Ocean temperatures on the atmosphere. Q. J. R. Meteorol. Soc., 98: 290–321.

Rowntree, P.R., 1976. Response of the atmosphere to a tropical Atlantic Ocean temperature anomaly. Q. J. R. Meteorol. Soc., 102: 607–623.

W.M.O., 1980. Programme on Short-, Medium- and Long-range Weather Prediction Research (PWPR). Report of the Informal Meeting of Experts on Long Range Forecasting, Geneva, 1-5 September 1980.

CHAPTER 27

MORPHOLOGY OF THE SOMALI CURRENT SYSTEM DURING THE SOUTHWEST MONSOON

MARK E. LUTHER, JAMES J. O'BRIEN and ALEX H. MENG

ABSTRACT

Results are presented from two cases of a model of the seasonal circulation in the Northwest Indian Ocean. The model is a nonlinear reduced gravity transport model, with realistic basin geometry and using observed winds as forcing. One case is forced by the monthly mean of the FGGE winds while the other is forced by the monthly mean climatological winds. The two cases are compared and inferences are made as to the importance of the different mechanisms at work in the generation and decay of the Somali Current system during the southwest monsoon. The Southern Hemisphere currents are shown to reverse before the local winds, due to the relaxation of the northeast monsoon winds. The southern gyre of the two-gyre system responds primarily to the Southern Hemisphere tradewinds, while the northern gyre forms north of 4°N in response to the development of the Findlater jet and its associated wind stress curl. The two-gyre system collapse is highly correlated with a decrease in the westerly component of the equatorial wind stress. The circulation patterns are strongly influenced by the gradient of the wind stress curl, as well as by the curl itself. The transition from southwest to northeast monsoon conditions depends on the remnants of the previous season's circulation patterns.

INTRODUCTION

Recent observations have revealed a great deal of information on the development of the Somali Current system with the onset of the southwest (SW) monsoon. The picture that has emerged is that of a complicated multiple-gyre circulation pattern, with rapid transitions from one configuration to another. The Somali Current develops from a weak southward flow to a very strong northward current, with transports of more than 60×10^6 m^3 s^{-1}, in roughly four months. This is quite a spectacular change, especially when compared to the seasonal variations in the other oceans. Excellent reviews of the recent observational and theoretical advances in this area are given by Schott (1983) and Knox and Anderson (1985).

During northern winter, the winds over the northwestern Indian Ocean are northeasterly (the NE monsoon) and drive a southwestward boundary current along the coast of Somalia. This current meets the northward flowing East African Coastal Current (EACC), which is driven by the Southern Hemisphere tradewinds, off the coast of Kenya, and both flow offshore. As the NE monsoon relaxes, the EACC pushes farther northward, causing the coastal currents to reverse before the local wind reverses along the coast of Kenya and southern Somalia (Leetmaa, 1972; Anderson, 1981). By mid-April, the winds in the Southern Hemisphere reverse, turning first onshore and then northward, as the Southern Hemisphere tradewinds extend to the northwest. There is northward flow up to the equator, where it turns offshore and recirculates to the south forming a

southern gyre. There is weak northward flow north of 5°N, with southward flow between
the equator and 5°N. In May, the winds reverse all along the Somali coast. The winds are
actually a jet-like structure (Findlater, 1971), fed in the south by the Southern
Hemisphere tradewinds. This jet, known as the Findlater jet, is the atmospheric equivalent
of an oceanic western boundary current, confined in the west by the highlands of Kenya
and Ethiopia. Much of the jet's path is over land at very low latitudes. The Findlater
jet migrates northward as it intensifies during the SW monsoon, so that the wind reversal
occurs at different times at different latitudes. There is northward flow and upwelling
all along the Somali coast, but the southern gyre still turns offshore within a few degrees
north of the equator. In early June, the core of the Findlater jet crosses the coast at
about 8°N and becomes very strong. At this time a second gyre appears between 4° and
8°N. It was this gyre that was described by Findlay (1866) as the great whirl. By late
June, the two-gyre system is well established, with wedges of cold upwelled water
extending out from the coast just to the north of the offshore flow in both gyres. This
pattern remains quasi-stationary until late July or early August, when the winds begin
to weaken. At this time, the southern cold wedge migrates rapidly northward and merges
with the northern wedge, signaling the coalescence of the two gyres. The Somali Current
is now a more or less continuous boundary current from 10°S to 10°N. Sometime during
the height of the SW monsoon, another clockwise gyre appears to the east of the island
of Socotra, appropriately name the Socotra Eddy (Bruce, 1979). This eddy also appears
to be a regular feature of the seasonal cycle. Sometime in October, the winds again
reverse with the onset of the NE monsoon, and the southward Somali Current quickly
develops. Much less is known about this part of the cycle, as it is much less dramatic
and has not aroused as much attention.

The features described above have proven to be highly repeatable. The presence of
the two-gyre system was found in historical data for most of the years where there was
sufficient data by Swallow and Fieux (1982). Bruce (1973, 1979) found the two-gyre
system in 1970 and 1976, and it also was quite prominent in the 1979 INDEX
observations (Schott and Quadfasel, 1982; Leetmaa et al., 1982; Swallow et al., 1983).
Bruce (1979) also documented the presence of the Socotra Eddy in the four years 1975
to 78. The observations indicate that the great whirl forms north of 4°N (Schott and
Quadfasel, 1982), rather than being advected from the equator as some theories suggest
(e.g. Hurlburt and Thompson, 1976; Cox, 1976, 1979; Lin and Hurlburt, 1981). It thus
appears that different mechanisms are responsible for the development of the southern
and northern gyres. Anderson (1981) implies that the southern gyre is forced by the
Southern Hemisphere tradewinds, and the current crosses the equator due to inertial
overshoot as in Anderson and Moore (1979), while the great whirl is generated farther
to the north by the Findlater jet as it leaves the coast around 9°N. Lin and Hurlburt
(1981) have also shown that the great whirl can be generated in this manner. Schott
and Quadfasel (1982) suggest that the local curl of the wind stress in the developing
Findlater jet may also be important in the generation of the great whirl.

Several authors have tackled the question of the relative importance of remote versus
local forcing in the generation of the Somali Current system (see Knox and Anderson,
1985, for a review). Most conclude that local forcing effects, i.e., those within the
boundary region, are important for the early stages of the current spin-up, but that
remote forcing effects become important at a later time. These models usually included a

very idealized wind stress forcing, with the region of wind stress curl located far away from the boundary, and neglected the spatially and temporally very complicated structure of the wind field over the Indian Ocean. In Luther and O'Brien (1985) we used observed climatological winds in a model with a realistic geometry, and were able to reproduce most of the observed features described above. The presence of a large, time-varying region of wind stress curl within and just offshore of the boundary region was a key factor in the development of the Somali Current system in that model. The fact that the model included the entire seasonal cycle was also shown to be important in accurately simulating the observed circulation patterns.

In this paper we will use the model described in Luther and O'Brien (1985) driven by the monthly mean of the winds from the First GARP Global Experiment (FGGE) of 1979, and compare those results to those from the model driven by the monthly mean climatology (MMC) winds. We will concentrate on the development and decay of the two-gyre system in these two cases. It will be shown that the great whirl develops at about 5°N under the influence of the local or very nearby wind stress curl, while the southern gyre responds to the local winds and the Southern Hemisphere tradewinds. The two-gyre system collapses when the westerly component of the winds at the equator begins to decrease. Not only is the local wind stress curl important, but also the strong gradient in the wind stress curl associated with the Findlater jet greatly influences the development of the Somali Current system through a differential Ekman pumping.

We will briefly describe the model in the next section, and then compare and contrast the two different wind fields used to drive the model. We next will describe and compare the model results in the two cases before finally interpreting these results in the context of previous theoretical work.

THE MODEL

We use the nonlinear reduced gravity transport equations to model the upper ocean response to an applied wind stress. Due to the latitudinal extent of the model, we use spherical coordinates, with ϕ (longitude) increasing eastward and θ (latitude) increasing northward. If we define the eastward and northward components of the upper layer transport as $U = uH$ and $V = vH$, respectively, where (u, v) are the depth-independent (ϕ, θ) velocity components in the upper layer and H is the thickness of the upper layer, the equations of motion are:

$$\frac{\partial U}{\partial t} + \frac{1}{a \cos \theta} \frac{\partial}{\partial \phi} \left(\frac{U^2}{H} \right) + \frac{1}{a} \frac{\partial}{\partial \theta} \left(\frac{UV}{H} \right) - (2\Omega \sin \theta) V = \frac{-g'}{2a \cos \theta} \frac{\partial H^2}{\partial \phi} + \frac{\tau^{(\phi)}}{\rho_1} + A\nabla^2 U \quad (1a)$$

$$\frac{\partial V}{\partial t} + \frac{1}{a \cos \theta} \frac{\partial}{\partial \phi} \left(\frac{UV}{H} \right) + \frac{1}{a} \frac{\partial}{\partial \theta} \left(\frac{V^2}{H} \right) + (2\Omega \sin \theta) U = \frac{-g'}{2a} \frac{\partial H^2}{\partial \theta} + \frac{\tau^{(\theta)}}{\rho_1} + A\nabla^2 V \quad (1b)$$

$$\frac{\partial H}{\partial t} + \frac{1}{a \cos \theta} \left[\frac{\partial U}{\partial \phi} + \frac{\partial}{\partial \theta} (V \cos \theta) \right] = 0 \quad (1c)$$

where $g' = \dfrac{(\rho_2 - \rho_1)}{\rho_2} g$ is the reduced gravitational acceleration, a is the Earth's radius, Ω is the Earth's rotation rate, and A is a kinematic eddy viscosity. The wind stress, $\tau = [\tau^{(\phi)}, \tau^{(\theta)}]$, is applied as a body force over the upper layer (Charney, 1955). The transport

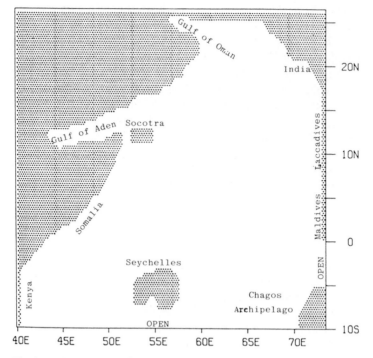

Fig. 1. Model geometry. Shading indicates land boundaries. The islands of Socotra and the Seychelles and their surrounding shallow banks are represented as land boundaries. The southern boundary and a portion of the eastern boundary are open.

form of the reduced gravity equations has the advantage that the continuity eqn. (1c) is linear. It also has the advantage that the discretization of the advective terms in eqns. (1a) and (1b) involves spatial averaging of the dependent variables, thus improving the numerical stability of the solution.

For the model geometry, we simulate the coastline of the Arabian Sea from $40°$ to $73°$E and from $10°$S to $25°$N. The model geometry is shown in Fig. 1. The boundary conditions along all solid (land) boundaries are the no-slip conditions:

$$u = v = 0$$

The southern boundary along $10°$S and a portion of the eastern boundary, from $5°$S to the equator, are open boundaries, the boundary condition there being a variation of the Sommerfeld radiation condition developed by Hurlburt (1974) and described in Camerlengo and O'Brien (1980). The eastern boundary from the equator to the Gulf of Khambat is closed, and the no-slip boundary condition applies there. This simulates the Laccadive and Maldive islands which close off the eastern side of the Arabian Sea. For computational convenience, we ignore the deep passages through these islands. The southeastern corner of the model basin is closed off by the Chagos Archipelago, and again, the no-slip boundary condition applies along the coast of this island.

Equations (1) are solved numerically on a 135×74 finite difference mesh. The

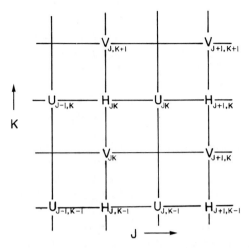

Fig. 2. Staggered mesh used in finite difference approximation, showing relative location of U, V and H points. J is the index in the zonal (ϕ) direction and K is the meridional (θ) index.

solution mesh is staggered in space as shown in Fig. 2. The model resolution is $1/8°$ in the zonal direction ($\Delta\phi$) and $1/4°$ in the meridional direction ($\Delta\theta$). The equations of motion are integrated in time using a leapfrog finite difference scheme, with a forward time difference used every 99th time step to eliminate the computational mode. The model time step is 30 min. The advective terms are computed by first averaging adjacent U, V and H values in space to form the desired product at the appropriate meshpoint and then forming the standard, second-order accurate, centered finite difference approximation. The spatial averaging helps suppress nonlinear growth of numerical noise in the model.

In the linearized form of the reduced gravity equations, one must prescribe the linear phase speed for the particular baroclinic mode one wishes to model: $C^2 = g'H_0$. In the nonlinear form, there is no analogous phase speed parameter, only g'; however, in the numerical solution of eqn. (1) one must prescribe the initial upper layer thickness, H_0. This is analogous to prescribing an initial phase speed, but this does not remain constant as the model fields evolve, as the variations in H are large compared to H_0. The average thickness of the upper layer is not constant, due to inflow and outflow at the open boundaries during the seasonal cycle. The other free parameter in the model is the kinematic eddy viscosity, A. This frictional term is required to damp out the grid-scale noise in the model to prevent this noise from growing through nonlinear interactions.

For the results presented here, we set $H_0 = 300\,\mathrm{m}$, $A = 1.4 \times 10^3\,\mathrm{m^2\,s^{-1}}$ and the reduced gravity $g' = 0.03\,\mathrm{m\,s^{-2}}$. The model is integrated from rest, beginning at 0000 GMT on 1 January, using an exponential taper with an e-folding time of 20 days to reduce the initial transients. For simplicity, the model year has 360 days, with each month having 30 days.

THE WIND FORCING

Two different wind data sets are used to derive the forcing fields for the model. For one case, we use the dynamically assimilated FGGE Level III-b wind data, obtained from the National Center for Atmospheric Research. These data consist of global analyses on a 1.875° latitude by 1.875° longitude mesh at each of 15 standard levels every 12 hours (every 6 hours during the winter and summer special observing periods) from 00h GMT on 1 December 1978 to 12 h GMT on 30 November 1979. We extract the winds only at the 1000 mbar level, and form a pseudostress from each observation. The pseudostress is defined as the wind vector W multiplied by its magnitude: $W|W|$. We then compute the monthly mean pseudostress from these data to facilitate comparison with the MMC data.

The MMC data were obtained from the United States National Climate Center in Ashville, North Carolina, from their Global Marine Sums TD-9757 data set. These data consist of standard meteorological observations compiled on 1° Marsden squares from over 60 years of ship reports. This data set was derived from the TD-11 data, that were also used by Hastenrath and Lamb (1979). The wind observations consist of an averaged eight-point wind rose on each 1° square for each month of the year. From this wind rose, we compute a pseudostress vector for each of the eight compass directions, and then average these vectors to get the monthly mean pseudostress on each square. The data are smoothed by a Hanning filter in both directions, additionally weighted by the number of observations on each square. The resultant fields compare well to those computed by Bruce (1983) and by Hellerman and Rosenstein (1983) for this area.

Once we have a monthly mean pseudostress for both wind data sets, they are treated identically, allowing, of course, for the different spatial resolutions. To interpolate from the monthly mean to the model time step, we compute the mean and first five Fourier harmonics at each point, and use these to construct daily values for the year. We then interpolate linearly from the daily values to the model time step. To interpolate in space from the wind data mesh to the model mesh, we use the natural bicubic spline interpolant. This results in a complete annual cycle of wind pseudostress for both the FGGE and MMC cases that is continuous in space and time and has continuous first and second derivatives in space. These pseudostress fields are converted to wind stress by the bulk aerodynamic formula:

$$\tau = \rho_a C_D W |W| \tag{2}$$

where ρ_a is the density of air and C_D is a constant drag coefficient. For the results presented here, we use the values $\rho_a = 1.2 \, \text{kg m}^{-3}$ and $C_D = 3.75 \times 10^{-3}$. The large value for C_D is used because we are projecting the wind stress on such a deep upper layer. The upper layer transports driven by the wind stress values thus obtained agree well with observations. For example, Leetmaa et al. (1982) report a transport in the upper 300 m just south of the offshore turning in the southern gyre of $36 \times 10^6 \, \text{m}^3 \, \text{s}^{-1}$ in late June, while the model gives a value of $35 \times 10^6 \, \text{m}^3 \, \text{s}^{-1}$ in the FGGE case and $28 \times 10^6 \, \text{m}^3 \, \text{s}^{-1}$ in the MMC case for the same time and location.

As stated above, we will concentrate here on the period March to October, which includes the SW monsoon and both transitions. Figure 3 shows the model wind stress for March from the FGGE case. This is the transition period from the NE monsoon to SW monsoon conditions. The wind stress field from the MMC case at this time is very similar. The Northern Hemisphere winds are still dominated by the NE monsoon, while the Southern Hemisphere winds are beginning to break up, and have an onshore component along the coast at 5°S. There is a convergence zone at 8–10°S, from 50° to

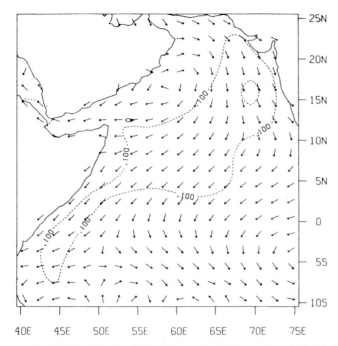

Fig. 3. Model wind stress derived from the FGGE wind data for 16 March. Arrows indicate direction, while contours give magnitude. The winds are weak and variable, but with some southwestward wind stress remaining from the NE monsoon. Contour interval is $0.1\,\mathrm{N\,m^{-2}}$.

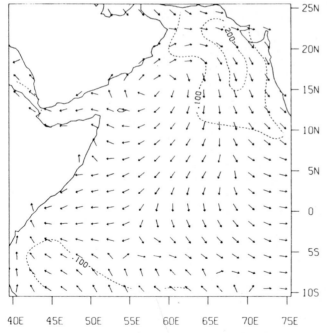

Fig. 4. Model wind stress derived from the FGGE wind data for 16 April. The winds along the southern Somali coast reverse in early April, due to a westward extension of the Southern Hemisphere tradewinds, while north of the equator the winds are still light and variable. Contour interval is $0.1\,\mathrm{N\,m^{-2}}$.

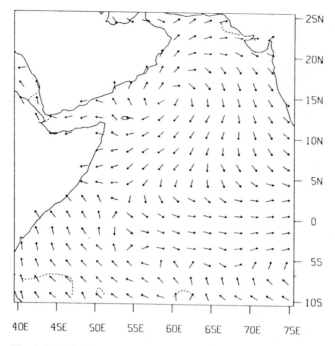

Fig. 5. Model wind stress derived from the MMC case for 16 April. Winds are very light, with some weak northward winds beginning to form along the Somali coast south of the equator. The westward extension of the Southern Hemisphere trades is not as strong in this case, possibly because of the sparse number of observations along 10°S. Contour interval is 0.1 N m^{-2}.

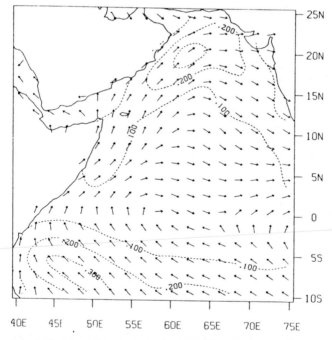

Fig. 6. Model wind stress from the FGGE case for 16 May. Northeastward winds are now found all along the coasts of Africa and Arabia. The maximum wind stress occurs in the Southern Hemisphere as a northwestward extension of the tradewinds. Simultaneously, another area of large wind stress builds over the northern Arabian Sea, forming the beginnings of the Findlater jet. There is a belt of weak convergent winds along the equator. Contour interval is 0.1 N m^{-2}.

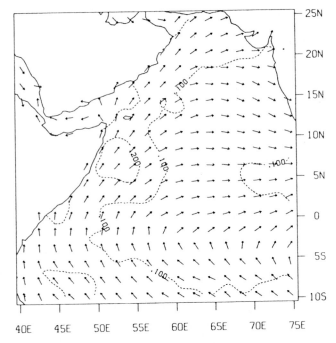

Fig. 7. Model wind stress from the MMC case for 16 May. North to northeastward winds are also found along the African and Arabian coasts, although not as strong as in the FGGE case, particularly in the Southern Hemisphere tradewind extension and in the extreme northern Arabian Sea. Contour interval is $0.1 \, \mathrm{N \, m^{-2}}$.

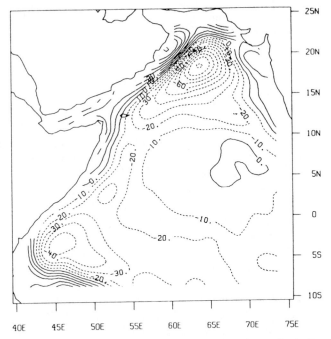

Fig. 8. Wind stress curl from the FGGE case for 16 May. Dashed contours indicate negative values. A large patch of negative curl has developed over the northern Arabian Sea, associated with the developing Findlater jet. Another less intense area of negative curl has formed off the southern Somali coast, associated with the extension of the Southern Hemisphere trades. From the equator to $10°\mathrm{N}$, there is almost no curl to the wind stress. Contour interval is $1.0 \times 10^{-7} \, \mathrm{N \, m^{-3}}$. Labels are in units of $10^{-8} \, \mathrm{N \, m^{-3}}$.

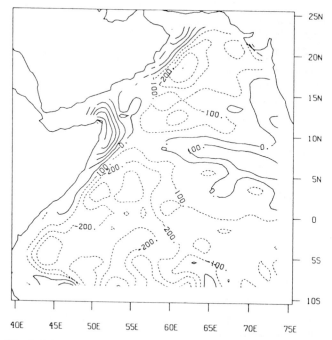

Fig. 9. Wind stress curl from the MMC case for 16 May. The patch of negative curl over the northern Arabian Sea is much weaker than in the FGGE case, because the Findlater jet is not as well developed at this time in this case. The negative curl in the Southern Hemisphere is almost as strong as in the FGGE case, even though the winds are much weaker. This is due to the northeastward turn the winds must make as they approach the African continent. Contour interval is 1.0×10^{-7} N m^{-3}. Labels are in units of 10^{-9} N m^{-3}.

70°E. In early April in the FGGE case, the winds along the southern Somalia and Kenya coasts reverse as the Southern Hemisphere tradewinds extend toward the northwest (Fig. 4). The convergence zone is now at about 7°S. The winds in the Northern Hemisphere have turned to onshore along the Somali coast. The northern Arabian Sea is dominated by a weak anticyclonic circulation, with stronger winds to the east along the Indian coast. The equatorial winds are easterly in the western part of the basin, and westerly in the eastern part. The patterns are somewhat similar in the MMC case (Fig. 5). The strength of the Southern Hemisphere tradewind extension is not as great. Westerly winds cover much more of the equatorial region in this case, and the convergence zone is slighly farther to the north. The winds along the Indian coast are also much weaker.

In May (Figs. 6 and 7) the winds become south-southwesterly all along the coast of Somalia and Kenya. The Findlater jet can now be seen in both cases, although it is more intense in the FGGE case. The Southern Hemisphere tradewinds have penetrated much farther north in the FGGE case, and consequently the convergence zone is farther north. The jet pushes farther inland over Kenya and Somalia near the equator in the FGGE case, resulting in a minimum in the alongshore wind stress at the equator. In the FGGE case, the jet is strongest in the Southern Hemisphere trades, and in the extreme northern Arabian Sea, while in the MMC case the strongest winds occur off the northern Somali coast. Westerly winds are found along the equator east of 58°E in both cases.

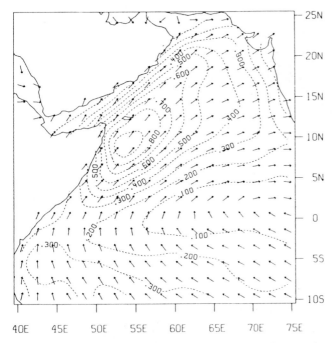

Fig. 10. Wind stress from the FGGE case for 16 June. The area of maximum wind stress is now south of the island of Socotra, and the magnitude of the stress has increased dramatically. There is still a relative minimum in the stress along the equator, and the extension of the Southern Hemisphere trades has strengthened. Contour interval is 0.1 N m^{-2}.

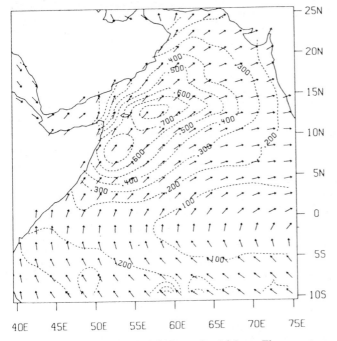

Fig. 11. Wind stress from the MMC case for 16 June. There are two relative maxima in the Northern Hemisphere, one to the east of Socotra, and another to the south. The minimum stress is again found along the equator, and the extension of the Southern Hemisphere trades has strengthened. Contour interval is 0.1 N m^{-2}.

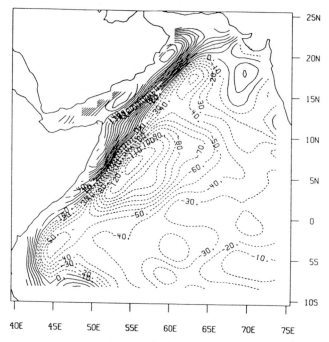

Fig. 12. Wind stress curl from the FGGE case for 16 June. There are now three relative maxima in negative curl, with the most intense patch of negative curl centered at 5°N, 53°E, just off the Somali coast. There is a strong gradient of wind stress curl that roughly follows the core of the Findlater jet. This curl gradient causes a differential Ekman pumping, with downward Ekman pumping on the eastern side of the jet core, and upward Ekman pumping on the western side. This differential Ekman pumping can greatly influence the wind-driven ocean circulation. Contour interval is 1.0×10^{-7} N m^{-3}. Labels are in units of 10^{-8} N m^{-3}.

The wind stress curl distribution is very different in the two cases, as can be seen from Figs. 8 and 9. Most of the differences can be attributed to differences in the strength and location of the Findlater jet. Since the jet is more intense in the FGGE case, the horizontal shears are larger and the curl is therefore greater. Much of this can be attributed in turn to the effect of the climatological averaging – the interannual variability in the position of the jet tends to make it more diffuse in the mean. The curl field in the FGGE case is of course much smoother, because the data were dynamically assimilated using a global general circulation model, which tends to produce relatively smooth curl fields. The gross features, however, are similar in some respects. There are large areas of negative curl to the right of the core of the jet (facing downstream), and a line of zero curl that roughly follows the core of the jet. The areas of negative curl are found off the southern Somali coast where the Southern Hemisphere tradewinds turn northward to feed the jet, and over the northern Arabian Sea, where the jet is most intense, and where it bends eastward toward the Indian subcontinent. An area of near-zero curl is found along and to the north of the equator east of 55°E.

In early to mid-June (Figs. 10 and 11) the "main onset" of the SW monsoon occurs, and the strength of the wind increases dramatically. Maximum wind stress values are found just off the coast of northern Somalia where the Findlater jet leaves the coast. In

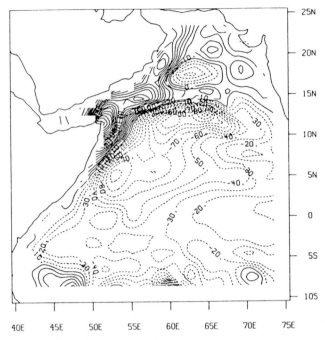

Fig. 13. Wind stress curl from the MMC case for 16 June. The curl pattern is quite different from the FGGE case, but we still see a relative maximum in negative curl centered at 5°N, 53°E. There is another relative maximum in negative curl to the east of Socotra, at about 12°N. There is also a strong gradient in the wind stress curl, but this gradient divides into two branches at 13°N. One branch is oriented east–west along 13°N, and the other, weaker branch trends northeast–southwest, roughly paralleling the Arabian coast. Contour interval is 1.0×10^{-7} N m^{-3}. Labels are in units of 10^{-8} N m^{-3}.

the MMC case there are two relative maxima, one to the east and one to the south of Socotra, while in the FGGE case there is only one to the south of Socotra. Again the jet is more intense in the FGGE case for the reasons mentioned above. The convergence zone is found at about 2°N in the FGGE case and at about 2°S in the MMC case, resulting in easterly winds along the equator in the FGGE case, while the MMC winds are still westerly there. A large area of negative wind stress curl develops between the equator and 12°N, with maximum negative values occurring at 5°N in both cases (Figs. 12 and 13). In the MMC case, there is another relative maximum, stronger than the one at 5°N, located along 12°N. In both cases there is a very strong gradient in the wind stress curl that follows the core of the jet. In the MMC case (Fig. 13) this gradient splits at 13°N, with one branch oriented to the northeast and the other to the east along 13°N. This is an indication that the jet itself splits, as reported by Findlater (1971). One would expect that these strong gradients in the wind stress curl will greatly influence the ocean's response, since it will drive a corresponding gradient or differential Ekman pumping, i.e., upward Ekman pumping beneath the positive curl and downward pumping beneath the negative curl, resulting in a tilting of the pycnocline. We can see from Figs. 12 and 13 that the curl and its gradients are stronger in the FGGE case.

The height of the monsoon occurs in mid-July in both cases (Figs. 14 and 15). In

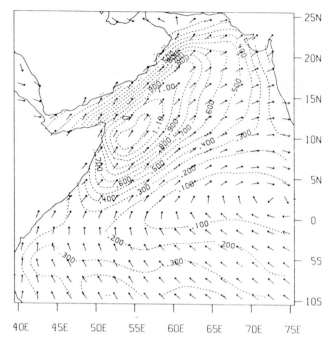

Fig. 14. Wind stress from the FGGE case for 16 July. The winds reach their maximum intensity in mid- to late July. The strongest winds occur to the south and east of Socotra, and the weakest winds occur along the equator. Contour interval is 0.1 N m^{-2}.

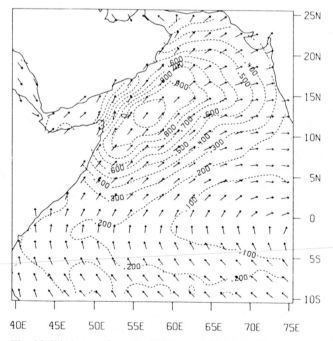

Fig. 15. Wind stress from the MMC case for 16 July. The strongest winds occur in mid- to late July in this case also, because of the Fourier harmonics interpolation scheme that was used in both cases. The pattern of the wind stress for this case is very similar to that for the FGGE case at this time, but the peak magnitude is about 25% lower. This is due to the effects of long-term averaging and inter-annual variability. Contour interval is 0.1 N m^{-2}.

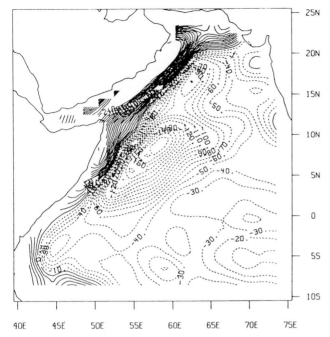

Fig. 16. Wind stress curl from the FGGE case for 16 July. A very large patch of negative wind stress curl has formed off the Somali coast from south of the equator to 20°N, with the maximum negative values occurring at 8°N, 58°E. The gradient of wind stress curl has intensified, but is in the same position as in the preceeding month. Contour interval is 1.0×10^{-7} N m^{-3}. Labels are in units of 10^{-8} N m^{-3}.

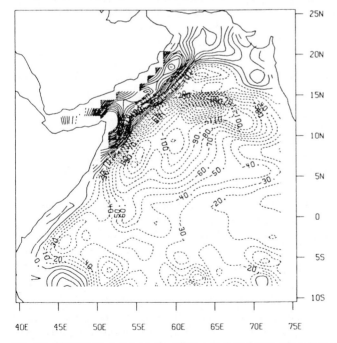

Fig. 17. Wind stress curl from the MMC case for 16 July. There is strong negative curl offshore of the Somali and Arabian coasts, with maximum negative values occurring at 13°N, 63°E, and secondary relative maxima occurring at 6°N, 55°E and at 9°N, 59°E. The gradient of the curl, with its two branches, has intensified, but remains in approximately the same location. The maximum negative values of the curl are the same for this case as for the FGGE case, although their locations differ. Contour interval is 1.0×10^{-7} N m^{-3}. Labels are in units of 10^{-8} N m^{-3}.

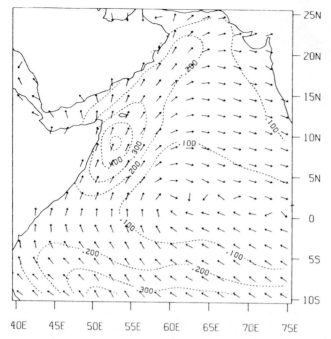

Fig. 18. Wind stress from the FGGE case for 16 September. Beginning in late July and continuing through August and early September, both the wind stress and wind stress curl decrease in magnitude very rapidly in the FGGE and in the MMC cases. By mid-September, the maximum wind stress is only 30% of its peak magnitude in the FGGE case. At this time, the wind stress fields are essentially the same in both cases. Contour interval is $0.1 \, \text{N m}^{-2}$.

the FGGE case, the maximum stress occurs southeast of Socotra, while in the MMC case, a single maximum now occurs to the east of Socotra. The maximum stress in the MMC case is about 25% lower than the FGGE case. In the FGGE case, the convergence zone is at 3°N, while the MMC case it is still just south of the equator. The curl patterns for both cases (Figs. 16 and 17) are similar to those of the preceeding month, but the curl and its gradients have become much stronger. Negative curl covers almost the entire basin, with the exception of the Arabian coastal region. The maximum negative curl is about the same in both cases, but occurs in different locations. Both cases have relative negative maxima in the curl at 7°–8°N, but in the MMC case the largest value of negative curl occurs along 13°N. It is important to note that in both cases there are very large values of negative curl very near the Somali coast boundary region.

In mid- to late July the winds begin to relax. This relaxation is very rapid in August and September. The patterns in the stress and its curl remain much the same as those for July in both cases, but their magnitudes decrease substantially. By mid-September, (Fig. 18) the wind stress values are about 30% of their July values in both cases. The Southern Hemisphere tradewinds are retreating to the south. By late October, the Northern Hemisphere winds begin to reverse, and in mid- to late November, the northeasterly winds push across the equator, along the southern Somali and Kenya coasts. The NE monsoon is fully developed by late December.

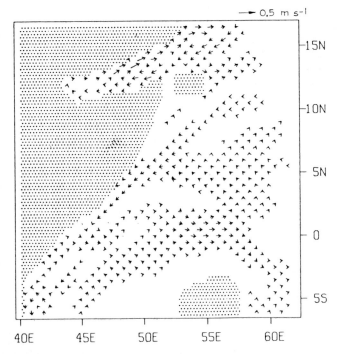

Fig. 19. Upper layer velocity from the FGGE case for 7 March for the Somali coast region. Arrows indicate speed and direction. Every fourth mesh point is shown. Scale is in upper right-hand corner. Vectors larger than $1.0 \, \text{m s}^{-1}$ are truncated. Velocities less than $0.03 \, \text{m s}^{-1}$ are not shown. Northward flow is just beginning along the southern Somali coast, even though the local winds do not reverse for another month. Eastward flow is present along a portion of the equator. There is a band of onshore flow at $5°N$, which splits northward and southward at the coast.

RESULTS

The model is integrated from rest for a period of three years for each case, with the annual wind cycle repeating each year. After the first year of integration, a regular seasonal cycle is established in the model fields. We disregard the first two years of model integration to be sure the seasonal cycle is completely spun up, and present the results from the third year. We concentrate here on the details of the development and decay of the Somali Current system during the SW monsoon in the two cases. Many features in the two cases are similar, and when there are differences, they must of course be related to differences in the two wind fields.

During the NE monsoon, the model Somali Current flows southward along the coast from Socotra to $5°S$, where it meets a weak northward flowing current, and both turn offshore and meander into the interior. The northward flowing current in the model is much weaker than its real world counterpart, the EACC, because the model stops at $10°S$ and therefore only includes a small portion of the Southern Hemisphere tradewinds, which drive the EACC. Therefore, the mechanism of Anderson and Moore (1979) and of Anderson (1981) of inertial overshoot of the EACC causing the early reversal of the Somali Current is missing in this model. Still, as early as March 7, in both cases, several

422

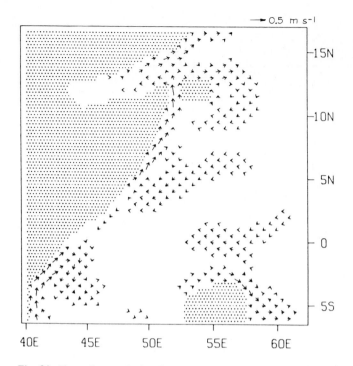

Fig. 20. Upper layer velocity from the FGGE case for 16 April, as in Fig. 19. The winds have just begun to reverse south of the equator, but are primarily onshore north of the equator. There is a gyre-like circulation pattern forming along the coast south of the equator. The onshore flow from the previous month has moved to the south, and feeds a northward flow along the Somali coast between 4°N and the Horn of Africa, with some recirculation at 6°N.

weeks before the local winds reverse, there is evidence of northward flow along the coast between 4°S and the equator (Fig. 19). This is due to the relaxation of the alongshore pressure gradient that formed in response to the NE monsoon winds, which reached their peak in mid-January. The pressure gradient force set up by the cross-equatorial wind stress along the boundary opposes the flow just south of the equator due to advective effects, but is in the same direction as the flow to the north of the equator, so that a relaxation of the wind stress causes a prompt reversal of the flow just south of the equator, but has little effect on the flow to the north of the equator. Similar behavior was reported in a model by Philander and Delecluse (1983).

Also in Fig. 19, we see an onshore flow at 5°N, that splits at the coast with part flowing to the south and a very weak part to the north. This flow is also found in the MMC case at about 7°N, and is a remnant of the source region of the Somali Current during the NE monsoon, which is moving southward as the wind stress curl patterns change.

By mid-April (Fig. 20), as the winds begin to reverse, a clockwise gyre begins to form south of the equator in the FGGE case, driven by the northwestward extension of the Southern Hemisphere tradewinds. In the MMC case (Fig. 21) the alongshore flow does not recirculate south of the equator, but feeds into an eastward equatorial current. This equatorial current is driven by the westerly winds along the equator seen in Fig. 5. Both

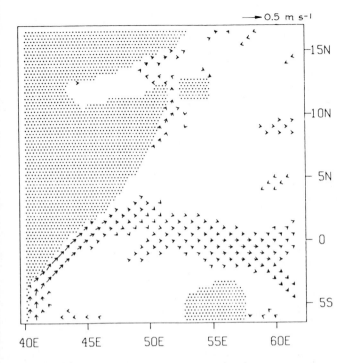

Fig. 21. Upper layer velocity from the MMC case for 16 April, as in Fig. 19. There is northward flow along the Kenya and Somali coasts, but instead of recirculating south of the equator as in the FGGE case, this flow turns eastward along the equator into a broad meandering equatorial current, that is not present in the FGGE case. There is northward flow along the Somali coast between 7° and 12°N, but it is much weaker than in the FGGE case.

the eastward current and the westerly winds are absent in the FGGE case. A jet similar to this equatorial current was first reported by Wyrtki (1973) during the monsoon transitions and explained theoretically by O'Brien and Hurlburt (1974) as a transient response to the onset of westerly winds. The predicted width of this jet is twice the baroclinic equatorial radius of deformation, or about 500 km. This is in agreement with equatorial current in the MMC case.

The currents north of the equator also differ at this time. In the FGGE case, the winds have a northward component north of 5°N, and drive a northward flow there, while south of 5°N, the winds still have a southward component and drive a weak southward flow there. In the MMC case, only the northward flow is found.

In May, northward flow is found all along the Kenya-Somali coast (Figs. 22 and 23). The Southern Hemisphere flow turns offshore at the equator and recirculates to the south in both cases. This southern gyre is narrower and stronger in the MMC case because the nearshore winds are stronger even though the offshore winds are weaker than in the FGGE case (see Figs. 6 and 7). The equatorial jet in the MMC case is dissipating. There are second offshore turning flows developing in both cases to the north of the equator. In the FGGE case, this flow is between 3° and 5°N, while the MMC case, it is between 6° and 8°N. These latitudes coincide in both cases with the latitude where the Findlater jet moves back offshore, as can be seen in Figs. 6 and 7. Anderson (1981) and Lin and

424

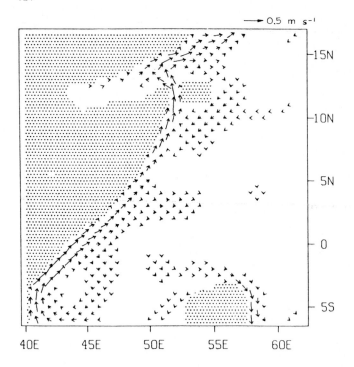

Fig. 22. Upper layer velocity from the FGGE case for 16 May, as in Fig. 19. The gyre south of the equator is now well developed. There is another weak offshore flow developing between 3° and 5°N. There is northward flow through the channel between Socotra and the Horn of Africa and across the Gulf of Aden, very similar to that inferred from satellite images at this time of the year.

Hurlburt (1981) have suggested that this northern part of the Findlater jet is responsible for generation of the great whirl. As we will see, this offshore flow does develop into the great whirl, but the local (or nearly local) curl of the wind stress is also important, as suggested by Schott and Quadfasel (1982), and by Figs. 8 and 9.

In the FGGE case, the flow along the northern Somali coast passes through the channel between Socotra and the Horn of Africa (Ras Aser or Cape Guardafui, as it is also known) and meanders west and then eastward across the mouth of the Gulf of Aden. Similar flow occurs in the MMC case, but it is much weaker. This behavior has been inferred from satellite infrared imagery and in situ data by B.J. Cagle (pers. commun., 1983) in early May of 1980. This current appears to be driven by the local wind stress, and controlled by the coastline geometry.

In June the winds increase considerably, and the Findlater jet leaves the coast at 9°N in the FGGE case and 8°N in the MMC case, which is where the maximum wind stress values occur (Figs. 10 and 11). The offshore flow, however, recirculates in a closed gyre centered at 5°N in both cases to form the great whirl (Figs. 24 and 25). This is also the latitude of the relative maximum in negative wind stress curl in both cases (Figs. 12 and 13). The center of these patches is located only 2° to the east of the center of the great whirl, so that the lag time between changes in the curl and changes in the great whirl is very short, especially at this low latitude, and is on the order of a few days.

The southern gyre now turns offshore at 2°N and recirculates southward across the

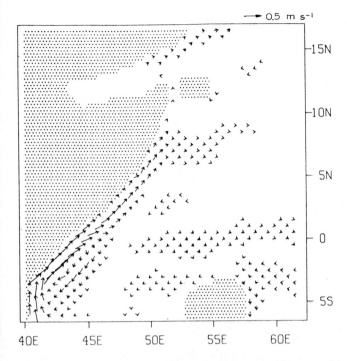

→ 0.5 m s⁻¹

Fig. 23. Upper layer velocity from the MMC case for 16 May, as in Fig. 19. The Southern Hemisphere gyre is tighter, with a larger transport than in the FGGE case. There is a weak offshore flow developing between 6° and 8°N, with some onshore flow beginning to appear at 3°N, 52°E. This flow will later become the great whirl. There is very little flow through the channel in this case.

equator. This recirculation is much stronger in the FGGE case because there is much more negative vorticity input by the wind stress curl in this case. In both cases, there is a Sverdrup-like southwestward flow farther offshore that crosses the equator and enters the southern gyre. This flow is stronger in the FGGE case because the negative wind stress curl is stronger.

In the MMC case, outflow from the great whirl passes to the southeast of Socotra and feeds an eastward jet along 13°N (Fig. 25). This jet follows the strong gradient in the wind stress curl along 13°N seen in Fig. 13, and is maintained by the differential Ekman pumping and induced pycnocline slope associated with that gradient (Fig. 26). The curl gradient in the FGGE case (Fig. 12) parallels the Arabian coast and also drives a northeastward current, part of which can be seen at the top of Fig. 24.

By early July, the two-gyre system is fully developed. In the FGGE case, the equatorial winds change from westerly to easterly to the east of 60°E (Fig. 10) in early June, and by late June the recirculation region from the southern gyre increases its transport and accelerates northward (Fig. 27). A similar response occurs in the MMC case, but not until mid-July and is triggered by a decrease in the westerly component of the equatorial wind stress. In Fig. 28, the recirculation region is still south of 2°N in the southern gyre on 1 July.

By late July in the FGGE case (Fig. 29), the southern gyre has reached 6°N and is

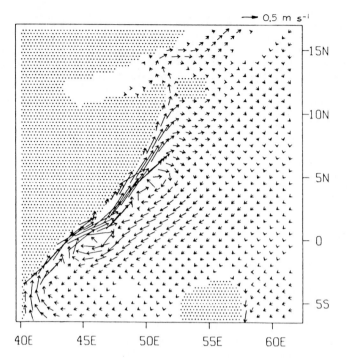

Fig. 24. Upper layer velocity from the FGGE case for 16 June, as in Fig. 19. The offshore flow that was between 3° and 5°N now extends from 5° to 10°N. Part of this flow has formed a closed eddy centered at 5°N, to form the beginning of the great whirl of the two-gyre circulation. The southern gyre now lies south of 2°N, with a strong recirculation stradling the equator. There is a broad south-westward Sverdrup-like flow farther offshore that crosses the equator and feeds into the southern gyre.

beginning to interact with the great whirl, which now between 7° and 12°N. The Northern Hemisphere winds are just beginning to decrease at this time. There is still very little flow from the great whirl passing through the channel.

In the MMC case at this time (Fig. 30), the southern gyre has just begun its north-ward migration, triggered, it appears, by the decrease in the westerly component of the near-equatorial winds. Another eddy has formed just south of the equator in the southern gyre. To the east of Socotra, another eddy is forming in the outflow from the great whirl and will become the Socotra Eddy.

Between 25 July and 4 August in the FGGE case, we see the southern gyre and the great whirl coalesce (Fig. 31). After the coalescence, there is a more or less continuous boundary current all along the east African coast with some smaller eddies found on its offshore side. The flow through the channel steadily increases, forming an eddy to the north of Socotra. To the east and southeast of Socotra, outflow from the great whirl forms large meanders that will later close into a larger clockwise eddy with a smaller counterclockwise eddy between it and the great whirl.

The coalescence of the southern gyre and the great whirl occurs in the MMC case between 19 and 28 August (Fig. 32), in much the same manner as in the FGGE case. A prominent eddy forms at the equator in this case, with significant offshore flow just

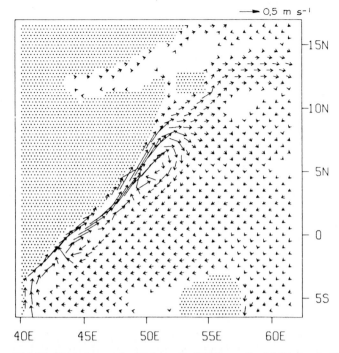

Fig. 25. Upper layer velocity from the MMC case for 16 June, as in Fig. 19. The two-gyre circulation pattern develops similarly to the FGGE case, but the current speeds are not as strong. The great whirl forms at about 5°N in both cases. The boundary current separates at a lower latitude in the MMC case, and the flow through the channel is in the opposite direction. In this case, an eastward jet forms along 13°N, fed by outflow from the great whirl. This jet is not seen in the FGGE case.

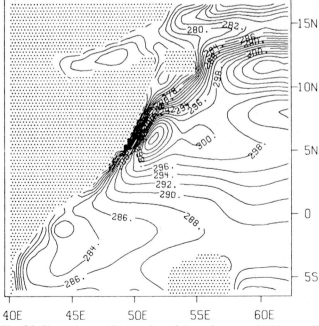

Fig. 26. Upper layer thickness for 16 June from the MMC case. Note slope along 13°N. Contour interval is 2 m.

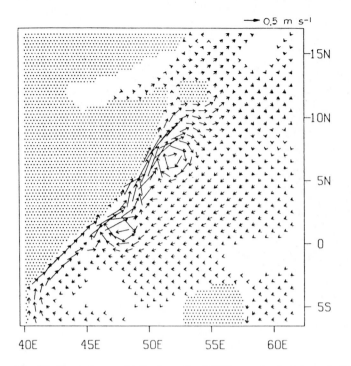

Fig. 27. Upper layer velocity from the FGGE case for 1 July as in Fig. 19. The velocity scale has changed and velocities less than 0.05 m s⁻¹ are not shown. The great whirl is fully developed. Flow through the channel has decreased considerably. The reciruclation region in the southern gyre is intensifying and is accelerating northward.

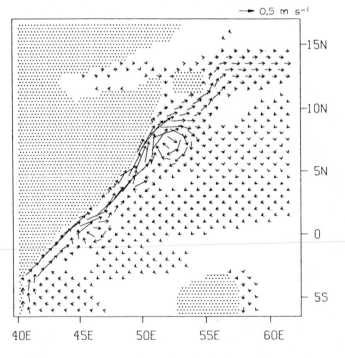

Fig. 28. Upper layer velocity from the MMC case for 1 July as in Fig. 27. The great whirl is well developed and is centered at 7°N. The recirculation region in the southern gyre is not as strong in this case and has not yet begun to accelerate northward, and there is flow from the southern gyre into the great whirl. The eastward jet at 13°N is still present, with no meandering evident.

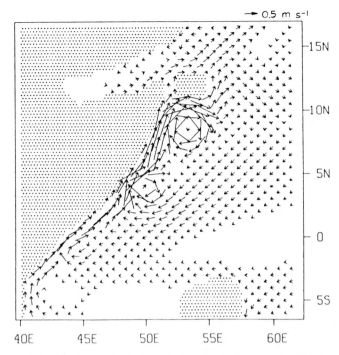

Fig. 29. Upper layer velocity from the FGGE case for 19 July, as in Fig. 27. As the wind stress begins to relax, both the great whirl and the southern gyre are moving northward. There is evidence of a new eddy forming in the southern gyre at the equator.

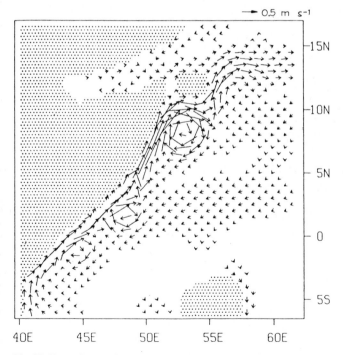

Fig. 30. Upper layer velocity from the MMC case for 19 July, as in Fig. 27. As the winds begin to relax, the recirculation region in the southern gyre intensifies and moves northward. A second eddy has formed in the southern gyre, just south of the equator. The great whirl is now centered at 8° to 9°N. The outflow from the great whirl is beginning to meander to form the Socotra eddy. Weak eddies are

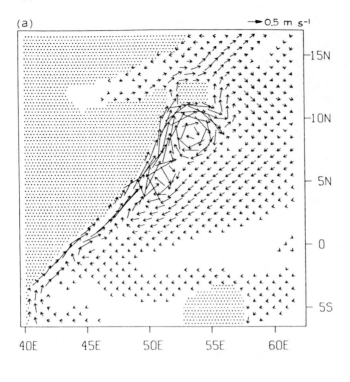

(a)

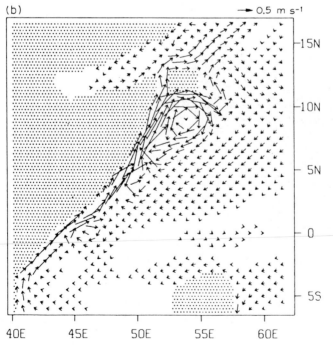

(b)

(c) → 0.5 m s⁻¹

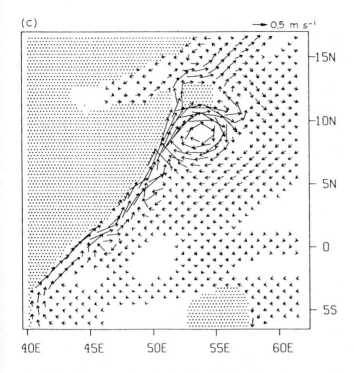

Fig. 31. Collapse of the two-gyre system in the FGGE case. (a) Upper layer velocity on 25 July. The recirculation region from the southern gyre is beginning to interact with the great whirl. Flow from the great whirl is beginning to pass through the channel. Outflow from the great whirl to the southeast of Socotra is beginning to meander. (b) Upper layer velocity for 1 August. The recirculation region from the southern gyre is coalescing with the great whirl. The flow through the channel and along the north side of Socotra is increasing, and small eddies are forming to the north and west. The outflow from the great whirl to the southeast of Socotra is meandering intensely. (c) Upper layer velocity for 4 August. The recirculation region from the southern gyre has completely merged with the great whirl. Closed eddy circulations are forming in the meandering outflow from the great whirl to the southeast of Socotra. The eddy at the equator is stengthening, as are the eddies to the north and west of Socotra.

north of the equator. In the FGGE case, only a small eddy forms at the equator at this time, but it is mostly confined to the offshore edge of the current. This difference can be explained by the difference in strength of the Southern Hemisphere winds, which are much weaker in the MMC case. The boundary current is not as strong, and does not have enough inertia to overcome its tendency to turn offshore north of the equator.

As the wind stress curl gradient relaxes in August and September, the pressure gradient that was maintained by the differential Ekman pumping becomes unbalanced. As this forcing is removed, more and more of the great whirl flow passes through the channel, until by October, there is a clockwise flow around the island of Socotra (Fig. 33). The southern part of this flow will become the source of the NE monsoon Somali Current. Southward flow steadily overrides the northward flow, pushing it to 5°S by late November or early December.

432

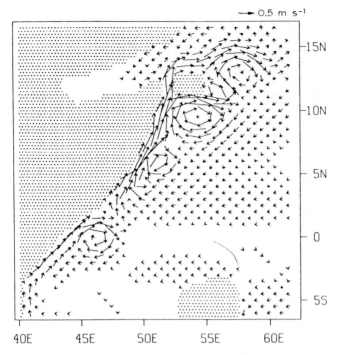

Fig. 32. Upper layer velocity from the MMC case for 19 August. The coalescence of the great whirl and the southern gyre is just beginning at this time, much later than in the FGGE case. The Socotra eddy is fully developed to the east of Socotra, with a small cyclonic circulation between it and the great whirl. There is a large eddy located at the equator, with considerable offshore flow to the north of it, so that there is not a continuous boundary current all along the coast. This situation persists until the onset of the northeast monsoon. Otherwise, the collapse of the two-gyre system in this case proceeds much the same as in the FGGE case.

DISCUSSION AND CONCLUSION

By comparing the results of this model for these two separate cases, we can draw some conclusions about the relative importance of the different forcing mechanisms in the generation and decay of the two-gyre system in the Somali Current. Figure 34 (a and b) summarizes the development and collapse of the two-gyre system by showing the north-ward component of transport across $2°$ zonal sections following the coast with time and distance. The minimum in northward transport along lines A and B indicates the separation between the southern gyre and the great whirl. The southern gyre recirculation region forms in early May in each case. Its center is indicated by the relative maximum along lines C and D. It moves slowly northward through May and June. During this period, the separation region remains at about $2°$N. In early June, the great whirl forms between $4°$ and $8°$N, indicated by line E, and steadily increases its transport. In late-June in the FGGE case and mid-July in the MMC case, the separation region abruptly begins to move northward, as the recirculation dramatically increases its transport and its northward speed. Both move northward at an average speed of 25 km/day in the FGGE case (18 km/day in the MMC case). The northward migration of the separation

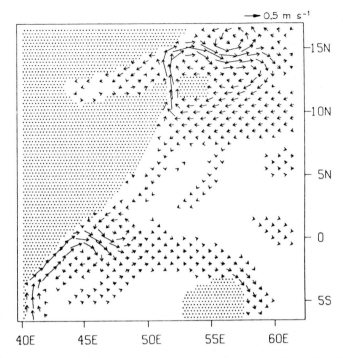

Fig. 33. Upper layer velocity from the MMC case for 16 October. The great whirl flow has been completely forced through the channel, due to the relaxation of the wind stress curl gradient, so that there is now a clockwise circulation around the island of Socotra. The southern part of this circulation becomes the source of the southward flowing winter Somali current.

region (the relative minimum along line B) can be interpreted as the northward movement of the cold wedge observed in satellite images by Brown et al. (1980), Evans and Brown (1981) and Swallow et al. (1983). As the southern gyre approaches the great whirl, it also moves northward, although much more slowly, as it is confined by Socotra to the north. In late July in the FGGE case (late August in the MMC case) the southern gyre catches the great whirl, and the two coalesce. This is observed in satellite images as a coalescence of the southern cold wedge with a northern cold wedge.

The question remaining is what triggers the northward migration of the southern gyre? The large increase in transport occurs only in the recirculation region as it moves away from the equator and not to the south. As its transport increases, its self-advection speed also increases. This transition occurs at northward transport values of about 25×10^6 m^3 s^{-1} in the FGGE case and 20×10^6 m^3 s^{-1} in the MMC case. This seems to rule out the idea that the steadily increasing transport reaches some critical value after which it becomes unstable and advects northward, as was suggested in several other models (e.g., Hurlburt and Thompson, 1976; Cox, 1976, 1979; Lin and Hurlburt, 1981). The northward movement coincides with a decrease in the westerly component of the winds along the equator in both cases, although this occurs at different times in each case. This certainly points to something in the wind field triggering the northward movement. If the winds along the equator have a westerly component that decreases or even

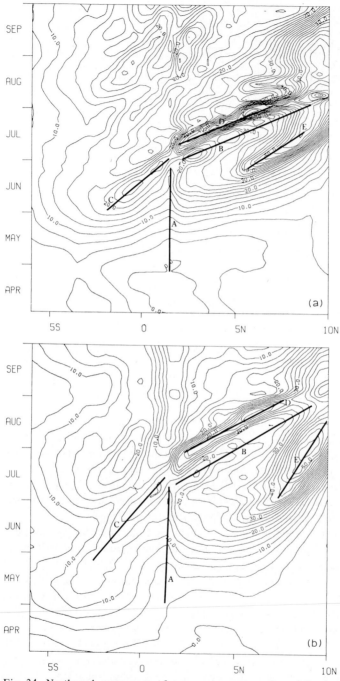

Fig. 34. Northward component of transport across sections following the coast versus time and latitude, for (a) the FGGE case and (b) the MMC case. Lines A and B follow the separation region between the southern and northern gyre. Lines C and D follow the center of the southern gyre recirculation region. Line E follows the northward movement of the great whirl. Units are 10^6 m^3 s^{-1}. Contour interval is 2.5×10^6 $m^3 s^{-1}$.

changes to easterly, this will send a stream of downwelling Rossby wave energy westward that will quickly reach the western boundary and cause a convergence of mass there. This convergence would intensify the recirculation region of the southern gyre and cause it to accelerate northward through nonlinear self advection. In the FGGE case, the convergence zone in the winds crosses the equator around 1 June, causing the equatorial winds to shift from westerly to easterly to the east of $60°E$. It would take a first mode equatorial Rossby wave approximately 20 days to travel from $60°E$ to the Somali coast, which is consistent with timing of the northward acceleration of the southern gyre. In the MMC case, the winds all across the equator have a westerly component. As these westerly winds decrease in mid-July, downwelling Rossby-wave energy almost immediately reaches the Somali coast, since it is generated all along the equator. This again is consistent with the increase in transport of the southern gyre occurring in mid-June. Swallow et al. (1983) observed a convergence at the equator with westward flow and thermocline deepening just prior to the migration of the southern gyre in 1979, which is consistent with the arrival of downwelling Rossby-wave energy. They, however, suggest that changes in the offshore curl pattern are responsible for the migration. Also from Fig. 34 we can see that the southern gyre in the FGGE case moves northward faster because its transport is about 25% larger than in the MMC case. Its translation speed increases slowly as its transport increases.

We have shown that the spatial and temporal inhomogeneity of the winds play a crucial role in the development of the Somali Current system. The pre-monsoon onset conditions are important in determining the early current reversals. The relaxation of the NE monsoon winds is sufficient to cause the currents to reverse south of the equator, so that the mechanism of early forcing by the Southern Hemisphere tradewinds, which are not included in this model, is not necessary. This is not to say that forcing by the Southern Hemisphere tradewinds is not important. Our EACC is very weak or non-existent at times because of their absence. The tradewinds do come far enough north later in the SW monsoon that they become important in our model. Anderson and Moore (1979) showed that it is the curvature of the wind near the equator that is important in determining the separation point of the southern gyre, but that this is modified by the inertial effects. Perhaps our southern gyre would penetrate farther north if we had resolved more of the Southern Hemisphere tradewinds.

It appears that the wind stress curl is as important in the development of the great whirl as is the local wind stress. As Anderson (1981) and Lin and Hurlburt (1981) showed, the great whirl can be generated by the alongshore winds associated with the Findlater jet as it crosses the Somali coast at about $9°N$, without the presence of a wind stress curl. Anderson (1981) found that the center of this northern gyre, and hence the maximum alongshore velocities, were located at the same latitude as the maximum longshore winds, while Lin and Hurlburt (1981) found that the gyre forms just south of the wind maximum. Their model winds also included a significant curl in the vicinity of the gyre development. In our model the center of the great whirl is located at the same latitude as the maximum negative wind stress curl throughout June and early July, while the maximum wind stress is located farther to the north. The negative vorticity input by the wind stress curl appears to control the location of the great whirl in both cases.

The mechanism of forcing by the differential Ekman pumping associated with the strong gradients in wind stress curl has not, to our knowledge, been reported elsewhere, and requires further investigation. It appears to be important in the formation of both the great whirl and the jet at 13°N. Compare for instance Figs. 13 and 26. The strong gradients in upper layer thickness closely parallel the gradients in wind stress curl. The relaxation of this forcing is also important in the formation of the Socotra eddy and in the movement of the great whirl flow through the channel.

The results of this model have been very encouraging thus far. We are in the process of extending the model to include the entire Indian Ocean north of 25°S so that we may include the effects of the Southern Hemisphere tradewinds and the equatorial wave guide east of 73°E. We are also studying the differential Ekman pumping mechanism. How sensitive is it to the parameterization of the wind stress as a body force over the upper layer? How does the relaxation of this forcing generate the eddies observed in the model? Might we expect this mechanism to be important in other areas of the world where the curl gradients are not as strong? We hope to report progress in these areas in the near future.

ACKNOWLEDGEMENTS

This research was supported by the Office of Naval Research. Programming assistance was provided by Mr. James D. Merritt and Mr. Chen-Tan Lin. The manuscript was typed by Mrs. Helen McKelder. The authors would also like to thank the other members of the Mesoscale Air–Sea Interaction Group and their other colleagues who have provided helpful discussion and comments during the course of this work.

REFERENCES

Anderson, D.L.T., 1981. The Somali Current. Ocean Modelling (unpublished manuscript).

Anderson, D.L.T. and Moore, D.W., 1979. Cross-equatorial jets with special relevance to very remote forcing of the Somali Current. Deep-Sea Res., 26: 1–22.

Brown, O.B., Bruce, J.G. and Evans, R.H., 1980. Evolution of sea surface temperature in the Somali Basin during the southwest monsoon of 1979. Science, 209: 595–597.

Bruce, J.G., 1973. Large scale variations of the Somali Current during the southwest monsoon, 1970. Deep-Sea Res., 20: 837–846.

Bruce, J.G., 1979. Eddies off the Somali coast during the southwest monsoon. J. Geophys. Res., 84 (C12): 7742–7748.

Bruce, J.G., 1983. The wind field in the western Indian Ocean and the related ocean circulation. Mon. Weather Rev., 111: 1442–1452.

Camerlengo, A.L. and O'Brien, J.J., 1980. Open boundary conditions in rotating fluids. J. Comput. Phys., 35: 12–35.

Charney, J.G., 1955. The generation of ocean currents by the wind. J. Mar. Res., 14: 477–498.

Cox, M.D., 1976. Equatorially trapped waves and the generation of the Somali Current. Deep-Sea Res., 23: 1139–1152.

Cox, M.D., 1979. A numerical study of Somali Current eddies. J. Phys. Oceanogr., 9: 311–326.

Evans, R.H. and Brown, O.B., 1981. Propagation of thermal fronts in the Somali Current system. Deep-Sea Res., 28: 521–527.

Findlater, J., 1971. Mean monthly airflow at low levels over the western Indian Ocean. Geophys. Mem., No. 115, 53 pp.

Findlay, A.G., 1866. A Directory for the Navigation of the Indian Ocean. Richard Holmes Laurie, London, 1062 pp.

437

Hastenrath, S. and Lamb, P.J., 1979. Climate Atlas of the Indian Ocean. University of Wisconsin Press, Madison, Wisc.

Hellerman, S. and Rosenstein, M., 1983. Normal monthly wind stress over the world ocean with error estimates. J. Phys. Oceanogr., 13: 1093–1104.

Hurlburt, H.E., 1974. The influence of coastline geometry and bottom topography on the eastern ocean circulation. Ph.D. dissertation, Florida State University, Tallahassee, Fla., 104 pp.

Hurlburt, H.E. and Thomson, J.D., 1976. A numerical model of the Somali Current. J. Phys. Oceanogr., 6: 646–664.

Knox, R.A. and Anderson, D.L.T., 1985. Recent advances in the study of the low-latitude ocean circulation. Prog. Oceanogr., 14: 259–318.

Leetmaa, A., 1972. The response of the Somali Current to the southwest monsoon of 1970. Deep-Sea Res., 20: 319–325.

Leetmaa, A., Quadfasel, D.R. and Wilson, D., 1982. Development of the flow field during the onset of the Somali Current, 1979. J. Phys. Oceanogr., 12: 1325–1342.

Lin, L.B. and Hurlburt, H.E., 1981. Maximum simplification of nonlinear Somali Current dynamics. In: M.J. Lighthill and R.P. Pearce (Editors), Monsoon Dynamics. Cambridge University Press, Cambridge.

Luther, M.E. and O'Brien, J.J., 1985. A model of the seasonal circulation in the Arabian Sea forced by observed winds. Prog. Oceanogr., 14: 353–385.

O'Brien, J.J. and Hurlburt, H.E., 1974. Equatorial jet in the Indian Ocean: Theory. Science, 184: 1075–1077.

Philander, S.G.H. and Delecluse, P., 1983. Coastal currents in low latitudes. Deep-Sea Res., 30: 887–902.

Schott, F., 1983. Monsoon response of the Somali Current and associated upwelling. Prog. Oceanogr., 12: 357–382.

Schott, F. and Quadfasel, D.R., 1982. Variability of the Somali Current system during the onset of the southwest monsoon, 1979. J. Phys. Oceanogr., 12: 1343–1357.

Swallow, J.C. and Fieux, M., 1982. Historical evidence for two gyres in the Somali Current. J. Mar. Res., 40 (suppl.): 747–755.

Swallow, J.C., Molinari, R.L., Bruce, J.G., Brown, O.B., and Evans, R.H., 1983. Development of near-surface flow pattern and water mass distribution in the Somali Basin, in response to the southwest monsoon of 1979. J. Phys. Oceanogr., 13: 1398–1415.

Wyrtki, K., 1973. An equatorial jet in the Indian Ocean. Science, 181: 262–264.

CHAPTER 28

A MODEL FOR THE SEA-SURFACE TEMPERATURE AND HEAT CONTENT IN THE NORTH ATLANTIC OCEAN

H. LE TREUT, J.Y. SIMONOT and M. CRÉPON

ABSTRACT

A simple model of the sea-surface temperature and of the thermocline depth is presented. It consists of a mixed layer which is advected by surface currents and by diffusion processes in order to simulate the large-scale heat flux. The results are consistent with the available observations.

INTRODUCTION

On an annual-term basis, it is found that the radiation budget of the atmosphere is negative and that the radiation budget of the ocean is positive; the atmosphere loses heat by radiation and the ocean gains heat by radiation. Since the heat budget of the globe is very close to equilibrium, the ocean must heat the atmosphere. This is done through sensible and latent heat fluxes across the sea surface. These two fluxes are controlled by the sea-surface temperature and by the heat content of the upper layers of the ocean. Hence, detailed climate studies require a model to describe the thermodynamics of the upper ocean. Since the radiative budget of the Earth is positive at low latitudes and negative in the polar regions, the atmosphere and the oceans must transport heat polewards. According to Oort and Vonder Haar (1976), the annual mean atmospheric and oceanic heat transports are almost equal. This implies that a realistic model of the ocean must incorporate meridional heat fluxes.

Two modelling strategies for these fluxes are possible. The first approach is to use oceanic general circulation models. But these models are difficult to construct and use and they require a large amount of computer time. It is thus necessary to develop also simpler and computationally efficient models, which can be restricted to the upper layers of the ocean. It is well known that the seasonal temperature variability of the ocean is limited to the upper three hundred meters (Oort and Vonder Haar, 1976). Therefore, the main idea of the second approach is to use a mixed layer model to parameterize the vertical exchanges and to embed it in an advective–diffusive process which accounts for the horizontal heat fluxes.

The thermodynamics of the upper ocean is controlled by the vertical exchanges of energy through the sea surface and by the horizontal heat transport. The variability in the heat transport is due to variations in atmospheric forcing, to large-scale advection and to eddy fluxes. In our model, eddy fluxes are represented by a diffusive process and large-scale advection is assumed to depend mainly on the variability of temperature gradients in the advection term rather than on the variability of the general circulation. The latter assumption is based on the fact that at middle latitudes, the dynamical response of the ocean to atmospheric forcings is very slow and in a first approximation one may assume that the ocean currents are fixed in time.

In the present paper, we focus our attention on the thermodynamics of the North Atlantic Ocean. This ocean is well documented and it will permit the careful validation of our model.

THE MIXED LAYER MODEL

In this section we describe the simple representation of the oceanic upper layers which has been used in our simulations. We have chosen a model whose discretization in the vertical allows us to record in memory the temperature profile under the thermocline. But from the surface down to the thermocline we may assume a perfectly mixed layer, the depth of which is computed by the model. In its present version the model has 20 vertical layers: one surface layer, 15 m deep, then nine layers of 5 m and finally ten layers of 20 m. Each layer is warmed by solar radiation and the temperature of the first layer also depends on the net infrared flux, on the sensible heat flux and on the evaporation. The surface fluxes have been determined by using observed sea-surface temperatures (SSTs) rather than predicted ones. This means that the atmospheric forcing on the model is the net energy flux at the surface and that there is no feedback of the SST on the fluxes. The reason for this is that the air temperature or air humidity adapt themselves very quickly to the oceanic conditions and cannot be used as a given forcing for an oceanic model in the absence of an atmospheric model. The decrease with depth of the solar absorption has been represented by the simple law:

$$R_s(z) = R_s(0)e^{-kz}$$

where $R_s(z)$ is the incoming solar flux at depth z.

This law neglects the spectral dependence of the solar absorption, but is sufficient at the level of approximation of our model.

The temperature profile may also be changed by two kinds of mechanisms: (1) adiabatic adjustment, whenever the vertical profile is not stable; and (2) deepening of the mixed layer induced by the wind.

Two formulations have been tested in order to represent this deepening:

(1) We have defined a bulk Richardson number R_i of the form:

$$R_i = g\frac{\Delta b(H^2/2)}{mu^{*3}}$$

where Δb is the variation of the buoyancy across the thermocline, H is the depth of the mixed layer, u^* the friction velocity of the wind and g the acceleration of gravity.

The term mu^{*3} represents very crudely the kinetic energy of the current generated by the wind in the mixed layer, the shear at the bottom of the mixed layer being responsible for the deepening. To conserve the total momentum when there is deepening, the coefficient m is chosen to decrease linearly with depth. We test downward and perform a mixing of the whole column until we find a level where R_i is less than a critical value s.

(2) We have also used the prognostic equation:

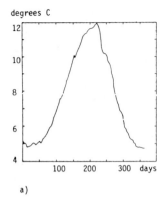

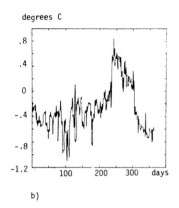

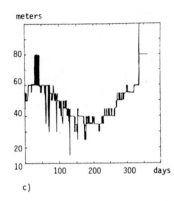

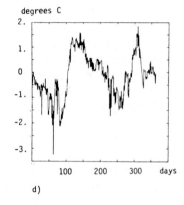

Fig. 1. (a) Simulated SST (in °C) at point "Papa"; (b) difference between the simulated and the observed SST (in °C) at point "Papa"; (c) simulated thermocline depth (in meters) at point "Papa"; and (d) difference between the simulated and the observed SST at point "Romeo". The abcissa are in days, starting on the 1st of March.

$$\frac{dH}{dt} = \frac{Cu^{*3} \exp(-yH)}{\Delta bH + q^2}$$

The numerator represents the flux of turbulent kinetic energy, the effect of dissipation is parameterized by the term $\exp(-yH)$, and q^2 represents a minimal value for the turbulent kinetic energy. This formulation is a simplified version of that given by Niller and Kraus (1977).

The unknown coefficients in these semi-empirical equations have been determined by fitting the results of the model to the data of weather ship stations at point "Papa" (50°N, 145°W; 1 yr of data) and "Romeo" (47°N, 16°W; 2 yrs of data). Since the model has been found sensitive to the coefficient k representing the absorption of solar radiation in the water, we have chosen values corresponding to the observations collected by Jerlov (1968): $k = 0.08 \, m^{-1}$ at point "Papa" and $k = 0.04 \, m^{-1}$ at point "Romeo". It has then been possible to compute values for the coefficient m and s in case (1), or for the coefficients C and y for case (2) that fit accurately the seasonal cycle at both stations. In Fig. 1 we give results for the formulation (1), which has been used in the experiments reported in the following section. The error is always less than 1°C at station "Papa" and less than 1.5°C at station "Romeo", the depth of the termocline going from 40 m in summer to more than 100 m in winter at station "Papa" and from 20 m to more than 100 m at station "Romeo".

MODELLING OF AN UPPER OCEAN BASIN

The aim of this work is to investigate whether a simple mixed-layer model may reproduce the observed SST's of an entire ocean basin.

The atmospheric forcing has been obtained from the daily E.C.M.W.F. analyses for the year 1981. They give the sensible heat flux, the latent heat flux, the net radiative budget and the wind at the surface. Due to the absence of archives for the solar flux, we were obliged to reconstruct it by using the analysed cloudiness and an empirical formula due to Perrin de Brichambaut (1975). It is also important to note that E.C.M.W.F. has calculated the fluxes using observed sea-surface temperatures. This forcing is in good agreement with the values of Esbensen and Kushnir (1981), both on a monthly and on an annual basis. Other boundary conditions for the model are the temperatures at the bottom of the thermocline which have been estimated from the Robinson et al. (1979) atlas and the solar absorption coefficient k which was obtained from Jerlov (1968, see above). The experiments begin on the 1st of March because the temperature profile is vertically homogeneous at that time of the year and changes rapidly afterwards. Initial conditions were derived from the atlas of Robinson et al. (1979).

We have first made experiments without imposing any advection: in this case we adjusted the fluxes at each grid point so as to have a zero-mean budget over the year. The results (not shown here) were incorrect in terms of SST, with temperatures higher off the African coast than near the Gulf of Mexico, in contradiction with observations. This implies that it is necessary to represent the advection of temperature by the general circulation and by the eddies. For this purpose we used the observed currents obtained from the Pilot Charts, which have been analyzed to retain only the divergent part (with a condition of zero velocity perpendicular to the coast). These currents represent the sum of an Ekman component and of a quasi-geostrophic component, the former being more important in the central part of the ocean, the latter in the western boundary current zone. We have therefore used them to advect the heat content over a depth corresponding to an Ekman layer (50 m), scaled by a fractional coefficient to account for the decrease of the currents from the surface to the bottom of the Ekman layer. We have also applied a diffusion operator to simulate the eddy fluxes with a viscosity coefficient equal to 10^5 m^2 s^{-1}.

The total heat transport has been redistributed through the vertical with a linear law that keeps the bottom temperature and the thermocline depth unchanged. This formulation constitutes one of the simplest ways of introducing observed currents in a mixed layer model.

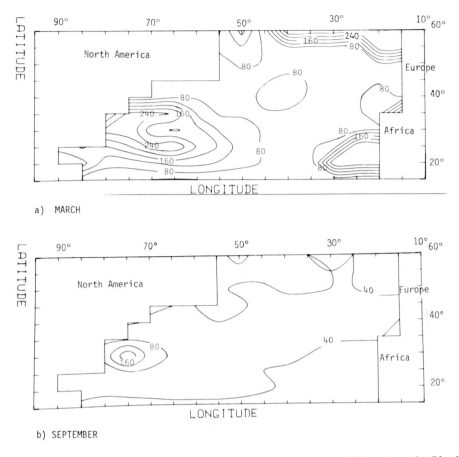

a) MARCH

b) SEPTEMBER

Fig. 2. Thermocline depth (in meters) for the fourth year at two different months (March and September). Contours every 40 m.

COMPARISON WITH DATA

In the conceptual phase of the problem, the aim was to produce a realistic north–south temperature gradient, in equilibrium with the north–south heat flux that it generates. This mean equilibrium has not been reached after five years of simulation. The temperature field is still evolving. This is due to the long thermal memory of the ocean: the diffusive times scale L^2/v, where L is the horizontal length scale and v the diffusion coefficient, is about 2.5×10^3 days, i.e. ≈ 6.5 yrs, and the advective time scale L/U, where U is the horizontal velocity scale, is about 5×10^2 days, i.e. ≈ 1.5 yrs. Longer computations are necessary therefore to reach a complete equilibrium.

In Figs. 2 and 3, we present a sequence of results extracted from a four-year simulation. They are compared with the climatological maps given by Robinson et al. (1979) for the North Atlantic Ocean or by Levitus (1982) over the world ocean. In October the thermocline depth (Fig. 2) is almost everywhere between 40 and 80 m. It is deeper in the Gulf Stream area and nearer the surface off the African coast. This is

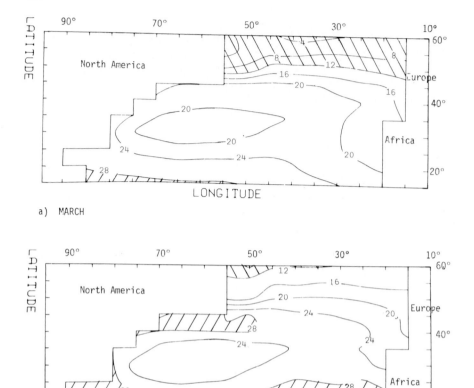

a) MARCH

b) SEPTEMBER

Fig. 3. SST in °C for the fourth year in the same months as in Fig. 2. Contours every 4°C. Shaded when temperature is less than 12°C or greater than 28°C.

in broad agreement with the observations, although the determination of the thermocline depth from observed data may be subject to uncertainties (Levitus, 1982). In March, the thermocline depths simulated by the model show a realistic pattern in the western part of the ocean basin although the deepening is somewhat insufficient. In the eastern part the thermocline is too shallow in spite of the cold SSTs caused by advection. This may be due to the difference between the wind series used to calibrate the model and the daily E.C.M.W.F. wind analyses and also possibly to the fact that our advection scheme keeps the thermocline unchanged. The SST fields (Fig. 3) present a correct east—west contrast. In particular at 20°N, cooler temperatures are obtained off Africa (where the thermocline is higher in summer) and warmer in the Gulf of Mexico area.

CONCLUSIONS

The results suggest that this model is an interesting working tool for a first modelling approach of the ocean—atmosphere interactions. It qualitatively reproduces the main patterns of observed SST throughout the North Atlantic basin, and is computationally efficient. We intend to develop our future research in the following directions. First, we will improve the validity of the model by adding the vertical advection due to divergence of the Ekman drift (De Szoeke, 1980). At mean latitudes, this effect is proportional to the curl of the wind stress (Ekman pumping). Then the variability of the general circulation will be included by using a Rossby wave perturbation approach similar to that described by Stevenson (1983). Finally, we will apply this technique to obtain a thermodynamic model of the surface layer of the world ocean. For this purpose, the above model will be coupled to a dynamical model at equatorial latitudes and to a sea-ice model at high latitudes. For longer time scales, it would be also necessary to represent the deep circulation of the ocean.

ACKNOWLEDGMENTS

The authors wish to thank P. Bertand and J.Y. Lebras who contributed to this work. A. Gerard from the French Météorologie Nationale and P. Gaspard from the University of Louvain la Neuve (Belgium) kindly provided us with the data. M. Ghil gave helpful comments on the manuscript. Financial support was provided by the french institutions CNRS, CNEXO and CNES.

REFERENCES

De Szoeke, R.A., 1980. On the effect of horizontal variability of wind stress on the dynamics of the mixed layer. J. Phys. Oceanogr., 10: 1439–1454.

Esbensen, S.K. and Kushnir, K., 1981. The Heat Budget of the Global Ocean. Clim. Res. Inst., Oregon State University, Corvallis, Oreg.

Jerlov, N.G., 1968. Optical Oceanography. Elsevier, Amsterdam, 194 pp.

Levitus, S., 1982. Climatological Atlas of the World Ocean. NOAA Prof. Pap. 13, U.S. Govt. Printing Office, Washington, D.C., 173 pp.

Niiler, P.P. and Kraus, E.B., 1977. One-dimensional models of the upper ocean. In: E.B. Kraus (Editor), Modelling and Prediction of the Upper Layers of the Ocean. Pergamon, New York, N.Y., pp. 143–172.

Oort, A.H. and Vonder Haar, T.H., 1976. On the observed annual cycle in the ocean–atmosphere heat balance over the Northern Hemisphere. J. Phys. Oceanogr., 6: 781–800.

Perrin de Brichambaut, C., 1975. Estimation des ressources énergétiques solaires en France. Supplément aux cahiers A.F.E.D.E.S., no. 1.

Robinson, M.K., Bauer, R.A. and Shroeder, E.H., 1979. Atlas of North Atlantic–Indian ocean monthly mean temperatures and mean salinities of the surface layer. Department of the Navy, Publ. no. 18.

Stevenson, J.W., 1983. The seasonal variation of the surface mixed layer response to the vertical motions of linear Rossby waves. J. Phys. Oceanogr., 13: 1255–1278.

CHAPTER 29

THE ROLE OF THE OCEAN IN CO_2-INDUCED CLIMATE CHANGE: PRELIMINARY RESULTS FROM THE OSU COUPLED ATMOSPHERE–OCEAN GENERAL CIRCULATION MODEL

MICHAEL E. SCHLESINGER, W. LAWRENCE GATES and YOUNG-JUNE HAN

ABSTRACT

The OSU coupled global atmosphere–ocean general circulation model (GCM) has been used to investigate the role of the ocean in CO_2-induced climate change. Two 16-year simulations have been made, a $1 \times CO_2$ simulation with a CO_2 concentration of 326 ppmv, and a $2 \times CO_2$ simulation with a CO_2 concentration of 652 ppmv. The results of the simulations are presented in terms of the $2 \times CO_2 - 1 \times CO_2$ differences in temperature.

The evolution of the global-mean temperature differences displays a rapid warming of the atmosphere followed by a more gradual warming of the ocean and atmosphere. The annual-mean zonal-mean warming for year 16 shows that the atmospheric warming at the surface increases from the tropics to the subtropics, decreases towards the middle latitudes, and increases towards high latitudes. The warming increases with altitude in the tropics and subtropics, and decreases with altitude elsewhere. The oceanic surface warming increases from the tropics towards the mid-latitudes of both hemispheres. This is similar to the latitudinal distribution of the excess ^{14}C over the pre-nuclear value that is observed at the sea surface. The oceanic warming penetrates to a greater depth in the subtropics and mid-latitudes than in the equatorial region. This also is similar to what is shown by the observed penetration depths of excess ^{14}C. The warming decreases with depth everywhere equatorward of 60 degrees latitude in each hemisphere, and extends from the subtropical surface water downward towards high latitudes. Particularly interesting features of the temperature change at depth are the maxima located near $65°N$ and $60°S$ in the 250–750 m ocean layer. The geographical distributions show that the latter feature is located in the vicinity of the Ross Ice Shelf.

Since the 16-year simulations are not of sufficient duration for the equilibrium change to have been attained, an analysis of the coupled GCM results is performed with an energy balance climate/box ocean model. From this analysis it is estimated that the gain (sensitivity) of the coupled GCM is $0.72°C (W m^{-2})^{-1}$, the global-mean air–sea heat transfer coefficient is $8.0 W m^{-2} °C^{-1}$, and the effective oceanic thermal diffusivity κ is $3.2 cm^2 s^{-1}$ at 50 m depth, $3.8 cm^2 s^{-1}$ at 250 m, and $1.5 cm^2 s^{-1}$ at 750 m. The mass-averaged $\kappa = 2.25 cm^2 s^{-1}$ is in agreement with the best estimate based on the observed penetration of bomb-produced tritium and ^{14}C into the ocean. Furthermore, a box-diffusion climate model with $\kappa = 2.25 - 2.50 cm^2 s^{-1}$ is successful in reproducing the evolution of the $2 \times CO_2 - 1 \times CO_2$ differences in the surface air and ocean surface layer temperatures simulated by the coupled GCM. Consequently, it appears that the coupled GCM transports heat from the surface downward into the ocean at a rate which is commensurate with the rate observed for the downward transport of tritium and ^{14}C.

A projection of the GCM results by the simpler climate model indicates that the time required for the ocean to reach 63% of its equilibrium warming is 75 years. The possible implications of this memory of the climate system are discussed in terms of the detection of a climate change and its attribution to CO_2.

1. INTRODUCTION

Measurements taken at Mauna Loa, Hawaii, show that the CO_2 concentration has increased from 316 ppmv in 1959 to 338 ppmv in 1980 (Keeling et al., 1982), a 7% increase in 21 years. A variety of direct CO_2 measurements and indirect reconstructions indicate that the pre-industrial CO_2 concentration during the period 1800–1850 was 270 ± 10 ppmv (WMO, 1983). A study by Rotty (1983) reports that the CO_2 concentration increased from 1860 to 1973 due to the nearly constant $4.6\%\,yr^{-1}$ growth in the consumption of fossil fuels (gas, oil, coal), and has continued to increase since 1973 due to the diminished $2.3\%\,yr^{-1}$ growth in fossil fuel consumption. A probabilistic scenario analysis of the future usage of fossil fuels predicts about an 80% chance that the CO_2 concentration will reach twice the pre-industrial value by 2100 (Nordhaus and Yohe, 1983). Computer simulations of the climate change induced by a doubling of the CO_2 concentration have been made with a hierarchy of mathematical climate models and give a warming of $1.3°$–$4.2°C$ in the global-mean surface air temperature (Schlesinger, 1984a; Hansen et al., 1984). Since such a global warming represents about 25–100% of that which is estimated to have occurred during the 10,000-year transition from the last ice age to the present interglacial (Gates, 1976a, b; Imbrie and Imbrie, 1979), there is considerable interest in the identification of a CO_2-induced climatic change, and in the potential impacts of such a change on the spectrum of human endeavors.

The evolution of the global-mean surface temperature, $T(t)$, measured from its value $T(t_{eq})$ at a time t_{eq} of climatic equilibrium, can be expressed as:

$$T(t) - T(t_{eq}) = \int_{t_{eq}}^{t} G \frac{dQ}{d\eta} F_c(t - \eta) d\eta \qquad (1.1)$$

where the change in Q represents the thermal forcing due to changes in the "external" climatic quantities such as the solar constant or CO_2 concentration, $G = dT/dQ$ is the gain or sensitivity of the climate system which includes the amplification and damping of the positive and negative climatic feedback processes, and $F_c(t - \eta)$ is a "climate response function" which measures the time lag between the actual response to a small-step function change in thermal forcing and that which would occur if the climate system had no thermal inertia. The temperature change measured from some time $t_0 > t_{eq}$ is by eqn. (1.1):

$$T(t) - T(t_0) = \int_{t_0}^{t} G \frac{dQ}{d\eta} F_c(t - \eta) d\eta + \int_{t_{eq}}^{t_0} G \frac{dQ}{d\eta} [F_c(t - \eta) - F_c(t_0 - \eta)]\, d\eta \qquad (1.2)$$

Thus, unless $F_c \equiv 1$, the change in temperature from t_0 to t depends on Q before, as well as after, t_0. If t_0 is taken as the time when CO_2 was first anthropogenically added to the atmosphere (ca. 1800–1850), the subsequent change in temperature depends not only on the rate of CO_2 increase since t_0, but also on the rate of heating by all the other "external" quantities, both before and after t_0. The actual CO_2-induced temperature change since the industrial revolution, and in the future, is thus seen to depend on the "memory" of the climate system through the climate response function F_c.

The majority of the simulations of CO_2-induced climate change have been performed to determine the gain of the climate system from the equilibrium warming resulting from an abrupt increase in CO_2 (see Schlesinger, 1984a). While these simulations are time dependent, their temperature evolution cannot be used to determine F_c because their treatment of the oceans was simplified for computational economy to enable equilibrium to be attained in as short a time as possible. For example, simulations of CO_2-induced climatic change have been carried out with fixed sea-surface temperatures (and sea ice), with "swamp" ocean models having zero heat capacity and zero heat transport, and with fixed-depth mixed layer models having heat capacity but no sub-mixed layer heat transport. In the last several years, however, simplified climate models with a variety of oceanic treatments have been applied to determine the characteristics of the climate response function. The characteristics and results of selected models are presented in Table 1 and are discussed below.

To fix ideas and introduce terminology we first consider a simplified energy balance climate model with a fixed-depth oceanic mixed layer governed by:

$$\rho c h \frac{\mathrm{d}\Delta T}{\mathrm{d}t} + \frac{\Delta T}{G} = \Delta Q \tag{1.3}$$

where ΔT is the change in temperature induced by ΔQ, and ρ, c and h are the mixed layer density, heat capacity and depth, respectively. For the case of constant ΔQ resulting from an abrupt CO_2 change, the solution of eqn. (1.3) is:

$$\Delta T(t) = \Delta T_{eq}[1 - \exp(-t/\tau_m)] \tag{1.4}$$

where $\Delta T_{eq} = G \Delta Q$ and $\tau_m = \rho c h G$. Thus the climate response function is a "complementary" exponential function characterized by a single parameter, the e-folding time for the thermally isolated mixed layer, τ_m. For the case of an exponential CO_2 increase, $C(t) = C_0 \exp(t/\tau_c)$, which is a reasonable approximation to the increase since 1800–1850, $\Delta Q = Rt/\tau_c$ where R is a "radiative-forcing" parameter (Augustsson and Ramanathan, 1977); in this case the solution of eqn. (1.3) is:

$$\Delta T(t) = \Delta T_{eq}(t)\left[1 - \frac{1 - \exp(-t/\tau_m)}{t/\tau_m}\right] \tag{1.5}$$

with ΔT_{eq} and τ_m defined as above. The time lag L between $\Delta T(t)$ and $\Delta T_{eq}(t)$ can be defined by $\Delta T(t) = \Delta T_{eq}(t - L)$, hence by eqn. (1.5):

$$L(t) = \tau_m[1 - \exp(-t/\tau_m)] \tag{1.6}$$

Thus, for a fixed-depth mixed layer that does not exchange heat with the deeper ocean, the lag L for an exponential CO_2 increase asymptotically approaches τ_m with an e-folding time of τ_m, and is independent of the rate of CO_2 increase, i.e. τ_c. For a global-mean h of 68 m (Manabe and Stouffer, 1980), the asymptotic lag is $L_{asy} \simeq \tau_m = 9$ G yrs. Thus, for $G \sim 0.325\text{--}1.05°C\,(W\,m^{-2})^{-1}$ corresponding to $\Delta T_{eq} \sim 1.3\text{--}4.2°C$ for $\Delta Q = 4$ $W\,m^{-2}$ due to a CO_2 doubling, the e-folding time and asymptotic lag are both about 3–10 yrs.

As shown in Table 1, Hunt and Wells (1979) investigated the climate response function using a radiative-convective model (RCM)* of the climate and a 300 m deep variable-depth mixed layer/thermocline ocean model. The time lag L for exponentially increasing

---

*For a review of RCMs see Ramanathan and Coakley (1978).

TABLE 1

Selected climate model studies of the climate response function $F_c(t)$

Study	Climate model[a] Gain G in °C (W m^{-2})$^{-1}$	Ocean model[b]	CO_2 increase[c]	τ_e, or L after t years (years)
Hunt and Wells (1979)	RCM (unknown)	Variable-depth mixed layer/thermocline model, 300 m total depth, $\kappa = 0$	Exponential with $\tau_c = 81, 104, 147$	$L = 8$ for $t = 56, 72, 102$ corresponding to different τ_c, respectively
Cess and Goldenberg (1981)	EBM (0.455 and 0.794)	Box-diffusion model consisting of a 70 m mixed layer and a semi-infinite, diffusive deep ocean with $\kappa = 1.3$	Exponential with $\tau_c = 104$	$L = 20$ and 28 for $G = 0.455$ and 0.794, respectively, both for $t = 72$
Hansen et al. (1984)	EBM (0.465, 0.698, 0.977)	Box-diffusion model consisting of a 110 m mixed layer and a 900 m diffusive deep ocean with $\kappa = 1$	Switch-on with $r = 2$	$\tau_e = 27, 55$ and 102 for $G = 0.465$, 0.698 and 0.977, respectively
Wigley and Schlesinger (1985)	EBM (arbitrary)	Box-diffusion model consisting of a mixed layer with arbitrary depth and a semi-infinite, diffusive deep ocean with arbitrary κ	Exponential with arbitrary τ_c, switch-on with arbitrary r	$\tau_e \approx 94\gamma^2 \kappa G^2$, $\quad L = L(t, \tau)$ – see text
Bryan et al. (1982) Spelman and Manabe (1984)	GCM (~ 0.7)[d]	Oceanic general circulation model extending from equator to pole over 60° longitude sector with uniform depth of 5000 m	Switch-on with $r = 5$	$\tau_e \simeq 25$

[a] RCM = radiative–convective model; EBM = energy balance model; and GCM = general circulation model. G is predicted in RCMs and GCMs, but is prescribed in EBMs.

[b] κ = effective coefficient of heat transfer to the deeper ocean in cm^2 s^{-1}.

[c] Exponential: $CO_2(t) = CO_2(0) \exp(t/\tau_c)$, τ_c in years. Switch-on: $CO_2(t) = [1 + (r - 1)H(t)](CO_2)_{ref}$ where H is the Heavyside step function.

[d] Estimated from 5.5°C global-mean surface air temperature change and an assumed heating of $\Delta Q = 8$ W m^{-2} for quadrupling the CO_2 concentration.

CO_2 was found to be eight years at the time when the CO_2 concentration reaches twice its initial value, regardless of the rate of CO_2 increase. These results are in general agreement with those above for the thermally isolated mixed layer*, and Hunt and Wells (1979) concluded that "Inclusion of mixed layer ocean in this problem does not significantly alter the results compared with previous, land only studies, although this finding might be modified for the real ocean because of heat transport by its currents".

The effect of the vertical transport of heat from the mixed layer to the deeper ocean was investigated by Cess and Goldenberg (1981) with an energy balance model (EBM)** of the climate and a box-diffusion ocean model. The latter consists of a fixed-depth mixed layer (the box) overlying a semi-infinite deeper ocean in which (vertical) heat transport is parameterized by diffusion with a prescribed constant diffusivity κ. The lag for the exponentially increasing CO_2 concentration at the time of CO_2 doubling was found to increase with G from 20 to 28 yrs for $G = 0.455$ and 0.794, respectively (Table 1). Cess and Goldenberg (1981) attributed their longer lag compared to that of Hunt and Wells (1979) to the sub-mixed layer heat transport.

Another approach to the determination of the climate response function has been used by Hansen et al. (1984) with an EBM/box-diffusion ocean model. These investigators performed a "switch-on" experiment (Manabe, 1983) in which the CO_2 concentration was increased (doubled) abruptly and then held constant. Although the resulting climate response function is not the complementary exponential function given in eqn. (1.5), the response was nevertheless characterized by the "e-folding time" defined as the time τ_e when $\Delta T/\Delta T_{eq} = 1 - e^{-1} \sim 0.63$. Hansen et al. (1984) found that τ_e rapidly increased with G for fixed $\kappa (= 1 \text{ cm}^2 \text{ s}^{-1})$ with $\tau_e = 27, 55$ and 102 yrs for $G = 0.465, 0.698$ and $0.977°C (W m^{-2})^{-1}$, respectively.

Recently Wigley and Schlesinger (1985) have obtained an analytic solution to the equations governing the EBM/box-diffusion ocean model which shows that:

$$\tau_e \simeq 94\gamma^2 \kappa G^2 \tag{1.7}$$

and:

$$L_{asy}(t) \simeq \left[1 - \frac{3/2}{1+\sqrt{t/\tau_e}}\right]\tau_m + \left[\sqrt{t/\tau_e} - 1 + \frac{1}{1+\sqrt{t/\tau_e}}\right]\tau_e \tag{1.8}$$

where γ is a factor that includes the ocean fraction (see Appendix) and t, τ_e and L_{asy} are in years, κ in $cm^2 s^{-1}$, and G in $°C (W m^{-2})^{-1}$. It can be seen that the e-folding time τ_e depends linearly on the heat transfer coefficient κ and quadratically on the gain of the climate system G. The asymptotic lag L_{asy} depends on two terms. The first involving τ_m represents the lag of the thermally isolated mixed layer modified by non-zero κ, while the second term involving τ_e represents the contribution to the lag by the heat transport from the mixed layer to the deeper ocean. It is seen that this lag, unlike that due to τ_m, is a function of time and increases as $\sqrt{t}$, and that $L_{asy} \sim \sqrt{\tau_e}$.

The numerical and analytical results from the EBM/box-diffusion ocean model show

---

*A complete comparison cannot be made since Hunt and Wells (1979) did not publish the heating ΔQ for their CO_2 doubling experiment, thereby precluding a determination of $G = \Delta T/\Delta Q$, where $\Delta T = 1.81°C$ is their CO_2-induced surface temperature change.
**For a review of EBMs see North et al. (1981).

that the climate response function depends strongly on the gain of the climate system and on the vertical transport of heat from the mixed layer to the deeper ocean. But diffusion is a crude parameterization of the many processes that can ultimately contribute to the vertical transport of heat in the real ocean.

A more physically comprehensive simulation of such heat transport processes has been performed with the GFDL* coupled atmosphere–ocean general circulation model (GCM) by Bryan et al. (1982) and by Spelman and Manabe (1984)**. For computational economy in these simulations the geographical domain was restricted to a 120°-longitude sector that extended from equator to pole, with the western half of the sector occupied by land at zero elevation, and the eastern half by ocean with a uniform depth of 5000 m. In these studies the coupled GCM was first integrated asynchronously in time to rapidly attain the equilibrium climates for two simulations, one with a normal (unspecified) CO_2 concentration, and the other with four times that concentration***. The coupled GCM was then integrated synchronously in time for both the normal ($1 \times CO_2$) and quadrupled ($4 \times CO_2$) simulations. The resultant evolution of the difference in the surface air temperature, $4 \times CO_2$ minus $1 \times CO_2$ (from the synchronous simulations), reached 63% of the equilibrium temperature difference (from the asynchronous simulations) in about 12 years over land and 25 years over ocean. This latter value is about half the e-folding time obtained by Hansen et al. (1984) and given by eqn. (1.7) for G estimated as $\sim 0.7°C(Wm^{-2})^{-1}$ $[= 5.5°C(8\,Wm^{-2})^{-1}]$ and $\kappa = 1\,cm^2\,s^{-1}$, and suggests that either the effective value of κ for the coupled GCM is about $0.45\,cm^2\,s^{-1}$ or the dependence of τ_e (and L_{asy}) for the coupled GCM is different from that for the EBM/box-diffusion ocean model.

More recently Bryan et al. (1984) used an uncoupled GCM of the world ocean driven by prescribed fluxes of momentum, heat and water to investigate the transient response to uniform temperature anomalies imposed at the sea surface. Four 50-year anomaly integrations were performed with somewhat distorted physics for computational economy, three with annual-mean forcing for a tracer and temperature anomalies of $\pm 0.5°C$, and one with seasonal forcing for the warm anomaly. It was found that the penetration depths of the tracer and positive anomaly were similar, while that of the negative anomaly was 25% greater. The ocean GCM results were reproduced by both a two-box ocean model and a box-diffusion ocean model. The former model displayed an initial rapid adjustment time scale of five years followed by a slow time scale of 100 years, while the latter model required $\kappa = 3\,cm^2\,s^{-1}$ for the warm anomaly and tracer, and $\kappa = 5\,cm^2\,s^{-1}$ for the cold anomaly. These large values were interpreted as the result of the pycnocline depth of the ocean GCM control integration being almost twice that observed.

The analytical and numerical studies described above, as well as those of Thompson and Schneider (1979), Hoffert et al. (1980), Dickinson (1981), Schneider and Thompson

---

*Geophysical Fluid Dynamics Laboratory/NOAA, Princeton University.
**An error in the code used by Bryan et al. (1982) was found after publication, hence the simulation was repeated, after correcting the error, by Spelman and Manabe (1984).
***The effective time integration periods for the atmosphere, upper ocean and deep ocean were 7.8, 850 and 23,000 years for the normal CO_2 simulation, and 11.8, 1290 and 35,000 years for the quadrupled CO_2 simulation.

(1981), Siegenthaler and Oeschger (1984) and others, have shown that the climate response function depends on the gain of the climate system and the rate at which heat is transported from the ocean surface downward into the interior of the ocean. However, these studies have been performed with models of the climate and ocean in which a variety of simplifications have been made. This includes the studies of Bryan et al. (1982) and Spelman and Manabe (1984) which, although performed with the most comprehensive model to date, the GFDL coupled atmosphere–ocean GCM, excluded two-thirds of the Northern Hemisphere and all of the Southern Hemisphere, did not include the orography of the continents and that of the ocean bottom, and prevented the interaction of clouds. It is likely that these simplifications significantly influence the simulated climate response function through their effects on the oceanic heat transports and climate gain.

An objective of the present study, therefore, is to investigate the characteristics of the climate response function with a global coupled atmosphere–ocean GCM which has realistic geography, realistic land and ocean-bottom topographies, and interactive clouds. A description of this model and the simulations is given in the next section. Selected results are presented in Section 3, and an analysis of the model's climate response function is given in Section 4. A summary and concluding remarks are presented in Section 5.

2. DESCRIPTION OF THE MODEL AND SIMULATIONS

2.1. Component atmosphere and ocean models

The atmospheric component of the coupled model is basically the same as the atmospheric general circulation model described by Schlesinger and Gates (1980, 1981), and documented by Ghan et al. (1982). This is a two-layer primitive equation GCM formulated using normalized pressure (sigma, σ) as the vertical coordinate, with the top at 200 mb and surface orography as resolved by a four degree by five degree latitude–longitude grid. The model predicts the atmospheric velocity (wind), temperature, surface pressure and water vapor, the surface temperature, snow mass, soil water and clouds, and includes both the diurnal and seasonal variations of solar radiation.

A number of relatively minor changes were made in the atmospheric GCM prior to its coupling with the ocean model. These include a slight reduction in cloud albedo (to reduce the erroneously high planetary albedo), a decrease in the emissivity of glaciated clouds (to reduce the excessively warm polar temperatures), and the formation of low-level clouds under stable stratification (to lower the excessive surface flux of solar radiation over the eastern equatorial oceans). Another change made in an effort to reduce what appeared to be excessively large surface winds was the restriction that the surface wind (found by extrapolation from the model's two tropospheric levels at approximately 400 and 800 mb) not exceed the wind at the lower model level. While these changes have in some cases increased the local accuracy of the coupled model, they have not significantly changed the model's overall performance (Gates et al., 1985).

The oceanic component of the coupled model is basically the same as that described by Han (1984a, b). This is a six-layer primitive equation model of the world ocean that includes realistic lateral and bottom topography as resolved by the four degree by five degree latitude–longitude grid. In distinction from the oceanic GCM described by

454

Han (1984a, b), however, the model version used in the present coupled simulation has been extended to include the Arctic Ocean. The model predicts the oceanic velocity (current), temperature and salinity (under the constraint of a prescribed surface concentration), and the formation and melting of sea ice.

2.2. Domain and solution procedure

The vertical structure of the coupled model is shown in Fig. 1 along with the primary quantities predicted during the course of a simulation. Here u and v are the horizontal components of the wind or current, T the temperature, q the atmospheric water vapor mixing ratio, s the oceanic salinity, and $\dot{\sigma}$ and w the atmospheric and oceanic vertical velocities, respectively. The surface boundary condition for the atmosphere is $\dot{\sigma} = 0$ at $\sigma = 1$, while that for the ocean is $w = 0$ at $z = 0$. At the ocean bottom the condition $w = V \cdot \nabla h$ is imposed where V is the horizontal velocity and h is the ocean depth, while $\dot{\sigma} = 0$ at the top of the model atmosphere at $\sigma = 0$ (200 mb). Figure 2 shows the horizontal resolution of the model and the orographies of the continental surface and layers 1–3, 4–5 and 6 of the ocean model. (The orographies of individual layers are shown in Figs. 6 and 7.) The coincidence of the horizontal grid in both the oceanic and atmospheric components of the model is computationally convenient, however, the $4° \times 5°$ grid is too coarse to effectively resolve oceanic boundary currents (Han, 1984a, b).

The coupling of the atmospheric and oceanic models is synchronous, that is, both component models simulate the same period of time. The atmospheric model is integrated forward in time one hour subject to the sea-surface temperature and sea-ice thickness fields predicted by the oceanic model, and the latter is integrated forward in time one hour subject to the net surface heat flux and surface wind stress fields calculated by the atmospheric model. This synchronous exchange of heat and momentum at the sea surface with no acceleration of one model with respect to the other, and no time smoothing or averaging, permits the coupled model's full realization of both diurnal and seasonal forcing.

2.3. Simulations

Two 16-year simulations have been performed with the coupled atmosphere–ocean model that differ only in their prescribed CO_2 concentrations. In the "reference" or "$1 \times CO_2$" simulation the CO_2 concentration was taken to be constant in space and time and equal to 326 ppmv*. In the "experiment" or "$2 \times CO_2$" simulation the concentration was doubled to 652 ppmv. Each simulation was started from the same initial conditions. For the atmospheric component of the model the initial conditions were taken as those on 1 November of year 1 of a 10-year atmospheric GCM integration that was itself initialized from an earlier model simulation. For the oceanic component of the model the initial conditions were taken as those on 1 November of year 9 of an 11-year oceanic

*Preliminary results of the global climate of this simulation have been given by Gates et al. (1985), and a critical evaluation of the model's performance in terms of the simulated air–sea–ice interaction has been presented by Han et al. (1985).

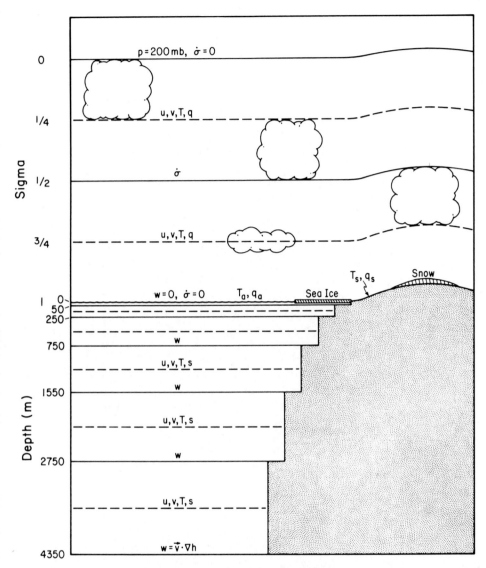

Fig. 1. The vertical structure of the coupled model, together with the primary dependent variables and boundary conditions in the atmosphere and ocean. The atmospheric GCM uses normalized pressure (sigma, $\sigma = (p - p_t)/(p_s - p_t)$, where p is pressure, p_t the 200 mb pressure of the top of the model, and p_s the variable surface pressure) as vertical coordinate and determines the horizontal velocity components u and v, the temperature T, and the water vapor mixing ratio q at two tropospheric levels $\sigma = 1/4$ ($p \sim 396$ mb) and $\sigma = 3/4$ ($p \sim 788$ mb), the vertical velocity $\dot{\sigma}$ at $\sigma = 1/2$ ($p \sim 592$ mb), the temperature T_a and water vapor mixing ratio q_a of the surface air, the temperature T_s of the earth's non-water surfaces, the soil water q_s, the snow mass, and the cloudiness. The z-coordinate oceanic GCM determines the horizontal velocity components u and v, the temperature T and the salinity s at six levels intermediate to those at which the oceanic vertical velocity w is determined. The boundary condition $\dot{\sigma} = 0$ is imposed at the model top and at the earth's surface, while the conditions $w = 0$ and $w = V \cdot \nabla D$ are imposed at the ocean surface and ocean bottom, respectively, where V is the horizontal velocity and D is the ocean depth.

456

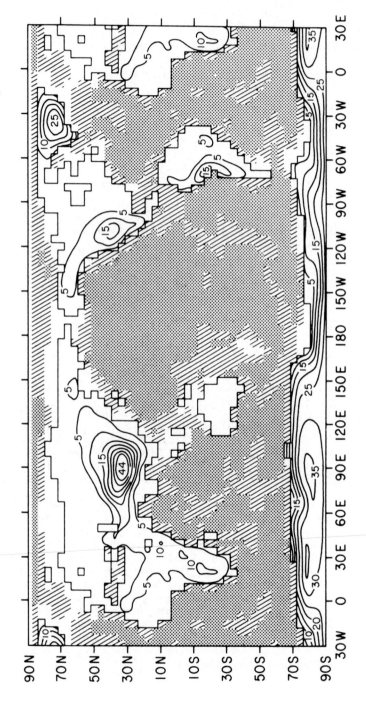

Fig. 2. The global domain of the coupled model, showing the continental outline and orography (in 10^2 m), and the oceanic depth resolved by the model's four degree by five degree latitude–longitude grid. Here the unshaded oceanic area is less than 750 m depth, the hatched area is between 750 and 2750 m depth, and the shaded area is between 2750 and 4350 m depth. (The orographies of the upper four layers of the ocean model are shown in Figs. 6 and 7.)

GCM integration with prescribed monthly atmospheric forcing that was itself initialized from an earlier 40-year simulation with annually averaged atmospheric forcing with initial conditions from the observed ocean climatology of Levitus (1982). In the following we shall identify the starting time for both the $1 \times CO_2$ and $2 \times CO_2$ simulations as 1 November of year 0. In this chronology each 16-year simulation terminates at the end of 31 October of year 16. In the next section we present the temporal and spatial distributions of the differences between the temperatures simulated by the coupled model for the $1 \times CO_2$ and $2 \times CO_2$ concentrations. For convenience, we shall describe these differences as warming when they are positive, and cooling when they are negative.

3. RESULTS

3.1. Evolution of the global-mean temperature change

The evolution of the change in global-mean temperature induced by the doubled CO_2 concentration is shown in Fig. 3 in terms of the vertical distribution of monthly-mean $2 \times CO_2 - 1 \times CO_2$ temperature differences for the atmosphere and ocean. The temperature changes shown have been filtered using a symmetric 12-pole, low-pass recursive filter with a 24-month cutoff period (Kaylor, 1977) to remove the large-amplitude, high-frequency oscillations which occur primarily in the atmosphere, thereby allowing the evolution to be seen more clearly.

The top panel of Fig. 3 shows a rapid and vertically uniform warming of the atmosphere by about 0.4°C during the initial two months of the $2 \times CO_2$ simulation compared with the $1 \times CO_2$ simulation. This warming is similar to, but somewhat larger than, the $2 \times CO_2$-induced warming simulated by the atmospheric model with prescribed monthly varying sea-surface temperature and sea-ice distributions (Gates et al., 1981). Unlike those earlier simulations, however, the present simulations with predicted ocean temperatures and sea ice show a continued atmospheric warming, but with a rate that decreases with time. For example, the initial 0.4°C warming occurs over two months, the second 0.4°C warming (from 0.4° to 0.8°C) requires two years at the surface and one year at the upper level, the third 0.4°C warming (from 0.8° to 1.2°C, shown stippled) requires four years at the surface and five years at the upper level, and the fourth 0.4°C warming (from 1.2° to 1.6°C) requires nine years at the upper level and has not yet occurred at the surface during the last ten years of the simulations.

The bottom panel of Fig. 3 shows that during the initial two months the warming of the ocean is about 0.3°C smaller than the warming of the atmosphere and, unlike the atmosphere, is restricted to the ocean surface layer. However, like the atmosphere, the rate at which the ocean surface warms decreases with time. For example, the initial 0.5°C warming of the upper oceanic layer occurs over 2.5 yrs, and the second 0.5°C warming (from 0.5°C to 1.0°C, shown stippled) occurs over 11 years (from year 2.5 to 13.5). This decrease in the warming rate is due in part to the transport of heat from the upper oceanic layer downward into the deeper ocean. Figure 3 shows that the warming has just penetrated into the fourth ocean layer with depth of 750–1550 m (Fig. 1) by the end of the 16th year.

458

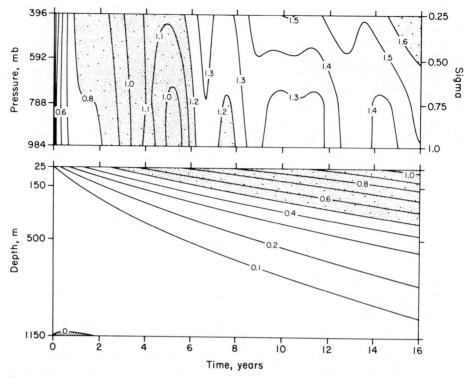

Fig. 3. Time-vertical distribution of the $2 \times CO_2 - 1 \times CO_2$ difference in monthly-mean global-mean temperatures (°C) of the atmosphere (above) and ocean (below). The temperature differences have been filtered using a symmetric 12-pole, low-pass recursive filter with a 24-month cutoff period. Atmospheric warming between $0.8°-1.2°C$ and larger than $1.6°C$ is indicated by light stipple, as is oceanic warming of $0.5°-1.0°C$, while oceanic cooling is shown by heavy stipple.

A comparison of the warming of the atmosphere and upper ocean layer shows that the $0.3°C$ warming differential established by the rapid early warming of the atmosphere is maintained throughout the 16-year period. The post-initial warming of the atmosphere therefore follows the warming of the upper ocean layer.

The temperature changes shown in Fig. 3, when normalized by the equilibrium temperature change, define the climate response function $F_e(t)$ for the global-mean temperature. Clearly, the present 16-year synchronously coupled $1 \times CO_2$ and $2 \times CO_2$ simulations are not of sufficient duration for the equilibrium change to have been attained, a result which was anticipated from the analysis of the EBM/box-diffusion ocean model summarized in Section 1. One way to determine the equilibrium change is to couple the amospheric and oceanic models asynchronously so that simulations extending to equilibrium become computationally possible. As described in Section 1, this is the approach taken by Bryan et al. (1982) and Spelman and Manabe (1984) with the simplified GFDL coupled sector atmosphere–ocean model. We have not taken this approach here. Instead, we shall estimate the equilibrium change using a one-dimensional model whose parameters are determined such that the evolution of its temperature changes matches that of the coupled atmosphere–ocean GCM. However, before presenting this model and its results, it is of interest to examine the global distribution of the temperature changes.

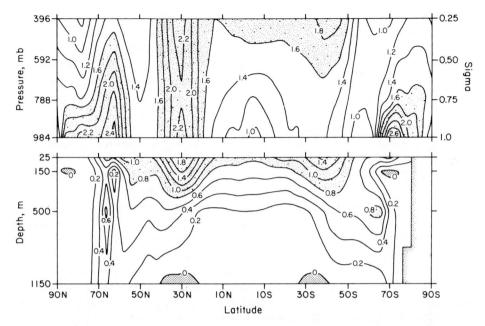

Fig. 4. Latitude-vertical distribution of the $2 \times CO_2 - 1 \times CO_2$ difference in the annual-mean zonal-mean temperatures (°C) of the atmosphere (above) and ocean (below) for year 16 (1 November–31 October). Atmospheric warming larger than 1.6°C is shown by light stipple, as is oceanic warming larger than 0.8°C, while oceanic cooling is indicated by heavy stipple.

3.2. Global distributions of the CO_2-induced temperature change

In this section we present first the latitude-vertical distributions of the zonal-mean temperature differences, and then the geographical distributions of the temperature differences whose global means are displayed in Fig. 3.

3.2.1. Zonal-mean temperature changes

The latitude-vertical distributions of the zonal-mean $2 \times CO_2 - 1 \times CO_2$ temperature differences are shown in Fig. 4 for the atmosphere and ocean in terms of the annual mean over year 16 (1 November–31 October). The zonal-mean temperature change shown for the ocean includes only those longitudes where there is ocean, while that for the atmosphere is the mean over all longitudes.

Figure 4 shows that the annual-mean zonal-mean atmospheric temperatures for year 16 of the $2 \times CO_2$ simulation are everywhere warmer than those of the $1 \times CO_2$ simulation, with a minimum difference of about 1°C and a maximum of 2.6°C. The zonal-mean ocean temperatures for $2 \times CO_2$ are generally warmer than those for $1 \times CO_2$ with a maximum warming of 1.8°C, but regions of slightly colder temperature are found in the 50–250 m ocean layer near the North Pole and Antarctica, and in the 750–1550 m ocean layer near 30 degrees latitude in each hemisphere.

The atmospheric warming at the surface increases from the tropics to the subtropics where temperature differences of 2.2° and 1.5°C are found in the Northern and Southern

Hemispheres, respectively. The warming decreases towards the middle latitudes where minimum temperature changes in the Northern and Southern Hemispheres are about 1.5° and 1.0°C, respectively. The temperature difference increases toward high latitudes with maximum values of 2.4°C located near 65°N and 2.6°C near 70°S. The warming increases with altitude in the tropics and subtropics, and decreases with altitude elsewhere. These results are qualitatively similar to those for the equilibrium $2 \times CO_2$-induced temperature changes simulated by the OSU atmospheric GCM/swamp ocean model (Schlesinger, 1984b), and by other atmospheric GCMs with either a swamp or fixed-depth mixed layer ocean model (see Schlesinger, 1984a; Schlesinger and Mitchell, 1985).

The oceanic surface warming simulated by the coupled atmosphere–ocean GCM increases from about 1°C in the tropics to 1.8°C at 30°N and 1.4°C at 40°S, and decreases towards the mid-latitudes of both hemispheres. This latitudinal distribution of the ocean surface layer temperature difference is similar to that for the surface air temperature difference. It is also very similar to the latitudinal distribution of the excess $\Delta^{14}C$ over the pre-nuclear value that is observed at the sea surface (Broecker et al., 1980). The ocean surface temperature change decreases poleward of about 70 degrees latitude in both hemispheres as does the surface air temperature change. The oceanic warming penetrates to a greater depth in the subtropics and midlatitudes than in the equatorial region. This also is very similar to what is shown by the observed penetration depths of excess ^{14}C (Broecker et al., 1980). The warming decreases with depth everywhere equatorward of 60 degrees latitude in each hemisphere, and extends from the subtropical surface water downward towards high latitudes. Particularly interesting features of the temperature change at depth are the maxima located near 65°N and 60°S in the 250–750 m ocean layer. Poleward of 60 degrees latitude in each hemisphere the warming decreases with depth in the upper ocean and is uniform below. In the next section we present the geographical distributions of the temperature differences whose zonal means are shown in Fig. 4.

3.2.2. Geographical distribution of temperature change

The geographical distributions of the $2 \times CO_2 - 1 \times CO_2$ difference in the annual-mean temperatures for year 16 (1 November – 31 October) are shown in Fig. 5 for the surface air temperature and the Earth's surface temperature, in Fig. 6 for the 0–50 and 50–250 m ocean layer temperatures, and in Fig. 7 for the 250–750 and 750–1550 m ocean layer temperatures.

An examination of Fig. 5 shows, not surprisingly, that the pattern of the $2 \times CO_2 - 1 \times CO_2$ differences in surface air temperature and that of the temperature differences of the Earth's surface are similar. In general, the temperature changes are positive, but regions of small temperature decreases are found in the northwestern United States, central North Atlantic Ocean, northern Brazil, near Antarctica at the dateline and, at the surface in parts of Africa, India, Southeast Asia and Australia. It is likely that these small temperature decreases do not represent statistically significant CO_2-induced changes but are instead manifestations of the natural year-to-year variability of temperature.

Figure 5 also shows that the $2 \times CO_2 - 1 \times CO_2$ temperature differences over most of the world ocean are smaller than the temperature differences over the land. The warming of the surface and the surface air exceeds 2°C over much of the land areas, particularly those outside the tropics, and over the Arctic Ocean and at particular longitudes over the oceans near 30 degrees latitude in each hemisphere and in the high latitudes of the Southern Ocean. Warming in excess of $4°C$ is found in central Europe, the Bering Strait, the Greenland Sea, and along the coast of Antarctica in the Amundsen and Weddell seas. In these high-latitude regions of both hemispheres the warming of the surface exceeds $8°C$, and similar surface air temperature increases are also found in many of the same regions. It is likely that these maximum temperature increases result from the CO_2-induced reduction in the extent of sea ice in these locations.

The changes in the surface temperature of the ocean are shown again in Fig. 6, but with a contour interval half that of Fig. 5, together with the temperature changes in the 50–250 m ocean layer*. It is seen that the $2 \times CO_2 - 1 \times CO_2$ differences in the 0–50 m ocean layer temperature are positive everywhere with the exception of regions of small decreases in the Arctic and North Atlantic oceans, and near the coast of Antarctica. Temperature increases larger than $1.5°C$ are found predominately in the subtropical and mid-latitude oceans, in the eastern Mediterranean, Black, Caspian and Bering seas, and in Hudson Bay. In general, the warming decreases from the 0–50 m ocean layer to the 50–250 m layer as can be seen from a comparison of the upper and lower panels of Fig. 6, and as was already shown in Fig. 4. The regions of maximum temperature difference in the 50–250 m layer off the east coast of Asia and North America, and in the eastern North Pacific Ocean, are located beneath the maxima of the 0–50 m layer. In the Southern Hemisphere the maximum temperature changes in the 50–250 m layer in the southern Indian Ocean are located about five degrees poleward of the maxima in the 0–50 m layer. This results in the poleward displacement of the isopleths with depth in the southern subtropics shown by Fig. 4. It can also be seen in Fig. 6 that the cooling of the 50–250 m layer in polar latitudes found in Fig. 4 is the result of widespread but small cooling in the Arctic Ocean, and larger but more localized cooling in the Amundsen and Bellingshausen seas.

The geographical distribution of the $2 \times CO_2 - 1 \times CO_2$ difference in the annual-mean temperatures for year 16 for the 250–750 and 750–1550 m ocean layers is shown in Fig. 7. In the 250–750 m layer it can be seen that regions of warming in excess of $1°C$ are located off the east coasts of North America and Asia between $30°$ and $60°N$ latitudes, and near the Antarctic coast from $120°E$ to $75°W$ longitudes. The extensive warming in the latter region was already in evidence in the zonal-mean temperature differences presented in Fig. 4. A comparison of the top panel of Fig. 7 with the bottom panel of Fig. 6 shows that the northern hemisphere warming maxima in the 250–750 m layer are located poleward of the regions of maximum warming in the 50–250 m layer. On the other hand, the Antarctic warming maximum in the 250–750 m layer is located about 150 degrees of longitude to the east of the region of maximum warming in the 50–250 m layer, but is located poleward of a region of increased 50–250 m layer temperatures centered near $135°W$ longitude. This poleward displacement of the 250–750 m layer warming centers relative to those of the 50–250 m layer results in the poleward slope of the isopleths with depth shown in the cross section of zonal-mean temperature differences

---

*The temperature differences in high latitudes shown in the lower panel of Fig. 5 and the upper panel of Fig. 6 differ because the latter is solely for the ocean while the former includes the surface temperature of sea ice when it exists.

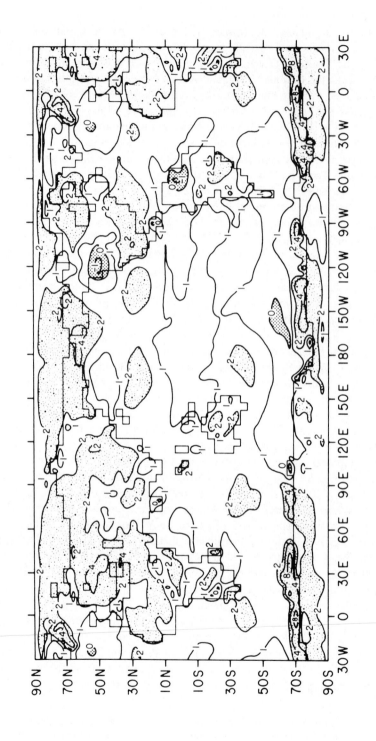

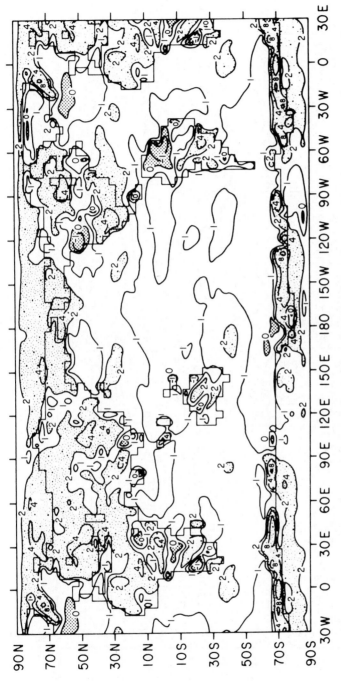

Fig. 5. Geographical distribution of the $2 \times CO_2 - 1 \times CO_2$ difference in the annual-mean temperatures (°C) of the surface air (above) and the earth's surface (below) for year 16 (1 November–31 October). Warming larger than 2°C is indicated by light stipple and cooling by heavy stipple.

464

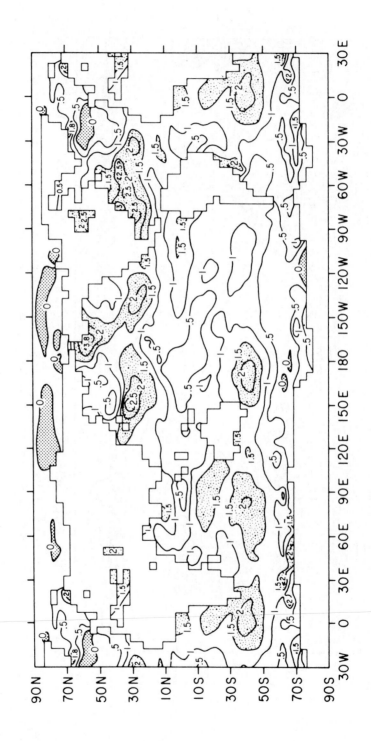

465

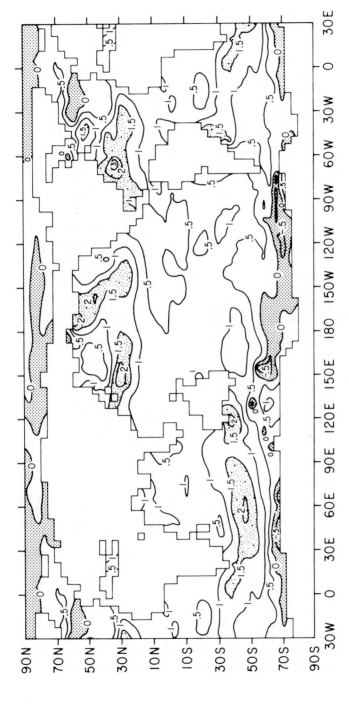

Fig. 6. Geographical distribution of the $2 \times CO_2 - 1 \times CO_2$ difference in the annual-mean temperatures (°C) of the $0-50$ m ocean layer (above) and the $50-250$ m layer (below) for year 16 (1 November– 31 October). Warming larger than $1.5°C$ is shown by light stipple and cooling by heavy stipple.

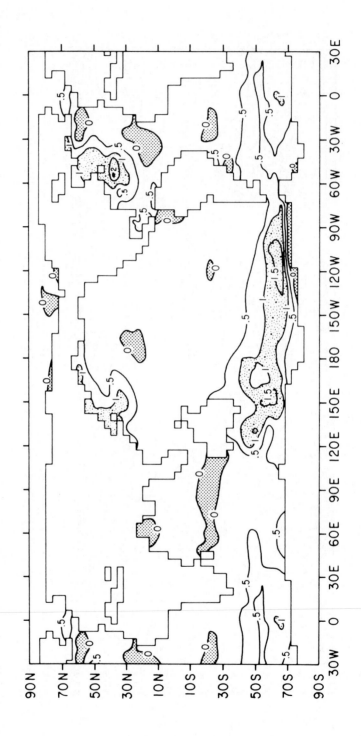

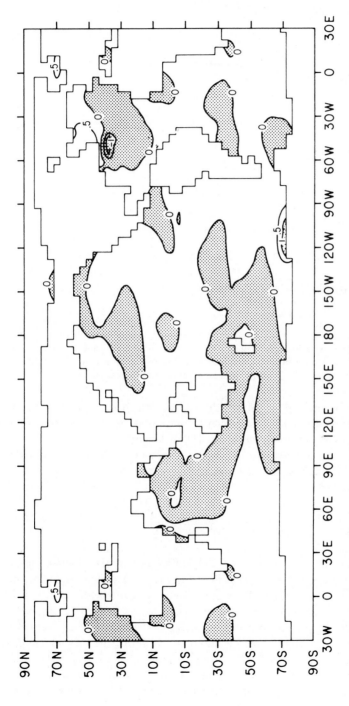

Fig. 7. Geographical distribution of the $2 \times CO_2 - 1 \times CO_2$ difference in the annual-mean temperatures (°C) of the 250—750 m ocean layer (above) and the 750—1550 m layer (below) for year 16 (1 November—31 October). Warming larger than 1°C is indicated by light stipple and cooling by heavy stipple.

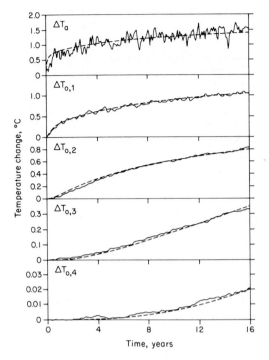

Fig. 8. Energy balance climate/box ocean model representation of the evolution of the $2 \times CO_2 - 1 \times CO_2$ differences in global-mean surface air temperature (ΔT_a) and ocean layer temperature $(\Delta T_{o,k}, \; k = 1, 2, 3, 4;$ dashed line) compared with the evolution of temperature differences simulated by the coupled atmosphere–ocean GCM (solid line) through year 16.

(Fig. 4). The lower panel of Fig. 7 shows that by the 16th year the warming has penetrated downward as far as the 750–1550 m layer off the east coast of North America near 60°N latitude and adjacent to West Antarctica at 120°W longitude.

In the following section we reconsider the time evolution of the global-mean temperatures of the atmosphere and ocean to determine the characteristics of the climate response function.

4. ANALYSIS OF THE CLIMATE RESPONSE FUNCTION

The evolution of the $2 \times CO_2 - 1 \times CO_2$ difference in the global-mean surface air temperature is shown in Fig. 8 (solid line) together with the corresponding differences for the area-mean temperatures of the upper four layers of the ocean model. This figure clearly shows the large-amplitude, high-frequency oscillations in the atmosphere which were filtered for the presentation in Fig. 3. As in that figure, Fig. 8 shows the rapid initial warming of the atmosphere by about 0.7°C during the first nine months, and a subsequent warming of about 0.7°C during the following 15 years. The ocean surface layer also displays two adjustment time scales, an initial rapid warming of about 0.3°C during the first year, and a gradual warming of about 0.7°C during the following 15 years. Similar dual time scales for the thermal adjustment of the atmosphere and upper

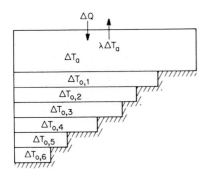

Fig. 9. Schematic representation of the global-mean energy balance climate/box ocean model.

ocean towards equilibrium have been obtained with energy balance climate/box ocean models by Thompson and Schneider (1979) and Dickinson (1981). Figure 8 shows that below the surface layer the adjustment of the ocean temperature towards equilibrium does not exhibit a rapid initial phase. Instead, the deeper ocean warms only slowly by an amount that decreases with depth. After 16 years the 0–50 m surface layer ($\Delta T_{0,1}$) has warmed by slightly more than 1°C, the 50–250 m layer ($\Delta T_{0,2}$) by 0.8°C, the 250–750 m layer ($\Delta T_{0,3}$) by slightly more than 0.3°C, and the 750–1550 m layer ($\Delta T_{0,4}$) by only 0.02°C. A comparison of the warming of the atmosphere and ocean indicates that the former is delayed by the transport of heat from the ocean surface layer progressively downward into the deeper ocean.

The temperature adjustments of Fig. 8, when normalized by the equilibrium temperature change ΔT_{eq}, define the climate response function $F_c(t)$ for the global-mean temperature. To estimate ΔT_{eq} we employ the energy balance climate/box ocean model shown schematically in Fig. 9. Here ΔT_a is the change in the global-mean surface air temperature, and $\Delta T_{0,k}$ the change in the temperature of ocean layer k, induced by a change in thermal forcing ΔQ due to the instantaneous doubling of the CO_2 concentration. The climate feedback parameter $\lambda = G^{-1}$, where G is the gain of the climate system. This model is similar to that employed by Dickinson (1981).

Mathematically the model is defined by:

$$C_a \frac{\mathrm{d}\Delta T_a}{\mathrm{d}t} = \Delta Q - \lambda \Delta T_a - \sigma_a \lambda_{a,0}(\Delta T_a - \Delta T_{0,1}) \tag{4.1}$$

$$C_{0,1} \frac{\mathrm{d}\Delta T_{0,1}}{\mathrm{d}t} = \lambda_{a,0}(\Delta T_a - \Delta T_{0,1}) - \sigma_1 \lambda_1 (\Delta T_{0,1} - \Delta T_{0,2}) \tag{4.2}$$

$$C_{0,k} \frac{\mathrm{d}\Delta T_{0,k}}{\mathrm{d}t} = \lambda_{k-1}(\Delta T_{0,k-1} - \Delta T_{0,k}) - \sigma_k \lambda_k (\Delta T_{0,k} - \Delta T_{0,k+1})$$

$$k = 2, \ldots, K-1 \tag{4.3}$$

$$C_{0,K} \frac{\mathrm{d}\Delta T_{0,K}}{\mathrm{d}t} = (\Delta T_{0,k-1} - \Delta T_{0,K}) \tag{4.4}$$

where C_a is the bulk heat capacity of the atmosphere, $C_{0,k}$ the bulk heat capacity of

TABLE 2

Characteristics of the energy balance climate/box ocean model

Reservoir	Layer thickness (mb for atmosphere, m for ocean)	Heat capacity C_a and $C_{0,k}$ (W m^{-2} yr K^{-1})	Area ratio σ_a and σ_k	λ, $\lambda_{a,0}$ and λ_k (W m^{-2} K^{-1})	κ_k (cm^2 s^{-1})
Atmosphere	784	0.25	0.71	1.39, 8.00	–
Ocean layer $k = 1$	50	6.25	0.97	10.84	3.23
2	200	26.57	0.98	4.60	3.84
3	500	66.44	0.97	0.98	1.52
4	800	106.30	0.93	–	–
5	1200	159.45	0.75	–	–
6	1600	212.60	–	–	–

ocean layer k, σ_a the ratio of the area of the ocean surface to the area of the atmosphere, σ_k the ratio of the area of ocean layer $k + 1$ to that of layer k, λ_{ao} the coefficient of heat transfer between the atmosphere and ocean, and λ_k the coefficient of heat transfer between ocean layers k and $k + 1$. Physically, ΔQ is the constant change in thermal forcing due to doubling CO_2 which perturbs the climate system from its initial conditions:

$$\Delta T_a(0) = \Delta T_{0,k}(0) = 0 \qquad k = 1, \ldots, K \tag{4.5}$$

$\lambda \Delta T_a$ is the adjustment of the system through the emission of longwave radiation to space and any climate system feedbacks towards the equilibrium:

$$\Delta T_{eq} = \frac{\Delta Q}{\lambda} \tag{4.6}$$

$\lambda_{a,0}(\Delta T_a - \Delta T_{0,1})$ is the change in heat transfer between the atmosphere and ocean, and $\lambda_k(\Delta T_{0,k} - \Delta T_{0,k+1})$ is the change in heat transfer between ocean layers k and $k + 1$. The change in heat transfer is assumed to be zero at the base of layer K. The values of the parameters of the model are given in Table 2.

To estimate ΔT_{eq} from eqn. (4.6) requires knowledge of ΔQ and λ. The former has been determined from the $2 \times CO_2 - 1 \times CO_2$ difference of the net radiation at the top of atmosphere at the initial time of the coupled atmosphere–ocean GCM simulations, and is $\Delta Q = 3.92$ W m^{-2}. The value of λ can be determined from eqns. (4.1–4) and the data of Fig. 8 as follows. First, eqn. (4.4) is integrated from the initial condition eqn. (4.5) to time t and solved for λ_{K-1} to yield:

$$\lambda_{K-1}(t) = \frac{C_{0,K}\Delta T_{0,K}(t)}{\int_0^t(\Delta T_{0,K-1} - \Delta T_{0,K})dt} \tag{4.7}$$

Then eqn. (4.3) is integrated and solved for λ_{k-1} sequentially from $k = K - 1$ to $k = 2$ to obtain:

$$\lambda_{k-1}(t) = \frac{C_{0,k}\Delta T_{0,k}(t) + \sigma_k\lambda_k\int_0^t(\Delta T_{0,k} - \Delta T_{0,k+1})dt}{\int_0^t(\Delta T_{0,k-1} - \Delta T_{0,k})dt} \qquad k = K-1, \ldots, 2 \tag{4.8}$$

In the identical manner $\lambda_{a,0}$ is determined from eqn. (4.2) to give:

$$\lambda_{a,0}(t) = \frac{C_{0,1}\Delta T_{0,1}(t) + \sigma_1\lambda_1\int_0^t(\Delta T_{0,1} - \Delta T_{0,2})dt}{\int_0^t(\Delta T_a - \Delta T_{0,1})dt} \tag{4.9}$$

and, finally, λ is obtained similarly from eqn. (4.1) as:

$$\lambda(t) = \frac{C_a\Delta T_a(t) + \sigma_a\lambda_{a,0}\int_0^t(\Delta T_a - \Delta T_{0,1})dt - t\Delta Q}{\int_0^t(-\Delta T_a)dt} \tag{4.10}$$

$K = 4$ was chosen in applying these equations to the coupled model results because the global-mean temperature changes in the fifth (1550–2750 m) and sixth (2750–4350 m) ocean model layers were negligibly small (7×10^{-4} and -1.5×10^{-3}°C, respectively) after only 16 years of simulation.

The λ's given by eqns. (4.7–10) and the data of Fig. 8 are time dependent. However, after an initial large and rapid adjustment, the λ's become quasi-constants. The values of the λ's at the end of year 16 are shown in Table 2. The climate feedback parameter $\lambda = 1.39\,\mathrm{W\,m^{-2}\,K^{-1}}$ corresponds to a climate system gain $G = \lambda^{-1} = 0.72\,\mathrm{K\,(W\,m^{-2})^{-1}}$ and, by eqn. (4.6) with $\Delta Q = 3.92\,\mathrm{W\,m^{-2}}$, $\Delta T_{eq} = 2.82$°C. The latter value lies in the middle of the 1.3–4.2°C range of the global-mean surface air temperature warming simulated by a hierarchy of mathematical climate models (Schlesinger, 1984a; Hansen et al., 1984). The coefficient of heat transfer between the atmosphere and ocean is $\lambda_{a,0} = 8.00\,\mathrm{W\,m^{-2}\,K^{-1}}$, a value which is only about 20% of that given by Dickinson (1981). This smaller value is due in part to the use of the global-mean surface air temperature in the first term of the right-hand side of eqn. (4.2), rather than the temperature averaged only over the oceans, and to the use of the area-averaged ocean surface layer temperature, rather than the area-averaged sea-surface temperature.

The oceanic heat transfer coefficients are shown in Table 2 together with the corresponding diffusivities calculated from:

$$\kappa_k = \tfrac{1}{2}(\Delta_{k+1}z + \Delta_k z)\lambda_k/\rho c \qquad k = 1, 2, 3 \tag{4.11}$$

where $\Delta_k z$ is the thickness of ocean layer k. The representation of the heat transfer by the Fourier law gives a diffusivity that increases from 3.23 cm^2 s^{-1} at 50 m to 3.84 cm^2 s^{-1} at 250 m, and then decreases to 1.52 cm^2 s^{-1} at the 750 m level. This is in contrast to the single value of κ assumed in box-diffusion ocean models. Broecker et al. (1980) determined values of κ required for a box-diffusion model to reproduce the observed penetration of bomb-produced tritium and ^{14}C into the ocean. These authors found that $1.7 \leqslant \kappa \leqslant 3.3$ cm^2 s^{-1} with a best estimate of $\kappa = 2.2$ cm^2 s^{-1}. The maximum and minimum values of κ for the coupled GCM are larger and smaller than the corresponding extreme values of Broecker et al. (1980), but the mass-weighted value of $\kappa = 2.25$ cm^2 s^{-1} is in agreement with the best estimate of Broecker et al. (1980). Furthermore, a box-diffusion climate model (see Appendix) with $\kappa = 2.25$–2.50 cm^2 s^{-1} is successful in reproducing the evolution of the $2 \times CO_2 - 1 \times CO_2$ differences in the surface air and ocean surface temperatures simulated by the coupled GCM. Consequently, it appears that in the global mean the coupled GCM transports heat from the surface downward into the ocean at a rate which is commensurate with the rate observed for the downward transport of tritium and ^{14}C.

To test the validity of the energy balance climate/box ocean model (EBM), eqns. (4.1–4) with the λ's shown in Table 2 were integrated in time from the initial condition

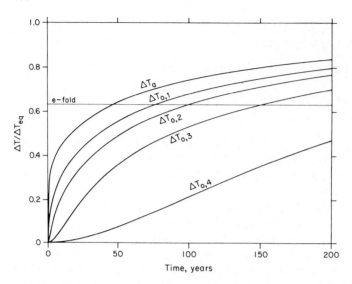

Fig. 10. Energy balance climate/box ocean model projection of the evolution of the $2 \times CO_2 -$ $1 \times CO_2$ difference in global-mean surface air temperature (ΔT_a) and ocean layer temperature ($\Delta T_{0,k}$, $k = 1, 2, 3, 4$) for years 16–200 as a fraction of the equilibrium temperature difference (ΔT_{eq}). The horizontal line labelled e-fold indicates the level at which $\Delta T/\Delta T_{eq} = 1 - 1/e \sim 0.63$.

eqn. (4.5) by the fourth-order Runge-Kutta method with a time step of 0.1 year. A comparison of the results of this integration with those from the coupled GCM are presented in Fig. 8 and shows that the EBM is successful in representing the secular part of the evolution of the $2 \times CO_2 - 1 \times CO_2$ temperature differences of the coupled GCM. Therefore, to estimate the characteristics of the climate response function, the EBM integration was extended to 200 years.

The results of the 200-year projection with the EBM, when normalized by the estimated equilibrium temperature change $\Delta T_{eq} = 2.82°C$, are presented in Fig. 10. This figure shows that the surface air warms to about 30% of the equilibrium value in less than a year. Thereafter, the rate of atmospheric warming greatly decreases and follows that of the ocean surface (0–50 m) layer. While the ocean surface layer begins to warm immediately, the warming of the deeper ocean layers is delayed until the warming above them is transported downward. This slow downward heat transport acts to sequester heat into an increasing volume of the ocean and, thereby, reduces the rate of warming of the upper ocean layers and the atmosphere. As a result, the ocean surface layer reaches the e-folding temperature of about 1.8°C in 75 years while the atmosphere, owing to its initial rapid adjustment, reaches the e-folding point about 25 years earlier.

5. SUMMARY AND CONCLUDING REMARKS

In this study the OSU coupled atmosphere–ocean general circulation model has been used to investigate the role of the ocean in CO_2-induced climate change. The evolution of the simulated $2 \times CO_2 - 1 \times CO_2$ differences in the global-mean surface air and ocean temperatures has been represented by an energy balance climate/box ocean model to determine the characteristics of the climate response function of the coupled GCM. This representation enables the estimation of the GCM's climate gain, the air–sea heat exchange coefficient, the ocean thermal diffusivity profile and the equilibrium temperature change induced by the doubled CO_2 concentration. The estimated equilibrium temperature change of $2.82°C$ is near the mid-point of the values simulated by the hierarchy of climate models (Schlesinger, 1984a), and the mass-averaged ocean thermal diffusivity of $\kappa = 2.25 \text{ cm}^2 \text{ s}^{-1}$ is in agreement with the best estimate based on the value required by a box-diffusion ocean model to reproduce the observed penetration of bomb-produced radionuclides into the ocean (Broecker et al., 1980). In fact a box-diffusion climate model with $\kappa = 2.25\text{–}2.50 \text{ cm}^2 \text{ s}^{-1}$ is successful in reproducing the $2 \times CO_2 - 1 \times CO_2$ differences in surface air and ocean surface layer temperatures simulated by the coupled GCM. These comparisons of the GCM results with other models and with observations suggest that such a model can be useful in discerning the ocean's role in CO_2-induced climate change.

The representation of the GCM results by the energy balance climate/box ocean model was therefore extended to 200 years to estimate the e-folding time of the GCM's response to the doubled CO_2 concentration. This projection shows that the ocean surface layer reaches 63% of its equilibrium response in about 75 years. Interestingly, this is within ten years of the e-folding time $\tau_e = 85$ yrs given by eqn. (1.7) for the EBM/box-diffusion ocean model with $\gamma = 0.88$ (see Appendix). Considering that the e-folding time for an ocean with a mixed layer that does not transport heat downward is about 3–10 years (see Section 1), it is seen that the ocean's role in CO_2-induced climate change is to delay the warming of the ocean mixed layer and atmosphere through the downward transport and sequestering of heat into the ocean's interior. For the case of an instantaneous CO_2 doubling the delay is about 65–70 years for the mixed layer and 40–45 years for the atmosphere.

The CO_2 concentration, however, is not changing by abrupt large increases such as an instantaneous CO_2 doubling, but rather is increasing in a continuous albeit accelerating manner. In this case an e-folding time cannot be defined, however, the climate response function can be characterized by its lag time $L(t)$ defined by $\Delta T(t) = \Delta T_{eq}(t - L)$. Since the EBM/box-diffusion ocean model provides a good representation of the GCM results for the case of CO_2 doubling, we can perhaps use the EBM to estimate what the GCM would simulate for the case of a continuous increase in CO_2. Because the actual increase in CO_2 concentration since the industrial revolution and into the future can be approximated by piecewise exponential increases, we estimate $L(t)$ from eqn. (1.8) which is independent of the initial CO_2 level and the time scale for the CO_2 increase (Wigley and Schlesinger, 1985). Asymptotically $L(t) \rightarrow \sqrt{t\tau_e}$. Consequently, the response of the climate system continues to fall further and further behind its equilibrium warming for the exponentially increasing CO_2. This has direct implications for the detection of a CO_2-induced warming.

The detection and attribution problem is further confounded by the fact that the temperature change depends not only on the rate of CO_2 increase since the beginning of the industrial revolution, but also on the thermal forcing by all other "external" quantities, both before and after that time. Since the e-folding time for abrupt changes in thermal forcing is $\tau_e \sim 50\text{–}75$ yrs, the memory of the climate system is about 3 $\tau_e \sim 150\text{–}225$ yrs. From this viewpoint the climate system may never really be in a state

474

of equilibrium. If so, this is a consequence of the large thermal inertia of the ocean.

While the discussion above has focused on the global-mean characteristics of the climate system, the results of the coupled GCM simulations show that the warming is geographically differentiated both at the Earth's surface and within the ocean. At the surface of the Earth the warming is larger over land than over the ocean. This has implications for the rate at which heat is transported between the ocean and land. An estimate of this rate can be obtained from a generalization of the energy balance model used here (see Appendix) to have separate boxes for the atmosphere over the ocean and over the land.

From the viewpoint of dynamical oceanography it is fundamental to inquire by what pathways and through which physical processes does the simulated ocean general circulation produce the penetration of CO_2-induced warming into the ocean. It is also of potential glaciological interest concerning the stability of the West Antarctic ice sheet to understand how the coupled model produces a considerable warming at depth in the vicinity of the Ross Ice Shelf. These inquiries can be made only through a time-dependent heat budget analysis for the world ocean. Such a study is now in progress.

ACKNOWLEDGEMENTS

We would like to thank Robert L. Mobley and William McKie for their management of the integration of the coupled GCM and their assistance in its analysis, Dean Vickers for his programming assistance with the energy balance model analysis, John Stark for drafting the figures, and Leah Riley for typing the manuscript. We also thank NCAR for the provision of the necessary time on their CRAY-1 computer which was accessed via a remote terminal at the OSU Climatic Research Institute. This research was supported by the National Science Foundation and the U.S. Department of Energy under grant ATM 8205992.

APPENDIX

Energy balance climate/box-diffusion ocean model

The model is described mathematically by :

$$C_a \frac{d\Delta T_a}{dt} = \Delta Q - \lambda \Delta T_a - \sigma_a \lambda_{a0}(\Delta T_a - \Delta T_m) \tag{A.1}$$

$$C_m \frac{d\Delta T_m}{dt} = \lambda_{a0}(\Delta T_a - \Delta T_m) + \frac{C_m}{h}\kappa \left(\frac{\partial \Delta T_0}{\partial z}\right)_{z=0} \tag{A.2}$$

where ΔT_a and ΔT_m are the changes in the temperatures of the atmosphere and mixed layer, with bulk heat capacities C_a and $C_m = \rho ch$, respectively, ΔQ is a change in thermal forcing, λ is the climate feedback parameter, σ_a is the ratio of the area of the ocean to the area of the atmosphere, λ_{a0} is the coefficient of heat transfer between the atmosphere and ocean, h is the depth of the mixed layer, ρ and c are its density and heat capacity per unit volume, κ the thermal diffusivity, and $\Delta T_0(z, t)$ the change in temperature of the deep ocean governed by:

$$\frac{\partial \Delta T_0}{\partial t} = \kappa \frac{\partial^2 \Delta T_0}{\partial z^2} \tag{A.3}$$

in the domain $0 \leqslant z \leqslant D$. Since the relaxation time of the atmosphere $\tau_a = C_a/\lambda \sim 2$ months (see Table 2) is much shorter than the relaxation time of the mixed layer, we assume here that the atmosphere is in equilibrium with the mixed layer. Thus from eqn. (A.1) with $d\Delta T_a/dt = 0$:

$$\Delta T_a = \frac{\Delta Q + \sigma_a \lambda_{a0} \Delta T_m}{\lambda + \sigma_a \lambda_{a0}} \tag{A.4}$$

Substitution of eqn. (A.4) into (A.2) then gives:

$$\gamma C_m \frac{d\Delta T_m}{dt} = \Delta Q - \lambda \Delta C_m + \gamma \frac{C_m}{h} \kappa \left(\frac{\partial \Delta T_0}{\partial z} \right)_{z=0} \tag{A.5}$$

where:

$$\gamma = \sigma_a + \frac{\lambda}{\lambda_{a0}} \tag{A.6}$$

The solution of eqns. (A.5) and (A.3) is obtained numerically. The domain is sub-divided into L layers that are denumerated downward by the index l, with $l = 1$ representing the mixed layer and $l = L$ the bottom layer. Letting ΔT_l, ΔF_l and $\Delta_l z$ denote the change in temperature of layer l, the change in downward heat flux at its base and its thickness, respectively, eqns. (A.5) and (A.3) are written using centered differences in the vertical and backward implicit time integration as:

$$\gamma \rho c \Delta_l z \frac{\Delta T_l^{(n+1)} - \Delta T_l^{(n)}}{\Delta t} = \begin{cases} \Delta Q^{(n)} - \lambda \Delta T_1^{(n+1)} - \Delta F_1^{(n+1)} & l = 1 \\ \Delta F_{l-1}^{(n+1)} - \Delta F_l^{(n+1)} & l = 2, \ldots, L-1 \\ \Delta F_L^{(n+1)} & l = L \end{cases} \tag{A.7}$$

with:

$$\Delta T_l^{(0)} = 0 \qquad l = 1, \ldots, L \tag{A.8}$$

and

$$\Delta F_l^{(n+1)} = \begin{cases} \gamma \rho c \kappa \dfrac{\Delta T_l^{(n+1)} - \Delta T_{l+1}^{(n+1)}}{0.5 (\delta_l \Delta_l z + \Delta_{l+1} z)} & l = 1, \ldots, L-1 \\ 0 & l = L \end{cases} \tag{A.9}$$

where:

$$\delta_l = \begin{cases} 0 & l = 1 \\ 1 & l \neq 1 \end{cases} \tag{A.10}$$

Substituting eqn. (A.9) into (A.7) gives after some manipulation:

$$A_1 \Delta T_1^{(n+1)} + B_1 \Delta T_2^{(n+1)} = D_1$$

$$A_l \Delta T_{l-1}^{(n+1)} + B_l \Delta T_l^{(n+1)} + C_l \Delta T_{l+1}^{(n+1)} = D_l \qquad l = 2, \ldots, L-1 \qquad (A.11)$$

$$A_L \Delta T_{L-1}^{(n+1)} + B_L \Delta T_L^{(n+1)} = D_L$$

TABLE A.1

Coefficients of eqns. (A.8)

Coefficient	Layer	
	$l = 1$	$l = 2, \ldots, L$
A_l	$1 + \dfrac{2\kappa \Delta t}{\Delta_1 z \Delta_2 z} + \dfrac{\lambda \Delta t}{\gamma \rho c \Delta_1 z}$	$-\dfrac{2\kappa \Delta t}{\Delta_l z (\delta_{e-1} \Delta_{l-1} z + \Delta_l z)}$
B_l	$-\dfrac{2\kappa \Delta t}{\Delta z_1 \Delta z_2}$	$1 + \dfrac{2\kappa \Delta t}{\Delta_l z}[(\Delta_l z + \Delta_{l+1} z)^{-1}\delta_{L+1-l} + (\delta_{l-1}\Delta_{l-1} z + \Delta_l z)^{-1}]$
C_l	$-$	$-\dfrac{2\kappa \Delta t}{\Delta_l z (\Delta_l z + \Delta_{l+1} z)}\delta_{L+1-l}$
D_l	$\Delta T_1^{(n)} + \dfrac{\Delta Q^{(n)} \Delta t}{\gamma \rho c \Delta_1 z}$	$\Delta T_l^{(n)}$

where the coefficients are presented in Table A.1. The tridiagonal matrix equations (A.11) with initial condition (A.8) are solved at each time step $n = 0, 1, \ldots$ using the method of Lindzen and Kuo (1969). For the computations reported herein the parameters were $\rho = 1.0 \, \mathrm{g \, cm^{-3}}$, $c = 1.0 \, \mathrm{cal \, g^{-1} \, K^{-1}}$, $\Delta t = 0.1 \, \mathrm{yr}$, $L = 40$, $\Delta_1 z = h = 50 \, \mathrm{m}$, $\Delta_l z = 100 \, \mathrm{m}$ for $l \neq 1$, $\lambda = 1.39 \, \mathrm{W \, m^{-2} \, K^{-1}}$, $\Delta Q^{(n)} = 3.92 \, \mathrm{W \, m^{-2}}$, and $\gamma = 0.71 + (1.39/8.00) = 0.88$ (see Table 2).

REFERENCES

Augustsson, T. and Ramanathan, V., 1977. A radiative–convective model study of the CO_2 climate problem. J. Atmos. Sci., 34: 448–451.

Broecker, W.S., Peng, T.-H. and Engh, R., 1980. Modeling the carbon system. Radiocarbon, 22: 565–598.

Bryan, K., Komro, F.G., Manabe, S. and Spelman, M.J., 1982. Transient climate response to increasing atmospheric carbon dioxide. Science, 215: 56–58.

Bryan, K., Komro, F.G. and Rooth, C., 1984. The ocean's transient response to global surface temperature anomalies. In: J.E. Hansen and T. Takahashi (Editors), Climate Processes and Climate Sensitivity. (Maurice Ewing Series, 5) Am. Geophys. Union, Washington, D.C., 29–38.

Cess, R.D. and Goldenberg, S.D., 1981. The effect of ocean heat capacity upon the global warming due to increasing atmospheric carbon dioxide. J. Geophys. Res., 86: 498–502.

Dickinson, R.E., 1981. Convergence rate and stability of ocean–atmosphere coupling schemes with a zero-dimensional climate model. J. Atmos. Sci., 38: 2112–2120.

Gates, W.L., 1976a. Modeling the ice-age climate. Science, 191: 1138–1144.

Gates, W.L., 1976b. The numerical simulation of ice-age climate with a global general circulation model. J. Atmos. Sci., 33: 1844–1873.

Gates, W.L., Cook, K.H. and Schlesinger, M.E., 1981. Preliminary analysis of experiments on the climatic effects of increased CO_2 with an atmospheric general circulation model and a climatological ocean. J. Geophys. Res., 86: 6385–6393.

Gates, W.L., Han, Y.-J. and Schlesinger, M.E., 1985. The global climate simulated by a coupled atmosphere—ocean general circulation model: Preliminary results. In: J.C.J. Nihoul (Editor), Coupled Ocean—Atmosphere Models. (Elsevier Oceanography Series, 40) Elsevier, Amsterdam, pp. 131—151 (this volume).

Ghan, S.J., Lingaas, J.W., Schlesinger, M.E., Mobley, R.L. and Gates, W.L., 1982. A documentation of the OSU two-level atmospheric general circulation model. Report No. 35, Climatic Research Institute, Oregon State University, Corvallis, Oreg., 395 pp.

Han, Y.-J., 1984a. A numerical world ocean general circulation model, Part I. Basic design and barotropic experiment. Dyn. Atmos. Oceans, 8: 107—140.

Han, Y.-J., 1984b. A numerical world ocean general circulation model, Part II. A baroclinic experiment. Dyn. Atmos. Oceans, 8: 141—172.

Han, Y.-J., Schlesinger, M.E. and Gates, W.L., 1985. An analysis of the air—sea—ice interaction simulated by the OSU-coupled atmosphere—ocean general circulation model. In: J.C.J. Nihoul (Editor), Coupled Ocean—Atmosphere Models. (Elsevier Oceanography Series, 40) Elsevier, Amsterdam, pp. 167—182 (this volume).

Hansen, J., Lacis, A., Rind, D., Russell, G., Stone, P., Fung, I., Ruedy, R. and Lerner, J., 1984. Climate sensitivity: Analysis of feedback mechanisms. In: J.E. Hansen and T. Takahashi (Editors), Climate Processes and Climate Sensitivity. (Maurice Ewing Series, 5) Am. Geophys. Union, Washington, D.C., pp. 130—163.

Hoffert, M.I., Callegari, A.J. and Hsieh, C.-T., 1980. The role of deep sea storage in the secular response to climate forcing. J. Geophys. Res., 85: 6667—6679.

Hunt, B.G. and Wells, N.C., 1979. An assessment of the possible future climate impact of carbon dioxide increases based on a coupled one-dimensional atmospheric—oceanic model. J. Geophys. Res., 84: 787—791.

Imbrie, J. and Imbrie, K.P., 1979. Ice Ages, Solving the Mystery. Enslow, Short Hills, N.J., 224 pp.

Kaylor, R.E., 1977. Filtering and decimation of digital time series. Inst. Phys. Sci. Technol., Technical Note BN 850, University of Maryland, College Park, Md., 42 pp.

Keeling, C.E., Bacastow, R.B. and Whorf, T.P., 1982. Measurements of the concentration of carbon dioxide at Mauna Loa Observatory, Hawaii. In: W.C. Clark (Editor), Carbon Dioxide Review: 1982. Oxford University Press, London, pp. 377—385.

Levitus, S., 1982. Climatological Atlas of the World Ocean, NOAA Prof. Pap. No. 13, U.S. Govt. Printing Office, Washington, D.C., 173 pp.

Lindzen, R.S. and Kuo, H.-L., 1969. A reliable method for the numerical integration of a large class of ordinary and partial differential equations. Mon. Weather Rev., 97: 732—734.

Manabe, S., 1983. CO_2 and climatic change. In: B. Saltzman (Editor), Advances in Geophysics, 25. Academic Press, New York, N.Y., pp. 39—82.

Manabe, S. and Stouffer, R.J., 1980. Sensitivity of a global climate model to an increase of CO_2 concentration in the atmosphere. J. Geophys. Res., 85: 5529—5554.

Nordhaus, W.D. and Yohe, G.W., 1983. Future paths of energy and carbon dioxide emissions. In: Changing Climate. National Academy of Sciences, Washington, D.C., pp. 87—153.

North, G.R., Cahalan, R.F. and Coakley, J.A., 1981. Energy balance climate models. Rev. Geophys. Space Phys., 19: 91—121.

Ramanathan, V. and Coakley, J.A., 1978. Climate modeling through radiative—convective models. Rev. Geophys. Space Phys., 16: 465—489.

Rotty, R.M., 1983. Distribution of and changes in industrial carbon dioxide production. J. Geophys. Res., 88: 1301—1308.

Schlesinger, M.E., 1984a. Climate model simulations of CO_2-induced climatic change. In: B. Saltzman (Editor), Advances in Geophysics, 26. Academic Press, New York, N.Y., pp. 141—235.

Schlesinger, M.E., 1984b. Mathematical modeling and simulation of climate and climate change. Contribution no. 41, Institut D'Astronomie et de Géophysique George Lemaitre, Université Catholique de Louvain, Louvain-la-Neuve, 87 pp.

Schlesinger, M.E. and Gates, W.L. 1980. The January and July performance of the OSU two-level atmospheric general circulation model. J. Atmos. Sci., 37: 1914—1943.

Schlesinger, M.E. and Gates, W.L., 1981. Preliminary analysis of the mean annual cycle and interannual variability simulated by the OSU two-level atmospheric general circulation model. Report No. 23, Climatic Research Institute, Oregon State University, Corvallis, Oreg., 47 pp.

Schlesinger, M.E. and Mitchell, J.F.B., 1985. Model projections of equilibrium climate response to increased CO_2. In: M.C. MacCracken and F.M. Luther (Editors), U.S. Department of Energy (in prep.).

Schneider, S.H. and Thompson, S.L., 1981. Atmospheric CO_2 and climate: Importance of the transient response. J. Geophys. Res., 86: 3135–3147.

Siegenthaler, U. and Oeschger, H., 1984. Transient temperature changes due to increasing CO_2 using simple models. Ann. Glaciol., 5: 153–159.

Spelman, M.J. and Manabe, S., 1984. Influence of oceanic heat transport upon the sensitivity of a model climate. J. Geophys. Res., 89: 571–586.

Thompson, S.L. and Schneider, S.H., 1979. A seasonal zonal energy balance climate model with an interactive lower layer. J. Geophys. Res., 84: 2401–2414.

Wigley, T.M.L. and Schlesinger, M.E., 1985. The response of global mean temperature to changing carbon dioxide levels. Nature (submitted).

WMO, 1983. Report of the WMO (CAS) Meeting of Experts on the CO_2 Concentrations from Pre-Industrial Times to I.G.Y. World Climate Programme, WCP-53, WMO/ICSU, Geneva, 34 pp.

CHAPTER 30

ATMOSPHERIC RESPONSE OF A GENERAL CIRCULATION MODEL FORCED BY A SEA-SURFACE TEMPERATURE DISTRIBUTION ANALOGOUS TO THE WINTER 1982–83 EL NIÑO

YVES TOURRE, MICHEL DÉQUÉ and JEAN FRANÇOIS ROYER

ABSTRACT

The response of a low-resolution spectral general circulation model (SGCM) to a large winter equatorial sea-surface temperature anomaly (SSTA) in the Pacific is presented. The response is also compared to the observation during winter 1982–83 mapped by the Climate Analysis Center (CAC). The statistical significance of the local response (Student t-test) for parameters such as zonal wind, vertical velocity, precipitation rate is also shown. The SGCM demonstrates skill in simulating the atmospheric response to a SSTA over the Pacific area and Equatorial regions during the same winter season.

1. INTRODUCTION

A warm southward current flows along the Ecuador–Peru coast (Wyrtki, 1975) near the end of each year. Every few years, the warming of the surface waters is not only stronger than usual (Wyrtki, 1979) and confined to the South American coastal regions, but extend westward along the equator towards the dateline. This is a typical "El Niño" phenomenon (Ramage, 1975; Weare et al., 1976). It is well-documented that "El Niño" is also associated to characteristic interannual variations of the surface pressure (Berlage, 1957; Troup, 1965; Bjerknes, 1969) and is now referred to as the El Niño/Southern Oscillation (ENSO) phenomenon. During the second stage of a typical ENSO phenomenon (Rasmusson and Carpenter, 1982), the Sea-Surface Temperature Anomalies (SSTA) expand westward. This was not the case during the late 1982–83 ENSO. The SSTA actually started in the western and central Pacific during mid-82, so that by December 1982, the SSTA exceeded 4°C over a vast area of the Equatorial Pacific. It was during that time of the year one of the largest positive anomalies ever observed over that region (Quiroz, 1983). Consequently, it was decided to perform a sensitivity experiment using the Spectral Global Circulation Model (SGCM) of the French Meteorological Service, forced by a large SSTA of approximately the same magnitude as the one observed during winter 1982–83, and analogous to the one used by Shukla and Wallace (1983; Fig. 1).

2. METHOD AND DATA

For the control experiment, the SST monthly climatology from Alexander and Mobley (1976) is used.

The SGCM for these numerical experiments (validation and anomaly), which is

480

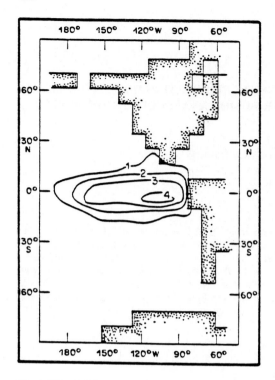

Fig. 1. The El Niño sea-surface temperature anomaly (°C) used in this simulation.

described in Royer et al. (1983) and in Déqué and Royer (1984), uses primitive equations of atmospheric dynamics. The fields are represented by spherical harmonics (Rochas, 1983), up to wavenumber 10 for the longitudinal modes and wavenumber 13 for the zonal modes. The associated global grid has 20 latitude circles and 32 points equidistant in longitude. The horizontal resolution is thus quite low, but sufficient to simulate fairly well the mean features of the atmospheric circulation. The vertical discretization uses ten levels. The main physical processes are parameterized: radiation transfer with diurnal and annual variation of insolation, interactive cloudiness, large-scale and convective precipitation, boundary layer representation, snow cover, and soil water content budget. The time step for the numerical integration is 60 minutes.

The control (validation) experiment is a simulation of five successive annual cycles. The "15 October" situation of the second year of this simulation was used to start the simulation with the anomalous SST boundary condition.

In order to assess the statistical significance of the model response to the SSTA we have calculated at each grid point the signal to noise ratio (difference perturbation minus control over standard deviation computed from ten simulated years of unperturbed climate). Absolute values of this ratio greater than 2.4 or 3.4 will be shown by hatching and double hatching, respectively, the sign of the response being indicated by the direction of the hatching. According to the methodology proposed by Chervin and Schneider (1976), these areas can be interpreted as significant at the 95% (resp. 99%) level for a local Student t-test. Such a collection of univariate tests cannot be used to

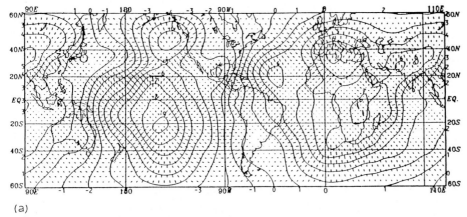

(a)

Fig. 2. a. Mean velocity potential at 200 mb during winter 1982–83 from CAC ($\times$ 10^6 m^2 s^{-1}); contour interval 10^6 m^2 s^{-1}. (*Continued on next page.*)

test the global response of the model (Storch, 1982). For that we need multivariate tests (Hannoschöck and Frankignoul, 1984). However, the Chervin and Schneider methodology is useful to study the local response of the model and has been commonly used to deal with such problems (e.g., Blackmon et al., 1983).

Following Helmholtz's theorem we partition the horizontal wind vector into its divergent and rotational components, and express them in terms of a velocity potential χ and streamfunction ψ according to the relation:

$$V = \nabla X + k \wedge \nabla \Psi$$

where k is vertical unit vector and Λ denotes vector products.

Such decomposition is motivated by the fact that the divergent and rotational parts (being respectively associated with energy generation by baroclinic and barotropic processes) play different roles in atmospheric dynamics. Theoretical studies have shown that available potential energy is first converted into kinetic energy of divergent flow (Chen and Wiin-Nielsen, 1976; Chen, 1980).

In this paper we compare the isopleths of several model-computed parameters (anomalies and means) to observed anomalies and means during the winter 1982–83 obtained from the Climate Analysis Center (CAC) in Washington D.C. They are respectively: (1) velocity potential at 200 mb; (2) stream function at 200 and 850 mb (and zonal wind); and (3) precipitation rate (SGCM) and OLR anomaly (CAC) (and vertical velocity at 500 mb).

3. RESULTS

200-mb velocity potential

We shall first look at the velocity potential at 200 mb which is an efficient parameter to investigate the atmospheric circulation in the tropics (Krishnamurti et al., 1973).

D8—D4   DJF

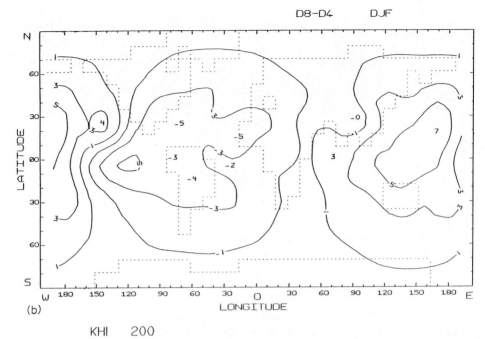

KHI   200

Fig. 2 (*continued*). b. Mean velocity potential at 200 mb during winter 1982–83 from SGCM ($\times$ 10^6 m^2 s^{-1}); contour interval 10^6 m^2 s^{-1}.

D8   DJF

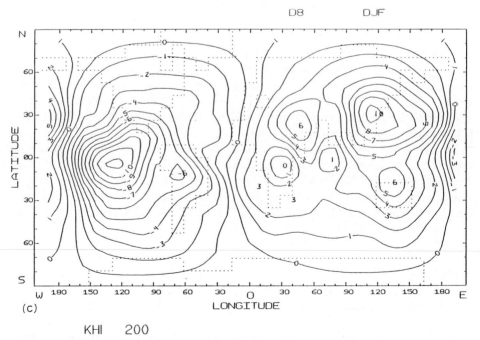

KHI   200

Fig. 2 (*continued*). c. Velocity potential anomalies from the SGCM ($\times$ 10^6 m^2 s^{-1}) during winter; contour interval 2 $\times$ 10^6 m^2 s^{-1}.

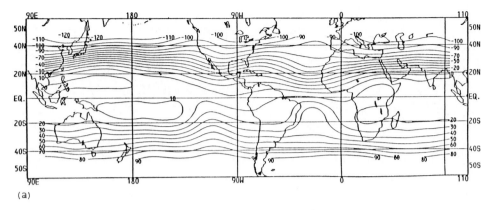

Fig. 3. a. Mean streamfunction at 200 mb during winter 1982–83 from CAC ($\times 10^6$ m² s^{-1}); contour interval 10×10^6 m² s^{-1}. (*Continued on next page.*)

The observed and simulated fields for winter 1982–83 (Fig. 2a, b) show a similar structure with a main center of divergence south of the equator in the Central Pacific and several centers of convergence mainly over eastern Asia, northern and Equatorial Africa. The shortcomings of the model simulations are the absence of a secondary divergence zone in the middle latitudes of the North Pacific, the absence of a convergence zone in the tropical North Atlantic and the presence of a convergence zone above North Australia. However, the response of the model to the equatorial SSTA (Fig. 2c) shows a simple pattern with increased divergence in the eastern Equatorial Pacific and increased convergence in the western Pacific. The anomalous divergent wind field produced in the model by SSTA at 200 mb is thus essentially an increased east–west circulation along the Equatorial Pacific and a smaller increase of westerly circulation over the Equatorial Atlantic, Africa and the Indian Ocean. In the eastern Pacific we can see on each side of the equator substantial zonal gradient which reflects an increased poleward component of the flow. Another significant feature of this figure is the extension of the divergence zones (low-velocity potential values) from the main center just above the anomaly in the northeast direction in the Northern Hemisphere. One can emphasize that the simulated response of this field is nearly symmetric in the two hemispheres in spite of their very different circulation due to differences in season and orography.

As a direct consequence of the continuity equation the regions of divergence at 200 mb are generally associated with rising motion in the middle troposphere and convergence in the lower troposphere.

In the velocity potential field 850 mb (not shown) we find just south of the equator at 150°W a maximum of convergence which can be unambiguously associated with the rising branch of the Walker circulation. Compared with its normal position this branch is displaced eastward in accordance with the eastward migration of the SST maximum during the winter 1982–83 (Philander and Rasmusson, 1984).

The response of the velocity potential field at 850 mb in the model is very similar, but with opposite sign, to the 200 mb response and show the same phenomena which can be further interpreted by looking at the anomalous vertical velocity field (see Fig. 6b).

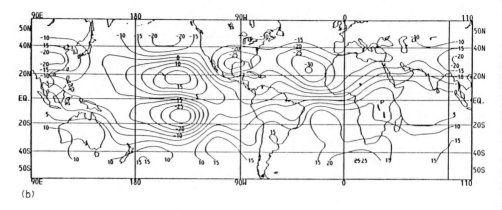

(b)

Fig. 3 (*continued*). b. Streamfunction anomalies at 200 mb during winter 1982–83 from CAC ($\times 10^6$ m^2 s^{-1}); contour interval 5×10^6 m^2 s^{-1}.

D8–D4   DJF

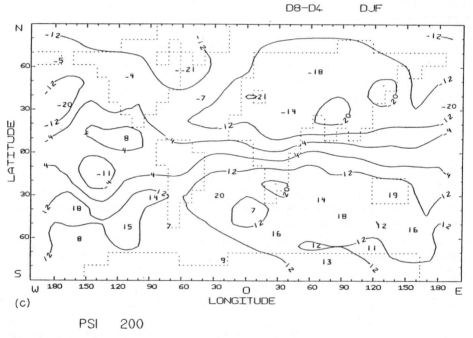

(c)

PSI   200

Fig. 3 (*continued*). c. Streamfunction anomalies at 200 mb from SGCM ($\times 10^6$ m^2 s^{-1}) during winter 1982–83; contour interval 4×10^6 m^2 s^{-1}.

Streamfunction at 200 and 850 mb

To complete the picture it is necessary to consider as well the response of the rotational part of the wind.

In Fig. 3a we display the mean streamfunction at 200 mb during the winter 1982–83. Figure 3b shows the anomaly of this field relative to the normal climatology and Fig. 3c the response of this field to the SSTA simulated by the model.

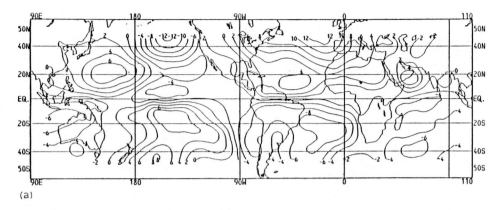

(a)

Fig. 4. a. Streamfunction anomalies at 850 mb during winter 1982–83 from CAC ($\times 10^6$ m^2 s^{-1}); contour interval 2×10^6 m^2 s^{-1}.

D8–D4   DJF

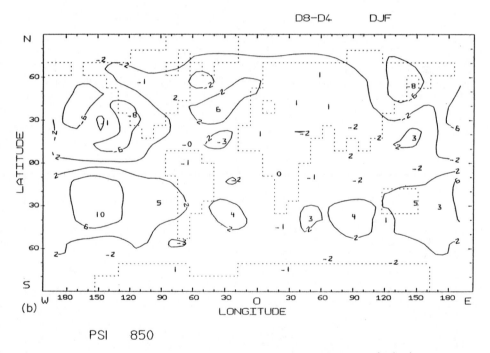

(b)

PSI   850

Fig. 4 (*continued*). b. Streamfunction anomalies at 850 mb from SGCM ($\times 10^6$ m^2 s^{-1}) during winter 1982–83; contour interval 4×10^6 m^2 s^{-1}.

The structure of the anomaly consists of an anticyclonic dipole straddling the equator in the Central Pacific and cyclonic circulations at higher latitudes (near 45°N and 45°S).

Over the Atlantic the anomaly is mainly a mirrored cyclonic dipole centered near 30°. The model appears quite successful in reproducing this structure although it generally underestimates somewhat the magnitude of the response.

In the lower troposphere (Fig. 4a) the anomaly shows two cyclonic cells in the Central

486

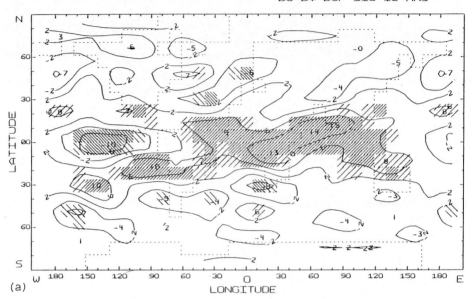

(a)

Fig. 5. a. SGCM zonal wind anomalies at 200 mb (m s^{-1}) during winter. Dense hatching corresponds to 99% significance level ($/\!/ = > 0$; $\backslash\!\backslash = < 0$). Light hatching corresponds to the 95% significance level.

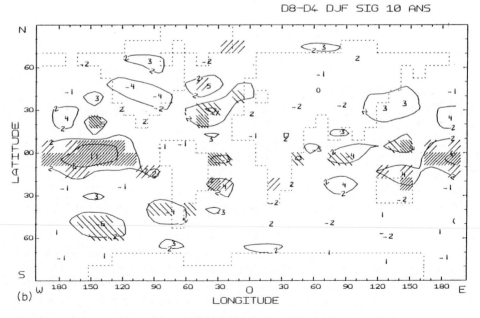

(b)

Fig. 5 (*continued*). b. SGCM zonal wind anomalies at 850 mb (m s^{-1}) during winter. Same statistical and sign conventions as Fig. 5a.

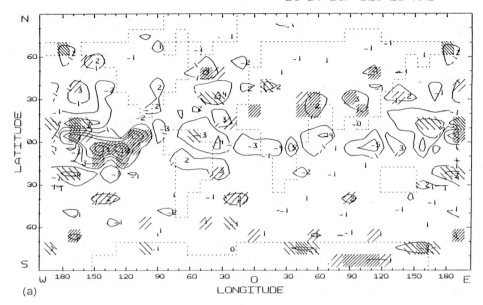

Fig. 6. a. SGCM precipitation anomalies (mm per day) during winter. Same statistical and sign conventions as Fig. 5.

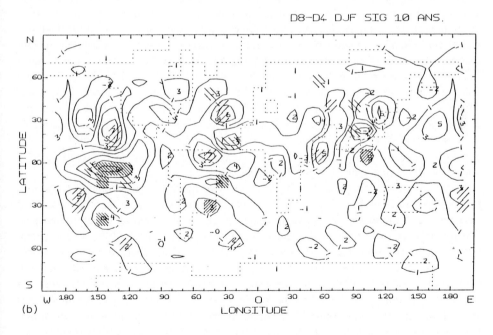

Fig. 6 (*continued*). b. SGCM vertical velocity anomalies cpa s^{-1}) during winter. Same statistical and sign conventions as Fig. 5.

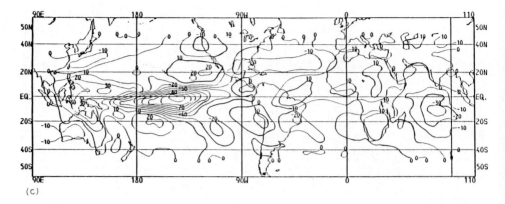

(c)

Fig. 6 (*continued*). c. Outgoing longwave radiation anomalies (w m^{-2}) during winter 1982–83 from CAC.

Pacific extending to higher latitudes and several anticyclonic cells extending from the Atlantic to the eastern Pacific. The model gives only a good simulation of the Pacific cyclonic cells, the simulated anticyclonic cells being of too small amplitude and insufficient spatial extent (Fig. 4b).

Zonal wind at 200 and 850 mb

The anomaly fields of the streamfunction show stronger gradients in the meridional direction. The anomalous flow will therefore appear especially prominent for the zonal component of the wind. In Fig. 5a and b we have plotted the response to the zonal wind U at 200 and 850 mb simulated by the model. In the lower troposphere we see that the two cyclonic cells in the Central Pacific are associated to a significant increase of the westerly component of the wind near the equator on the western side of the SSTA. The anticyclonic cells, even though not so well defined in the model than in the observed fields, lead nonetheless to small increase of the easterly component along the equator over the eastern Pacific and the other oceans. The equatorial response at 200 mb is opposite to the one in the lower troposphere (Fig. 5b). There is an increase of the easterly component just above the SSTA (from 90° to 180°W) and an increase in the westerly component on the remaining part of the equatorial belt.

In the Central Pacific (near 150°W) there are two maxima of increased westerly component near 20°N and 30°S which can be readily explained by an increased poleward transport of absolute angular momentum by the anomalous thermal Hadley circulation that develops on both sides of the SSTA. As a result of the increase of the westerly component of the wind on large equatorial areas there is a 13.5% increase in the total relative angular momentum of the simulated atmosphere. This result is consistent with the conclusion of Stefanick (1982) who showed a strong similarity between the observed temporal variations of the length of day (as an index of atmospheric angular momentum) and a Southern Oscillation Index. It was then inferred that the fluctuations of equatorial temperatures associated with ENSO were the cause of the fluctuations of the zonal wind.

Precipitation, vertical velocity (model) and OLR (observation) anomalies

The alternation of maxima and minima of precipitation and vertical velocity at 500 mb over the Pacific Ocean is not only well correlated with the observed Outgoing Long-wave Radiation OLR pattern, but the level of significance is quite high (Fig. 6). Basically there are large zones of upward motion over the SSTA. These zones also extend poleward and eastward in a Y-shaped fashion. One branch of the Y could be associated to the northeastward displacement of the South Pacific Convergence Zone (Heddinghaus and Krueger, 1981).

Another significant pattern is the rainfall deficit (subsident anomalies) over north-east Brazil due to the non-penetration of the ITCZ over that region (recall the easterly anomaly in the zonal wind at 850 mb; Tourre and Rasmusson, 1984). This is also the area where there is an actual increase in OLR of $10 \, \text{W m}^{-2}$.

CONCLUSION

The response of the model to the SSTA can be qualitatively interpreted as an east-ward displacement of the east—west overturning. There is a simultaneous increase of the Hadley circulation in the Central Pacific near 150°W. Westerlies do not dominate the subtropics of the Northern Hemisphere, but are concentrated over a narrower equatorial band, with a disruption of the classical three-wave structure of the subtropical jet-stream (Krishnamurti et al., 1973).

There is a fairly good response over the Pacific area and equatorial regions. This confirms the fact that, by and large, the cause of the equatorial circulation can be connected to ENSO. The response pattern is quite similar to the one given by simple linear models (Webster, 1981).

Introduction of other simultaneous/lagged SSTA, which will certainly have other local effects, should improve the model output results. This will be the main ingredient of our future work.

ACKNOWLEDGEMENTS

The authors would like to thank the CAC and Drs. Rasmusson and Arkin in particular for providing to us the necessary climatological data. The attendance by one of the authors at the Liège colloquium was supported by ORSTOM.

This work was supported in part by PNEDC, and by EEC (contract No. STi-006-JC-CD).

REFERENCES

Alexander, R.C. and Mobley, R.L., 1976. Monthly average sea surface temperature and ice pack limits on a 1° global grid. Mon. Weather Rev., 104: 143—148.
Berlage, H.P., 1957. Fluctuations in the general atmospheric circulation of more than one year, their nature and prognostic values. K. Ned. Meteorol. Inst. Meded. Verh., 69, 152 pp.

490

Bjerknes, J., 1969. Atmospheric teleconnections from the equatorial Pacific. Mon. Weather Rev., 97: 163–172.

Blackmon, M.L., Geeisler, J.E. and Pitcher, E.J., 1983. A general circulation model of January climate anomaly patterns associated with interannual variation of equatorial Pacific Sea surface temperatures. J. Atmos. Sci., 40: 1410–1425.

Chen, T.C., 1980. On the energy exchange between the divergent and rotational components of atmospheric flow over the tropics and subtropics at 200 mb during two Northern summers. Mon. Weather Rev., 108: 896–912.

Chen, T.C. and Wiin-Nielsen, A.C.; 1976. On the kinetic energy of the divergent and non-divergent flow in the atmosphere. Tellus, 28: 486–497.

Chervin, R.M. and Schneider, S.H., 1976. On determining the statistical significance of climate experiments with general circulation models. J. Atmos. Sci., 33: 391–404.

Déqué, M. and Royer, J.F., 1984. Présentation et validation du modèle climat de l'EERM. La Météorologie, in press.

Hannoschöck, G. and Frankignoul, C., 1984. Multivariate statistical analysis of a sea surface temperature anomaly experiment with the GISS general circulation model. J. Atmos. Sci., in press.

Heddinghaus, T.R. and Krueger, A.F., 1981 Annnual and interannual variations in outgoing longwave radiation over the tropics. Mon. Weather Rev., 109: 1208–1218.

Krishnamurti, T.N., Kanamitsu, M., Koss, W.J. and Lee, J.D., 1973. Tropical East–West circulations during the Northern winter J. Atmos. Sci., 30: 780–787.

Philander, G. and Rasmusson, E.M., 1984. On the evolution of El Niño. Trop. Ocean Atmosphere Newslett., 24, p. 16.

Quiroz, R., 1983. The climate of the "El Niño" winter of 1982–83. A season of extraordinary climatic anomalies. Mon. Weather Rev., 111: 1685–1706.

Ramage, C.S., 1976. Preliminary discussion of the meteorology of the 1972–73 El Niño. Bull. Am. Meteorol. Soc., 56: 234–242.

Rasmusson, E.M. and Carpenter, T.H., 1982. Variations in tropical sea surface temperature and surface wind fields associated with the Southern Oscillation/El Niño. Mon. Weather Rev., 110: 354–384.

Rochas, M., 1983. A note of the computation of grid point values of the wind components in spectral models. Mon. Weather Rev., 111, p. 1707.

Royer, J.F., Déqué, M. and Pestiaux, P., 1983. Orbital forcing the inception of the Laurentide ice sheet. Nature, 304: 43–46.

Shukla, J. and Wallace, J.M., 1983. Numerical simulation of the atmospheric response of equatorial Pacific sea surface temperature anomalies. J. Atmos. Sci., 40: 1613–1630.

Stefanick, M., 1982. Interannual atmospheric angular momentum variability. 1963–1973 and the Southern Oscillation. J. Geophys. Res., 87: 428–432.

Tourre, Y.M. and Rasmusson, E.M., 1984. The tropical atlantic region during the 1982–83 Equatorial Pacific warm event. TOAN., 25: 1–2.

Troup, A.J., 1965. The Southern Oscillation. Q. J. R. Meteorol. Soc., 91: 490–506.

Von Storch, H., 1982. A remark on Chervin-Schmeider's algorithm to test significance of climate experiments with GCMs. J. Atmos. Sci., 39: 1877–1890.

Weare, B.C., Navato, A.R. and Newell, R.E., 1976. Empirical orthogonal analysis of Pacific sea surface temperature anomalies. J. Phys. Oceanogr., 6: 671–678.

Webster, P.S., 1981. Mechanisms determining the atmospheric response to sea surface temperature anomalies. J. Atmos. Sci., 38: 554–571.

Wyrtki, K., 1975. El Niño — The dynamic response of the equatorial Pacific Ocean to atmospheric forcing. J. Phys. Oceanogr., 5: 572–584.

Wyrtki, K., 1979. El Niño. La Recherche, 10: 1212–1220.

CHAPTER 31

THE RESPONSE OF AN EQUATORIAL PACIFIC OCEAN MODEL TO WIND FORCING

M.A. ROWE and N.C.WELLS

ABSTRACT

Results are presented from preliminary experiments with an ocean general circulation model of the equatorial Pacific Ocean. The model has been developed to be suitable for investigating the response of the equatorial Pacific Ocean to seasonal and interannual variations in atmospheric forcing. A comparison is made of the model responses during the first eighty days of integration forced by: (1) a purely zonal idealised wind-stress distribution; and (2) the annual-mean wind-stress field. This study shows the ability of the real wind field to excite equatorial waves from within the basin interior and the role of the horizontal advection in redistributing temperature. Significant differences between the frictionally dominated surface circulations in the two experiments are attributed to features of the real wind-stress distribution and in particular the existence of cross-equatorial components.

1. INTRODUCTION

The primary motivation of this work has been to develop an ocean general circulation model, suitable for investigations into the response of the equatorial Pacific Ocean, to seasonal and interannual variations in atmospheric circulation.

In recent years, a number of studies have considered important aspects of this problem. Of particular note are the investigations by Philander and Pacanowski (1980, 1981), who have studied the response of a general circulation model in an idealised geometry, to simple wind distributions, and the work of Busalacchi and O'Brien (1981), and Busalacchi et al. (1983) who have applied observed wind distributions to a linear reduced gravity model of the equatorial Pacific Ocean. Schopf and Cane (1983) have incorporated thermodynamics and mixed-layer physics into a two-layer model, to investigate the response of the sea-surface temperature and mixed-layer depth to simple wind-stress distributions in an idealised geometry. More recently, with the availability of fast vector processing machines, it has become practicable to use realistic ocean geometries, with high-resolution general circulation models (Cox, 1980; Philander and Seigel, 1985; and Latif, 1985).

In this paper, preliminary results from a general circulation model of the equatorial Pacific Ocean, which has been forced by two different wind distributions, are presented. In the first experiment an idealised zonal wind stress, varying only with latitude is used, whilst in the second experiment, the annual-mean wind stress derived from a 20-year climatological data set (Weare et al., 1980) is used. The primary discussion is directed to a consideration of the temperature and velocity fields, at the surface and in the equatorial thermocline, during the first 80 days of the integration. In addition, a diagnosis of the thermal budget in the surface layer at $(0.6°S, 157°W)$ is presented.

2. THE MODEL

The model used in this study is based upon the primitive equation model of Bryan (1969), but was coded by the authors. The model uses a staggered finite difference grid (Arakawa B grid) on which values of the three velocity components (u, v, w), temperature (T) and density (ρ) are stored. The equations of the model are as follows:

$$\frac{\partial u}{\partial t} + \frac{1}{r \cos (\phi)} \frac{\partial (u^2)}{\partial \lambda} + \frac{1}{r \cos (\phi)} \frac{\partial [uv \cos (\phi)]}{\partial \phi} + \frac{\partial (uw)}{\partial z} - fv - \frac{uv \tan (\phi)}{r} =$$

$$= -\frac{1}{\rho_0 r \cos (\phi)} \frac{\partial P}{\partial \lambda} + \frac{\partial}{\partial z} \left(\kappa_M \frac{\partial u}{\partial z} \right) + A_M \left| \frac{1}{r^2 \cos^2 (\phi)} \frac{\partial^2 u}{\partial \lambda^2} \right.$$

$$\left. + \frac{1}{r^2 \cos (\phi)} \frac{\partial}{\partial \phi} \left[\cos (\phi) \frac{\partial u}{\partial \phi} \right] + \frac{[1 - \tan^2 (\phi)]}{r^2} u - \frac{2 \sin (\phi)}{\cos^2 (\phi)} \frac{\partial v}{\partial \lambda} \right\} \tag{1}$$

$$\frac{\partial v}{\partial t} + \frac{1}{r \cos (\phi)} \frac{\partial (vu)}{\partial \lambda} + \frac{1}{r \cos (\phi)} \frac{\partial [v^2 \cos (\phi)]}{\partial \phi} + \frac{\partial (vw)}{\partial z} + fu + \frac{u^2 \tan (\phi)}{r} =$$

$$= -\frac{1}{\rho_0 r} \frac{\partial P}{\partial \phi} + \frac{\partial}{\partial z} \left(\kappa_M \frac{\partial v}{\partial z} \right) + A_M \left| \frac{1}{r^2 \cos^2 (\phi)} \frac{\partial^2 v}{\partial \lambda^2} + \frac{1}{r^2 \cos (\phi)} \frac{\partial}{\partial \phi} \left[\cos (\phi) \frac{\partial v}{\partial \phi} \right] \right.$$

$$\left. + \frac{[1 - \tan^2 (\phi)]}{r^2} v - \frac{2 \sin (\phi)}{\cos^2 (\phi)} \frac{\partial u}{\partial \lambda} \right\} \tag{2}$$

$$\frac{1}{r \cos (\phi)} \frac{\partial u}{\partial \lambda} + \frac{1}{r \cos (\phi)} \frac{\partial [v \cos (\phi)]}{\partial \phi} + \frac{\partial w}{\partial z} = 0 \tag{3}$$

$$\frac{\partial P}{\partial z} = -\rho g \tag{4}$$

$$\rho = \rho_0 (1 - \alpha T) \tag{5}$$

$$\frac{\partial T}{\partial t} + \frac{1}{r \cos (\phi)} \frac{\partial (uT)}{\partial \lambda} + \frac{1}{r \cos (\phi)} \frac{\partial [vT \cos (\phi)]}{\partial \phi} + \frac{\partial (wT)}{\partial z} = \frac{\partial}{\partial z} \left(\kappa_T \frac{\partial T}{\partial z} \right)$$

$$+ A_T \left\{ \frac{1}{r^2 \cos^2 (\phi)} \frac{\partial^2 T}{\partial \lambda^2} + \frac{1}{r^2 \cos (\phi)} \frac{\partial}{\partial \phi} \left[\cos (\phi) \frac{\partial T}{\partial \phi} \right] \right\} \tag{6}$$

To improve the efficiency of the model, the barotropic calculation is neglected permitting the use of a longer time step (1/4 day). The barotropic component of the velocity field is known to be small in equatorial regions and this assumption may therefore be expected to have little effect upon the surface temperature distributions obtained. The wind stress forcing is applied as a body force on the surface model level and in the series of experiments presented here there is no surface heat flux. Other boundary conditions on the model are: The rigid lid ($w = 0$ at $z = 0$) and no-slip and no-normal flow at lateral boundaries [(u, v) = 0 on lateral boundaries].

The ocean basin extends from 130°E to 70°W and from 22.5°S to 22.5°N, and has a coastline as shown in Fig. 3. In the vertical, the model has a constant depth of 3650 m represented by sixteen model levels, six of which are within 600 m of the surface. The

TABLE 1
Model vertical structure

Level	Level base (m)	Level centre (m)	Eddy viscosity (m² s⁻¹)	Temperature (°C)
1	25	12.5	5×10^{-4}	25
2	75	50	5×10^{-4}	24.5
3	150	112.5	4.5×10^{-4}	22
4	250	200	1.5×10^{-4}	10
5	400	325	1.3×10^{-4}	9
6	600	500	1.0×10^{-4}	8

Below 600: Level thickness = 300 m; eddy viscosity = 10^{-4}; Initial temperature decreases linearly to 3°C

vertical eddy viscosity (κ_M) is a function of depth in order to represent surface mixing processes. However, to prevent diffusive spreading of the thermocline, in the absence of a surface heat flux, the vertical eddy diffusivity (κ_T) is zero at all depths. The values of vertical eddy viscosity, initial temperature and level depths are tabulated in Table 1. The horizontal eddy diffusivity (A_T) is 10^4 m² s⁻¹ and the eddy viscosity (A_M) is 10^5 m² s⁻¹. The high value of the latter is required by the coarse (1.25° latitude, 2.5° longitude) grid. The horizontal eddy viscosity is increased to prevent the propagation of Kelvin waves along the northern and southern closed boundaries.

The response of the model, when subjected to: (1) idealised wind stress [eqn. (7) and Fig. 1a]; and (2) an annual-mean wind stress (Fig. 1b), is discussed for temperature in Section 2.1 and for velocity in Section 2.2. The annual-mean wind stress was obtained by interpolation of the Weare data set, from a 5° × 5° grid over the tropical Pacific Ocean. The idealised wind-stress distribution is given by:

$$\tau_x = -0.065 + 0.035 \cos(12\phi)$$
$$\tau_y = 0$$

(7)

In Section 3.3 an analysis of the time-dependent response of the zonal velocity and temperature on the equator is discussed.

2.1. The temperature distribution

The temperature at 50 and 200 m after 80 days is depicted in Figs. 2 and 3, for the idealised wind (a) and the realistic wind (b). The corresponding vertical and horizontal velocity fields are shown in Figs. 4, 5 and 6, respectively.

At 50 m, in both experiments, there is a clearly defined cold tongue along the equator and along the eastern boundary. For the idealised wind, the cooling is most pronounced off the South American coast, though the winds are symmetrical about the equator, and therefore geometry appears to play a part in the induction of upwelling at the eastern boundary. The asymmetry between the two hemispheres, however, becomes more pronounced with realistic winds. In this case the temperature minimum at the eastern boundary is now displaced from the equator to 5°S. Another feature is the development of a second temperature minimum at 150°W on the equator. This occurs in the zone of the maximum easterly wind stress, where the southeast trades cross the equator (Fig. 1b).

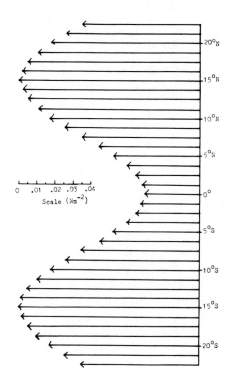

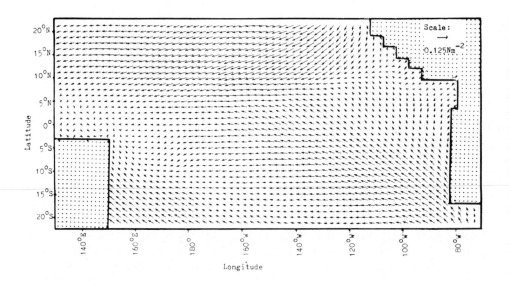

Fig. 1. (a) Idealised wind stress (eqn. 7). (b) Wind-stress forcing for real wind experiment.

The zonal component of the trade winds induces a frictional Ekman flow away from the equator, which in turn causes upwelling in the upper levels of the model. Comparison between Figs. 2 and 4 clearly shows that the maximum upwelling $(3 \times 10^{-5} \, \mathrm{m \, s^{-1}})$ is coincidental with the temperature minimum on the equator.

The break-up of the cold tongue into two distinct minima is related to the change in the direction of the surface winds from southeasterly in the Central Pacific, to southerly in the eastern Pacific. With a southerly wind stress, the frictional divergence is antisymmetric about the equator, with upwelling south of the equator and downwelling to the north. Hence, the cold tongue tends to bow southwards in the eastern basin. Immediately, poleward of the upwelling zone (Fig. 4) is a compensating region of downwelling between $3°$ and $7°$ in both hemispheres. In the realistic wind case, the most intense downwelling zones $(1.2 \times 10^{-5} \, \mathrm{m \, s^{-1}})$, are located poleward of the major upwelling region, though the downwelling region is located further east in the Northern Hemisphere than in the Southern Hemisphere. At the western boundaries, at $9°$N and $9°$S, two cold gyres are apparent, which do not appear in the idealised case, and also a cold tongue is evident at $15°$N in the eastern Pacific. Both these features are associated with Ekman upwelling.

At 200 m, which corresponds to the lower part of the thermocline, most of the features described above are in evidence. In addition the temperature field now clearly shows the regions of warming, as the result of downwelling from the surface mixed layer. In both wind cases, warming occurs along both zonal boundaries as the result of Ekman convergence in the surface layers, and some warming is also apparent at the western boundaries. Again the cold gyres are present in both hemispheres in the western Pacific. The equatorial downwelling regions are evident, though comparison with the surface downwelling patterns shows an important difference. The warmest water is located further east in the South Pacific than in the North Pacific, though the surface downwelling pattern shows the opposite behaviour. This is possibly the result of the propagation of a baroclinic Rossby wave from the eastern boundary, which would be expected from the geometry to be further advanced in the Northern Hemisphere than in the Southern Hemisphere.

Along the equator, the temperature pattern at 200 m is quite different to that at 50 m. First, the minimum temperature is located at the eastern boundary, in both the realistic and idealised wind cases. Second, the maximum longitudinal temperature gradient occurs in the Central Pacific for the realistic wind case, where the easterly wind stress reaches a maximum. Whilst, with a zonally uniform wind stress, the zonal temperature gradient is nearly constant with longitude. Superimposed on this basic set-up state, longitudinal variations in temperature are present in both cases, as the result of long equatorial waves propagating along the thermocline. These equatorial waves will be discussed in Section 3.3.

2.2. The current distribution

The surface current distribution is dominated by the wind-stress pattern. On the equator, westward zonal currents of up to $70 \, \mathrm{cm \, s^{-1}}$ are induced by the idealised wind stress, whilst with a realistic wind stress a maximum of $100 \, \mathrm{cm \, s^{-1}}$ is found. Either side of the equator, the turning of current vectors is evident, and beyond $10°$ latitude, the current vector is perpendicular to the wind stress, as expected for a steady state Ekman

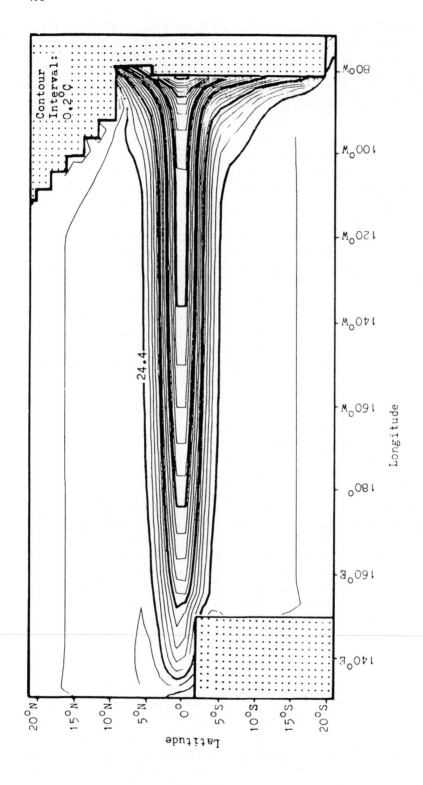

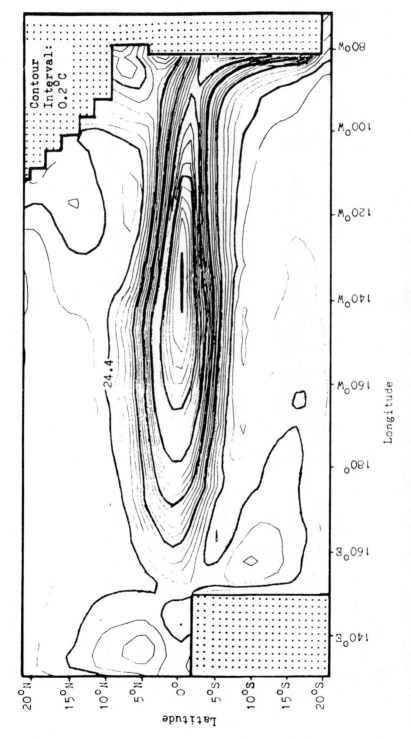

Fig. 2. Temperature at 50 m on day 80 of: (a) idealised wind experiment; and (b) real wind experiment.

498

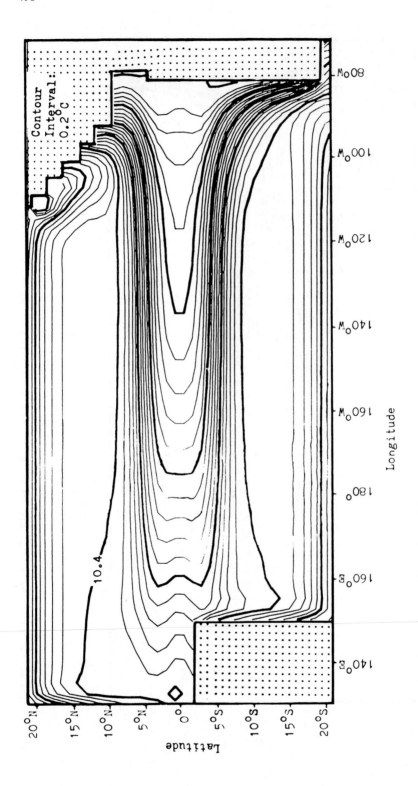

499

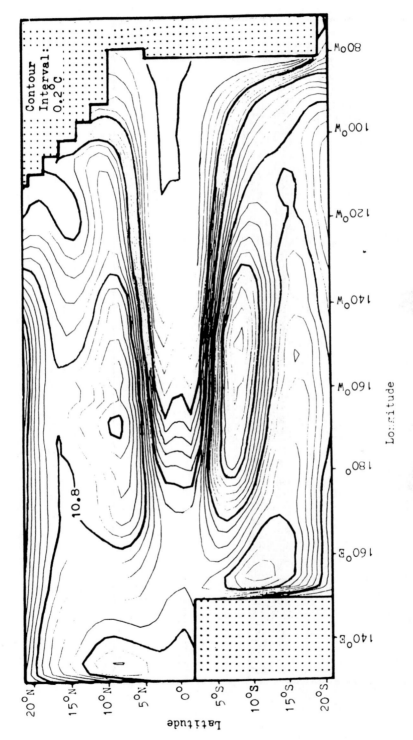

Fig. 3. Temperature at 200 m on day 80 of: (a) idealised wind experiment; and (b) real wind experiment.

500

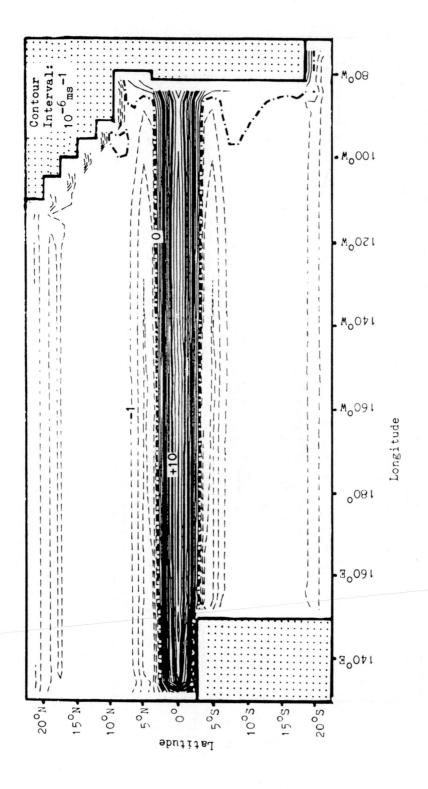

501

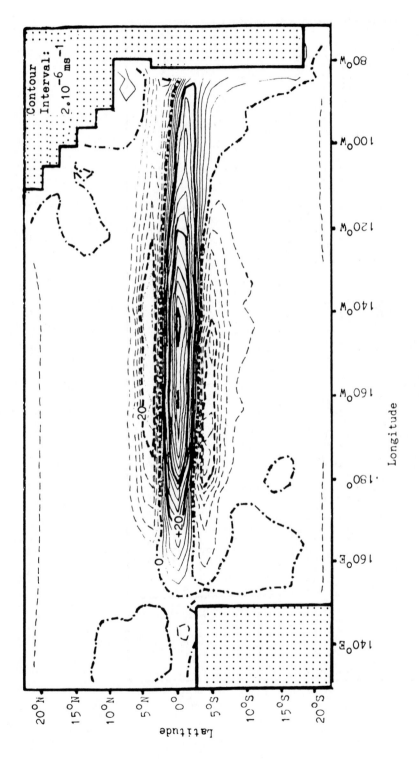

Fig. 4. Vertical velocity at 25 m on day 80 of: (a) idealised wind experiment; and (b) real wind experiment.

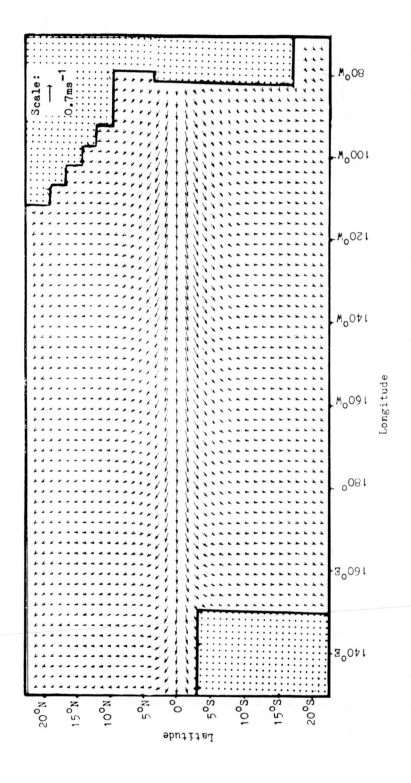

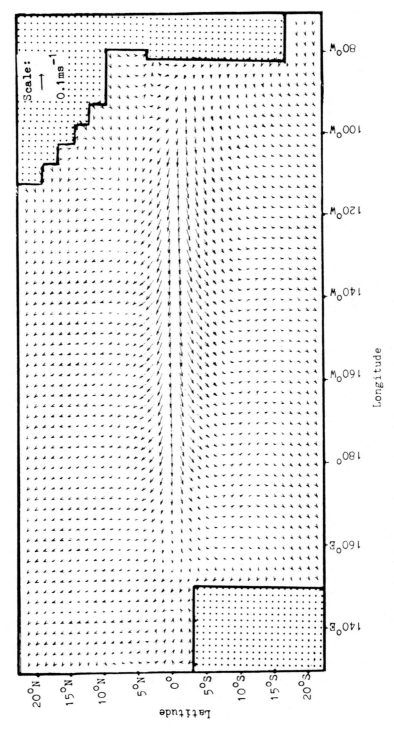

Fig. 5. Horizontal velocity at 12.5 m on day 80 of: (a) idealised wind experiment; and (b) real wind experiment.

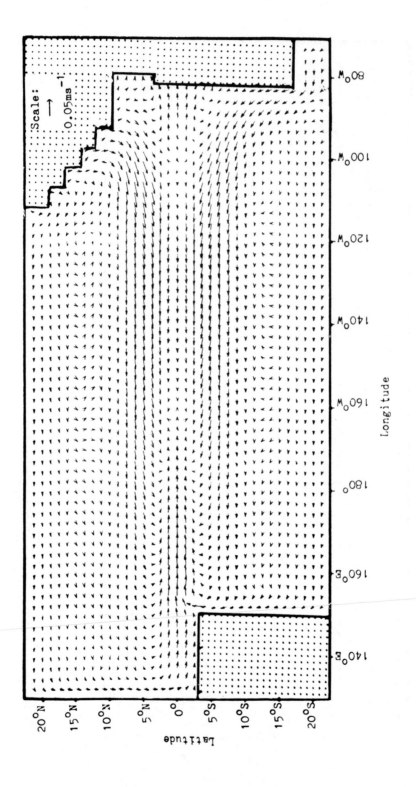

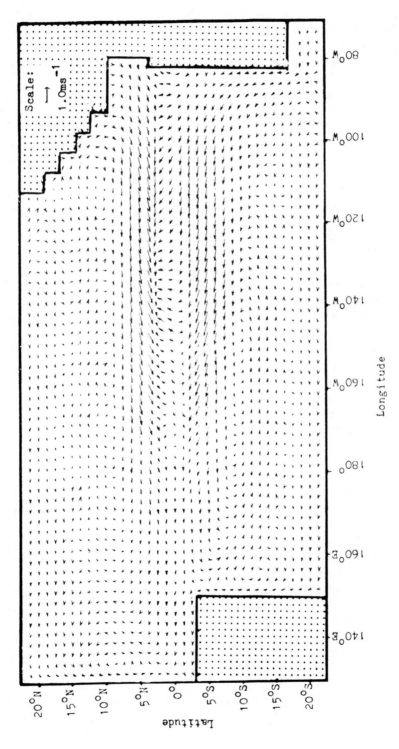

Fig. 6. Horizontal velocity at 200 m on day 80 of: (a) idealised wind experiment; and (b) real wind experiment.

Scale:
1.0ms^{-1}

Longitude

Latitude

layer. In the idealised wind case, the frictional divergence induces upwelling along the length of the equatorial zone, whilst in the realistic wind case, the major upwelling zone is located in the Central Pacific. In the extreme west of the Pacific, a westerly wind stress (Fig. 1), induces a frictional convergence on the equator and downwelling. Comparison of Figs. 5a and b also indicates important differences in the eastern Pacific. With a realistic wind stress, an anticyclonic circulation is induced at 10°N, 100°W, and close to the eastern boundary, between 3° and 7°N, the surface flow is towards the equator. At 90°W the southerly wind stress induces a clockwise rotation in the surface current as it crosses the equator.

In the thermocline, the current patterns, both for idealised and realistic wind distributions, are complicated by the appearance of equatorial Kelvin and Rossby waves, produced by the impulsive wind stress at the beginning of the integrations. For idealised winds, on the equator, there is evidence for two current reversals, with an eastward undercurrent in the western and eastern basin, and a westward current in the central basin. For realistic winds, the current is generally westward everywhere. Beyond 5° of the equator, the patterns between the two cases are more consistent, with general westward flow. However there is evidence of an eastward countercurrent in both cases at about 12°N and S, with the stronger countercurrents for the realistic winds. These countercurrents are on the poleward side of the warm-water gyres at 9°N and S, which is consistent with geostrophic motion. Furthermore, there is evidence of cyclonic geostrophic circulation around the cold water gyres in the western Pacific, in the realistic wind case. The strongest flow, in both cases, occurs along the boundary with the cold equatorial tongue, where the westward currents reach up to $10 \, \mathrm{cm \, s^{-1}}$. With realistic winds, there is convergence of the predominantly westward flow, towards the equator (between $\pm 6°$ latitude). This convergence is caused by the longitudinal equatorial temperature gradient which produces a meridional component of geostrophic flow towards the equator. It is not as pronounced in the idealised wind case because of the weaker longitudinal temperature gradient present.

2.3. Time-dependent response

Figure 7a shows the time-dependent response of the zonal current at 200 m on the equator. For the idealised wind, the response is similar to that described by many other authors (e.g., Philander and Pacanowski, 1981). There is a steady increase in zonal velocity which reaches a peak between 50 and 60 days. This steady acceleration is arrested by the arrival of an equatorial Kelvin wave from the west, and an equatorial Rossby wave from the east. The response is, however, complicated by the fact that there are two Kelvin waves, one originating from the western boundary in the Northern Hemisphere, and the other originating from the Southern Hemisphere, each having different arrival times because of the geometry. The Kelvin wave has a propagation speed lying between the theoretical first and second baroclinic mode speeds (Table 2). In the wake of the Kelvin wave, an eastward undercurrent flow develops, and continues to grow in amplitude for the remainder of the integration. A weaker eastward flow also develops in the extreme east of the basin, behind the westward propagating Rossby wave. For realistic winds (Fig. 7b) the time-dependent response is more complicated than for the idealised case. However, there are two important observations. First, the main westward

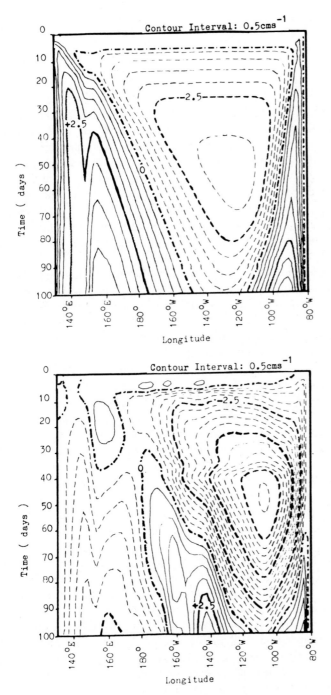

Fig. 7. Zonal velocity at 200 m as a function of longitude and time on the equator, (a) idealised wind experiment; and (b) real wind experiment.

TABLE 2
Model equatorial wave characteristics

Mode	H (m)	$c = \sqrt{gH}$ (m s^{-1})	$L = \sqrt{c/\beta}$ (km)
1	0.58	2.39	323
2	0.31	1.75	277

velocity occurs both earlier (between day 45 and day 50) and further eastward than in the previous case. Either the Kelvin wave speed has increased in the real wind case, or the Kelvin wave has not originated from the western boundary. Though there are variations in the slope of the Kelvin wave front, the evidence for a large increase in phase speed is not strong. However, the origin of the Kelvin wave appears to be further east for the real wind than for the idealised wind. It is known that in the real wind case longitudinal variations in wind stress can initiate equatorial Kelvin waves, away from western boundaries. The most likely region of origin would be between the position of maximum easterly wind stress and the position of zero wind stress (i.e. between the date-line and the western boundary).

The second observation is that no eastward currents form behind the Kelvin wave fronts, though there is a weak eastward flow in the centre of the basin. It is clear that the model is not in equilibrium and equatorial waves are still actively adjusting the pressure gradient to the imposed wind stress.

The time-dependent response of the temperature field on the equator is less complicated than that for the velocity (Fig. 8). For the idealised wind stress, the temperature decreases steadily, until the arrival of the eastward propagating Kelvin wave. Behind the Kelvin wave, the temperature changes only very slowly with time and by day 100, over the western half of the basin, the longitudinal temperature gradient is close to equilibrium. However, in the eastern basin, the temperature continues to decrease throughout the period, though the rate of cooling does decrease significantly between day 80 and 100. In the realistic wind case, the adjustment in the western half of the basin is more rapid, though there is a small steady increase in temperature behind the Kelvin wave. An interesting difference between the ideal and real wind is that in the latter case the coldest water occurs in the centre of the basin, and propagates eastward during the 100 days, whilst in the former case the coldest water is always located at the eastern boundary. Again, at the eastern boundary, the temperature continues to decrease throughout the integration.

An analysis of the terms in the temperature equation [eqn. (6)] in the surface level of the model is presented in Fig. 9 for the point 0.5°S, 156°W. In the absence of surface heating and vertical diffusion, the budget is a balance between the local change in temperature, the advection of temperature, and horizontal temperature diffusion.

The principal features of the budget, for both cases, is the primary role of vertical advection in establishing the temperature field in the first 40 days. However, as the horizontal temperature field develops, there is an increasing importance of both meridional and zonal advection in augmenting the decrease in the upper-level temperature. Because of the lack of surface heating and vertical diffusion, the advection of heat has to be counter-balanced by horizontal diffusion in a steady state. For the idealised wind, a near-balance between vertical advection, horizontal advection, and horizontal

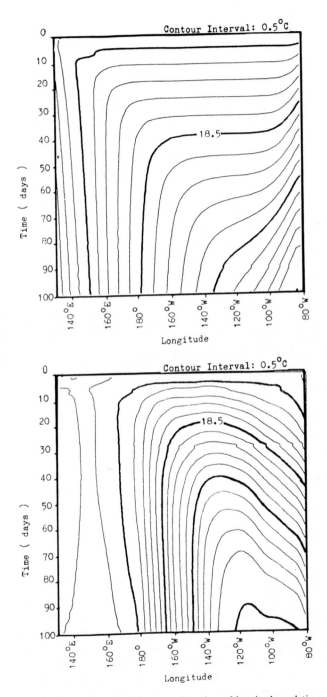

Fig. 8. Temperature at 112 m as a function of longitude and time on the equator, (a) idealised wind experiment; and (b) real wind experiment.

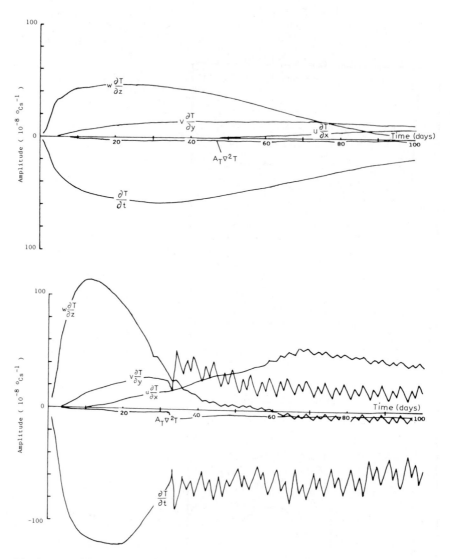

Fig. 9. Thermal budget for 12.5 m at 0.6°S, 156°W for: (a) idealised wind experiment; and (b) real wind experiment.

diffusion is established after 100 days. With realistic winds, the terms in the budget are significantly larger and these fields are noisier after 30 days. This is numerical noise induced by convective adjustment of the upper levels. The convection is caused by advection of cold upwelled water over warmer deep water. In contrast to the ideal wind case, the zonal advection of temperature dominates all other advective terms, in the last 40 days of the integration. The dominance of zonal advection is related to the presence of a large zonal temperature gradient at this point (Fig. 3b), and a strong westward surface flow. The contribution of vertical advection is small, because the convective adjustment scheme removes the vertical gradient of temperature at this level, and does not therefore imply an absence of upwelling. The convective adjustment scheme, is responsible for an upward flux into the surface layer. Again, horizontal diffusion is necessary to balance the advection and convective processes.

3. CONCLUSIONS AND SUMMARY

The purpose of this paper has been to compare the response of an equatorial model to a simple latitudinally varying wind-stress distribution and a realistic wind-stress distribution for the Pacific Ocean. The principal results are:

(1) The surface circulation is dominated by frictionally induced Ekman currents caused by the local wind stress. The surface temperature distribution is in turn dominated by the upwelling distribution, along the equator and the eastern boundary. However, an analysis of the thermal budget on the equator indicates that horizontal advection is a significant factor for redistributing temperature in the latter stages of the integration.

(2) The horizontal temperature distribution in the thermocline again reflects the pattern of Ekman upwelling and downwelling poleward of $5°$ of the equator. In the equatorial zone, both baroclinic Kelvin and Rossby waves are in evidence, in adjusting both velocity and pressure fields to the wind forcing. The Kelvin and Rossby waves can be identified, though the response is complicated by irregular boundaries and longitudinal wind variations. The most significant feature in the real wind case at 200 m is the propagation of cold upwelled water, initially in the centre of the basin, to the eastern boundary by a Kelvin wave.

There are some deficiencies in the model. First, the model is not in equilibrium and therefore further integration of the model will be required to achieve a final steady state.

A more serious deficiency is the absence of surface heating and vertical diffusion of temperature. The purpose of this study, however, has been to investigate the dynamically induced temperature changes near the surface. Whilst the surface heating is recognised as giving rise to temperature changes of similar magnitude, it is assumed that on the time scales considered the thermal structure is not sufficiently changed to alter the dynamical response. In the absence of surface heating which maintains the stratification, it is necessary to remove the processes which destroy the stratification, namely those processes parameterised by the vertical diffusion of temperature. In future experiments it is hoped to include both a parameterisation of surface heat flux and the vertical diffusion of temperature. The latter parameterisation should include both the effects of surface and wind mixing, and sub-grid scale mixing processes in the equatorial thermocline, such as the breaking of internal waves and fine structured turbulence (Crawford and Osborn, 1980).

ACKNOWLEDGEMENTS

The provision of a studentship by the University of Southampton in support of the above work, and the travel grant from the National Environmental Research Council given to M.A. Rowe to attend the Liège Colloquium are gratefully acknowledged.

REFERENCES

Bryan, K., 1969. A numerical method for the study of the world ocean. J. Comp. Phys., 3: 347–376.

Busalacchi, A.J. and O'Brien, J.J., 1981. Interannual variability of the Equatorial Pacific in the 1960's. J. Geophys. Res., 86C: 10901–10907.

Busalacchi, A.J., Takeuchi, K. and O'Brien, J.J., 1983. Interannual variability of the Equatorial Pacific revisited. J. Geophys. Res., 88C: 7551–7562.

Cox, M.D., 1980. Generation and propagation of 30 day waves in a numerical model of the Pacific. J. Phys. Oceanogr., 10: 1168–1186.

Crawford, W.R. and Osborn, T.R., 1980. Energetics of the Atlantic Equatorial undercurrent. In: W. Düing (Editor), GATE supplement to Deep-Sea Res. 26A, Pergamon Press, Oxford, pp. 309–323.

Latif, M., Maier-Reimer, E. and Olbers, D., 1985. Climate variability studies with a primitive equation model of the equatorial Pacific. In: J.C.J. Nihoul (Editor), Coupled Ocean–Atmosphere Models. (Elsevier Oceanography Series, 40) Elsevier, Amsterdam, pp. 63–81 (this volume).

Philander, S.G.H. and Pacanowski, R.C., 1980. The generation of Equatorial currents. J. Geophys. Res., 85C: 1123–1136.

Philander, S.G.H. and Seigel, A.D., 1985. Simulation of El Niño of 1982–1983. In: J.C.J. Nihoul (Editor), Coupled Ocean–Atmosphere Models. (Elsevier Oceanography Series, 40) Elsevier, Amsterdam, pp. 517–541 (this volume).

Schopf, P.S. and Cane, M.A., 1983. On equatorial dynamics, mixed layer physics, and sea surface temperature. J. Phys. Oceanogr., 13: 917–935.

Weare, B.C., Strub, P.T. and Samuel, M.D., 1980. Marine Climate Atlas of the Tropical Pacific Ocean. Dept. of Land, Air and Water Resources, Univ. Calif., Davis, Calif.

CHAPTER 32

A SIMPLE WIND- AND BUOYANCY-DRIVEN THERMOCLINE MODEL

PETER D. KILLWORTH

EXTENDED ABSTRACT

The ocean is driven primarily by processes occurring at its surface. Wind stress drives horizontal fluxes in the surface mixed layer and produces Ekman divergence at its base. Surface buoyancy fluxes, produced both by heating/cooling and evaporation/precipitation, yield varying densities in the upper layer of the ocean, which drive motions through the resulting pressure imbalance.

Whereas motions on short (days to months) timescales are dominated by wind driving, those on longer timescales are produced by an intricate and highly nonlinear interaction between wind and buoyancy effects; an interaction we do not fully understand.

Past attempts to study the dynamics of the large-scale, quasi-steady ocean circulation have centered on the "thermocline equations": conservation of mass and buoyancy (perhaps with mixing terms for the latter), plus the geostrophic hydrostatic balances, with surface boundary conditions involving specified surface density and Ekman pumping. Solutions to date have either involved the simplification of similarity solutions (Welander, 1971) or of slab-like dynamics (Luyten et al., 1983). Since by their nature the equations cannot apply to western boundary layers, solutions have been limited in their spatial extent or applicability.

The model discussed here serves two distinct functions. First, it is an attempt to circumvent some of these difficulties (or at least replace them) by reducing the vertical resolution drastically, to two vertical levels, topped by an Ekman layer whose sole function is to provide specified Ekman pumping

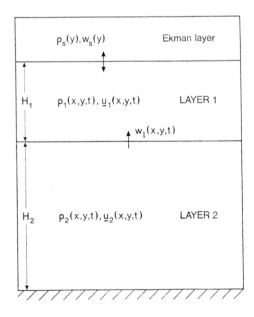

Fig. 1. The two-level model geometry used.

514

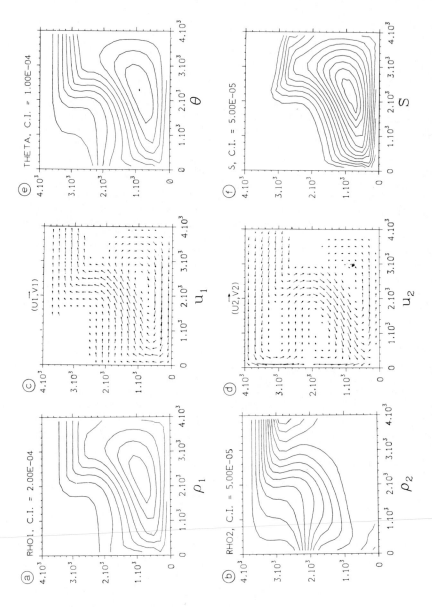

Fig. 2. Typical solutions on a 20 × 20 grid for a two-gyre forcing, of total size 4000 km. The Ekman pumping is sinusoidal north–south, of magnitude 10^{-4} cm s^{-1}. The applied surface density varies like a single cosine north–south, being negative (lighter) in the south and positive (denser) in the north. Its total variation is 4×10^{-3} g cm^{-3} north–south. Shown are: (a) level 1 density; (b) level 2 density; (c) level 1 velocities; (d) level 2 velocities (note change of scale); (e) average density; and (f) stratification.

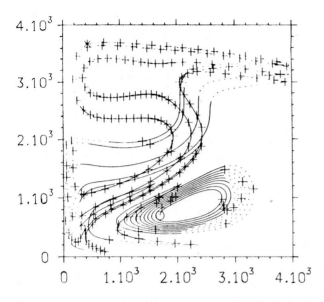

Fig. 3. A typical particle trajectory for solution in Fig. 2. Initial position marked by circle. Firm line indicates downward motion, dotted line upward. A cross is marked every year. Abrupt direction changes indicate transition between the two levels. Length of trajectory is several hundred years. As a rough guide, slow speeds indicate lower level, rapid speeds upper level.

and surface density (i.e., to maintain the same boundary conditions as in classical thermocline solutions; Fig. 1). Second, the model works toward the development of a simple numerical model which can permit rapid, cheap evaluation of the ocean circulation which could be used for climate studies. It turns out, however, that the model is anything but simple in its outcome; simple modelling of its results proves difficult.

In each level of the model the motion is geostrophic, plus a small linear drag which allows a Stommel-like western boundary layer. The motion is mass-conserving, and buoyancy-conserving save for a small horizontal diffusion necessary to close the physics. Inclusion of vertical diffusion has little effect on the results. In order to avoid the analytic version of a numerical mode in the vertical, one-sided differences are necessary for vertical advection of buoyancy (which induce an effective vertical diffusivity anyway). Sidewall conditions are not well understood; in this model the (almost certainly over-specified) conditions are of no normal flow and no normal buoyancy flux at all sidewalls.

The depth-integrated flow is known from the specified Ekman pumping, so that the only unknown flow is the (single) baroclinic mode, which can be derived from the thermal wind equations if the density field is known. Steady solutions may then be found by time integrations of the equations for each level. The time taken to a steady state is a few hundred years for a two-gyre basin of side 4000 km.

Despite the apparent simplicity of the model, the solution is fairly realistic and quite complicated (Fig. 2). The solution requires convective adjustment in the northern (cool) part of the basin, at least between the Ekman layer and top level. There is a strong western boundary current which separates rather south of its observed North Atlantic latitude and flows towards the NE corner of the basin, where there is strong downwelling as the flow is returned in the lower level. The average of the level densities serves as an approximate streamfunction for the flow. This average density spins-up initially like a long Rossby wave response of an ocean to wind forcing; in the later stages the response is still that of a Rossby wave, but now one with spatially varying stratification. The relevant timescale becomes that for vertical advective effects to reach the entire basin. Analysis of the steady state is difficult because of the variations induced by the one-sided differencing used.

516

To examine the ventilation of the lower subtropical level of the ocean, trajectories were examined for particles emitted from the (downwelling) Ekman pumping at the surface. The idea is to see how much of the lower subtropical water has ever been in contact with the atmosphere, and how much remains a "pool" region as discussed by Luyten et al. (1983) for their layer model. Particles released in the southern half of the subtropical gyre have quite complex tracks, as Fig. 3 shows. There is a tendency for anticyclonic circulation for several years, followed by a cross-gyre movement to the subpolar gyre. Here, the particle may surface, or sink near the northern boundary to return to the deep subtropics by this somewhat circuitous route. There seems to be only small direct ventilation by such particles. Particles released nearer the gyre boundary also show little tendency to direct ventilation of the lower level. Tests which add random walks to the particle tracks, in simulation of the diffusivity used, show that apart from the expected 'smearing out' of the tracks, there is little qualitative difference induced by diffusion.

REFERENCES

Luyten, J., Pedlosky, J. and Stommel, H., 1983. The ventilated thermocline. J. Phys. Oceanogr., 13: 292–309.
Welander, P., 1971. The thermocline problem. Philos. Trans. R. Soc. London, Ser. A, 270: 415–421.

CHAPTER 33

SIMULATION OF EL NIÑO OF 1982–1983

S.G.H. PHILANDER and A.D. SEIGEL

ABSTRACT

A general circulation model of the ocean simulates El Niño of 1982–1983 with reasonable success and provides the following results. The massive eastward transfer of warm surface waters from the western to the eastern Pacific was accomplished by unusual eastward surface currents which, by November 1982, extended from 9°S to 9°N across 120°W. Further east, the persistence of the southeast trades over the eastern tropical Pacific inhibited eastward surface flow at, and to the south of the equator at that time, but the eastward flow between 3° and 8°N penetrated right to the coast of Central America. The relaxation of the trades and the changes in the curl of the windstress, that caused the redistribution of heat in the upper ocean, occurred so gradually between June and November 1982 that the response of the ocean was approximately an equilibrium one. The zonal pressure gradient along the equator and the intensity of the Equatorial Undercurrent, for example, decreased gradually as the trade winds weakened. In December 1982, the anomalous eastward winds west of the dateline suddenly changed to northerly winds. The westward pressure force which the eastward winds had established was left unbalanced. This excited an eastward travelling equatorial Kelvin wave which elevated the thermocline and accelerated the equatorial currents westward. The wave front dispersed downwards as it propagated eastward and there is no evidence of its reflection at the South American coast affecting the surface layers of the ocean. Eastward winds in the eastern equatorial Pacific Ocean in March and April interrupted the recovery from El Niño and generated an intense local eastward surface jet. The reappearance of the tradewinds in May 1983 signalled the end of El Niño and the gradual return to normal conditions.

INTRODUCTION

Though El Niño of 1982–1983 is far better documented than any previous El Niño, the available data are too fragmented to provide a complete picture of the development of this El Niño. Simulation of El Niño therefore serves two purposes. It is a test of the realism of the model and, should the model be reasonably successful in reproducing the available measurements, then the model results from regions where data are unavailable can be used to paint a coherent picture of the evolution of the event. This paper describes a simulation of El Niño of 1982–1983 with a general circulation model of the tropical Pacific Ocean. Unfortunately, the winds used to drive the model — 1000 mb monthly mean winds provided by the National Meteorological Center — are known to have flaws (E. Rasmusson and J. O'Brien, pers. commun., 1984). It is, therefore, unclear whether discrepancies between the measurements and the results from the model are attributable to the poor wind data, or to deficiencies of the model. The same model has performed successfully in a simulation of the well-documented seasonal cycle of the tropical Atlantic Ocean (Philander and Pacanowski, 1984; Katz, 1984). We are therefore inclined to attribute discrepancies between the model's results and the measurements primarily to

518

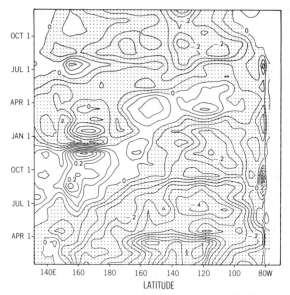

Fig. 1. Zonal wind stresses averaged between 5°N–5°S for 1982–1983. Shading represents westward direction. Contour interval is 0.05 dynes cm^{-2}.

inaccuracies in the wind fields. These discrepancies are fortunately not too serious, so that the results from the model can be used to obtain a coherent picture of how El Niño developed.

This paper is organized as follows: Section 2 describes the winds that force the model, and briefly summarizes the measurements obtained in 1982 and 1983; Section 3 is a description of the model; Section 4 is a presentation of the results; and Section 5 is a summary.

2. THE MEASUREMENTS

El Niño of 1982–1983 evolved in three stages: Up to November 1982, there was a relaxation of the trade winds that led to the appearance of eastward winds west of the dateline; between December 1982 and May 1983, the winds over the western Pacific were cross-equatorial and had only a small zonal component near the equator, while the easterly winds over the eastern Pacific weakened and became westerly; the final stage of the event started in May 1983 with the return of the tradewinds, first in the central Pacific, then elsewhere. These changes are evident in Figs. 1–3.

The relaxation of the trade winds during the first phase of El Niño resulted in a massive transfer of warm surface waters from the western to the eastern tropical Pacific Ocean. The available measurements give an incomplete picture of how this occurred. Sea-level measurements (Wyrtki, 1984a, b) reveal that the thermocline first started to rise in the region west of the dateline, between the equator and 15°N approximately. During the subsequent months (until December 1982) this region in which the thermocline shoaled, expanded eastward to about 160°W, and also expanded southward across the equator. Geostrophic calculations based on XBT data from tracks between New

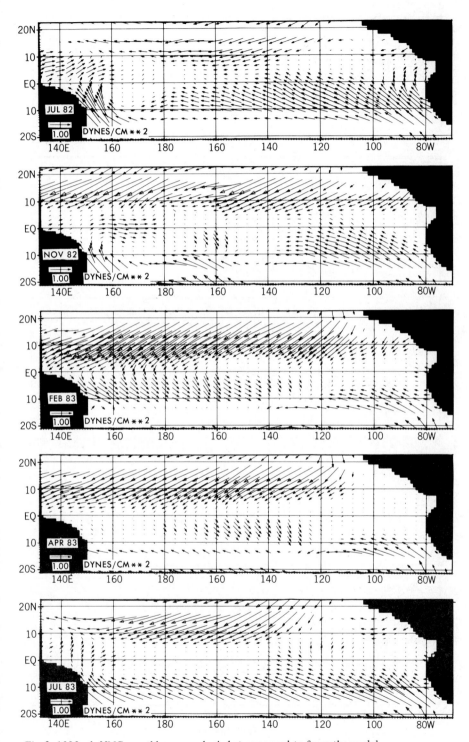

Fig. 2. 1000 mb NMC monthly averaged wind stresses used to force the model.

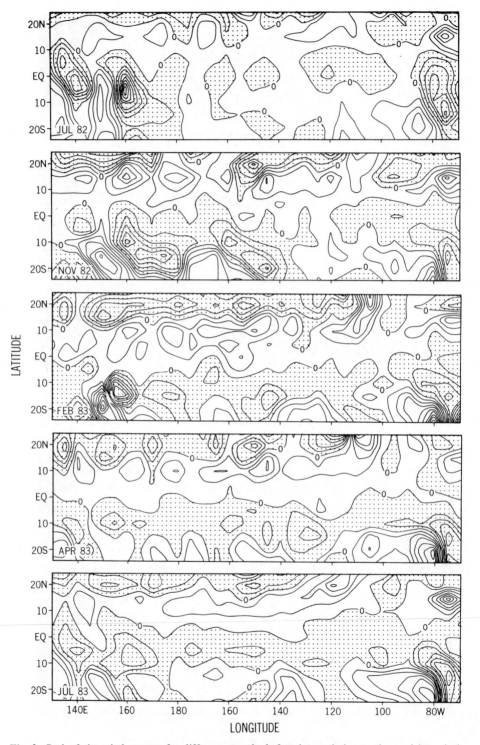

Fig. 3. Curl of the wind stresses for different months before interpolation to the model resolution. Contour interval is 1.0×10^{-8} dynes cm^{-3}. The curl is negative in shaded areas.

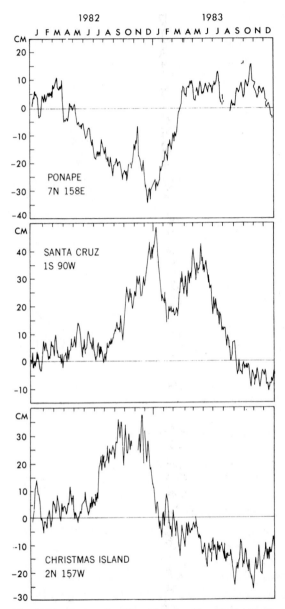

Fig. 4. Observed sea level at various locations in the Pacific (after Wyrtki, 1984a).

Caledonia and Japan (Meyers and Donguy, 1984) and between New Zealand and California (Kessler, pers. commun., 1984) show that an intensification of the North Equatorial Countercurrent (between 3° and 9°N approximately) transferred the warm surface waters eastward as the thermocline in the west rose. (Direct current measurements are unavailable for the region west of 160°W.) Sea-level measurements at different meridians document the eastward march of the increase in heat content of the water column. In Fig. 4, the increase in sea level at Christmas Island is seen to preceed that

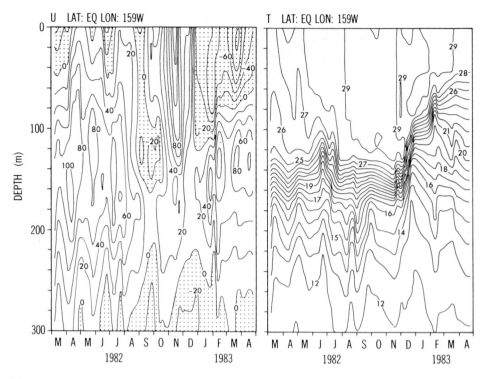

Fig. 5. The observed zonal velocity fields (cm s⁻¹) and temperature (in °C) at 159°W and the equator during 1982–1983 as measured by Firing et al. (1983). Shaded areas denote westward flow.

at the Galapagos Islands by several weeks. The complex vertical structure of this change in the heat content of the ocean, as observed at 0°N 159°W and 0°N 95°W (Figs. 5 and 6) precludes an interpretation of the measurements in terms of a simple eastward travelling vertical mode of the ocean. At 159°W on the equator, there was an increase in the temperature of the surface layers but not a deepening of the thermocline. At 95°W, on the other hand, the thermocline deepened steadily during 1982. At both locations the Equatorial Undercurrent is seen to decelerate during the first phase of El Niño. The flow in the surface layers is not the usual swift westward drift, but is sporadically eastward and includes an intense eastward equatorial jet at 159°W in November 1982. By this time, a sufficient amount of warm surface waters had been transferred from the western to the eastern Pacific to eliminate the usual zonal slope of the thermocline (Firing et al., 1983).

The second phase in the development of El Niño starts in December 1982 with the deceleration of the eastward equatorial jet at 159°W. This coincided with a rise of the thermocline and a fall in sea level. In the western Pacific to the south of the equator, sea level dropped steadily during the next six months (Wyrtki, 1984a) but in the eastern Pacific, developments were more complex. At first, in January and February 1983, sea level fell, the thermocline shoaled, the Equatorial Undercurrent disappeared and the surface flow was westward (Fig. 6; Hansen, 1984). In April and May, however, sea level rose again because of an increase in sea-surface temperature rather than a deepening of

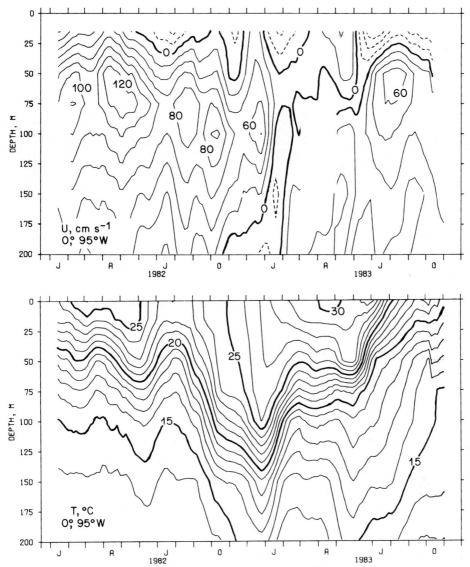

Fig. 6. The observed zonal velocity fields and temperature at 95°W and the equator during 1982–1983 as measured by Halpern (1984). Dashed contours indicate westward flow.

the thermocline (Fig. 6), and an eastward jet appeared in the surface layers. Finally, in June 1983, the return of the trade winds started the restoration of normal conditions.

The oceanic changes during El Niño are in response to the changes in the surface winds but with the available measurements it is impossible to establish which wind variations caused specific changes in the ocean. The relaxation of the tradewinds during the first phase of the event presumably resulted in the eastward transfer of heat in the upper ocean, but the pattern of currents that effected this redistribution is unclear. The eastward phase propagation of the rise in sea level near the equator suggests that a Kelvin

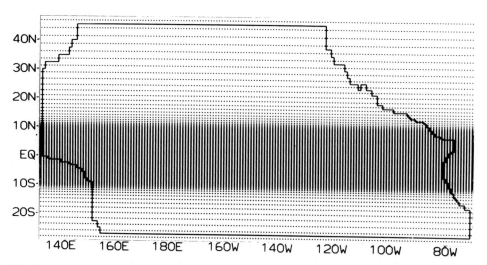

Fig. 7. Coastlines and grid-spacing of the Pacific Ocean model. Highest resolution is between 10°S and 10°N latitude.

wave was involved but the considerable differences in the vertical structures at 159° and 95°W indicate that this wave was not involved in a simple manner. The sudden fall of the sea level near the equator at the end of 1982 is a puzzle, as is its renewed rise in the east in April 1983. We next turn to a model for more information about the response of the ocean to the winds.

3. THE MODEL

The Pacific Ocean model extends from 130°E to 70°W longitude and 28°S to 50°N latitude. The coastlines and grid-point distribution are shown in Fig. 7. The longitudinal resolution is a constant 100 km, but the latitudinal distance between grid points is 33 km between 10°S and 10°N and increases gradually poleward of this region. The spacing at 25°N is 200 km. The flat-bottom ocean is 4000 m deep. There are 27 levels in the vertical; the upper 100 m have a resolution of 10 m.

The primitive equations are solved numerically by means of finite differencing methods discussed in Bryan (1969). The use of Richardson number dependent vertical-mixing coefficients is explained in detail in Pacanowski and Philander (1981). The coefficient of vertical eddy viscosity is assigned a constant value of $10 \, \text{cm}^2 \, \text{s}^{-1}$ in the upper 10 m of the model to compensate for mixing by the high-frequency wind fluctuations which are absent from the monthly mean winds. The coefficient of horizontal eddy viscosity has the constant coefficient $2 \times 10^7 \, \text{cm}^2 \, \text{s}^{-1}$ equatorward of 10° latitude and elsewhere varies inversely as the latitudinal resolution to a value of $50 \times 10^7 \, \text{cm}^2 \, \text{s}^{-1}$ at 50°N.

Poleward of 20°S and 30°N the heat equation for the temperature T gains a term $\gamma(T - T^*)$, where T^* is the prescribed monthly mean climatological temperature for the region under consideration, and γ is a Newtonian cooling coefficient. Its value is 1/(2 days) near the zonal boundaries and decreases to a value of zero equatorward of 30°N and 20°S. This device mitigates the effect of the artificial zonal walls along the southern and northern boundaries of the ocean and forces the solution towards the climatology in these regions. There are similar terms in the horizontal momentum equations.

The heat flux across the ocean surface is:

$$Q = SW - LW - QS - QE$$

The solar short wave heating SW is taken to be 500 ly per day equatorward of 20° latitude and decreases linearly to 300 ly per day between 20° and 45° latitude. The long wave back radiation LW has the constant value of 115 ly per day. The sensible heat flux is:

$$QS = \rho C_D C_p V(T_0 - T_A)$$

and the evaporation is:

$$QE = \rho C_g L V [\exp(T_0) - \gamma \exp(T_A)] (0.622/p)$$

Here $\rho = 1.2 \times 10^{-3}$ g cm^{-3}; $L = 595$ cal g^{-1}; $C_D = 1.4 \times 10^{-3}$; $p = 1013$ mb; $C_p = 0.24$ cal g^{-1} °C^{-1}; T_0 is the sea-surface temperature in degrees Kelvin; T_A is the atmospheric temperature at the surface; V is the surface wind speed; and the relative humidity γ is assigned the constant value 0.8. No provision is made for clouds. The sensible heat flux is found to be of secondary importance so that variations in heat flux are primarily a consequence of change in the evaporation. For the El Niño simulations the atmospheric temperature T_A is assumed to be 1°C less than the predicted sea-surface temperatures. The results to be shown are for this value. The sea-surface temperatures are found to be unrealistically high and can exceed 32°C. An experiment in which $T_0 - T_A = 2$°C gives exactly the same sea-surface temperature patterns except that the values of the contours are 3°C less. This suggests that $T_0 - T_A = 1.5$°C would have been the best parameterization. Fields other than sea-surface temperature were unaffected by the change in the value for $T_0 - T_A$. Since evaporation depends on the wind speed, avoidance of excessively high temperatures in regions of weak winds required that the wind speed not be less than 4.8 m s^{-1}. This minimum parameterizes evaporation caused by high-frequency wind fluctuations that are absent from the mean monthly winds.

The initial conditions for the model are zero currents and the climatological temperature field (Levitus, 1982). Month averaged climatological winds then force the model for three years by which time the model has an equilibrium seasonal cycle. After this stage the model was forced with monthly mean winds for 1982 and 1983, provided by the National Meteorological Center.

There are several reasons for expecting discrepancies between the results from this model and the measurements. The parameterization of sub-grid scale mixing processes, the artificial zonal boundaries along the northern and southern extremes of the basin, the absence of all islands, including the Galapagos, and the incorrect initial conditions – conditions in January 1982 differed from those in the model at that time – all contribute to the errors. Probably of most importance are errors in the surface boundary conditions, especially the surface winds which were available on a very coarse spatial and temporal grid. In spite of these handicaps, the model performed surprisingly well. Changes in the heat content of the ocean, a variable that is correlated with sea level, are in good agreement with the measurements in Fig. 4. It can be argued that this is not a severe test for the model because the heat content, being a (vertical) integral, is a relatively simple variable and can readily be simulated with linear one-level models (Busalacchi and O'Brien, 1981). Simulation of changes in the vertical structure of the flow is a more stringent test for the model. Figures 8 and 9 demonstrate the ability of the model to reproduce such changes on the equator at 160° and 95°W. The considerable differences in the variability at these locations (Figs. 5 and 6), to be discussed in more detail later, are reproduced fairly well. [Keep in mind that spatial gradients are large near 0°N 95°W which is close to the Galapagos Islands (at 0°N 90°W).] Serious flaws appear in the model primarily towards the end of 1983 when isotherms fail to rise to the surface at 0°N 95°W, for example. Comparison with measurements in other regions reveal that the simulation of the recovery from El Niño after July 1983 was in general poor in all parts of the tropical Pacific Ocean. The following discussion of results is therefore limited to the period through July 1983. The extent to which errors in the wind field caused the discrepancies between the measurements and the model will be assessed once better wind data are available.

4. RESULTS

Figure 10 shows the simulated surface currents at various states of El Niño. Advection by these currents altered the sea-surface temperature patterns as shown in Fig. 11, and redistributed the warm waters of the upper ocean as shown in Fig. 12. By July 1982, there was intense eastward flow in the western equatorial Pacific while the southeast trades maintained westward flow in the eastern equatorial Pacific. During the subsequent months the region of unusual eastward flow expanded latitudinally — the region extends from 9°S to 9°N approximately — and eastward. The persistence of the southeast trade-winds inhibited the eastward flow at and south of the equator from penetrating much beyond 120°W by November 1982, but the North Equatorial Countercurrent had reached the coast of Central America by that time. This intensified the anti-clockwise circulation around the Coasta Rica Dome (Fig. 12), a feature observed by Barberan et al. (1984) and also resulted in the southward advection of warm surface waters across the equator in the region east of the Galapagos Islands (Figs. 13 and 11). The increase in the heat content of the eastern tropical Pacific after July 1982 was at the expense of the off-equatorial regions to the west of the dateline where the thermocline shoaled (Fig. 12).

Figure 14 shows the development of the surface and subsurface flow along the equator. The eastward winds west of the dateline between July and November 1982 (Fig. 1) are seen to drive intense eastward surface jets in the ocean. The jet of November 1982 is exceptionally intense and penetrates far east. It was this one that Firing et al. (1983) observed at 159°W (Fig. 5). Below the surface the Equatorial Undercurrent decelerated steadily after July 1982. This happened because the zonal pressure gradient that drives this current decreased steadily as more and more warm surface water was advected eastward (Fig. 15). By November 1982 the slope of the thermocline to the west of 160°W had actually reversed (Fig. 16).

The changes during the first phase of El Niño, up to December 1982, are in response to the relaxation of the tradewinds. Studies of the response to idealized changes in the wind indicate that the time-scale that characterizes the wind variations is a very important

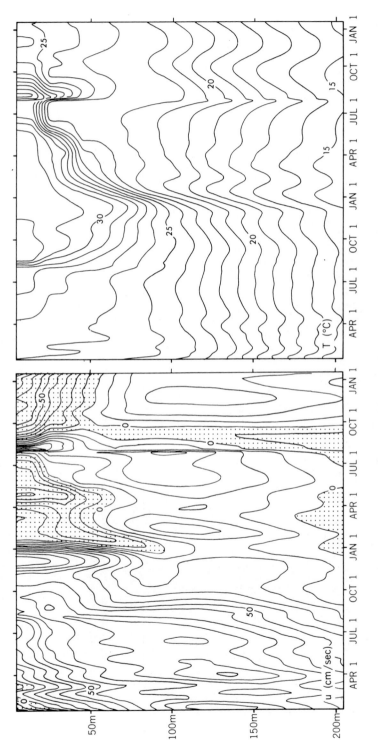

Fig. 8. Zonal velocity (contour interval = 10 cm s^{-1}) and temperature (contour interval = 1°C) fields at the equator and 159.5°W during 1982–1983 in the model. Shaded areas denote westward flow.

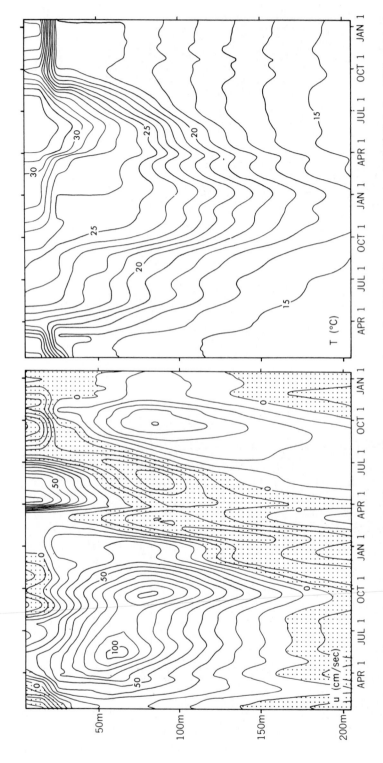

Fig. 9. Zonal velocity (contour interval = 10 cm s⁻¹) and temperature (contour interval = 1°C) fields at the equator and 94.5° W during 1982–1983 in the model. Shaded areas denote westward flow.

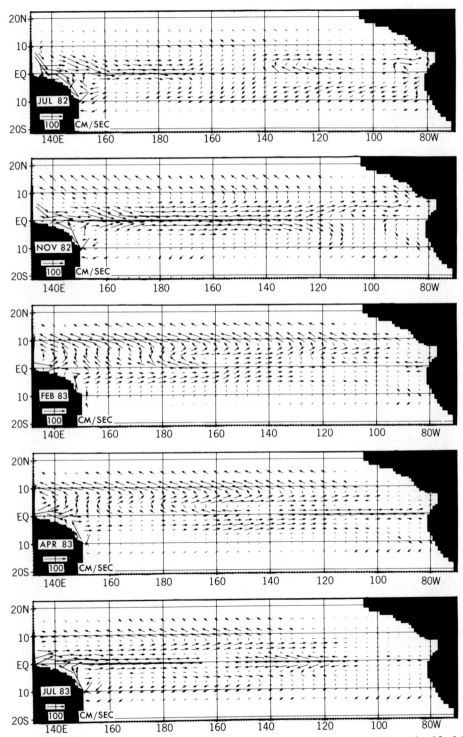

Fig. 10. Horizontal velocity vectors at a depth of 5 m. These are instantaneous values at day 15 of the indicated months.

530

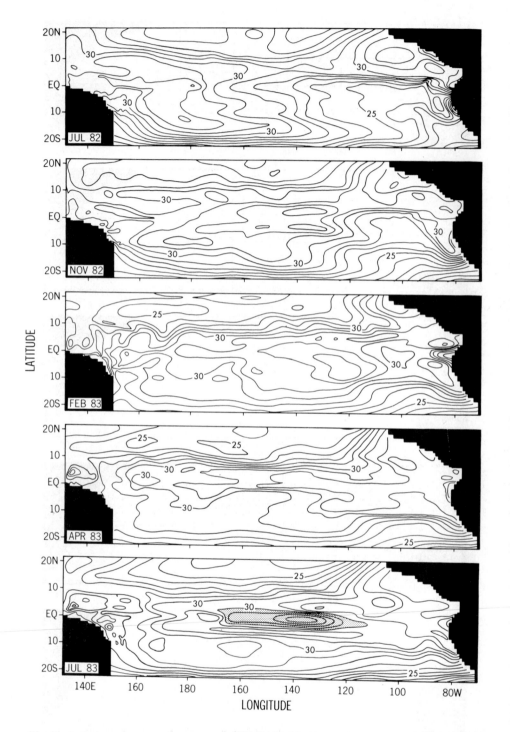

Fig. 11. Instantaneous maps of temperature (°C) at day 15 of several months at 5 m depth. Contour interval is 1°C. In the map for July 1983 the temperature in the shaded equatorial region is less than 30°C.

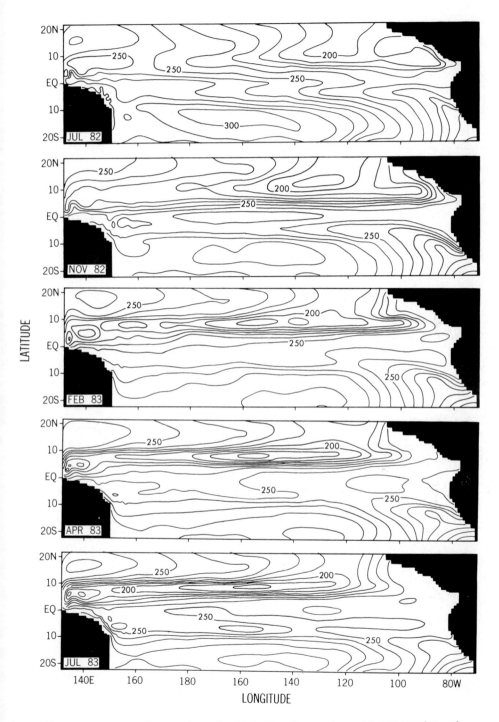

Fig. 12. Heat storage maps integrated to a depth of 317 m. Contour interval is 1.0×10^{-5} J cm^{-2}.

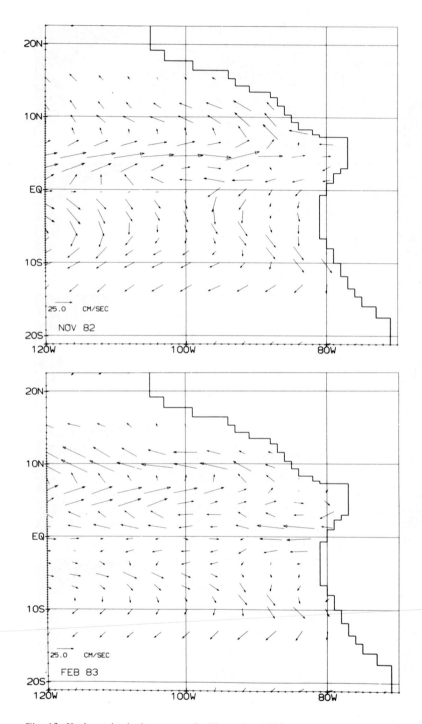

Fig. 13. Horizontal velocity vectors for November 1982 and February 1983 shown on a larger scale than Fig. 10 for the eastern Pacific.

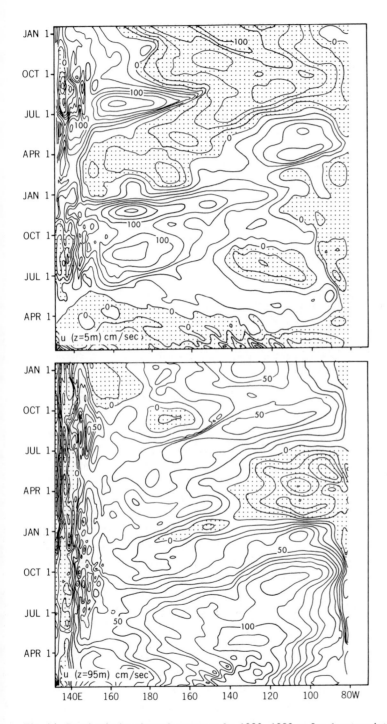

Fig. 14. Zonal velocity along the equator for 1982–1983 at 5 m (contour interval = 20 cm s^{-1}) and 95 m (contour interval = 10 cm s^{-1}). Westward flow is shaded.

534

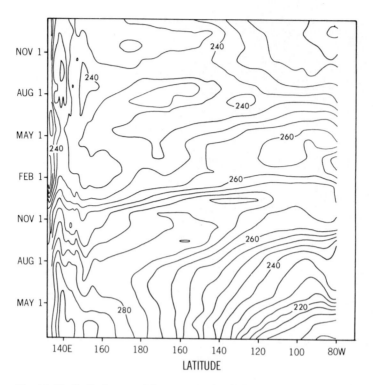

Fig. 15. Vertically integrated heat storage (to 317 m) along the equator during 1982–1983. Contour interval is 5.0×10^4 J cm^{-2}.

factor (Philander, 1981). In 1982 there were examples of both a gradual change in the winds (between June and November 1982) and an abrupt change, in December 1982 (Fig. 1). The response to a gradual change is one in which the ocean is at each moment practically in equilibrium with the wind. The gradual decrease of the zonal pressure gradient and the gradual deceleration of the Equatorial Undercurrent between June and November 1982 are examples of such a response. Kelvin waves undoubtedly play a role during this period, but they are not particularly prominent. [The highly variable eastward phase propagation evident in Fig. 14 is in part attributable to the eastward propagation of the weakening of the trades (Fig. 1).] Kelvin waves are far more prominent in the oceanic response to the abrupt relaxation of the tradewinds west of the dateline during December 1982. The eastward winds had, by that time, established a westward pressure force in the western Pacific where the thermocline sloped down to the east (Fig. 16). The abrupt relaxation of the winds left this pressure force unbalanced. It accelerated the surface waters westward and caused an elevation of the thermocline. Figures 14 and 15 show Kelvin waves propagating this effect eastward, as in the model of McCreary (1976). The eastward phase propagation is at a speed in the neighbourhood of 160 cm s^{-1}, considerably faster than any speed that can be inferred from Fig. 14 for the months before December 1982. This speed implies an equivalent depth of 25 cm, a value close to that of the second baroclinic mode. However, inspection of the vertical structure of the flow reveals that far more than a single vertical mode was involved. The abrupt rise of the thermocline in December 1982 affected the ocean at 160°W from the surface

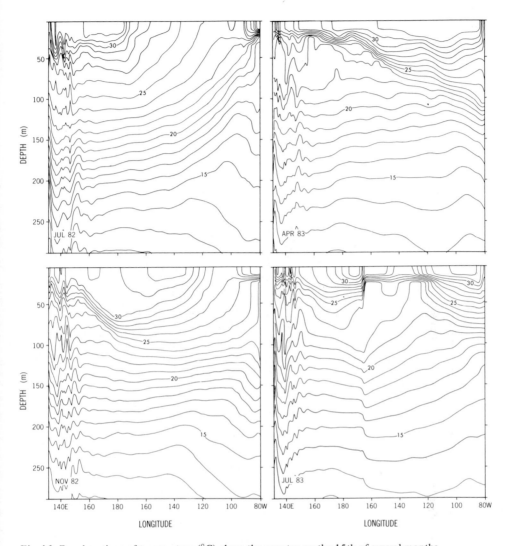

Fig. 16. Zonal sections of temperature (°C) along the equator on the 15th of several months.

to a depth of 200 m at least (Fig. 8), but further east, at 95°W, the thermal field of the upper 75 m of the ocean was unaffected (Fig. 9). This suggests that the Kelvin wave excited in the western Pacific in early December 1982 travelled eastward *and* downward. This is confirmed by Fig. 17 which shows that, at a depth of 55 m the rise of the thermocline in December 1982 peters out in the neighbourhood of 140°W, but at a depth of 95 m this rise is evident much further east. Note that none of Figs. 14, 15 and 17 shows westward phase propagation that could suggest reflection of Kelvin waves off the South American coast. The reason presumably is that a downward propagating Kelvin wave reflects as a Rossby wave that continues to propagate further downward. In response to periodic forcing, there is relatively little vertical dispersion as beams travel eastward and downward (McCreary, 1984) but wind pulses, such as that of November 1982 will cause considerable vertical dispersion. This interesting topic will be pursued on another

536

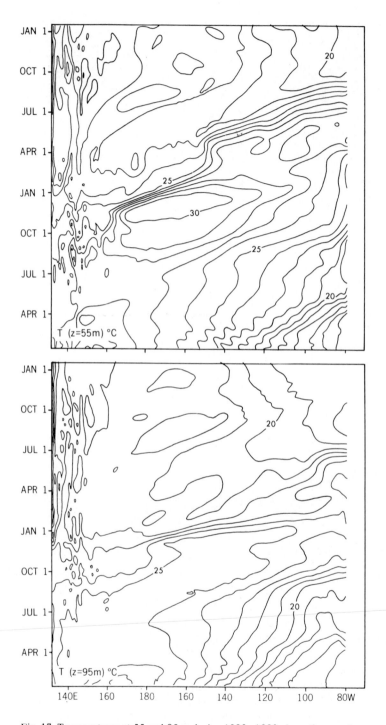

Fig. 17. Temperatures at 55 and 95 m during 1982–1983 along the equator.

occasion. Here, we note that one-level models that ignore the continuous stratification of the ocean could greatly exaggerate the importance of reflected waves on the upper ocean.

After the relaxation of the eastward winds west of the dateline in December 1982, the trades in the eastern Pacific also relaxed (Fig. 1). The response to no zonal wind is a horizontal thermocline, but the adjustment to this state was interrupted by the appearance of eastward winds in the eastern equatorial Pacific in March and April 1983 (Fig. 1). This generated an intense eastward surface jet near the equator (Fig. 10) and caused the thermocline to slope downwards to the east at all meridians (Fig. 16), a complete reversal of the situation in July 1982.

The return of the tradewinds in May 1983 signaled the end of El Niño. Sea-surface temperatures started to decrease first in the central Pacific (Fig. 11) where the winds quickly succeeded in reestablishing the usual slope of the thermocline, downward to the west (Fig. 16). Eastward winds apparently appeared to the west of the dateline in July 1983 (Fig. 1) but they did not expand and merely interrupted the recovery in that region.

Figures 18 and 19 shows the changes in the off-equatorial currents during 1982 and 1983. Note that eastward surface flow across 160°W extends from 6°S to 10°N by October 1982. After the deceleration of the eastward equatorial jet early in 1983, and the appearance of westward motion near the equator, the eastward North Equatorial Countercurrent, and an eastward current centered on 6°S, continue to persist. The westward North Equatorial current in the neighbourhood of 10°N is also seen to intensify. El Niño clearly is a phenomenon with a considerable latitudinal extent and involves far more than the equatorial wave-guide.

5. SUMMARY

A simulation, with a general circulation model of the tropical Pacific Ocean, of El Niño of 1982–1983 succeeds reasonably well in reproducing the few available measurements. This establishes confidence in the results in regions where measurements are unavailable. The model shows that the eastward transfer of heat from the western to the eastern tropical Pacific was affected by unusual eastward surface currents which, by November 1982, extended from 9°S to 9°N across 120°W. Further east the persistence of the southeast tradewinds inhibited eastward surface flow at and to the south of the equator at that time, but the eastward North Equatorial Countercurrent between 3° and 8°N penetrated right to the coast of Central America. The intensification of the North Equatorial Countercurrent affected the sea-surface temperature field significantly (Fig. 11), especially to the east of the Galapagos Islands where warm surface waters flowed southward across the equator. The pattern of changes in the heat content of the upper 300 m of the ocean (Fig. 12) differs from that of the sea-surface temperature and reflects the importance of the equatorial wave-guide where there were oceanic changes to a considerable depth. The oceanic response to the gradual relaxation of the trade winds between June and November 1982 was approximately an equilibrium so that the zonal pressure gradient, which the trades had maintained, and the intensity of the Equatorial Undercurrent, which the pressure gradient had maintained, decreased with the winds.

u (y,z)   T (y,z)

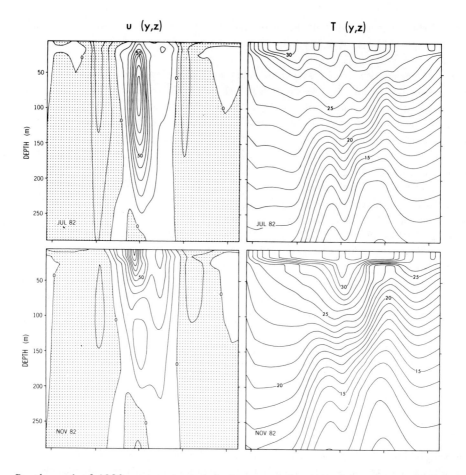

By the end of 1982, eastward winds had reversed the sign of the pressure gradient and eliminated the Equatorial Undercurrent (Fig. 14). Kelvin waves undoubtedly played a role in the adjustment of the ocean to the weakening of the trades, but they were not particularly prominent until December 1982. An abrupt weakening of eastward winds to the west of the dateline during that month caused a rapid rise of the thermocline and a westward acceleration of the surface currents right across the equatorial Pacific Ocean (Figs. 14 and 15). The Kelvin waves that induced these changes dispersed downward while propagating eastward. Waves excited by reflection at the coast of South America apparently continued to propagate downwards because there is no evidence of such reflected waves affecting the upper ocean.

The initial recovery from El Niño in early 1983 was interrupted by the appearance of eastward winds near the equator in the eastern half of the basin in March and April 1983 (Fig. 1). By this time the slope of the thermocline all along the equator, was downwards to the east, the opposite of what it had been in July 1982 (Fig. 16). The return of the tradewinds in May 1982 signalled the end of El Niño. A fall in sea-surface temperatures and reestablishment of the normal slope of the thermocline first occurred in the central equatorial Pacific Ocean.

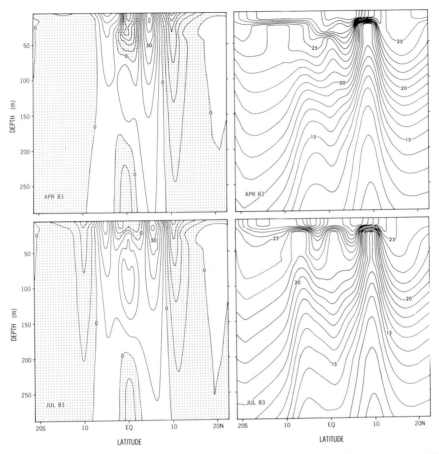

Fig. 18. Meridional sections of zonal velocity and temperature along 159.5W on the 15th of different months. Contour intervals are 1°C and 10 cm s⁻¹, respectively. Shaded regions indicate westward flow.

The success of the model is surprising because it is widely believed (as stated earlier) that the winds that drive the model are not very accurate. The monthly mean winds suppress information about high-frequency fluctuations. Such fluctuations must have caused the abrupt changes in sea level that were seen at Christmas Island in July 1982 (Fig. 4) but which are absent from the model. The grid of 5° longitude by 5° latitude on which the data are available precludes accurate estimates of the curl of the wind which must have been an important factor in the intensification of the off-equatorial currents. Time-series measurements of the winds at 0°N 110°W and 0°N 95°W (Halpern, 1984) suggest that the region of considerable wind variations during 1982 did not extend as far eastward as is shown in Fig. 1. Given the inaccuracies in the wind field, it is premature to make a detailed comparison between the results from the model and the measurements. Such comparisons, and more detailed analyses of the model results, will be made once the calculations have been repeated with an improved wind data set.

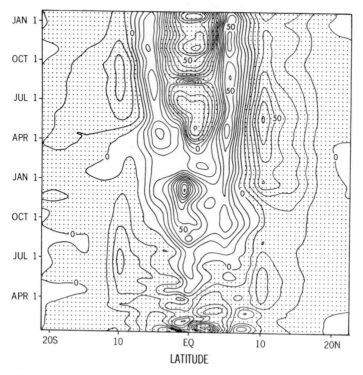

Fig. 19. Zonal velocity (cm s^{-1}) along 155.5°W at the surface during 1982–1983. Shaded areas are regions of westward flow.

ACKNOWLEDGEMENTS

We are indebted to D. Hansen for valuable comments on an earlier draft of the manuscript, to R. Pacanowski for assistance with the computations, and to P. Tunison and J. Pege for assistance in the preparation of the paper.

REFERENCES

Barberan, J., Gallegus, A. and Padilla, A., 1984. The Costa Rica Dome during the onset of the 1982–1983 El Niño. Trop. Ocean–Atmos. Newsl., 24: 13–14.
Bryan, K., 1969. A numerical method for the study of the world ocean. J. Comput. Phys., 4: 347–376.
Busalacchi, A.J. and O'Brien, J.J., 1981. Interannual variability of the equatorial Pacific in the 1960's. J. Geophys. Res., 86: 10901–10907.
Firing, E., Lukas, R., Sades, J. and Wyrtki, K., 1983. Equatorial undercurrent disappears during 1982–1983 El Niño. Science, 222: 1121–1123.
Halpern, D., 1984. Upper ocean current and temperature observations along the equator west of the Galapagos Islands before and during the 1982–83 ENSO event. El Niño/Southern Oscillation Workshop, Climate Res. Comm., Natl. Res. Counc., Miami, Fla.
Hansen, D., 1984. Surface drifter measurements in the eastern equatorial Pacific during El Niño of 1982–1983. (In prep.)

Katz, E.J., 1984. Basin wide thermocline displacements along the equator of the Atlantic in 1983. Geophys. Res. Lett., 11: 729–732.

Levitus, S., 1982. Climatological atlas of the world ocean. NOAA Prof. Pap. 13, 173 pp., U.S. Govt. Printing Office, Washington, D.C.

McCreary, J.P., 1976. Eastern tropical ocean response to changing wind systems: with application to El Niño. J. Phys. Oceanogr., 6: 632–645.

McCreary, J.P., 1984. Equatorial beams. J. Mar. Res., 42: 395–430.

Meyers, G. and Donguy, J.R., 1984. The North Equatorial Countercurrent and heat storage in the western Pacific Ocean during 1982–83. Nature, in press.

Pacanowski, R. and Philander, S.G.H., 1981. Parameterization of vertical mixing in numerical models of tropical oceans. J. Phys. Oceanogr., 11: 1443–1451.

Philander, S.G.H., 1981. The response of equatorial oceans to a relaxation of the trade winds. J. Phys. Oceanogr., 11: 176–189.

Philander, S.G.H. and Pacanowski, R.C., 1984. Simulation of the seasonal cycle in the tropical Atlantic Ocean. Geophys. Res. Lett., 11: 802–804.

Wyrtki, K., 1984a. Pacific-wide sea level fluctuations during the 1982–1983 El Niño. In: Galapagos 1982–1983: A Chronicle of the Effects of El Niño. Charles Darwin Research Station, Guayaguil, Ecuador, in press.

Wyrtki, K., 1984b. Monthly maps of sea level in the Pacific during El Niño of 1982 and 1983. In: Time Series of Ocean Measurements, Vol. 2. I.O.C. Tech. Ser. No. 25, UNESCO, Paris.

CHAPTER 34

THE PHYSICS OF THERMOCLINE VENTILATION

J.D. WOODS

ABSTRACT

The potential vorticity profile in the seasonal pycnocline can be predicted from the surface buoyancy and momentum fluxes by means of the Lagrangian correlation of seasonally varying mixed-layer depth and density. Water flows geostrophically from the seasonal pycnocline into the permanent pycnocline, through the sloping surface of depth D defined by the annual maximum depth of the mixed layer. Potential vorticity flows into the permanent pycnocline where $U(D) \cdot \nabla D \leqslant W(D)$ and vice versa. A theory is given for the regional variation of D, and methods of determining D from hydrographic data are reviewed. This physical understanding of potential vorticity sources and sinks makes it possible to reformulate ventilated thermocline models in terms of flux rather than density boundary conditions, and guides the design of coupled models of ocean–atmosphere circulation. It leads to formulae for water-mass formation and the nutrient balance in the seasonal boundary layer.

INTRODUCTION

This paper is concerned with the ventilation of the permanent pycnocline of the ocean. The problem is to predict the flux of potential vorticity into the pycnocline where it circulates in response to Ekman pumping as described by the thermocline equations (Robinson and Stommel, 1959; Welander, 1959). The aim is to relate the regional distribution of that influx of potential vorticity to the regional variation of the flux of buoyancy through the sea surface. That involves developing a theory for the sources and sinks of potential vorticity in the seasonal boundary layer, which lies on top of the permanent pycnocline.

Processes occurring above the permanent pycnocline are not considered explicitly in classical models of thermocline circulation (Luyten et al., 1982). They avoid the problem by prescribing the distribution of density and vertical velocity at the top of the permanent pycnocline, which is assumed to coincide with the bottom of the Ekman layer. The models implicitly incorporate Iselin's (1939) notion of water being pumped down isopycnically from the Ekman layer. That approach is unsatisfactory for two reasons: first because Iselin's conceptual model of the boundary layer is wrong, and second because prescribing the surface density prevents us from applying the model to climate problems in which the surface density distribution is unknown and changing.

The problem with Iselin's conceptual model of the subduction of surface water into the permanent pycnocline is that it does not take account the regional variation of the depth of the pycnocline. The top of the pycnocline lies by definition at the annual maximum depth of the mixed layer (D). The upper boundary conditions of models of circulation in the permanent pycnocline apply at depth $z = D$, not at the base of the

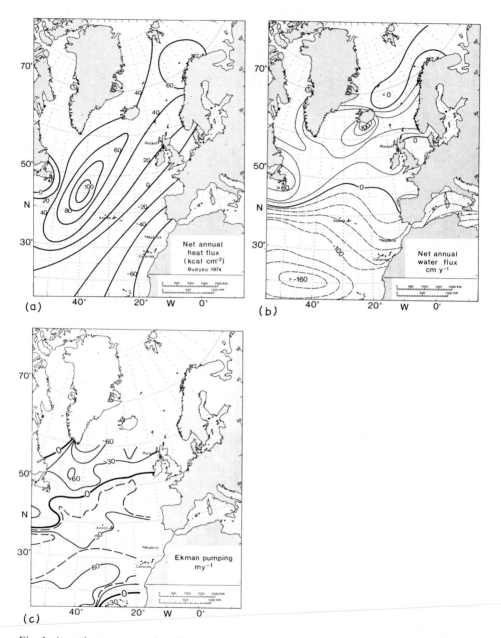

Fig. 1. Annual mean net surface fluxes of (a) heat (contour interval $20 \, \mathrm{KCal \, cm^{-2} \, yr^{-1}}$), (b) water (contour interval $20 \, \mathrm{cm \, yr^{-1}}$ and (c) annual vertical displacement by Ekman pumping in the NE Atlantic (from Woods, 1984b).

Ekman layer $(z = H_E)$. Iselin was wrong to assume that water is pumped down from the surface into the pycnocline: vertical mixing is an order of magnitude faster at most extra-tropical locations. (Compare the Ekman pumping displacement per year, Fig. 1c,

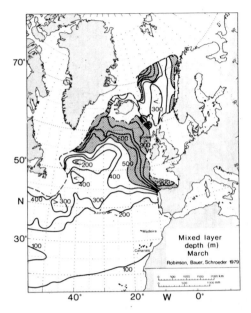

Fig. 2. The mean depth of the mixed layer for March in the Northeast Atlantic, which is one of the best available indicators of the regional variation of D, the annual maximum depth of winter convection (from Woods, 1984b; based on Robinson et al., 1979).

with the annual descent of the mixed layer, Fig. 2.) The vertical velocity is known (from the wind stress) at the base of the Ekman layer, which is much shallower than the top of the permanent pycnocline. Thermocline dynamics apply in the seasonal pycnocline between the Ekman layer and the permanent pycnocline ($H_E \leqslant z \leqslant D$), so water is not pumped down from H_E to D, but flows geostrophically in the seasonal pycnocline until it passes nearly horizontally through the sloping surface marking the top of the permanent pycnocline at depth D. According to this revised conceptual model, Ekman pumping drives lateral circulation: it is only indirectly related to the vertical motion in the pycnocline, which can better be explained in terms of adiabatic stretching of vortex tubes as the circulation carries them to different latitude. The new conceptual model is compared with Iselin's in Fig. 3. Sarmiento (1983b) found that the former produces better GCM simulation of tritium distribution in the Atlantic Ocean.

It is believed that the large thermal capacity of the permanent pycnocline will delay climate response to CO_2 pollution in the atmosphere (Charney, 1983). It has been conjectured (Bretherton, 1982) that the ventilation of the CO_2-induced thermal anomaly into the permanent pycnocline might be treated as though it were a passive scalar, like the tritium studied by Sarmiento (1983). If that were so the ventilation and circulation predicted by pycnocline models based on prescribed surface density might be relevant to the CO_2 problem. However, one has to be cautious because the heat inflow is caused by a change of a few $W\,m^{-2}$ in the surface net IR flux. Computer studies have shown that the seasonal variations of mixed layer depth and temperature are sensitive to such changes in the surface buoyancy flux (Woods and Barkmann, 1985). If the changes in the mixed layer affect the source of potential vorticity in the seasonal pycnocline that

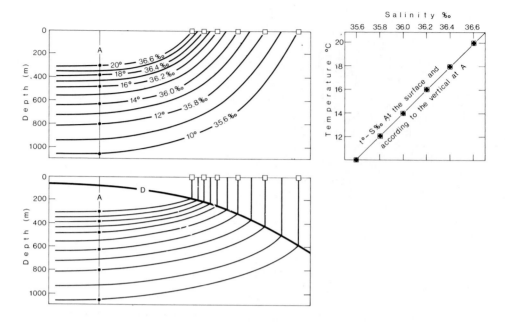

Fig. 3. *Upper*: Iselin's (1939) conceptual model of the generation of the density profile in the permanent pycnocline by the pumping down of water isopycnically from the Ekman layer. *Lower*: The new conceptual model in which water is mixed in winter down to a depth D. D is then much greater than the annual displacement by Eckman pumping. (The geostrophic circulation in the seasonal pycnocline, $z \leq D$, has been omitted for simplicity.)

will in turn change the flux of potential vorticity into the permanent pycnocline leading to a change in its circulation. Eventually that might show up as a reduction in the depth of the permanent pycnocline (D), which would further alter the ventilation rate. Bryan and Spelman (1984) have reported such changes in GCM experiments designed to investigate the role of the pycnocline in the CO_2 problem. More fundamentally the purpose of such models is to predict the rate of change of sea-surface temperature, so it is illogical to use a model based on prescribed surface temperature. The extra complication of using the surface buoyancy flux seems justified.

This paper investigates the problem of constructing a model of the pycnocline ventilated from the seasonal boundary layer in response to the surface fluxes. The aim is to develop a conceptual framework for the design of process models and GCMs, and experiments to test them.

THE SEASONAL BOUNDARY LAYER

This section briefly reviews the main features of the seasonal boundary layer of the ocean, which extends down to the top of the permanent pycnocline at depth D. It will be shown later that a quantitative description of the seasonal boundary layer can best be obtained by considering changes occurring along particle trajectories, i.e. from the Lagrangian viewpoint. However, it is easier to introduce the subject from the Eulerian

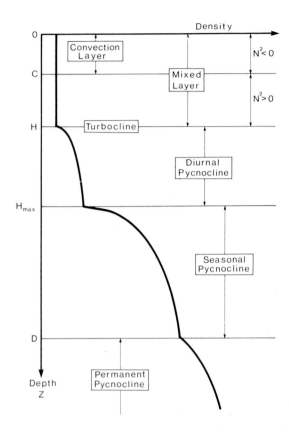

Fig. 4. Layers in the upper ocean. The convection layer is statically unstable ($N^2 \leqslant 0$), the remainder of the mixed layer is weakly stable ($N^2 \geqslant 0$). Turbulence in the mixed layer contains eddies with vertical scale up to H; the largest eddies in the diurnal pycnocline have scales much smaller than its thickness ($H_{max} - H$). The flow is predominantly laminar in the seasonal and permanent pycnoclines.

viewpoint at a fixed location in the ocean. To begin it is convenient to neglect advective changes, although later it will be shown that these dominate the seasonal evolution of the mixed layer. Advection will be neglected in describing those features that it only affects quantitatively, but cannot be avoided in discussing D, which depends fundamentally on the circulation of potential vorticity.

The description is based on Eulerian integration of a one-dimensional boundary layer model forced by the astronomical cycle of solar elevation and climatological mean seasonal variation of the atmosphere and surface fluxes derived from Bunker's (Bunker and Goldsmith, 1979) monthly mean data set. The illustrations are taken from integrations at a site (41°N, 27°W) where the net annual heat and water fluxes are close to zero (Fig. 1) and the currents are weak (Fig. 14), so the influence of advection is likely to be small. The predicted seasonal variation of mixed layer depth and temperature are sufficiently close to the observed values for the present purpose. A full account is available in Woods and Barkmann (1985). The water column is divided into the convection and mixed layers, and the diurnal, seasonal and permanent pycnoclines, as shown

548

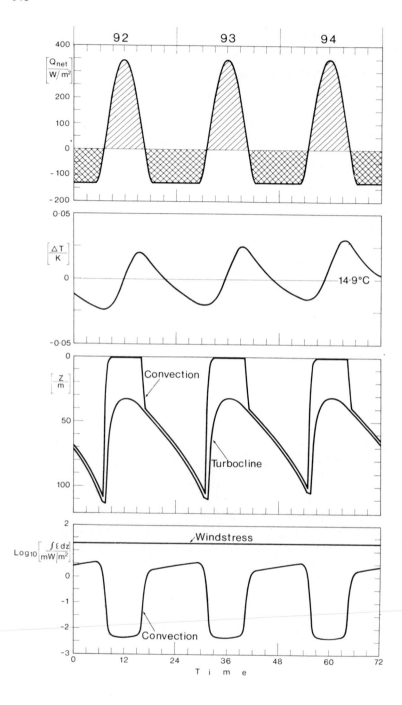

Fig. 5. Diurnal variation of surface net buoyancy flux, mixed-layer temperature, convection layer and mixed layer depths, and the power supply to turbulence in the mixed layer from convection and the wind stress (from Woods and Barkmann, 1985).

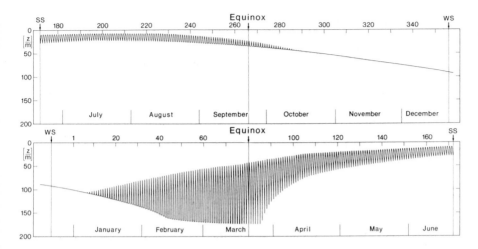

Fig. 6. Seasonal modulation of the mixed-layer depth H, showing the diurnal minima and maxima (from Woods and Barkmann, 1985).

in Fig. 4. The convection layer is unique in being statically unstable; the upward flux of buoyancy supplies the loss at the surface due to heat and water flow into the atmosphere. The density gradient is stable in the lower portion of the mixed layer $(C \leqslant Z \leqslant H)$ and in the pycnoclines. The climatological mean diurnal and seasonal variations in those layers are described below. No account will be taken here of changes induced by the weather, which modulates the climatological mean cycles.

Diurnal variation

The diurnal variation of convection depth and mixed-layer depth and temperature are illustrated in Fig. 5. During the day solar heating quenches convection, reducing the convection depth to less than one metre while the solar heat input exceeds twice the heat loss to the atmosphere (Woods, 1980b). Turbulence in the mixed layer loses nearly half its power input when convection weakens during the day, and loses 15% of the remainder to work against the Archimedes force as solar heating stabilizes the water below the convection layer. As the result, the turbocline marking the bottom of the mixed layer (across which the viscous dissipation rate drops from $mW\,m^{-3}$ to $\mu W\,m^{-3}$) rises during the day. The convection layer deepens again as the sun goes down in the afternoon, until the convection depth equals the turbocline depth $(C = H)$ about an hour before sunset. The mixed layer is then statically unstable from top to bottom, and remains in that state throughout the night as the turbocline descends to a diurnal maximum H_{max} about an hour after sunrise, when convection is again quenched.

Seasonal variation

Astronomical variation of day length and noon solar elevation combines with the seasonal variations of cloud cover and surface fluxes to produce the modulation of mixed-layer depth shown in Fig. 6. The seasonal variations of diurnal maximum mixed-layer

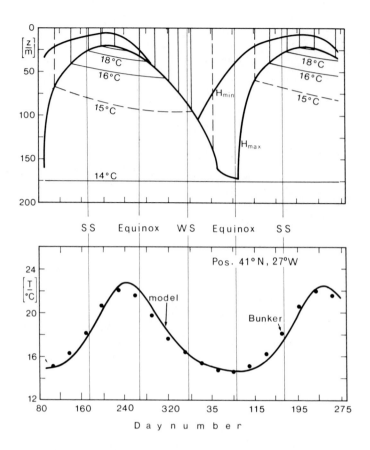

Fig. 7. Seasonal variation of the diurnal minimum (H_{min}) and maximum (H_{max}) of the mixed-layer depth, and the depths of isotherms. During spring the isotherms rise vertically through the mixed layer on the day they are subducted into the seasonal thermocline. Their subsequent deepening is due to solar heating. During the autumn and winter the isotherms are re-entrained into the mixed layer. Note the mixed layer acquires their temperature (vertical isotherm) some days after they are entrained; a consequence of turbulent penetration of the seasonal thermocline. The mixed-layer temperature predicted by the model is compared with Bunker's observed monthly mean values in the lower panel.

depth (H_{max}) and of the depths of isotherms is illustrated in Fig. 7. The descent of isotherms in the seasonal thermocline arises solely from solar heating as advection is not included in the model. The seasonal changes of temperature, salinity and density profiles are shown in Fig. 8. Note that seasonal storage of heat and water occurs mainly in the top 50 metres. That is determined by the penetration depth of solar heating during spring, which is considerably less than the annual maximum depth of the extra-tropical mixed layer D (Fig. 9).

The lower limit of the seasonal boundary layer

The seasonal boundary layer is defined as the water column that is changed at least once per year by diabatic processes related directly to the local surface energy and

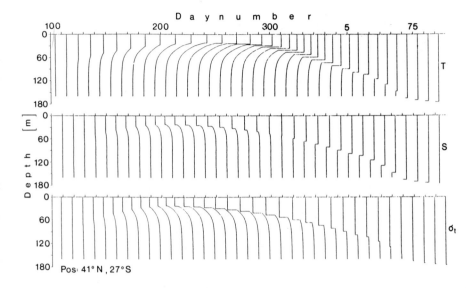

Fig. 8. Climatological variation of temperature, salinity and density profiles calculated by Woods and Barkmann (1985) using a one-dimensional model forced by the climatological mean seasonal variation of surface heat and water fluxes at 41°N, 27°W where the annual net heat and water fluxes are both close to zero. At this site the flux divergence of heat and water carried by ocean currents is nearly zero. Note that almost all of the seasonal storage of heat and water occurs in the top 50 m.

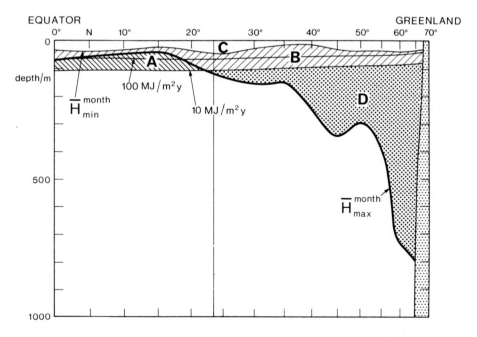

Fig. 9. A comparison of the depth of solar heating and the annual maximum depth of the mixed layer along a meridional section at 30°W (from Woods et al., 1984).

552

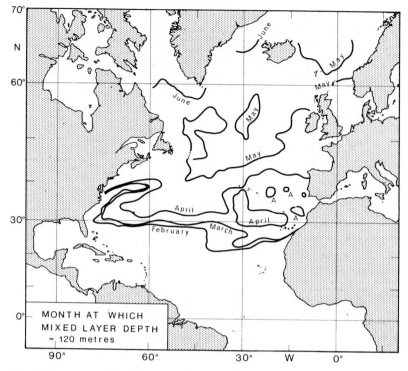

Fig. 10. The regional variation of the spring month in which the mixed-layer depth on average rises to 120 m.

water fluxes. The permanent pycnocline lying directly below the seasonal boundary layer is defined by the absence of diabatic processes related directly to the local surface fluxes. The surface separating the seasonal boundary layer and the permanent pycnocline is defined (in the one-dimensional model with no advection) in the tropics by the maximum penetration of solar heating, and at higher latitude by the annual maximum depth of the mixed layer, D (Fig. 9). In this paper we are concerned with the extra-tropical regime.

The effect of advection on D

During the extra-tropical winter, when the mixed layer is much deeper than the Ekman layer, its rate of descent is determined mainly by convection. (Turbulence generated by the wind stress is so weakened by dissipation in upper levels that it has little power available for entrainment of water from the seasonal thermocline when $H \gg H_E$.) The increase of H_{max} each day then depends upon the daily mean surface buoyancy flux and the density gradient in the top of the seasonal pycnocline. H_{max} reaches its annual maximum value D when the daily buoyancy flux changes sign at the beginning of spring. In regions where there is a net annual transfer of heat from the ocean to the atmosphere (e.g. north of a line from Europe to the Caribbean; see Fig. 1), the value of D calculated by a one-dimensional model without advection increases each year. If D is constant in nature the net annual heat buoyancy must be balanced by a

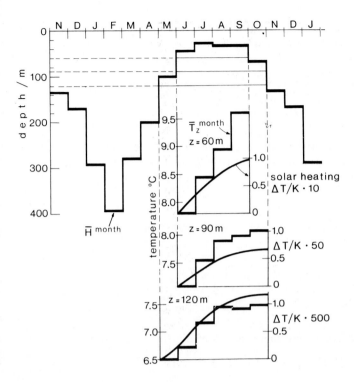

Fig. 11. The monthly mean depth of the mixed layer at OWS "C" according to Robinson et al. (1979) indicating $D = 400$ m. The lower three graphs show the monthly mean temperature inside the seasonal pycnocline at depths of 60, 90 and 120 m, indicating a mean rise of $\dot{T} = 1K/3$ months which varies little with depth. This is equivalent to a decrease in density of $\dot{\sigma}_t = 0.2$ kg m^{-3} over the three months (compare with Fig. 19).

divergence of the advective buoyancy flux. It is therefore impossible to predict D with a boundary layer model that does not include advection.

It is also to be expected that the differences at all seasons between the observed mixed-layer depth and temperature profile in the seasonal thermocline and those predicted by the one-dimensional model used to illustrate this section are partly due to neglect of advection. Seasonally varying advection dominates the seasonal cycle in the tropics. Its effects are also seen in strong steady currents at high latitudes. Note the effect of the Gulf Stream on the timing of the spring rise of the mixed layer in Fig. 10. The seasonal variation of temperature in the seasonal thermocline at OWS "C" (located in the North Atlantic current on the front between the anticyclonic and cyclonic gyres) is far larger than can be explained by solar heating; it reflects the advective heat flux divergence required to balance the 3 GJ m^{-2} yr^{-1} net heat loss to the atmosphere (Fig. 11).

Models of the seasonal boundary layer have hitherto ignored the contribution of advection. The main emphasis has been on checking whether they work tolerably well in regions of small mean surface buoyancy flux and weak currents (like OWS "P"). The main reason for ignoring advection has been the lack of an adequate description of the circulation. But the problem cannot be avoided if we want to predict the regional variation of D, which is controlled by advection. The resolution of this problem will be one of the major topics to be discussed below.

SOURCES AND SINKS OF POTENTIAL VORTICITY

Potential vorticity

Classical pycnocline theory (Robinson and Stommel, 1959; Welander, 1959; Luyten et al., 1982) expresses the density stratification (N) in terms of isopycnic potential vorticity (Q), which is assumed to be a Lagrangian conservative scalar:

$$N^2 = (g/\rho)\,\partial\rho/\partial z \tag{1}$$

where g is the acceleration of gravity, ρ is the seawater density and z the depth;

$$Q = (\xi + f)N^2/g \tag{2}$$

where ξ and f are the vertical components of relative and planetary vorticity. (Note that this is a simplified version of Ertel's potential vorticity suitable for use on a rotating sphere; in intense currents it is necessary to add terms neglected here, but they do not affect the discussion in this paper.)

For gyre-scale circulation, described by Sverdrup dynamics, the Rossby number is small:

$$Ro = (\xi/f) \ll 1 \tag{3}$$

permitting the use of the Sverdrupian potential vorticity Q_s;

$$Q \to Q_s = fN^2/g \tag{4}$$

The convection layer

Isopycnic potential vorticity cannot usefully be defined by eqn. (2) in a convection layer where N^2 is negative. Any convection layer therefore acts as a sink of (isopycnic) potential vorticity for water flowing into it and a source of potential vorticity for water flowing out. Here we are concerned with the upper ocean convection layer driven by the surface buoyancy flux (convection layers driven by double diffusion inside the pycnocline will be considered later).

In the last section it was shown that the depth of the convection layer varies diurnally from less than one metre in the day to the depth of the turbocline marking the base of the mixed layer at night. The depth of the turbocline H varies diurnally, reaching a maximum depth H_{max} each morning just before nocturnal convection is quenched by the rising sun, i.e. by the morning reversal of the surface buoyancy flux. The density of the water in the convection layer also varies diurnally, but the range seldom exceeds $0.1\ \mathrm{kg\,m^{-3}}$. Potential vorticity can be defined by eqn. (2) in the water below the surface convection layer because it is statically stable ($N^2 \geqslant 0$). It cannot be defined in the convection layer where $N^2 \leqslant 0$. The interface between these two regimes is not a material surface. Water flowing into the convection layer loses its potential vorticity; water emerging from the convection layer acquires potential vorticity.

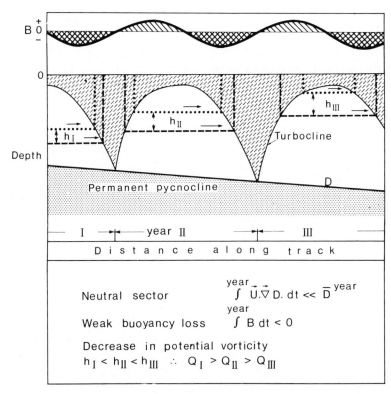

Fig. 12. The conceptual model of potential vorticity sources and sinks in the seasonal boundary layer of the ocean. The depths of the same pair of isopycnals are shown in three successive summers along the track of a column of water assumed to move barotropically across the ocean. The annual variation of the net daily buoyancy flux B is shown in the top panel, the water column suffers a small net annual buoyancy loss, giving a weak downstream increase in the depth D of the top of the permanent pycnocline. For simplicity it has been assumed that f is constant, and diabatic processes are negligible, so that the spacing between the pair of isopycnals remains constant during their residence in the seasonal pycnocline. The depth at which water of the same density is subducted is different each year. The spacing between the pair of isopycnals with the same densities (but containing different water) increases each year. The initial spacing and therefore the potential vorticity in the seasonal pycnocline can be calculated by Lagrangian integration of a one-dimensional model such as that used to produce Figs. 5–8. The non-linearity of the model makes it impossible to express $Q(\rho)$ as an analytical function of B.

In other words, if a water parcel lies in the convection layer its potential vorticity cannot usefully be defined by eqn. (2): if the water parcel lies in the thermocline, Q can be defined by eqn. (2), but the water cannot be cooling. This difficulty was avoided by Stommel (1979a), in his pioneering treatment of surface cooling in the β-spiral theory of flow in the seasonal boundary layer at OWS "J", by means of a parameterization in which Q decreases progressively during cooling.

Parameterizing diurnal variation

That is formally the physical mechanism of the source and sink of potential vorticity in the boundary layer of the ocean. In practice, however, it is not very convenient to work with the convection depth C because the water in the (weakly stable) depth range $C \leqslant Z \leqslant H_{max}$ is sufficiently turbulent to produce rapid change in Q. We therefore ignore the diurnal variation of both C and H, and assume that eqn. (2) is valid only for $Z \geqslant H_{max}$. We shall adopt that simplification throughout this paper and, unless specially stated to the contrary, mean the diurnal maximum depth and density of the convection layer when we refer to the mixed-layer depth and density.

The climatological mean seasonal variations of the mixed layer depth and density were described in the last section. It is important to resolve or parameterize the diurnal variation of convection depth in predicting the seasonal cycle, but provided that is done we can stick to our simplification of treating H_{max} as the effective source/sink boundary for Q. The justification is that the difference between successive values of H_{max} and ρ_{max} are small, so that there is little error in holding constant the correlation between mixed layer depth and density in the 24-hour interval between successive values of (H_{max}, ρ_{max})*. Furthermore, the distance moved by water during 24 h is small compared with the correlation scales of the surface fluxes. The substitution of H_{max} for C greatly simplifies modelling thermocline ventilation, by allowing us to relate it to the seasonal rather than diurnal variation of surface fluxes, as illustrated in Fig. 12.

Ventilation of the seasonal pycnocline

The rapid rise of the turbocline (H_{max}) in spring leaves behind a water column with a density profile $\rho(z)$ from which we can determine $Q_s(\rho)$ by eqn. (4). That is the mechanism of the source of potential vorticity to be discussed in this paper.

The rather slower descent of the turbocline (H_{max}) in autumn and winter entrains water from the underlying pycnocline, consuming its potential vorticity. The annual maximum depth of the mixed layer D depend on the profile of $Q(\rho)$ and the surface fluxes. That is the mechanism of the sink of potential vorticity to be discussed in this paper.

In many regions the distance moved by water in one year (Fig. 14) is not negligible compared with the downstream correlation scales of the surface fluxes (Fig. 1), so it is not possible to assume that the potential vorticity profile consumed in the autumn and winter is the same as that formed at the same location the previous spring. It is necessary to know the circulation and the regional variation of the surface fluxes if we are to predict the profile of potential vorticity in the seasonal pycnocline, $Q(\rho)$.

Sensitivity of ventilation to buoyancy flux and advection

The seasonal variation of mixed-layer depth in a model with no advection is very sensitive to the surface buoyancy flux B (Woods and Barkmann, 1985). So we expect ventilation to respond to small changes in the buoyancy flux climate. At sites where the net annual buoyancy flux is negative (e.g. where annual transfer of heat from the ocean to the atmosphere exceeds the heat input from the sun), the value of D increases

---

*This is the diurnal equivalent of Stommel's (1979b) argument that water subducted into the pycnocline under the ITCZ has the density of the mixed layer when it is deepest in the annual cycle. But Rayleigh's demon, rather than Ekman's, is responsible.

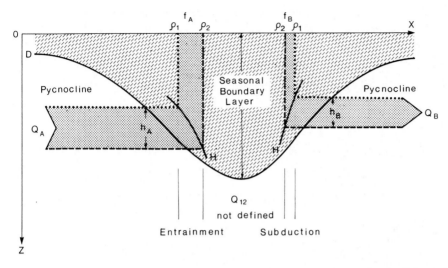

Fig. 13. The conceptual model of the flux of potential vorticity into and out of the permanent pycnocline from the seasonal boundary layer (shaded) by geostrophic flow through the sloping surface (D) between them. (For simplicity the seasonal cycle of mixed-layer depth has been omitted.) The lightly stippled layer represents water of potential vorticity Q_A in the pycnocline flowing into the boundary layer where the Coriolis parameter is f_A giving a separation h_A between the isopycnals ρ_1 and ρ_2. After passing into the boundary layer those isopycnals do not occur in the pycnocline until somewhere downstream (different) water flows out of the boundary layer into the pycnocline with the same densities ρ_1 and ρ_2. That occurs at a location where the Coriolis parameter is f_B and the spacing between the isopycnals is h_B, giving a Sverdrupian potential vorticity Q_B. The relationship between $(Q_B - Q_A)$ and the surface fluxes depends on boundary layer processes that can be modelled explicitly along individual trajectories (as in Fig. 12) given the Lagrangian variation of surface fluxes.

each year in such a model. In order to stabilize D, buoyancy loss through the surface must be balanced by a buoyancy gain from advection. That advective gain of buoyancy can be expressed in terms of the (Eulerian) change in potential vorticity profile of the seasonal pycnocline wrought by advection. The Gulf Stream causes the turbocline to ascend much earlier than in neighbouring regions with comparable surface fluxes. This influences the potential vorticity profile of water subducted that spring and subsequently the depth to which convection descends next winter a few megametres downstream. The effects of advection cause D (Fig. 2) to have a much smaller Lagrangian correlation scale than the surface buoyancy flux (Fig. 1).

Ventilation of the permanent pycnocline

The permanent pycnocline receives potential vorticity from and loses potential vorticity to the seasonal pycnocline as water flows geostrophically through the sloping surface defined by depth D. Potential vorticity is defined as entering the permanent pycnocline wherever D decreases downstream, and leaving it where D increases downstream.

i.e. At sources of Q, $\qquad U \cdot \nabla D \leqslant 0$

and at sinks of Q, $\qquad U \cdot \nabla D \geqslant 0$

(5)

558

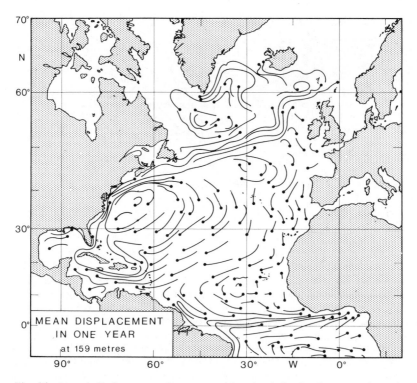

Fig. 14. Annual displacement of water particles circulating in the mean circulation calculated by Sarmiento (1983b) using the Princeton GCM, in which lateral mixing by quasi-geostrophic eddies is parameterized by a horizontal eddy-viscosity.

Figure 13 shows water with potential vorticity Q_A between isopycnals ρ_1 and ρ_2 passing from the permanent pycnocline into the seasonal pycnocline at a location where the planetary vorticity is f_A, so that the spacing between the isopycnals $h_A = [(\xi_A + f_A)/Q_A]$ $[2(\rho_2 - \rho_1)/(\rho_2 + \rho_1)]$. Provided the water spends at least one winter above the permanent pycnocline it will become blended into the mixed layer, and subsequently spread evenly through the seasonal pycnocline that forms next spring. There is therefore no continuity of water particles between water leaving the permanent pycnocline between ρ_1 and ρ_2 and water later entering the permanent pycnocline between a pair of isopycnals with the same densities. Those two isopycnals do not exist and Q_{12} is not defined in the permanent pycnocline between entrainment into the seasonal pycnocline and later subduction from it. The spacing h_B and the Coriolis parameter f_B determine the potential vorticity Q_B of the water entering the permanent pycnocline between ρ_1 and ρ_2 at that subduction site (assuming $\xi_B = 0$). In order to predict this source potential vorticity Q_B for the permanent pycnocline we have to trace the water back a few months to its subduction from the mixed layer into the seasonal pycnocline where it was stored until drifting to a site where D is shallower than the isopycnals.

Classical theories of the ventilated pycnocline (i.e. those based on a surface density boundary condition) tell us that the gyre circulation depends on the distribution of both the wind stress curl (Ekman pumping) and the source of potential vorticity. It has been shown that the source of Q depends on the surface fluxes during spring and the downstream variation of D. So accurate prediction of the regional variation of D will be an essential test of models of the ventilated pycnocline based on the surface buoyancy flux. Comparison of maps of the surface fluxes (Fig. 1) and D (Fig. 2) shows only a crude correlation. That leads us to the conclusion that D is very sensitive to the circulation of Q.

THE SUBDUCTION OF ISOPYCNALS

Eulerian models

Potential vorticity is defined by the vertical spacing between isopycnals. In order to quantify our conceptual model of potential vorticity injection it is necessary to predict the depths of isopycnals as they are subducted from the mixed layer during its spring ascent. It was shown earlier (Fig. 7) that the subduction depth of isopycnals can be predicted by a one-dimensional mixed-layer model. That illustration was based on Eulerian integration of the model, forced by the seasonally varying surface fluxes at a fixed site, with no allowance for changes due to advection. The new conceptual model of pycnocline ventilation focusses attention onto those regions where $|U \cdot \nabla D|$ is large, i.e. where currents are strong. It is therefore necessary to introduce advection explicitly into the boundary layer model before it can be used to predict the subduction depths of a set of isopycnals, and the profile of potential vorticity in the seasonal pycnocline. This might be achieved in principle by adding an Eulerian advection term $U \cdot \nabla \rho$ to the model, as in general circulation models of thermocline ventilation (Bryan et al., 1982). The success of that approach depends on the accuracy of velocity field predicted by the GCM. The results of existing GCMs are encouraging, but are they sufficiently accurate for the prediction of isopycnal subduction depth? How sensitive is the subduction depth to advection?

Diagnostic studies based on hydrographic data might help answer those questions if the empirical distributions of velocity and density they generate were sufficiently accurate. There has been some success in recent years in estimating the net buoyancy flux through sections crossing oceans from coast to coast (Bryden and Hall, 1980), but there is controversy about the details of circulation patterns on the scale of seasonal water displacement. It is concluded that diagnostic studies based on the correlation of fields of velocity and density derived from hydrographic data cannot help to clarify the role of advection on the density profile in the seasonal pycnocline.

The Lagrangian correlation of mixed-layer depth and density

A different approach is investigated below. The attempt to estimate advection terms in Eulerian integrations of the boundary layer model is abandoned in favour of a Lagrangian approach. The method is to use the correlation of the seasonally varying depth and density of the mixed layer to calculate the depth of each isopycnal as it is subducted at successive locations along the trajectory of a water column. To first approximation the potential vorticity profile remains constant during the spring

560

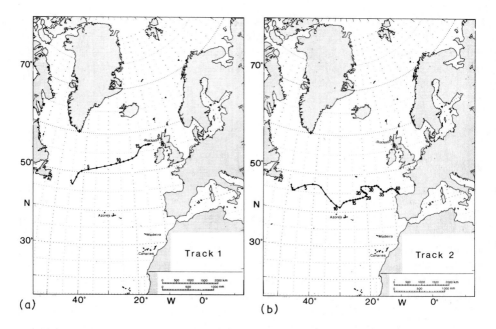

Fig. 15a. Track of the drifter used in the first case study. b. Track of the drifter used in the second case study.

subduction season, i.e. $\dot{Q}(\rho) = 0$*. (Processes that change $Q(\rho)$ will be discussed later.) It is possible in this way to predict the potential density profile of water in the seasonal pycnocline passing through a fixed site at any given instant, either by extracting from oceanographic data banks the observed climatological mean Lagrangian correlation of mixed layer and density at successive locations occupied by the water as it drifted along the trajectory that eventually carries it through the chosen site, or by computing it using a mixed layer model driven by the climatological mean surface fluxes at the same sequence of locations.

The process of potential vorticity injection into the seasonal pycnocline can be quantified in that way for individual trajectories defined by the tracks of drifters. Ignoring any slippage relative to the water, a drifter describes a trajectory that occurred once. The trajectory is one of many possibilities. If we ignore seasonal variation in the mean circulation (small outside the tropics) and in the eddy statistics (Dickson et al., 1982), there is an equal probability of a water particle following the chosen trajectory at any season (i.e. at any phase relative to the seasonal cycle) although a drifter was actually observed to pass along the track on only one occasion with a particular seasonal phase. We can contemplate a hypothetical set of particles following the same track at increasing phase lags relative to the seasonal cycle; increments of two weeks were used in the example that follows.

---

*The Newtonian overdot notation for Lagrangian rate of change is used throughout this paper.

POTENTIAL VORTICITY INJECTION

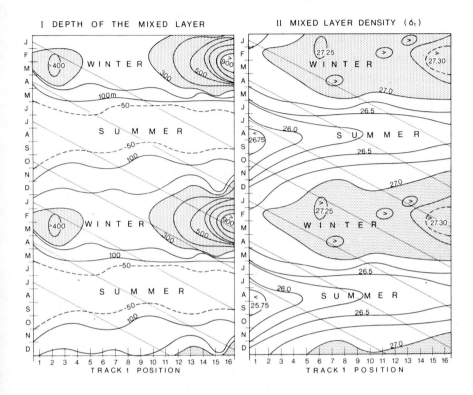

Fig. 16. a. Phase diagram for seasonal variation of mixed layer depth (m) along track 1. b. Phase diagram for seasonal variation of mixed-layer density (σ_t) along track 1.

A diagnostic study

The theory outlined above is illustrated by a diagnostic study in which the Lagrangian correlation of mixed-layer depth and density were computed along the tracks of two drifters with drogues at 100 m depth. It is assumed that the water in the seasonal pycnocline ($H_{max} \leqslant z \leqslant D$) moved barotropically along the tracks so that a water column remained vertical. (The effect of current shear will be discussed later.) Position numbers were marked along the buoy tracks at intervals of 15 days (Fig. 15). The synoptic variations of mixed-layer depth and density along the drogue tracks were extracted for each month of the year from the monthly mean maps in Robinson et al. (1979). They were then contoured with the month of year on the ordinate and the buoy positions on the abscissa (Figs. 16 and 17).

In the resulting phase diagrams, the synoptic variation along the track at a given time of year is aligned horizontally, seasonal variation at a fixed position along the track is aligned vertically, and the Lagrangian variation for a water column travelling barotropically along the track is indicated by the straight sloping phase lines. Temporal variation for a

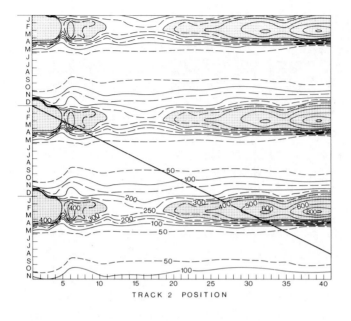

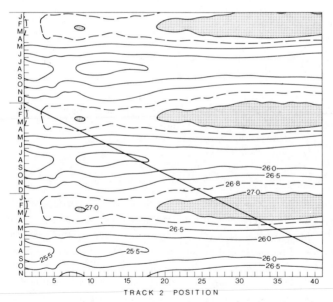

Fig. 17. a. Phase diagram for seasonal variation of mixed layer depth (m) along track 2. b. Phase diagram for seasonal variation of mixed-layer density (σ_t) along track 2.

water column starting along the track at any given day of the year can be extracted by choosing the correct seasonal phase for the sloping line. For example, a water column starting along track 1 in mid-February is seen to be mixed down to a maximum depth of 400 m between positions 2 and 3, and thereafter crosses the Atlantic while the mixed

563

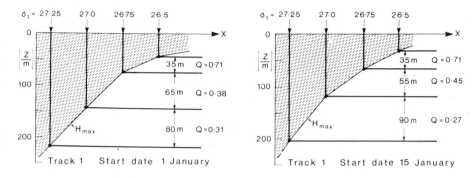

Fig. 18. The subduction depths of isopycnals and the potential vorticities (Q in radians per Gigametre second) along track 1. (a) Starting date 1 January; (b) Starting date 15 January.

layer depth is shallow in summer, arriving off Scotland before it is 100 m. Another water column starting along the same track in July is mixed to a depth of 900 m before reaching position 17. A third column starting at the end of November passes through the winter in a region where the maximum depth of mixing does not exceed 300 m. So examination of Fig. 16a shows that the Lagrangian variation of mixed-layer depth depends on the seasonal phase with which a water column flows through the seasonal and regional variation of mixed-layer depth. Large regional changes of mixed-layer climate are encountered during the nine months that the water takes to flow from end to end of track 1.

Seasonal stratification along track 1

The subduction depths of isopycnals emerging from the mixed layer along track 1 are determined by correlating mixed-layer depth and density along a common seasonal phase line in Fig. 16. The results are shown in Fig. 18 for two water columns starting on the 1st and 15th January, respectively. The Sverdrupian potential vorticity profile $Q_s(\rho)$ derived from the spacing between the isopycnals is indicated. The values are similar to those mapped in the same region (McDowell et al., 1982; Fig. 21). The difference in $Q_s(\rho)$ in the two water columns indicates the change that would be encountered in half a month at a fixed location along the track, due solely to the change in subduction depths of isopycnals emerging from the mixed layer at different seasonal phase along the same track. (Other causes of Eulerian change in $Q_s(\rho)$ will be discussed later.) The difference also indicates the downstream gradient of potential vorticity $\nabla_\rho Q(\rho)$ measured along isopycnals. Note the variation of $\nabla_\rho Q(\rho)$ with depth. (The role of $\nabla_\rho Q(\rho)$ in mesoscale frontogenesis will be discussed later.) The Eulerian development of the density profile near OWS "C", calculated from the variation of subduction depth with seasonal phase along track 1, is shown in Fig. 19. This variation can be compared with the climatological mean development of the temperature profile in the seasonal thermocline at OWS "C" if we assume the seasonal variation of salinity is negligible (Taylor and Stephens, 1980). The climatological mean rate of change predicted by Lagrangian correlation of mixed-layer depth and density along track 1 is approximately 50% greater than the climatological mean change observed at OWS "C". This is considered

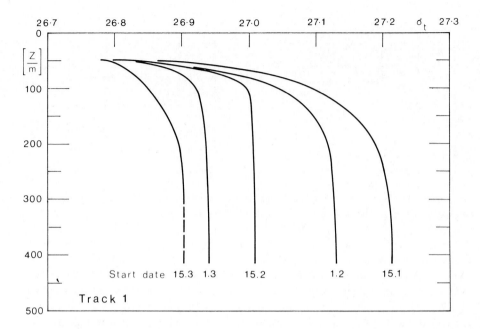

Fig. 19. Development of the density profile in the seasonal pycnocline at a fixed position near OWS "C" computed from the successive arrival of water columns that started flowing barotropically along track 1 at intervals of 15 days.

a remarkably close fit given the assumptions of the calculation, and the fact that track 1 passes to the south of OWS "C". It provides strong support for the Lagrangian theory of the seasonal pycnocline presented in this paper.

Seasonal stratification along track 2

The second track is longer and allows us to see subduction, re-entrainment and a second subduction during the twenty months that the water column takes to go from the Grand Banks to NW Spain (Fig. 20). The striking difference in the depths at which isopycnals are subducted into and later entrained from the seasonal pycnocline can be attributed to the neglect of vertical shear, which acts on the phase variation of subduction depth seen in Fig. 18. (This and other causes of change in $Q_s(\rho)$ will be discussed later.) The difference between the depths of isopycnals at entrainment and subsequent subduction represents the change wrought by the winter, as does the corresponding change of potential vorticity profile $\Delta Q(\rho)$. Note that $\Delta Q(\rho) \leqslant 0$ if the depth of subduction exceeds 50 m where convection controls mixed-layer depth, and $\Delta Q(\rho) \geqslant 0$ if subduction occurs in the top 50 m where the wind stress is more important.

Comment

It was pointed out earlier that there is no continuity of water particles between isopycnals entrained in winter and re-established in spring, so it would be incorrect to

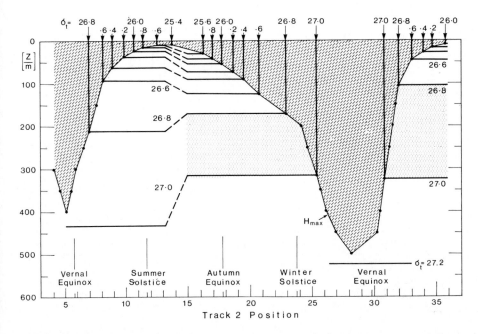

Fig. 20. Depths at which isopycnals are subducted, re-entrained and subducted again during the 20 months taken by water to flow along track 2 used in the diagnostic study. Note that Q is decreased by winter mixing if the subduction depth is greater than 50 m, and vice versa.

talk about a continuous change of Q due to winter cooling, as did Stommel (1979a) and Olbers et al. (1982). The difference in Q between the same isopycnals before and after a winter mixing event depends non-linearly on the surface fluxes, and, although it can, in principle, be calculated from Lagrangian integration of a one-dimensional mixed layer model, there is no simple functional relation between ΔQ and B, as assumed by Olbers and Willebrand (1984). Curiously, the magnitude of the change depends mainly on the correlation of the mixed-layer depth and density during the spring warming season, and not on the integrated heat loss during the winter. The cooling controls the regional variation of D rather than the downstream change in Q.

REGIONAL VARIATION OF D

The conceptual model (Fig. 3) underlying the present theory of potential vorticity injection states that lateral subduction by geostrophic flow through regions of decreasing D is more important than vertical subduction by Ekman pumping, i.e.:

$$\overline{U_g \cdot \nabla D}^{yr} \gg \overline{W_E}^{yr} \tag{6}$$

Inspection of maps of U_g, D and W_E suggests that eqn. (6) is satisfied in the NE Atlantic, (Woods, 1984b). In this section we consider two problems. The first is how to construct maps of D from oceanographic observations. The second is how to predict D.

566

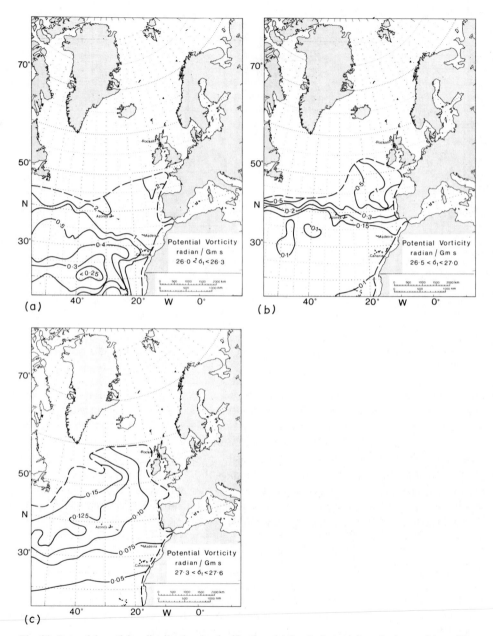

Fig. 21. Potential vorticity distribution in the Northeast Atlantic derived from hydrographic archives. (Redrawn from McDowell et al., 1982.)

Observations of D

Our knowledge of the regional variation of deep mixing in the extratropical winter has advanced rapidly in the past few years with the analysis of bathythermograph and

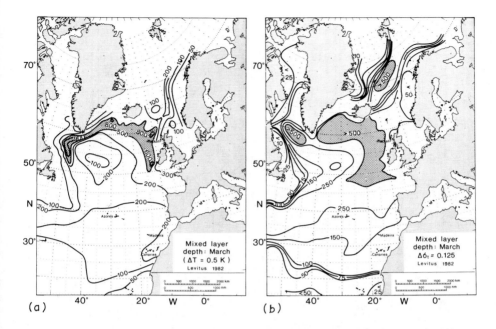

Fig. 22. March-mean depth of the mixed layer according to Levitus (1982) based on the following criteria: (a) $\Delta T = 0.5$ K; (b) $\Delta\sigma_t = 0.0125$ (from Woods, 1984).

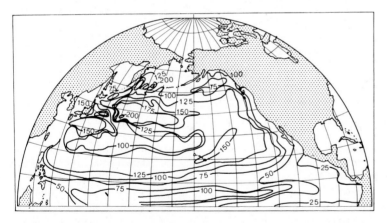

Fig. 23. The depth of winter mixing in the North Pacific deduced by Reid (1983) from the depth of the oxycline.

hydrographic archive data. Maps of monthly mean mixed-layer depth have been published by Robinson (1976) for the North Pacific and Robinson et al. (1979) for the North Atlantic (Fig. 10). Levitus (1982) gives world maps of the March and September mixed-layer depths based on the depth of the thermocline and, in an alternative version, the pycnocline (Fig. 22). Reid (1982) used oxygen saturation ratio to map the annual maximum depth of winter mixing in the Pacific (Fig. 23). Broecker and Peng (1982) have reviewed methods based on transient tracers, such as ^{3}He, which is formed by

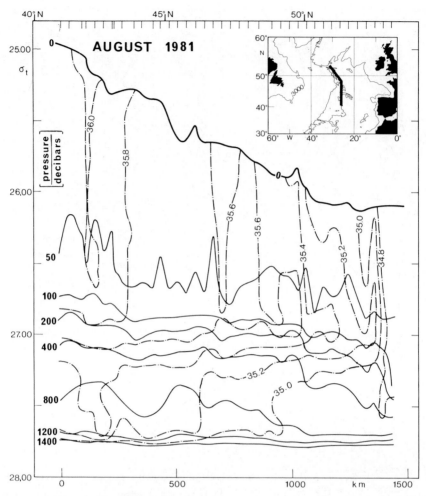

Fig. 24. The maximum depth of mixing in the previous winter revealed by the thermoclinicity elbow in a hydrographic section going north from the Azores (data from Fahrbach et al., 1983, replotted isopycnically by the present author).

radioactive decay in the interior and is mixed up to the surface where it outgases (Jenkins, 1980, 1982). Woods (1984a) exploited the fact that the seasonal pycnocline depends much more on temperature than on salinity change to determine the annual maximum depth of the mixed layer from hydrographic sections collected in summer, the "thermoclinicity elbow" method (Fig. 24). Spot values of the depth of the mixed layer have been estimated from hydrographic data either by the criterion of an adiabatic lapse rate or more simply by identifying it with a kink or elbow in the temperature and salinity profiles (Fuchs et al., 1984; Fig. 25). Advective restratification of the mixed layer after a deep mixing event can destroy the evidence in single profiles within a few weeks (Fig. 26), but the evidence from the oxygen, ^{3}He and thermoclinicity elbows persists through the summer. These methods therefore offer the practical advantage of using data collected during summer to estimate the maximum depth of mixing achieved during the preceding winter.

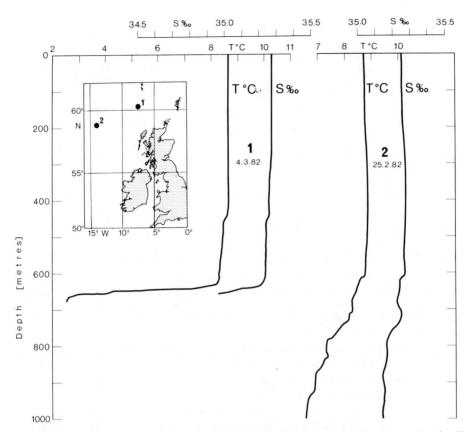

Fig. 25. The maximum depth of mixing in the previous winter revealed by a kink or by the elbow in temperature and salinity profiles from individual hydrographic stations (data kindly provided by Prof. Dr. J. Meincke).

In order to exploit that advantage it is necessary to make a correction for the horizontal displacement of water from its location at the end of winter mixing to its encounter some months later by a reseach ship making summer hydrographic stations. The tracks of surface drifters with drogues below the Ekman layer provide an indication of the displacement. Inspection of many drifter tracks (Richardson, 1983; Rossby et al., 1983; Krauss and Käse, 1984) shows that synoptic scale (quasi-geostrophic) motion produces considerable variability in the trajectories of water circulating in the upper ocean (Fig. 28). Although their mean stream function may resemble those deduced from general circulation models of the ocean (Fig. 14), the winter location of water sampled during a summer hydrographic station may have deviated by up to a megametre from that predicted according to the climatological mean or model-predicted trajectory. The eddy-induced uncertainty in the winter locations covers an area which spans large variations of D. The only way to reduce the uncertainty arising from this large seasonal catchment area is to collect the hydrographic data as early as possible in the heating season.

A theory for D

It was shown earlier that the annual maximum depth of convection cannot be predicted by Eulerian integration of a one-dimensional model forced by seasonal cycles of surface

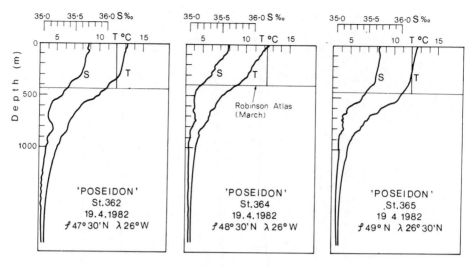

Fig. 26. In regions of strong vertical current shear the elbows created in temperature/salinity profiles by deep winter convection soon become undetectable. These profiles come from stations in the centre and 100 km to the north and south respectively of the North Atlantic current core. The Robinson et al. (1979) March-mean depths and temperatures of the mixed layer are indicated for comparison.

fluxes at the chosen location. Worthington (1976) argued that D increases downstream where there is a net annual surface buoyancy loss and vice versa. The diagnostic study of stratification in the seasonal pycnocline, reported above, revealed the importance of both seasonal and regional variation, and showed how they could be related to the surface fluxes by Lagrangian integration of a one-dimensional mixed-layer model. We begin to develop a theory for D by considering the same approach, namely Lagrangian integration of a one-dimensional mixed-layer model. It will be shown that D depends on the inflow of water from the permanent pycnocline in regions where $U \cdot \nabla D \geqslant 0$.

Figure 27 shows the seasonal variation of mixed layer along the trajectory of a particle lying initially in the permanent pycnocline below the reach of the mixed layer at all seasons (i.e. $Z_p \geqslant D$). Finally, in year IV, the particle reaches the entrainment zone and enters the mixed layer on the first night when $H_{max} \geqslant Z$. The actual data on which that occurs depends on the phase of the particle's position along its trajectory relative to the seasonal cycle of surface buoyancy flux. The particle that penetrates furthest in the pycnocline is entrained earlier in the winter. The particle that penetrates least ($Z_p = D$) is entrained last, as the daily net buoyancy flux changes sign around the vernal equinox. The density of the mixed layer equals that of the particle on those two dates. The geographical locations of the particle on those two dates provide two points in the winter, upstream migration of the isopycnal across a map of sea-surface density. The range of the migration of the isopycnal, measured along the particle trajectory, depends on the particle speed and the downstream gradient of D:

$$\Delta X_\sigma = \alpha U_\sigma \cdot \nabla D \tag{7}$$

where α is the fraction of the year for which the mixed-layer density is less than σ (i.e. $H \leqslant Z_p$).

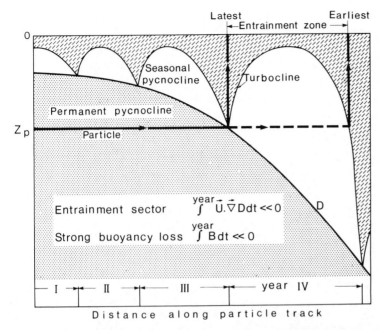

Fig. 27. The variation of turbocline depth (the climatological mean diurnal maximum H_{max} as defined by Woods and Barkmann, 1985) directly over a particle drifting adiabatically in the permanent pycnocline until entrained into the mixed layer at a location along its trajectory between the two limits marked "earliest" and "latest" during the winter of year IV. Note that by the time the turbocline descends in the winter of year IV at the site marked "earliest", the seasonal pycnocline below the particle depth (Z_p) comprises water from the permanent pycnocline that has not been changed by deep mixing during recent years.

The layer between the diurnal and annual maximum depths of the mixed layer ($H_{max} \leqslant Z \leqslant D$) is called the seasonal pycnocline. This definition is not altered by the advection of particles into that depth range from the permanent pycnocline, as shown in Fig. 27. However, it means that the density profile in the seasonal pycnocline at a fixed location changes because of the influx of particles from upstream, with water from the permanent pycnocline arriving at deeper levels. That is important for determining the value of D, because it means that during the winter, after the buoyancy stored during the spring and summer has been lost, further loss of buoyancy involves convective adjustment into a water column with the density profile of the permanent pycnocline. For climate problems that do not involve a change in that density profile, hydrographic archive data can be used to predict D. Otherwise it must be predicted by the circulation model.

The particle trajectory in Fig. 27 marks the critical depth Z_p for the location marked "earliest". At depths greater than Z_p the mixed layer descends into water that has the stratification of the permanent pycnocline as described above. At shallower depths it descends into the part of the seasonal pycnocline in which the density gradient was established by subduction of water from the mixed layer upstream as H_{max} decreased after the last winter. The process of subduction and re-entrainment is illustrated in Fig. 12, for a region where D increases downstream and there is a net buoyancy loss through

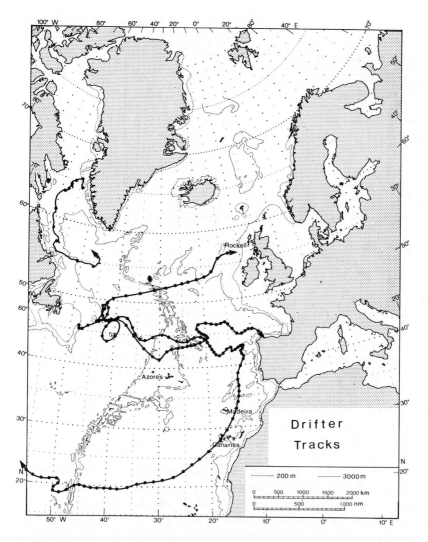

Fig. 28. Trajectories of surface drifters tracked by satellite, smoothed to reduce small-scale motion associated with transient, quasi-geostrophic eddies. The dots indicate positions every two weeks. (Redrawn from Richardson, 1983, and Lazier, pers. commun., 1984.)

the surface each year. The density gradient in the seasonal pycnocline at depths less than Z_p depends on the vertical spacing between the two isopycnals. Their initial spacing (h_s) is established in spring subduction. If the flow in the seasonal pycnocline is adiabatic and the Rossby number is small, then the Sverdrupian potential vorticity is conserved [eqn. (4)] and the spacing varies with latitude along the trajectory, so that the spacing at the moment of re-entrainment (h_e) is given by:

$$h_e = h_s \left(\frac{f_e}{f_s} \right) \tag{8}$$

Inspection of Argos drifter tracks (Fig. 28) shows that the change of f in one year is sometimes sufficient to alter the density profile in the seasonal thermocline significantly between subduction and re-entrainment. That is why we relate D to $Q(\rho)$ rather than $N(z)$.

To summarize, the deepening of the mixed layer in autumn and winter depends primarily on convective adjustment to the surface buoyancy flux through a potential vorticity profile $Q(\rho)$ that was established in the upper levels of the seasonal thermocline $(Z \leqslant Z_p)$ by subduction upstream the previous spring, and in the lower levels $(Z_p \leqslant Z \leqslant D)$ by inflow of permanent pycnocline water.

This first order theory neglects the following processes, which will be discussed later: (1) changes in potential vorticity due to solar heating and diffusion; (2) vertical motion due to local Ekman pumping; (3) vertical motion due to vortex stretching; and (4) transient vertical motions due to eddies and internal waves.

The seasonal catchment area

Figure 27 showed schematically the seasonal changes of mixed-layer depth occurring above a particle flowing adiabatically in the permanent pycnocline at depth Z_p. The Lagrangian frame of reference allows us to compute the variation of $N(Z_p)$ up to the moment the particle is entrained into the mixed layer by exploiting the fact that $\dot{Q} = 0$. Our knowledge of large-scale particle trajectories in the ocean has been greatly influenced by observations of the motion of neutrally buoyant (SOFAR) floats in the permanent pycnocline (Rossby et al., 1983) and satellite-tracked surface drifters attached to drogues in the seasonal thermocline (Richardson, 1983; Krauss and Käse, 1984). They reveal large deviations from the idealized mean gyre circulation, as deduced from hydrographic data (Dietrich, 1969; Stommel et al., 1978) and from general circulation models (Sarmiento, 1983b). The dispersion of clusters of drifters have been interpreted in terms of horizontal eddy diffusivity (Rossby et al., 1983). The inverse problem, namely determining the upstream tracks of drifters that pass through a given location is more difficult to study statistically, but the evidence of individual drifters show that for each location there is an upstream catchment area, the dimensions of which increase with time lag. We are interested in the size of the catchment at intervals of one month up to one year. Although we lack a statistically significant sample, there is little doubt that OWS "C", lying at the intersection of the Mid-Atlantic Ridge and the front between the anticyclonic and cyclonic gyres, receives a mixture of water including samples from both the Gulf Stream and Labrador current.

A Lagrangian interpretation of summer hydrographic profiles

This leads to a Lagrangian model to explain the temperature and salinity profiles measured on a summer day at some location in the Northeast Atlantic. For simplicity we neglect vertical shear and diabatic processes below the mixed layer (they will be discussed later). The water sampled at the hydrographic station arrived there after following some complicated track extending upstream across the seasonal catchment area. The starting point of the track was the location of the water column at the last occasion when $H = D$, roughly the previous vernal equinox. The initial T–S profile

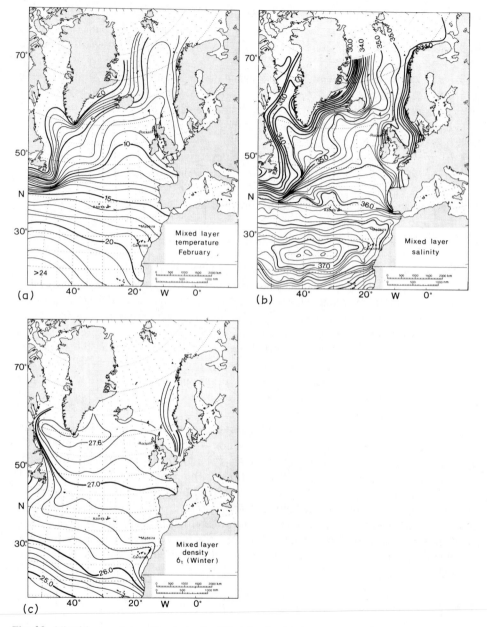

Fig. 29. Mixed-layer winter climate in the NE Atlantic. (a) Temperature; (b) salinity; (c) density (from Woods, 1984b).

was uniform down to depth D at that start location, and below it can be assumed to follow the T–S profile for the permanent pycnocline at that location, as (Levitus, 1982) given in atlases. The T–S profile then changes as the water column drifts from its starting location until it arrives some months later at the site of the hydrographic station. These seasonal changes, which only influence the water at depth $Z \leqslant H_{max}$, can be

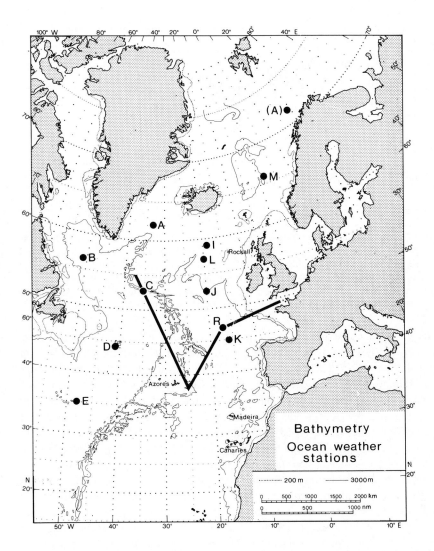

Fig. 30. The location of the Kiel batfish section used to collect data for statistical analysis of seasonal variations in temperature and salinity arising from multi-trajectory inflow from the broad seasonal catchment to the west. Each section, collected by an undulating CTD towed at 8 knots behind a research ship, comprises ca. 500 T, S profiles per degree of latitude.

calculated by Lagrangian integration of a mixed-layer model, i.e. a Woods-Barkmann model forced each day by the surface fluxes appropriate for the position that the water column has then reached along its track.

The dimensions of the seasonal catchment area for a site such as OWS "C" are indicated roughly by the trajectories in Figs. 14 and 28. Using those dimensions we can estimate the potential range of initial conditions for water that may pass through OWS "C" from the maps of winter temperature, salinity and density (Fig. 29). And we note from the corresponding maps of surface heat and water fluxes (Fig. 1) that there is a large regional variation across the catchment area. So the seasonal changes in temperature,

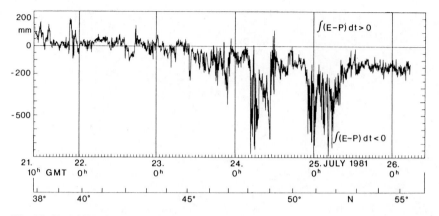

Fig. 31. Variability of seasonal water content sampled in one pass of the standard section from the Azores through OWS "C" to 55°N. Since the start of the seasonal storage: (1) evaporation exceeded precipitation along almost all trajectories arriving at the section south of 42°N; (2) precipitation exceeded evaporation along almost all trajectories arriving at the section north of 45°N; and (3) precipitation minus evaporation fluctuates about zero for trajectories between 42 and 45°N. Note that the zero line for annual mean water flux (Fig. 1b) crosses the line at 47° N.

salinity and density calculated in a Lagrangian integration of the Woods-Barkmann model will differ significantly from those calculated in an Eulerian integration at OWS "C", or some intermediate location. The differences will depend on the details of the trajectory of the water column. The tracks determined from a GCM (Fig. 14) do not take account of the lateral displacements and, more importantly, the capture of a water column by an eddy for up to a few weeks, giving a bias in favour of the surface fluxes at that location during that time of year. Tests have shown that surface drifters can follow these eddy circulations as well as the mean drift (Kraus and Meincke, 1982), at least in calm weather and up to the moment when the drogue becomes detached from the float.

It is possible to specify an inverse problem, namely to identify the initial position (when $H = D$) and the approximate trajectory of a water column, given data from a summer hydrographic station and climatological maps of winter temperature and salinity distribution, and monthly mean surface heat and water fluxes. We are undertaking such an analysis with batfish profiles collected at $1/2$ km intervals along the 2.5 Mm section starting at the Azores and going North through OWS "C" to 55°N (Fig. 30). The first results are encouraging. For example, Fig. 31 shows the water content anomalies of the 5000 profiles arising from the initial condition and net water flux histories integrated along the 5000 independent trajectories that fathered them. This new data set, which permits statistical analysis with 300 profiles per degree of latitude, is providing the first experimental test of the Lagrangian interpretation of summer hydrographic profiles.

Predicting D in ocean GCMs

The diagnostic studies reported above indicate the magnitude of the problem facing anyone seeking to modify OGCMs to predict D from the surface fluxes. (Sarmiento (1983b) had to assume the observed distribution of D in his pioneering use of a GCM to predict the distribution of tritium in the North Atlantic.) Predicting D lies beyond

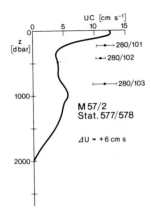

Fig. 32. The current profile near OWS "C" determined by the dynamic method compared with two-week mean samples from moored current meters (from Meincke and Sy, 1983).

the existing generation of models in which eddy motion is parameterized by lateral diffusivity. While that may be acceptable in an ensemble mean sense, it cannot be used to simulate changes that occur within a single year, because the number of eddies affecting a single trajectory is very small, less than five. Parameterization by a diffusivity cannot be justified in the sense of a time average over many years, because the slate is wiped clean each winter. The potential of eddy resolving circulation models to simulate particle trajectories explicitly is as yet not proven, although experiments with two-dimensional turbulence models are not encouraging (Haidvogel et al., 1983). One of the harshest tests of such models will be to generate realistic trajectories and seasonal catchment areas. The diagnostic studies suggest that it will be important to do so.

Lagrangian modelling based on observed trajectories allows us to explore the sensitivity of the seasonal pycnocline, D distribution, and potential vorticity injection to uncertainty in the trajectories and catchment area, in the absence of suitable velocity fields from EGCMs. My group is pursuing the Lagrangian approach both diagnostically, as described above, and prognostically, by integrating our model (Woods and Barkmann, 1985) of boundary layer processes following a drifter trajectory. We expect comparison of those two approaches to stimulate new parameterizations of the influence of eddies on deep convection in winter.

CHANGES IN THE POTENTIAL VORTICITY PROFILE

It is remarkable how little the values estimated from the Lagrangian correlation of mixed layer depth and density differ from the values in those maps derived from hydrographic archive data (McDowell et al., 1982; Sarmiento et al., 1982; Sarmiento, 1983a), despite the assumptions that the flow was barotropic and that the diabatic processes were negligible. Here we consider changes in $Q_S(\rho)$ due to shear and diabatic processes.

Vertical shear in the pycnocline

Vertical shear acting on the isopycnic gradient of potential vorticity $\mathbf{\nabla}_\rho Q_S(\rho)$ created by the variation of subduction depth with location and seasonal phase causes the potential

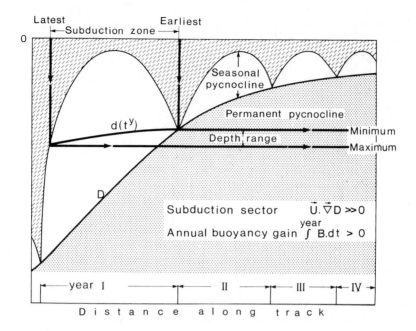

Fig. 33. The effect of seasonal variation of mixed-layer depth on subduction of an isopycnal into the permanent pycnocline, showing the annual variation of depth arising from the changing correlation of mixed-layer depth and density in the subduction zone.

vorticity profile in the seasonal pycnocline to change continuously. For the sake of simplicity, vertical shear in the pycnocline was neglected in the conceptual models and diagnostic studies presented above. In reality there is significant vertical shear, as we see in both moored current meter data and in geostrophic calculations based on hydrographic data (Meincke and Sy, 1983; Fig. 32). The cross-stream component can be calculated from the downstream baroclinicity created during subduction (Figs. 18 and 19), but without the cross-stream density gradient it is impossible to describe the geostrophic shear vector and so we remain ignorant of the velocity anomalies in the seasonal pycnocline resulting from regional variability in the subduction depths of isopycnals. Perhaps it is sufficient to assume that the low mode velocity profiles associated with the gyre and eddy circulations outcrop through the seasonal thermocline, so that the velocity shear can be estimated without regard to the baroclinicity anomalies created by the subduction process. We plan to use our batfish data to test that conjecture.

Seasonal variation of subduction depth

An isopycnal on a spring synoptic map of surface density defines the locus of sites at which water of that density is being subducted. Water in the depth range between successive downstream values of H_{max} is subducted with the density (and temperature, salinity, oxygen concentration, etc.) pertaining in the mixed layer as the turbocline H reaches H_{max}. On a spring day water of a particular density is being subducted at a variety of depths, defined by the values of H_{max} at the locations where the mixed layer has that density. The distribution of depths at which water of that density is being subducted $[H_{max}(\rho)]$ varies as the subduction locations migrate during the spring. The seasonal variation of subduction location and depth for water of the same density is illustrated in Fig. 33 for a single trajectory. The result is that an isopycnal emerges from the mixed layer over a range of source depths. The range increases with the downstream extent of the subduction zone, given by eqn. (9):

$$\Delta X(\rho) = |U(\rho) \cdot \nabla D| \cdot \Delta t \tag{9}$$

Relative vorticity

The potential vorticity of water between two isopycnals varies inversely with their spacing divided by the absolute vorticity, the sum of the vertical components of relative and planetary vorticity (eqn. 2). The Rossby number is the ratio of the relative and planetary vorticities (eqn. 3). The spectrum of ocean Rossby number is blue. At low wave number, large gyre circulation has negligible Ro, and we can use the Sverdrupian potential vorticity, in which the spacing between isopycnals varies only with latitude. At intermediate scales, small gyres circulating in topographically constrained seas (e.g. the Gulf of Lyons and Labrador Sea) can have sufficient vorticity for significant cyclonic uplifting of isopycnals which reduces the buoyancy content of the water column, preconditioning it for deep convection in winter (Killworth, 1982). However, the Rossby number is still much smaller than unity and fractional variation of h with gyre relative vorticity can normally be neglected in the seasonal pycnocline. That is still true in the case of the synoptic scale eddy motions with Rossby number approaching 0.1, i.e. giving at most a 10% transient modulation of spacing between isopycnals at constant latitude. The synoptic scale peak in the spectrum of oceanic kinetic energy is close to the Rhines scale L_{Rh} at which the eddy relative vorticity equals the change of planetary vorticity (Δt) experienced by particles circling the eddy:

$$L_{Rh} = \left(\frac{2U}{\beta}\right)^{1/2} \tag{10}$$

where U is the eddy circulation speed and β is the rate of change of f with distance from the equator.

The mesoscale waveband, which lies to the high-wavenumber side of the synoptic waveband in the energy spectrum, is characterized by high Rossby number and horizontal anisotropy. The kinetic energy is concentrated into jets several kilometres wide and tens of kilometres long. The relative vorticity approaches the limiting value of $(-f)$ on the anticyclonic side of the jet and often exceeds $(+f)$ on the cyclonic side. The quasi-geostrophic potential vorticity equation, which efficiently predicts the development of synoptic scale motion, cannot describe these mesoscale motions with Rossby number about unity. The structure of the jet changes significantly in a few days, which is about the time taken for the swiftest particles to flow through it from end to end. In a few days a little vortex tube drifting into, along and out of such a jet experiences large fractional changes in the spacing between its isopycnals.

The jets are formed by mesoscale frontogenesis (MacVean and Woods, 1980; Bleck et al., 1985) in confluences between synoptic scale eddies, which occur at intervals of a few times the Rossby radius of deformation for the permanent pycnocline. They are nourished in the seasonal pycnocline by potential vorticity gradients generated by the inhomogeneities of isopycnal subduction described earlier. Mesoscale jets therefore ubiquitously vein the seasonal pycnocline, as we see in satellite infrared images[*]. The first attempts to measure both spacing and relative vorticity in the seasonal pycnocline as reported by Woods et al. (1985), as a step towards mapping Q and so testing the prediction of the subduction model of potential vorticity injection presented in this paper.

Diabatic processes in the pycnocline

It is often assumed that $\dot{Q}$ is zero in the pycnocline. We now briefly examine five processes that cause the isopycnic potential vorticity of particles to change as they drift in the pycnocline: compression, solar heating, molecular diffusion, billow turbulence and double-diffusive convection.

Compression

Potential vorticity can be defined in terms of many different surfaces in the ocean (R. Hide, pers. commun., 1984). In this paper we have concentrated on the isopycnic potential vorticity $Q(\rho)$ defined in terms of the spacing between two surfaces $\rho \pm \delta\rho$. The density ρ is in practice normally taken to be the potential density referred to the sea surface, or to some intermediate pressure preferably at the mean depth of the isopycnal along its trajectory between subduction and entrainment. (Bray and Fofonoff, 1980). That corrects for first-order changes arising from changes of depth along the particle trajectory (due to gyre, eddy or internal wave dynamics). Second-order effects of depth rises from thermoclinicity (i.e. from variation of temperature and salinity along the isopycnal) due to the non-linearity of the equation of state $\rho = \rho(T, S)$. MacDougall (1984) has shown how to define thermodynamically neutral surfaces which differ slightly from isopycnals. In practice these compression effects are often neglected in calculating $Q(\rho)$, which is normally based on in situ density or potential density referred to the sea surface. The errors arising from that simplification are much smaller than the uncertainty in estimating the value of $Q(\rho)$ at the moment of subduction of a single water parcel and in estimating a climatological mean $\overline{Q}(\rho)$ for the ensemble of particles passing through a particular location from the large seasonal catchment area.

Solar heating

The rate of change of potential vorticity due to solar heating is given by:

$$\dot{Q} = \left[(\xi + f) \cdot \alpha \frac{\partial^2 I(z)}{\partial z^2} \right] \tag{11}$$

where α is the coefficient of thermal expansion and $I(z)$ is the solar irradiance at depth z. Solar heating can significantly change Q in the top 100 m of the seasonal pycnocline (Fig. 7) but becomes negligible thereunder (Woods et al., 1984).

---

[*]The surface temperature signature of mesoscale jets is predicted by the frontogenesis models of MacVean and Woods (1980) and Bleck et al. (1985).

Molecular diffusion

The fractional rate of change of potential vorticity due to molecular conduction of heat is given by eqn. (12):

$$\frac{\Delta Q}{Q} = \kappa \frac{\partial^3 \rho}{\partial z^3} \bigg/ \frac{\partial \rho}{\partial z} \, \Delta t \tag{12}$$

which is negligible over $\Delta t = 6$ months.

Billow turbulence

The flow in the seasonal thermocline is predominantly laminar. Turbulence occurs with overturning scales of a few decimetres, either as sets of isolated billows triggered by short internal waves (Woods, 1968) or close-packed into layers of a few metres thick maintained by inertial wave shear (Woods and Strass, 1985). The lifetime of individual billow turbulence events is typically 5 min, and the mean intermittency factor is 3%, yielding an average interval of $2\frac{1}{2}$ h between successive events. If the events are located randomly, it is reasonable to parameterize the ensemble of events occurring during one month by an eddy diffusivity:

$$K = \frac{1}{12} \, \bar{\epsilon} N^{-2} \tag{13}$$

where $\bar{\epsilon}$ is the mean power input to the turbulence (Garrett and Wunsch, 1982). The mean rate of change of potential vorticity is given by eqn. (14):

$$\frac{\dot{Q}}{Q} = \left(\frac{\partial^2}{\partial z^2} K \frac{\partial \rho}{\partial z} \right) \bigg/ \left(\frac{\partial \rho}{\partial z} \right) = \left[\frac{\partial^2}{\partial z^2} (\rho \epsilon) \right] \bigg/ (12 N^2) \tag{14}$$

The variation of ϵ with z or N is not known (Garrett and Wunsch, 1982). However, the mean value of energy input to the seasonal thermocline is of order $10 \, \text{mW m}^{-2}$ (Olbers, 1983), which is so small that the fractional change of potential energy ($\Delta Q/Q$) is negligible over the residence time of water in the seasonal pycnocline. (We are concerned with the monthly average Q, and not with transient finestructure of Q created and erased by vertically migrating layers of turbulence associated with inertial period shears.)

Double diffusive convection

Potential vorticity is not defined by eqn. (2) in convection layers where N^2 is negative, whether driven by buoyancy loss to the atmosphere at the sea surface or by double-diffusive buoyancy release inside the pycnocline. It has been shown that entrainment and subduction of isopycnals into the surface convection layer plays a crucial role in establishing the density structure of the pycnocline. One of the central points of the paper is that there is no simple function relationship $\dot{Q}(B)$ between the surface buoyancy flux, on the one hand, and the mean change of potential vorticity between entrainment and subduction. Lagrangian diagnostic studies reported above were designed to explore the possibility of constructing parameterizations of $\Delta Q(B)$ suitable for use in coupled ocean–atmosphere models. The problem seems to be one susceptible to solution on the basis of an explicit description of changes occurring along particle trajectories.

We must follow the same approach in parameterizing the effect of double-diffusive

(DD) convection layers inside the pycnocline. Empirical evidence concerning the occurrence of DD convection layers is still fragmentary. The DD process has been identified optically in the ocean (Williams, 1975), and inferred from hydrographic profiles (Tait and Howe, 1968). Laboratory experiments (Schmitt, 1979) have led to an empirical relationship between the rate of release of buoyancy during DD convection and the density parameter $R = \alpha(\partial T/\partial z)/\beta(\partial S/\partial z)$. That leads to parameterization of the net effect of an ensemble of DD convection events on the mean T, S profiles in terms of eddy diffusivities K_T, K_S that are related empirically to R. There has been no discussion in the literature of the effect of DD convection on potential vorticity, although the K parameterization suggests an approach similar to that used by Olbers et al. (1982) to parameterize $\dot{Q}(B)$ in the surface convection layer by a kind of negative solar heating. That approach becomes unreliable when only a few events occur during the time interval of interest. It was rejected as a method of parameterizing $\dot{Q}(B)$ in the North Atlantic Current because particles only experience two or three winters during their drift from America to Europe. How many DD convection events will particles experience during their residence in the pycnocline? To answer that question we need to have figures for the mean lifetime of individual DD convection events, and the interval between them.

The stepped T, S profiles characteristic of DD convection are found most frequently at fronts in the ocean (Turner, 1981). Onken and Woods (1985) have explained that observation in terms of the increase in $\nabla_\rho R$ that occurs as the result of ageostrophic circulation at thermoclinic mesoscale jets. This leads to a conceptual model in which DD convection events are the result of synoptic scale confluence acting on a combination of baroclinicity and thermoclinicity, which are created by spatial inhomogeneities in the spring subduction of isopycnals and isotherms (Woods, 1980a). The spacing of double diffusion convection events is controlled by that of confluences in the synoptical scale and meanders on the mesoscale jets at the confluences.

Water residing in the seasonal pycnocline for one summer experiences only two or three synoptic scale "weather" events between subduction and re-entrainment: the change of Q by double diffusion must be treated in terms of discrete events. But water residing in the permanent pycnocline for many years will pass through many oceanic storms between subduction and re-entrainment, and it may be appropriate to parameterize $\dot{Q}(B)$ in terms of R using a method similar to that of Olbers et al. (1982).

Schmitt (1981) has taken that approach to explain the observed constancy of the $R(\rho)$ profile at any location in the permanent pycnocline. He points out that the Iselin (1939) model (Fig. 3) leads to a vertical profile of R that reflects the meridional variation at the surface. That is not substantially changed by replacing Iselin's assumption that water is pumped down by the new idea that it is implanted at depth by spring subduction following deep winter convection. The point is that if we trace water encountered at a station in the Sargasso Sea back to its source in the mixed layer we find that the catchment area for water of each level covers a range of values of the source density parameter R_s. And if we consider just the mean $\overline{R_s}$ for water drawn from the catchment area for a particular isopycnal, then there is a systematic variation of source density parameter with density $R_s(\rho)$. Schmitt (1981) has argued convincingly that the transition from an inhomogeneous $R_s(\rho)$ to an almost constant $R(\rho)$ in the Sargasso Sea is the result of double diffusive convection. He shows that DD convection acts on average to reduce the amplitude of $R(\rho)$ finestructure.

This concept of a smoothing of the $R(\rho)$ profile has important consequences for the prediction of D. The slow elimination of $R_0(\rho)$ irregularities during the multiyear circulation of water around the anticyclonic gyre is accompanied by elimination of $Q_0(\rho)$ irregularities, so that water approaching a region of deep winter convection has a universal $Q(\rho)$ profile. The irregularities arising from spatial and temporal variation of $Q_0(\rho)$ over the extended catchment area that feeds water flowing into the entrainment site are eliminated by double diffusive convection. The prediction of D can be based on convective adjustment of the surface mixed layer into a well-conditioned $Q(z)$ profile.

DISCUSSION

Upwelling and downwelling in the seasonal pycnocline

It was pointed out earlier that Ekman pumping plays no role in the present theory of potential vorticity injection into the pycnocline. Injection is achieved by the subduction of successive isopycnals during the vernal ascent of the mixed layer. The role of Ekman pumping is limited to forcing the geostrophic circulation, as described by the thermocline equations. Here we consider that process in the seasonal pycnocline. The wind stress curl determines the Ekman pumping speed at the bottom of the Ekman layer. In summer that depth $[H_E]$ is limited by the turbocline, so (ignoring diurnal variation; Woods and Strass, 1985) the Ekman pumping velocity $[W_E]$ is then defined at the top of the seasonal pycnocline. Its amplitude decreases downwards through the seasonal pycnocline with a corresponding divergence in the geostrophic mass flux. The profile of vertical velocity changes seasonally with the windstress curl and with the depth of the top of the seasonal pycnocline. At locations where $D \geqslant H_E$ there will be some period of the year when $H \gg H_E$ with the result that the Ekman pumping velocity (calculated from the wind stress curl) occurs inside the mixed layer; the vertical velocity at the top of the seasonal pycnocline is then less than W_E.

Upwelling and downwelling is too feeble significantly to affect the spacing between isopycnals as they are subducted into the seasonal pycnocline. Consider for example the spacing between the isopycnals with densities 26.75 and $27.0 \, \text{kg m}^{-3}$ in Fig 18a. The time interval between their respective subduction is less than one month, so the lower surface cannot have moved vertically by more than the Ekman displacement in that month, which is approximately 3 m according to Bunker's climatological data (Isemer and Hasse, 1985). That is less than 4% of the 80 m spacing calculated from the Lagrangian correlation of mixed-layer depth and density. Furthermore the vertical displacement in the interval between subduction and re-entrainment is too small to explain the difference between the subduction and entrainment depths of the same isopycnal in Fig. 20. (The difference is due to vertical shear; see above.)

The existence of vertical motion does not affect the value of potential vorticity flowing into the permanent pycnocline in regions where $U \cdot \nabla D \leqslant 0$. However, more of the water subducted into the seasonal pycnocline from the mixed layer becomes reclassified as permanent pycnocline water in downwelling regions, as shown in Fig. 34. The thickness of the water column that escapes from the seasonal to the permanent

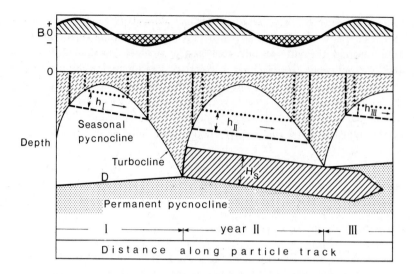

Fig. 34. The lowest H_S of the water column subducted from the mixed layer into the seasonal pycnocline flows geostrophically into the permanent pycnocline. The remainder is re-entrained into the mixed layer. This Lagrangian diagram represents a region in which there is downwelling in the seasonal pycnocline due to Ekman pumping, and a small net annual gain of buoyancy, making $U_g(D) \cdot \nabla D \leqslant W(D)$, as off North West Africa. The thickness H_S of the water mass subducted into permanent pycnocline in one annual cycle is given by the water mass formation equation (eqn. 15).

pycnocline in one year is given approximately by eqn. (15):

$$H_S = \int\limits^{\text{year}} [W(D) - U_g(D) \cdot \nabla D]\, \mathrm{d}t \qquad (15)$$

This formula can also be used to estimate the mass of water formed each year with the temperature and salinity of the mixed layer when its depth lies in the range D to $(D - H_S)$. Famous masses, such as the 18 degree water (Worthington, 1976), generally flow into the permanent pycnocline at locations where the Ekman term in eqn. (15) is small compared with the pycnocline slope term. The situation is reversed in the eastern Atlantic Ekman pumping region.

Water mass formation

The theory of potential vorticity flow into the permanent pycnocline presented in this paper can easily be extended to predict the source function of water types classified by temperature and salinity. The water acquires its T, S characteristic, as its density, on the day it is subducted from the mixed layer into the seasonal pycnocline. A theory for water mass formation therefore begins with the Lagrangian double correlation of mixed layer depth with its temperature *and* salinity, which requires the separation of surface buoyancy flux into heat and water fluxes. The mass of water in a chosen range of temperature and salinity is determined from the spacing between the bounding isothermal surfaces at the locations where they lie within the salinity range. The flux of water from the seasonal boundary layer into the permanent pycnocline is described by eqn. (15), which will therefore be named the "Water mass formation equation".

The source of thermoclinicity

Isothermal and isopycnic surfaces do not coincide because the surface heat and water fluxes are uncorrelated (Fig. 1) and because the coefficients of thermal expansion varies with temperature and salinity. The downstream component of source thermoclinicity $U \cdot \nabla_\sigma T/|U|$ can be calculated by the Lagrangian mixed-layer correlation method. Assuming statistical isotropy between down-stream and cross-stream components for the ensemble of trajectories crossing the seasonal catchment area, we can compute the annual rate of thermoclinic available potential vorticity (Woods, 1980a) flowing into the seasonal pycnocline, and, in source regions where $U \cdot \nabla D \leqslant W$, into the permanent pycnocline (there being little loss in one summer). That is the power supply for the mesoscale intrusion process that controls the rate at which double-diffusive convection eliminates inhomogeneities in R and Q originating in the seasonal boundary layer.

Annual storage of heat and water

The annual storage of heat and water is divided into two components. The surface fluxes create the first in the upper part of the boundary layer, mainly in the top 50 m. Advection creates the second throughout the water column $0 \leqslant Z \leqslant D$, drawing on an upstream catchment area several megametres long in some regions. In the North Atlantic current at OWS "C" the two contributions are similar, but in most extratropical places the surface fluxes make a larger contribution than advection to the annual range of heat storage. That encouraged Gill and Turner (1976) to use the annual cycle of the Eulerian correlation of heat content and mixed-layer temperature as a criterion for testing Eulerian integrations of mixed-layer models without advection (like Woods and Barkmann, 1985) against OWS observations. The method is less successful with the annual water content, because evaporation and precipitation nearly balance at all seasons (Taylor and Stephens, 1980).

Gill and Niiler (1973) argued that the error arising from neglecting advection becomes negligible when the annual cycles of heat and water content are averaged over a large area. The present theory for the seasonal boundary layer shows that the averaging area must be larger than the catchment area from which the seasonal pycnocline acquires its properties. Whatever success one can have with averaging away the effects of advection in calculating the annual cycles of heat and water content, that does not help solve the problem of thermocline ventilation which can only be done by explicit treatment of the balance between long-range advection and local vertical mixing in response to the surface fluxes.

Modelling climate response to carbon dioxide pollution

Models of pycnocline ventilation in terms of the surface buoyancy flux (rather than prescribed surface density) are vital for prediction of decadal climate change which is paced by the thermal response of the waters with decadal residence times, namely those in the upper gyres. Changes in the surface buoyancy flux due to CO_2 pollution in the atmosphere will alter the annual heat balance. Initially the solar heat input will be unchanged (until increasing sea-surface temperature changes the evaporation and cloud cover). Increasing pCO_2 will reduce oceanic cooling, weakening convection. D will tend to decrease (Bryan and Spelman, 1984) and the subduction depths of isopycnals will decrease. The injection of potential vorticity $Q_0(\rho)$ will also change with the climatological variation of net surface buoyancy flux (B). The geostrophic circulation around the gyre depends on $Q_0(\rho)$, and will therefore also respond to change of B. In summary, the heat storage, potential vorticity source term and the circulation will all respond to the change in surface buoyancy flux as the atmospheric pCO_2 increases. It is not correct to assume that the thermal anomaly will enter and circulate around the pycnocline as though it were a passive tracer.

Nutrients in the seasonal boundary layer

The new ventilation theory provides a physical model for discussing nutrient flow into the euphotic zone. Consider the seasonal cycle at a location where D increases downstream, starting at t_0 when $H = D$, with a uniform nutrient concentration N_0. During the year primary production exhausts the nutrients in the euphotic zone, leaving a nutricline at depth H_N. A fraction b of the nutrients consumed in primary production are lost from the seasonal boundary layer in particles that fall into the permanent pycnocline. If there were no mechanism to recharge the boundary layer, the concentration at the end of one year, when winter mixing has eliminated the nutricline, would be as follows:

$$N_1 = N_0(D - bH_N)/D \tag{16}$$

However, nutrient-rich (N_p) water from the permanent pycnocline replaced the seasonal pycnocline water (N_0) to depth $\Delta Z = D - (U \cdot \nabla D)\,\Delta t$ where Δt is circa 12 months (see Fig. 27). This increases the nutrient content of the water column by:

$$\Delta M = (N_p - N_0)\Delta Z \tag{17}$$

The value of N_0 can be deduced by equating the biological loss to the advective gain:

$$N_0 = N_p(D - U \cdot \nabla D \Delta t)/[bH_N + (D - U \cdot \nabla D \Delta t)] \tag{18}$$

Experimental errors

The new conceptual model of pycnocline ventilation presented in this paper is consistent with the available experimental evidence. However, the evidence is as yet fragmentary and not robust. It is easy to weaken the case made for the model by pointing out, for example, that the drifter tracks used in the Lagrangian diagnostic calculations may have deviated significantly from water trajectories in the pycnocline if the drogues had fallen off (a common occurrence according to Prof. J. Meincke). Equally the calculations can be criticized for mixing drifter tracks, which are single realizations, with monthly mean maps of mixed-layer depth and density assembled from data collected over many years. The only defence is that the results are rather convincing, in that the calculated change in density profile in the seasonal pycnocline near OWS "C" lies within 50% of the atlas mean value. It must be remembered that the purpose of the paper is to present a conceptual model to guide future research, both in model design and in experiments such as WOCE. The paper is a battle plan for the future, not a review of past victories. It must be expected that the plan will have to be changed as the battle proceeds.

CONCLUSION

Classical theories of circulation in the permanent pycnocline are based on the regional variations of wind stress curl (for Ekman pumping) and of surface density (for the source of potential vorticity). This paper investigates the possibility of using the surface buoyancy flux to determine the source of potential vorticity. That would open the way to modelling the response of the circulation to climatic changes of the surface buoyancy flux (due, for example, to changing atmospheric concentration of CO_2).

The upper boundary of the permanent pycnocline lies at depth D, the annual maximum depth of the mixed layer. The seasonal pycnocline lies above the permanent pycnocline. Potential vorticity flows from the seasonal pycnocline into the permanent pycnocline at locations where D decreases downstream, and vice versa. In order to predict the source of potential vorticity for the permanent pycnocline it is therefore necessary to have theories for (1) the source of potential vorticity for the seasonal pycnocline; (2) the regional variation of D and (3) processes that change $Q(\rho)$ in the pycnocline.

It has been shown that the source of potential vorticity for the seasonal pycnocline depends on the Lagrangian correlation of the seasonally varying mixed-layer depth and density. Realistic values of $Q_S(\rho)$ have been obtained by that method from a combination of climatological mean mixed-layer data and the tracks of surface drifters with drogues in the seasonal pycnocline. The results confirm the seasonal variation of $Q_S(\rho)$ predicted by the theory, and provide realistic estimates for the observed temporal change of density profile at OWS "C", a site in the North Atlantic current on the front between the cyclonic and anticyclonic gyres.

A theory for predicting D is presented in terms of convective adjustment to the surface buoyancy flux and the profile of potential vorticity arriving at a given site each day along a different trajectory crossing the broad seasonal catchment area. In regions where D increases downstream the seasonal pycnocline is invaded each year by water from the permanent pycnocline, so the value of D depends not only on potential vorticity injected into the seasonal pycnocline upstream during the previous spring, but also on potential vorticity that has circulated around the gyre while residing inside the permanent pycnocline for several years. Methods of estimating D from hydrographic data were reviewed critically.

Processes that change the vertical profile $Q_S(\rho)$ in the seasonal and permanent pycnocline were reviewed. It was concluded that the main problem is the change effected by vertical shear in the seasonal pycnocline acting on inhomogeneities in the depths at which isopycnals are subducted from the mixed layer. Changes due to diabatic processes can be parameterized, with the exception of double-diffusive convection in the seasonal pycnocline. The longer residence time of water in the permanent pycnocline permits double diffusive convection to be parameterized, simplifying the problem of predicting $Q_S(\rho)$ at locations where water leaves the permanent pycnocline, and therefore the problem of predicting D.

588

The main problem thrown up by this investigation of the possibility of using a buoyancy flux boundary condition for modelling pycnocline circulation is the large seasonal catchment area created by synoptic scale dispersion. It is not yet clear whether eddy resolving GCMs will be able to predict the statistics of particle trajectories sufficiently accurately to calculate catchment areas. Experimental determination of the catchment area using drifters is likely to be expensive. An attempt is being made to determine catchment areas by an inverse method using the large number of T–S profiles available in five megametre batfish sections spanning the North Atlantic current.

Although the investigation has raised a number of challenging problems, it has provided a physically realistic conceptual framework for future experimental and theoretical research on the ventilation of the permanent pycnocline and the circulation of the warm-water sphere.

REFERENCES

Bleck, R., Onken, R. and Woods, J.D., 1985. High Rossby number frontogenesis – A two-dimensional model. Deep-Sea Res. (in prep.)
Bray, N.A. and Fofonoff, N.P., 1981. Available potential energy for mode eddies. J. Phys. Oceanogr. 11: 30–117.
Bretherton, F.P., 1982. Ocean climate modelling. Prog. Oceanogr., 11: 93–129.
Broecker, W.S. and Peng, T.-H. 1982. Tracers in the Sea. Lamont-Doherty Geological Observatory, Columbia University, Pallisades, N.Y.
Bryan, K. and Spelman, M.J., 1984. The ocean as a climate buffer for CO_2-induced warming. (unpubl.)
Bryan, K., Komro, F.G., Manabe, S. and Spelman, M.J., 1982. Transient climate response to increasing atmospheric carbon dioxide. Science, 215: 56–58.
Bryden, H.L. and Hall, M.M., 1980. Heat transport by ocean currents across 25°N latitude in the Atlantic Ocean. Science, 207: 884–886.
Bunker, A.F. and Goldsmith, R.A., 1979. Archived time-series of Atlantic Ocean meteorological variables and surface fluxes. Tech. rep., WHOI-79-3, Woods Hole Oceanographic Inst.
Charney, J., 1983. Climate Change. (CO_2 assessment ctte.) Natl. Acad. Sci., Washington, D.C., 496 pp.
Dickson, R.R., Gould, W.J., Garbutt, P.A. and Killworth, P.D., 1982. A seasonal signal in ocean currents to abyssal depths. Nature, 295: 193–198.
Dietrich, G., 1969. Atlas of the hydrography of the Northern North Atlantic Ocean based on the polar front survey of the IGY winter and summer 1958. Cons. Int. Explor. Mer, Serv. Hydrogr., Charlottenlund Slot.
Fuchs, G., Jancke, K. and Meincke, J., 1984. Observations of winter convection in the Northeastern Atlantic. ICES Hydrogr. Committee CM 1984/C: 18.
Garrett, C. and Wunsch, C., 1982. Turning Points in Universal Speculation on Internal Waves. A Celebration in Geophysics and Oceanography – 1982 in Honor of Walter Munk. Scripps Inst. Oceanogr. Ref. Ser. No. 84-5, La Jolla, Calif.
Gill, A.E. and Niiler, P.P., 1973. The theory of the seasonal variability in the ocean. Deep-Sea Res., 20: 141–177.
Gill, A.E. and Turner, J.S., 1976. A comparison of seasonal thermocline models with observation. Deep-Sea Res., 24: 391–401.
Haidvogel, D.B., Robinson, A.R., and Rooth, C.G.H., 1983. Eddy-induced dispersion and mixing. In: A.R. Robinson (Editor), Eddies in Marine Science. Springer, Berlin, pp. 481–491.
Iselin, C.O'D., 1939. The influence of vertical and horizontal turbulence on the characteristics of waters at mid-depths. Trans. Am. Geophys. Union, 3: 414–417.
Isemer, H.J. and Hasse, L., 1985. The Bunker Climate Atlas of the Atlantic Ocean. Springer, Berlin (in prep.)

Jenkins, W.J., 1980. Tritium and ³He in the Sargasso Sea. J. Mar. Res., 38: 533–569.

Jenkins, W.J., 1982. On the climate of a subtropical ocean gyre: Decade time scale variations in water mass renewal in the Sargasso Sea. J. Mar. Res., 40: 265–290.

Killworth, P.D., 1982. Deep convection in the world ocean. In: Rev. Geophys. Space Phys., 21: 1–26.

Krauss, W. and Käse, R., 1984. Mean circulation and eddy kinetic energy in the East North Atlantic. J. Geophys. Res., 89 (C3): 3407–3415.

Krauss, W. and Meincke, J., 1982. Drifting buoy trajectories in the North Atlantic Current. Nature, 296: 737–740.

Levitus, S., 1982. Climatological Atlas of the World Ocean. NOAA Tech. Pap. No. 3, Rockville, Md., 173 pp.

Luyten, J., Pedlosky, J. and Stommel, H., 1982. The ventilated thermocline. J. Phys. Oceanogr., 13: 292–309.

MacDougall, T.J., 1985. The relative roles of diapycnal and epipycnal mixing on the variation of properties along isopycnals. J. Phys. Oceanogr. (submitted).

MacVean, M.K. and Woods, J.D., 1980. Redistribution of scalars during upper ocean frontogenesis: a numerical model. Q. J. R. Meteorol. Soc., 106: 293–311.

McDowell, S., Rhines, P. and Keffer, T., 1982. North Atlantic potential vorticity and its relation to the general circulation. J. Phys. Oceanogr., 12: 1417–1436.

Meincke, J. and Sy, A., 1983. Large-scale effects of the Mid-Atlantic Ridge on the North Atlantic Current. ICES Hydrogr. Committee, Cm 1983/C:8, 10 pp.

Olbers, D.J., 1983. Models of the oceanic internal wave field. Rev. Geophys. Space Phys., 21: 1567–1606.

Olbers, D.J. and Willebrand, J., 1984. The level of no motion in an ideal fluid. Reply. J. Phys. Oceanogr., 14: 203–212, 214.

Olbers, D., Willebrand, J. and Wenzel, M. 1982. The inference of ocean circulation parameters from climatological hydrographic data. Ocean Modell. 46: 5–9.

Onken, R. and Woods, J.D., 1985. Mesoscale preconditioning for double-diffusing interleaving. (in prep.).

Reid, J.L., 1982. On the use of dissolved oxygen as an indicator of winter convection. Naval Res. Rev., 3: 28–39.

Richardson, P.L., 1983. Eddy kinetic energy in the North Atlantic for surface drifters. J. Geophys. Res., 88 (C7): 4355–4367.

Robinson, M.K., 1976. Atlas of North Pacific Ocean monthly mean temperatures and mean salinities of the surface layer. U.S. Naval Oceanogr. Off. Ref. Publ., 2, 262 pp.

Robinson, A.R. and Stommel, H., 1959. The oceanic thermocline and the associated thermohaline circulation. Tellus, 11: 295–308.

Robinson, M.K., Bauer, R.A. and Schroeder, E.H., 1979. Atlas of North Atlantic – Indian Ocean monthly mean temperatures and mean salinities of the surface layer. U.S. Naval Oceanogr. Off. Ref. Publ. 18, Washington, D.C.

Rossby, H.T., Riser, S.C., and Mariano, A.J., 1983. The Western North Atlantic – A Lagrangian viewpoint. In: A.R. Robinson (Editor), Eddies in Marine Science. Springer, Berlin, pp. 67–91.

Sarmiento, J.L., 1983a). A tritium box model of the North Atlantic thermocline. J. Phys. Oceanogr., 13: 1269–1274.

Sarmiento, J.L., (1983b). A simulation of bomb tritium entry into the Atlantic Ocean. J. Phys. Oceanogr., 13: 1924–1939.

Sarmiento, J.L., Rooth, C.G.M. and Roether, W., 1982. The North Atlantic Tritium Distribution in 1972. J. Geophys. Res., 87: 8047–8056.

Schmitt, R.W., 1979. Flux measurements on salt fingers at an interface. J. Mar. Res., 37: 419–436.

Schmitt, R.W., 1981. Form of the temperature–salinity relationship in the central water: evidence for double-diffusive mixing. J. Phys. Oceanogr., 11: 1015–1026.

Stommel, H., 1979a. Oceanic warming of western Europe. Proc. Natl. Acad. Sci., U.S.A., 76: 2518–2521.

Stommel, H., 1979b. Determination of water mass properties of water pumped down from the Ekman layer to the geostrophic flow below. Proc. Natl. Acad. Sci. U.S.A., 76: 3051–3055.

Stommel, H., Niiler, P. and Anati, D., 1978. Dynamic topography and recirculation of the North Atlantic. J. Mar. Res., 36: 449–468.

Tait, R.I. and Howe, M.R., 1968. Some observations of thermo-haline stratification in the deep ocean. Deep-Sea Res., 15: 275–280.

Taylor, A.H. and Stephens, J.A., 1980. Seasonal and year-to-year salinity at the nine North Atlantic Ocean Weather Stations. Oceanol. Acta, 3: 421–430.

Turner, J.S., 1981. Small-scale mixing processes. In: B. Warren and C. Wunsch (Editor), Evolution of Physical Oceanography. MIT Press, Cambridge, Mass., pp. 236–263.

Welander, P., 1959. An advective model of the ocean thermocline. Tellus, 11: 309–318.

Williams, A.J., 3rd (1975). Images of ocean microstructure. Deep-Sea Res., 22: 811–829.

Woods, J.D., 1968. Wave-induced shear instability in the summer thermocline. J. Fluid Mech., 32: 791–800.

Woods, J.D., 1980a. The generation of thermohaline finestructure at fronts in the Ocean. Ocean Modelling, 32: 1–4.

Woods, J.D., 1980b. Diurnal and seasonal variation of convection in the wind-mixed layer of the ocean. Q. J. R. Meteorol. Soc., 106: 379–394.

Woods, J.D., 1984a. The upper ocean and air–sea interaction in global climate. In: J.T. Houghton (Editor), The Global Climate. Cambridge University Press, Cambridge pp. 141–187.

Woods, J.D., 1984b. The Warmwatersphere of the North East Atlantic. A miscellany. Inst. Meereskunde, Kiel, Berichte 128, 39 pp.

Woods, J.D. and Barkmann, W., 1985. The response of the upper ocean to solar heating. I. The mixed layer. Q. J.R. Meteorol. Soc. (submitted).

Woods, J.D. and Strass, V., 1985. The response of the upper ocean to solar heating. II. The wind-driven current. Q. J. Meteorol. Soc. (submitted).

Woods, J.D., Barkmann, W. and Horch, A., 1984. Solar heating of the oceans – diurnal, seasonal and meridional variation. Q. J. R. Meteorol. Soc., 110: 633–686.

Woods, J.D., Leach, H. and Fischer, J., 1985. Mapping the components of potential vorticity in the seasonal thermocline. (In prep.)

Worthington, L.V., 1976. On the North Atlantic Circulation. Johns Hopkins Unversity Press, Oceanogr. Stud., 6, 110 pp.

CHAPTER 35

PRELIMINARY EXPERIMENTS ON THE SENSITIVITY OF ATMOSPHERIC MONTHLY MEAN PREDICTION TO SEA-SURFACE TEMPERATURE SPECIFICATION

RÉJEAN MICHAUD, T.N. KRISHNAMURTI and ROBERT SADOURNY

ABSTRACT

A series of January 1983 forecasts is being performed to investigate the effect of specifying the sea-surface temperature in different ocean basins. Two different SST data sets are used. Preliminary results show that the tropical response of the atmosphere is reasonably coherent. The extratropical response is more complex and requires further experimentation.

INTRODUCTION

The warming of the sea-surface temperature (SST) in large areas of tropical oceans, in particular the extraordinary warming in the eastern tropical Pacific, has strongly modified the tropical circulation during the winter 1982–1983. For example, the winter monsoon circulation over Indonesia was found further east in the central Pacific, strong precipitations were recorded in the eastern and central equatorial Pacific and a drought took place in Indonesia. An anomalous circulation was also observed in the northern mid-latitudes: in January for instance, the West Pacific trough at 500 mb was shifted eastward and extended all over the North Pacific; the Northwest American ridge was similarly displaced eastwards, the Northeast Atlantic ridge was enhanced and also shifted eastwards to the British Isles. The flow did not change very much in February, except that the Northeast Atlantic ridge went back to 25°W with still a higher-than-normal amplitude.

Our purpose here is to present some preliminary results from a series of experimental forecasts of the January 1983 circulation and their sensitivity to the prescription of observed sea-surface temperature in different ocean basins.

THE MODEL

The model used in this study is a standard version of the LMD general circulation model (Sadourny and Laval, 1984). It is a sigma-coordinate gridpoint model with 64 points equally spaced in longitude and 50 points equally spaced in sine of latitude. The corresponding mesh size at the equator is 625 km in longitudinal direction, against 225 km in meridional direction; a square 400 km mesh is obtained in the vicinity of 50° N. The model has eleven layers with increased resolution near the surface. The horizontal momentum advection scheme is designed to conserve potential enstrophy

592

exactly for divergent barotropic flow (Sadourny, 1975). The vertical differencing is constructed to conserve total energy, and the two first moments of potential enthalpy exactly.

The computational scheme of the solar radiative heating (Fouquart and Bonnel, 1980), takes into account absorption by water vapour, carbon dioxide and ozone, as well as scattering by cloud particles. The cooling due to long-wave radiation is computed using a scheme originally due to Katayama (1972), modified to account for stratospheric processes; in this scheme the mean transmission function is calculated as the product of transmissions by carbon dioxide and water vapour.

Large-scale and small-scale condensation are modelled separately; the former takes place when the air is supersaturated and stably stratified, while the latter occurs under unstable conditions with respect to the moist adiabatic lapse rate. In this case, the moist convective adjustment technique (Manabe and Strickler, 1964) is used when the air is saturated; for unsaturated air, convection occurs in moisture convergence areas (Kuo, 1965). The clouds generated by the convection scheme interact with radiation according to Le Treut and Laval (1984).

NUMERICAL EXPERIMENTS

At this time two series of four one-month integrations have been completed. All start from a single initial state interpolated from the European Centre for Medium-range Weather Forecasts analysis of 12 GMT, 1 January 1983. Series 1 uses sea-surface temperature data prepared at Florida State University from the U.S. Navy operational data sets; series 2 uses instead the sea-surface temperatures produced by the NOAA Climate Analysis Centre. The four integrations within each series correspond to four different specifications of the SST field: January climatology + global January 1983 anomaly (G), climatology + Pacific anomaly alone (P), climatology + Indian Ocean anomaly alone (I), and climatology (C). In our labelling system, the experiment using observed global SST from NOAA data sets will be thus referred to as G2, etc. The Sunda Isles were selected as the artificial boundary separating the Indian Ocean from the Pacific Ocean. The climatological SSTs for series 1 was the RAND climatology (Schutz and Gates, 1971). The RAND climatology is warmer than the CAC climatology by around half a degree along the equator in central-to-eastern Pacific; it is colder by around half a degree around the dateline near 15 degrees south, in the Caribbean and in the South China sea. The major discrepancy between the two climatologies is near the north coast of Australia, where the RAND climatology is two degrees colder. For January 1983, NOAA data are warmer by half a degree to one degree in the equatorial eastern Pacific, by half a degree in the Caribbean and South China seas, and by up to two degrees off the Peruvian coast and the north coast of Australia; NOAA data, on the other hand, are colder by half a degree in tropical Northwest Pacific.

The sea-surface temperatures used in series 1 are shown in Fig. 1, which displays the observed, climatological and anomaly fields. Besides the warm positive anomaly in the eastern tropical Pacific, the dominant feature is a positive anomaly over the Indian ocean of the order of $1°C$ in the vicinity of the equator. In the January climatology, the warmest water is found near the Solomon Islands and the $28°C$ line is nearly confined

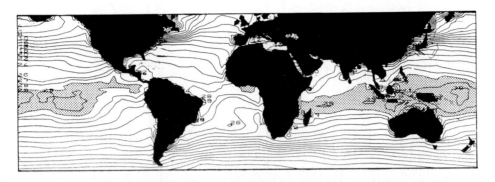

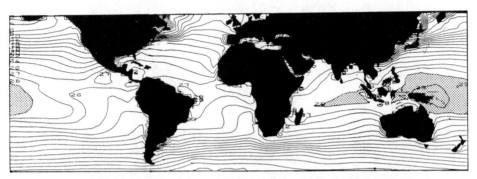

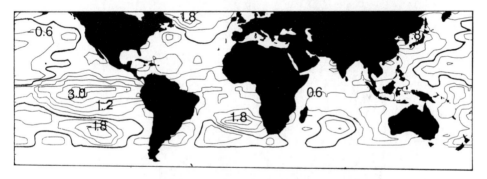

Fig. 1. Sea-surface temperatures used in series 1 forecasts. (a) Observed January 1983; (b) January climatology; (c) Anomaly a–b. Contour interval: 2 degrees, 1 degree above 26°C. Stippled areas indicate temperatures exceeding 28°C.

west of dateline. During January 1983, the warm water exceeding 28°C extends over the whole equatorial Pacific and Indian oceans, with a maximum of 30°C east of the Solomon Islands. Also in January 1983, the equatorial Atlantic is found warmer than normal.

The main objective of our long-term experimental strategy, of which the experiments just completed are but an initial phase, is to evaluate the impact on one-month forecasts, of SST specification in various geographical areas. Note that forecast experiments are, in this respect, fundamentally different from the usual climate sensitivity experiments,

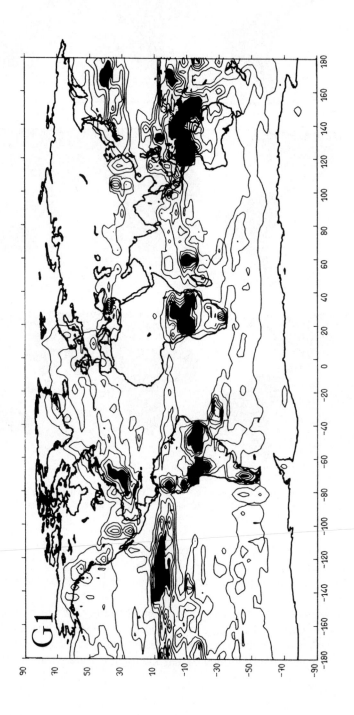

595

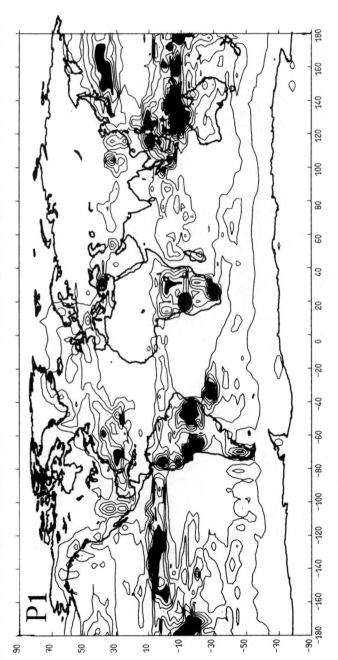

Fig. 2. *Caption on p. 597.*

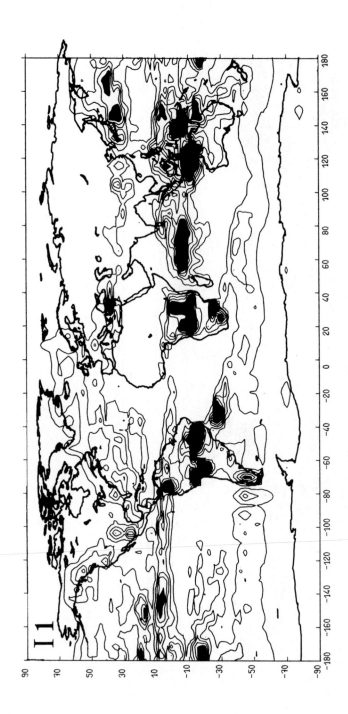

597

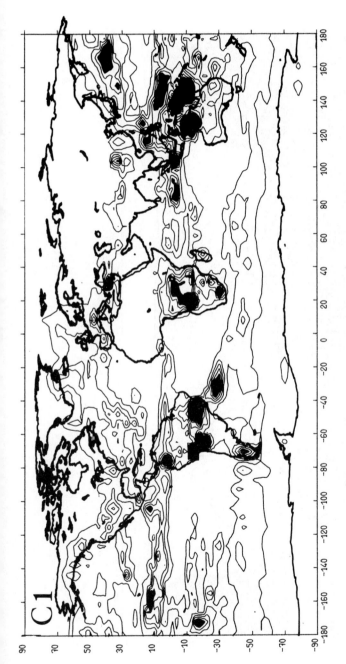

Fig. 2. January average precipitation in experiments G1, P1, I1, C1. Contour interval: 2 mm per day. Black areas indicate precipitation exceeding 10 mm per day.

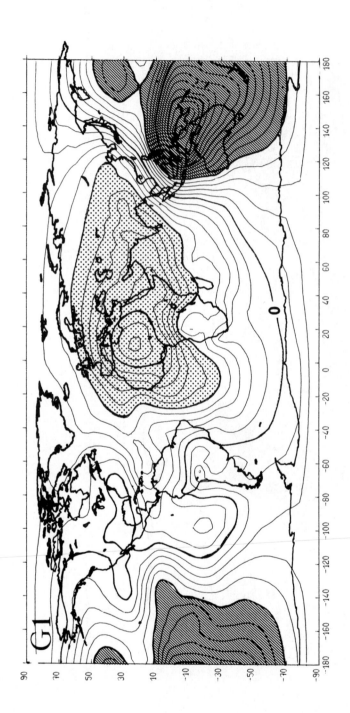

599

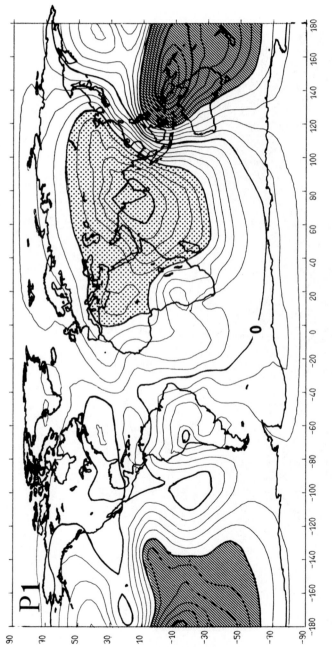

Fig. 3. *Caption on p. 601.*

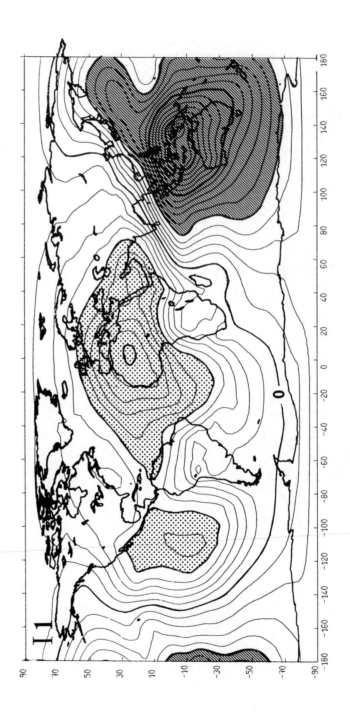

601

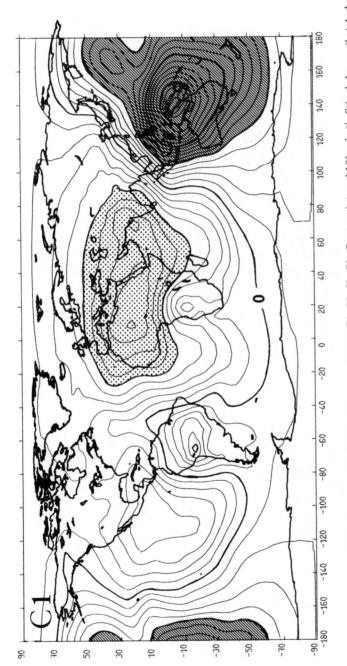

Fig. 3. January average velocity potential at 200 mb in experiments G1, P1, I1, C1. Contour interval 10^6 m^2 s^{-1}. Stippled areas (hatched areas) indicate regions of main confluence (main diffluence).

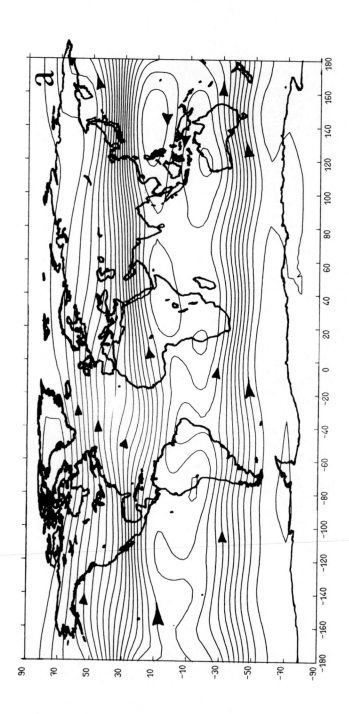

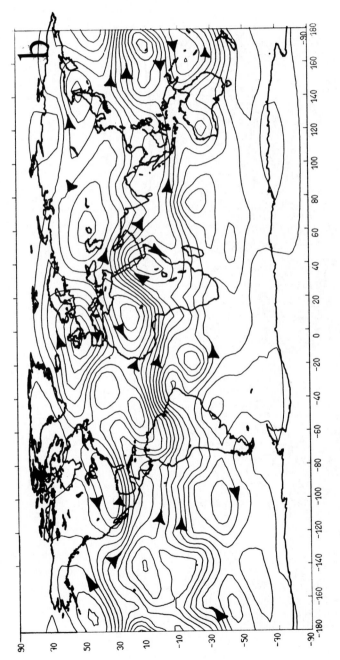

Fig. 4. *Caption on p. 605.*

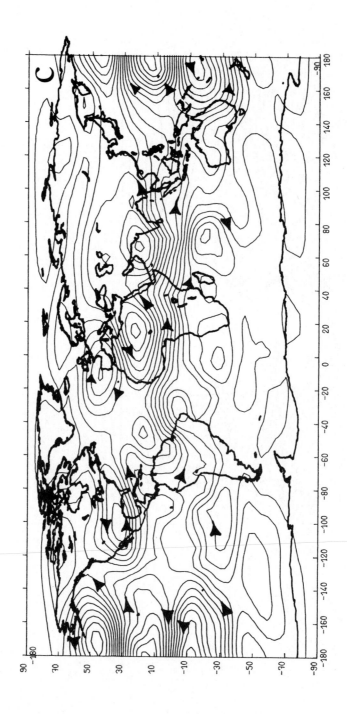

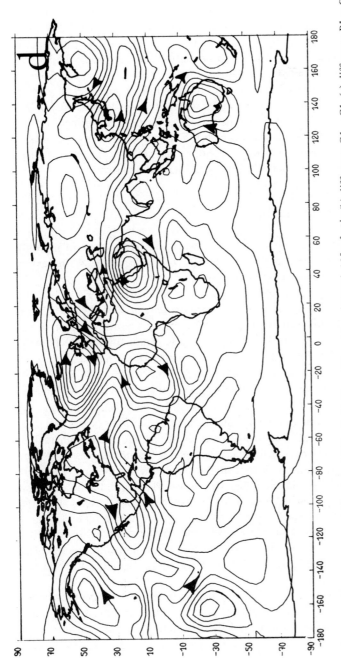

Fig. 4. January average streamfunction at 200 mb. (a) From G1, contour interval, 10^7 m^2 s^{-1}; (b) difference G1 − C1; (c) difference P1 − C1; (d) difference I1 − C1. Contour interval for b, c, d: 2×10^6 m^2 s^{-1}.

since an influence of the SST is already present in the initial conditions. Also, we are not interested in a (relatively weak) mean atmospheric response, but rather in a (likely more intense) transient response, strongly dependent on initial conditions and overbearing the mean response. Our experimentation being still in progress, the limited set of forecasts available at the moment (and enumerated above) is obviously insufficient to give a reliable answer to this difficult question — at least in the extra-tropics where the level of transient variance is high — for lack of adequate statistics. To assess the statistical significance of forecast skill estimates, a much larger set of forecasts is still needed, where both initial and boundary conditions are varied within reasonable limits. An appropriate choice for varying initial conditions would be a set of time-lagged initial states for the atmospheric fields (but soil moisture, for example, should also be varied); SST conditions, on the other hand, should be varied within the uncertainties due to data errors and retrieval algorithms. In a way, we have already begun to experiment on the latter aspect of the problem by using two distinct SST data sets; further experiments, involving time-lagged initial conditions will be performed in the near future. For these reasons, all conclusions drawn at this stage have to be considered tentative.

DISCUSSION: TROPICS AND SUBTROPICS

All results presented in this section and the following will be in terms of monthly (January) averages, i.e. averages from day 0 to day 30 of each integration. No discussion of the transient evolution will be attempted at this stage.

The monthly mean January precipitation fields for experiments G1, P1, I1, C1 are displayed in Fig. 2. One conspicuous feature is that the decrease in rainfall observed over Indonesia in winter 1983 is not well simulated in all cases; this result, however, is not inconsistent with the fact that all our SST fields, including or not the Pacific anomaly, have their maximum value in the western equatorial Pacific. On the other hand, the only experiments which succeed in maintaining the observed high precipitation rate in the eastern equatorial Pacific are those with the Pacific SST anomaly (G1 and P1). Similarly, only G1 and I1 produce high rainfall rates in the southern Indian Ocean. Looking at the tropical Pacific—Indian Ocean area as a whole, and comparing the four experiments, we observe that the Indian Ocean SST anomaly tends to attract the Indonesian rainfall towards the west, while the Pacific anomaly tends to attract it towards the east. Also, the Indian Ocean anomaly tends to attract African rainfall maxima towards the east. Therefore it seems plausible that higher SSTs in the Indian Ocean, associated with further eastward displacement of SST maxima in the equatorial Pacific, would produce drought situations in both Africa and Indonesia. A realistic feature of G1 — and, to a lesser extent, P1 — is the abnormally high rainfall rate over Florida and the southeastern North-American coast. Another realistic feature of G1 is the relatively southward position of the ITCZ over the gulf of Guinea (see Thépenier, 1984), an effect thus associated to local SST anomalies.

The response of Hadley-Walker circulations to changes in tropical forcing is described in Fig. 3, which shows monthly average velocity potential fields at 200 mb. The effect of the Pacific anomaly is to extend the West Tropical Pacific diffluence area eastward, deep into eastern Pacific (P1, G1); and to shift the main confluence center to the east, from Africa to India (P1). The effect of the Indian Ocean anomaly is to shift the diffluence area and the main confluence center westward (I1). For the global anomaly experiment, the response appears as a rather straightforward mixture of P1 and I1. Apparently, the divergent circulation over the Pacific Ocean is not very sensitive to Indian Ocean SST. We note, however, that high SSTs in the Indian Ocean tend to enhance the Walker cell subsidence in the eastern Pacific (compare G1 to P1 and I1 to C1), as the main subsidence cell over Africa and the Middle East is pushed westward.

The main (but tentative) conclusion we draw here is that the tropical forcing and the associated divergent circulation tend to respond to tropical SST specification in a relatively straightforward, and therefore probably to some extent predictable, manner. In tropical areas where SST gradients are weak, relatively small SST changes are able to produce long-distance shifts of SST maxima, which the forcing field and the corresponding Hadley-Walker ascending branches tend to follow. Thus the response of divergent circulation has a relatively large amplitude at planetary scales.

A similar coherence is found in the response of the rotational wind in the tropics. Figure 4 shows the monthly average streamfunction at 200 mb from experiment G1, together with streamfunction differences $G1 - C1$, $P1 - C1$ and $I1 - C1$. In all our simulations the anomalous circulation over the Pacific Ocean is not quite realistic: For instance, in G1, the equatorial easterlies are confined to the west of the dateline, contrary to observations (Arkin et al., 1983); they are better in P1 and worse in I1. These shortcomings are of course consistent with our above-mentioned forcing responses. It is interesting, however, to compare the $G1 - C1$, $P1 - C1$, $I1 - C1$ charts to the streamfunction anomaly actually observed in January 1983 (Arkin et al., 1983). The comparison shows a qualitative agreement of $G1 - C1$ with the 1983 anomalies over the equatorial Atlantic Ocean and Africa. Comparing now $G1 - C1$ to $P1 - C1$ and $I1 - C1$, we infer that the observed anomalous westerly flow over the equatorial Atlantic Ocean and west equatorial Africa might have been driven by the Pacific SSTs, while the observed anomalous southwesterly flow over Northeast Africa appears related to the influence of the Indian Ocean.

DISCUSSION: MID-LATITUDES*

A thorough discussion of the mid-latitude responses would require statistics based on a larger number of experiments. Here we shall restrict our discussion to an analysis of the 500 mb geopotential height, using the simplified diagnostics proposed in Von Storch and Kruse (1985): our basic variable from now on, z, is the 500 mb geopotential height, monthly averaged and meridionally averaged from $30°$ to $60°N$, for which we discuss the longitudinal distribution of troughs and ridges.

The background of Fig. 5 is the 95% significance band for January normality, and the observed distribution of z in January 1983, both according to Von Storch and Kruse (1985). The "climatology" of the LMD model for January is also indicated. This climatology is not entirely adequate in the present context, because it has been based on a 150-day mean from a perpetual January run. A more proper climatology, based on a

---

*This section has been added after the colloquium.

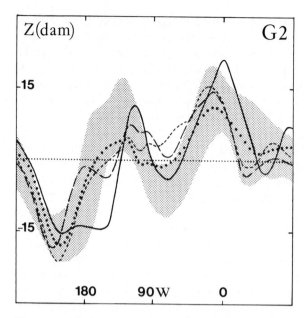

Fig. 5. Monthly average geopotential height in dam, meridionally averaged from 30° to 60° N, as a function of longitude. Stippled: 95% significance band for January normality, and continuous line: observations for January 1983, both from Von Storch and Kruse (1985). Dots: LMD model climatology; discontinuous line with short dashes: first month from G2; with long dashes: second month from G2.

series of January forecasts, is now being constructed. The present climatology shows that the model is reasonably successful in reproducing the observed troughs and ridges, except for a systematic underestimation of the North American ridge—trough system. On this background, Fig. 5 shows the performance of G2. G2, in fact, has been integrated up to two months, and the two monthly averages (January and February) are shown. The main points of interest are the following: (1) the slight eastward shift of the northwest Pacific trough is not predicted; the trough is deeper than observed; (2) the anomalous northeast Pacific trough at 150°W is only very slightly indicated during the first month, it becomes more marked — but still not deep enough — during the second month (note that this feature actually subsisted throughout February); (3) the model predicts an eastward shift of the northwest American ridge; the shift, however, is weaker than observed; (4) the northeast American trough tends to fill up unrealistically during the first month; it deepens again in February; (5) the observed strengthening of the northeast Atlantic ridge also occurs in the model; but it remains weaker than observed and the phase is not right (20°W in January, 10°W in February, against 0° and 25°W in observations); and (6) the occurrence in January of a ridge in Western Siberia is forecast by the model; but like the northeast Atlantic ridge, it is predicted 20 degrees west of its actual location.

We now proceed to discuss the effect of SST specification on this distribution of troughs and ridges. Figure 6 shows the same variable z for each of our eight January forecasts, together with the observed distribution. Series 2, in general, gives better results than series 1. We may describe some of the systematic effects associated with variations in SST specification as follows:

609

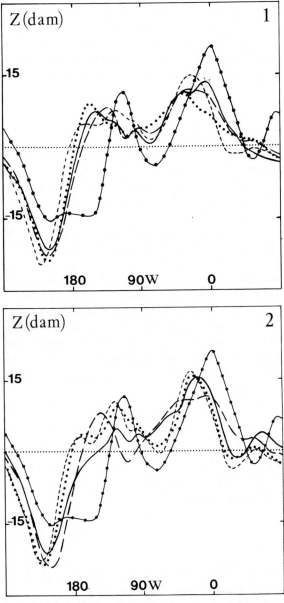

Fig. 6. Same as Fig. 5, for first month forecasts from series 1 and 2. Continuous line: G; discontinuous line with long dashes: P; with short dashes: I; dotted line: C. Observations from Von Storch and Kruse (1985): dots on continuous line.

(1) Higher SST in the Indian Ocean tend to pull the northwest Pacific trough westward, to enhance the northeast Atlantic ridge and to shift it also westward; these effects seem qualitatively consistent with the aforesaid displacement of the Asia–African subtropical subsidence towards the west (Fig. 3).

610

(2) The Pacific SST anomaly tends to produce an eastward shift of the West Pacific trough (contrary to the Indian Ocean SST), which is again qualitatively consistent with the eastward displacement of the Asian subtropical subsidence observed in Fig. 3; the effect of the Pacific SST anomaly at other longitudes is not clear: In particular, it seems to play no determining role in the strengthening of the northeast Atlantic ridge.

CONCLUSIONS AND PROSPECTS

We have performed a series of January 1983 forecasts to investigate the influence of various SST specifications. The two main shortcomings of the simulations are: (1) the inability of the forecasts to produce the observed marked decrease of rainfall in the Indonesian area; and (2) the inability of the forecasts to produce the right response pattern in northeast Pacific–North American longitudes.

These two effects are probably related, but (2) might also result in part from the systematic error of the model in its treatment of the quasi-stationary wave over the North American continent.

In general, we have noted a fairly coherent response of the tropical circulation to SST specifications; this is reasonable, in view of the direct link between Hadley-Walker circulations and tropical forcing. The response in mid-latitudes is less obvious, and requires a larger amount of statistics to be fully elucidated: Our conclusions therefore remain to a large extent tentative. The present experimentation is still in progress: part of this series of forecasts will be repeated starting from time-lagged initial conditions; one of our main concerns is also the removal of the mid-latitude systematic error, which appears as a crucial step for improving the simulation.

ACKNOWLEDGEMENTS

We thank the U.S. Navy for letting us use their operational SST data in the present experiments, and the European Centre for Medium Range Weather Forecasts, for providing the initial atmospheric fields analyses on January 1st, 1983.

REFERENCES

Arkin, P.A., Kopman, J.D. and Reynolds, R.W., 1983. 1982–1983 El Niño/Southern Oscillation Event Quick Look Atlas. NOAA/National Weather Service, Washington, D.C.
Fouquart, Y. and Bonnel, B., 1980. Computations of solar heating of the earth's atmosphere: a new parameterisation. Contrib. Atmos. Phys., 53: 35–62.
Katayama, A., 1972. A simplified scheme for computing radiative transfer in the troposphere. Tech. Rep. No. 6, Dept. of Meteorol., Univ. of Calif, Los Angeles, Calif.
Kuo, H.L., 1965. On the formation and intensification of tropical cyclones through latent heat release by cumulus convection. J. Atmos. Sci., 22: 40–63.
Le Treut, H. and Laval, K., 1984. The importance of cloud–radiation interaction for the simulation of climate. In: A.L. Berger and C. Nicolis (Editors), New Perspectives in Climate Modelling. (Developments in Atmospheric Science, 16) Elsevier, Amsterdam, pp. 199–221.
Manabe, S. and Strickler, R.F., 1964. On the thermal equilibrium of the atmosphere with convective

adjustment. J. Atmos. Sci., 21: 361–385.

Sadourny, R., 1975. The dynamics of finite difference models of the shallow water equations. J. Atmos. Sci., 32: 680–689.

Sadourny, R. and Laval, K., 1984. January and July performance of the LMD general circulation model. In: A.L. Berger and C. Nicolis (Editors), New Perspectives in Climate Modelling. (Developments in Atmospheric Science, 16) Elsevier, Amsterdam, pp. 173–197.

Schutz, C. and Gates, W.L., 1971. Global climatic data for surface, 800 mb, 400 mb. Rep. R. 915 – ARPA, Rand Corporation, Santa Monica, Calif.

Thépenier, R.M., 1984. Etude des fluctuations des zones convectives en Afrique, à partir des images Meteosat. Veille Climatique Satellitaire, Météorologie Nationale, No. 1. pp. 11–16.

Von Storch, H. and Kruse, H., 1985. The significant tropospheric midlatitudinal El Niño response patterns observed in January 1983 and simulated by a GCM. In: J.C.J. Nihoul (Editor), Coupled Ocean–Atmosphere Models. (Elsevier Oceanography Series, 40) Elsevier, Amsterdam, pp. 275–288 (this volume).

CHAPTER 36

THE PARAMETRIZATION OF THE UPPER OCEAN MIXED LAYER IN COUPLED OCEAN–ATMOSPHERE MODELS

C. GORDON and M. BOTTOMLEY

ABSTRACT

In an ocean model suitable for detailed climate studies it is necessary to represent the ocean temperature changes arising from both heat advection and the exchange of heat across the air–sea interface. The relative importance of these two effects in determining the large-scale sea-surface temperature field depends upon geographical location. In mid-latitude oceans, away from the boundary regions, local exchanges dominate the upper ocean heat budget so that a simple mixed layer model might be expected to predict the seasonal changes in sea-surface temperature reasonably well.

A number of experiments are described in which a simple mixed layer model is forced with both observed and atmospheric model fluxes. The experiments have a global domain, although particular attention is given to the simulation of the seasonal cycle in middle latitudes. Comparisons between experiments with observed and model fluxes indicate that care should be taken to match the detail of the mixed layer parametrization to the quality of the atmospheric model forcing.

1. INTRODUCTION

In the analysis of numerical experiments using atmospheric general circulation models (AGCMs) with climatologically prescribed sea-surface temperatures (SSTs) and sea-ice extents, little attention has been given to the simulation of the surface fluxes of heat and momentum over the oceans. This is because in such experiments the oceans act as an infinite heat source or sink. In a coupled atmosphere–ocean model the situation is very different since SSTs and sea-ice extents are dependent on the surface forcing. A useful preliminary to a coupled experiment is therefore to look in detail at the fluxes of heat and momentum from AGCM experiments with prescribed SSTs and sea-ice extents.

The major problem in attempting to verify the AGCM simulation of the surface fluxes is the lack of sufficiently accurate climatological data sets. This is particularly true over the global oceans where, in many areas, the heat and momentum fluxes across the surface are poorly known due to the lack of data coverage. The existing data sets of global heat budget components (Budyko, 1963; Schutz and Gates, 1971–1974; Esbensen and Kushnir, 1981) and momentum flux (Hellerman, 1967; Han and Lee, 1981; Hellerman and Rosenstein, 1983) have allowed modellers to assess the overall patterns simulated by their models. A point is inevitably reached, however, when it becomes impossible to say whether it is the models or the data which are in error. These problems are particularly severe in the Southern Hemisphere.

A similar problem occurs in the verification of the results from ocean general circulation models (OGCMs). If these models were provided with the best estimates of the surface forcing, how well would they simulate the observed ocean circulation and, most importantly in a coupled model, the spatial and temporal variations of SST? Again a point must be reached at which it is impossible to disentangle errors due to surface forcing from those due to inadequacies in the models.

In a coupled atmosphere–ocean GCM both of the above problems occur together so that, for example, errors in certain aspects of the atmospheric circulation may arise from errors in the SST field which, in turn, may be to some extent a consequence of errors in the surface forcing. The possibilities for getting the wrong answer, or even the right answer for the wrong reasons, are many and the complex interactions are such that it is very difficult to pin-point the source of a particular error.

It is obviously desirable to test both the oceanic and atmospheric components of the coupled system in an uncoupled mode in order to assess the models individually and to ensure that the complexity of the models is suitably matched. With the AGCMs this has been done in experiments with climatologically specified SSTs and sea-ice extents. The results of the various models are well documented.

When testing OGCMs, Haney forcing (Haney, 1971) has often been used to parametrize the net surface heat flux via the difference between an effective air temperature and the predicted SST. Such an approach is valuable in assessing the performance of some aspects of the model. However, it can say little about how the SST may be expected to respond in a coupled atmosphere–ocean experiment since the SST is essentially constrained to be close to the effective air temperature. The alternative is to use climatological estimates of the surface heat fluxes. The objection to this approach is that in reality the fluxes will change with the SST. This largely involves a negative feedback via the turbulent heat losses at the sea surface.

Ultimately, of course, it is necessary to use a coupled ocean–atmosphere model if the important feedback mechanisms are to be effectively represented. However, coupled experiments are expensive in terms of computer resources and the relatively cheap non-interactive tests using both Haney and heat flux forcing can be useful in providing some insight into the behaviour of the coupled system. Most of the experiments reported in this paper are non-interactive in the sense that the surface fluxes do not respond to the SST field. The exception is the last experiment reported using model data, in which the anomalous surface heat flux due to a SST anomaly is assumed to be linearly proportional to the temperature anomaly.

Results of experiments using both observed climatological and modelled surface fluxes to simulate the seasonal cycle of SST in middle latitudes will be discussed. Emphasis is given to temperature changes in middle latitudes for two reasons. Firstly, as indicated in Fig. 1, the seasonal range of SST peaks in the middle latitudes of both hemispheres where it is by far the most dominant signal. Secondly, there are good theoretical and observational reasons to expect the open ocean large-scale seasonal changes in SST to be determined by the seasonal variations in the heat and mechanical energy fluxes across the sea surface (Gill and Niiler, 1973; Barnett, 1981). In fact, the experiments with observed climatological data reported below may be taken as a further indication that this is the case. This means that a simple mixed layer model may well be able to produce a reasonable simulation of the seasonal changes in SST in these regions.

It should be emphasised that it is not being suggested here that a simple mixed-layer model provides a suitable representation of the ocean in a global atmosphere–ocean

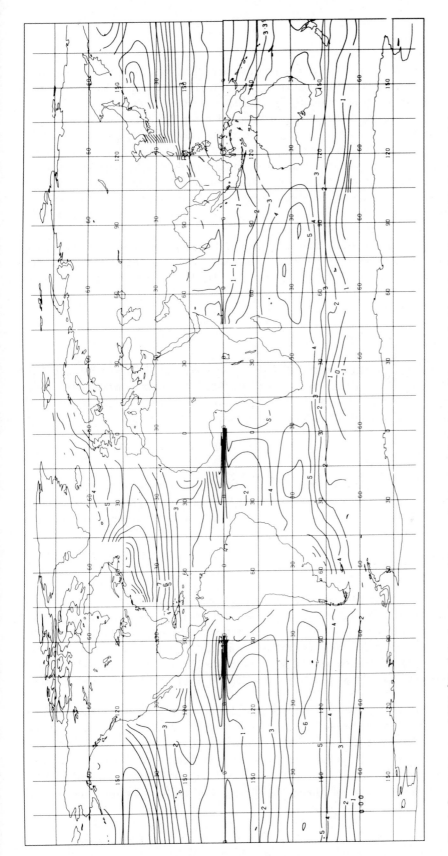

Fig. 1. Seasonal range of SST (°C), defined as the monthly mean for August minus the monthly mean for March in the Northern Hemisphere and the February value minus the September value in the Southern Hemisphere.

616

model. This is clearly not the case since advective effects are of primary importance in both the tropics and the regions of large ocean to atmosphere heat exchange associated with the western boundary currents. The mixed-layer model is being used rather as a diagnostic tool to assess the sensitivity of the predicted mid-latitude SSTs to the quality of the surface fluxes. If, using modelled fluxes, the seasonal changes in mid-latitude SSTs are predicted to the same accuracy as obtained using observed climatological fluxes (which may themselves provide a fairly poor simulation) then the modelled fluxes are as good as the climatological estimates. As will be shown below, this is not the case and most of the errors in the simulated SST using model fluxes can be directly linked with errors in these fluxes. In fact, the worst errors occur in precisely those regions where a simple mixed-layer model should work well.

This paper is organised as follows: Section 2 describes the mixed layer model and Section 3 an experiment using observed climatological surface fluxes. Section 4 describes experiments with surface fluxes derived from an AGCM and conclusions are drawn in Section 5.

2. THE MIXED-LAYER MODEL

The mixed-layer model used in this study is essentially that of Gill and Turner (1976) as developed by Mitchell (1977a) and Gordon (1985). This bulk model relates changes in the potential energy of a vertical water column to the rate of working by the wind at the sea surface. The rate of turbulent energy input at the surface is given by the wind-mixing power:

$$W = m_\sigma \rho_w v_*^3 \tag{1}$$

where v_* is the oceanic friction velocity, ρ_w is the density of seawater and m_σ is a scaling factor of order unity. This is decayed exponentially with depth to prevent over-deepening of the winter-time mixed layer. W can also be expressed in terms of the local wind speed at some level, so that:

$$W = \frac{m_\sigma (\rho_{air} C_L)^{3/2} U_L^3}{\rho_w^{1/2}} \tag{2}$$

where C_L is the drag coefficient appropriate to a level L in the atmosphere and U_L is the wind speed at that level. ρ_{air} is the density of air. This energy is used to mix the surface heat input throughout the mixed layer. Convection in the model is partly penetrative with 85% of convectively generated turbulent kinetic energy dissipated within the mixed layer. The penetrative short-wave flux is represented by a double exponential decay function (Paulson and Simpson, 1977). The effects of salinity on density are not included in the model.

The model parameter m_σ and the e-folding depth for the decay of windmixing were fixed so that, when the mixed-layer depth was of order 20–30 m, the total dissipation in the mixed layer was equivalent to that obtained by Davis et al. (1981) using data from the MILE experiment. Although fine tuning of these parameters could have improved the model simulation at a particular location, previous work has shown that it is not possible to choose one set of parameters which give a best fit at all locations (Mitchell,

1977b). With the MILE parameters the modelled seasonal range of SST at Ocean Weather Station "Papa" (50°N, 145°W) was within about 0.5°C of the observed value. The model also produces realistic variations in sub-surface temperatures.

Tests were performed to investigate the sensitivity of the model to both vertical resolution and timestep. In these experiments a control integration was run having a uniform 1 m vertical resolution to 200 m and a timestep of one hour. When compared with the control integration, a model with 6 layers in the upper 200 m (centred on 5, 17.5, 32.5, 55, 95 and 160 m, i.e. giving greater resolution near to the surface) and a timestep of 24 h, was found to be quite adequate to simulate the seasonal cycle of SST to an accuracy of 1°C. This version of the model was adopted for the experiments to be described later in Section 4.

3. EXPERIMENTS WITH CLIMATOLOGICAL FLUXES

Turbulent and radiative fluxes for this study were taken from the monthly data sets compiled by Esbensen and Kushnir (1981). At the time when these experiments were performed no suitable global data sets of the windmixing power were available. A number of tests were carried out to determine whether existing data sets of monthly mean wind stress (Han and Lee, 1981) and wind speed (Esbensen and Kushnir, 1981) could be used to provide a reasonable estimate of the windmixing over the globe. This was not the case since the cubic dependence of this quantity on the wind speed [see eqn. (2)] meant that the lack of wind variability in the existing mean data sets led to serious errors when the windmixing power was calculated. Although this can be compensated for, locally, by introducing an empirical enhancement factor it is not possible to do this globally. Tests with the mixed layer model showed that these errors are not insignificant in the simulation of the seasonal cycle of SST. A global data set of windmixing is currently being created and will subsequently be used to force the mixed layer model. Details of the tests referred to above can be found in a recent report by Grahame (1984a).

In the absence of a suitable windmixing data set Grahame (1984b) performed a series of experiments in which the mixed-layer depths were specified in various ways. When the daily mean heating at the sea surface is positive and there is insufficient windmixing to redistribute this heat throughout the existing mixed layer, the layer base will establish itself at a shallower depth (h). If the effects of penetrative radiation, decay of mixing energy with depth and heat advection are ignored this is given by the Monin-Obukhov depth (Niiler and Kraus, 1977):

$$h_* = \frac{2W}{\frac{\alpha g Q}{c}} \tag{3}$$

where W is the windmixing power, Q the net surface heating, α the expansion coefficient of seawater, g the acceleration due to gravity and c the specific heat capacity of seawater. The surface temperature, T, increases according to:

$$\frac{\partial T}{\partial t} = \frac{Q}{c \rho_w h} \tag{4}$$

In his experiments Grahame (1984b) determined the mixed-layer depths during the shallowing season in each hemisphere by interpolation from the Levitus and Oort (1977) hydrographic data set. The net surface heating from the Esbensen and Kushnir (1981) data set was distributed over the mixed layer at each ocean point on a 2.5° by 3.75° latitude–longitude grid i.e. the surface temperatures were updated using eqn. (4). It is clear from eqn. (3) that this procedure is loosely equivalent to specifying a windmixing power. At each point on the grid the seasonal range of SST in this experiment is what one might expect to obtain by running the full mixed-layer model described in Section 2 providing the full model reproduces the same time evolution of mixed layer depths as derived from the Levitus and Oort (1977) data set.

Problems arise, however, when eqn. (4) is used without modification. For example, in regions of the ocean where there is a net heat loss from the surface over the year the annual mean mixed-layer temperature will fall from year to year. In reality this heat loss is compensated for by warm advection due to ocean currents. In the absence of better estimates Grahame (1984b) approximated the effects of the oceanic heat flux convergence at each grid point by assuming that it leads to a constant heating rate in the mixed layer throughout the year which he derived from the annual mean heat exchange across the surface. This ensured that the annual mean heat content of the water column at each grid point does not change. Thus, when the mixed layer is shallowing eqn. (4) was replaced by:

$$\frac{\partial T}{\partial t} = \frac{Q}{c\rho_w h} - \frac{\bar{Q}}{c\rho_w \bar{h}} \tag{5}$$

where $\bar{Q}$ and $\bar{h}$ are the annual mean net surface heating and mixed-layer depths respectively. The last term in eqn. (5) is the heating rate corresponding to the assumed oceanic heat flux convergence. If the instantaneous mixed-layer depth was used in this term rather than the annual mean, then the heating rate would have a spurious seasonal signal associated with the seasonal changes in mixed-layer depth. Grahame (1984b) performed a number of experiments in which the annual mean heating was distributed in the vertical in a variety of ways (e.g., over the instantaneous mixed layer, over a 200 m column, etc.). It was found that in mid-latitude open ocean regions the details of how the annual mean heating is distributed in the vertical does not fundamentally alter the simulation of the seasonal cycle.

Figure 2 is the difference between the simulated and observed seasonal ranges in SST using the model described by eqn. (5). As already stated Q and $\bar{Q}$ were taken from Esbensen and Kushnir (1981) and h and $\bar{h}$ from Levitus and Oort (1977). The seasonal range of SST is defined as the monthly mean for August minus the monthly mean for March in the Northern Hemisphere and the February minus the September value in the Southern Hemisphere (which leads to the discontinuity in Figs. 1 and 2 at the equator). A detailed discussion of the results of this and a number of other experiments can be found in Grahame (1984b). It is evident from Fig. 2 that the seasonal range of SST is simulated to within 2°C over much of the world's mid-latitude oceans. This typical accuracy of 2°C should be kept in mind when assessing the simulation of the seasonal cycle of SST using AGCM fluxes.

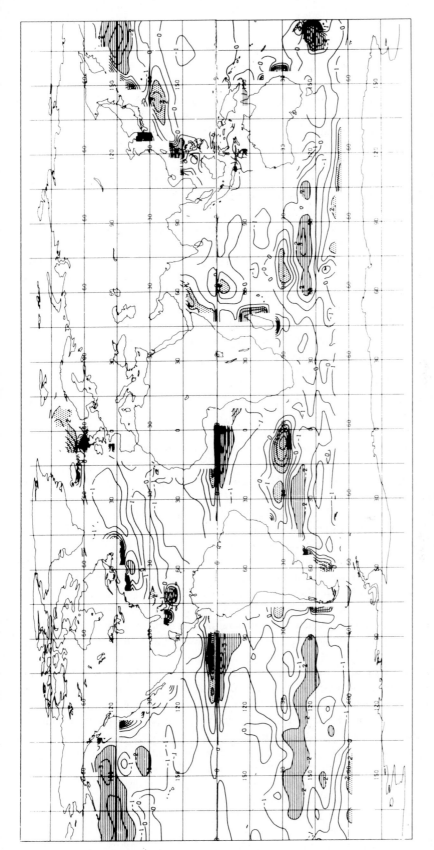

Fig. 2. Difference (°C) between the simulated and observed climatological seasonal ranges of SST. The simulated range was predicted using observed climatological heat fluxes and mixed layer depths, as discussed in Section 3. (Hatched areas indicate differences less than −2°C; stippled areas indicate differences greater than 2°C.)

4. EXPERIMENTS WITH MODELLED FLUXES

In the experiments described in this section the fluxes of heat and mechanical energy needed to force the mixed layer model described in Section 2 were extracted from a four-year integration of the Meteorological Office 11-layer AGCM. This is a global model, a limited area version of which was used for the GARP Atlantic tropical experiment (Lyne and Rowntree, 1983). It is a primitive equation model using σ (pressure/surface pressure) as a vertical co-ordinate, and a regular $2.5° \times 3.75°$ latitude—longitude grid. The seasonal and diurnal variations of solar radiation are represented, and the radiative fluxes are a function of temperature, water vapour, carbon dioxide and ozone concentrations, and prescribed zonally averaged cloudiness. SSTs and sea-ice extents are prescribed from climatology, and updated every five days.

Fluxes of turbulent and radiative heat at the sea surface and the windmixing power were accumulated every timestep of the atmospheric model and stored on tape once per model day. Subsequently, these fluxes were used to drive the mixed-layer model at each grid point.

A number of experiments were performed. In each experiment the temperatures in the upper 200 m of the ocean were initialised using values interpolated from the Levitus and Oort (1977) data set for "spring".

Preliminary experiments

An experiment in which no account was taken of the annual mean advective heat flux [see eqn. (5)] produced summer SSTs which were too high (by 5°C or more in places) in both the central North Pacific and the North Atlantic. In both regions the inclusion of the annual mean advective heat flux convergence would have made the simulation worse. These high temperatures were due to the overestimation by the AGCM of the absorbed short-wave flux at the surface. This is associated with the use of zonally meaned cloud amounts in the model. A further experiment in which the modelled short wave flux was replaced by the corresponding fields from the Esbensen and Kushnir (1981) data sets (interpolated to daily values) showed a considerable improvement in the simulation of the Northern Hemisphere summer SSTs. Subsequent experiments with the mixed-layer model were therefore carried out with the AGCM short wave flux replaced by the Esbensen and Kushnir (1981) value. In future experiments with the AGCM an interactive cloud scheme will be used in which cloud amounts are predicted within the model, thus allowing longitudinal variations in the modelled short-wave flux to be represented.

The annual mean advective heating correction discussed in the previous section was not applied in the above experiments. When using AGCM fluxes a problem arises in the implementation of this correction since the modelled annual mean mixed-layer depth is not known in advance. [See eqn. (5).] In order to circumvent this problem, values of Q were evaluated at each grid point from modelled turbulent and long-wave fluxes and the observed climatological short-wave flux. A heat flux equal to Q was then removed at each timestep from the instantaneous modelled mixed layer. The experiment was run for one year and the modelled annual mean mixed-layer depths obtained were used as an estimate of h in the calculation of the heating rate:

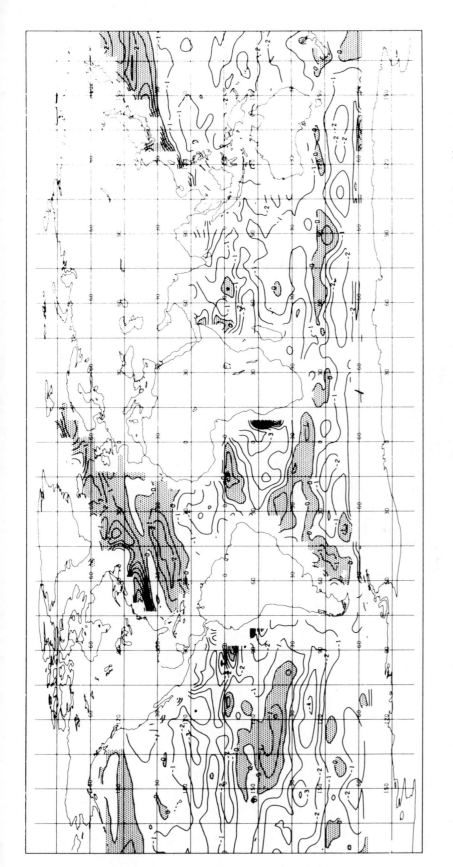

Fig. 3. Difference (°C) between the monthly mean SST for March as predicted by the full mixed-layer model and the observed climatological monthly mean value. (Stippled areas indicate differences that are positive.)

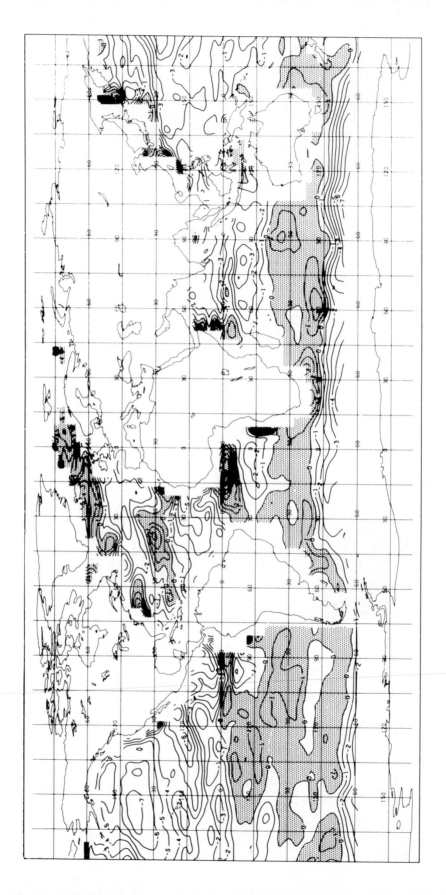

Fig. 4. As Fig. 3 but for September.

$$\left(\frac{\partial T}{\partial t}\right)_{\text{advection}} = \frac{-\bar{Q}}{c\rho_w \bar{h}} \tag{6}$$

This heating rate was used in all subsequent experiments to represent the annual mean advective heat flux convergence at each location.

Experiment 1

An integration was carried out using daily AGCM fluxes of long-wave heating, turbulent heating and windmixing and the Esbensen and Kushnir (1981) short-wave flux and with eqn. (6) applied to the instantaneous mixed layer. The experiment was run for a period of four years and investigates, primarily, the response of the mixed-layer model to the above AGCM fluxes.

Figures 3 and 4 show the global distribution of modelled minus observed monthly mean SSTs for March and September, respectively. The observed temperatures are from Esbensen and Kushnir (1981) and the modelled values are four-year means for these months. Figures 5 and 6 show the corresponding distributions of mixed-layer depths. The observed values are from Levitus (1982).

March simulation

Over most of the oceans the modelled March SSTs differ from climatology by less than $2°C$ (Fig. 3). The largest errors occur in the western boundary regions of the Northern Hemisphere and are probably associated with the simple treatment of the advective heat flux. The model net annual mean heating (Q) is large in these areas and the mixed layer model would not be expected to perform well.

The modelled mixed-layer depths in the North Pacific are of the right order but exhibit a larger east–west gradient than appears in the observations (Fig. 5). In much of the North Atlantic modelled mixed-layer depths are generally too shallow, especially between $20°$ and $45°N$. A relatively shallow tongue extends across the sub-tropical North Atlantic and leads to the relatively warm modelled SSTs in this region (Fig. 3). In the middle latitudes of the Southern Hemisphere the mixed-layer depths are generally too deep. The area in the region of $37°S$, $120°W$ shows simulated SSTs more than $3°C$ lower than climatology. This is associated with the relatively deep modelled mixed-layer depth (about $70–80\,m$ whereas the Levitus (1982) data set indicates a shallow mixed layer of about $25–50\,m$). It is also significant that climatological SSTs in this region have a large seasonal range, presumably associated with the shallow summer time mixed layer.

September simulation

In September the Northern Hemisphere oceans tend to be too cold, by as much as $7°C$ in the North Pacific (Fig. 4). The mixed layer depths are correspondingly too deep (Fig. 6), although this cannot account for the large magnitude of the error in the SSTs. The experiment using observed fluxes described earlier also underestimated the summer SSTs in this region, but with the maximum errors occurring somewhat further north (see Fig. 2).

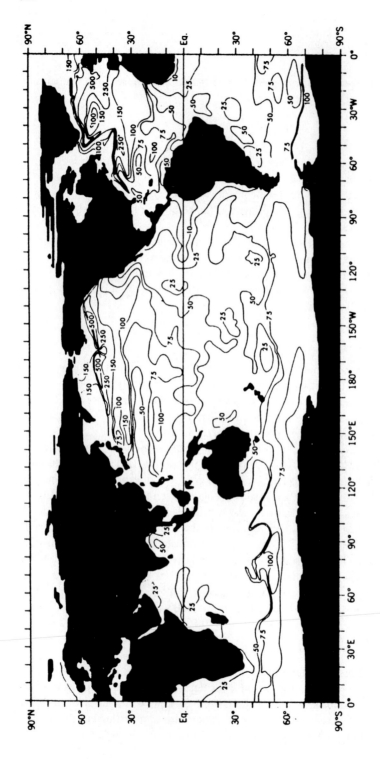

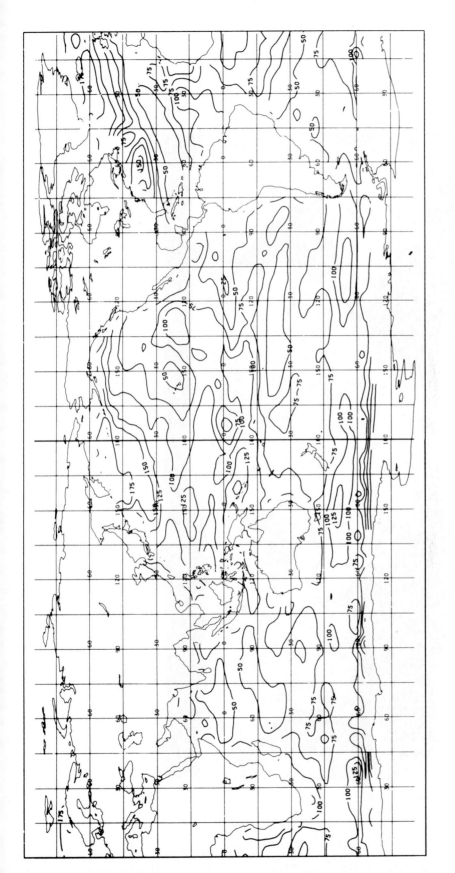

Fig. 5. a. Observed climatological distribution of monthly mean mixed-layer depths (m) for March (after Levitus, 1982). b. Modelled distribution of monthly mean mixed-layer depths (m) for March from Experiment 1.

626

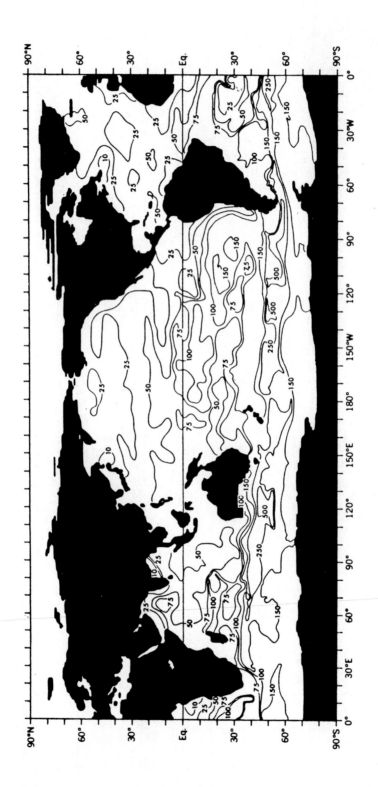

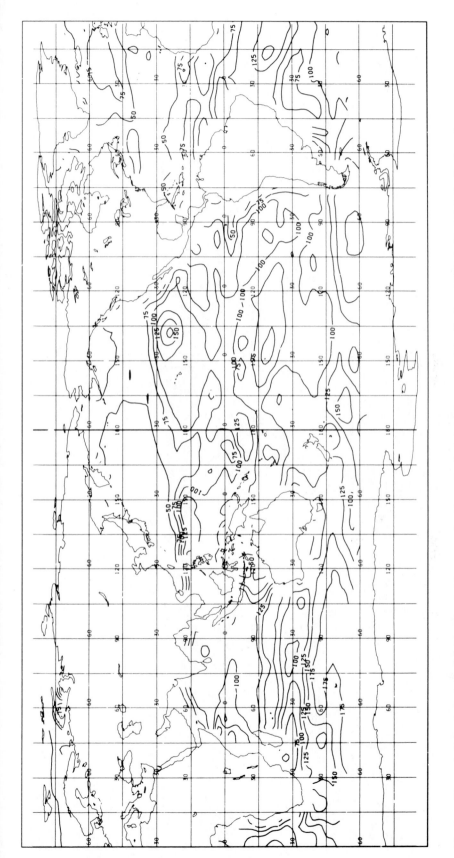

Fig. 6. a. Observed climatological distribution of monthly mean mixed-layer depths (m) for September (after Levitus, 1982). b. Modelled distribution of monthly mean mixed-layer depths (m) for September from Experiment 1.

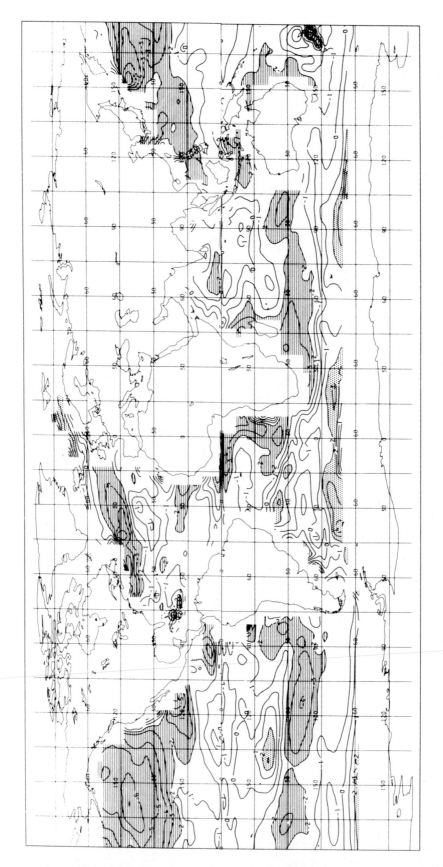

Fig. 7. Difference (°C) between simulated and observed climatological seasonal ranges of SST. The simulated range was predicted using AGCM heat and windmixing fluxes to force the full mixed-layer model. (Shading as in Fig. 2.)

One of the major errors in the AGCM integration is a systematic underestimation of the downward long-wave flux in the model (Rowntree, 1981). In the areas where SSTs are much too cold the modelled net downward long-wave flux is consistently of order $20-30\,\mathrm{W\,m^{-2}}$ lower than the estimates given by Esbensen and Kushnir (1981). The effects of this error are to some extent compensated for by the treatment of the annual mean advective heat flux convergence in the model. This puts heat back into the mixed layer so as to ensure that the net change in oceanic heat content at each location is zero over the year. However, this heat flux is applied as a constant heating rate over the whole mixed layer [see eqn. (6)], not as a surface flux, and no windmixing energy is used to mix the heat throughout the layer. The modelled mixed-layer depths are consequently too deep since in reality the instantaneous surface heating is greater and the net annual mean heating is smaller. Greater surface heating means that more windmixing energy must be used to mix the surface heat input throughout the mixed layer and less energy is therefore available for actually deepening the layer.

Mixed-layer depths in the Atlantic are also too deep and the SSTs too cold. In the Southern Hemisphere SSTs are generally simulated to within 1°C of climatology except along the ice edge where differences are larger.

Seasonal range simulation

Figure 7 shows the modelled minus climatological seasonal range of SST. The definition of seasonal range is the same as that used in Section 3. It is evident that over most of the North Pacific and much of the North Atlantic the seasonal range is underestimated. This is associated with the poor simulation of the summer SSTs as discussed above. In the Southern Hemisphere the simulation is generally better although it is clear that the largest errors tend to occur in those regions where the climatological range is greatest.

A comparison of Figs. 2 and 7 shows that the model simulation of seasonal range is considerably worse than that obtained using observed fluxes and mixed-layer depths.

Experiment 2

This experiment was identical to Experiment 1 except for the inclusion of a linear feedback term in the surface heat balance. The original purpose of the experiment was to investigate the variability of SST predicted by the mixed layer model when it was forced with daily AGCM fluxes. For this purpose a feedback is essential (Frankignoul and Hasselmann, 1977). The simulated anomalies will not be discussed here but the experiment also produced some interesting features in the mean fields which will be described.

The net heat flux over the ocean is a function of SST and various atmospheric variables such as air temperature, wind speed, cloudiness, etc. Denoting the latter collectively by x, the net surface heating is linearised around the flux obtained with the observed climatological SST, so that:

$$Q(T, x) \approx Q(T_c, x) - \lambda (T - T_c) \tag{7}$$

where:

$$\lambda = \left\langle -\frac{\partial Q}{\partial T}(T, x) \right\rangle_{T = T_c} \tag{8}$$

The angle brackets in eqn. (8) represent an ensemble mean over many realisations of the atmospheric variables x (Frankignoul, 1979). The net heat flux $Q(T_c, x)$ in eqn. (7) is the flux obtained with the climatological SST and is therefore the flux produced in the AGCM simulation. The last term is the anomalous heat flux due to the SST anomaly and is essentially equivalent to a local Haney-type forcing (Gill, 1979). The feedback parameter λ was assigned a globally constant value of $35\,\mathrm{W\,m^{-2}\,K^{-1}}$. In reality this parameter should vary with geographical location and the spatial scale of the anomaly so that it decreases to a few $\mathrm{W\,m^{-2}\,K^{-1}}$ for changes of SST on a global scale (Bretherton, 1982).

The difference between simulated and observed climatological SST range is shown in Fig. 8. Not surprisingly the differences are generally less than those obtained using the model without feedback (Fig. 7). Figures 9a and b show the mixed-layer depths from Experiment 2 for March and September respectively. In March in the North Pacific and sub-tropical North Atlantic the mixed-layer depths are generally deeper than those obtained in Experiment 1, whereas they are shallower in the South Pacific. In September the North Pacific depths are shallower and closer to climatology in this experiment and in the Southern Hemisphere they are deeper. Each of these changes can be understood in terms of the anomalous stabilising or de-stabilising effects of the anomalous heat flux associated with the feedback term.

In general the distribution of mixed-layer depths is closer to the Levitus (1982) climatology and, of course, the SSTs are also simulated considerably better than those in Experiment 1. Taken at face value the results of Experiment 2 would give a mis-leading indication of the ability of the mixed-layer model to simulate global SSTs. This illustrates the limitations of using a heat flux forcing which ensures that the SST remains close to its observed climatological value. If the spatial scale-dependence of the feedback parameter λ was introduced as suggested by Bretherton (1982), its value would decrease as the large scale SST differences evident from Experiment 1 started to develop. With this dependence included the simulated SSTs would be a better guide to how the SSTs might respond in a fully coupled integration. The results would, no doubt, exhibit many of the errors found in Experiment 1 but with reduced magnitude. [Ideally, non-local effects should also be taken into account in the manner outlined by Bretherton (1982).]

5. CONCLUSIONS

The simulation of the seasonal cycle of SST using fluxes derived from the AGCM must be considered to be rather poor. However, the systematic error in the downward long-wave flux at the surface, which was due to inadequacies in the treatment of water vapour absorption and emission in the radiation scheme, has now been corrected. An improve-ment in the simulation of the SST seasonal cycle is therefore expected in future experi-ments with the mixed-layer model.

The quality of the atmospheric model fluxes is highly dependent on the model in question. For example, previous experiments with the Meteorological Office 5-level

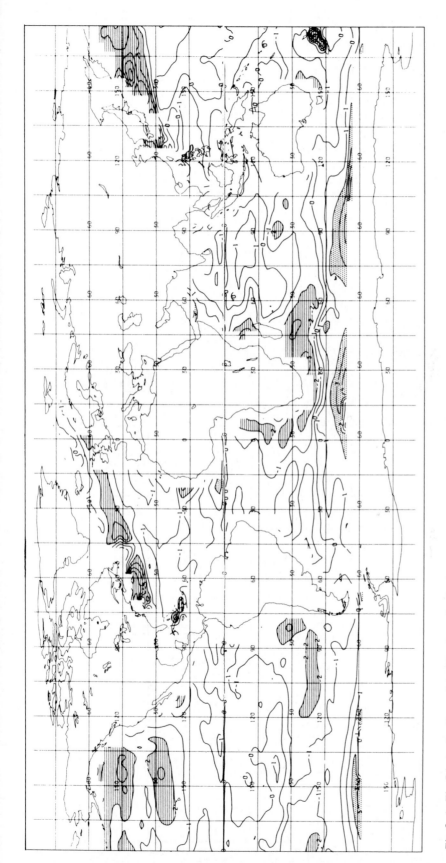

Fig. 8. As Fig. 7 but using the model with a linear feedback term.

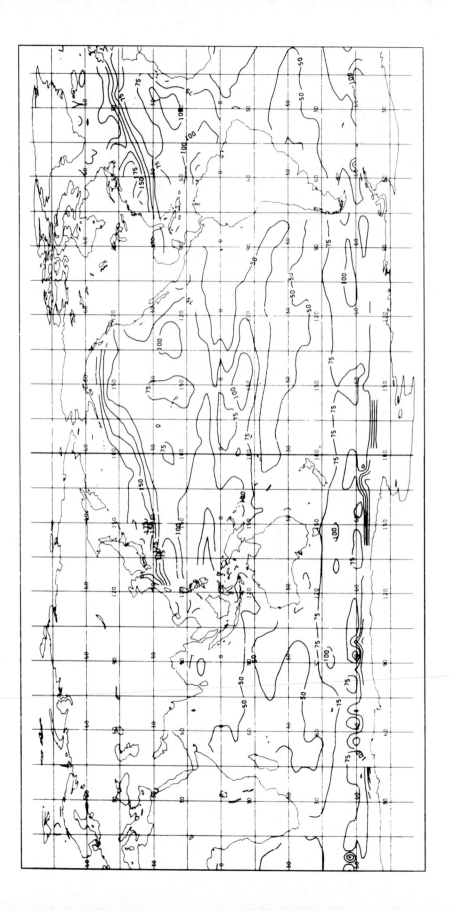

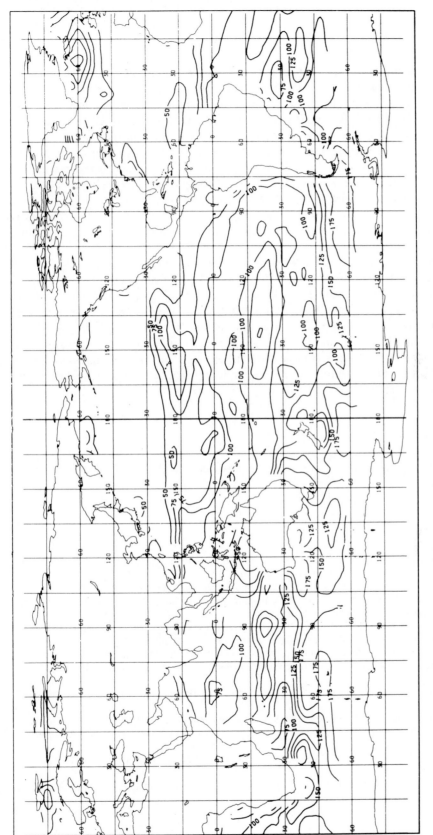

Fig. 9. a. Modelled distribution of monthly mean mixed-layer depth (m) for March from Experiment 2. b. As (a) but for September.

AGCM produced much too high summer SSTs in the North Pacific, partly because the summer time windmixing was much too low. The results presented here should not, therefore, necessarily be taken to be applicable to other AGCMs although analogous problems will occur in all models. Using the AGCM fluxes to force the mixed-layer model is a useful way to assess the significance of the errors in these fluxes in terms of predicting SSTs. In this context the mixed-layer model is a useful diagnostic tool.

Also relevant here is the question as to which upper ocean parametrization should be used in a coupled model. For example, if the AGCM produces a very poor simulation of the global windmixing field, a model with a fixed mixed-layer depth may well produce better results than a model which determines the mixed-layer depth. The complexity of the ocean model has therefore to be matched to the quality of the fluxes used to force it.

In the near future it is intended to repeat the experiments with climatological heat fluxes but using a global data set of climatological windmixing power to force the full mixed layer model. Simulations of the seasonal cycle will also be performed in which fluxes from an AGCM experiment with improved parametrizations of the downward long-wave flux and the surface drag coefficient (see Mitchell et al., 1985) as well as model predicted surface short wave heating produced by an interactive cloud scheme, will be used. In the longer term the mixed-layer model will be embedded into a global version of the Bryan/Semtner model currently being developed for coupling to the atmospheric model.

ACKNOWLEDGEMENTS

The authors gratefully acknowledge the assistance of Mr. N.S. Grahame in the preparation of this paper.

REFERENCES

Barnett, T.P., 1981. On the nature and causes of large scale thermal variability in the central North Pacific Ocean. J. Phys. Oceanogr., 11: 887–904.
Bretherton, F.P., 1982. Ocean climate modelling. Progr. Oceanogr., 11: 93–129.
Budyko, M.I., 1963. Atlas of the Heat Balance of the Earth. Akad. Nauk SSSR, Prezidum Mezhduvedomstennyi Geofiz. Komitet, 69 pp.
Davis, R.E., DeSzoeke, R. and Niiler, P., 1981. Variability in the upper ocean during MILE. Part II. Modelling the mixed layer response. Deep-Sea Res., 28A: 1453–1475.
Esbensen, S.K. and Kushnir, Y., 1981. The heat budget of the global ocean: An atlas based on estimates from surface marine observations. Report No. 29, Climatic Research Institute, Oregon State University, Corvallis, Oreg.
Frankignoul, C., 1979. Stochastic forcing models of climate variability. Dyn. Atmos. Oceans, 3: 465–479.
Frankignoul, C. and Hasselmann, K., 1977. Stochastic climate models, part II. Tellus, 29: 284–305.
Gill, A.E., 1979. Comments on stochastic models of climate variability. Dyn. Atmos. Oceans, 3: 481–483.
Gill, A.E., and Niiler, P.P., 1973. The theory of the seasonal variability in the ocean. Deep-Sea Res., 20: 141–178.
Gill, A.E. and Turner, J.S., 1976. A comparison of seasonal thermocline models with observation. Deep-Sea Res., 23: 391–401.

Gordon, C., 1985. Sensitivity tests with a simple mixed layer model. Dynamical Climatology Technical Note, Meteorological Office, Bracknell. (In prep.)

Grahame, N.S., 1984a. A technique for representing synoptic scale variability for calculations of monthly mean wind forcing for mixed layer experiments. Met 0 20 Technical Note II/221 (unpublished, copy available in the National Meteorological Library, Bracknell).

Grahame, N.S., 1984b. Simulation of the seasonal cycle of SST using globally specified climatological data to force simple ocean models. Dynamical Climatology Technical Note, No. 5 (unpublished, copy available in the National Meteorological Library, Bracknell).

Han, H.J., and Lee, S.W. 1981. A new analysis of monthly mean wind stress over the global ocean. Report No. 26, Climatic Research Institute, Oregon State University, Corvallis, Oreg.

Haney, R.L., 1971. Surface thermal boundary conditions for ocean circulation models. J. Phys. Oceanogr., 1: 241–248.

Hellerman, S., 1967. An updated estimate of the wind stress on the world ocean. Mon. Weather Rev., 95: 607–626.

Hellerman, S. and Rosenstein, M., 1983. Normal monthly wind stress over the world ocean with error estimates. J. Phys. Oceanogr., 13: 1093–1104.

Levitus, S., and Oort, A., 1977. Global analysis of oceanographic data. Bull. Am. Meteorol. Soc., 58: 1270–1284.

Levitus, S., 1982. Climatological atlas of the world ocean. NOAA Prof. Pap. 13, Geophysical Fluid Dynamics Laboratory, Princeton, N.J.

Lyne, W.H. and Rowntree, P.R., 1983. Forecast model. In: P.R. Rowntree and H. Cattle (Editors), The Meteorological Office Gate Modelling Experiment. Meteorological Office Scientific Paper No. 40, pp. 34–40.

Mitchell, J.F.B., 1977a. An oceanic mixed layer model for use in general circulation models. Met 0 20 Technical Note II/85 (unpublished, copy available in the National Meteorological Library, Bracknell).

Mitchell, J.F.B., 1977b. Some experiments using a simple oceanic mixed layer model. Met 0 20 Technical Note II/86 (unpublished, copy available in the National Meteorological Library, Bracknell).

Mitchell, J.F.B., Wilson, C.A. and Price, C., 1985. On the specification of surface fluxes in coupled atmosphere–ocean general circulation models. In: J.C.J. Nihoul (Editor), Coupled Ocean–Atmosphere Models. (Elsevier Oceanography Series, 40) Elsevier, Amsterdam, pp. 249–262 (this volume).

Niiler, P.P., and Kraus, E.B., 1977. One-dimensional models of the upper ocean. In: E.B. Kraus (Editor), Modelling and prediction of the Upper Layers of the Ocean. Pergamon, Oxford, pp. 143–172.

Paulson, C.A. and Simpson, J.J., 1977. Irradiance measurements in the upper ocean. J. Phys. Oceanogr., 7: 952–956.

Rowntree, P.R., 1981. A survey of observed and calculated longwave fluxes for the cloud-free tropical atmosphere. Met 0 20 Technical Note II/174 (unpublished, copy available in the National Meteorological Library, Bracknell).

Schutz, C. and Gates, W.L., 1971. Global climatic data for surface, 800 mb, 400 mb, January. R-915-ARPA, The Rand Corporation, Santa Monica, Calif.

Schutz, C. and Gates, W.L., 1972. Global climatic data for surface, 800 mb, 400 mb: July R-1029-ARPA, The Rand Corporation, Santa Monica, Calif.

Schutz, C. and Gates, W.L., 1973. Global climatic data for surface, 800 mb, 400 mb: April. R-1317-ARPA, The Rand Corporation, Santa Monica, Calif.

Schutz, C. and Gates, W.L., 1974. Global climatic data for surface, 800 mb, 400 mb: October. R-1425-ARPA, The Rand Corporation, Santa Monica, Calif.

CHAPTER 37

STABILITY OF A SIMPLE AIR–SEA COUPLED MODEL IN THE TROPICS

T. YAMAGATA

ABSTRACT

A simple theoretical discussion is given of the stability of air–sea coupled models in the Tropics. It is shown that growing perturbations similar to the oceanic Kelvin waves associated with Matsuno's (1966) circulation pattern in the atmosphere exist for the situation such that changes of atmospheric winds induce further release of latent heat. A net positive correlation between atmospheric winds and oceanic flows is a necessary condition for instability.

The relevance of the model results to the 1982/83 ENSO event is discussed. In spite of a simple assumption that the latent heat release is proportional to the depth changes of the thermocline, the model results are encouraging in explaining the temporal and spatial evolution of the anomalous air–sea conditions of the 1982/83 ENSO event.

1. INTRODUCTION

There is no doubt that air–sea interaction plays a key role during El Niño and Southern Oscillation events. Although Bjerkness (1966) pointed out almost intuitively that El Niño and Southern Oscillation are two aspects of the same phenomenon, almost all attempts have been so far to explain the response of one component of the coupled system to the other.

Recently, Philander (1983) pointed out with clarity that unstable interactions are possible in the tropical ocean–atmosphere system. The positive feedback of this instability is provided only when the release of latent heat by the ocean changes the winds in the atmosphere so that the further release of latent heat becomes possible (Philander et al., 1984).

The purpose of this paper is to provide a more complete analysis of the instability by use of a simple air–sea coupled system as an extension of the previous work (cf. Philander et al., 1984). The direct application of the model to realistic ENSO events is premature, because our simple model assumes that linear shallow water equations govern the oceanic and atmospheric motion. In addition, the procedure of relating SST changes to variations in depth of the thermocline (or in the equivalent depth) is quite naive. However, the model results simulate, to some extent, the evolution of anomalous conditions observed in the 1982/83 event. In particular, the fact that modest anomalies which first appeared in the western Pacific Ocean in May grew both spatially and temporally until they reached a maximum amplitude in October can be well explained within the framework of the present simple model. This suggests that the above process of unstable air–sea interactions may be modelled, to some extent, by our simple coupling parameterization.

The presentation of the present paper is the following. In Section 2, the details of the model equations are given and in Section 3 both analytical and numerical solutions are

638

discussed. The relevance of the results to numerical experiments is also discussed. Section 4 is devoted to summary and discussion.

2. THE MODEL

2.1. Atmosphere

In order to reduce an atmospheric model to its simplest form, we shall adopt the one-layer model originally developed by Matsuno (1966) for tropical waves. The crucial assumption of this model is that the vertical structure can be described by a single vertical mode. Geisler and Stevens (1982) demonstrated that the altitude dependence of the response amplitude in the troposphere is very close to that of the forcing for $\omega < A$ where ω is the frequency of the periodic forcing and A is the dissipation coefficient. Therefore, if the forcing is of a single dominant mode, the shallow-water equations for this mode can describe the response fairly well within the layer of forcing. The vertical distribution of the forcing used in several studies (Stevens et al., 1977; Geisler, 1981; Geisler and Stevens, 1982) is a reasonable fit to the first baroclinic mode with the equivalent depth of 400 m (cf. Gill, 1980). This is mainly because the heating has a single maximum at about 500 mb in the tropics (cf. Yanai et al., 1973).

The basic equations with the equatorial beta-plane approximation ($f = \beta y$) are:

$$U_t - fV + gH_x = -AU$$
$$V_t + fU + gH_y = -AV \tag{2.1}$$
$$H_t + D(U_x + V_y) = -BH - Q$$

where U, V, H and Q are the zonal velocity, the meridional velocity, the depth and the mass source or sink (which corresponds to the heating rate). A and B are inverse time scales for Rayleigh friction and Newtonian cooling. The atmosphere has an equivalent depth D, perturbations to which are measured by H. In the present study $D = 400$ m so that the inviscid equatorial radius of deformation is about $10°$ wide approximately. The atmospheric long gravity wave speed $C_a (= \sqrt{gD})$ is 63 m s^{-1}.

2.2. Ocean

We also adopt shallow-water equations for the ocean. Assuming the equatorial beta-plane approximation, the equations are written as:

$$u_t - fv + gh_x = -au + \tau^x$$
$$v_t + fu + gh_y = -av + \tau^y \tag{2.2}$$
$$h_t + d(u_x + v_y) = -bh$$

where u, v, h and (τ^x, τ^y) are the zonal velocity, the meridional velocity, the depth perturbation and a body force. In the present study the equivalent depth d is 20 cm so that the long gravity wave speed $C_0(= \sqrt{gd})$ is 140 cm s^{-1}. The oceanic motion is also damped by Rayleigh friction (the coefficient of which is a) and Newtonian cooling (the coefficient of which is *b*). The reduced gravity (or single mode) model adopted here has been widely used by oceanographers to simulate sealevel variations caused by wind variations with success in spite of its limited capability (Hurlburt et al., 1976; McCreary, 1976; Busalacchi and O'Brien, 1981; Gill, 1983).

2.3. Air–sea coupling

The manner of coupling between the atmosphere and ocean must be specified in order to solve the coupled model. We assume simply that changes in the depth of the thermocline correspond to SST changes. This assumption is relatively good in the central and eastern equatorial Pacific (Gill, 1982; Donguy et al., 1984), although there seems to be some inconsistency in the central equatorial Pacific (cf. Gill, 1983). We further assume that the SST changes, in turn, affect the evaporation which leads to the latent heat release at a suitable height (say, 500 mb) in the troposphere. In reality, the heating of the atmosphere occurs where moisture converges and condenses. Therefore if the mean wind converges, it may play an important role in the actual heating. Our latter assumption will be more appropriate to areas where seasonal changes of mean wind convergence are relatively small such as the central equatorial Pacific. Notice here that the big seasonal changes of the mean wind do not necessarily mean big changes of low-level convergence (cf. Horel, 1982).

Thus, the thermodynamical coupling is reduced to:

$$Q = -\alpha h \tag{2.3}$$

where α is the coefficient of coupling. We also assume the following simple form for the dynamical coupling, that is:

$$(\tau^x, \tau^y) = \gamma(U, V) \tag{2.4}$$

where γ is the coefficient of coupling. In our air–sea coupled system, the atmosphere is driven by the release of latent heat from the ocean and the ocean is driven by the surface winds. Rough estimates of the interaction coefficients α and γ are shown in the Appendix. All parameters are listed in Table 1 for reference. However, the reader should not take the above values too seriously because the parameterization itself is so simple.

2.4. A necessary condition for instability

Once the manner of coupling is specified, it is easy to derive a necessary condition for instability. An energy integral of the oceanic equations leads to:

$$\tfrac{1}{2} \langle d(u^2 + v^2) + gh^2 \rangle_t = -a \langle d(u^2 + v^2) \rangle - b \langle gh^2 \rangle + \gamma \langle d(uU + vV) \rangle \tag{2.5}$$

where $\langle \rangle$ denotes the integration with respect to x over a wavelength of the disturbance and y from $-\infty$ to $+\infty$. Since the first two terms on the right-hand side are negative definite, the positive correlation between atmospheric winds and oceanic flows is necessary to instability.

TABLE 1

List of parameters used in the analysis (see text for details)

Type of parameter	Parameter values	Nondimensional values
Atmospheric Rayleigh damping: A	2×10^{-1} days^{-1}	3×10^{-1}
Oceanic Rayleigh damping: a	10^{-2} days^{-1}	1.5×10^{-2}
Atmospheric Newtonian cooling: B	6.7×10^{-2} days^{-1}	5×10^{-5}
Oceanic Newtonian cooling: b	10^{-2} days^{-1}	1.5×10^{-2}
Thermodynamical coupling: α	10^{-2} s^{-1}	6.5×10^{-1}
Dynamical coupling: γ	5×10^{-7} s^{-1}	6.5×10^{-2}
Unit of length (oceanic equatorial Rossby radius): $\sqrt{C_0/(2\beta)}$	175 km	1
Unit of time: $1/\sqrt{2\beta C_0}$	1.5 days	1

3. SOME SIMPLE SOLUTIONS OF THE MODEL EQUATIONS

3.1. Non-rotating case

In order to have an insight into an equatorial case, it is very useful to consider the case in which neither the atmosphere nor the ocean is rotating. Since the atmospheric equivalent depth is much larger than the oceanic one, the restoring force (associated with the stratification) in the atmosphere is much stronger than in the ocean. Therefore, the time scale of the driving force, which is determined by the oceanic motion, in the atmospheric equations is large compared with the periods of natural oscillation in the atmosphere. Thus, the atmospheric motion is always in a quasi-steady state as far as the slow phenomenon of the oceanic time scale is concerned. This is the reason we neglect the acceleration terms of the atmospheric equations.

Assuming a wave-like disturbance of the form:

$$(u, v, h, U, V, H) = (u, v, h, U, V, H)\, e^{i(kx + ly - \omega t)} \tag{3.1}$$

we obtain the dispersion relation:

$$(\omega - ia)(\omega - ib) = C_0^2 \left(\frac{\kappa^2}{\kappa^2 + AB/C_a} \right) \cdot (\kappa^2 - \kappa_c^2) \tag{3.2}$$

and:

$$(\omega - ia) = 0 \tag{3.3}$$

where:

$$\kappa^2 = k^2 + l^2 \qquad \text{and} \qquad \kappa_c^2 = (\alpha\gamma - AB)/C_a^2$$

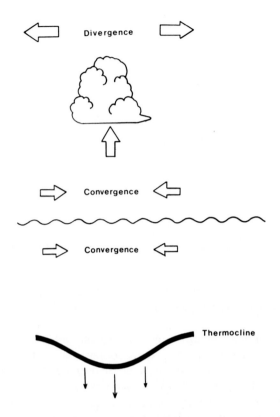

Fig. 1. Schematic picture of unstable air–sea interaction in non-rotating fluids.

We see that two of three eigenmodes are affected by the air–sea coupling. In an inviscid case, stationary unstable disturbances can exist for $\kappa < \kappa_c$. This unstable mode is reduced to a neutral oceanic gravity mode when $\alpha\gamma = 0$ and has a maximum growth rate in the long wavelength limit ($\kappa \to 0$) for the inviscid case. The physical reason for this instability is the following (see also Fig. 1). An initial perturbation, say, a depression of the thermocline causes a heating of the atmosphere, resulting in the convergent motion at a low level in the atmosphere. This convergent wind accelerates the convergent current in the ocean. Thus, the initial depression of the thermocline grows (see Section 2.4). The above positive feedback mechanism operates independently of a lateral scale of the initial disturbance. However, the gravitational restoring force associated with the stratification of the ocean provides a negative feedback mechanism. This restoring force decreases as the lateral scale of the disturbance increases. When the positive feedback overcomes the negative feedback, the instability occurs for $\kappa < \kappa_c$. Mathematically, the instability is due to the coupling between the two gravity modes which propagate in opposite directions.

The effect of the oceanic dissipation is not essential, because it decreases the growth rate by the amount of a (if $a = b$). The atmospheric dissipation, however, plays some significant roles. If we keep the term of Newtonian cooling in eqn. (2.1), the relative importance of the convergence terms in the mass balance decreases with increasing B (or with decreasing κ). Therefore, the positive feedback becomes less efficient. Also,

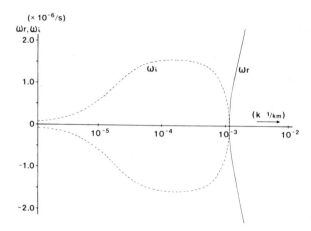

Fig. 2. Real (ω_r) and imaginary (ω_i) parts of frequencies as a function of the wavenumber k for the non-rotating case.

if Rayleigh damping is strong, velocity fields become weak so that the relative importance of the convergence terms in the mass balance decreases. In fact, if AB exceeds $\alpha\gamma$, no unstable mode exists.

Figure 2 shows the dispersion relationship and growth rates for parameters listed in Table 1. Most important is the existence of the large-scale stationary unstable mode. It should be noted here that this instability is essentially the same as demonstrated by Lau (1981). Although Lau (1981) claimed that the equatorial Kelvin waves are unstable based on the model without rotation, we need to analyse a more complete model with rotation in order to have a definite answer about the stability of equatorial waves. This is because the instability discussed here is due to the coupling between two gravity modes which propagate in opposite directions with phase speeds of equal magnitude.

3.2. Rotating case

When the Coriolis parameter f has a constant value, the dispersion relation becomes:

$$(\omega - ia)^3 + E_1(\omega - ia) + F_1 = 0 \tag{3.4}$$

where:

$$E_1 = -C_o^2\kappa^2 - f^2 + \frac{\alpha\gamma C_o^2\kappa^2}{C_a^2\kappa^2 + f^2 + A^2}$$

and:

$$F_1 = -i\frac{\alpha\gamma C_o^2\kappa^2 f^2}{A(C_a^2\kappa^2 + f^2 + A^2)}$$

If the air–sea interaction is weak enough, an expansion in powers of $\alpha\gamma$ gives:

$$\omega_I \sim \frac{-i\alpha\gamma C_o^2\kappa^2 f^2}{A(C_a^2\kappa^2 + f^2 + A^2)(C_o^2\kappa^2 + f^2)} - ia + 0(\alpha^2\gamma^2)$$

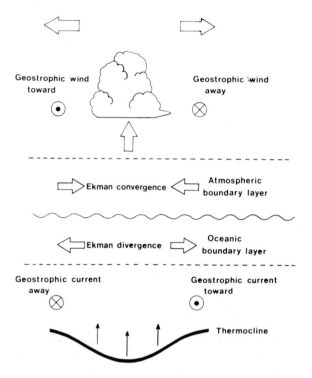

Fig. 3. Schematic picture of decaying air–sea interaction in rotating fluids.

$$\omega_{\mathrm{II}}^{\pm} \sim \pm\sqrt{C_o{}^2\kappa^2 + f^2} - ia + 0(\alpha\gamma) \qquad (3.5)$$

The mode with frequency given by ω_{I} corresponds to the geostrophic mode. For this mode a negative feedback mechanism exists in contrast to the gravity mode. The physical reason is as follows (see also Fig. 3). The thermocline depression generates convergent winds in the atmosphere. This wind, in the presence of a positive Coriolis parameter, is associated with a cyclonic circulation in the following manner:

$$U_x + V_y = \frac{A}{f}(U_y - V_x) \qquad (3.6)$$

The cyclonic circulation, in turn, generates divergent flows in the ocean at low frequencies provided that the oceanic dissipation time is long compared to the inertial period. Thus the initial depression of the thermocline is smeared out (see Section 2.4). In other words, convergent winds in the atmosphere, which are originally caused by convergent flows in the ocean, drive divergent flows in the ocean by an approximate relationship of the form:

$$u_x + v_y = -\frac{\alpha}{A}(U_x + V_y) \qquad (3.7)$$

The other modes correspond to the inertia-gravity waves affected by the air–sea interaction. In the limit of no rotation, these modes are reduced to the gravity modes. Since the time scale is relatively short, these modes can satisfy the necessary condition

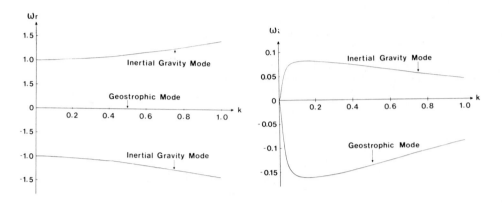

Fig. 4. As in Fig. 2 except for the rotating case.

for instability that convergent atmospheric winds drive convergent oceanic flows. Figure 4 shows the dispersion relationship and growth rates for parameters listed in Table 1.

3.3. The equatorial β-plane

The equatorial β-plane has properties of both the rotating case and the non-rotating case, since the Coriolis parameter vanishes just at the equator. Also, it acts as a wave-guide as discussed by Matsuno (1966). Therefore, it is not straightforward to see how air–sea coupling occurs.

Using the oceanic equatorial Rossby radius $\sqrt{C_0/(2\beta)}$ as a length scale and $1/\sqrt{2\beta C_0}$ as a time scale, we can write the nondimensional equations in the form:

$$\left. \begin{array}{l} -\tfrac{1}{2}yV = -H_x - A^*U \\[4pt] \tfrac{1}{2}yU = -H_y - A^*V \\[4pt] U_x + V_y = -B^*H - \alpha^*h \end{array} \right\} \tag{3.8}$$

and:

$$\left. \begin{array}{l} u_t - \tfrac{1}{2}vv = -h_x - a^*u + \gamma^*U \\[4pt] v_t + \tfrac{1}{2}yu = -h_y - a^*v + \gamma^*V \\[4pt] h_t + u_x + v_y = -b^*h \end{array} \right\} \tag{3.9}$$

where the nondimensional coefficients are defined as:
coefficient of atmospheric Rayleigh damping:

$$A^* = A/\sqrt{2\beta C_0} \tag{3.10}$$

coefficient of atmospheric Newtonian cooling:

$$B^* = (B/\sqrt{2\beta C_0})\,(d/D) \tag{3.11}$$

coefficient of oceanic Rayleigh damping:

$$a^* = a/\sqrt{2\beta C_0} \tag{3.12}$$

coefficient of oceanic Newtonian cooling:

$$b^* = b/\sqrt{2\beta C_o} \tag{3.13}$$

coefficient of thermodynamical coupling:

$$\alpha^* = (\alpha/\sqrt{2\beta C_o})(d/D) \tag{3.14}$$

coefficient of dynamical coupling:

$$\gamma^* = \gamma/\sqrt{2\beta C_o} \tag{3.15}$$

Notice here that the accelerations are neglected in the atmospheric equations. All non-dimensional coefficients are listed in Table 1. Although B^* is much smaller than A^* (mainly because the atmospheric equivalent depth is much larger than the oceanic one), it is of interest to discuss the special case of $A^* = B^*$ and $a^* = b^*$, since it permits a simple analytical treatment without losing the heart of the problem.

First, following Gill (1980), we introduce new variables q, r, Q and R, which are defined by:

$$q = h + u, \qquad r = h - u, \qquad Q = H + U \qquad \text{and} \qquad R = H - U \tag{3.16}$$

Expanding the variables in terms of parabolic cylinder functions $\{D_n\}$, viz.:

$$(q, r, v, Q, R, V) = \sum_{n=0}^{\infty} (q_n, r_n, v_n, Q_n, R_n, V_n) \cdot D_n \cdot e^{i(kx - \omega t)} \tag{3.17}$$

and substituting this series (after introducing the long-wave approximation and truncating the series at $n = 2$) into eqns. (3.8) and (3.9) by use of the orthogonality of the parabolic cylinder functions, we obtain the following simple equations governing symmetric modes:

$$\left.\begin{aligned}
(A^* + ik)Q_0 &= -\frac{\alpha^*}{2} q_0 - \alpha^* q_2 \\[2mm]
(3A^* - ik)Q_2 &= -\frac{\alpha^*}{2} q_0 - \frac{3}{2}\alpha^* q_2
\end{aligned}\right\} \tag{3.18}$$

and:

$$\left.\begin{aligned}
[-i(\omega - k) + a^*]q_0 &= \frac{\gamma^*}{2} Q_0 - \gamma^* Q_2 \\[2mm]
\left[-3i\left(\omega + \frac{k}{3}\right) + a^*\right]q_2 &= -\frac{\gamma^*}{2} Q_0 + \frac{3}{2}\gamma^* Q_2
\end{aligned}\right\} \tag{3.19}$$

The meaning of the above simplified system is the following. The atmospheric quasi-stationary Kelvin wave (Q_0) and the gravest Rossby wave (Q_2) are excited by the heat source provided by the oceanic Kelvin wave (q_0) and the gravest Rossby wave (q_2). The oceanic Kelvin wave (q_0) and the gravest Rossby wave (q_2) are, on the other hand, excited by the winds composed of the atmospheric quasi-stationary Kelvin wave (Q_0) and the gravest Rossby wave (Q_2).

The dispersion relationship is given by:

$$(\omega - ia^*)^2 + E_2(\omega - ia^*) + F_2 = 0 \tag{3.20}$$

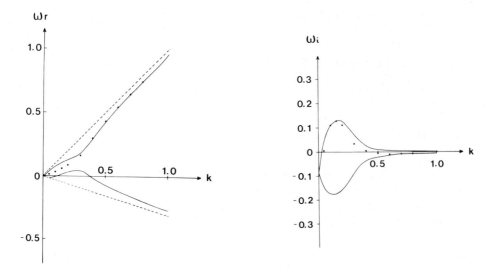

Fig. 5. As in Fig. 2 except for the equatorial beta-plane ($\alpha^*\gamma^* = A^* = B^* = 0.04$). Dots correspond to numerical results.

TABLE 2

Eigenvalues for $\alpha^*\gamma^* = A^* = B^* = 0.04$ and $k = 0.15$

Type of mode	Eigenvalues $(\omega = \omega_r + i\omega_i)$	Corresponding neutral wave
Gravity (antisymmetric)	1.5838 + 0.0443 i	1.6029
Gravity (symmetric)	1.2380 + 0.0491 i	1.2578
Yanai (anti)	0.7386 + 0.0635 i	0.7861
Kelvin (sym)	0.0658 + 0.1298 i	0.1500
Rossby (sym)	0.0002 − 0.0567 i	
Yanai (anti)	− 0.6556 + 0.0332 i	− 0.6361
Gravity (sym)	− 1.2063 + 0.0322 i	− 1.2085
Gravity (anti)	− 1.5837 + 0.0289 i	− 1.5732

where:

$$E_2 = \frac{2k^3 + 4iA^*k^2 + (6A^{*2} + \frac{1}{2}\alpha^*\gamma^* + \frac{1}{4}\alpha^{*2}\gamma^{*2})k - \frac{3}{2}i\alpha^*\gamma^*A^*}{-3(k - iA^*)(k + 3iA^*)}$$

and:

$$F_2 = \frac{k^4 + 2iA^*k^3 + (3A^{*2} - \frac{7}{2}\alpha^*\gamma^*)k^2 + \frac{i}{2}\alpha^*\gamma^*A^*k + \frac{1}{16}\alpha^{*2}\gamma^{*2}}{-3(k - iA^*)(k + 3iA^*)}$$

Figure 5 shows the dispersion diagrams for $\alpha^*\gamma^* = A^* = B^* = 0.04$. It is clear that the large-scale unstable mode exists for a certain range of zonal wavenumber k. The growth rate is sensitive to the magnitudes of $\alpha^*\gamma^*$ and A^*; it increases with increasing $\alpha^*\gamma^*$ and decreases with increasing A^*.

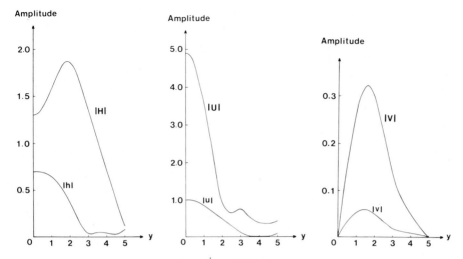

Fig. 6. Amplitudes (H, h), (U, u) and (V, v) of the gravest growing mode for $k = 0.15$ and $\alpha^* \gamma^* = A^* = B^* = 0.04$.

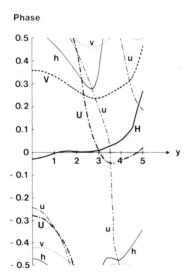

Fig. 7. Phases (H, h), (U, u) and (V, v) of the gravest growing mode for $k = 0.15$ and $\alpha^* \gamma^* = A^* = B^* = 0.04$ ± 0.5 cycle corresponds to $\pm 180°$.

In order to check truncation errors we solved eqns. (3.8) and (3.9) by a numerical variational method described by Orlanski (1968) in detail. We assumed boundaries at $y = \pm 5$. The interval was divided into 100 intervals. The eigenfrequencies thus computed are shown in Table 2 for $\alpha^* \gamma^* = A^* = B^* = 0.04$ and $k = 0.15$. It can be seen that the mode which corresponds to the Kelvin wave in the limit of no air–sea interaction has the largest growth rate among various growing modes. Since we are interested in the low-frequency regime, unstable Yanai and gravity modes are neglected hereafter. Dots in

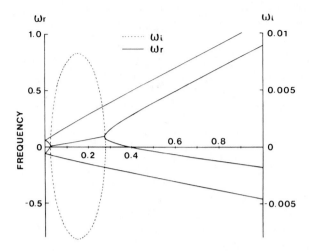

Fig. 8. As in Fig. 5 except for the inviscid time-dependent atmosphere and ocean.

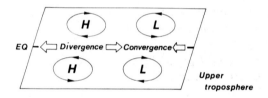

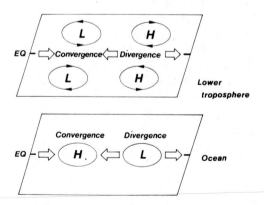

Fig. 9. Schematic picture of unstable air−sea interaction in the equatorial beta-plane.

Fig. 5 are based on the numerical analysis. One sees that the growth rates obtained analytically are relatively good in spite of the severe truncation.

The amplitude and phase of variables for a typical growing mode are shown in Figs. 6 and 7. It is interesting to note that the gravest growing mode is, on the ocean side,

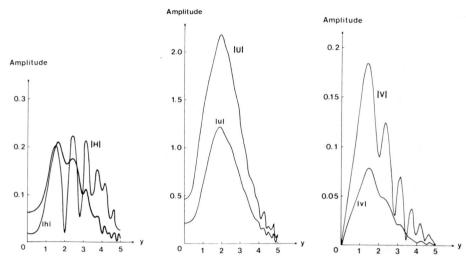

Fig. 10. As in Fig. 6 except for the decaying mode ($k = 0.3$).

basically a Kelvin wave modified a little by Rossby waves. As the wavenumber decreases, this modification increases because Kelvin and Rossby waves can be coupled more closely with decreasing frequency gap. The inviscid results of a truncated D_n analysis with a time-dependent atmosphere shows clearly the importance of this coupling for the instability (Fig. 8). Partly because of this coupling and partly because of the asymmetric response of the atmosphere to the heating (cf. Matsuno, 1966; Gill, 1980), the zonal velocity fields in the atmosphere and ocean can be cooperative so that the instability can occur (Fig. 9). This situation is, therefore, similar to that depicted in Fig. 1 except for that east–west asymmetry of the atmospheric zonal wind which appears as a meridional phase tilt of the zonal wind in the present, zonally periodic model (see Fig. 7). The air–sea coupled Kelvin mode progresses eastward with a speed less than the long gravity wave speed C_0. This is partly because the atmospheric wind introduces the westward phase propagation by inducing downwelling to the west and upwelling to the east of the warm anomaly (cf. Horel, 1982) and partly because the oceanic Kelvin and Rossby waves are coupled to each other.

The computed amplitudes and phases of variables for a typical decaying mode are shown in Figs. 10 and 11. In contrast to the growing mode, the corresponding velocity fields in the atmosphere and ocean are almost out of phase (Fig. 11). This situation is similar to that discussed in Section 3.2. One sees that the contribution from a Kelvin wave is much smaller than that from Rossby waves. Since eigenfrequencies of the original neutral Rossby modes are so close to each other compared to the interaction frequency roughly given by $\sqrt{\alpha^* \gamma^*}$, many modes can interact so that the structure of decaying eigen modes is very complicated.

In reality, the assumption of $A^* = B^*$ breaks down (see Table 1). Therefore it is of interest to discuss more realistic cases in detail. Although Zebiak (1982) discussed the nature of the atmospheric response for general heating distribution, the sensitivity to changes in the parameters A^* and B^* is left to be answered.

First, it is useful to introduce a similarity law governing a steady atmospheric response.

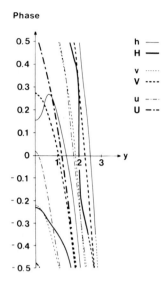

Phase

Fig. 11. As in Fig. 7 except for the decaying mode ($k = 0.3$).

If we adopt the long-wave approximation and introduce the transformation of variables and forcing function such as:

$$
\begin{Bmatrix} x \\ y \\ U \\ V \\ H \\ \alpha^* h \end{Bmatrix} = \begin{Bmatrix} (A^*/B^*)^{1/2}X \\ (A^*/B^*)^{1/4}Y \\ (A^*/B^*)^{1/2}U \\ (A^*/B^*)^{1/4}V \\ (A^*/B^*)H \\ Q \end{Bmatrix} \tag{3.21}
$$

Equation (3.8) can be reduced to:

$$
\left. \begin{aligned} -\tfrac{1}{2}YV &= -H_X - A^*U \\ \tfrac{1}{2}YU &= -H_Y \\ U_X + V_Y &= -A^*H - Q \end{aligned} \right\} \tag{3.22}
$$

Therefore, as far as the long-wave approximation is valid, the difference of magnitude between Rayleigh damping and Newtonian cooling disappears ostensibly. The long-wave approximation is valid if the zonal scale is large compared with the meridional scale and if the nondimensional value of Rayleigh damping coefficient is small. These conditions are satisfied for observed warming events. Once we know a solution of eqn. (3.22) such as shown by Matsuno (1966), Gill (1980) and Zebiak (1982), it is easy to derive the family of solutions of eqn. (3.8) by changing the magnitude of B^*. Also, it is easy to prove that the above similarity does not exist if we retain the short wave dynamics such as the damped Yanai wave, damped gravity waves and damped short Rossby waves.

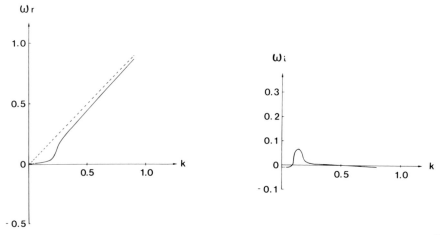

Fig. 12. As in Fig. 5 except for $A^* = 0.4$ and $B^* = 0.04$. Boundaries are assumed at $y = \pm 10$. The interval was divided into 200 intervals for the present case.

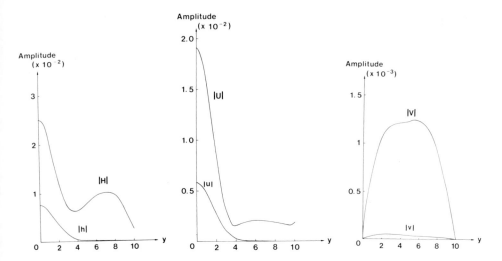

Fig. 13. As in Fig. 6 except for $k = 0.3$, $\alpha^*\gamma^* = 0.04$, $A^* = 0.4$ and $B^* = 0.04$.

If the meridional extent of the heating field is of order unity in the y coordinate, it is of order $(B^*/A^*)^{1/4}$ in the Y coordinate. Therefore, if B^* is much smaller than A^*, the extent is smaller than unity in the Y coordinate. In other words, the effect of the Coriolis force is much reduced as far as the atmospheric response is concerned.

Figure 12 shows the dispersion relationship and the growth rates obtained by the numerical variational method for the case $A^* = 0.4$ (other parameters are kept the same as those of Fig. 5). Since the value of A^* is ten times larger than that of Fig. 5, both growth rates and unstable regime are much suppressed. Figures 13 and 14 show the amplitudes and phases of variables for a weakly growing mode with the wavenumber $k = 0.3$. The situation is schematically shown in Fig. 15. The prominent feature is the

652

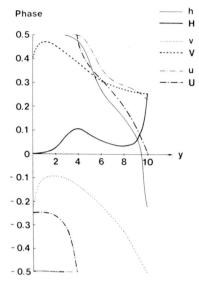

Fig. 14. As in Fig. 7 except for $k = 0.3$, $\alpha^* \gamma^* = 0.04$, $A^* = 0.4$ and $B^* = 0.04$.

existence of an anomalous atmospheric low just above the oceanic warm anomaly. The response of this type is reminiscent of the non-rotating case (cf. Fig. 1). However, the zonal velocity fields of the atmosphere and ocean are almost a quarter cycle out of phase for the present weakly growing mode, since the oceanic warm anomaly progresses almost as an oceanic Kelvin wave in contrast to the non-rotating case. Therefore, the net work done by the convergent winds is small in the present example. It is important that the phase of the atmospheric zonal wind lags as one proceeds northward (or southward) from the equator. This is mainly because the stationary Rossby waves are generated as a response to the heating. Even if the magnitude of these Rossby waves is small, they provide a positive correlation between the oceanic zonal flow and the atmospheric zonal winds. For a more strongly growing mode, the phase lag is mainly provided by the coupling among oceanic waves.

Figure 5 of Philander et al. (1984) shows the evolution of an initial depression of the thermocline with a Gaussian shape given by:

$$h = 0.01 \exp \left[-(x^2 + y^2)/8.16 \right] \tag{3.23}$$

All parameters used in the experiment are the same as listed in Table 1. It was shown that the atmospheric and oceanic disturbances may grow at the same time with the growth rate of about 1/20 per day. In particular, it was demonstrated that the depression of thermocline may expand eastward with a speed less than the oceanic Kelvin wave speed C_0. The cooperative nature of the zonal velocity fields both in the atmosphere and the ocean is provided partly by the asymmetric response of the atmosphere (see fig. 5 of Philander et al., 1984) to the heating and partly by the coupling among oceanic Kelvin and Rossby waves. Those oceanic Rossby waves can be detected in fig. 6 of Philander et al. (1984), where the evolution of spatial structures of oceanic u and h fields were shown.

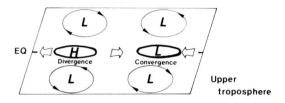

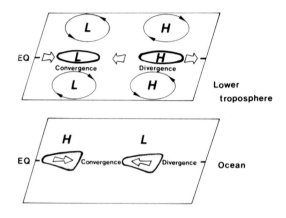

Fig. 15. Schematic picture of unstable air–sea interaction in a more realistic situation of the tropics.

4. SUMMARY AND DISCUSSION

We have studied the stability of a simple air–sea coupled model using analytical and numerical methods. Our simple model assumes that linear shallow water equations govern the oceanic and atmospheric motion. The atmospheric heating is assumed to be proportional to variations in the depth of the thermocline. The ocean is assumed to be driven by the winds.

The energy integral of the oceanic equations show that the necessary condition for the instability is the net positive correlation between atmospheric winds and oceanic flows. This positive correlation is possible in at least two ways. One is provided by the phase lag caused by a coupling among oceanic modes. A typical example is the non-rotating case, where two gravity modes propagating in opposite directions are coupled when the positive feedback overcomes the restoring force due to the stratification. The other is provided by an asymmetric response of the atmosphere to the heating. As Gill (1980) noted using a simple atmospheric model as adopted here, the response of the tropical atmosphere to diabatic heating has a significant east–west asymmetry even if the heating distribution is symmetric. The more intense westerly winds to the west of the heating are due to the stationary long Rossby waves. The less intense easterly winds to the east of the heating are due to the stationary Kelvin wave. We note here that the CISK type process (Charney and Eliassen, 1964) may enhance the above east–west asymmetry. As described in terms of the similarity law in Section 3.3, however, the atmospheric response becomes more similar to the non-rotating case with increasing A^* (non-dimensional Rayleigh friction) or decreasing B^* (non-dimensional Newtonian cooling). Thus, the asymmetry of the atmospheric response diminishes with decreasing B^*/A^*. In reality, the above quantity is extremely small. As a consequence, the atmospheric response is similar to the non-rotating case near the equator. Away from the equator, however, the response is similar to the case discussed by Gill (1980) so that a meridional phase tilt of the zonal wind can contribute to the net positive correlation between atmospheric winds and oceanic flows. Although this correlation is small, an unstable mode can exist even for the extreme case $B^*/A^* \ll 0\,(1)$.

A major defect of our model is of course the naive way of the thermodynamical coupling, since we oversimplified the oceanic mixed layer physics and the mechanism of the atmospheric diabatic heating. Nevertheless, our simple model turned out to be quite encouraging. The evolution of the outgoing long-wave radiation anomaly, 850 mb stream function anomaly and 200 mb stream function anomaly observed in the 1982/83 ENSO event was recently shown by Arkin et al. (1983). It is now quite clear that the modest anomaly which first appeared in the western tropical Pacific grew both spatially and temporally. In particular, the anomaly, associated with Matsuno (1966)'s circulation pattern in the atmosphere, progressed eastward with a speed less than a typical oceanic Kelvin wave speed. It is amazing how this actual evolution pattern looks like Fig. 15.

It is interesting to note here that the location of the maximum OLR anomaly corresponds well with the location of the maximum actual SST (see fig. 1 of Philander et al., 1984). This is because the evolution of the net convergence leading to the latent heat release in the lower troposphere is determined by the evolution of the actual SST, not by the evolution of the anomaly of SST (cf. Philander and Rasmusson, 1984). In the western and central tropical Pacific, the seasonal changes of the net divergence are relatively small in the lower troposphere (cf. Horel, 1982). This may partly explain why our anomaly model looks successful especially in the 1982/83 event. As discussed in Philander et al. (1984), the evolution of the anomalous conditions near the eastern tropical Pacific is expected to be more regulated by the seasonal cycle because of the large seasonal changes of the net atmospheric convergence. The difference in the evolution of the anomalous conditions between the 1972 event and the 1982/83 event may be explained by the difference in the timing of penetration of interannual Los Niños into the more seasonally regulated Los Niños in the eastern tropical Pacific.

APPENDIX

Estimation of α

First we assume the air is always saturated by the water vapor supplied from the ocean surface. The latent heat release per one gram of dry air is given by $-L_c \cdot \Delta q_s$, where L_c denotes the latent heat and Δq_s is the change of the saturated mixing ratio q_s^*. From the equation of state for the ideal gas, the quantity Δq_s is related to the change of the saturated water vapor pressure e_s in the following way:

---

*The mixing ratio is defined by: $q \equiv \rho_v/\rho_d$, where ρ_v is the density of the water vapor and ρ_d is the density of the dry air.

$$\Delta q_s \doteq \frac{m_v}{m_d} \cdot \frac{e_s}{p} \tag{A1}$$

where $m_v (= 18.00)$ is the molecular weight for the water and $m_d (= 28.97)$ is the molecular weight for the dry air. Here we have assumed that the change of the total pressure is small. The temperature dependence of e_s is given by the Clausius-Clapeyron relation:

$$\frac{d}{dT} e_s = m_v L_c e_s \frac{1}{R^* T^2} \tag{A2}$$

where R^* is the gas constant. Since $e_s = 6.11$ mb at $T = 273°$K, the integral of the above relation is given by:

$$e_s = 6.11 \cdot e^{b[(1/273)-(1/T)]} \tag{A3}$$

where $b = m_v L_c / R^{*2}$

The latent heat release is then:

$$-L_c \Delta q_s \sim -L_c \cdot \frac{m_v}{m_d} \cdot \frac{\Delta e_s}{p}$$

$$\sim -6.11 \cdot L_c \cdot \frac{m_v}{m_d} \cdot \frac{b \Delta T}{p T^2} \cdot e^{b[(1/273)-(1/T)]} \tag{A4}$$

The quantity b is a constant which is approximately 5400K^*. Then the above relation is reduced to:

$$-L_c \Delta q_s \sim -120 L_c \frac{\Delta T}{T^2} \tag{A5}$$

where T is assumed to be $300°$K (which corresponds to the mean SST). A heat source Q which appears in eqn. (2.1) is thus related to the SST.

The relation between ΔT and the depth perturbation h of the thermocline is not simple. Here ΔT is simply assumed to be linearly related to h (cm), giving:

$$\Delta T \sim h \tag{A6}$$

Therefore Q is given by:

$$Q \doteq \frac{R}{g} \cdot \frac{L_c \Delta q}{C_p \Delta t} \sim 9.5 \times 10^3 \cdot \frac{h}{\Delta t} (\text{cm s}^{-1}) \tag{A7}$$

where Δt is the characteristic time of latent heat release, R is the gas constant for air and C_p the specific heat at constant pressure. If we tentatively adopt $\Delta t = 10^6$ s (which gives diabatic heating of about 1K per day in the mature stage of warming events), we obtain:

$$Q \sim 10^{-2} h \tag{A8}$$

This finally yields:

$$\alpha \sim 10^{-2} (\text{s}^{-1}) \tag{A9}$$

---

*Using $m_v = 18$ g, $L_c = 597$ cal g^{-1} and $R^* = 8.32 \times 10^7$ erg $\cdot$ mol$^{-1} \cdot$ K^{-1} yields $b \sim 5400$ K.

Estimation of γ

If a linear drag law is introduced, then a body-force (τ^x, τ^y) in eqn. (2.2) is written in the form:

$$
\begin{Bmatrix} \tau^x \\ \tau^y \end{Bmatrix} = \frac{\rho_{air}}{\rho_{water}} C_D U_{typ} \frac{1}{H_{mix}} \begin{Bmatrix} U \\ V \end{Bmatrix}
\tag{A10}
$$

where H_{mix} corresponds to the mixed-layer depth. For $C_D = 1.5 \times 10^{-3}$, $\rho_{air}/\rho_{water} = 1.2 \times 10^{-3}$, $U_{typ} = 5 \text{ m s}^{-1}$ and $H_{mix} = 20 \text{ m}$, we obtain:

$$
\gamma = 4.5 \times 10^{-7} \text{ s}^{-1}
\tag{A11}
$$

There is another approach which leads to almost the same value of γ. It is to assume that a steady wind of 2 m s^{-1} that persists for 10 days accelerates an equatorial jet to 1 m s^{-1}. The above estimate will give the uppermost value of γ.

REFERENCES

Arkin, P.A., Kopman, J.D. and Reynolds, R.W., 1983. 1982–1983 El Niño/Southern Oscillation Event Quick Look Atlas. Climate Analysis Center, National Meteorological Center, NOAA National Weather Service, Washington, D.C.
Bjerkness, J., 1966. A possible response of the atmospheric Hadley circulation to equatorial anomalies of ocean temperature. Tellus, 18: 820–829.
Busalacchi, A.J. and O'Brien, J.J., 1981. Interannual variability of the equatorial Pacific in the 1960's. J. Geophys. Res., 86: 10901–10907.
Charney, J.G. and Eliassen, A., 1964. On the growth of the hurricane depression. J. Atmos. Sci., 21: 68–75.
Donguy, J.R., Dessier, A., Eldin, G., Morliere, A. and Meyers, G., 1984. Wind and thermal conditions along the equatorial Pacific. J. Mar. Res., 42: 103–121.
Geisler, J.E., 1981. A linear model of the Walker circulation. J. Atmos. Sci., 38: 1390–1400.
Geisler, J.E. and Stevens, D.E., 1982. On the vertical structure of damped steady circulation in the tropics. Q. J. R. Meteorol. Soc., 108: 87–93.
Gill, A.E., 1980. Some simple solutions for heat-induced tropical circulation. Q. J. R. Meteorol. Soc., 106: 447–462.
Gill, A.E., 1982. Changes in thermal structure of the equatorial Pacific during the 1972 El Niño as revealed by bathythermograph observations. J. Phys. Oceanogr., 12: 1373–1387.
Gill, A.E., 1983. An estimation of sea-level and surface-current anomalies during the 1972 El Niño and consequent thermal effects. J. Phys. Oceanogr., 13: 586–606.
Horel, J.D., 1982. On the annual cycle of the tropical Pacific atmosphere and ocean. Mon. Weather Rev., 110: 1863–1878.
Hurlburt, H.E., Kindle, J.C. and O'Brien, J.J., 1976. A numerical simulation of the onset of El Niño. J. Phys. Oceanogr., 6: 621–631.
Lau, K.M.W., 1981. Oscillations in a simple equatorial climate system. J. Atmos. Sci., 38: 248–261.
Matsuno, T., 1966. Quasi-geostrophic motions in the equatorial area. J. Meteorol. Soc. Japan, 44: 25–43.
McCreary, J., 1976. Eastern tropical ocean response to changing wind system: with application to El Niño. J. Phys. Oceanogr., 6: 632–645.
Orlanski, I., 1968. Instability of frontal waves. J. Atmos. Sci., 25: 178–200.
Philander, S.G.H., 1983. El Niño Southern Oscillation phenomena. Nature, 302: 245–301.
Philander, S.G.H. and Rasmusson, E.M., 1984. On the evolution of El Niño. Trop. Ocean–Atmos. Newslett., 24: p. 16.

Philander, S.G.H., Yamagata, I. and Pacanowski, R.C., 1984. Unstable air–sea interactions in the Tropics. J. Atmos. Sci., 41: 604–613.

Stevens, D.E., Lindzen, R.S. and Shapiro, L.J., 1977. A new model of tropical waves incorporating momentum mixing by cumulus convection. Dyn. Atmos. Oceans, 1: 365–425.

Yanai, M., Esbensen, S. and Chu, J.-H., 1973. Determination of bulk properties of tropical cloud clusters form large-scale heat and moisture budgets. J. Atmos. Sci., 30: 611–627.

Zebiak, S.E., 1982. A simple atmospheric model of relevance to El Niño. J. Atmos. Sci., 39: 2017–2027.

CHAPTER 38

NUMERICAL EXPERIMENTS ON A FOUR-DIMENSIONAL ANALYSIS OF POLYMODE AND "SECTIONS" PROGRAMMES – OCEANOGRAPHIC DATA

A.S. SARKISYAN, S.G. DEMYSHEV, G.K. KOROTAEV and V.A. MOISEENKO

ABSTRACT

The main correlations for a multi-element four-dimensional analysis of ocean fields are presented. Numerical model experiments are carried out to estimate the effect of density and flow velocity fields on the analysis of observed data. Calculations performed for the physical-geographical conditions of the Polymode area have revealed the representativeness of the data assimilation algorithm. Preliminary numerical experiments on the assimilation of the sea-surface and deep-layers temperature data of the "Sections" programme have been conducted for the North Atlantic region with the coordinates $35.5°-55.5°$ N and $30.5°-70.5°$ W.

INTRODUCTION

The Polymode programme has presented interesting results but insufficient information for the large-scale ocean circulation study. The intensive observation phase of the multinational "Sections" programme started in January 1981 for a decade. According to this programme, oceanographic, meteorological, upper-air and satellite observations are carried out three or four times a year for five energetically active zones (EAZO) of the Atlantic and Pacific oceans. The main aim of the "Sections" programme is to study the influence of characteristic oceanographic anomalies on short-range climate change. Within the framework of this programme 5000 deep oceanographic stations, 2000 aerological stations, and 10,000 meteorological observations have been conducted for three EAZO of the Atlantic Ocean during the period from January 1981 to December 1983 (Itogi nauki i tekhniki, 1984). From the beginning of 1981 to the end of 1984 the "Sections" programme remained the largest multinational programme on the atmosphere and ocean research.

The active observation phase of the large-scale international programme Toga starts at the beginning of 1985. Within the framework of Toga a large amount of oceanographic information is to be collected. Thus, the necessity of elaboration and testing of four-dimensional analysis models for the oceanographic information is obvious.

The methods for reconstructing hydrophysical fields from measurement data should allow for specific features of the acquisition of oceanographic information, namely, the absence of a stationary network of observation stations, the small number of deep-sea measurements, a large variety of the relevant parameters to be measured, and significant unsynchronized and non-uniform experimental data now available. The present paper describes one of the possible approaches to solving this urgent problem – a four-dimensional analysis of hydrophysical ocean fields (Knysh et al., 1980; Sarkisyan, 1982).

660

FOUR-DIMENSIONAL ANALYSIS SCHEME

Let the state vector of the ocean $U(x,t)$ have the following components: $u(x,t)$ and $v(x,t)$, the horizontal projections of the velocity vector to the Cartesian axes x,y directed to the east and north, respectively; $\rho(x,t)$, the anomaly of sea-water density; $\zeta(x,y,t)$, the conventional sea level, then:

$$U(x,t) = \|u(x,t), v(x,t), \rho(x,t), \zeta(x,y,t)\|^T, \qquad x = (x,y,z) \tag{1}$$

where the vertical axis z points downward. Let the results of measuring $Z(x_r,t_l)$ of the state vector [eqn. (1)] be known at N points x_1, x_N at discrete moments $t_l, l = 1, \ldots, L$. Our aim is to determine the evolution of the main hydrophysical fields of the ocean in a certain time interval $[t_0, T]$, proceeding from the information on the state of the ocean at moments $t_l \in [t_0, T]$. An optimal estimate of eqn. (1) is defined as a vector $U(x,t)$ for which the mean-square error of the estimate is minimal (this estimate is sought in the class of linear functions of measurement data; Sakawa, 1972; Padmanabhan and Colantuoni, 1974).

In the Boussinesq and β-plane approximation the system of equations for optimal estimates of the components of eqn. (1) in a small time interval $t_n \leq t \leq t_{n+1}$ is of the form (Sarkisyan, 1977; Demyshev and Knysh, 1981; the arguments are omitted for brevity):

$$\frac{\partial \hat{u}}{\partial t} - f\hat{v} - \nu \frac{\partial^2 \hat{u}}{\partial z^2} = -\rho_0 g \frac{\partial \hat{\zeta}}{\partial x} - g \int_0^z \frac{\partial \hat{\rho}}{\partial x} d\mu + \tilde{f}_1 \tag{2}$$

$$\frac{\partial \hat{v}}{\partial t} + f\hat{u} - \nu \frac{\partial^2 \hat{v}}{\partial z^2} = -\rho_0 g \frac{\partial \hat{\zeta}}{\partial y} - g \int_0^z \frac{\partial \hat{\rho}}{\partial y} d\mu + \tilde{f}_2 \tag{3}$$

$$\frac{\partial \hat{\rho}}{\partial t} + \hat{u}\frac{\partial \hat{\rho}}{\partial x} + \hat{v}\frac{\partial \hat{\rho}}{\partial y} + \hat{W}\frac{\partial \hat{\rho}}{\partial z} = \kappa_z \frac{\partial^2 \hat{\rho}}{\partial z^2} + \kappa_l \Delta\rho + \tilde{f}_3 \tag{4}$$

$$\frac{gH}{f^2}\frac{\partial \Delta\hat{\zeta}}{\partial t} + \frac{g}{2af}\Delta\hat{\zeta} + \frac{g}{f^2}\beta H\frac{\partial \hat{\zeta}}{\partial x} - \frac{g}{f}J(H,\hat{\zeta}) =$$

$$= \frac{1}{\rho_0 f}\left(\frac{\partial \tau_x}{\partial y} - \frac{\partial \tau_y}{\partial x}\right) + \frac{\beta}{\rho_0 f^2}\tau_x - \frac{g}{2a\rho_0 f}\int_0^H \Delta\hat{\rho}\,dz - \frac{g}{\rho_0 f}\int_0^H J(H,\hat{\rho})\,dz -$$

$$- \frac{\beta g}{\rho_0 f^2}\int_0^H (H-z)\frac{\partial \hat{\rho}}{\partial x}dz + \frac{1}{f}\left[\frac{\partial}{\partial y}\int_0^H \left(\frac{\partial \hat{u}^2}{\partial x} + \frac{\partial \hat{u}\hat{v}}{\partial y}\right)dz - \right. \tag{5}$$

$$\left. - \frac{\partial}{\partial x}\int_0^H \left(\frac{\partial \hat{u}\hat{v}}{\partial x} + \frac{\partial \hat{v}^2}{\partial y}\right)dz\right] + \tilde{f}_4$$

The vertical component of the stream velocity estimate $\hat{W}$ is calculated by the formula (Demyshev and Knysh, 1980):

$$\hat{W} = \frac{1}{f}\left[\beta \int_0^z \hat{v}d\mu + \frac{1}{\rho_0}\left(\frac{\partial \tau_x}{\partial y} - \frac{\partial \tau_y}{\partial x}\right) + \int_0^z \left(\frac{\partial \hat{\zeta}}{\partial t} + \frac{\partial \hat{u}\hat{\zeta}}{\partial x} + \frac{\partial \hat{v}\hat{\zeta}}{\partial y}\right)d\mu\right] \tag{6}$$

where $\hat{\xi} = (\partial\hat{v}/\partial x) - (\partial\hat{u}/\partial y)$. The following notation has been used in eqns. (2)–(6): $f = f_0 + \beta y$ is the Coriolis parameter; $\beta = df/dy$; g is the acceleration due to gravity; τ_x, τ_y are the components of the tangential wind stress; H is the ocean depth; ν, κ_z, κ_l are the coefficients of turbulent vertical momentum exchange and turbulent diffusion in ther vertical and horizontal directions, respectively; Δ is the 2D-Laplacian; J is the Jacobian; and $\alpha = (f/2\nu)^{1/2}$.

In the case of discrete data acquisition the components $\hat{f}_i$ in eqns. (2)–(5) will be of the form:

$$\tilde{f}_i(x, t_l) = F_i(x, t_l)\,\delta(t - t_l)$$

where $\delta(t - t_l)$ is a Dirac delta function. The procedure for four-dimensional analysis will consist of two consequent stages (Sakawa, 1972; Padmanabhan and Colantuoni, 1974): (1) determination of prediction values of $\hat{U}(x, t_l^-)$ by integrating eqns. (2)–(6) for $\tilde{f}_i = 0$, $i = 1,\ldots,4$, on the interval $[t_n, t_l)$; and (2) an analysis of observation data at moment $t = t_l$ by the formula:

$$\hat{U}(x, t_l^+) = \hat{U}(x, t_l^-) + F(x, t_l) \tag{7}$$

where $F = \|F_1 F_2 F_3 F_4\|^T$ and the components F_i are calculated from the relations (Demyshev and Knysh, 1981):

$$F_i(x, t_l) = \sum_{r=1}^{N} [\Delta_r^{iu}(x, t_l^-)\,\delta u(x_r, t_l) + \Delta_r^{iv}(x, t_l^-)\,\delta v(x_r, t_l) + \Delta_r^{i\rho}(x, t_l^-)\,\delta\rho(x_r, t_l) +$$

$$+ \Delta_r^{i\xi}(x, t_l^-)\,\delta\xi(x_r, y_r, t_l)], \qquad i = 1,\ldots,4$$

$x_r\,(r = 1,\ldots,N)$ are the points of an area where observation data exist at moment t_l; the upper index "–" means that the corresponding values are calculated regardless of measurement data acquired by this moment. Prediction errors for the state vector [eqn. (1)] are written as:

$$\delta U(x_r, t_l) = Z(x_r, t_l) - \hat{U}(x_r, t_l^-)$$

To determine the weight coefficients one should know the estimate errors covariance matrix:

$$P = \begin{Vmatrix} P_{uu}^{pr} & P_{vu}^{rp} & P_{\rho u}^{rp} & P_{\xi u}^{rp} \\ P_{uv}^{pr} & P_{vv}^{pr} & P_{\rho v}^{rp} & P_{\xi v}^{rp} \\ P_{u\rho}^{pr} & P_{v\rho}^{pr} & P_{\rho\rho}^{pr} & P_{\xi\rho}^{rp} \\ P_{u\xi}^{pr} & P_{v\xi}^{pr} & P_{\rho\xi}^{pr} & P_{\xi\xi}^{pr} \end{Vmatrix} \tag{8}$$

where $P_{\varphi\psi}^{rp} = P_{\varphi\psi}(x_r, x_p, t_l^-)$ (Demyshev and Knysh, 1981). The weight factors can be calculated by the relations:

$$\Delta^{1u}(x, t_l^-) = \|P(x_r, x_p, t_l^-) + R(x_r, x_p, t_l)\|^{-1} P_{uu}(x, x_r, t_l^-)$$
$$\Delta^{2u}(x, t_l^-) = \|P(x_r, x_p, t_l^-) + R(x_r, x_p, t_l)\|^{-1} P_{vu}(x, x_r, t_l^-)$$
$$\Delta^{3u}(x, t_l^-) = \|P(x_r, x_p, t_l^-) + R(x_r, x_p, t_l)\|^{-1} P_{\rho u}(x, x_r, t_l^-) \tag{9}$$
$$\Delta^{4u}(x, t_l^-) = \|P(x_r, x_p, t_l^-) + R(x_r, x_p, t_l)\|^{-1} P_{\xi u}(x, x_r, t_l^-)$$

where $R(x_r.x_p,t_l)$, the covariance error matrix for measurements of the state vector is of the form:

$$R(x_r, x_p, t_l) = \left\| \begin{array}{cccc} R_{uu}^{pr} & & & \\ & R_{vv}^{pr} & 0 & \\ 0 & & R_{\rho\rho}^{pr} & \\ & & & R_{\zeta\zeta}^{pr} \end{array} \right\|$$

$$\Delta^{iu}(x.\,\bar{t_l}) = \|\Delta^{iu}(x,\bar{t_l})\Delta^{iv}(x,\bar{t_l})\Delta^{i\rho}(x,\bar{t_l})\Delta^{i\zeta}(x,\bar{t_l})\|^T \qquad i = 1,\ldots,4$$

$$P_{uu}(x,x_r,\bar{t_l}) = \|P_{uu}(x,x_r,\bar{t_l})P_{uv}(x,x_r,\bar{t_l})P_{u\rho}(x,x_r,\bar{t_l})P_{u\zeta}(x,x_r,\bar{t_l})\|^T$$

The expressions for P_{vu}, $P_{\rho u}$, and $P_{\zeta u}$ have a similar form. Using the geostrophic approximation for calculating δu, δv and the dynamic method for determining $\delta\zeta$, one can express the remaining components of the covariance matrix P in terms of $P_{\rho\rho}$ (Knysh et al., 1980). The evolution equation for the covariance matrix $P_{\rho\rho}$ is of the form:

    (1) *at the prediction stage* (Sakawa, 1972),

$$\frac{\partial P_{\rho\rho}}{\partial t} = L_x[P_{\rho\rho}] + L_{x'}[P_{\rho\rho}] + Q_{\rho\rho} \tag{10}$$

where:

$$L_x = -\hat{u}\frac{\partial}{\partial x} - \hat{v}\frac{\partial}{\partial y} - \hat{W}\frac{\partial}{\partial z} + \kappa_z\frac{\partial^2}{\partial z^2} + \kappa_l\Delta$$

$Q_{\rho\rho}$ is the covariance error matrix for hydrothermodynamic modelling of the density field; $x' = (x',y',z')$ is a point of three-dimensional space;

    (2) *at the analysis stage*,

$$P_{\rho\rho}(x,x',t_l^+) = P_{\rho\rho}(x,x',\bar{t_l}) - \sum_{r=1}^{N}[\Delta_r^{1\rho}x,t_l)P_{\rho u}(x',x_r,\bar{t_l}) +$$

$$+ \Delta_r^{2\rho}(x,\bar{t_l})P_{\rho v}(x',x_r,\bar{t_l}) + \Delta_r^{3\rho}(x,\bar{t_l})P_{\rho\rho}(x',x_r,\bar{t_l}) +$$

$$+ \Delta_r^{4\rho}(x,\bar{t_l})P_{\rho\zeta}(x',x_r,\bar{t_l})] \tag{11}$$

    The boundary and initial conditions are of the form:

at $z = 0$, $\nu\dfrac{\partial\hat{u}}{\partial z} = -\dfrac{T_x}{\rho_0}, \nu\dfrac{\partial\hat{v}}{\partial z} = -\dfrac{T_y}{\rho_0}, \kappa_z\dfrac{\partial\hat{\rho}}{\partial z} = -\gamma_1(T_A + \gamma_2\hat{\rho})$ (12)

at $z = H$, $\hat{u} = \hat{v} = 0, \hat{\rho} = \rho(x,y,H,t_0)$ (13)

on side boundaries, $\hat{\rho} = \hat{\rho}(x,t_0), \hat{\zeta} = -\dfrac{1}{\rho_0}\displaystyle\int_0^H \hat{\rho}^*\,dz$ (14)

In time intervals between the measurement data acquisition, $P_{\rho\rho}$ at the side boundaries was specified according to the formula (Knysh et al., 1980; Demyshev and Knysh, 1981):

$$P_{\rho\rho}(x, x, t_{n+1}) = P_{\rho\rho}(x, x, t_0) + [P_{\rho\rho}(x, x, t_n) - P_{\rho\rho}(x, x, t_0)] \cdot$$
$$\cdot \exp[-\alpha(t_{n+1} - t_n)], \text{ where } \alpha = 0.05 \text{ s}^{-1}$$

The initial conditions are of the form:

$$(\hat{u}, \hat{v}, \hat{\rho}, \hat{\zeta})|_{t=t_0} = (u^0, v^0, \rho^0, \zeta^0), \qquad P_{\rho\rho}|_{t=t_0} = P_{\rho\rho}^0 \qquad (15)$$

Equation (12) at $z = 0$ was obtained with the assumption that the density has a linear dependence on temperature. The constants γ_1 and γ_2 are determined from average climatic data (Haney, 1971; Demyshev and Knysh, 1981), T_A is the temperature at the lower atmospheric boundary, and $\rho^* = \rho - \rho_s$, where ρ_s is an average value of density of the upper ocean layer.

The procedure of calculating optimal estimates is as follows. Using eqns. (2)–(6) and (10) with boundary (12)–(14) and initial (15) conditions, we predict the fields $\hat{u}, \hat{v}, \hat{\rho}, \hat{\zeta}$, $P_{\rho\rho}$, and W before the moment t_l of data acquisition. Calculating the covariance estimate error functions $P_{\varphi\psi}^{rp}$, one determines the weight factors at moment t_l by eqn. (9). The fields $\hat{u}, \hat{v}, \hat{\rho}, \hat{\zeta}$, and $P_{\rho\rho}$ are then corrected by eqns. (7) and (11). The fields thus obtained are used as initial ones to make predictions at the next moment. Then, the procedure is repeated. The numerical algorithm of the dynamic-stochastic model (on which the four-dimensional analysis is based) is founded on implicit calculation schemes and has a second-order accuracy with respect to the spatial variables (Demyshev and Knysh, 1980).

MODEL NUMERICAL EXPERIMENTS

Using the algorithm just described, we performed three series of numerical experiments. In the first series, the effectiveness and authenticity of the four-dimensional analysis was tested with the aid of model experiments. Calculations were carried out for a grid region corresponding to the central part of the Polymode area of the size 13×15 points with a grid step $\Delta x = \Delta y = 3.1484 \times 10^4$ m. In the vertical direction, we studied nine levels with a grid step of $\Delta z = 200$ m. The main parameters of the model had the following values: $\kappa_1 = 10^2$ m^2 s^{-1}, $\kappa_z = 10^{-4}$ m^2 s^{-1}, $\nu = 10^{-3}$ m^2 s^{-1}, $\beta = 0.2 \cdot 10^{-10}$ m^{-1} s^{-1}, $\gamma_1 = 0.712 \cdot 10^{-5}$ m s^{-1}, $\gamma_2 = 0.735 \cdot 10^{-1}$ kg K^{-1}, $\Delta t = 8.64 \cdot 10^4$ s. The values of τ_x, τ_y, T_A, and H were determined from observed data. Below the depth $H_1 = 1000$ m, the density anomaly was extrapolated by the formula:

$$\hat{\rho}(x, y, z, t) = \hat{\rho}(x, y, H_1, t) \exp[-2\pi(z - H)/(H - H_1)] \qquad (16)$$

with the aim of considerable reduction of the eqn. (10) dimension, the covariance matrix $P_{\rho\rho}$ was approximated by the following expression (Knysh et al., 1980):

$$P_{\rho\rho}(x, x', t) \approx D(x, t) D(x', t) P_0(x - x', y - y') \varphi(z, z') \qquad (17)$$

where $P_0(x - x', y - y')$ and $\varphi(z, z')$ are covariance functions determined from observed data (Knysh and Yarin, 1978); $D(x, t)$ is the variance of the density field estimates errors. The prognostic equation for $D(x, t)$ is obtained by substituting eqn. (17) into (10) (see Knysh et al., 1980). In the first and second series of experiments the measurement errors were neglected. The model time duration of both prediction and analysis stages was five days in the first series. The second density survey of the Polymode, referred to August 1, 1977 (Demyshev and Knysh, 1981) and interpolated by objective analysis into the computational grid points served as the initial field.

The values of u, v, ρ calculated by eqns. (2)–(6) with boundary conditions (12)–(14) have been used as control fields. Observed data were sampled from this control state each day. The density fields, obtained by the smoothing of the initial field ρ was taken as the initial one for all further numerical experiments of the first series. The first series contains nine experiments.

To compare further calculated results, we performed a preliminary experiment (PE), which was, in fact, an ordinary hydrodynamic prediction of the fields with the smoothed initial data.

In experiment I of the first series we carried out a four-dimensional analysis of hydrophysical fields in which observed data for u and v were assimilated each day. The location points and the sequence of assimilation are shown in Fig. 1a. In experiment II, unlike experiment I, density estimates were corrected with data of the velocity field. In experiment, III, the measured values of the density field were assimilated [the explicit effect of the observed ρ values on the components of the state vector (eqn. (1)) was neglected]. Experiment IV differed from III by considering the influence of the measured density values on v and u. In experiment V both the data of velocity and density fields were assimilated. In experiment VI (Fig. 1b) the order of density data assimilation was reversed. In experiment VII, as compared to VI, it was assumed that the field of the density estimate errors is homogeneous and isotropic relative to the covariance function. The presence of two research ships in the test area was imitated in experiment VIII (Fig. 1c). In the last experiment (IX) stream velocity data were assimilated on the third to the fifth day of the model time, when observation stations were stationary near the test area boundaries (as it is marked by stars in Fig. 1b).

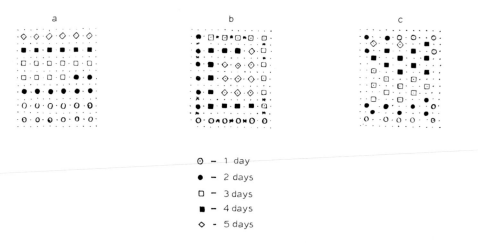

Fig. 1. The order and number of observation stations for density and current-velocity fields on each day of model time for numerical experiments of the first series: (a) experiments I–V; (b) experiments VI, VII, IX; (c) experiment VIII.

TABLE 1

The decrease (in percent) of the mean ($\bar{\delta}$), root mean-square (σ), and maximal errors (δ_{max}) of u, v, and σ_t in numerical model experiments of the first series relevant to the same errors of the preliminary experiment. Negative values mean the increase of the errors

Experiment	Corrected components	Measured components	Position of observation stations	$\bar{\delta}$			σ			δ_{max}		
				u	v	σ_t	u	v	σ_t	u	v	σ_t
I	u, v	u, v	Fig. 1a	3.8	8.3	−1.3	4.1	−0.5	−2.2	−2.2	0	−0.4
II	u, v, ρ	u, v	Fig. 1a	5.7	11	4.9	3.5	5.9	5	0.9	1.3	1.1
III	ρ	ρ	Fig. 1a	28.8	26.9	38.6	32.5	32.1	36.3	43.4	56.7	18.8
IV	ρ	ρ	Fig. 1a	30	28.1	38.9	32.6	32.8	36.4	42.1	55.4	17.6
V	u, v, ρ	u, v, ρ	Fig. 1a	33.1	33.1	39.1	36.1	36.7	37	41.4	55.8	21.1
VI	ρ	ρ	Fig. 1b	20.3	21.4	46.0	21.8	28.8	38.8	23.0	55.5	42.9
VII	ρ	ρ	Fig. 1b	16.5	24.1	34.4	19.5	30.2	31.3	23.8	55.6	36.4
VIII	ρ	ρ	Fig. 1c	32.1	30.2	50.7	27.9	22.7	49.7	32.5	8.7	25.0
IX	u, v	u, v	Fig. 1b	12.4	22.7	−1.5	13.8	22.0	−0.6	28.5	41.2	0.4

Let us introduce the notation:

$$\bar{\delta}(z) \ = \ \frac{1}{\Sigma} \int\limits_{S} |\varphi_i - \varphi_1| \, \mathrm{d}S, \qquad \delta_{\max}(z) \ = \ \max |\varphi_i - \varphi_1|,$$

$$\sigma(z) \ = \ \left\{ \frac{1}{\Sigma} \int\limits_{S} [|\varphi_i - \varphi_1| - \bar{\delta}(z)]^2 \, \mathrm{d}S \right\}^{1/2} \tag{18}$$

where S and Σ are the surface and area of the test region, φ_i is the control value of the field, φ_1 is the value obtained in experiment. Table 1 lists the calculation results and the decrease (in percent) of the errors $\bar{\delta}$, $\delta_{\max}$, and σ relative to the corresponding errors in the preliminary experiment (the decrease was calculated for the final analysis moment at the depth $z = 400$ m). It can be seen that assimilation of velocity data (experiments I and II) improves little the analysis results for u and v. However, correction of the density field ρ should not be neglected in the assimilation of data for u and v. The errors were found to decrease significantly in experiments III–V. Hence, assimilation of density data seems to contribute most of all to the four-dimensional analysis of hydrophysical fields (Demyshev and Knysh, 1981; Carpenter and Lowther, 1982).

Comparison of the results of experiments III, VI and VIII leads to the following conclusions. Reversal of the order of observation stations for the density field leads to essential change of analysis results. With the same amount of measurement data, the effectiveness of the four-dimensional analysis depends on a proper position of the observation stations. Apparently, there exists an optimal version of positioning of these stations. Comparison of the results obtained in experiments VI and VIII shows that it is very important to take into account inhomogeneity and anisotropy of the density estimates error field with respect to the covariance function (in experiment VII the errors $\bar{\delta}$ and σ for the field ρ increased by 12 and 8%, respectively). Finally, experiment IX clearly demonstrates a higher effectiveness of the stationarity of velocity observation stations position as compared to experiments I and II.

NUMERICAL EXPERIMENTS WITH THE ASSIMILATION OF POLYMODE PROGRAMME DATA

In the second series of numerical experiments hydrophysical fields were predicted and analysed on the Polymode area for the period from August 1 to 15, 1977. The density field described above was taken as the initial one. The first experiment was designed as an ordinary hydrodynamic prediction, with the stationary boundary conditions. The second experiment differed from the first one in the following condition for the density on the side boundaries:

$$\rho(x, t) \ = \ \rho(x, t_0) \frac{T - t}{T - t_0} + \rho(x, T) \frac{t - t_0}{T - t_0} \tag{19}$$

where $\rho(x, T)$ are density values on the side boundaries obtained in the grid points by an objective analysis of measurement data of the third density survey performed in the test area on August 8–18, 1977, and referred to August 15, 1977. The third numerical experiment represented a four-dimensional analysis of hydrophysical fields with sequential assimilation of density data obtained on August 8–15, 1977.

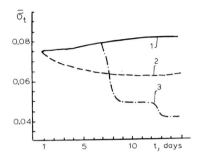

Fig. 2. Evolution curves of the mean (with respect to volume) prediction error of conventional density for the second series of experiments: 1 = ordinary hydrodynamical prediction; 2 = prediction with boundary conditions being changed with time; 3 = prediction with the assimilation of density-field data beginning from the 8th day.

Figure 2 shows the time behaviour of the mean (with respect to volume) prediction error for conventional density $\bar{\sigma}_t$ in comparison with measurement data obtained on August 19–29, 1977. It can be seen that in experiment II the mean error of the conventional density (Fig. 2, curve 2) decreases almost 2 times as compared with $\bar{\sigma}_t$ in experiment I. Comparison with actual observations (Bulgakov, 1978) shows, however, that the hydrological structure of the fields in the test area is far from the real structure. Comparison of the curves of Fig. 2 shows that in this series of experiments the best result is obtained in the case of a four-dimensional analysis of ocean fields with density data assimilation. This is also confirmed by the results of in-situ measurements (Bulgakov, 1978). Since we compared calculation results with observation data which are "future" relative to the final moment of prediction and analysis, the experiments revealed the representativeness of the analysis scheme on the synoptic scale.

NUMERICAL EXPERIMENTS WITH THE ASSIMILATION OF THE "SECTIONS"
PROGRAMME DATA

In the third series of numerical experiments we used the following version of the dynamic-stochastic model. Optimal estimates of the stream velocity between the times of measurements were calculated by the geostrophic relation, the tangential wind stress was neglected; the Dirichlet condition for the density was assumed to be valid on the ocean surface. The nonlinear terms from eqns. (5) and (6) were neglected. Boundary conditions on side boundaries are as follows (Fig. 4):

$$\frac{\partial \hat{\rho}}{\partial n} = \left(\frac{\partial \rho}{\partial n}\right)_c \qquad \text{on section ABCDE} \qquad (20)$$

$$\hat{\rho} = \rho_c \qquad \text{on section AGE} \qquad (21)$$

$$\frac{\partial \hat{\zeta}}{\partial x} = -\frac{1}{H\rho_0} \int_0^H z \left(\frac{\partial \rho}{\partial x}\right)_c dz \qquad \text{on sections AB and CD} \qquad (22)$$

$$\hat{\zeta} = -\frac{1}{\rho_0} \int_0^H \hat{\rho}^* dz \qquad \text{on the rest of boundary} \qquad (23)$$

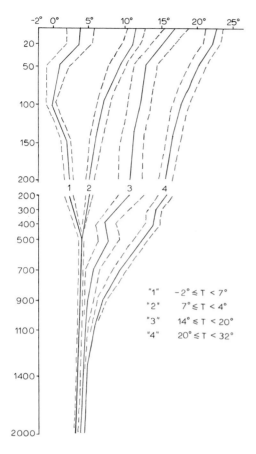

Fig. 3. Sea-surface temperature variability intervals, and the corresponding model profiles $T_j(z)$ °C, and $\sigma_i(z)$ °C.

Index "c" indicates the the corresponding values in eqns. (20)–(23) are determined from climatic values of the density field.

It is known that the assimilation of surface temperature data without corresponding approximation of the vertical temperature profile can disturb density stratification, that leads to the fictitious appearance of heat fluxes. We suggest a procedure of approximation of the vertical temperature profile, which uses the property of temperature changes correlation on the sea surface and in the depth. From measuring sea-surface temperature we passed to the "effective" temperature data $T^d(x, t)$, and to the vertical profile of the temperature measurement error variances $R^d(x, t)$. With the aim of constructing the model profiles we chose four preliminary intervals of surface-temperature variations for the region under consideration. For these intervals we plotted the profiles of mean and root mean-square deviation. These curves and intervals of the surface-temperature variations are represented in Fig. 3. The values of the model-temperature profile were calculated by the formula:

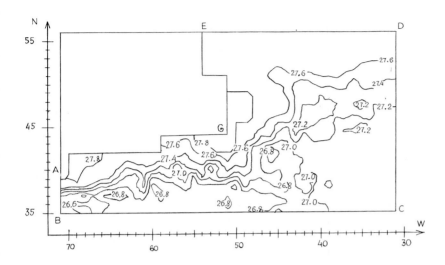

Fig. 4. Climatic density field calculated by temperature and salinity fields (data of Princeton University, USA) at $z = 500$ m.

$$\bar{T}_j^m(x, y, z, t_i) = \bar{T}_j(z) + [T(x, y, t_i) - \bar{T}_j(0)] \frac{\sigma_j(z)}{\sigma_j(0)} \qquad (24)$$

where $\bar{T}_j(z)$ is the mean temperature profile, $\sigma_j(z)$ is the profile of root mean-square deviation of temperature for the jth water type, $T(x, y, t_i)$ are sea-surface temperature data. The "effective" temperature data, used for the assimilation, are the sum of the model profile [eqn. (24)] and the "effective" measurement error $\delta_j^d(x, t)$:

$$T_j^d(x, t) = \bar{T}_j^m(x, t) + \delta_j^d(x, t) \qquad (25)$$

The effective measurement error value is defined by two factors: (1) sea-surface temperature measurement errors $\delta_0(x, y, t)$; and (2) model approximation errors $\delta_j^m (x, t)$:

$$\delta_j^d(x, t) = \frac{\sigma_j(z)}{\sigma_j(0)} \delta_0(x, y, t) + \delta_j^m(x, t) \qquad (26)$$

According to above speculations $R_j^d(x, t)$ will be of the form:

$$R_j^d(x, t) = \sigma_0(x, y, t) \frac{\sigma_j(z)}{\sigma_j(0)} + \sigma_j^m(x, t) \qquad (27)$$

where $\sigma_0(x, y, t)$ is surface measurements error variance, which one can consider to be known; $\sigma_j^m(x, t)$ is model approximation error variance, which is preliminarily determined from the ocean deep-layer measurements. The values of density field for the assimilation were obtained from temperature field measurements by the formula (Mamaev, 1970):

$$\rho = [28.152 - 0.0735\,T - 0.00469\,T^2 + (0.802 - 0.002\,T) \cdot (S_c - 35)] \cdot 10^{-3}$$

where S_c is the climatic salinity field.

To carry out numerical calculations we considered the region with the coordinates 35.5°–55.5°N and 289.5°–330.5°E. The step in the horizontal direction was 1°. In the vertical direction we considered 16 layers: $z = 0, 20, 50, 100, 150, 200, 300, 400, 500, 700, 900, 1100, 1500, 2000, 3000,$ and 4000 m. Coefficients of turbulent diffusion were: $\kappa_z = 10^{-4}$, $\kappa_e = 10^3\ \mathrm{m^2\,s^{-1}}$.

We performed two numerical experiments, each of them was carried out in two stages. The first stage of both experiments contains ordinary hydrodynamic prediction (according to semi-diagnostic method) for 40 days of model time with the aim of obtaining the adjusted fields of density, current velocities and sea-surfce topography (Sarkisyan and Demin, 1983). Figures 4 and 5 show the initial (climatic) and adjusted density fields. One can see that adjustment leads to disappearance of small inhomogeneities of synoptic scale in density field; the ρ field has more smooth structure. By the final moment of adjustment the velocities in upper layers of the Gulfstream run up to 60–80 cm s^{-1}. The Gulfstream flow is of regular character, the flow parallel to the Newfoundland bank is expresssed distinctly. One can trace velocities of the Gulfstream flow of 20–30 cm s^{-1} in near-surface layers of the open ocean area.

We assumed the density and sea-surface topography fields, obtained at the adjustment stage, as initial fields for the second stage of numerical experiments. In the first numerical experiment we performed an ordinary hydrodynamic prediction for two months with the aim of comparison of calculation results with the assimilation of measurement data. The time step in this experiment was 1 day. In the second numerical experiment we carried

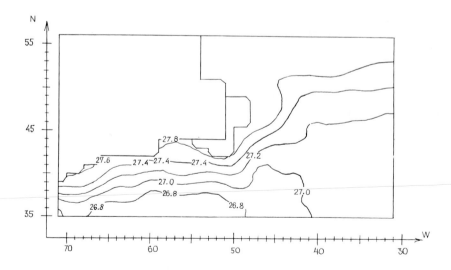

Fig. 5. Density field at $z = 500$ m after adjustment.

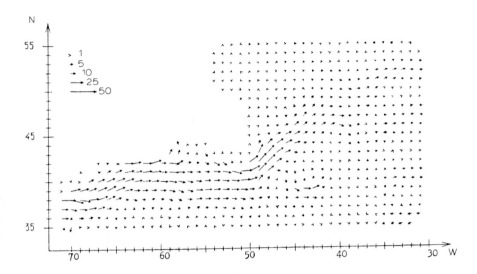

Fig. 6. Currents field at $z = 150$ m after 52 model days of the assimilation of the data of programme "Sections".

out a four-dimensional analysis of the fields with the assimilation of surface and deep-layer temperature data. In the calculation scheme of optimal estimates in this experiment there was an additional adjustment procedure, which was performed after analysis. With this aim we carried out an ordinary hydrodynamic prediction for five days with the step of four hours.

Using results of the second experiment, we plotted a number of charts of sea-surface topography, density and current velocities. Figure 6 shows the current charts by the finite moment of analysis (the 52nd day) in the layer of 150 m. Comparison of results of the experiment shows that assimilation of measurement data, unlike an ordinary hydrodynamic prediction, leads to increasing velocity in the upper Gulfstream layers by 5–7 cm s^{-1}. The mass flux in a quasistationary meander, which is situated on the right

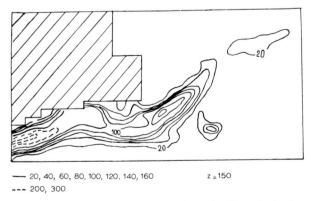

—— 20, 40, 60, 80, 100, 120, 140, 160 $z = 150$
--- 200, 300

Fig. 7. Kinetic energy after 52 model days of ordinary hydrodynamical prediction.

of the flow jet, increased. It should be noted that the jet-like flow divides into two weak currents in the open ocean. This is shown in the charts of kinetic energy (Figs. 7 and 8).

CONCLUSIONS

The four-dimensional analysis scheme we have considered in the present paper, has in its basis the Kalman filter for the systems with distributed parameters (Sakawa, 1972; Padmanabhan and Colantuoni, 1974). The characteristic features of this four-dimensional analysis scheme are: (1) mutual dependence of dynamic and stochastic parts of a model; (2) the possibility of optimal estimation in the case when not all the components of state vector are measured, and/or not in all points of space under consideration the data are available; (3) the possibility of usage of the same variable measurement results being carried out by different techniques; and (4) measurement errors filtering.

An algorithm of a four-dimensional analysis of observed data under concrete physical-geographical conditions of Polymode experiment and Newfoundland energetically active zone, has been performed. The former contains: (1) a procedure of the estimates' hydro-dynamic adjustment to a prognostic model; (2) calculation of state vector prognostic estimates; and (3) complex assimilation of measurement data.

Numerical calculations with complex assimilation of measurement data have shown that the most informative are the density-field observations. The realization by this model, performed for the physical-geographical conditions of the Polymode area, has revealed the representativeness of the four-dimensional data assimilation algorithm.

The measurement assimilation procedure suggested above is of principle character from the point of view of further use of information about the ocean surface, since this procedure allows one to solve the ocean monitoring problem on synoptic and large scales.

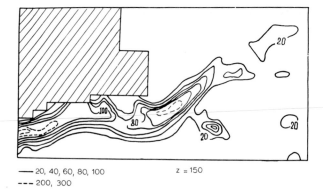

——— 20, 40, 60, 80, 100   z = 150
- - - 200, 300

Fig. 8. The same as in Fig. 7, with the assimilation of the data of programme "Sections".

REFERENCES

Bulgakov, N.P., 1978. General characteristics of synoptic vortices. In: Eksperimentalnye Issledovaniya po Mezhdunaorodnoi Programme Polymode (Results of the 16th Cruise of the Research vessel "Akademik Vernadsky"). Marine Hydrophysical Institute of the Ukranian Academy of Sci., Sevastopol, pp. 44–54 (in Russian).

Carpenter, K.M. and Lowther, L.R., 1982. An experiment on the initial conditions for a mesoscale forecast. Q. J. R. Meteorol. Soc., 108: 643–660.

Demyshev, S.G. and Knysh, V.V., 1980. Computer simulation of synoptic and large-scale oceanic currents. In: Struktura, Kinematika, i Dinamika Sinopticheskikh Vikhrei. Marine Hydrophysical Institute of the Ukranian Academy of Sci., Sevastopol, pp. 46–58 (in Russian).

Demyshev, S.G. and Knysh, V.V., 1981. Model numerical experiments on estimating the reliability of the multielement four-dimensional analysis of the major physical oceanic fields. In: Teoriya Okeanicheskikh Protsessov. Marine Hydrophysical Institute of the Ukranian Academy of Sci., Sevastopol, pp. 61–69 (in Russian).

Haney, R.L., 1971. Surface thermal boundary condition for ocean circulation models. J. Phys. Oceanogr., 1(4): 241–248.

Itogi nauki i tekhniki, 1984. Ser. Atmosphere, Ocean, Space – Programme "Sections". Moscow, Vol. 1, 60 pp.

Knysh, V.V. and Yarin, V.D., 1978. Certain hydrodynamical characteristics and statistical structure of temperature and density fields on a test area. In: Eksperimentalnye Issledovaniya po Mezhdunorodnoi Programme Polymode (Results of the 17th Cruise of the Research Vessel "Akademic Vernadsky" and of the 33d Cruise of the Research Vessel "Mikhail Lomonosov"). Marine Hydrophysical Institute of the Ukranian Academy of Sci., Sevastopol, p. 74–81 (in Russian).

Knysh, V.V., Moiseenko, V.A., Sarkisyan, A. S. and Timchenko, I. E., 1970. Complex use of measurement data obtained on hydrophysical oceanic test areas in a four-dimensional analysis. Dokl. Akad. Nauk SSSR, (4): 832–836 (in Russian).

Mamaev, O.N., 1970. T-S Water Analysis of the World Ocean. Gidrometeoizdat, Leningrad, 364 pp. (in Russian).

Padmanabhan, L. and Colantuoni, G., 1974. Sequential estimation in distributed systems. Int. J. Syst. Sci., 5 (10): 937–986.

Sakawa, Y., 1972. Optimal filtering in linear distributed parameter systems. Int. J. Control, 16 (1): 115–127.

Sarkisyan, A.S., 1977. Numerical Analysis and Prediction of Sea Currents. Gidrometeoizdat, Leningrad, 179 pp. (in Russian).

Sarkisyan, A.S., 1982. Monitoring large-scale ocean circulation with the aid of time series. WCRP papers, Time series (Tokyo, 11–15 May, 1981), pp. 13–22.

Sarkisyan, A.S. and Demin, Yu.L., 1983. A semidiagnostic method of sea current calculations. In: Large-Scale Oceanographic Experiments in the WCRP. WCRP Publ. Ser., 1, Vol. II, pp. 201–214.

CHAPTER 39

EFFECTS OF VARYING SEA-SURFACE TEMPERATURE ON 10-DAY ATMOSPHERIC MODEL FORECASTS

PETER H. RANELLI, RUSSELL L. ELSBERRY, CHI-SANN LIOU and SCOT A. SANDGATHE

ABSTRACT

A 10-day prediction with the U.S. Navy Operational Global Atmosphere Prediction System that has fixed sea-surface temperatures (SST) is compared with a hindcast in which the observed SST fields at each 12 h are supplied as a lower boundary condition. A period during Spring 1983 is selected to determine the atmospheric response to SST changes. The maximum changes in surface latent heat flux are about 30% of the mean values when the SST is fixed. The maximum surface sensible heat flux changes are also about 30% larger than in the fixed SST case. Changes in six maritime cyclone developments are studied using the Systematic Error Identification System. Five of these cyclones occurred in regions of small SST changes. Two of the cyclone predictions were improved in the time-dependent SST case. The degradations in the other four cyclone predictions were less than the departures of the predictions from the verifying analyses.

INTRODUCTION

It is well known that important interactions between the atmosphere and the ocean exist on time scales of a month or longer. The interaction between the atmosphere and the ocean on shorter time scales is less well understood. However, heat fluxes from the ocean into the atmosphere are believed to play an important role in many atmospheric circulations.

Rapid advancements in the last three decades have greatly improved both the quality and the speed of numerical weather prediction models. The present model forecasts are, in general, no better than climatology after five or six days. To improve model forecasts beyond this present limitation, some type of feedback between the atmosphere and the ocean may have to be included.

This work is the first in a series of case studies designed to study the necessity and feasibility of coupling an atmospheric model and an ocean model. In this first step, we examine the response of 10-day atmospheric forecasts to the *observed* sea-surface temperature (SST) changes during the forecast period. We concentrate our analysis on the effects of SST changes on the model predictions of midlatitude cyclone development.

There are many oceanic processes that can cause significant changes in the SST on ten-day time scales that may result in a response in the atmosphere. Large changes in SST can occur during the spring and autumn transition periods in the ocean. During this period, there are warm and shallow ocean mixed layers that may deepen and cool rapidly in response to atmospheric forcing (Camp and Elsberry, 1978; Elsberry and Camp, 1978; Elsberry and Raney, 1978). During winter, the ocean mixed layer is deeper and the

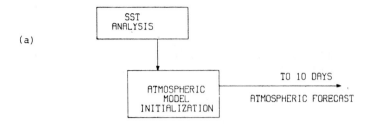

(a)

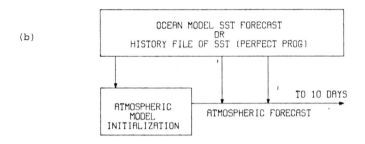

(b)

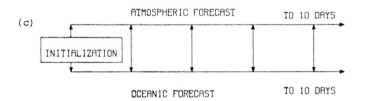

(c)

Fig. 1. Methods for coupling atmospheric and oceanic models: (a) minimal feedback; (b) non-synchronous; and (c) synchronous.

changes in SST due to winter storms are smaller. Summer storms are less intense than winter storms and do not force a large change in the SST.

The SST changes alter the boundary forcing of the atmospheric forecast model via surface sensible and latent heat fluxes, and possibly changes in momentum fluxes due to static stability modifications. Both surface fluxes depend on SST and also on PBL status, and on atmosphere conditions above the PBL in the case of surface latent heat flux. The deviations in surface sensible heat flux will be directly felt by the atmosphere, whereas the deviations in surface latent heat flux will affect the atmosphere after condensation has occurred. The changes in cloud cover and water vapour content also affect the radiative transfers. Therefore, the impact of SST changes on atmospheric circulations is related to the magnitude and also the location of the SST changes. This nonlinear character of the SST influence demands a full forecast atmospheric model be used to study this problem, and the response may be highly case-dependent.

There are three methods (Elsberry et al., 1982) for coupling an atmospheric model to an oceanic model. Ranked in terms of increasing sophistication, the methods are: (1) minimal feedback or weak coupling; (2) non-synchronous coupling; and (3) synchronous coupling (Fig. 1). For the weak coupling scheme (Fig. 1a), the analyzed SST field at the initialization time of the atmospheric model is used as input for the model initial conditions. The SST then remains constant for the entire model run. This type of coupling is used in present operational models.

For the non-synchronous coupling scheme (Fig. 1b), a SST prediction for the entire forecast period is assumed to be available and is used as input during the appropriate times in the atmospheric model runs. The time-dependent SST can be provided from either an ocean model forecast that has been run independently of the atmospheric model, or from analyzed SST fields in a "perfect-prog" hindcast. The non-synchronous method may be impractical for operational use if an independently run ocean prediction develops biases which would overwhelm a forecast by the atmospheric model during a 10–15 day forecast. However, this type of coupling is an excellent research tool when using the "perfect-prog" SST fields (Rosmond et al., 1983).

The final and most sophisticated scheme is the fully sychronous coupling (Fig. 1c). The two models are run concurrently and provide feedback to each other at the appropriate points in the model integration. In this way, the SST used in the atmospheric model and the atmospheric forcing (surface wind, surface heat fluxes, precipitation) for the ocean model, are being continually updated. While this is obviously the most complicated of the three schemes, it should also provide the best forecast.

This study was conducted using a non-synchronous coupling with "perfect-prog" SST to isolate the atmospheric response. The purpose of studying the response in this type of model is to understand the role of air–sea interaction. This better understanding of the model air–sea processes should ultimately lead to a fully synchronous coupled model.

We investigate the atmosphere model response to SST changes by integrating the Navy Operational Global Atmospheric Prediction System (NOGAPS) to ten days. In the control run, the analyzed SST field from the Fleet Numerical Oceanography Center (FNOC) at the initial time is the lower boundary condition throughout the forecast (Fig. 1a). In the test case (called the SST run), the analyzed SST fields during the ten-day forecast period were inserted each 12 h. A period during the spring was selected when the SST changes might be large enough to cause significant differences in the model predictions. A period between 1800 GMT 26 May 1983 and 1800 GMT 3 June 1983 was chosen. The resulting forecasts are analyzed by the Systematic Error Identification System (SEIS) to trace the changes in cyclone developments due to the time-dependent SST.

MODELS AND EXPERIMENTAL PROCEDURES

Navy Operational Global Atmospheric Prediction System (NOGAPS)

NOGAPS is the Navy's state-of-the-art atmospheric prediction model. The model was made available by Dr. T.E. Rosmond of Navy Environmental Prediction Research Facility (NEPRF). The version of NOGAPS used in this experiment contains all modifications made to the system through July 1983.

The NOGAPS forecast model is a six-layer, sigma coordinate, primitive equation model (Rosmond, 1981). The finite difference of staggered grid C is used with a spatial resolution of 2.4° latitude by 3° longitude covering the whole global area. The model is based upon the UCLA general circulation model (GCM), described by Arakawa and Lamb (1977). The diabatics of the model are of full GCM sophistication. NOGAPS includes the parameterization of the planetary boundary layer (PBL) after Deardorff (1972) and Randall (1976); cumulus convection using the Arakawa-Schubert (1974) scheme, as described by Lord (1978); and radiation as described by Katayama (1972) and Schlesinger (1976). It should be noted that NOGAPS differs from the UCLA GCM in that the PBL is not allowed to exceed the first sigma level. This effectively limits the PBL to the bottom 200 mb of the atmosphere.

The extremely complete package of diabatic processes used in the model was felt to be an important consideration in the selection of NOGAPS for this experiment. The full parameterization of both the PBL and the cumulus convection was necessary for the effects of the changing SST to be felt in the rest of the model. Without full diabatics, the response to the SST would have been diminished.

Sea-surface temperature analysis

The time-dependent SST used in this experiment was obtained from FNOC. FNOC performs SST analyses twice daily at 0000 and 1200 GMT. The analysis is an integral part of the Thermal Ocean Prediction System-Expanded Ocean Thermal Structure (TOPS-EOTS). TOPS is an upper ocean forecast model that includes the Mellor and Yamada (1974) level-2 turbulence parameterization scheme and advection by instantaneous wind drift and climatological geostrophic current (Clancy and Martin, 1979). EOTS is an ocean thermal analysis procedure which uses information blending techniques (Holl and Mendenhall, 1971) to blend XBT and surface ship reports to a three-dimensional grid. Twenty-six ocean parameters including primary layer depth, temperatures and vertical temperature derivatives are analyzed in the upper 400 m represented on 18 vertical levels. The first guess fields are the previous 24 h TOPS forecasts. Satellite-derived SST reports are not presently used in the analysis. The combined TOPS-EOTS had only been in an operational status a few months when the NOGAPS initial conditions and SST were captured. However, in a four-month study the TOPS-EOTS combination had less noise in the daily analysis than the conventional EOTS (Clancy and Pollack, 1983).

Systematic Error Identification System (SEIS)

SEIS is a tool to objectively analyze numerical model predictions and produce error statistics for use by operational forecasters (Harr et al., 1983). SEIS operates in a quasi-Lagrangian frame with the reference center located at the center of the storm. The primary algorithm within SEIS is the vortex tracking program (VTP) after Williamson (1981). The purpose of VTP is to track Synoptic-scale features and produce a listing of operationally relevant parameters following the feature. This program fits each vortex by six parameters at a time to approximate the shape and location of the vortex. The parameters chosen include amplitude (A), ellipticity (ϵ), radius (R), orientation (α), and position of the feature. Amplitude is the magnitude of the vortex central pressure relative to the zonal-mean pressure. Ellipticity is a measure of the deviation of the shape of the storm from circular. It is computed as the square of the ratio of the semi-major and semi-minor axes. Orientation is the angle between the X-axis and the semi-major axis, measured counterclockwise from the positive X-axis. Position is specified as either the model grid position or the geographical position.

Initial conditions

The time period was chosen to coincide with the occurrence of the spring transition period in the ocean. During the transition, increases in the SST are associated with a rapid shallowing of the mixed layer. Passage of a storm may then eliminate this warm layer and rapidly decrease the SST. Thus, it was desirable to select a period before the seasonal thermocline had become very strong. The atmospheric analyses were monitored for occurrences of cyclogenesis and a storm track across a large portion of the Pacific Ocean. If this occurred, the increased mixing due to the increased surface wind stress could act to mix through the incipient seasonal thermocline and rapidly deepen the mixed layer. The increased mixing would reduce the SST due to the entrainment of cold water into the mixed layer. In the atmospheric model runs with the time-dependent SST, the reduced SST should act to impede the cyclogenesis compared to the control run.

Given these constraints, the most favorable conditions for a model run appeared in the Pacific Ocean in late May. The initial conditions captured were for 1800 GMT May 1983. This was NOGAPS 6-h update and not an actual forecast. Since this was a full initialization for a NOGAPS forecast, this should not have caused a problem in the experimental forecasts.

In capturing the SST fields for this ten-day period, the TOPS-EOTS analysis was not available from the 0000 GMT 3 June analysis to the end of the ten-day forecast at 1800 GMT 5 June 1983. The lack of a changing SST for the last three days of the forecast period may cause some differences in the overall final forecast. However, we feel that the lack of the last three days of SST changes may reduce the differences between the two model runs but should not cause a significant effect on the model forecasts.

Experimental procedures

In the control run from 1800 GMT 26 May 1983, the SST was held fixed at the initial values, as is presently done in the operational forecasts. The model was integrated to ten days with no changes in any of the input initial fields (Fig. 2a). A complete history tape was written every 6 h during the model run for future analysis. These fields include the winds, heights, humidities and temperatures for several levels. Various PBL parameters were output as well, including the total heat flux, moisture (latent heat) flux, sensible heat flux and long- and short-wave radiative heat fluxes. Precipitation fields associated with cumulus convection and large-scale lifting were also ouput.

The second model run, designated the SST run, was made using the "perfect-prog" time-dependent SST. The new SST was input every 12 h at 0000 and 1200 GMT during the forecast (Fig. 2b). No time interpolation of the SST fields to smooth the effect of the change was performed. The last of the changing SST fields was input at 162 h and held constant for the remainder of the integration. A similar history tape was generated from the SST run as for the control run.

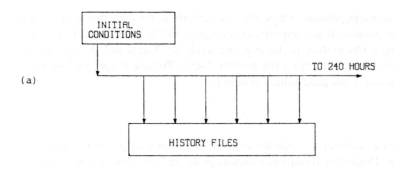

(a)

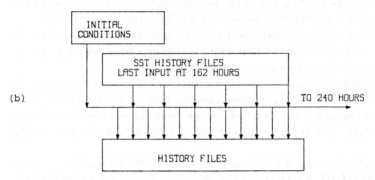

(b)

Fig. 2. Schematic of experiment design for (a) control run and (b) SST run. SST fields are input every 12 h. History files are output every 6 h.

SST FIELDS AND SURFACE FLUXES

SST fields

The initial SST fields (Fig. 3) are from the TOPS-EOTS analysis at 1800 GMT 26 May 1983. This field has a predominant north—south gradient with very little east—west structure, except along the coastal regions. The warmest areas of 26°C are found in the southwestern corner of the basins. The Pacific Ocean SST field shows the Kuroshio current as a strong gradient along the east coast of Japan. Along the west coast of North America, a plume of warm water extends northward into the Gulf of Alaska. In the Atlantic Ocean, the predominant gradient is associated with the Gulf Stream and is oriented NW—SE over most of the ocean north of 40°N.

To summarize the time-dependent SST fields, a simple sum of seven daily SST departures from the initial value was calculated (Fig. 4). For both basins, maximum of the *accumulated* SST changes during this period of spring are about 10°C. In general, the cumulative temperature departures over the Pacific Ocean have positive values of 2.5°C over much of the basin, and negative values are seen along the coasts of both

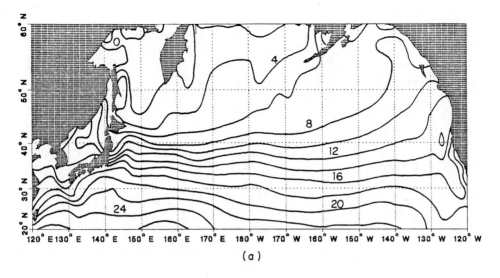

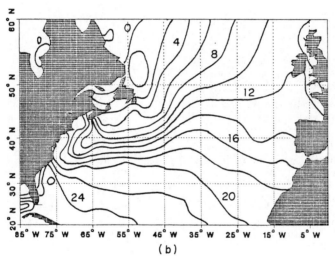

Fig. 3. (a) Pacific and (b) Atlantic sea-surface temperature fields used as initial conditions in the model run. Contour interval is 2°C.

continents and in a large region of the eastern Pacific. The Atlantic Ocean has a similar change, except the negative values are concentrated along the western boundary.

Surface heat fluxes

The surface heat fluxes (sensible, latent and total) are responsive to changes in the SST. The surface sensible heat flux directly affects the atmosphere flow in PBL. The surface latent heat flux only modifies the moisture balance in PBL. Subsequent changes in the atmospheric circulation will depend on how efficiently these fluxes are transported from

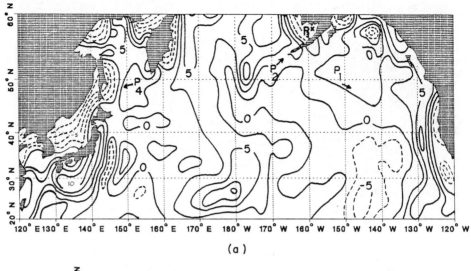

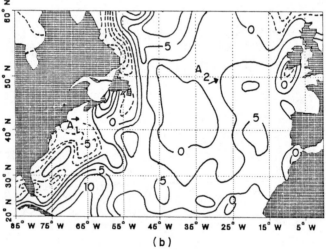

Fig. 4. The cumulative change in the (a) Pacific and (b) Atlantic SST field summed over seven 24-h intervals. Contour interval is 2.5°C. Solid (dashed) lines are positive (negative) changes. Initial positions of six maritime cyclones that occurred during the forecast period are indicated by arrows.

the PBL to the free atmosphere and where the condensation occurs. In this regard, a model with higher vertical resolution than the six-layer version of NOGAPS used in this study will probably improve the model response to a time-dependent SST. The total surface heat flux (sensible heat plus latent heat and radiation) is more relevant to the changes in oceanic circulations than atmospheric circulations.

The analysis consisted of determining the means and standard deviations at each grid point for the sensible, latent and total heat fluxes. The means and standard deviations were computed over the 40 six-hourly calculations. The difference of the two model runs was then determined to illustrate the change in model response. In the following

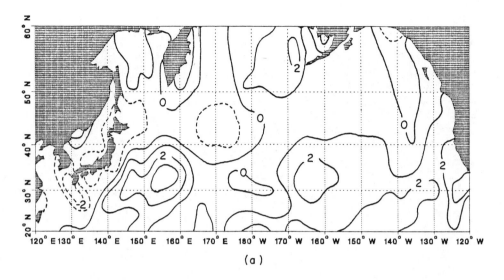

(a)

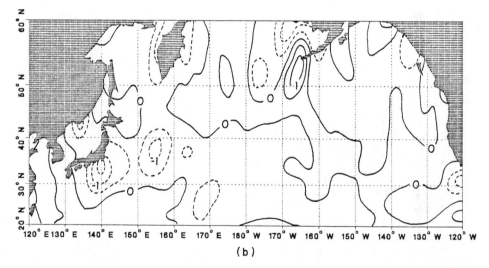

(b)

Fig. 5. (a) Mean sensible heat flux for the Pacific Ocean, SST run. Contour interval is 1 cal cm^{-2} h^{-1}. Solid lines are positive (upward) heat flux. Dashed lines are negative (downward) heat flux. (b) Differences in the sensible heat flux, control run minus SST run. Positive differences (solid lines) indicate less energy available to the atmosphere in the SST run. Contour interval is 0.5 cal cm^{-2} h^{-1}.

figures, upward heat flux (from the ocean to the atmosphere) is positive. A positive difference value indicates less heat flux was available to the atmosphere in the SST run.

Sensible heat flux

The sensible heat fluxes for the Pacific Ocean (Fig. 5) and the Atlantic Ocean (Fig. 6) show upward sensible heat flux over most of the ocean basins for both model runs. Areas of downward heat flux (indicating the air is warmer than the ocean surface) are found

684

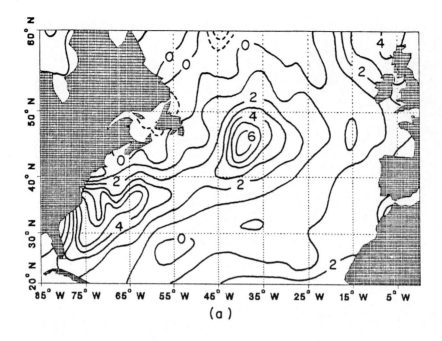

(a)

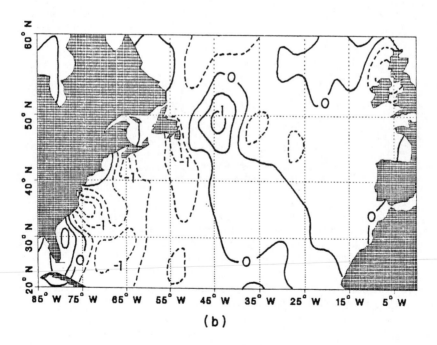

(b)

Fig. 6. As in Fig. 5, except for the Atlantic Ocean.

685

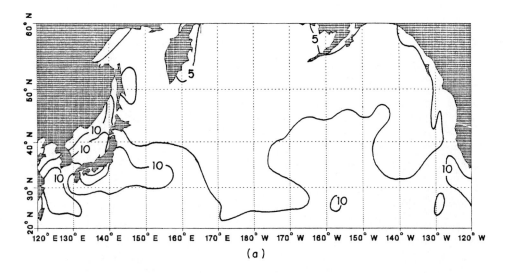

(a)

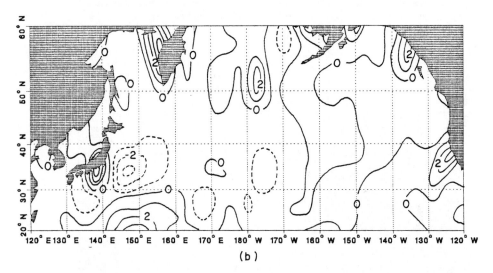

(b)

Fig. 7. As in Fig. 5, except for the latent heat flux. (a) Contour interval is 2.5 cal cm^{-2} h^{-1}; (b) contour interval is 1 cal cm^{-2} h^{-1}.

along the western edge of the Pacific Ocean extending southeastward from Kamchatka as far as 30°N, and in the northwestern portion of the Atlantic Ocean. The effect of the Gulf Stream is very evident as the maximum upward flux extends from the east coast of the U.S. to the northeast and into the central Atlantic. Fluxes in the Atlantic are larger than 6 cal cm^{-1} h^{-1}* as opposed to the largest flux of 3 cal cm^{-2}h^{-1} in the Pacific.

The difference in the flux generally follows the change in SST (Fig. 4). Areas of higher SST have resulted in a positive change in the flux and large changes in the SST correlate

*Note, cal cm^{-2} h^{-1} means gram-calorie per square centimeter per hour.

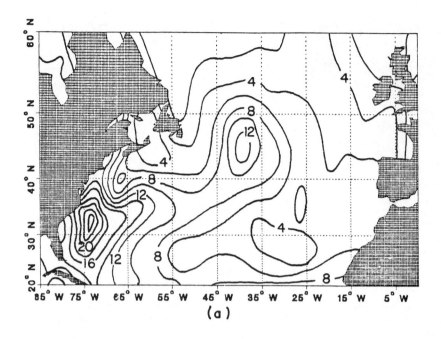

(a)

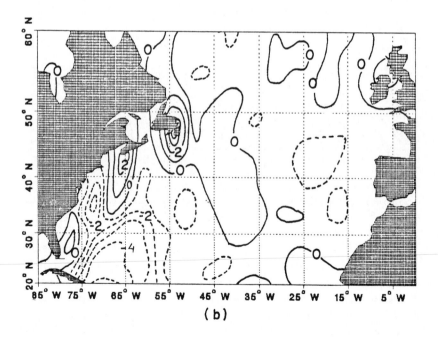

(b)

Fig. 8. As in Fig. 5, except for the latent heat flux in the Atlantic Ocean. (a) Contour interval is $2\ cal\,cm^{-2}\,h^{-1}$; (b) contour interval is $1\ cal\,cm^{-2}\,h^{-1}$.

with large changes in flux. However, there are several exceptions at western boundaries and middle of the two basins where SST changes result in opposite changes in surface sensible heat flux. For both basins, the largest change in sensible heat flux was about 25–30% compared to the mean values of the SST run. Largest changes of $1.0 \, cal \, cm^{-2} \, h^{-1}$ in the Atlantic Ocean and $1.5 \, cal \, cm^{-2} \, h^{-1}$ in the Pacific Ocean occurred along the east coast of continents.

Latent heat flux

The mean fields of the surface latent heat flux (Figs. 7 and 8) show a general trend of increasing flux from the north to the south. This pattern corresponds generally to the SST field. The highest fluxes near the western boundary currents are on the order of $20 \, cal \, cm^{-2} \, h^{-1}$ in the Atlantic and $15 \, cal \, cm^{-2} \, h^{-1}$ in the Pacific. The effect of the Gulf Stream can be seen more than halfway across the Atlantic and as far as $50°N$.

The changes in surface latent heat flux for the SST experiment were relatively larger than for surface sensible heat flux. The differences in the fluxes and the SST change were concentrated near the western boundaries of the two basins. Over most of both oceans, there was an increase in the latent heat flux to the atmosphere in the SST run compared to the control run. The largest changes were about 30% mean values of the SST run. The Kuroshio had increases of $3 \, cal \, cm^{-2} \, h^{-1}$ over a small region to the east of Japan. The increase over the Gulf Stream region was as large at $5 \, cal \, cm^{-2} \, h^{-1}$ and occurred over a larger area than associated with the Kuroshio.

Total surface heat flux

The total heat flux is the sum of the latent, sensible, solar (shortwave) and back (long-wave) radiation. The total heat flux is expected to have a larger diurnal component than the sensible and latent heat fluxes because of the strong downward component during the day.

The long-wave radiation is the heat energy loss by the ocean to the atmosphere or space. However, back radiation from the sea surface may be absorbed by clouds or water vapour and reradiated. The effective back radiation is the net long-wave radiation loss from the sea surface. Since the SST is relatively constant, the controlling factors are the amount of water vapour in the atmosphere and the cloud amount. The solar radiation is also affected by the amount of water vapour and clouds.

The mean fields of the total heat flux (Figs. 9 and 10) show downward heat flux over most of the ocean. Exceptions are found near the Kuroshio and the Gulf Stream and in a region in the central Atlantic. In these regions, the upward surface heat fluxes exceed the solar flux. The largest magnitudes of $10 \, cal \, cm^{-2} \, h^{-1}$ occur near the Gulf Stream and the central part of the two oceans.

The differences in total heat flux between the two model runs are relatively small, which suggests the total heat flux is a result of processes not strongly dependent on the SST. The mean and variation of the total heat flux during this time period are more dependent on the solar radiation than on a time-dependent SST.

STORM TRACK COMPARISON

To investigate the effects of varying SST on the forecast, the storm track comparisons are made only over the oceans. Six maritime cyclones appeared during the forecast period,

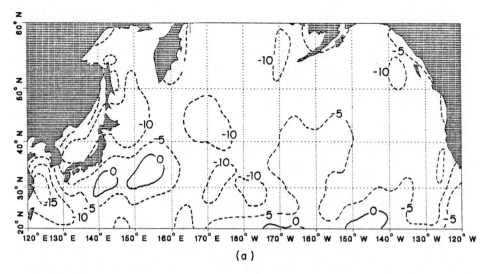

(a)

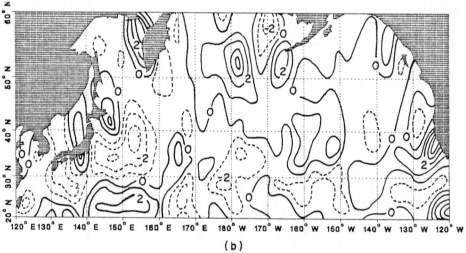

(b)

Fig. 9. As in Fig. 5, except for the total heat flux. (a) Contour interval is 5 cal cm^{-2} h^{-1}; (b) contour interval is 1 cal cm^{-2} h^{-1}.

four over the Pacific Ocean and two over the Atlantic Ocean (Fig. 4). These storms are identified with a letter for the ocean and a number that indicates the sequence in the development. For example, A1 indicates the first storm that developed in the Atlantic Ocean. The storms of most significance were the ones that occurred over the large surface flux change areas and developed late in the forecast period, because these storms will have been exposed to the largest SST changes.

Storm A1 was the first detected by SEIS at 102 h (Fig. 11a). It began as a small low over the middle Atlantic states and traveled to the northeast. When it crossed into the Gulf of St. Lawrence at 162 h, the additional energy available to the storm from the surface heat flux resulted in a rapid deepening of the storm by 10 mb in 24 h. In both

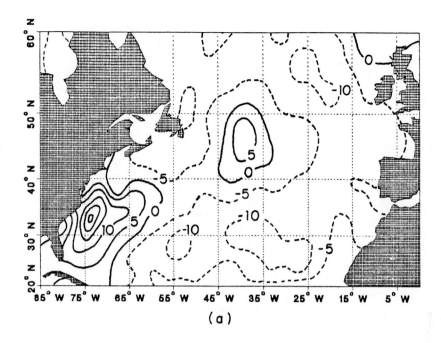

(a)

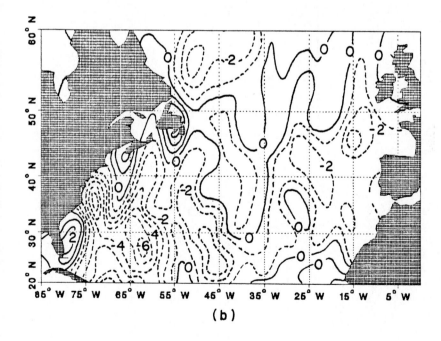

(b)

Fig. 10. As in Fig. 5, except for the total heat flux in the Atlantic Ocean. (a) Contour interval is 5 cal cm^{-2} h^{-1}; (b) contour interval is 1 cal cm^{-2} h^{-1}.

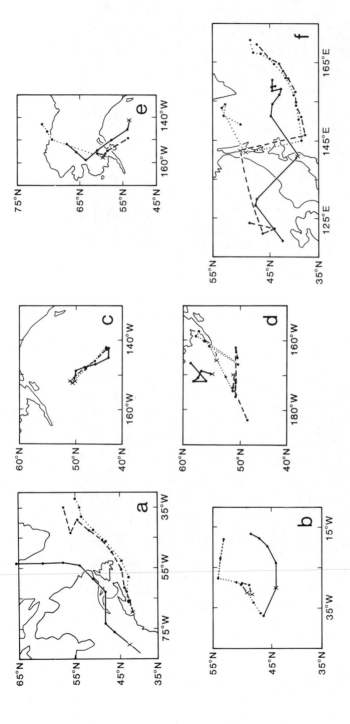

Fig. 11. Storm tracks from analysis (solid), control (dashed) and SST (dotted) runs for (a) Atlantic 1 or A1; (b) A2; (c) P1; (d) P2; (e) P3; and (f) P4. The x in each figure denotes a common time. Thus x corresponds to 126, 138, 6, 66, 90 and 126 in panels (a) through (f), respectively.

model forecasts this storm developed over the ocean just off Cape Cod 24 h later. The model storm tracks parallel the actual track, which indicates that after the initial error in the development position, the storms were correctly moved by the model.

Storm A2 developed at 114 h in the middle of the Atlantic Ocean (Fig. 11b) and tracked eastward before dying at 186 h. The two model runs handled this storm differently. The SST run forecast deepening 12 h earlier than the actual storm, moved the center to the northeast, and maintained it until 186 h. The control run developed this storm 24 h later and forecast it to remain active until 174 h. The SST run was a better forecast in terms of the forecast development time and life cycle (duration) of this storm.

Storm P1 was present in the Gulf of Alaska at the initial time. The storm track (Fig. 11c) was to the southwest before ending at 54 h, while the forecasts terminated the storm at 42 h. The agreement between the two model runs was very good, although both were ahead of the analyzed position. The SST for the control run was lower than the SST run at 30 h. The total heat flux was initially lower for the SST run, but then was almost the same in the two forecasts. The small changes in these inputs resulted in only small changes in the forecast storm. Both the radius and the central pressure were not changed. As expected, the time-dependent SST did not cause a change in the model response early in the model run.

Storm P2 developed after the 66 h into the forecast. At the time SEIS began to tract storm P2, it was already in a mature stage and began to occlude and fill. Storm P2 was followed until 126 h (Fig. 11d). During this time it was almost stationary in the Bering Sea. Both model runs developed this storm 24 h earlier and the storm centers were to the southeast of its observed location. The track for the control run began to turn to the southeast while the track for the SST run moved into the vicinity of the actual storm. The SST run continued the storm to the same time as the analysis, whereas the control run ended the storm 24 h early.

Storm P3 developed at 90 h of the forecast (Fig. 11e). This storm in the Gulf of Alaska moved to the north across Alaska. The two forecast storm tracks seem close to this path, but with very different timing. The control run forecast storm development 24 h early and dissipation 12 h after the storm actually started. The SST run began the storm at the same time as the analysis but had predicted the end of the storm 36 h too soon.

Storm P4 developed at 90 h over Asia and continued until the end of the forecast. The SST model run developed this storm 12 h earlier. In both runs and analysis the storm was very weak and extended over considerable distance. The SEIS program had difficulty fitting a regular pattern to this feature which resulted in the large variability in the early storm tracks (Fig. 11f). The storm began to organize and develop between 126 and 138 h when it first crossed from the Asian continent into the Pacific Ocean. After 138 h, the forecast tracks were to the south and lag the actual storm track, which indicated the model was slow in the movement of this storm. The track for the SST run was marginally closer than the control run to the actual track position over the ocean.

The biggest improvement in the SST run was that the evolution of the storm was closer to the actual life of the storm. This change was most apparent for storm A2, but could be seen for other storms, particularly P2. Storm P2 was the only one located near a center of SST variation and large surface sensible heat flux changes. Storm A1 passed by a region of large surface latent heat flux changes. All other maritime cyclones in this case

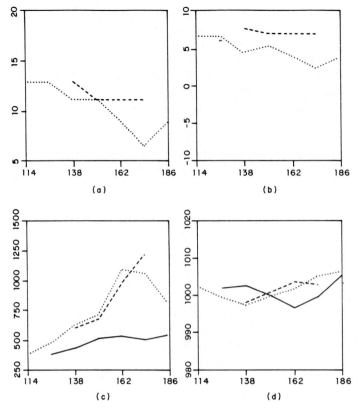

Fig. 12. Storm parameters for the analysis (solid), control (dashed) and SST run (dotted) for the storm A2. (a) SST (°C) averaged over nine grid points following the storm center; (b) heat flux (cal cm^{-2} h^{-1}) averaged over nine grid points following the storm center; (c) radius of storm in km; (d) central pressure in mb.

study occurred in regions of small SST changes and small surface flux changes. For storms A2 and P2, the time-dependent SST results in a better prediction of cyclone evolution. For all other cyclones, the differences in predicted cyclone positions between the two runs are degraded. In general, all differences are less than the departures from the verifying analyses, except for storm P4 at its early stage when the SEIS program had difficulty in tracking the storm. Although storm A1 passed by an area of large latent heat flux changes, the lack of improvement in the SST run indicated that the surface latent heat flux difference was not effectively felt by the storm during the forecast. Storm A2 did not pass over an area with large SST variations or surface flux changes. However, the improved prediction of the storm in the SST run may have been due to the long exposure to·an air stream that had been modified by the SST variations since the storm developed downstream from an area of large surface flux changes.

The latitudinal differences in Fig. 11 indicate the SST run consistently placed the storm farther to the north than in the control run. This may be due to the warming of the sea surface due to the time-dependent SST. The longitude error shows storms in the SST run were generally to the east of the storms in the control run. The combined latitude

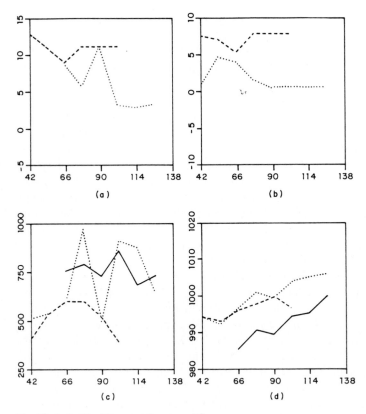

Fig. 13. As in Fig. 12 except for storm P2.

and longitude differences indicate that the storms in the SST run moved faster than in the control run. When the two model-run positions are compared with the analysis, the positions from the control-run forecasts are better.

The comparison of storm radius, storm center pressure, a nine point average of SST and of surface heat flux centered on the storm centers for the improved predictions associated with storms A2 and P2 are shown in Figs. 12 and 13. In general, both the control run and the SST run underforecast the central pressure relative to the analysis, but the error was smaller for the control run than for the SST run. The storm radii have large standard deviations and often depart significantly from the analysis. In almost all cases, the storm radius forecast in the SST run was closer to the analysis than was the control run.

SUMMARY AND CONCLUSIONS

Even after five days, each observed cyclone development was predicted fairly well by NOGAPS. Maximum accumulated 24-h SST changes during this period of spring was about $10°C$. Six maritime cyclones were identified during the ten-day forecast and were objectively tracked using SEIS. All cyclones in this case study occurred in regions of small

SST changes, except for cyclone P2. All cyclones in this study occurred in regions of small surface flux change, except for cyclones P2 and A1. Under these synoptic conditions, the results of varying SST are the following: (1) maximum surface sensible heat flux changes due to time-dependent SST fields were about $1.5 \, \mathrm{cal \, cm^{-2} \, h^{-1}} \cong 30\%$; (2) maximum surface latent heat flux changes due to time-dependent SST fields were about $5 \, \mathrm{cal \, cm^{-2} \, h^{-1}} \cong 30\%$; (3) surface flux changes in coastal areas may have different signs that would be expected from the SST changes; (4) for cyclone P2, the time-dependent SST resulted in a better prediction of cyclone evolution and cyclone size; (5) for cyclone A1, the time-dependent SST did not change the cyclone prediction significantly, which indicated the surface latent heat difference did not affect a deep layer of the atmosphere above it; (6) even though cyclone A2 did not pass by a region of large surface flux changes, the cyclone developed late in the forecast period and was located downstream of surface flux change centers, therefore, the time-dependent SST resulted in a better prediction of the cyclone evolution; and (7) for all other cyclones, the differences in the cyclone predictions between the control and the time-dependent SST integrations were less than their departures from the verifying analyses.

This study was limited in its application and results. The initial conditions were chosen during the spring transition in the ocean, when a decrease in the SST that is expected in the wake of a cyclone might have decreased the amount of cyclogenesis in the SST run. Rather, a general increase in SST was observed and this hypothesis could not be tested. Additionally, SST fields were only available for the first 7.25 days of the model run. This may have reduced the size of the differences between the two model runs.

In this single experiment, even though SST variation during the forecast period may cause 30% changes in surface sensible heat flux and surface latent heat flux, none of the cyclones are directly affected by the centers of largest heat flux changes. P2 is the only cyclone directly experiencing relatively large SST changes, and the SST run predicted a better cyclone evolution and cyclone size. Additional spring cases in which cyclones pass through regions with significant surface flux variations should be studied. Studies are now being made of the impact of SST changes on atmospheric predictions during the autumn, when larger SST variations occur.

ACKNOWLEDGEMENTS

The time and effort that Dr. T.E. Rosmond of the Naval Environmental Prediction Research Facility (NEPRF) always made available to assist in the running of NOGAPS was absolutely essential to this research. Dr. T. Tsui and Mr. P. Harr of NEPRF provided considerable aid, expertise and support in using SEIS. This research was funded by NEPRF under Program Element 62759N, Project WF59-551, "Numerical modelling of unique tropical phenomena". The Fleet Numerical Oceanography Center provided computer resources for running NOGAPS and the Church Computer Center of the Naval Postgraduate School supplied graphics support. The manuscript was expertly prepared by Mrs. M. Marks.

REFERENCES

Arakawa, A. and Lamb, V., 1977. Computational design of the basic dynamical processes of the UCLA general circulation model. Methods Comput. Phys., 17: 173–265.

Arakawa, A. and Schubert, W.H., 1974. Interaction of a cumulus cloud ensemble with the large scale environment, Part I. J. Atmos. Sci., 31: 674–701.

Camp, N.T. and Elsberry, R.L., 1978. Oceanic thermal response to strong atmospheric forcing II. The role of one-dimensional processes. J. Phys. Oceanogr., 8: 215–224.

Clancy, R.M. and Martin, P.J., 1979. The NORDA/FLENUMOCEANCEN thermodynamical ocean prediction system (TOPS): A technical description. NORDA Tech. Note 54, NORDA, NSTL Station MS, 28 pp.

Clancy, R.M. and Pollack, K.D., 1983. A real time synoptic ocean thermal analysis forecast system. Progr. Oceanogr., 12: 383–424.

Deardorff, J.W., 1972. Parameterization of the planetary boundary layer for use in general circulation models. Mon. Weather Rev., 100: 93–106.

Elsberry, R.L. and Camp, N.T., 1978. Oceanic thermal response to strong atmospheric forcing. Part I. Characteristics of forcing events. J. Phys. Oceanogr., 8: 206–214.

Elsberry, R.L. and S.D. Raney, 1978. Sea-surface temperature response to variations in atmospheric wind forcing. J. Phys. Oceanogr., 8: 881–887.

Elsberry, R.L., Haney, R.L., Williams, R.T., Bogart, R.S., Hamilton, H.D. and Hinson, E.F., 1982. Ocean/troposphere/stratosphere forecast systems: a state-of-the-art review. Tech. Rep. CR 8204, Systems and Applied Sciences Corporation, Monterey, Calif., 79 pp.

Harr, P.A., Tsui, T.L. and Brody, L.R., 1983. Model verification statistics tailored for the field forecaster. Preprint volume, Seventh Conference on Numerical Weather Prediction, Omaha, NE, Am. Meteorol. Soc., Boston, Mass., pp. 241–246.

Holl, M.M. and Mendenhall, B.R., 1971. Fields by information bending, sea level pressure version. Tech. Rep. M167, Meteorol. Int. Inc., Monterey, Calif., 71 pp.

Katayama, A., 1972. A simplified scheme for computing radiative transfer in the troposphere. Tech. Rep. No. 6, Dept. of Meteorology, UCLA.

Lord, S.J., 1978. Development and observational verification of a cumulus cloud parameterization. Ph.D. Thesis, Dept. of Atmos. Sci., UCLA.

Mellor, G.L. and T. Yamada, 1974. A hierarchy of turbulence closure models for planetary boundary layers. J. Atmos. Sci., 31: 1791–1806.

Randall, D.A., 1976. The interaction of the planetary boundary layer with large scale circulations. Ph.D. Thesis, Dept. of Atmos. Sci., UCLA.

Rosmond, T.E., 1981. NOGAPS: Navy operational global atmospheric prediction system. Preprint volume, Fifth Conference on Numerical Weather Prediction, Monterey, CA., Am. Meteorol. Soc., Boston, Mass., 7479 pp.

Rosmond, T.E., Weinstein, A.L. and Piacsek, S.A., 1983. Coupled ocean–atmosphere modelling for 3–15 day numerical prediction: A workshop report. NEPRF Tech. Rep. TR 8305, NEPRF, Monterey, Calif., 81 pp.

Schlesinger, M.E., 1976. A numerical simulation of the general circulation of the atmospheric ozone. Ph.D. Thesis, Dept. of Atmos. Sci., UCLA.

Williamson, D.L., 1981. Storm track representation and verification. Tellus, 33: 513–530.

CHAPTER 40

INTERANNUAL CLIMATE VARIABILITY ASSOCIATED WITH THE EL NIÑO/ SOUTHERN OSCILLATION

EUGENE M. RASMUSSON and PHILLIP A. ARKIN

ABSTRACT

The Southern Oscillation is the dominant global climate signal on time scales of a few months to a few years. Its relative contribution to interannual climate variability is largest in the tropical Pacific where at irregular intervals of 2–10 years it enters an extreme "low-index" phase now referred to as an El Niño/Southern Oscillation (ENSO) episode.

The characteristic events and evolution of an ENSO episode, which are intimately tied to the seasonal cycle, are reviewed with emphasis on observational evidence of large-scale sea–air interaction in the tropical Pacific. The 1982–83 ENSO episode was one of the strongest of the past 100 years. Its intensity and peculiar time history have stimulated important new developments in the description and understanding of this phenomenon.

1. INTRODUCTION

On the monthly–seasonal time scale, the atmospheric circulation in the tropics is dominated by quasi-stationary heat sources associated with large-scale circulation features – monsoons, convergence zones, and upper tropospheric anticyclones – whose space–time scales are much larger than those of transient tropical disturbances. Tropical spectra show a significantly higher variance at lower frequencies, i.e. are redder than those from the extratropics, reflecting the dominance of the relatively slow fluctuations of these planetary scale features. Empirical studies and model experiments, e.g. Manabe and Hahn (1981), Charney and Shukla (1981) and Shukla (1984) support the hypothesis that slowly varying boundary conditions e.g. sea-surface temperature (SST), soil moisture and land vegetation are an important factor in this low-frequency variability.

The circulation systems of the atmosphere and the ocean are coupled through interactions at the air–sea interface. However, it is only in the deep tropics, where the time scale of the oceans is relatively short, and that of the atmosphere relatively long, that there is a close dynamic coupling between the large scale circulation features of the two media on time scales shorter than one year. The dynamically distinct region of the ocean within about five degrees of the equator is of particular significance. In addition to the well-known equatorial upwelling associated with the local wind forcing, intense zonal geostrophic currents are easy to generate near the equator and equatorially trapped waves which are possible at all frequencies, can modify the thermocline and SST significantly (Philander, 1979). Such wave-induced changes are not correlated with local wind forcing because the waves are not excited locally, and thus represent an oceanic counterpart to the atmospheric teleconnection phenomenon.

An examination of monthly–interannual variability in the global atmosphere reveals one clearly dominant signal associated with Walker's Southern Oscillation (SO) (Walker, 1924; Walker and Bliss, 1932). Its relative importance is greatest over the tropical Pacific, where its amplitude may equal or exceed that of the annual cycle. The SO signal also extends deep into the extratropics of both hemispheres (Van Loon and Madden, 1981; Van Loon, 1984), where it is associated with major wintertime circulation anomalies at locations as widely separated as Canada and New Zealand.

While atmospheric SO fluctuations have been documented for over 80 years, a fairly clear understanding of their relationship to tropical ocean variability has only developed during the past two decades. Using data from the 1957, 1963 and 1965 Pacific warm episodes, Bjerknes (1966, 1969, 1972) was able to uncover a number of basic connections between the SO, the eastern Pacific El Niño phenomenon, SST and atmospheric circulation anomalies across a broad stretch of the equatorial Pacific, and wintertime circulation anomalies over the North Pacific–North America sector. These and related studies clearly showed that the SO and the equatorial Pacific SST fluctuations associated with El Niño events are part of a single global climate fluctuation, which is now referred to as El Niño/Southern Oscillation (ENSO).

The deluge of empirical studies and model experiments since 1970 has significantly expanded our knowledge and understanding of ENSO. In particular, each additional ENSO episode since that time has resulted in important new insights. For example, analysis of data from the major 1972 episode led to the identification of remote ocean forcing as an important process in the development of eastern equatorial Pacific SST anomalies (Wyrtki, 1975). Availability of global outgoing long-wave radiation data from satellites during 1976–77 resulted in the first comprehensive description of the Pacific basin-wide rainfall anomaly pattern during an ENSO episode (Heddinghaus and Krueger, 1981; Leibmann and Hartmann, 1982). The accumulation of data from ENSO episodes during the period 1951–79 allowed construction of composite ENSO surface anomaly fields over the Pacific (Rasmusson and Carpenter, 1982; Van Loon, 1984), and statistical descriptions of the ENSO signal in the upper troposphere (Horel and Wallace, 1981; Arkin, 1982). The 1982–83 ENSO episode provided another natural experiment for detailed study. It is by far the best documented to date and because of its remarkable intensity and unusual timing has already provided important new insights into the fundamental nature of the phenomenon (Rasmusson and Wallace, 1983; Cane, 1983; Philander, 1983; Gill and Rasmusson, 1983; Tropical Ocean–Atmosphere Newsletter, 1983a, 1983b; Philander and Rasmusson, 1984).

In this paper we describe some of the salient aspects of the ENSO phenomenon in the atmosphere. The paper is not intended to be an exhaustive review of ENSO in general, or the 1982/83 episode in particular as there is already an extensive literature on both topics (see, for example, previously cited references and the earlier ENSO-related references in Rasmusson and Carpenter, 1982).

This paper is focused on three aspects of the phenomenon: (1) the global nature of ENSO in both ocean and atmosphere; (2) the central role of the annual cycle in the timing and evolution of the ENSO episode; and (3) the nature of the anomalous atmospheric forcing in the tropical Pacific during the 1982/83 episode. Particular attention is paid to questions which the new results raise concerning the conceptual framework which has developed from previous empirical studies and simple model experiments.

Data and analyses are discussed briefly in Section 2. Certain aspects of the tropical annual cycle which appear to be particularly relevant to ENSO are discussed in Section 3, and aspects of ENSO in the tropical Pacific are reviewed in Section 4. In Section 5 the focus is on the description of the 1982/83 episode, with emphasis on the evolution of and relationship between the SST, convection, and 850 and 200 mb circulation anomalies. Section 6 consists of some further observations and concluding remarks.

2. DATA AND ANALYSES

The data and analyses used in this study were derived from the Climate Diagnostics Data Base (CDDB) and supporting climatological summaries of the NOAA Climate Analysis Center (CAC). The CDDB summaries were, in turn, derived from NOAA operational data and analyses. The climatology from which the SST anomalies were derived was taken from Reynolds (1982). The monthly and seasonal SST analyses for the 1982/83 episode were prepared at the CAC from ship and buoy observations using the analysis scheme described by Reynolds and Gemmill (1984). Tropical Pacific SST fields for the period prior to 1977 were derived from the data of Rasmusson and Carpenter (1982).

The outgoing long-wave radiation (OLR) data, which span the period June 1974 to the present, are derived from a variety of window channel measurements from U.S. polar orbiter satellites, using a non-linear regression scheme based on model calculations (Krueger and Gruber, 1984). The data have been subjected to a retrospective correction to remove discontinuities introduced by changes in instrumentation. The day and night values have been averaged to remove the influence of the diurnal cycle insofar as possible. These data are used to infer changes in the distribution and intensity of tropical convective rainfall. This approach is based on the hypothesis that, in the tropics, decreased OLR corresponds to increased coverage of cold (high) cloud tops, which indicates increased convective rainfall. The results of Arkin (1979), Heddinghaus and Krueger (1981), Leibmann and Hartmann (1982) and Lau (1985) support these assumptions. The OLR values are used as an index of the spatial and temporal variability of tropical condensation heating.

The wind fields at 850 and 200 mb are derived from the U.S. National Meteorological Center (NMC) operational global optimum interpolation (OI) analyses, using as a first guess the previous 12 h forecast. Analyses were performed twice daily, about 11 h after the 0000 and 1200 GMT observation times on a 2.5° latitude–longitude grid. The OI analyses are modified by an adiabatic non-linear normal mode initialization prior to use as initial conditions for the NMC forecast cycle. The characteristics of this analysis scheme, which became operational in September 1979, have been described by Bergman (1978) and Kistler and Parrish (1982).

The time-averaged stream function (ψ) defining the rotational part of the flow, and velocity potential (χ), defining the divergent part were computed by relaxation of the vorticity and divergence of the initialized OI analyses. The iterative technique of Dey and Brown (1976), which requires boundary conditions only at the poles, was used for the relaxation. The relationship among these fields is given by:

$$V = \nabla\chi + k\nabla\psi = V_d + V_r$$

One should not confuse the planetary scale pattern of the χ field with the field of

divergence $\nabla^2\chi$, which has a much smaller spatial scale, and is more appropriately compared with the tropical OLR patterns. In terms of divergence, the χ field is an integral
quantity, i.e. the average divergence over an area enclosed by a χ contour is the line
integral of V_d around the contour divided by the enclosed area. The divergent wind field
derived from the χ field is useful in describing the large-scale direct thermal circulations
associated with the non-uniform surface boundary conditions in the tropics.

The degree to which the OI/normal mode initialized analysis captures the divergent
component of flow is still a matter of conjecture. The most significant effect of the
normal mode initialization appears to be a smoothing of smaller scale features of the
divergent flow and a reduction in the amplitude of the larger scale features (Rosen and
Salstein, 1985). Many of the "climatological" features exhibited by the five-year base
period mean fields as well as the major year to year variations appear quite plausible
when viewed in the context of the OLR fields. In our judgment the most suspect tropical
area is the African–southern Indian Ocean region, where data have been inadequate in
the past.

In summary, the smaller scale features of the χ field are likely to be attenuated and the
intensity of the large-scale divergent flow underestimated, but we believe that the χ
analyses provide a description of the planetary scale features of the 200 mb divergent
flow which is adequate within the scope of the subsequent discussions.

3. MEAN ANNUAL CYCLE

Over portions of the equatorial Pacific the ENSO signal equals or exceeds the amplitude of the mean annual cycle. For the most part, however, ENSO represents a relatively
modest perturbation of the mean annual cycle. Even in those regions where the signal is
relatively strong, the evolution of anomalies over the 18–24-month lifetime of a typical
episode shows a strong tendency for phase locking with the annual cycle. Indeed, one
way to interpret the appearance, evolution and decay of ENSO anomalies over the various
regions of the world is in terms of a phase shift or amplitude modulation of some feature
of the annual cycle.

Figure 1 shows the mean SST distribution for the two solstice seasons December–
February (DJF) and June–August (JJA). The longitudinal variations in the tropics are
striking, particularly the temperature difference between the upwelling region of the
eastern equatorial Pacific and the huge "warm pool" which overlies the west Pacific–
eastern Indian Ocean monsoon region.

Figure 2 shows the OLR distribution for the comparable seasons. Areas in the tropics
with values less than $240\,W\,m^{-2}$ are shaded, indicating regions of heavy convective
rainfall and mean upward vertical motion. Two of these are located over summer
hemisphere continental regions (Africa, South America/Central America and adjacent
waters). The third and most extensive area of tropical convection is centered over the
warm waters and adjacent land areas of the Indonesian–Australian–East Asian monsoon
region, with eastward extensions along the warm SST axes marking the Intertropical
Convergence Zone (ITCZ) north of the equator, and the South Pacific Convergence
Zone (SPCZ) in the southwest Pacific.

The location and seasonal migration of the oceanic convective zones and regions of

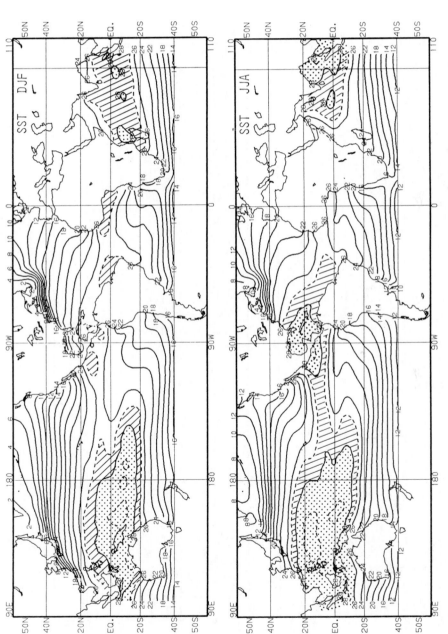

Fig. 1. Mean SST for December–February (upper) and June–August (lower). Areas where SST is greater than 28°C are stippled; areas where SST is between 27° and 28°C are hatched. Contour interval (solid lines) 2°C.

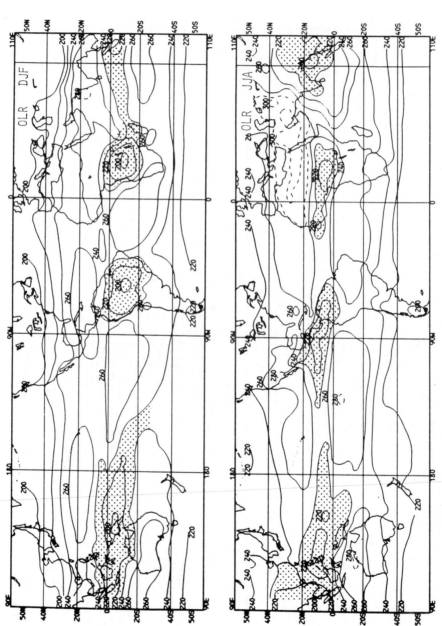

Fig. 2. Mean outgoing long-wave radiation (OLR) for December–February (upper) and June–August (lower). Areas where OLR is less than 240 W m^{-2} are stippled. Contour interval 20 W m^{-2}.

high SST largely coincide. Throughout the year the tropical oceanic regions of OLR less than 240 W m^{-2} remain almost entirely within areas enclosed by the 27°C SST isotherm, and primarily within the 28°C isotherm (compare shaded or hatched areas on Figs. 1 and 2). The modeling of the interactions associated with the annual migration of the convection and SST warm pool in the Indian Ocean—eastern Pacific region appears to be an important step in the modeling and understanding of ENSO.

West of the dateline, equatorial Pacific SST anomalies are usually less than 1°C and year-to-year changes rarely exceed 2°C, but since the zonal gradient is small, this is sufficient to produce interannual east—west excursions of several tens of degrees longitude in the 28°C SST isotherm (Fig. 4). These SST anomalies are associated with comparable zonal shifts in the West Pacific convection as will be described in Section 5.

The 200 mb divergent flow provides a picture of the upper branches of the planetary scale thermal circulations. Figure 3 shows the mean χ (5 yrs) and ψ (16 yrs) fields as derived from the NMC analyses, for DJF and JJA. The divergent flow is directed normal to the contours of χ from low values toward high, while the rotational circulation is parallel to the ψ contours, clockwise about maxima and counterclockwise about minima. Both components of the flow show large longitudinal asymmetries related to those in the OLR. In general, the χ charts show a planetary-scale 200 mb cross-equatorial divergent flow from the convective regions of the tropics to the radiative sink regions of the subtropics, and flow to the relatively cold water regions of the eastern tropical Pacific. The ψ fields show anticyclonic circulations in each hemisphere in the longitudes of the maxima in condensation heating.

Regions of tropical convection are characterized by low-level convergence and upper troposphere divergence, with maximum vertical motion and latent heat release in the middle troposphere. Temperatures are relatively high throughout the troposphere. The climatological convective regions are capped by mean upper level anticyclonic flow. The circulation asymmetries arising from the longitudinal asymmetries in latent heating were studied by Webster (1972) with the aid of a relatively simple model.

The strongest upper level anticyclones are part of the Australasian monsoon circulation system. During JJA, the 200 mb anticyclone centered over southern Asia dominates the Northern Hemisphere circulation over approximately 180° of longitude. The tropical easterly jet (TEJ; Koteswaram, 1958; Flohn, 1964), with its core around 150 mb, lies on the southern flank of the anticyclone. On the corresponding χ chart, the upper level outflow over the Pacific region of high SST lies in the entrance region of the TEJ, while the Southern Hemisphere region of inflow centered near 20°S is in the longitudes of the TEJ exit region. The largest values of V_d are near the axis of the TEJ, which reaches maximum values near 10°N, 75°E.

There is a companion 200 mb anticyclonic circulation in the Southern Hemisphere monsoon region. During the Southern Hemisphere summer the anticyclones are centered further east, over the Pacific—Australian region where the upper-level inter-hemispheric monsoon circulation is most vigorous. The major features of the DJF 200 mb χ field are associated with the outflow from the convective region extending from Indonesia southeastward along the SPCZ, and the upper level inflow and subsidence over the cold land surfaces of southern and eastern Asia. Comparison of JJA and DJF χ fields shows a shift in direction of the major cross-equatorial divergent flow at 200 mb which in both seasons is the reverse of the low-level cross-equatorial monsoon flow. This seasonal

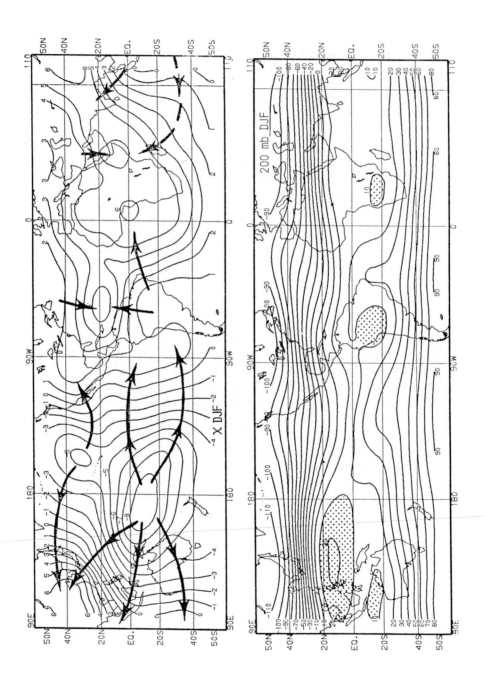

705

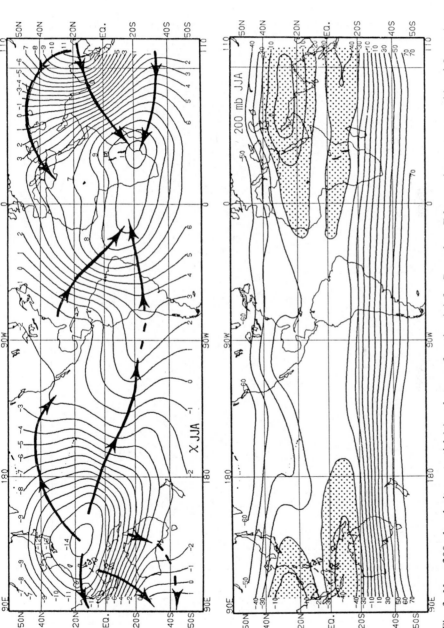

Fig. 3. Mean 200 mb velocity potential (χ) and stream function (ψ) for December–February (upper two panels) and June–August (lower two panels). χ contour interval is 10×10^6 m² s⁻¹. ψ contour interval is 10^6 m² s⁻¹. χ mean is for 5 yrs; ψ mean is for 16 yrs.

706

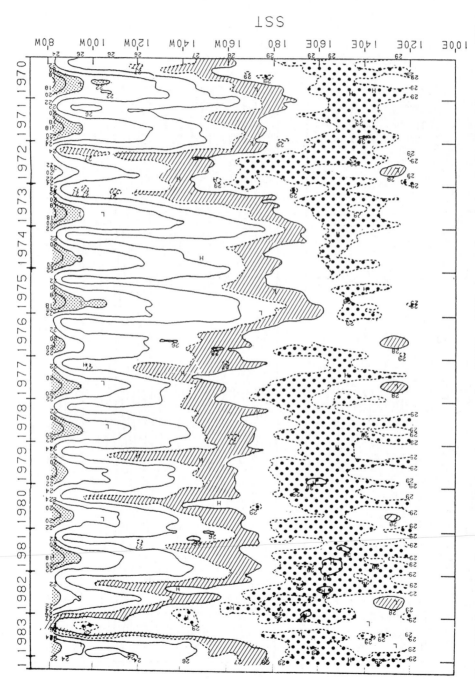

Fig. 4. SST time section. Section extends along the equator from 120°E to 95°W, then southeastward to the South American coast at 8°S. Contour interval (solid lines) 2°C.

change is consistent with and reflected in the seasonal cycle of the zonally averaged Hadley circulations (Oort and Rasmusson, 1971).

The upper level divergent flow from the convective region in the West Pacific to the subsidence region over the cold waters of the eastern tropical Pacific is of particular relevance to ENSO, since it relates to the much discussed Walker Circulation (Bjerknes, 1969). As previously noted by Krishnamurti (1971), this east–west divergent flow is of broad latitudinal extent, rather than being narrowly confined to the equatorial zone, where the dynamics described by Bjerknes (1969) are valid. Furthermore, except for the western Pacific during DJF, the divergent component of the flow is typically a small fraction of the total flow. The NMC analyses rarely show 200 mb monthly values of V_d which exceed $5 \, \mathrm{m \, s^{-1}}$, although actual values are probably somewhat larger as previously discussed.

These data clearly show the important longitudinal asymmetries of the planetary scale mean flow and their relationship to the asymmetries in surface temperature and precipitation. They also show pronounced differences between the hemispheres. Despite the approximate symmetry of the 200 mb anticyclonic circulations in the longitudes of the maxima in convection, it is clear from the OLR and χ fields that the summer hemisphere anticyclone is characterized by extensive cloudiness and associated upper level outflow which contributes to local anticyclonic vorticity generation. In contrast, relatively high values of OLR are observed over most of the area of the winter hemisphere anticyclonic circulation, and the upper level divergent flow is more difficult to characterize. The maintenance of anticyclonic vorticity in the two hemispheres appears to involve somewhat different processes.

4. EVOLUTION OF ENSO

The effectiveness of an anomalous SST boundary forcing in changing the atmospheric circulation hinges on its ability to produce a deep heat source, and the propagation of this influence away from the source (Webster, 1981; Shukla, 1984). Boundary forcing from SST's is generally not sufficient in itself to produce these changes. The effectiveness of an anomalous forcing depends on the existence of a favorable dynamical environment through which the forcing can be transformed into a three-dimensional heat source. For example, a positive SST anomaly in a region of weak low-level convergence and upward motion will further strengthen the atmospheric heat source through enhanced vertical motion and convection, until a new equilibrium state is reached. In a region of low-level divergence, the anomaly may be of insufficient strength to reverse the downward motion and initiate the required positive feedback process (Webster, 1981). Therefore, the effect of an SST anomaly may be very different depending on geographical location and the time of year, as well as on the amplitude of the anomaly.

The major sea–air forcing associated with ENSO takes place in the equatorial Pacific. Here, as elsewhere in the tropics, the low-level convergence regions closely coincide with the regions of highest SST (Figs. 1 and 2). A relatively small SST anomaly in a climatologically favorable warm region, such as the central equatorial Pacific, may well produce a more pronounced atmospheric response than a much larger anomaly in a dynamically unfavorable cold region. For example, although the largest ENSO SST anomalies occur

708

over the relatively cold waters of the eastern equatorial Pacific, the major atmospheric effects are associated with eastward and equatorward shifts of the convective regions located on the climatologically warm fringes of the anomaly area.

The pattern of interannual SST variability in the equatorial Pacific and its relationship to the annual cycle is illustrated in Fig. 4, which shows an SST time section spanning the

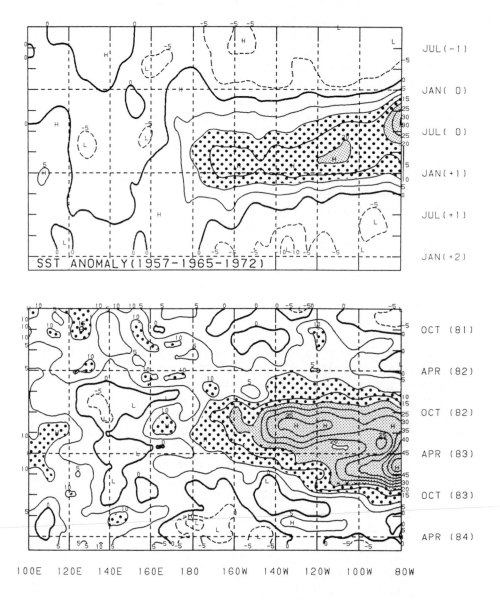

Fig. 5. SST anomaly time sections (along the equator from 120°E to 95°W, then southeastward, to the South American coast at 8°S. Upper: Composite for 1957, 1965, 1972 episodes; lower: 1982/83 episode. Contour interval 0.5°C × 10.

period 1970–83. A pronounced seasonal cycle appears in the eastern Pacific, distorted but rarely overwhelmed by ENSO events. The annual cycle becomes progressively weaker towards the west (Horel, 1982), and without multi-year averaging is almost indistinguishable west of 160°W. From there to Indonesia the fluctuations are largely interannual in nature, reflecting changes in the location, extent and intensity of the West Pacific warm pool.

The major longitudinal excursions of the warm pool, as delineated by the shaded 27°–28°C isotherm band on Fig. 4, are associated with major swings in the Southern Oscillation (Rasmusson and Wallace, 1983). The pronounced eastward excursions of this isotherm band correspond to ENSO episodes and associated El Niño conditions in the eastern equatorial Pacific.

There are two time scales associated with the ENSO phenomenon, the recurrence interval and the lifetime of an individual event. The recurrence interval is irregular, generally ranging between two and seven years. In contrast, individual episodes follow a rather similar evolution over a period of 18–24 months, depending on one's choice of starting and ending conditions. The evolution of the SST anomaly field during a typical episode has been illustrated and discussed by Rasmusson and Carpenter (1982). Comparisons of the unusual but not unprecedented evolution of the 1982/83 episode with the more typical pattern have been made by Rasmusson and Wallace (1983), Cane (1983), Philander (1983) and others. During most episodes, El Niño conditions develop near the South American coast early in the year with SST anomalies typically peaking during April–June (Fig. 5). The most rapid developments in the central equatorial Pacific usually take place during mid-year, when SST anomalies are diminishing in the eastern Pacific. The large-scale anomaly pattern peaks around the end of the year, during the "mature" stage of the episode, when El Niño conditions may briefly reappear in the eastern Pacific. The anomalies then enter a period of decay which usually spans several months.

During the 1982–83 episode, the first major developments took place during mid-year in the western and central Pacific, giving the appearance that the early-year El Niño warming phase in the eastern Pacific had been bypassed (Fig. 5). El Niño conditions did appear in the eastern Pacific late in the year, consistent with a mature phase pattern. Again contrary to the normal sequence of events, this was followed during the first half of 1983 by the development of intense El Niño conditions east of 120°W, at the same time that the Pacific-scale SST anomaly pattern was in decline. All in all, the evolution of the SST anomaly field had the appearance of a reversal of the normal sequence of events in the eastern Pacific, with the El Niño warming coming at the end, rather than near the beginning of the episode. Locally, however, the anomalies still showed a rather typical phasing with the annual cycle. In the atmosphere, the reversal of the anomaly sequence appeared to be associated with an extension of the pronounced DJF mature phase anomaly pattern into the next season.

5. 1982/83 ENSO ANOMALIES

The relatively long lifetime and consistent pattern of evolution of an ENSO episode is viewed as strong evidence of the key role of the tropical Pacific in the evolution of the

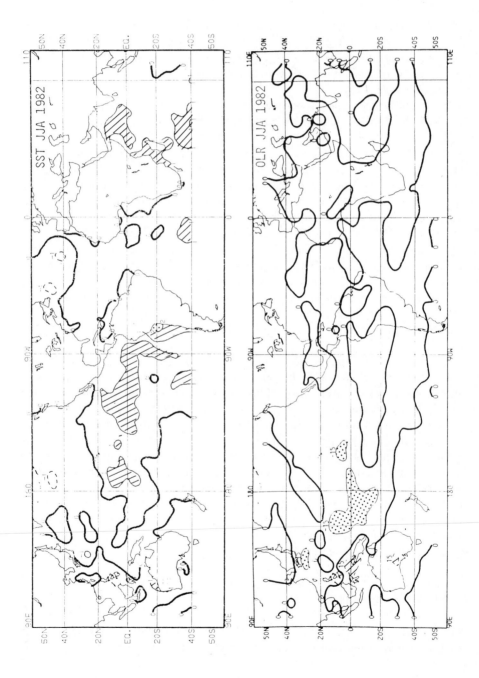

711

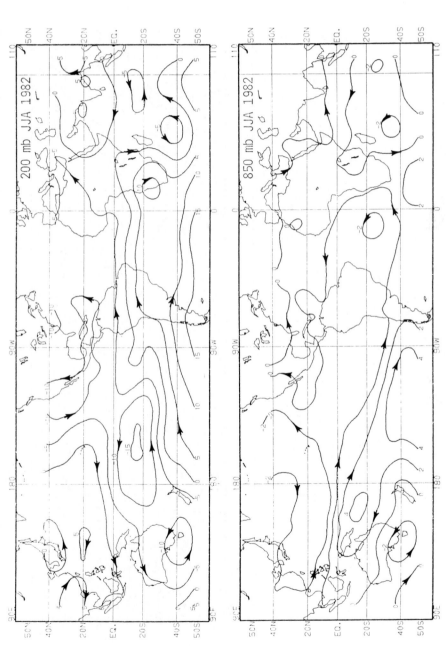

Fig. 6. Three-month averaged anomalies for June–August 1982. SST: Contour interval 1°C; areas with anomalies greater than 1°C hatched. OLR: Contour interval 15 W m^{-2}. Areas less than −15 W m^{-2} stippled; areas greater than +15 W m^{-2} hatched. 200 mb stream function; contour interval 5 × 10^6 m^2 s^{-1}. 850 mb stream function; contour interval 2 × 10^6 m^2 s^{-1}.

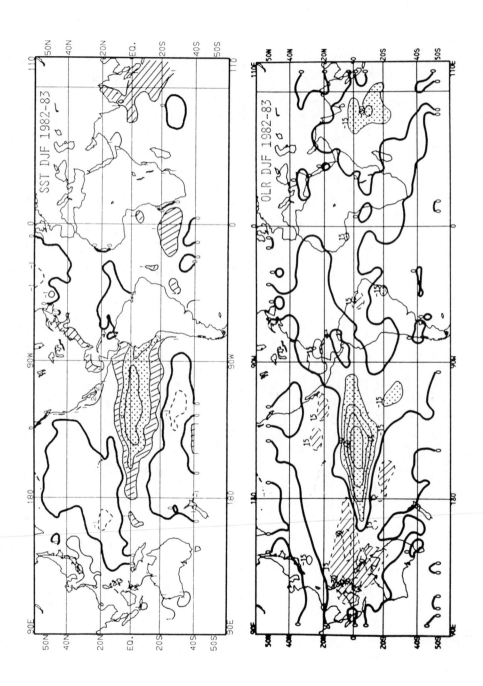

713

Fig. 7. Same for December–February 1982/83.

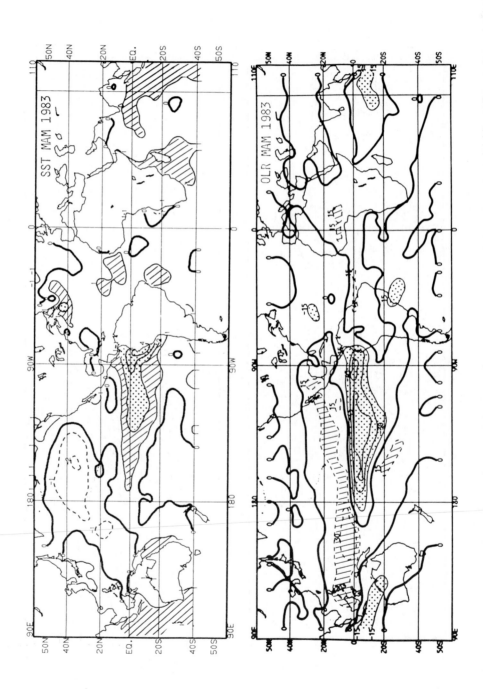

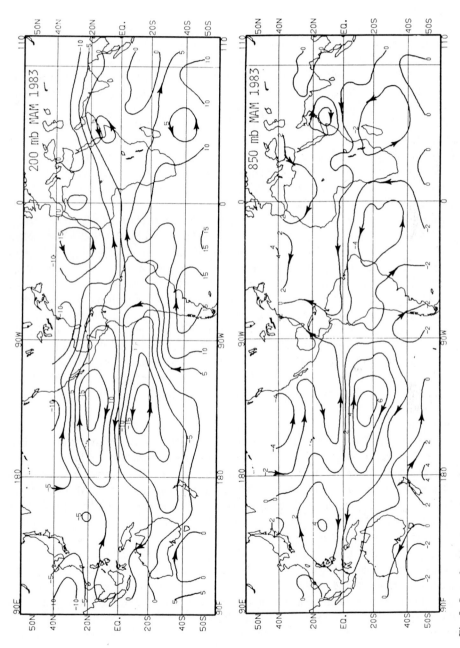

Fig. 8. Same for March–May 1983.

phenomenon. The sequence of events during the 1982/83 episode strongly support this view. Perhaps the most striking evidence was the simultaneous eastward migration along the equator of the west Pacific warm water as tracked by the 28°C SST isotherm, the region of enhanced rainfall, as tracked by the negative OLR anomalies, and the low-level westerly and upper-level easterly wind anomalies (Quiroz, 1983; Gill and Rasmusson, 1983; Philander and Rasmusson, 1984).

This steady eastward migration, across more than 100 degrees of longitude over a period of almost one year, is truly a remarkable relationship between ocean and atmosphere parameters, and provides compelling evidence of the close coupling of the two media. Some features of the evolution appear to be explainable in terms of ocean advection (Gill, 1983; Gill and Rasmusson, 1983), and the pattern of anomalies in the equatorial plane is broadly consistent with results from relatively simple linear models of the atmospheric response to thermal forcing (Gill, 1980).

While revealing a salient feature of the ENSO 1982/83 episode, the equatorial time sections fail to display important three-dimensional aspects of the evolution. For example, they are inappropriate for examining the evolution of rainfall anomalies associated with the equatorward displacement of the ITCZ. More generally, they fail to bring out the marked asymmetries about the equator of the anomalous forcing and atmospheric circulation anomalies.

Figures 6–8 show the three-month averaged anomaly fields for SST, OLR, and the 850 and 200 mb streamfunction for the seasons JJA 1982, DJF 1982/83 and MAM 1983 respectively. The well-documented pattern of the eastward extension of the West Pacific monsoon convection and the equatorward displacement of the ITCZ is apparent in the evolution of the OLR anomalies. The ITCZ-related negative OLR anomalies appear as an east–west band north of or near the equator, while the anomaly associated with the eastward displacement of the SPCZ and monsoon convection is centered south of the equator. These two features are clearly distinguishable as separate entities on the JJA chart, but later in the episode the two patterns tend to merge with further eastward migration of the SPCZ. The DJF pattern is dominated by the eastward moving OLR minimum. The resulting Southern Hemisphere bias is accentuated during MAM as a pronounced band of positive OLR anomalies spreads across the Northern Hemisphere subtropics.

Much of the analysis of the 1982/83 episode, and most of the atmospheric GCM modelling to date has focused on conditions during DJF. Particular attention has been given to the relationship between the anomalous atmospheric forcing and the upper-level anticyclonic couplet which straddles the equator directly north and south of the OLR anomaly minimum. The couplet pattern became reasonably well organized during SON, 1982, although an earlier stage of development appears on the JJA 200 mb chart. At this time an anomalous ridging developed in the 200 mb flow field over the western North Pacific, part of an anticyclonic anomaly which extends westward to Arabia. In this connection, note that the seasonally averaged circulation anomalies in the vicinity of the TEJ were rather unimpressive during this episode. The 200 mb anomaly pattern reflects a southward and eastward expansion of the Northern Hemisphere monsoon anticyclone, which extends eastward into the Pacific along about 20°N, north of the region of enhanced convection in the western Pacific.

A pronounced 200 mb anticyclonic anomaly also appears in the Southern Hemisphere, but it is centered at 15°S, 150°W, far to the east of the Northern Hemisphere ridging, in the longitudes of enhanced convection north of the equator. The anomalous ridging extends southwestward across Australia, where it reflects an equivalent barotropic blocking ridge that persisted throughout most of the three-month period. The ridging in the westerlies was strongly reflected in the 850 mb anomaly pattern, as was the pronounced downstream anomalous troughing in the southwest Pacific, and the entire ridge-trough anomaly pattern in the lower troposphere was associated with the sharp negative swing in the Tahiti minus Darwin SO Index. The anomalous southerly flow on the east side of the ridge extended to the equator where it turned eastward, becoming the equatorial westerly anomalies associated with the enhanced convection (Harrison, 1984). The region of anomalous convection directly to the south of the westerly anomalies was associated with a local troughing in the low-level anomalous flow, but except for these features the 850 mb tropical wind anomalies were not particularly noteworthy, and were, in general, difficult to relate to the 200 mb anomaly pattern.

Distinct 200 mb anticyclonic anomaly centers developed in SON in both hemispheres, directly north and south of the OLR anomaly minimum. Both circulation anomalies were centered around 20° latitude but the Southern Hemisphere anomalous anticyclone remained the more intense, and extended eastward to the region where the anomaly center was located in JJA.

The well-known DJF 200 mb pattern (Fig. 7) consists of anticyclonic anomalies of roughly equal strength, even though the anomalous forcing is clearly biased towards the Southern Hemisphere. The circulation anomalies reflect eastward extensions of the climatological monsoon highs. The anomalous westerlies on the poleward flanks of the anticyclones reflect an associated eastward extension of the strong north–south gradient on the poleward side of the North Pacific high associated with the West Pacific jetstream. Thus the eastward extension of the West Pacific jetstream, which is related to wintertime climate anomalies over the northwest Pacific and North America, is clearly tied to the eastward migration of the low-latitude forcing through the associated 200 mb anticyclonic anomaly. The 1982/83 situation was extreme, both in terms of the intensity of the anticyclonic anomaly, and the eastward displacement of the convection.

Contrary to the situation at 200 mb, the low-level circulation pattern reflects the Southern Hemisphere forcing bias. The Southern Hemisphere cyclonic anomaly at 850 mb is by far the stronger of the two which straddle the anomalous convection, and is centered directly below the 200 mb anticyclonic anomaly. In contrast, the weaker Northern Hemisphere circulation is centered about 10° equatorward of the upper-level anticyclonic anomaly. The MAM pattern of anomalies shows the same qualitative relationships (Fig. 8). Beginning in June, 1983, SST in the eastern equatorial Pacific began a rapid fall due to the combined effects of decreasing anomalies and the normal seasonal cooling. The atmospheric circulation anomalies decayed even more rapidly. The eastward shift of the tropical Pacific convection was accompanied by diminished convection over Indonesia, Australia, and the West Pacific, the development over the western equatorial Pacific of low-level easterly anomalies, upper-level westerly anomalies and a weak upper-level cyclonic couplet. This pattern persisted through DJF. However, by this stage of the episode, the upper-level anomalies were only part of a belt of equatorial westerly anomalies which extended around most of the world. The low-level easterly

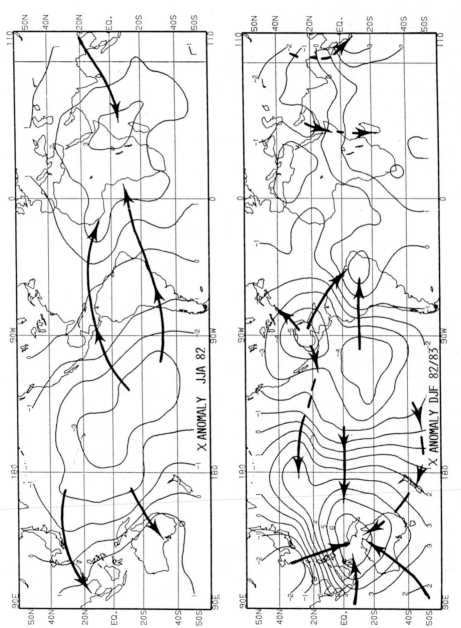

Fig. 9. 200 mb velocity potential (χ) anomaly for June−August 1982 (upper) and December−February 1982/83. Contour interval 10^6 m^2 s^{-1}.

anomalies continued over Indonesia during MAM, but the upper-level cyclonic couplet disappeared as rainfall in the area tended towards more normal values.

The anomalies in the divergent component, like the anomalies in the rotational component discussed above, were surprisingly large and resulted in the eastward migration of the planetary-scale features of the flow. The 200 mb χ anomaly field was primarily a wave number-one feature covering the Indian Ocean–Pacific region and exhibiting a pattern reminiscent of the SO surface pressure seesaw (Fig. 9). Like the seesaw, and the equatorial Pacific anomalous forcing, the 200 mb χ anomaly field exhibited a Southern Hemisphere bias throughout the lifetime of the episode.

The 200 mb divergent flow which developed during the early stages of the event clearly reflected the eastward shift of the West Pacific convection and the associated heavy rainfall over the central equatorial Pacific and drought conditions over Indonesia, eastern Australia and the southwest Pacific. An anomalous divergent flow extended from a 200 mb outflow region over the Central and South-central Pacific westward to Australia and Indonesia and eastward to North and South America and the Atlantic.

The wave number-one pattern migrated eastward, so that by DJF the major features over the Pacific were shifted far east of their normal climatological positions (Fig. 10). The anomaly χ field had intensified to such a degree that it was now comparable in magnitude to the five-year mean field. In addition to the Pacific features, the anomaly map also shows an inflow region over northern South America consistent with evidence of subnormal rainfall over the Amazon Basin derived from the OLR positive anomalies over the area (Fig. 7).

The DJF χ anomaly pattern appears to reflect several regional rainfall anomalies in addition to the major anomalies over and in the neighbourhood of the Pacific Basin. These include dry conditions over southern India and Southwest Africa, and heavy rainfall over the southeastern United States, Cuba and the Gulf of Mexico. By JJA 83 the 200 mb χ field had returned to near normal, reflecting the decay of the ENSO episode.

6. CONCLUDING REMARKS

In the atmosphere, the year-to-year variability of seasonal means is determined by the combined effects of internal atmospheric dynamics and the more slowly varying surface boundary conditions. Since the oceans constitute 71 percent of the Earth's surface, the role of upper-ocean thermal anomalies in short-term climate variability is an important issue for climate prediction for seasons of longer (Rasmusson, 1982). The ENSO phenomenon now occupies a central role in research on short-term climate variability. On a purely scientific level, this elegant system of ocean/atmosphere interactions offers an opportunity for significant expansion of our understanding of climate interactions in general. At the applied level, it offers a potentially important element of predictability on the seasonal–interannual time scale.

We have focused our analysis on the evolution of the atmospheric circulation anomalies most directly associated with the anomalous SST and forcing in the tropical Pacific. We have attempted to place these anomalies in the context of the mean annual

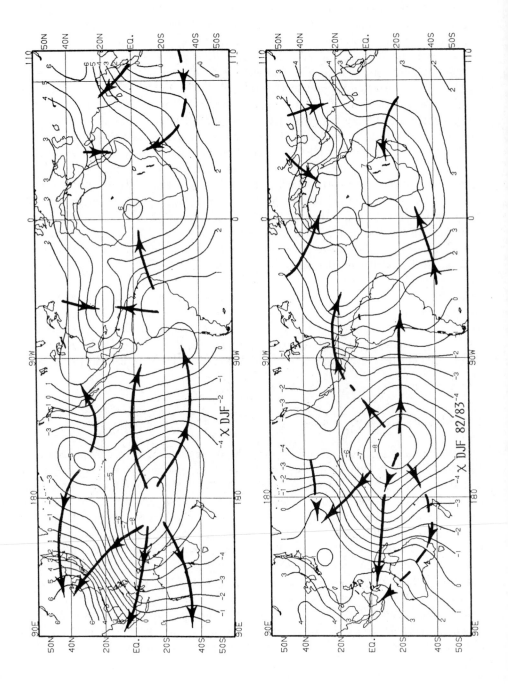

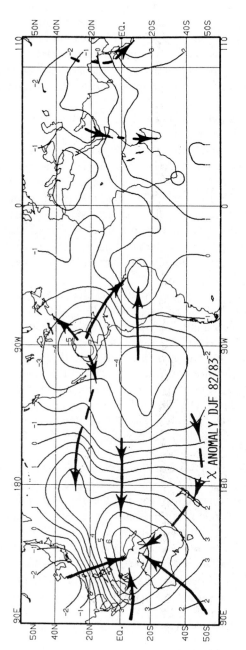

Fig. 10. 200 mb velocity potential (χ). Upper: December–February five-year mean. Middle: December–February 1982/83. Lower: December–February 1982/83 anomaly. Contour interval 10^6 m^2 s^{-1}.

cycle, which appears to be a dominant factor in the phásing and evolution of an ENSO episode. A better understanding of the annual cycle, particularly the ocean–atmosphere processes which link such features as the seasonal migration of the high SST/convective region of the western Pacific, is required if we are to ultimately understand the year-to-year variability associated with ENSO.

The region of enhanced convection in the tropical Pacific, as defined by the area of negative OLR anomalies, generally corresponds to the area where SST anomalies are positive and SST is greater than $27°–28°C$. During the 1982/83 episode this region showed a pronounced Southern Hemisphere bias. The low-level circulation anomalies also reflected this asymmetry with 850 mb cyclonic anomaly centers poleward of the convection in each hemisphere, but more nearly symmetric about the latitude of maximum convection than about the equator, and with the Southern Hemisphere cyclonic anomaly the stronger of the two.

A qualitatively different anomaly pattern developed at 200 mb. Anomalous anticyclonic circulations of roughly equal strength developed between 15–20 degrees latitude in each hemisphere directly north and south of the anomalous forcing. The Southern Hemisphere anticyclonic anomaly was directly over the low-level cylonic anomaly, but the comparable Northern Hemisphere anomaly was centered north of the region of enhanced convection, over a low-level anomaly ridge. Thus the striking reversal of the anomalous circulation between the upper and lower troposphere observed throughout the episode on and south of the equator was not present a few degrees north of the equator. The implication of these asymmetries in terms of the vorticity dynamics and the validity of simple models of the atmospheric response to thermal forcing has been discussed by Sardeshmukh and Hoskins (1985).

The discussion of SST anomalies has centered on the evolution of the pattern in the primary region of sea–air interaction in the tropical Pacific. During 1982/83, the global SST anomaly pattern evolved in a manner characteristic of previous ENSO episodes. By the mature phase (Figs. 7 and 8), the dominant positive anomaly area in the eastern equatorial Pacific was flanked by weaker negative anomalies in the North and South Pacific and positive anomalies extending poleward along the west coast of the Americas. The evolution of this North Pacific anomaly pattern has been analyzed and described by Reynolds and Rasmusson (1982). Over the tropics the SST anomalies are generally positive during the mature phase of ENSO, as are surface temperatures over land (Rasmusson et al., 1983). This feature is clearly reflected in the composite zonal averages of Pan and Oort (1983), and in Figs. 7 and 8.

This global pattern of SST anomalies as well as the rainfall anomalies over the Amazon Basin and other regions remote from the equatorial Pacific raise complex questions concerning secondary effects and feedbacks, since at least some of these anomalies represent significant changes in atmospheric forcing and possibly in surface boundary conditions. Are they merely passive reflections of atmospheric anomalies remotely forced from the tropical Pacific, or are they active ENSO elements which lead to important downstream effects, and exert a significant influence on the evolution of the episode in their own right? These questions clearly reflect the need for a global perspective of the phenomenon, and may be difficult to answer without the aid of experiments with fully coupled global atmosphere/ocean/land surface GCM's.

Model experiments with atmospheric GCM's, using specified SST anomalies in the equatorial Pacific, have reached a high level of sophistication as displayed by the results of Lau and Oort (1985) and the simulation of the 1982/83 mature phase anomalies described at this colloquium. A realistic companion simulation of the 1982/83 equatorial Pacific Ocean fluctuations using an ocean GCM driven by relatively crude estimates of surface wind has been described by Philander (1985). These developments bring closer the day when realistic simulations of the evolution of the complete ocean/atmosphere ENSO episode become commonplace.

REFERENCES

Arkin, P.A., 1979. The relationship between fractional coverage of high cloud and rainfall accumulations during GATE over the B-scale array. Mon. Weather Rev., 107: 1382–1387.
Arkin, P.A., 1982. The relationship between interannual variability in the 200 mb tropical wind field and the Southern Oscillation. Mon. Weather Rev., 110: 1393–1404.
Bergman, K.H., 1978. Multivariate analysis of temperature and winds using optimum interpolation. Mon. Weather Rev., 107: 1423–1444.
Bjerknes, J., 1966. A possible response of the atmospheric Hadley circulation to equatorial anomalies of ocean temperature. Tellus, 18: 820–829.
Bjerknes, J., 1969. Atmospheric teleconnections from the equatorial Pacific. Mon. Weather Rev., 97: 163–172.
Bjerknes, J., 1972. Large-scale atmospheric response to the 1964–65 Pacific equatorial warming. J. Phys. Oceanogr., 2: 212–217.
Cane, M.A., 1983. Oceanographic events during El Niño. Science, 222: 1189–1194.
Charney, J.G. and Shukla, J., 1981. Predictability of monsoons. In: J. Lighthill and R.P. Pearce (Editors), Monsoon Dynamics. Cambridge Univ. Press, Cambridge, pp. 99–110.
Dey, C. H. and Brown, J. A., 1976. Decomposition of a Wind Field on the Sphere. NOAA Tech. Memo., NWS NMC59, 13 pp.
Flohn, H., 1964. Investigations on the tropical easterly jet. Bonn. Meteorol. Abh., 4: 1–83.
Gill, A.E., 1980. Some simple solutions for heat induced tropical circulation. Q. J. R. Meteorol. Soc., 106: 447–462.
Gill, A. E., 1983. An estimation of sea level and surface current anomalies during the 1972 El Niño and consequent thermal effects. J. Phys. Oceanogr., 13: 586–606.
Gill, A. E., and Rasmusson, E.M., 1983. The 1982–83 climate anomaly in the equatorial Pacific. Nature, 306: 229–234.
Harrison, D.E., 1984. The appearance of sustained equatorial surface westerlies during the 1982 Pacific warm event. Science, 224: 1099–1102.
Heddinghaus, T. R. and Krueger, A.F., 1981. Annual and interannual variations in outgoing longwave radiation over the tropics. Mon. Weather Rev., 109: 1208–1218.
Horel, J.D., 1982. On the annual cycle of the tropical Pacific atmosphere and ocean. Mon. Weather Rev., 110: 1863–1878.
Horel, J.D. and Wallace, J.M., 1981. Planetary scale atmospheric phenomena associated with the Southern Oscillation. Mon. Weather Rev., 109: 813–829.
Kistler, R.E. and Parrish, D.F., 1982. Evolution of the NMC data assimilation system: September 1978–January 1982. Mon. Weather Weather Rev., 110: 1335–1346.
Koteswaram, P., 1958. The easterly jet stream in the tropics. Tellus, 10: 43–57.
Krishnamurti, T.N., 1971. Tropical east–west circulation during the northern summer. J. Atmos. Sci., 28: 1342–1347.
Krueger, A.F. and Gruber, A., 1984. The status of the NOAA outgoing longwave radiation data set. Bull. Am. Meteorol. Soc. (submitted).
Lau, K.-M., 1985. Subseasonal scale oscillation, bimodal climate state and the El Niño/Southern Oscillation. In: J.C.J. Nihoul (Editor), Coupled Ocean–Atmosphere Models. (Elsevier Oceanography Series, 40) Elsevier, Amsterdam, pp. 29–40 (this volume).

Lau, N.-C. and Oort, A.H., 1985. Response of a GFDL general circulation model to SST fluctuations observed in the tropical Pacific Ocean during the period 1962–1976. In: J.C.J. Nihoul (Editor), Coupled Ocean–Atmosphere Models. (Elsevier Oceanography Series, 40) Elsevier, Amsterdam, pp. 289–302 (this volume).

Leibmann, B. and Hartmann, D. L., 1982. Interannual variations of outgoing IR associated with tropical circulation changes during 1974–78. J. Atmos. Sci., 39: 1153–1162.

Manabe, S. and Hahn, D.G., 1981. Simulation of atmospheric variability. Mon. Weather Rev., 109: 2260–2286.

Oort, A.H. and Rasmusson, E.M., 1971. Atmospheric Circulation Statistics. NOAA Professional Paper 5. Dept. of Commerce, 323 pp.

Pan, Y.H. and Oort, A.H., 1983. Global climate variations connected with sea surface temperature anomalies in the eastern equatorial Pacific Ocean for the 1958–73 period. Mon. Weather Rev., 111: 1244–1258.

Philander, S.G.H., 1979. Variability of the tropical oceans. Dyn. Atmos. Oceans, 3: 191–208.

Philander, S.G.H., 1983. El Niño Southern Oscillation phenomena. Nature, 302: 295–301.

Philander, S.G.H. and Rasmusson, E.M., 1985. The Southern Oscillation and El Niño. Adv. Geophys., (in press).

Philander, S.G.H. and Seigel, A.D., 1985. Simulation of El Niño of 1982–1983. In: J.C.J. Nihoul (Editor), Coupled Ocean–Atmosphere Models. (Elsevier Oceanography Series, 40) Elsevier, Amsterdam, pp. 517–541 (this volume).

Quiroz, R.S., 1983. The climate of the El Niño winter of 1982–83: a season of extraordinary anomalies. Mon. Weather Rev., 111: 1685–1706.

Rasmusson, E.M., 1982. Ocean effects. Proc. WMOCAS/JSC Expert Study Meeting on Long-Range Forecasting, World Meteorological Organization, pp. 97–122.

Rasmusson, E.M. and Carpenter, T.H., 1982. Variations in tropical sea surface temperature and surface wind fields associated with the Southern Oscillation/El Niño. Mon. Weather Rev., 110: 354–384.

Rasmusson, E.M. and Wallace, J.M., 1983. Meteorological aspects of the El Niño/Southern Oscillation. Science, 222: 1195–1202.

Rasmusson, E.M., Reynolds, R.W. and Carpenter, T.H., 1983. A global view of the Southern Oscillation/El Niño signal in precipitation and surface temperature. Paper presented at the Second Conf. on Climate Variations, New Orleans, La., 10–14 January 1983.

Reynolds, R.W., 1982. A monthly averaged climatology of sea surface temperature. NOAA Tech. Rep. NWS 31, U.S. Dept. of Commerce, Washington, D.C., 35 pp.

Reynolds, R.W. and Gemmill, W.H., 1984. An objective global monthly mean sea surface temperature analysis. Trop. Ocean–Atmosphere Newslett., 23: 4–5.

Reynolds, R.W. and Rasmusson, E.M., 1982. The North Pacific sea surface temperature associated with El Niño events. Proc. Seventh Annual Climate Diagnostic Workshop, October 18–22, 1982, Boulder, Colo., pp. 298–310.

Rosen, R.D. and Salstein, D.A., 1985. Effect of initialization on diagnoses of NMC largescale circulation statistics. Mon. Weather Rev., (submitted).

Sardeshmukh, P.D. and Hoskins, B.J., 1984. Vorticity balances in the tropics during the 1982–83 El Niño–Southern Oscillation event. Q.J.R. Meteorol. Soc., 111 (in press).

Shukla, J., 1984. Predictability of monthly means, Part II. Influence of the boundary forcings. In: D.M. Burridge and E. Kallen (Editors), Problems and Prospects in Long and Medium Range Weather Forecasting. Springer, Berlin, 274 pp.

Tropical Ocean–Atmosphere Newsletter, 1983a. The 1982 Equatorial Pacific Warm Event. Special Issue, 16, 20 pp.

Tropical Ocean–Atmosphere Newsletter, 1983b. 1982–83 Equatorial Pacific Warm Event. Special Issue II Update, 21, 34 pp.

Van Loon, H., 1984. The southern Oscillation, Part III: Association with the trades and with the trough in the westerlies in the South Pacific Ocean. Mon. Weather Rev., 112: 947–954.

Van Loon, H. and Madden, R.A., 1981. The Southern Oscillation, Part I: Global associations with pressure and temperature in northern winter. Mon. Weather Rev., 109: 1150–1162.

Walker, G.T., 1924. Correlation in seasonal variations of weather IX: A further study of world weather. Mem. India Meteorol. Dept., 24(9): 275–332.

Walker, G.T. and Bliss, E.W., 1932. World Weather V. Mem. R. Meteorol. Soc., 4: 53–84.

Webster, P.A., 1972. Response of the tropical atmosphere to local, steady forcing. Mon. Weather Rev., 100: 518–541.

Webster, P.A., 1981. Mechanisms determining the atmospheric response to sea surface temperature anomalies. J. Atmos. Sci., 38: 554–571.

Wyrtki, K., 1975. El Niño – the dynamic response of the equatorial Pacific Ocean to atmospheric forcing. J. Phys. Oceanogr., 5: 572–584.

CHAPTER 41

MODELING TROPICAL SEA-SURFACE TEMPERATURE: IMPLICATIONS OF VARIOUS ATMOSPHERIC RESPONSES

PAUL S. SCHOPF

ABSTRACT

Models of tropical oceans have been found to be rather sensitive to the assumptions made about the atmospheric response to SST anomalies. The changes induced in the ocean model response affect not only the mean state of the model's "climatology", but also alter the character of anomalous warming events. This is so much so that under certain different assumptions about the atmosphere, the generation of anomalies can arise from entirely different processes. A model for the coupled ocean–atmosphere thermodynamics is constructed which examines the role of radiation, moisture, advection, and diffusion in the atmosphere and ocean. Linearization of the surface heat balance and radiative transfer equations leads to a set of equations parameterizing the atmosphere for purposes of determining the damping of surface temperature anomalies. The change from a pre-specified atmosphere to no atmosphere (black body radiation from the sea surface) changes the anomaly time-scale from about 1 month to 2 years. Diffusive and advective effects provide scale dependent damping timescales that lie between these two extremes – with short wavelength anomalies being damped rapidly. Thus processes which tend to produce larger-scale anomalies will enjoy a more favorable environment for growth.

INTRODUCTION

Much of the recent work on coupled ocean–atmosphere models has focused on the influence of the sea-surface temperature (SST) anomalies on the surface wind. This influence comes from the heating of the atmosphere over warm water, which drives the atmosphere through the modification of the pressure forces. In the studies to date, however, this heating has been treated somewhat sketchily and the consequence of the surface heat flux on the anomaly of SST itself is not often considered.

In this study, the dynamical response and the feedback through the surface winds will be ignored, and the thermal effects of the interaction between ocean and atmosphere will be examined. This approach is not taken in the belief that the dynamical effects are unimportant (or even dominant), but to clarify the heat balance considerations which arise in coupled ocean–atmosphere systems.

The problem originates in the specification of the surface heat flux from regions of anomalously warm (or cold) water in a predictive sense. While the heat flux can be estimated reasonably well from observations, the prediction of these fluxes requires the prediction of surface air temperature, humidity and wind speed (or their parameterization). A model for the evolution of anomalies in the coupled system must therefore include a model of these variables.

In an earlier attempt at describing the surface flux, Haney (1971) proposed a linearization of the surface heat balance in which the surface heating over an SST anomaly was proportional to the strength of the anomaly:

$$Q' = K^* T' \tag{1}$$

where T' is the SST anomaly, Q' is the surface heating anomaly and K^* is a constant. Arguments have been raised in the past as to what value to use for K^*, since it can embody so much physics. It has been fit to the drag laws (with constant wind speed assumed) and to radiative equilibrium models. The two types give much different values. In the first case the air is assumed unchanged over the SST anomaly, while in the later the air is assumed in a radiative balance.

This treatment has the advantage that time-dependent experiments on the ocean thermal response can be made, somewhat at the expense of a detailed description of the underlying physics. On the other hand, Sarachik (1978) has examined a radiative-convective equilibrium model which includes evaporation and a moisture budget with radiative considerations for the atmosphere. In Sarachik's discussion, however, the equilibrium nature of the solution implies that we have little information on the temporal evolution of the state. While this may be so, several important mechanisms arise in the model that are useful to consider:

The radiative-convective model introduces a second physical mechanism to the heat balance. If one takes the view that the Haney formulation parameterizes evaporation, the additional balance describes how the latent heat that is put into the atmosphere is radiated to space (maintaining a balanced atmospheric heat budget). Or, if the view of Schopf (1983) is taken that K^* embodies the radiative relaxation of the atmosphere, the convective assumption completes the moisture budget.

Sarachik's model can be extended to the non-equilibrium case, in which the requirement of zero net heating of the ocean can be replaced with the specification of the SST. The zero heat flux condition is a necessary condition for equilibrium, but not for the solution of the model equations. If it is waived, the resulting solution is no longer in a locally balanced state, but if the heat capacity of the ocean is sufficiently large, the atmosphere can be considered in local thermal balance, as will be discussed below. If this approach is taken, the net surface heating can be determined as a function of SST and incoming solar radiation. A linearization of the resulting curve gives a coupling constant K as would be appropriate in a Haney-type model of the surface flux.

Without going into the detail associated with such a radiative-convective "semi-equilibrium" model, we can construct a simple model for the ocean–atmosphere system which includes the effects of radiation and convection in a time-dependent (and therefore predictive) way. In the process, we may add the effects of horizontal heat fluxes in the ocean and atmosphere.

A SIMPLE THERMODYNAMIC MODEL

The model envisioned is shown schematically in Fig. 1. In this case the atmospheric temperature of interest is the radiating temperature, lying well up into the troposphere. In the tropics, however, strong heating from below drives convection quite vigorously, which keeps the troposphere quite near the moist adiabatic lapse rate. Thus, both the mid-tropospheric temperature and the near-surface temperature are strongly linked. The dependence of the lapse rate on the air temperature can be introduced by modeling the lower troposphere response as some fraction of the mid-tropospheric response:

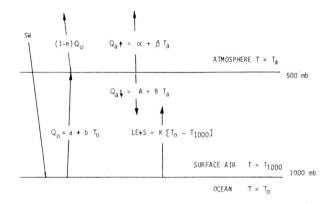

Fig. 1. Schematic of the coupled ocean–atmosphere model. Incoming short wave radiation (*SW*) is absorbed by the ocean mixed layer at temperature T_o. The ocean emits longwave radiation $Q_o\uparrow$, a fraction (*e*) of which is absorbed by the atmosphere. Latent plus sensible heating (*LE + S*) are linearized about the difference in surface air and sea temperatures. The mid-troposphere emits upward ($Q_a\uparrow$) and downward ($Q_a\downarrow$) longwave radiation from its mean temperature T_a. The effects of convection are embodied in the treatment of the lapse rate as along a moist adiabat, so that the surface air temperature changes are a fraction $(1 - \delta)$ of the mid-tropospheric temperature changes.

$$T_{1000} = T_a(1 - \delta) \tag{2}$$

The lower troposphere temperature is important for determining the surface fluxes of water and sensible heat. In this view, we will parameterize the atmospheric planetary boundary layer (pbl) implicitly, and make the evaporative fluxes of water and sensible heat proportional to the difference between the SST and the surface air temperature. This approach assumes either a fixed relative humidity at the ground or a constant depth pbl which is saturated at its top. Embodied in the linearization are the variations in the wind speed, the Clausius-Claperyon relation and drag coefficients. A similar treatment of the sensible heating allows the combination of latent plus sensible heat fluxes into:

$$LE + S = K[T_o - T_a(1 - \delta)] \tag{3}$$

All radiation laws will also be linearized for discussion purposes, so that the upward longwave from the ocean surface will be:

$$Q_o\uparrow = a + bT_o \tag{4}$$

a fraction (*e*) of which is absorbed by the atmosphere, the remainder going to space as window radiation. The downward and upward *IR* from the atmosphere are given by:

$$Q_a\uparrow = \alpha + \beta T_a \tag{5}$$

$$Q_a\downarrow = A + BT_a \tag{6}$$

where the different coefficients reflect slightly different emission levels (and hence mean temperatures).

So far, this model represents a linearized version of the physics included in Sarachik (1978). For simple spatial patterns, we can add horizontal transport operators in the ocean and atmosphere: L_o and L_a, which operate on their respective temperatures. Such processes as advection by a steady mean wind and diffusion are envisioned.

EIGENMODES AND EIGENVALUES

For the atmosphere–ocean system envisioned by the above, the thermodynamic equations may be written:

$$C_a \partial T_a / \partial t = - [\beta + B + K(1 - \delta) + L_a] T_a + (b + K) T_o + F_a$$

$$C_o \partial T_o / \partial t = [B + K(1 - \delta)] T_a - (b + K + L_o) T_o + F_o$$

where F_a and F_o contain the constants A, a, etc. With the linearizations introduced, these give a simple coupled set which may be treated in terms of their eigenmodes and eigenvalues. The gravest mode represents a situation in which the atmosphere and ocean warm and cool together, while the second mode represents (as will be seen) temperature changes in the atmosphere alone with nearly zero temperature change in the ocean.

This behavior of the modes derives from the small heat capacity of the atmosphere in relation to that of the ocean's mixed layer: for a 50-m ocean mixed layer, the atmospheric heat capacity represents 4% of the ocean's. On longer time scales, the oceanic thermal inertia is considerably larger, since heat from the surface can be mixed to greater depths, but for the purposes of this scale argument, we shall ignore the variation in mixed-layer depth.

To represent the eigenvalue as a decay scale, we consider the eigenvalue problem presented by:

$$\partial X / \partial t = AX + F$$

$$AX = - \lambda X$$

The ratio of the two heat capacities will be defined as $\epsilon = (C_a / C_o)$, giving the eigenvalue equation

$$\epsilon \lambda^2 - [\beta + B + K(1 - \delta) + L_a + \epsilon(b + K + L_o)] \lambda / C_o + (\beta + L_a)(b + K + L_o)/C_o^2 +$$

$$+ [B + K(1 - \delta)] [b(1 - e) + L_o]/C_o^2 = 0$$

The two eigenvalues give a slow mode:

$$\lambda_1 = \frac{(eb + K)(\beta + L_a)}{C_o [\beta + B + K(1 - \delta) + L_a]} + \frac{(1 - e)b + L_o}{C_o}$$

plus a fast mode (which is order $(1/\epsilon)$ faster then the slow one):

$$\lambda_2 = [\beta + B + K(1 - \delta) + L_a]/C_a$$

The eigenmodes have the structure given by:

$$T_{a1}/T_{o1} = (eb + K)/[\beta + B + K(1 - \delta) + L_a]$$

$$T_{o2}/T_{a2} = 0(\epsilon)$$

from which it can be seen that the fast mode essentially represents a change in air temperature (although roughly equal changes in heat content of the ocean and atmosphere).

Before examining the nature of these modes as they may be realized in the tropical ocean–atmosphere system, some limits on the slow mode can be considered:

No atmosphere. In the case of no atmosphere, the ocean equilibrates through long-wave radiation to space directly plus horizontal re-distribution through the transport flux:

$$\lambda_1 = (b + L_o)/C_o$$

Greenhouse. If the atmosphere is treated as a greenhouse, with no evaporative or sensible coupling between the ocean and atmosphere (and ignoring horizontal transports) the usual greenhouse solution is derived:

$$\lambda_1 = b/C_o(b + \beta)$$

which can be seen to give roughly half the damping rate as the transparent atmosphere case when the window radiation vanishes.

Strongly coupled. If the coupling constants for evaporation and sensible heating are much larger than the long-wave sensitivity $(K \gg b)$, then the ocean and atmosphere are strongly linked, and the relaxation process occurs through the atmospheric long-wave emission, with:

$$\lambda_1 = \beta/C_o(1 - \delta)$$

This case looks quite like the case without the atmosphere, except that the ocean cools with the black-body sensitivity appropriate for the top of the atmosphere (β) rather than that for the ocean surface (b). Note that the adjustment of the lapse rate as a function of the temperature can cause increased damping because of the fact that a $1°C$ SST change can give a larger temperature change at height, thereby enhancing the heat loss to space. This increased heat flux to space is readily taken out of the ocean by the strong evaporative coupling.

Strong transport. If the atmosphere can transport heat horizontally much more effectively than the other processes, an interesting balance derives in which heat is communicated from the ocean to atmosphere, and then transported away. The limiting factor in this case is how fast the heat can be put into the air, for once it reaches the atmosphere, the strong transport removes it. In this case the decay is given by:

$$\lambda_1 = (b + K + L_o)/C_o$$

where the ocean transport L_o/C_o is an additional damping effect. This case has the interesting property that the air temperature change is essentially zero:

$$T_{a1}/T_{o1} \simeq (eb + K)/L_a \ll 1$$

while in the others the air and water temperature changes are of the same order. This is the case most equivalent to specifying the atmosphere as unchanged in the face of SST anomalies, and may well be the most appropriate approximation at small scales.

On the global scale, it seems appropriate that the radiative-convective model for the ocean–atmosphere system is probably appropriate. In the present context this means a case in which the coupling between the atmosphere and ocean is accomplished by radiation, evaporation and sensible heating, and the essential limiting process is the rate of longwave loss from the top of the atmosphere to space. The time constant for this is essentially C_o/β, perhaps modified by a factor of 2 due to greenhouse effects or lapse rate considerations.

On the smallest scales, however, the radiation balance seems inappropriate. The horizontal transports by both the ocean and atmosphere are active enough to provide effective damping through horizontal redistribution. The difference between these types is significant and gives quite different decay times for the two extremes. On the large scale the anomalies can last for very long times, since the radiative feedback is weak, while small-scale features can be erased in the matter of days.

SCALING OF THE THERMAL DAMPING FACTORS

To examine the transition from radiative to transport regimes, a model for the various processes must be constructed. The radiative coupling parameters B and β are linearizations of the black-body law for the atmosphere, and lie in the range of $4-5 \, \mathrm{W \, m^{-2}}$ while the emission from the surface of the ocean is treated with $b \approx 6 \, \mathrm{W \, m^{-2}} \, K$, the evaporative and sensible coupling coefficient is a linearization of a much more complex system, and should be regarded as much more uncertain. Experiments with GCMs and estimate based on bulk formulae give estimates of K in the range of $10-100 \, \mathrm{W \, m^{-2}}$, depending on exactly how the surface moisture balance is maintained and what wind speed is chosen. For our purposes, we shall consider a range of values and examine the nature of the results as a function of this parameter.

The transport processes can be divided into two types: advective and diffusive. In the diffusion model we will assume a sinusoidal anomaly structure:

$$T(x) = T e^{ikx}$$

so that:

$$L_o T_o = -k^2 C_o D_o T_o$$
$$L_a T_a = -k^2 C_a D_a T_a$$

where D_a, D_o are the diffusivities of the atmosphere and ocean. Rough estimates give $D_a \approx 3 \times 10^6 \, \mathrm{m^2 \, s^{-2}}$, and $D_o \approx 10^3 \, \mathrm{m^2 \, s^{-2}}$. With these values for the diffusivities, the atmospheric transport damps at a rate that varies from $30{,}000 \, \mathrm{W \, m^{-2}} \, K^{-1}$ at a length scale of 200 km to $3 \, \mathrm{W \, m^{-2}} \, K^{-1}$ at a length scale of 20,000 km. Clearly, the case of transport-dominated regime obtains at the smaller length scales. The ocean transport varies from 200 to $0.02 \, \mathrm{W \, m^{-2}} \, K$ over the same range of length scales, and gives less of a contribution to the transport equation, although it can be the dominant mechanism through the following: When the atmospheric transport is highly effective the limiting process on the transport through the atmopshere is how fast the heat can be transported up into the air. This occurs through evaporation and radiation, and is therefore limited by the value of K. When the ocean transport factor is stronger than K, it will dominate the heat flow, regardless of the high efficiency of the atmospheric flux.

An advective model can also be constructed which may be appropriate at coasts. Consider $T = T \exp(-kx)$, with flow to the west (away from the coast). The advective damping will occur linearly in $U_a k C_a$ and $U_o k C_o$.

Figure 2 gives the decay time for thermal anomalies as a function of their length scales for both the advective and diffusive cases. For the longest scales, the weakness of the radiative damping leaves anomalies with several hundred day decay times. The sensitivity

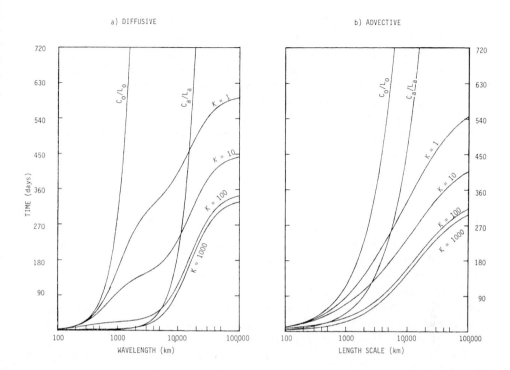

Fig. 2. Decay times vs. length scales and strength of the evaporative coupling (K) for diffusive (a) and advective (b) transport mechanisms.

to the value of K is seen and is related to the difference between the greenhouse ($K = 0$) and strongly coupled ($K \gg b$) cases. At this scale, the limiting process is that of radiation from the atmosphere.

At the shorter scales, the decay time is very short by this model, and is dominated by the transport processes. At this range, the dependence of the decay on K is again apparent, as the ocean transport takes over. Between these two extremes lies the length range of most interest, from the eddy scale (200 km or so) up to a scale corresponding to that of the Pacific basin. Here, the atmospheric transport is relatively important, and the essential limiting factor is how fast the heat can be put into the air, particularly in the diffusive case. The difference between $K = 100$ and $K = 10$ can be seen to be the difference between a 100 day damping time and a 10 day damping time. Anomalies which can grow through their feedback with the atmospheric wind systems will behave much differently when damped with a 10 day time scale as opposed to damped with 100 days.

This simple scaling serves to draw attention to some important unresolved issues in the coupled ocean—atmosphere problem. The evaporative flux has been linearized here, and linearized with respect to air—sea temperature difference. The sensitivity of the results to this parameter in the length-scale range of interest raises an immediate concern as to the nature of the evaporation process, particularly as to how the surface humidity is maintained.

The second issue arises from the development of anomalies through feedback processes, and the mechanisms that act in the ocean and atmosphere to produce the growth. A perturbation to the ocean which tends to cause a large-scale SST change should grow more efficiently than one which leads to local response only. For example, an anomalous advection of water along a gradient takes on a space scale corresponding to the space scale of the mean field itself. In the case of the Pacific, this means that the SST anomaly can have a basin-wide scale even when driven by relatively localized winds. A mechanism which starts with a localized warm pool of water must overcome strong initial damping in order to grow.

CONCLUSIONS

This study has sought to examine some of the factors arising in the thermodynamics of a coupled ocean—atmosphere model. It amounts to a scaling consideration of the thermal equations, and through the linearizations allows for the derivation of varying regimes of ocean—atmosphere heat paths.

The important questions raised are that the thermal damping of anomalies is significantly scale dependent. We can expect that in fully coupled, complex GCM experiments, spontaneous generation of strong anomalies from spatially limited SST perturbations will not be as important as larger scale processes.

The linearizations taken here may represent rather liberal parameterizations of processes which are at present poorly understood. Particularly important is the description of the surface evaporative flux. Of most concern here are the processes which control the surface humidity.

REFERENCES

Haney, R.A., 1971. Surface thermal boundary condition for ocean circulation models. J. Phys. Oceanogr., 1: 241–248.
Sarachik, E.S., 1978. Boundary layers on both sides of the tropical ocean surface. In: Review Papers of Equatorial Oceanography FINE Workshop Proceedings, Nova/N.Y.I.T. University Press, 73 pp.
Schopf, P.S., 1983. On equatorial waves and El Niño: II. Effects of air—sea thermal coupling. J. Phys. Oceanogr., 13: 1878–1893.

CHAPTER 42

MODELS OF INTERACTIVE MIXED LAYERS

ERIC B. KRAUS

ABSTRACT

Governing equations for interactive mixed layers above and below the air–sea interface are listed and discussed.

It is shown that the atmospheric layer depth affects the magnitude of the stress and the Ekman mass transport in both fluids. A correction formula to compensate for an arbitrary choice of the atmospheric layer depth is developed.

Although the tropical ocean supplies much more latent than sensible heat to the atmosphere, the thickness of the atmospheric mixed layer is predominantly a function of the sensible heat flux. This amplifies the energy amount which the atmospheric layer contains and transports. It makes the development very sensitive to relatively small changes in sea-surface temperature.

The interactive evolution of the oceanic and atmospheric boundary layers in the stratus-covered source regions of the trades is discussed. Sensitivity tests show the development to be strongly affected by the upstream boundary conditions and by the advection velocities in the two fluids.

INTRODUCTION

The near-surface mixed layers of the ocean and the marine atmosphere are obviously boundary layers in the sense that their very existence is caused by their proximity to the air–water boundary. In contrast, however, to the common definition of boundary layers as regions in which changes normal to the boundary are very much larger than changes along it, they are characterized by a uniform or quasi-uniform distribution of various conservative properties in the normal direction. Mixed layers are formed by mechanical and convective stirring. Changes in the mixed-layer thickness are determined by the interplay between turbulence-generating processes and the stabilizing influences of a downward buoyancy flux. The latter is equivalent to a lifting of the center of gravity of the fluid column.

Mixed-layer models have to allow for the exchange of mass and momentum across their outer boundary. The relevant data for conditions in the interior of the two fluids have to be supplied by observations or by large-scale circulation models. So far, there have been few attempts to incorporate mixed-layer physics explicitly into large-scale models. A paper by Schopf and Cane (1983) is a noticeable exception. However, layers are well-mixed by definition in all layered models. The depth of the layers adjacent to the surface is usually fixed arbitrarily or determined by the position of an isobaric or isopycnal surface. Such an arbitrary determination of the depth of the surface layers can affect the character of the fluxes across the interface and hence the subsequent development of the system.

THE MODEL EQUATIONS

Ocean mixed layers are commonly characterized by a vanishing or negligibly small vertical gradient of the temperature T and the salinity s (see Appendix). A uniform distribution of mean horizontal momentum implies an absence of shear within the layer and hence horizontal displacement like a solid slab. It is possible for some properties to be well mixed throughout the layer while others exhibit vertical gradients.

Atmospheric mixed layers tend to be topped by inversions. They can be stirred convectivly by heating from the sea surface and also by radiative cooling from their upper boundary. The humidity mixing ratio and the potential temperature are conserved by mixing in the absence of condensation. However, the higher strata of marine atmospheric mixed layers often contain clouds. The properties which are conserved in these circumstances are the total water content σ and the isobaric equivalent-potential temperature:

$$\xi = \theta + Lq/c_p \tag{1}$$

where θ is the potential temperature referred to the local sea-level pressure and not to the standard 1000 mb.

The mixed-layer equations can now be written in the following form:

$$-\rho c(h\, dT/dt + w^* \Delta T) - S = Q + I(0) = \rho_a c_p (H d\xi/dt + W^* \Delta\xi) + I(H) \tag{2}$$

$$\rho(h\, ds/dt + w^* \Delta s) = E = \rho_a (H d\sigma/dt + W^* \Delta\sigma) \tag{3}$$

$$\rho(h\, d\mathbf{v}/dt + hf\mathbf{n} \times \Delta\mathbf{v} + W^* \Delta\mathbf{v}) = \tau = -\rho_a (H d\mathbf{V}/dt + Hf\mathbf{n} \times \mathbf{V} + W^* \Delta\mathbf{V}) \tag{4}$$

The change in the layer depths depends on the available power for turbulence generation. The entrainment process always requires work to lift denser water from below into the ocean mixed layer or to draw relatively tenuous air from above into the atmospheric layer. Either process raises the center of gravity of the fluid column and hence requires work. When the buoyancy flux is downward — e.g. when the water is heated from above or the air is cooled from below — and when the wind stirring is weak or decreasing, there may not be enough power available for entrainment. In these circumstances, the mixed layer can not be affected by conditions in the interior ocean or the free atmosphere and its depths become proportional to the so-called Monin-Obukhov length. This means that it can be determined diagnostically from the ratio of the turbulent kinetic energy generation, which is proportional to the third power of the friction velocity, and the work needed to balance the downward buoyancy flux.

$$h \propto -u_*^3/b \qquad H \propto -U_*^3/B \tag{5}$$

The proportionality factors are different in the two media.

When the buoyancy flux is upward ($w^* > 0$, $W^* > 0$), the change in layer depth is given by:

$$\partial h/\partial t + \nabla \cdot h\mathbf{v} + w^* = 0 = \partial H/\partial t + \nabla \cdot H\mathbf{V} + W^* \tag{6}$$

The specific buoyancy fluxes immediately below and above the sea surface can be expressed in the form:

$$b = g[\alpha(S + Q)/c + \beta sE]/\rho \tag{7}$$

$$B = g[(Q-LE)/c_p\Theta + 0.6\,E]/\rho_a \tag{7a}$$

The computations are not very sensitive to resonable values of the reference potential temperature Θ and an arbitrary choice $\Theta = 290\,K$ will not cause significant errors.

To close the preceding system of equations, it is necessary to represent the entrainment velocities w^*, W^* and the surface flux terms Q, E, τ as functions of the external forcing and of the bulk variables. The oceanic entrainment velocities can be derived from the turbulence-energy equation:

$$w^* = \Lambda(\psi)/g'h \tag{8}$$

where g' is the reduced gravity at the base of the mixed layer:

$$g' = g\Delta\rho/\rho = g(\alpha T - \beta s) \tag{9}$$

and the Heavyside function $\Lambda(\psi)$ is equal to the its argument ψ when ψ is positive and equal to zero otherwise.

$$\psi = mu_*^3 + hb(0.5+r) - |hb|(0.5-r) \tag{10}$$

The empirical factor $m \approx 3.0$ parameterizes the efficiency of kinetic energy generation by the wind stress and the associated dissipation. The parameter $r \approx 1.0$ plays the same role for the convectively generated energy. It has no effect when kinetic energy is being consumed by a downward buoyancy flux ($b < 0$).

The corresponding equations for entrainment across the inversion are more complicated, because the vertical gradients of potential density, temperature and mixing ratio change at cloud base. It is convenient to introduce two new variables a and z':

$$a \equiv a(q) = (1 + L^2 q/c_p R_v\Theta^2)^{-1} = (1 + 155.3q)^{-1} \tag{11}$$

where R_v is the gas constant for water vapour and

$$z' = (z_c/H + 0.5) - |z_c/H - 0.5| \tag{12}$$

The quantity $z' = 1$ in a cloud free mixed layer when $z_c > H$; when the cloud base is below the inversion $z' = z_c/H$.

The general expression for entrainment across the inversion has the same form as the corresponding oceanic expression:

$$W_* = \Lambda(\Psi)/g_a'H \tag{8a}$$

where:

$$g_a' = [(a - az' + z'^2)\Delta\xi - (Lz'^2/c_p)\Delta q]/\Theta \tag{9a}$$

and:

$$\Psi = mU_*^3 + HB_*(0.5+r) - |HB_*|(0.5-r) \tag{10a}$$

The quantity

$$B_* = z'(2-z')B + [a(1-z')^2 Q + (a - az'^2 + z'^2)I(H)]\,g/c_p\rho_a\Theta \tag{13}$$

An explicit derivation of the preceding expressions can be found in a paper by Kraus and Leslie (1982). The same paper also lists a set of empirical formulas for the flux terms

Q, E and τ. The entrainment process has been considered in a more sophisticated manner by Deardorff (1979). More detailed considerations of the flux terms can be found in papers by Busch (1977), Sarachik (1978) or Smith (1980).

The set of eqns. (2)–(6) does not admit a non-trivial stationary solution. If the surface flux terms are finite, the total time derivatives of the mixed-layer depths and heat or water content can not all be zero. Mixed layers evolve either in time or advectively downstream. In the following sections, simplified or reduced forms of the complete mixed-layer equations are applied to some particular problems.

THE EFFECT OF THE ATMOSPHERIC MIXED-LAYER DEPTH ON THE SURFACE WIND STRESS

It is customary in oceanographic and atmospheric models to assume that the wind stress acts like a body force, which is distributed uniformly over a layer adjacent to the interface. This implies horizontal movement of the layer like a solid slab. The present section deals specifically with the magnitude and direction of the Ekman transports in both media, which would be consistent and compatible with such an assumption. It will be seen that the mass transports and their divergence are sensitive to changes in the real or assumed thickness of the atmospheric boundary layer.

In reality, momentum does not mix at the same rate as heat. Mixed layers of quasi-constant temperature in the ocean or equivalent-potential temperature in the atmosphere are rarely completely free of shear. However, apart from special cases like the layer above the equatorial undercurrent, the shear wthin these mixed layers is often significantly smaller than the shear at their base in the ocean or at the topping inversion in the atmosphere. This provides some justification for the assumption that to first order of approximation, temperature mixed layers can be considered also as layers of vertically quasi-uniform, mean horizontal momentum. Combinations of one-dimensional mixed-layer physics and slab dynamics, have been used by Schopf and Cane (1983), De Ruyter (1983) and many others before. The arguments below are applicable, however, not only to these cases but also to layers of arbitrary depth, provided the stress is modelled to act like a body force.

To treat the interactive dynamics of the two boundary layers rigorously, one has to know the actual surface drift as a reference velocity. This is the velocity which might be realized by a monomolecular layer or a thin oil patch in the absence of wave action. It is the velocity which is "seen" by each fluid as a boundary value. Under steady-state conditions, with no slip at the interface, the surface drift velocity can be derived from the equation (Kraus, 1977):

$$v_s = \frac{\rho_a H V_g + \rho h v_g}{\rho_a H + \rho h} \approx \frac{\rho_a H}{\rho_w h} V_g + v_c \tag{14}$$

The last term in this equation represents the effect of a surface slope on the geostrophic current:

$$v_c = v_g - \frac{\rho_a}{\rho_w} V_g \tag{15}$$

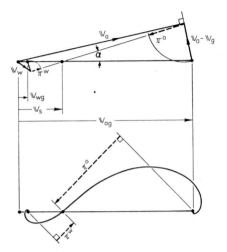

Fig. 1. Velocity distributions above and below the interface, for slab layers (upper diagram), hodograph of Ekman spirals (lower diagram). The wind and current velocities are represented here by v_a, v_w; geostrophic velocities by v_{ag}, v_{wg}; the surface drift by v_s. The angle between the actual and the geostrophic wind is α; the stress τ is not in scale.

As it affects the computations only as an additive vector, it is convenient in a general treatment to consider this velocity zero. In other words, the sea surface is assumed to be horizontal. With $v_c = 0$, eqn. (14) shows the surface drift to have the direction of the geostrophic wind, as in the classical double Ekman spiral illustrated in the lower diagram of Fig. 1.

The surface velocity is usually small compared to the wind speed. It is therefore usually possible to neglect it in wind-stress computations – though its effect can be noticeable in special circumstances, e.g. when a wind blows against the Florida current as compared to one which blows downstream. In the kinematics of the ocean surface layer v_s has to be considered as a relatively large quantity and it becomes an essential parameter in any physical discussion of the two-fluid problem.

The stress can be expressed in parametric form by:

$$\tau = C\rho_a |V_a - v_s| (V_a - v_s) \approx C\rho_a V_g |V_g| \qquad (16)$$

For the present demonstration it is sufficient to use a constant drag coefficient C. Denoting the velocity components by u, v, etc., and chosing a coordinate system with the X-axis in the direction of the geostrophic wind ($U_g = |V_g|$, $V_g = 0$), one can introduce the new non-dimensional variables:

$$U' = (U - u_s)/(U_g - u_s) \qquad V' = V/(U_g - u_s) \qquad \Pi = C(U_g - u_s)/fH \qquad (17)$$

Introduction of (16) and (17) into the second equality (4) with $dv/dt = W^* = 0$, yields, after simple algebraic manipulation:

$$U' = [(1 + 4\Pi^2)^{1/2} - 1]/2\Pi^2 \qquad V' = \Pi U^{3/2} \qquad (18)$$

The angle between the actual layer wind and the geostrophic wind is given by $\tan \psi = V'/U' = \Pi\sqrt{U'}$.

The preceding expressions show that the wind, stress and mass transport vectors, can

740

all be expressed as a function of the geostrophic wind V_g and of the non-dimensional number Π which is inversely proportional to the layer depth. The geometry of the situation is illustrated by the upper diagram in Fig. 1, which is based on an assumed density ratio of 1:20 between the two fluids. The real air/water density ratio of about 1:800 involves velocity differences which are too large to be representable on the same diagram. For the same reason, the ratio between the layer thicknesses had to be kept relatively small. Stipulating $H = 600$ m, $h = 120$ m and $U_g = 12\ \mathrm{m\,s^{-1}}$, one gets then $u_g = 0.6\ \mathrm{m\,s^{-1}}$, $u_s = 3\ \mathrm{m\,s^{-1}}$, $\Pi = 0.3$ and a stress angle $\psi = 16°$.

The ageostrophic mass transports are equal and opposite in the two fluids. The ageostrophic wind component is represented by the normal from the end point of the geostrophic wind vector to the stress line; the actual mixed-layer wind vector v_a is given by a line from the origin to the foot point of that normal. In the lower fluid, the current velocity is given similarly by a line from the origin to the intersection of the normal from the geostrophic current to the stress line.

For comparison's sake, the hodogram of a double Ekman spiral with scale depths equal to the mixed-layer depths H, h is shown below the mixed-layer diagram. The vertical integral of the ageostrophic wind component divided by the scale depth H can be represented again by the normal from the endpoint of the geostrophic wind vector to the stress line. In a spiral this distance is much larger than in the slab case. The difference is due to the stress angle now being $45°$ and also to the fact that the influence of surface friction extends well above the scale depth in the Ekman spiral.

Figure 2 provides a dimensional representation of the y-component of the Ekman mass transport in either the ocean or the atmosphere. It has been constructed with assumed values of $C = 1.3 \times 10^{-3}$ and $f = 0.65 \times 10^{-4}\ \mathrm{s^{-1}}$, which makes the ratio $C/f = 20$ s. In reality, this ratio will be usually smaller except in low latitudes. The range of mixed-layer heights and geostrophic winds is realistic. The figure illustrates that changes

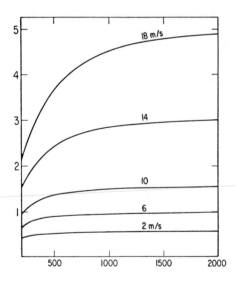

Fig. 2. Ekman mass transport component (in $10^3\ \mathrm{kg\,m^{-1}\,s^{-1}}$) at right angle to geostrophic wind as a function of atmospheric mixed-layer depth and geostrophic wind strength.

in layer height can produce relatively large changes in the Ekman transport. Deep layers are often the result of static instability and an upward buoyancy flux. The computed stress becomes large in these circumstances, even if the value of the drag coefficient C is held constant. This deduction confirms orthodox − if unquantified − nautical opinion, that "there is more weight in the wind when it is colder than the sea". For a given geostrophic wind, the surface stress and the ageostrophic transport component tend to be more than three times larger when the atmospheric mixed layer is 2500 m deep, than when it is 500 m deep.

The relevant angles are indicated in Fig. 3. According to classical Ekman theory, the angle between the geostrophic wind and the difference between the near-surface wind and the surface drift should be exactly 45°. On the other hand, the angle observed from a ship's mast is often considered to be of order 16°. Figure 3 shows that this last value is compatible with the present results for the angle between the mixed-layer wind velocity and the geostrophic wind. However, this angle is not a constant. Even at the same latitude, it can vary easily between about 10° and 25° as a function of wind velocity and inversion height. This corresponds to a change in the ageostrophic velocity component by a factor of more than 2.5.

In a truly well mixed layer the functional relationship, between the computed surface stress and the layer depths, expresses a physical reality. On the other hand, with H fixed arbitrarily as a model parameter, the computation of the stress and of the associated ageostrophic mass transport in the ocean will be affected by an arbitrary factor which depends on the particular chosen value of H. It is possible to compensate for this arbitrariness by appropriate adjustments in the value of the drag coefficient C. To do that, the drag coefficient has to be adjusted as a function of the layer depth $[C = C(H)]$, so as to make the stress and the mass transport independent of H.

The magnitude of the ageostrophic mass transport MT in either fluid is determined by

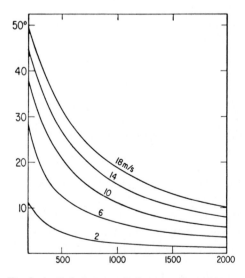

Fig. 3. Angle between actual near-surface and geostrophic wind as a function of atmospheric mixed-layer depth and geostrophic wind strength.

$$MT = \rho_a C(U^2 + V^2)/f = \rho_a CU' U_g^2/f \equiv \rho_a C_o U_o' U_g^2/f \tag{19}$$

The last equality represents the stipulation that the transport should be independent of H and that for a given geostrophic wind, it should always be equal to some standard transport characterized by a nondimensional velocity U_o' and hence a height H_o. It implies that $CU' = C_o U_o'$. Introduction of the expressions (17) and (18) into this equality yields:

$$C\{[(fH)^2 + (2CU_g)^2]^{1/2} - fH\}/H^3 = C_o\{[fH_o)^2 + (2C_o U_g)^2]^{1/2} - fH_o\}/H_o^3$$

This is an implicit equation for C as a function of H, H_o and C_o. With arbitrary $U_g = 10\,\mathrm{m\,s}^{-1}$, $C_o = 1.3 \times 10^{-3}$, $f = 0.65 \times 10^{-4}\,\mathrm{s}^{-1}$ and $H_o = 800\,\mathrm{m}$ one gets:

$$\frac{C}{C_o} = \frac{0.944\,(H/H_o)^{3/2}}{H/H_o - 0.056} \approx \mathrm{const.} \times \sqrt{H}/H_o \tag{20}$$

The last relation indicates that it would be appropriate in a layered atmospheric model to adjust the value of the drag coefficient, by a factor which is proportional to the square root of the depth of the model boundary layer. The proportionality factor depends on the values of C_o, H_o and hence U_o which characterize the standard state. The values chosen here for illustrative purposes were quite arbitrary. We know little about the actual relation between a mixed-layer wind and the surface drag. Further observational studies and experimentation would be useful for this purpose.

Boundary layer meteorologists are accustomed to think of a decrease of the drag coefficient with height. In the preceding computation it was found that its value should be increased if the height of the layer increases. The apparent paradox is easily explained. When the momentum is well mixed, the wind does not change with height in the layer. The same must then apply to the drag coefficient. However, in a deeper layer the wind is closer to geostrophic equilibrium and hence faster. This increases the surface stress. The Ekman transport increases at a corresponding rate. As a note of caution it is mentioned once more that the adjustment equation (20) is meant to correct for the virtual change in the Ekman transport and the stress, which may be caused by the arbitrary choice of the depth of the frictional layer in a model. In a real mixed layer, the transport and the stress will in fact vary as a function of both the geostrophic velocity and the depth. The adjustment is then irrelevant.

THE EFFECT OF SEA-SURFACE TEMPERATURE CHANGES ON THE ATMOSPHERIC MIXED-LAYER DEPTH AND MOISTURE CONTENT

The transport of energy by the trades towards the Intertropical Convergence Zone is sensitive to relatively small changes in sea-surface temperature. This sensitivity is due to a combination of two processes: firstly increased surface saturation vapour pressure and a faster evaporation rate, secondly an increased buoyancy flux and a correspondingly deeper atmospheric mixed layer. The deeper layer can contain and transport more energy towards the convergence zone.

The effect is illustrated schematically in Fig. 4. When the tropical sea-surface temperature is increased by one degree, the equivalent-potential temperature $\xi(0)$ of the

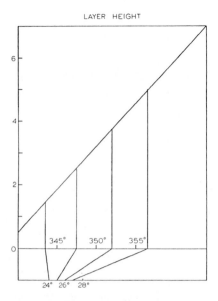

Fig. 4. Schematic diagram of atmospheric mixed-layer depth H and equivalent potential temperature ξ as a function of surface temperature.

air in immediate contact with the surface is increased by about four degrees. Assume that the equivalent-potential temperature ξ_u in the upper air above the mixed layer increases linearly with the height z:

$$\xi_u(z) = \xi_u(0) + \lambda z \tag{21}$$

It is also assumed — naively — that there is no entrainment and that the equivalent potential temperature is continuous ($\xi = \xi_u$) at H. The mixed layer is then determined diagnostically from:

$$H = [\xi_u(H) - \xi_u(0)]/\lambda = [\xi - \xi_u(0)]/\lambda \tag{22}$$

The figure suggests then that a one degree increase in sea-surface temperature could be associated with a deepening of the layer by about 1000 m.

The downstream evolution of the layer depends upon the upward energy flux from the surface. In parameterized form, with the infrared radiation balance assumed zero and with sea-surface conditions denoted by $\xi(0)$ and $q(0)$:

$$Q = \rho_a C |V| \{ c_p [T(0) - \Theta] + L [q(0) - q] \} = \rho_a C |V| c_p [\xi(0) - \xi] \tag{23}$$

Introduction (22) and (23) into the steady-state form of the second equality (2), with the indicated neglect of W^* and I, yields after division by $\rho_a C_p |V|$:

$$\partial \xi / \partial x = C \lambda [\xi(0) - \xi] / [\xi - \xi_u(0)] \tag{24}$$

where x represents now the distance along a trajectory of the air column or the stream-line of V. We stipulate $\xi \equiv \xi_o$ at the initial or entry point where $x = 0$. Integration of the last equation yields then the following transcendental equation for ξ:

TABLE 1

Change of air column heat content in 10^6 J m^{-2} as a function of the sea-surface temperature T (in °C) and the downstream distance x (in km)

T	Downstream evolving H			Constant $H = H_0 = 1500$ m		
	$x = 1000$	2000	3000	1000	2000	3000
24	11.6	18.9	24.4	8.3	11.9	15.7
25	17.0	28.6	37.8	12.7	18.2	23.9
26	22.7	39.1	52.0	14.9	21.4	28.2

$$\xi - \xi_o + [\xi(0) - \xi_u(0)] \ln \{ [\xi(0) - \xi]/[\xi(0) - \xi_o] \} = -\lambda C x \tag{25}$$

The change in heat content which is associated with the downstream evolution is $\rho_a c_p$ times:

$$H\xi - H_o\xi_o - \int_{H_0}^{H} \xi_u dz = (\xi - \xi_o) [(\xi + \xi_o)/2 - \xi_u(0)]/\lambda \tag{26}$$

The preceding equations can be used to compute heat content changes during the downstream displacement of the air column. Some characteristic values are listed in Table 1, which is based on assumed $\xi_u(0) = 332°$, $\lambda = 2°/1000$ m, $\xi_o = 335°$ and, therefore, an initial layer depth $H_o = 1500$ m. For comparison, Table 1 lists also the change which would have occurred if the layer depth would have been constrained to remain constant $(H \equiv H_o)$.

In the construction of Fig. 4 and in the calculation of the numbers in Table 1, entrainment, infrared cooling and shear were disregarded. This is obviously unrealistic. As far as the numbers go, they do indicate that a two-degree increase in the tropical sea surface would more than double the energy which is being transported downstream by the atmosphere. With entrainment, this increase should have been even larger. The table also shows that the change of energy with temperature and distance is much larger when the depth of the boundary layer is allowed to adjust itself, than when it is constrained to remain constant.

THE INTERACTIVE EVOLUTION OF THE OCEANIC AND ATMOSPHERIC MIXED LAYERS IN THE SOURCE REGION OF THE TRADES

None of the preceding discussions dealt with truly interactive developments — in the sense of an active feedback between the two fluids. A particular example of an investigation in which these processes had been considered can be found in a paper by Kraus and Leslie (1982), which deals specifically with conditions in and above the eastern regions of the subtropical Pacific and South Atlantic. These regions are characterized by extensive low stratus covers. The cloud decks tend to form over cold water; their presence then helps to keep the water cold.

We have investigated this phenomenon in the framework of a steady-state model which was based on eqns. (1)–(13). Except for the pressure gradient, all gradients normal to the trajectories of the mixed layer air and water columns were considered zero. This reduced the treatment to two dimensions which involved only the vertical fluxes and the downstream advection. The integration was continued downstream until convective instability developed $\Delta\xi = [\xi - \xi_u(H)] > 0$ or, failing that, for a distance of 2000 km. Radiation, upstream boundary and interior conditions as well as the advection velocities in both media were specified. All the following figures are reproduced from the original paper, which also provides additional references and more information about details of the treatment.

Figure 5 illustrates a particular integration. The abscissa represents the potential temperature in either fluid. (In the water this is practically equal to the actual temperature.) The height scale is different above and below the interface. At the upstream boundary, the water and surface air temperatures were both 17°C ($T = \theta = 290$ K). It was assumed also that the relative humidity of the surface air at the upstream boundary was 50% and that the advection velocities were 6 m s^{-1} for the atmosphere and 0.2 m s^{-1} for the ocean mixed layer.

The figure shows that during the early stages of the development, the surface air becomes warmer by entrainment of potentially warmer air from above. The resulting downward conduction of heat also warms the water layer. Further downstream the entrainment of cold water from below becomes the dominant factor and both layers become colder, although warm air is still entrained from above. The cooling and the growing humidification of the air layer causes a gradual lowering of the cloud base, that is of the mixing condensation level z_c which is represented by the lowest kink in atmospheric potential temperature curve. The depth of both layers increases downstream. It should be noted that the temperature changes in the two layers are of the same order, in spite of the much greater heat capacity of the water column. This can be explained by the unequal advection velocities. The water is affected by the same heat exchanges and interactions as the air; but it is affected for a 30 times longer time span.

The sensitivity of the evolution to the initial conditions at the upstream atmospheric boundary is indicated by Fig. 6. It shows that relatively small changes in the initial values

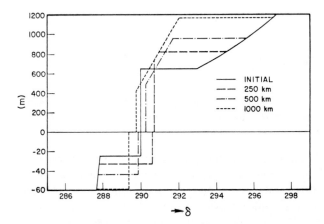

Fig. 5. Computed interactive downstream evolution of oceanic and atmospheric mixed layers ($|V| = 6$ m s^{-1}, $|v| = 0.2$ m s^{-1}, δ = potential temperature in either fluid).

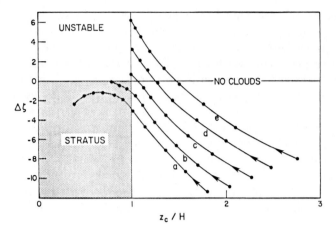

Fig. 6. Evolution of the equivalent potential temperature jump $\Delta\xi$ at the inversion as a function of the initial value of the ration z_c/H. The evolution proceeds from right to left; dots mark distances of 50 km.

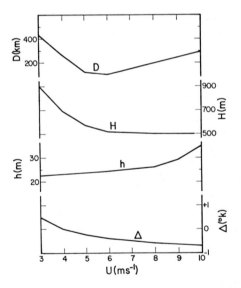

Fig. 7. Distance D of cloud extent, final mixed-layer thicknesses H, h and air–sea temperature difference Δ as functions of wind strength U with $|v| = 0.2\,\mathrm{m\,s^{-1}}$ and initial conditions $T_o = 17°$, $\theta_o = 290°$ and q_o = half saturation mixing ratio for T_o.

of the layer hight and of the humidity or the mixing condensation level z_c can lead either to the development of a stratus deck, or unstable cumulus clouds, or no clouds at all.

The distance D to which the stratus reaches along the trajectory before break-up is also a function of the wind velocity and this is illustrated in Fig. 7. At low wind velocities, evaporation is not very vigorous. The air can travel therefore a relatively long distance D before ξ has increased to its critical value and conditions become convectively unstable at the inversion. The weak evaporation and wind stirring, combined with continuing solar irradiation, actually causes the sea surface to remain warmer than the air above for winds of less than $4 \, \mathrm{m \, s^{-1}}$. Evaporation increases and the destabilization process becomes more effective with increasing wind velocity up to about $6 \, \mathrm{m \, s^{-1}}$ and this results in a relatively early break-up. At still higher wind velocities, the roiling of the water and the associated deepening and cooling of the oceanic mixed layer causes the surface vapour pressure and evaporation, as well as the model inversion height H, to become lower again. This permits the stratus to persist for a greater distance D downstream before break-up occurs. In tests with initially slightly deeper oceanic mixed layers, the stratus persisted beyond the 2000 km limit for all winds in excess of $9 \, \mathrm{m \, s^{-1}}$.

SUMMARY

Each of the applied sections in this paper indicated the relative importance of changes in the depth of the boundary or mixed layer. Variations in this depth can exert a controlling influence upon the values of the surface stress and heat flux. They can affect therefore the large scale oceanic and atmospheric development. The effect of mixed-layer depth changes can be modelled in various ways — some of them very different from those illustrated here. However, they have to be allowed for somehow in any realistic model of the atmosphere—ocean system. A compensatory change in the drag coefficient, when the depth of the lowest atmospheric layers is determined arbitrarily, has been suggested at the end of the third section.

The sensible heat flux from the ocean surface is generally much smaller than the latent heat flux, particularly in low latitudes. However, the sensible heat flux has a relatively large influence upon the buoyancy flux and hence upon the height of the atmospheric mixed layer. It acts therefore as a trigger or amplifying factor. This was illustrated in the fourth section.

In an interactive mixed-layer model, initial upstream conditions can have a large effect for a long distance downstream — in other words, advection in the boundary layers is not negligible. The effect is enhanced by a positive feedback in the case of the sub-tropical stratus layers. Temperature changes in the atmospheric and oceanic surface layers tend to be of the same order during the downstream evolution. This is a result of the unequal advection rates in the two fluids. The local heat loss from the water column, which is associated with an upward flux of equivalent-potential temperature, exceeds the corresponding local gain in the atmosphere by a factor which is equal to the wind velocity divided by the surface current velocity.

ACKNOWLEDGEMENTS

The work of the author is supported by the National Science Foundation under grants No. OCE-8109022 and ATM-8315320 and by the National Oceanic and Atmospheric Administration under the EPOCS program.

APPENDIX

b, B	buoyancy flux immediately below and above the interface
c, c_p	specific heat of sea water and air at constant pressure
D	Drag coefficient
E	evaporation
f	Coriolis parameter
g	gravitational acceleration
h, H	depths of oceanic and atmospheric mixed layers
$I(O), I(H)$	net infrared radiation flux at surface and inversion level
l	concentration or mixing ratio of liquid water in clouds
L	latent heat
n	vertical unit vector
q	vapour mixing ratio
Q	turbulent heat flux across the sea surface
S	solar irradiation
s	salinity
T	ocean mixed-layer temperature
u_*, U_*	friction velocity in water $(|\tau|/\rho)^{1/2}$ air $(|\tau|/\rho_a)^{1/2}$
v, V	horizontal velocity vectors in water and air
w^*, W^*	entrainment velocities into the oceanic and atmospheric mixed layers
z_c	mixing condensation level
α	coefficient of thermal expansion $(\partial \ln \rho / \partial T)$
β	coefficient of haline contraction $(\partial \ln \rho / \partial S)$
Δx	discontinuity of property x at outer mixed-layer boundary
θ, Θ	Actual and reference potential (air) temperature
ξ	(isobaric) equivalent-potential temperature
ρ, ρ_a	density of water, air
σ	total water mixing ratio $(q + 1)$
$\boldsymbol{\tau}$	wind stress vector

REFERENCES

Busch, N., 1977. Fluxes in the surface boundary layer over the sea. In: E.B. Kraus (Editor), Modelling and Prediction of the Upper Layers of the Ocean. Pergamon Press, New York, N.Y., pp. 72–91.

Deardorff, J.W., 1979. Prediction of convective mixed layer entrainment for realistic capping inversion structure. J. Atmos. Sci., 36: 424–436.

De Ruyter, W.P.M., 1983. Frontogenesis in an advective mixed layer model. J. Phys. Oceanogr., 13: 487–495.

Kraus, E.B., 1977. Ocean surface drift velocities. J. Phys. Oceanogr., 7: 606–609.

Kraus E.B. and Leslie, L.D., 1982. The interactive evolution of the oceanic and atmospheric boundary layers in the source region of the trades. J. Atmos. Sci., 39: 2760–2772.

Sarachik, E.S., 1978. Boundary layers on both sides of the tropical ocean surface. Review Papers on Equat. Oceanogr. Proc. FINE Workshop, Ft Lauderdale, Nova/N.Y.I.T. Univ. Press.

Schopf, P.S. and M.A. Cane, 1983. On equatorial dynamics, mixed layer physics and sea surface temperature. J. Phys. Oceanogr., 13: 917–935.

Smith, S.D., 1980. Wind stress and heat flux over the ocean in gale force winds. J. Phys. Oceanogr., 10: 709–726.

CHAPTER 43

A GCM STUDY OF THE ATMOSPHERIC RESPONSE TO TROPICAL SST ANOMALIES

MAX J. SUAREZ

ABSTRACT

The response of the UCLA GCM to the sequence of tropical sea-surface temperatures anomalies that occurred during 1982/83 is examined and compared to a multi-year control experiment. Preliminary results are reported here. Both the tropical and extratropical responses are of interest. Results are presented for Northern Hemisphere fall and winter averages. The extratropical response is found to be qualitatively different in the two seasons.

INTRODUCTION

The response of the atmosphere to anomalies in equatorial sea-surface temperatures (SSTs) has received a great deal of attention in recent years. Both statistical studies using observational data (Bjerknes, 1966, 1969; Krueger and Winston, 1974; Horel and Wallace, 1981; Rasmusson and Carpenter, 1982) and general circulation model experiments (Rowntree, 1972; Julian and Chervin, 1978; Keshavamurty, 1982; Shukla and Wallace, 1983; Blackmon et al., 1983) have clearly established that the ocean temperature perturbations occurring during warm El Niño episodes cause a discernible climate signal over much of the globe. A body of theory has also been developed that attempts to explain the mid-latitude portion of the response in terms of global teleconnection patterns (Wallace and Gutzler, 1981; Horel and Wallace, 1981; Simmons et al., 1983) and propagating Rossby waves forced by the anomalous atmospheric heating (Hoskins and Karoly, 1981).

The most recent (1982/83) warm episode is of particular interest because of the exceptional strength and highly unusual sequence of the SST anomalies (Cane, 1983; Rasmusson and Wallace, 1983; Quiroz, 1983).

The purpose of the work presented here is to understand the evolution of the atmospheric anomalies associated with the most recent warm episode by the use of simulation studies with the UCLA general circulation model (GCM). Our approach is to integrate the model using the observed sequence of SST anomalies during 1982/83, and compare it with a control run in which all boundary conditions vary from month to month as in the climatology.

The UCLA GCM incorporates sophisticated parameterizations of cumulus convection and planetary boundary layer processes (Arakawa and Schubert, 1974; Lord et al., 1982; Suarez et al., 1983). Since these processes are intimately involved in transforming the SST anomaly into diabatic heating, and hence forcing of the free atmosphere, their parameterizations are particularly important to the current study. A brief description of the model and some features of its climatology relevant to our problem are presented in the next section.

THE GCM AND ITS CLIMATOLOGY

The UCLA model uses finite-difference discretization on latitude–longitude coordinates in the horizontal and on a modified sigma coordinate in the vertical. The version used here has a resolution of $4°$ latitude by $5°$ longitude, with nine layers between the surface and a rigid lid at 50 mb. Its physical parameterizations include Arakawa–Schubert convection and a mixed-layer formulation of the planetary boundary layer. Details of the latter and a brief description of the model may be found in Suarez et al. (1983). Further details may be obtained from references therein.

To give some idea of the model's simulated climatology we show in the first figures boreal fall (SON) and winter (DJF) averages of total precipitation and 200 mb zonal wind from the control run. The averaged precipitation in Figs. 1 and 2 are compared with observations by Jaeger (1976). In both seasons the Pacific ITCZ and the South Pacific convergence zone (SPCZ) are fairly well simulated, though the ITCZ is not as narrow as one might wish. The zonal wind at 200 mb is shown in Figs. 3 and 4, together with the observed climatology used at the Climate Analysis Center (Arkin et al., 1983). The position and orientation of the jet axes are well-simulated in both seasons, although their strength is exaggerated. In the tropics the model shows a definite westerly bias, though the positions of easterly maxima, being associated with zonal asymmetries of convective heating, are well simulated. An exception is the absence of easterlies over South America during DJF, a serious weakness of the simulation.

Figure 5 shows the DJF average of the 200 mb eddy height field simulated by the model. (It may be compared with observations presented in Wallace, 1983.) The overall pattern is in excellent agreement with observation. In particular the model captures very well the nearly nodal transition at $\sim 35°N$ and the position of all major features. However, it overestimates the strength of the highs over western Europe and North America, while stretching the East Asian low clear around to Africa.

The vertical structure of the stationary wave pattern is shown in Fig. 6. (For comparison see observations in Lau, 1979, or Wallace, 1983.) At $45°N$ the pattern differs significantly from the observations only in the low over the Eastern Hemisphere, which is far too broad and does not decay above 300 mb, as is observed. At high latitudes the strength of both highs in the Western Hemisphere is exaggerated as already noted, and the whole pattern appears more equivalent barotropic than the observed. The strong phase tilts near the surface, however, are well simulated. The $180°$ phase change between middle and low latitudes is evident in the two lower panels. At low latitudes the amplitude simulated in the upper troposphere is roughly twice the observed.

In general we feel the model produces a very good climatology, adequate to study the effects of tropical sea-surface temperature anomalies. However, deficiencies such as those we have noted here may affect the model's response. The preliminary results shown in the following section are in fact suggestive of considerable sensitivity of the response – particularly in the extratropics – to the model's climatology.

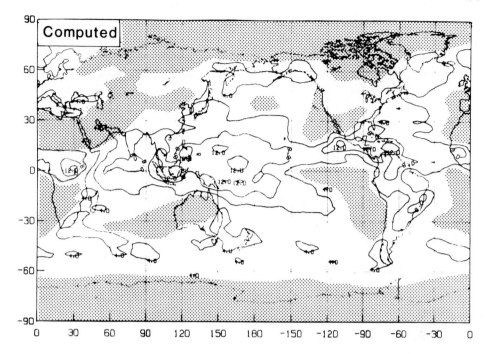

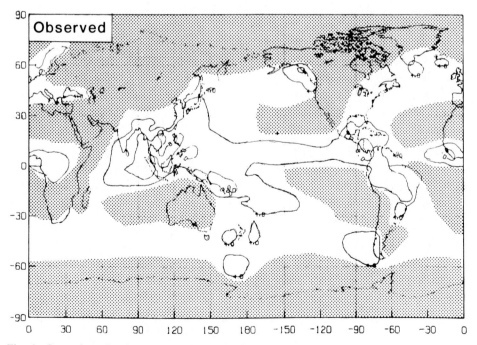

Fig. 1. September–October–November average of observed (Jaeger, 1976) and computed precipitation. Shading indicates values less than 2 mm per day.

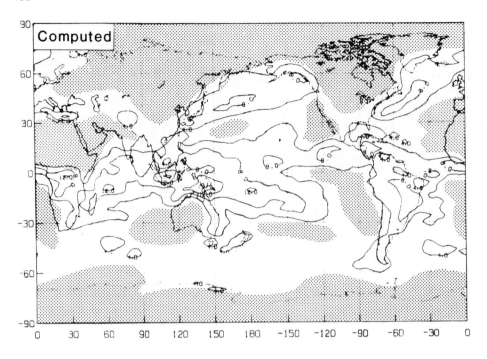

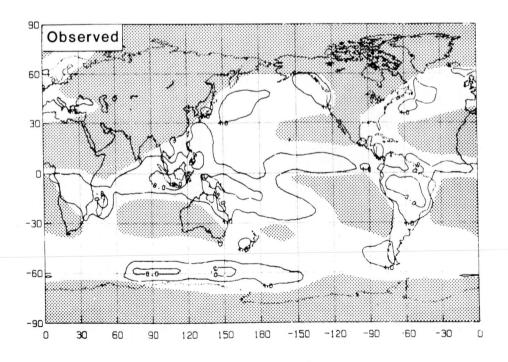

Fig. 2. As in Fig. 1, for December—January—February average.

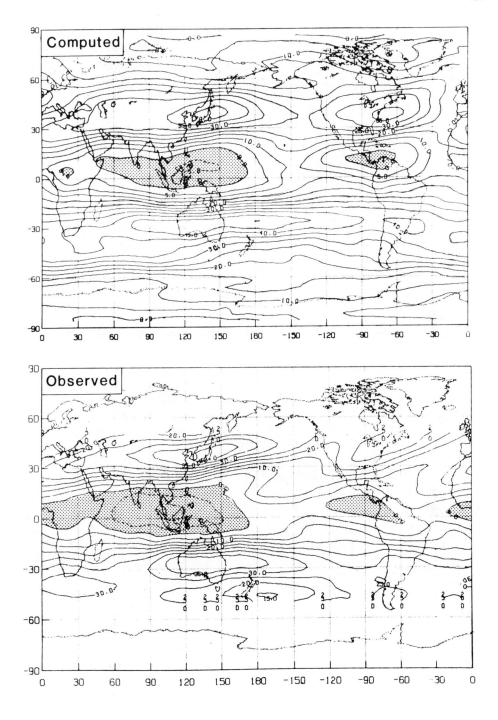

Fig. 3. September–October–November average of zonal wind component of 200 mb height. Regions of easterly wind are shaded. The observations are reproduced from Arkin et al. (1983).

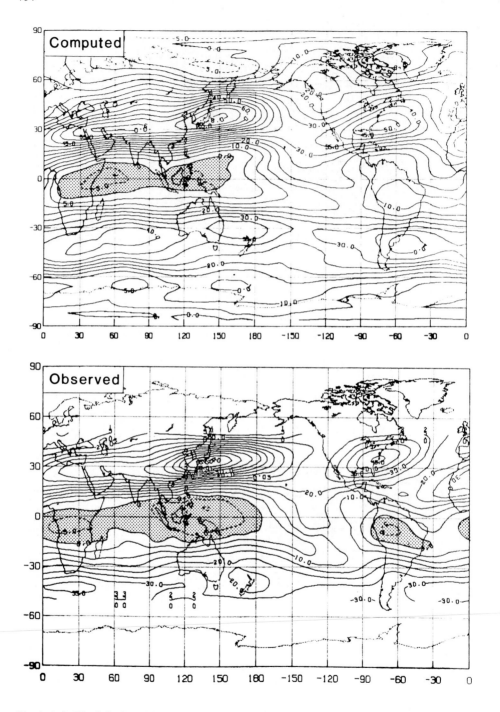

Fig. 4. As in Fig. 3, for December–January–February average.

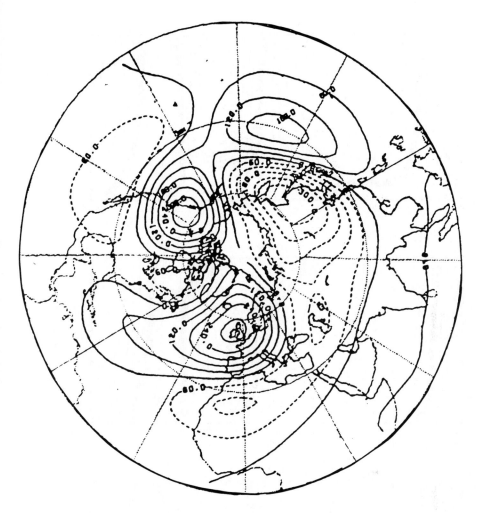

Fig. 5. Zonal deviations of the December–January–February average of 200 mb height. The contour interval is 60 m.

RESULTS

The control for the experiment was a three-year simulation using seasonally varying climatological SSTs. The anomaly calculation was initialized from 15 June of the first year of the control. From 15 June to 1 July, the SST was gradually modified by the anomalies observed during June–July of 1982. From 1 July on, the run was continued using the control's SST plus the 1982–83 anomalies. SSTs were varied daily, interpolating between monthly means. This second integration was carried to the end of February of the second year (1983 in the anomaly). Only anomalies over the tropical Pacific were used.

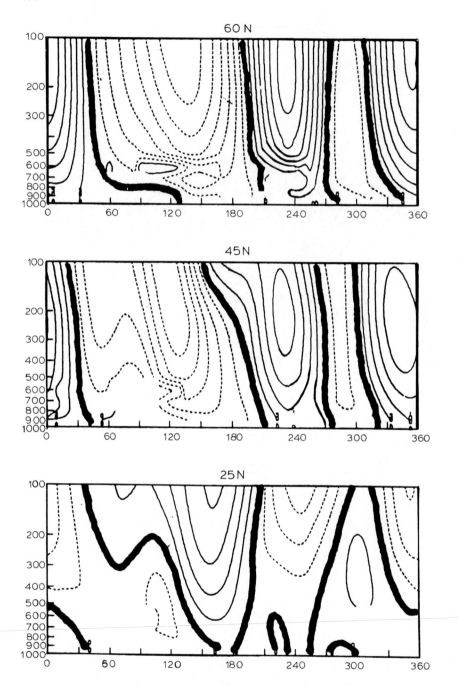

Fig. 6. Vertical structure of zonal deviations in the height field averaged for December–January–February average. The contour interval is 50 m.

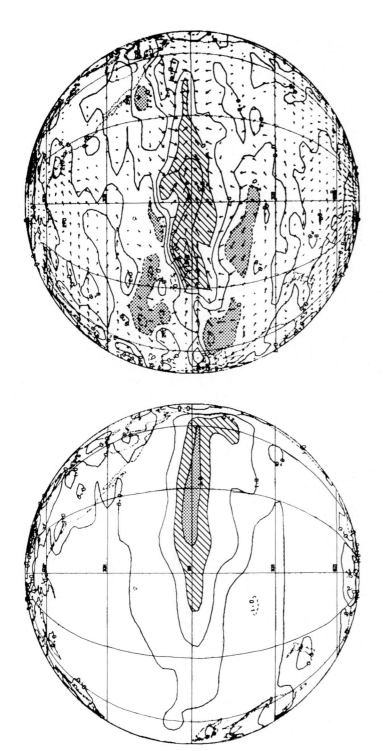

Fig. 7. Anomalies averaged over September–October–November. The surface temperature anomaly (left) is prescribed over the oceans. In the precipitation anomalies (right), regions of less than 2 mm per day are stippled and regions of more than 4 mm per day are hatched. Arrows are anomalous winds in the model's planetary boundary layer.

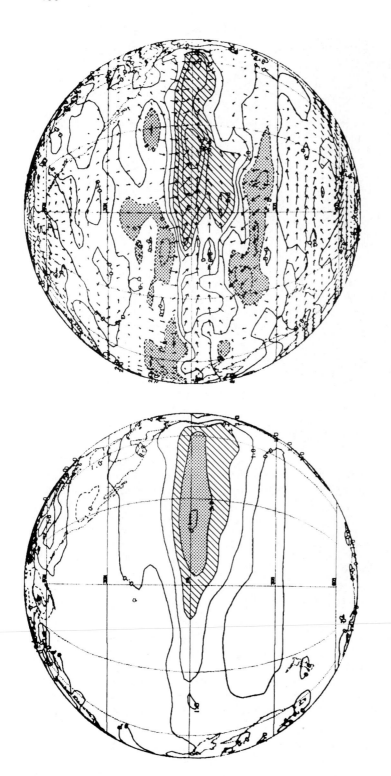

Fig. 8. As in Fig. 7 for December—January—February.

Figures 7–11 summarize the model's response. Seasonally averaged SST anomalies are shown together with the resulting precipitation and surface wind anomalies in Figs. 7 and 8. During SON, precipitation has increased over the equator between the dateline and 120°W with the greatest anomalies just east of the dateline. The precipitation anomaly is thus well west of the SST anomaly. By DJF the response has shifted east some 30° and anomalies of ~ 6 mm per day extend to the South American coast. As we will see below, this eastward progression of the region of anomalous convective activity was observed during the 1982–83 event. The total rainfall averaged over the tropical Pacific changes little. Positive anomalies over the equator are compensated by negative anomalies over a large horseshoe-shaped region poleward and westward of the increased rainfall. The major effect of the SST anomaly is thus a redistribution of the precipitation over the tropical Pacific. The anomaly in the surface wind, which must be doing the bulk of the anomalous transport of water vapor, is also shown on the figure. Over the western half of the region of maximum forcing (180°–150°W during SON and 150°–120°W during DJF) the result is qualitatively as one would expect from the simplest linear theory (cf. Matsuno, 1966; Webster, 1972; Gill, 1980): Cyclonic flow north and south of the equator just west of the maximum forcing, and strong westerly anomalies over the equator. East of the maximum rainfall anomaly, however, the response is very different from what one would expect from linear theory. The model produces a strong meridional convergence with particularly strong northerly flow just north of the equator. The linear response about a state of rest would lead to either flow away from the equator or, far east of the forcing, no meridional component.

We have already mentioned the eastward progression of the rainfall anomaly. Since, as we will see below, the model's extratropical response changes dramatically from SON to DJF, suggesting some sensitivity to the longitude of the tropical forcing, we compare in Fig. 9 the modelled and observed regions of anomalous convective activity for the two seasons. For the observed we take as proxy the outgoing infrared radiation anomalies reported by Arkin et al. (1983). Since outgoing infrared anomalies are mostly the result of changes in high cloudiness, negative values should be associated with regions of increased convective rainfall. During the fall the greatest changes in convective activity are near the dateline in both model and observations. The outgoing longwave anomalies, however, indicate a predominately east–west shift in the precipitation patterns, with large positive anomalies (negative in convective activity) over the maritime continent and negative anomalies over much of the Indian Ocean. There is also some indication of a decrease in strength of the SPCZ. In the model there is little sign of a major east–west shift in rainfall. In winter the maximum longwave anomaly has moved to 150°W; the model's rainfall anomaly, however, has moved to nearly 120°W, and large changes extend to the South American coast. As in the fall the model produces less of an east-west shift in precipitation than appears in the longwave anomaly; however, the pattern of decreased precipitation is in better agreement with the observation than in the previous three months. There was no significant response over the Indian Ocean in either season.

Turning now to the extratropical response, Fig. 10 shows the anomaly of the stationary wave pattern in the 200 mb height field. During SON, with the anticyclonic pair in the tropics centered at 150°W, the Northern Hemisphere response is very reminiscent of the PNA pattern described by Wallace and Gutzler (1981). In DJF the response is quite different, suggestive of a wavetrain emanating at the tropical anticyclone

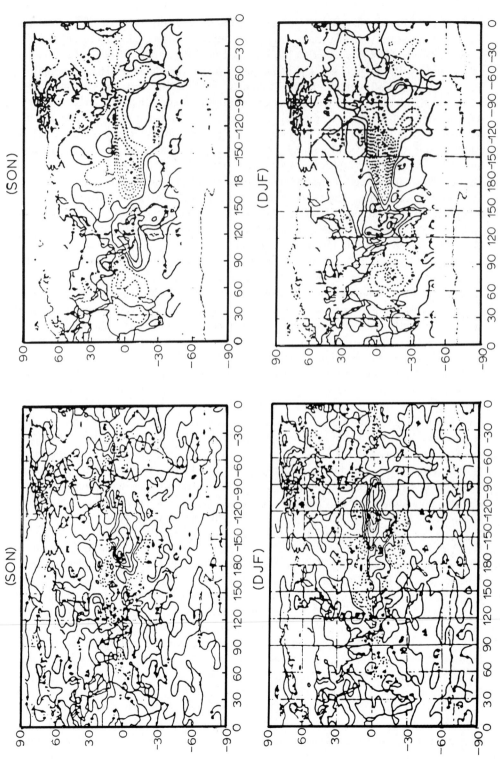

Fig. 9. Model's precipitation (left) and observed outgoing longwave anomalies (right) for the two seasons. The observed is reported from Arkin et al.

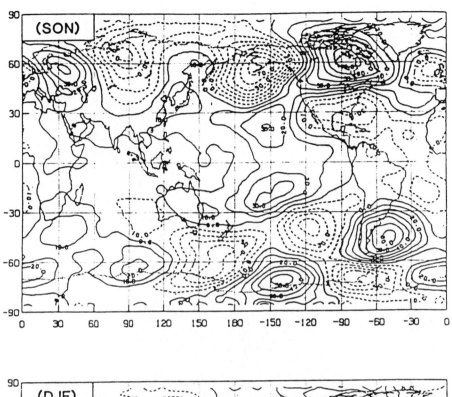

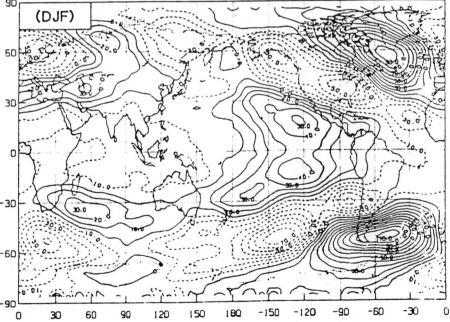

Fig. 10. The zonal deviation of 200 mb height anomalies produced by the GCM during the two seasons.

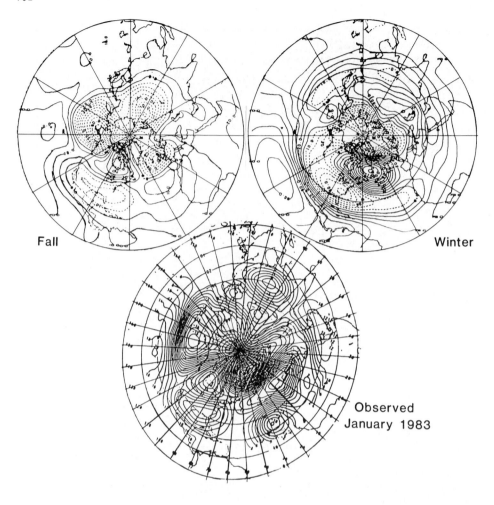

Fall

Winter

Observed
January 1983

Fig. 11. Northern Hemisphere 200 mb height anomalies produced by the GCM during fall and winter. Observed anomalies are from Arkin et al. (1983) for January 1983. The contouring interval in the GCM results is half that in the observed.

now located much further east, at 120°W, and disappearing over northern Europe. Remarkably, the response shows a strong symmetry about the equator, even in details far in the Eastern Hemisphere that one might not otherwise consider significant. In Fig. 11, these two patterns are compared with the January 1983 anomalies produced by Arkin et al. (1983). The model's anomalies are much weaker than the observed, but note the strong resemblance of the PNA pattern between the model's SON anomaly and the observed January, both of which have the tropical anticyclone at ~ 150°W.

Assuming these extratropical patterns represent a true response to the anomalies rather than a statistical residual of the model's natural variability, and if we view the extratropical response as an equivalent barotropic extension of the tropical response in the upper troposphere (Simmons, 1982), it is important to understand why the model's tropical anticyclone pair is consistently east of the observed, and why the extratropical response is so sensitive to the longitude of the tropical pattern aloft.

REFERENCES

Arakawa, A. and Schubert, W.H., 1974. Interaction of a cumulus cloud ensemble with the large-scale environment. Part I. J. Atmos. Sci., 31: 674–701.

Arkin, P.A., Dipman, J.D. and Reynolds, R.W., 1983. The 1982/83 El Niño/Southern Oscillation Event Quick-Look Atlas. Climate Analysis Center, NOAA, Washington, D.C.

Bjerknes, J., 1966. A possible response of the atmospheric Hadley circulation to equatorial anomalies of ocean temperature. Tellus, 18: 820–829.

Bjerknes, J., 1969. Atmospheric teleconnections from the equatorial Pacific. Mon. Weather Rev., 97: 163–172.

Blackmon, M.L., Geisler, J.E. and Pitcher, E.J., 1983. A general circulation model study of January climate anomaly patterns associated with interannual variation of Equatorial Pacific sea surface temperatures. J. Atmos. Sci., 40(6): 1410–1425.

Cane, M.A., 1983. Oceanographic events during El Niño. Science, 222: 1189–1195.

Gill, A.E., 1980. Some simple solutions for heat-induced tropical circulation. Q. J. R. Meteorol. Soc., 106: 447–462.

Horel, J.D. and Wallace, J.M., 1981. Planetary-scale atmospheric phenomena associated with the Southern Oscillation. Mon. Weather Rev., 109: 813–829.

Hoskins, B.J. and Karoly, D.J., 1981. The steady linear response of a spherical atmosphere to thermal and orographic forcing. J. Atmos. Sci., 38: 1179–1196.

Jaeger, L., 1976. Monatskarten des Niederschlags für die ganze Erde. Ber. Dtsch. Wetterdienstes, 18, No. 139, Offenbach, 38 pp.

Julian, P.R. and Chervin, R.M., 1978. A study of the Southern Oscillation and Walker circulation phenomenon. Mon. Weather Rev., 106: 1433–1451.

Keshavamurty, R.N., 1982. Response of the atmosphere to sea surface temperature anomalies over the equatorial Pacific and the teleconnections of the Southern Oscillations. J. Atmos. Sci., 39: 1241–1259.

Krueger, A.F. and Winston, J.S., 1974. A comparison of the flow over the tropics during two contrasting circulation regimes. J. Atmos. Sci., 31: 358–370.

Lau, N.C., 1979. The observed structure of tropospheric stationary waves and the local balances of vorticity and heat. J. Atmos. Sci., 36: 996–1016.

Lord, S.J., Chao, W.C. and Arakawa, A., 1982. Interaction of a cumulus cloud ensemble with the large-scale environment. Part IV: The discrete model. J. Atmos. Sci., 39: 104–113.

Matsuno, T., 1966. Quasi-geostrophic motions in the equatorial area. J. Meteorol. Soc. Jpn., 44: 25–42.

Quiroz, R.S., 1983. Seasonal climate summary: The climate of the "El Niño" winter of 1982–83 – A season of extraordinary climatic anomalies. Mon. Weather Rev., 111: 1685–1706.

Rasmusson, E.M. and Carpenter, T.M., 1982. Variations in tropical sea surface temperature and surface wind fields associated with the Southern Oscillation/El Niño. Mon Weather Rev., 110: 354–384.

Rasmusson, E.M. and Wallace, J.M., 1983. Meteorological aspects of the El Niño/Southern Oscillation. Science, 222: 1195–1202.

Rowntree, P.R., 1972. The influence of tropical east Pacific ocean temperatures on the atmosphere. Q. J. R. Meteorol. Soc., 98: 290–321.

Shukla, J. and Wallace, J.M., 1983. Numerical simulation of the atmospheric response to equatorial Pacific sea surface temperature anomalies. J. Atmos. Sci., 40: 1613–1630.

Simmons, A.J., 1982. The forcing of stationary wave motion by tropical diabatic heating. Q. J. R. Meteorol. Soc., 108: 503–534.

Simmons, A.J., Wallace, J.M. and Branstator, G.W., 1983. Barotropic wave propagation and instability and atmospheric teleconnection patterns. J. Atmos. Sci., 40: 1363–1392.

Suarez, M.J., Arakawa, A. and Randall, D.A., 1983. The parameterization of the planetary boundary layer in the UCLA general circulation model: formulation and results. Mon. Weather Rev., 111: 2224–2243.

Wallace, J.M., 1983. The climatological mean stationary waves: observational evidence. In: B.J. Hoskins and R.P. Pearce (Editors), Large Scale Dynamical Processes in the Atmosphere. Academic Press, New York, N.Y., 397 pp.

Wallace, J.M. and Gutzler, D.S., 1981. Teleconnections in the geopotential height field during the Northern Hemisphere winter. Mon. Weather Rev., 109: 784–812.

Webster, P.J., 1972. Response of the tropical atmosphere to local steady forcing. Mon. Weather Rev., 100: 518–541.

LIST OF PARTICIPANTS

JSC/CCCO 16th International Liège Colloquium

"Coupled Ocean–Atmosphere Models"

7–11 May, 1984, University Campus, Sart Tilman, Liège, Belgium

ANDERSON D., Prof. Dr., University of Oxford, Oxford, UK
ALDERSON St., Mr., Institute of Oceanographic Sciences, Wormley, UK
BAH A., Dr., Université Laval, Québec, Canada
BECKERS Ph., Mr., University of Liège, Liège, Belgium
BERGER A., Prof. Dr., Université Catholique de Louvain, Louvain-la-Neuve, Belgium
BIGG G.R., Dr., University of Cambridge, Cambridge, UK
BLACKMON M.L., Dr., National Center for Atmospheric Research, Boulder, Colorado, USA
BOER G.J., Dr., Canadian Climate Centre/CCRN, Downsview, Ontario, Canada
BORMANS M., Mrs., Dalhousie University, Halifax, Nova Scotia, Canada
BOUKARI S., Mr., University of Niamey, Niamey, Niger
BRETHERTON F.P., Dr., National Center for Atmospheric Research, Boulder, Colorado, USA
BRICKMAN D., Mr., Dalhousie University, Halifax, Nova Scotia, Canada
BRYAN K., Dr., Princeton University, Princeton, New Jersey, USA
BYE J.A.T., Dr., The Flinders University of South Australia, Australia
CANE M., Dr., Lamont-Doherty Geological Observatory, Palisades, New York, USA
CLEMENT F., Mr., University of Liège, Liège, Belgium
COOPER N., Dr., University of Oxford, Oxford, UK
COREN Cl., Miss, University of Liège, Liège, Belgium
CRÉPON M., Dr., Museum d'Histoire Naturelle, Paris, France
CUBASH U., Mr., European Centre for Medium Range Weather Forecasts, Reading, UK
CUSHMAN-ROISIN B., Dr., Florida State University, Tallahassee, Florida, USA
DAVEY M.K., Dr., University of Cambridge, Cambridge, UK
DELECLUSE P., Dr., Museum d'Histoire Naturelle, Paris, France
DELEERSNIJDER E., Ing., University of Liège, Liège, Belgium
DE RUIJTER W.P.M., Dr., Rijkswaterstaat/Deltadienst, The Hague, The Netherlands
DISTECHE A., Prof. Dr., University of Liège, Liège, Belgium
DJENIDI S., Ing., University of Liège, Liège, Belgium
DORE A., Mme, Editions Dunod-Bordas-Gauthier-Villars, Paris, France
DYMNIKOV V., Dr., USSR Academy of Sciences, Moscow, USSR
ENDOH M., Dr., Meteorological Research Institute, Ibaraki, Japan
ESBENSEN S.K., Prof. Dr., Oregon State University, Corvallis, Oregon, USA
EVERBECQ E., Ing., University of Liège, Liège, Belgium
FENNESSY M.J., Mr., The University of Maryland, College Park, Maryland, USA
FRANKIGNOUL Cl., Prof. Dr., Université Pierre et Marie Curie, Paris, France
GALLARDO Y., Mr., Centre Océanologique de Bretagne, Brest, France
GASPAR Ph., Ing., Université Catholique de Louvain, Louvain-la-Neuve, Belgium
GATES W.L., Prof. Dr., Oregon State University, Corvallis, Oregon, USA
GHAZI A., Dr., Commission des Communautés Européennes, Bruxelles, Belgium
GILL A.E., Dr., University of Oxford, Oxford, UK
GILLOT R.H., Dr., Commission des Communautés Européennes, Ispra, Italy
GONELLA J., Prof. Dr., Museum d'Histoire Naturelle, Paris, France
GORDON C., Dr., Meteorological Office, Bracknell, UK

GUO Y.-F., Mr., Academia Sinica, Beijing, China
HACKER P.W., Dr., National Science Foundation, Washington, D.C., USA
HAN Y.-J., Dr., Oregon State University, Corvallis, Oregon, USA
HAPPEL J.J., Ing., University of Liège, Liège, Belgium
HEBURN G.W., Dr., NORDA, NSTL Station, Massachusetts, USA
HIRST A.C., Mr., University of Wisconsin, Madison, Wisconsin, USA
JAMART B., Dr., University of Liège, Liège, Belgium
JENSEN T.G., Mr., University of Copenhagen, Copenhagen, Denmark
JOHNSON J.A., Dr., University of East Anglia, Norwich, UK
KILLWORTH P.D., Dr., University of Cambridge, Cambridge, UK
KITAIGORODSKII S., Prof., Dr., The Johns Hopkins University, Baltimore, Maryland, USA
KRAUSE E., Prof. Dr., University of Colorado at Boulder, Boulder, Colorado, USA
KRUSE A.H., Dr., Universität Hamburg, Hamburg, F.R.G.
KURBATKIN G.P., Dr., USSR Academy of Sciences, Novosibirsk, USSR
LATIF M., Mr., Universität Hamburg, Hamburg, F.R.G.
LAU W.K.-M., Dr., NASA/Goddard Space Flight Center, Greenbelt, Maryland, USA
LEBON G., Prof., Dr., University of Liège, Liège, Belgium
LE TREUT H., Dr., Ecole Normale Supérieure, Paris, France
LIOU C.-S., Dr., Naval Postgraduate School, Monterey, California, USA
MALVESTUTO V., Dr., CNR/Istituto di Fisica dell'Atmosfera, Roma, Italy
McCREARY J., Dr., Nova Oceanographic Laboratory, Dania, Florida, USA
McINTYRE A., Prof. Dr., Lamont-Doherty Geological Observatory, Palisades, New York, USA
McPHADEN M., Dr., University of Washington, Seattle, Washington, USA
MERLE J., Dr., Université Pierre et Marie Curie, Paris, France
MERZI N., Mr., Ecole Poytechnique Fédérale de Lausanne, Switzerland
MICHAUD R., Mr., Ecole Normale Supérieure, Paris, France
MIRALLES L., Dr., Instituto de Investigaciones Pesqueras de Barcelona, Barcelona, Spain
MITCHELL J.F.B., Dr., Meteorological Office, Bracknell, UK
MOLCARD R., Dr., UNESCO, Paris, France
MOLIN A., Mr., Université Pierre et Marie Curie, Paris, France
MONREAL A., Mrs., CICESE, Ensenada, Mexico
MUIR D.J., Dr., University of Exeter, Exeter, UK
NEMRY B., Ing., University of Liège, Liège, Belgium
NEVES R., Ing., Instituto Superior Tecnico, Lisboa, Portugal
NEWMAN M.R., Mr., Meteorological Office, Bracknell, UK
NEWSON R.L., Dr., World Meteorological Organization, Geneva, Switzerland
NIILER P., Prof. Dr., Scripps Institution of Oceanography, La Jolla, California, USA
NIHOUL J.C.J., Prof. Dr., University of Liège, Liège, Belgium
OBERHUBER J.M., Mr., Universität Hamburg, Hamburg, F.R.G.
O'BRIEN J.J., Prof. Dr., The Florida State University, Tallahassee, Florida, USA
OLBERS D.J., Dr., Universität Hamburg, Hamburg, F.R.G.
OORT A.H., Prof. Dr., Princeton University, Princeton, New Jersey, USA
OPSTEEGH Th., Dr., K.N.M.I., De Bilt, The Netherlands
PALMER T.N., Dr., Meteorological Office, Bracknell, UK
PAN Y., Ing., Second Institute of Oceanography, Hangzhou, China
PHILANDER G., Dr., Princeton University, Princeton, New Jersey, USA
PICAUT J., Dr., Université de Bretagne Occidentale, Brest, France
PIERINI S., Dr., Istituto Universitario Navale, Napoli, Italy
PIRLET A., Ing., University of Liège, Liège, Belgium
POULAIN P.M., Mr., University of Liège, Liège, Belgium
PRANGSMA G.J., Dr., K.N.M.I., De Bilt, The Netherlands
PREISENDORFER R.W., Dr., PMEL/NOAA, Seattle, Washington, USA
PRICE P., Mr., World Meteorological Organization, Geneva, Switzerland
PU S., Ing., First Institue of Oceanography, Qingdac, China
RAMMING H.-G., Dr., Aumühle, F.R.G.

767

RASMUSSON E.M., Dr., Climate Analysis Center, Washington, D.C., USA
REBERT J.P., Mr., Office de la Recherche Scientifique et Technique d'Outre-Mer, Centre de Nouméa, New Caledonia
RENNICK M.A., Dr., Naval Postgraduate School, Monterey, California, USA
ROGERS C., Dr., Institute of Oceanographic Sciences, Wormley, UK
RONDAY F.C., Dr., University of Liège, Liège, Belgium
ROWE M., Mr., The University of Southampton, Southampton, UK
SALAS DE LEON D., Mr., CICESE, Ensenada, Mexico
SARKYSIAN A., Dr., USSR Academy of Sciences, Moscow, USSR
SAUVEL J., Mr., Service Hydrographique et Océanographique de la Marine, Paris, France
SCHLESINGER M.E., Prof. Dr., Oregon State University, Corvallis, Oregon, USA
SCHOPF P., Dr., NASA/Goddard Space Flight Center, Greenbelt, Maryland, USA
SEMTNER A.J., Dr., National Center for Atmospheric Research, Boulder, Colorado, USA
SERVAIN J., Dr., Université de Bretagne Occidentale, Brest, France
SHUKLA J., Prof. Dr., University of Maryland, College Park, Maryland, USA
SILVER D., Dr., The Johns Hopkins University, Laurel, Maryland, USA
SMITH N.R., Dr., The University of Southampton, Southampton, UK
SMITZ J., Ing., University of Liège, Liège, Belgium
SPITZ Y., Miss, University of Liège, Liège, Belgium
SUAREZ M. Dr., NASA/Goddard Space Flight Center, Greenbelt, Maryland, USA
TAYLOR A.H., Mr., Institute of Marine Environmental Research, Plymouth, UK
THOMPSON B.J., Mr., UNESCO, Paris, France
THOMPSON K., Dr., Dalhousie University, Halifax, Nova Scotia, Canada
TIDMARSH E., Mrs., Dalhousie University, Halifax, Nova Scotia, Canada
TINTORE J., Dr., Instituto de Investigaciones Pesqueras de Barcelona, Spain
TOURRE Y., Dr., Université Pierre et Marie Curie, Paris, France
VAN YPERSELE J.P., Mr., National Center for Atmospheric Research, Boulder, Colorado, USA
VON STORCH H., Dr., Universität Hamburg, Hamburg, F.R.G.
WEBSTER F., Dr., University of Delware, Lewes, Delaware, USA
WELLS N., Dr., The University of Southampton, Southampton, UK
WILLEBRAND J., Prof. Dr., Universität Kiel, Kiel, F.R.G.
WOODS J.D., Prof. Dr., Universität Kiel, Kiel, F.R.G.
YAMAGATA T., Dr., Kyushu University, Kasuga, Japan
YEPDJUO E., Ing., Service Météorologique National, Douala, Cameroun
ZICARELLI M., Dr., Istituto Universitario Navale, Napoli, Italy
ZIMMERMANN D., Mr., University of Liège, Liège, Belgium